AF479091

Mycobacteria

Molecular Biology and Virulence

Mycobacteria
Molecular Biology and Virulence

EDITED BY

Colin Ratledge
Department of Biological Sciences
University of Hull
Hull

AND

Jeremy Dale
School of Biological Sciences
University of Surrey
Guildford

Blackwell
Science

© 1999
Blackwell Science Ltd
Editorial Offices:
Osney Mead, Oxford OX2 0EL
25 John Street, London WC1N 2BL
23 Ainslie Place, Edinburgh EH3 6AJ
350 Main Street, Malden
  MA 02148-5018, USA
54 University Street, Carlton
  Victoria 3053, Australia
10, rue Casimir Delavigne
  75006 Paris, France

Other Editorial Offices:
Blackwell Wissenschafts-Verlag GmbH
Kurfürstendamm 57
10707 Berlin, Germany

Blackwell Science KK
MG Kodenmacho Building
7–10 Kodenmacho Nihombashi
Chuo-ku, Tokyo 104, Japan

First published 1999

Set by Excel Typesetters Co., Hong Kong
Printed and bound in Great Britain
by MPG Books Ltd, Bodmin, Cornwall

The Blackwell Science logo is a
trade mark of Blackwell Science Ltd,
registered at the United Kingdom
Trade Marks Registry

A catalogue record for this title
is available from the British Library

ISBN 0-632-05304-6

Library of Congress
Cataloging-in-Publication Data

Mycobacteria: molecular biology and
  virulence/edited by Colin
  Ratledge and Jeremy Dale.
      p.   cm.
    Includes bibliographical references
    and index.
    ISBN 0-632-05304-6
    1. Mycobacterial diseases —
    Molecular aspects.
    2. Mycobacteria.
  I. Ratledge, Colin.   II. Dale, Jeremy.
  QR201.M96M95   1999
  579.3'74 — dc21   99-24452
                                     CIP

DISTRIBUTORS

Marston Book Services Ltd
PO Box 269
Abingdon, Oxon OX14 4YN
(*Orders*: Tel: 01235 465500
            Fax: 01235 465555)

USA
Blackwell Science, Inc.
Commerce Place
350 Main Street
Malden, MA 02148-5018
(*Orders*: Tel: 800 759 6102
            781 388 8250
            Fax: 781 388 8255)

Canada
Login Brothers Book Company
324 Saulteaux Crescent
Winnipeg, Manitoba R3J 3T2
(*Orders*: Tel: 204 837-2987)

Australia
Blackwell Science Pty Ltd
54 University Street
Carlton, Victoria 3053
(*Orders*: Tel: 3 9347 0300
            Fax: 3 9347 5001)

For further information on
Blackwell Science, visit our website:
www.blackwell-science.com

Contents

List of contributors

B. G. BARRELL *The Sanger Centre, Wellcome Trust Genome Campus, Hinxton CB10 1IL, UK*

C. E. BARRY, III *Tuberculosis Research Section, Laboratory of Host Defences, National Institute of Allergy and Infectious Diseases, National Institutes of Health, Rockville, MD 20852, USA*

A. R. BAULARD *U447, Institut Pasteur de Lille, 1 rue du Professeur Calmette, BP245 59019 Lille Cedex, France*

G. S. BESRA *School of Microbiological, Immunological and Virological Sciences, The Medical School, University of Newcastle upon Tyne, Framlington Place, Newcastle upon Tyne NE2 4HH, UK*

G. H. BOTHAMLEY *East London Tuberculosis Services, Department of Respiratory Medicine, Homerton Hospital, Homerton Row, London E9 6SR, UK*

P. J. BRENNAN *Department of Microbiology, Colorado State University, Fort Collins, CO 80523-1677, USA*

R. BROSCH *Unité de Génétique Moléculaire Bactérienne, Institut Pasteur, 28 rue Dr Roux, 75724 Paris Cedex 15, France*

D. CATTY *Departments of Immunology and Infection, The University of Birmingham Medical School, Edgbaston, Birmingham B15 2TT, UK*

R. CLIFTON-HADLEY *Central Veterinary Laboratory, New Haw, Addlestone, Surrey KT15 3NB, UK*

S. T. COLE *Unité de Génétique Moléculaire Bactérienne, Institut Pasteur, 28 rue Dr Roux, 75724 Paris Cedex 15, France*

D. M. COLLINS *AgResearch, Wallaceville Animal Research Centre, PO Box 40063, Upper Hutt, New Zealand*

M. J. COLSTON *The Division of Mycobacterial Research, The National Institute for Medical Research, The Ridgeway, Mill Hill, London NW7 1AA, UK*

A. M. COOPER *Mycobacteria Research Laboratories, Department of Microbiology, Colorado State University, Fort Collins, CO 80523-1677, USA*

R. A. COX *The Division of Mycobacterial Research, The National Institute for Medical Research, The Ridgeway, Mill Hill, London NW7 1AA, UK*

J. DAVIES *Department of Microbiology and Immunology, University of British Columbia, Room 300, Wesbrook Building, 6174 University Blvd, Vancouver, BC V6T 1Z3, Canada*

K. EIGLMEIER *Unité de Génétique Moléculaire Bactérienne, Institut Pasteur, 28 rue Dr Roux, 75724 Paris Cedex 15, France*

K. D. EISENACH *University of Arkansas for Medical Sciences, 4301 W. Markham Street, Little Rock, AR 72205, USA*

J. O. FALKINHAM, III *Fralin Biotechnology Center, Virginia Tech Blacksburg, VA 24061-0346, USA*

T. GARNIER *Unité de Génétique Moléculaire Bactérienne, Institut Pasteur, 28 rue Dr Roux, 75724 Paris Cedex 15, France*

B. GICQUEL *Unité de Génétique Mycobactérienne, Institut Pasteur, 25 rue du Dr Roux, 75015 Paris, France*

M. GOMEZ *Public Health Research Institute, 455 First Avenue, New York, NY 10016, USA*

S. V. GORDON *Unité de Génétique Moléculaire Bactérienne, Institut Pasteur, 28 rue Dr Roux, 75724 Paris Cedex 15, France*

F. GRIFFIN *Deer Research Laboratory, Department of Microbiology, University of Otago, PO Box 56, Dunedin, New Zealand*

C. GUILHOT *Unité de Génétique Mycobactérienne, Institut Pasteur, 25 rue du Dr Roux, 75015 Paris, France*

G. F. HATFULL *Pittsburgh Bacteriophage Institute and Department of Biological Sciences, University of Pittsburgh, Pittsburgh, PA 15260 USA*

P. VAN HELDEN *Director, MRC Centre for Molecular & Cellular Biology, Department of Medical Biochemistry, Faculty of Medicine, University of Stellenbosch, PO Box 19063, Tygerberg, 7505, South Africa*

G. HEWINSON *Central Veterinary Laboratory, New Haw, Addlestone, Surrey KT15 3NB, UK*

N. HONORÉ *Unité de Génétique Moléculaire Bactérienne, Institut Pasteur, 28 rue Dr Roux, 75724 Paris Cedex 15, France*

M. JACKSON *Unité de Génétique Mycobactérienne, Institut Pasteur, 25 rue du Dr Roux, 75015 Paris, France*

R. JANSSEN *Department of Infectious Diseases & Microbiology, Imperial College School of Medicine, Norfolk Place, London, UK*

J. LIU *Department of Medical Genetics and Microbiology, University of Toronto, 4382 Medical Science Building, 1 Kings College Circle, Toronto, Ontario M5S 1A8, Canada*

D. B. LOWRIE *The Division of Mycobacterial Research, National Institute for Medical Research, The Ridgeway, Mill Hill, London NW7 1AA, UK*

J. MCFADDEN *School of Biological Sciences, University of Surrey, Guildford, Surrey GU2 5XH, UK*

H. NIKAIDO *Department of Molecular and Cell Biology, 229 Stanley Hall, University of California, Berkeley, CA 94720-3206, USA*

I. M. ORME *Mycobacteria Research Laboratories, Department of Microbiology, Colorado State University, Fort Collins, CO 80523, USA*

J. POLLOCK *Department of Agriculture for Northern Ireland, Veterinary Sciences Division, Stoney Road, Stormont, Belfast BT4 3SD, UK*

C. RATLEDGE *Department of Biological Sciences, University of Hull, Hull HU6 7RX, UK*

G. A. W. ROOK *Department of Bacteriology, Windeyer Institute of Medical Sciences, Royal Free and University College Medical School, 46 Cleveland street, London W1P 6DB, UK*

D. G. RUSSELL *Molecular Microbiology, Washington University School of Medicine, 660 South Euclid Avenue, St. Louis, MO 63110, USA*

I. SMITH *Public Health Research Institute, 455 First Avenue, New York, NY 10016, USA*

J. L. STANFORD *Department of Bacteriology, Windeyer Institute of Medical Sciences, Royal Free and University College Medical School, 46 Cleveland Street, London W1P 6DB, UK*

G. R. STEWART *School of Biological Sciences, University of Surrey, Guildford, Surrey GU2 5XH, UK*

J. THOLE *Department of Infectious Diseases & Microbiology, Imperial College School of Medicine, Norfolk Place, London, UK*

J. TIMM *Public Health Research Institute, 455 First Avenue, New York, NY 10016, USA*

V. WEBB *Lookfar Solutions, Inc., P.O. Box 811, Tofino, BC, VOR 2ZO Canada*

D. YOUNG *Department of Infectious Diseases & Microbiology, Imperial College School of Medicine, Norfolk Place, London, UK*

Preface

Tuberculosis still kills more people than any other single infective agent. The scourge of the White Death, as tuberculosis has been called, is still with us and the spectre of it re-emerging as a major killer in developed countries, as well as continuing to wreak havoc in developing ones, is continually before us. Tuberculosis is a problem that has never gone away. Western societies have become complacent about its ravages over the past 40 years as antibiotics and new drugs have kept the disease more or less under control. However, we have now run out of new antibiotics, and new chemotherapeutic agents are not materializing as rapidly as had been anticipated. Moreover, like all bacteria, the tubercle bacillus has learnt how to defend itself against the existing drug regimens and the emergence of multiple drug-resistant strains of *Mycobacterium tuberculosis* is now a very real threat to world health.

Now, more than ever, it is imperative to understand this enemy. Only by understanding the organism, and the disease it causes, do we stand any chance of conquering it. This can only be achieved by determined efforts in the research laboratories of the world, but mycobacterial research has always run at a different pace from other bacterial pathogens. The slow growth of *Mycobacterium tuberculosis*, coupled with many other quirks, make it a frustrating organism to deal with, while laboratory culture of the other major pathogen, *Mycobacterium leprae*, remains an elusive dream. The application of the techniques of molecular biology, starting about 15 years ago, held out a promise that at last we would be able to get to grips with many aspects of the fundamental biology of these organisms that had proved so difficult to study by other means. Counterbalancing that promise was an implicit (and sometimes explicit) assumption that mycobacteria are so different from other bacteria that their molecular biology would be difficult to interpret—or even impossible to study with conventional techniques.

How times have changed! The molecular biologists soon provided valuable tools for the study of certain aspects of mycobacteriology, first with the identification and cloning of genes for apparently dominant protein antigens, and then with methods for characterizing the molecular epidemiology of tuberculosis and the development of molecular diagnostics. More recently, the modes of action of antimycobacterial antibiotics and the major resistance mechanisms have been elucidated. And now the publication of the complete genome sequence of *M. tuberculosis* H37Rv, followed by the sequence of a second strain CSU#93, and with the *M. leprae* sequence not far behind, means that we are entering a whole new world of comparative genomics, enabling a detailed inventory of the metabolic potential of these organisms. We can now appreciate, as indeed we should have appreciated (and sometimes did) all along, that at a molecular genetic level mycobacteria are not so different from other bacteria. They behave differently because they use the information in different ways. The difference is in the detail. The view expressed by Frank Winder, one of the most eminent of mycobacterial biochemists, that 'Mycobacteria are nothing more than *E. coli* wrapped up in a fur coat!' can now be seen to have some force, if we allow for a degree of poetic licence.

Coupling this existing and forthcoming information on genome structure with molecular techniques for analysing gene expression, especially for intracellular organisms, as well as with the recent advances in obtaining gene knock-outs through homologous recombination (another technique that was thought at one stage to be impossible), will enable us to understand much more fully the unique characteristics of mycobacterial physiology. This applies especially to features such as cell envelope structure and synthesis, and to the basis of the slow growth and dormancy of these organisms—and flowing from that, the molecular basis of the virulence of mycobacteria. We can expect rapid advances to be made from the integration of molecular biology studies with other techniques such as microbial physiology and immunology.

And yet, there is considerable room for humility. In our frankest and most self-critical moods, we can and should ask ourselves how much impact any of these advances have yet made towards the control of tuberculosis. Target sites have been identified that may prove useful in the development of new antimycobacterial agents, and molecular techniques are available that will assist in testing candidate agents—but

these new agents are not yet with us. Techniques have been developed that may prove useful in the development of new TB vaccines, but we do not yet understand why BCG does not work fully as a vaccine (or indeed whether it works at all), which means that the nature of an effective vaccine remains uncertain. We can identify local outbreaks of TB with a greater degree of certainty (which tells us that it is not a good idea to put a TB patient into an AIDS ward!) but the impact of these advances on population-based epidemiology and control strategies is limited.

Undoubtedly we will see, over the next few years, major advances in our understanding of tuberculosis and other mycobacterial infections, and we hope that this review of the progress that has been made in recent years will help to stimulate and inform those advances. But the real prize awaits those who can employ this knowledge to achieve effective control of this disease.

Colin Ratledge
Jeremy Dale
November 1998

Chapter 1 / Recombination

GRAHAM R. STEWART & JOHNJOE McFADDEN

1 Introduction

Homologous recombination (HR) is ubiquitous among living organisms. Much of what we understand about the mechanisms of recombination has been gained from studies on *Escherichia coli* in which mutants can be readily obtained and recombination easily quantified. However, until recently, it has proved difficult to demonstrate this fundamental of genetics in mycobacteria, particularly in the slow-growing mycobacteria. At its simplest, recombination is the interaction between two lengths of DNA resulting in the exchange, deletion, duplication, insertion or translocation of sequence. General or homologous recombination involves interaction between two identical or near-identical sequences. HR is involved in the repair of DNA, in the segregation of chromosomes into daughter cells and in the maintenance of genetic diversity. HR is utilized by bacteria in such a variety of functions and on such a multitude of substrates that it should be of no surprise that the enzymatic pathways by which it can be achieved are several, involving a complexity of enzymes with a high degree of overlap and interchangability.

The mechanisms of recombination and DNA repair in mycobacteria are of particular interest and importance because the pathogenic species *Mycobacterium tuberculosis* and *Mycobacterium leprae* have evolved to exist in environments that may require unique properties from their recombination machinery. Both these bacteria are intracellular pathogens which can survive and replicate inside phagosomal compartments of macrophages where the bacteria are exposed to DNA-damaging agents such as reactive oxygen and nitrogen intermediates. It is also a quality of these pathogens that they can exist for long periods in an apparent state of dormancy, which may require an unusually fine control of replication. Interference with the processes of maintaining genetic integrity or replication thus provide potential targets for control measures.

A number of approaches can be taken to identify the function of a gene or a gene product. However, the definitive method for assigning functionality and/or contribution to virulence is to create a null mutant which can be complemented by reintroduction of the gene. In many organisms null mutants are easily produced by HR-driven allelic exchange, but this has yet to be routinely achieved in slow-growing mycobacteria. This chapter provides an overview of HR in both fast- and slow-growing mycobacteria. In addition to examining the technical difficulties involved in producing gene knock-outs in mycobacteria, it is attempted to relate experimental observations with information from the *M. tuberculosis* genome project to produce putative biochemical pathways for recombination.

1

2 Genes and enzymes involved in recombination

2.1 Homologous recombination

HR can occur between any two lengths of DNA that carry regions of homology, whether these regions are present on the same molecule or different molecules and irrespective of the source of DNA, e.g. chromosomes, plasmids and phages. In *E. coli,* the mechanisms involved in HR have been split into three pathways: RecBCD, RecF and RecE (Kowalczykowski *et al.* 1994). However, many of the enzymes involved have overlapping functions and the exact pathway of recombination utilized may depend largely on the nature of the DNA substrate. All models of HR require some initial processing of one of the dsDNA molecules to produce a region of ssDNA which is then paired with the dsDNA homologous partner and used to form a heteroduplex joint, which comprises a dsDNA region with one strand donated from each parent molecule. The crossover connection can then migrate along the parent molecules until breaks in the crossover region allow resolution of the junction into recombinant progeny molecules. Whatever the pathway utilized, all general recombination relies on the ability to pair homologous DNA strands and exchange DNA. The RecBCD and RecF pathways rely on RecA protein to promote homologous pairing and exchange of DNA strands. The RecE pathway is RecA independent, relying on the pairing activity of RecT (Kowalczykowski *et al.* 1994).

2.2 Recombination enzymes of mycobacteria

With the information from the *M. tuberculosis* genome project and the ongoing *M. leprae* genome project (Cole *et al.* 1998; see also Chapter 5) it has been possible to identify many mycobacterial homologues of enzymes involved in recombination in *E. coli*. A list of these enzymes and their databank accession numbers is given in Table 1.1. A list of mycobacterial genes involved in both DNA repair and recombination is also available (Mizrahi & Andersen 1998). The most obvious difference is the absence of RecE and RecT homologues in mycobacteria, which is perhaps not surprising as in *E. coli* these enzymes are encoded by the cryptic Rac prophage (Gillen *et al.* 1977). There are, however, homologues of RecBCD and RecF in *M. tuberculosis* and the inference from this is that all recombination in mycobacteria is dependent on RecA-mediated homologous pairing and strand exchange.

2.2.1 RecA

In *E. coli* and other bacteria, RecA is the key enzyme involved in HR but it is also involved in DNA repair, chromosomal replication, mutagenesis, induction of phage λ lysogens and induction of the SOS response (Kowalczykowski *et al.* 1994). It is a complex protein that binds adenosine triphosphate (ATP) and DNA, polymerizing in a helical filament along the DNA and actively performing the homology search and DNA strand exchange (Kowalczykowski & Eggleston 1994). The RecA protein was the first recombination enzyme to be identified in *M. tuberculosis* (Davis *et al.* 1991; Nair & Steyn 1991) and it is also the recombination gene/gene product about which we know most. The structure of the gene was shown to be quite unusual and different from that of *E. coli* in that the coding sequence contained an in frame insertion of coding sequence unrelated to *recA* (Davis *et al.* 1991). This central region or intein is post-translationally excised from the precursor protein and the remaining amino- and carboxy-terminal regions are spliced together to form the mature RecA protein (Davis *et al.* 1991, 1992). Expression of wild-type and mutant *M. tuberculosis recA* genes in *E. coli* demonstrated, firstly, that splicing of the precursor can occur in *E. coli* and, secondly, that only the post-translationally spliced protein was capable of complementing *E. coli recA* mutants in HR and repair of UV-damaged DNA (Davis *et al.* 1992).

While there are a number of examples of inteins in eukaryotic and prokaryotic organisms, they remain a relatively rare occurrence. Inteins from *Saccharomyces cerevisiae* and *Thermococcus littoralis* have been identified as site-specific endonucleases which cut intein-

Table 1.1 Mycobacterial homologues of enzymes involved in homologous recombination.

Enzyme	Activity	*M. tuberculosis* accession no.	*M. leprae* accession no.
RecA	Promotes DNA strand transfer, DNA renaturation, DNA-dependent ATPase, DNA- and ATP-dependent coprotease	Z77165	Z94723 U00019
Exonuclease V subunits: RecB; RecC; RecD	DNA helicase, ATP-dependent dsDNA and ssDNA exonuclease, ATP-stimulated ssDNA endonuclease, DNA-dependent ATPase	Z92772	
RecF	Activities largely unknown, dsDNA, ssDNA and ATP binding	X92504	Z70722 L39923
RecR	Interaction with RecF	AL022121	
RecG	DNA helicase, DNA-dependent ATPase, branch migration of crossover connection	Z83018	
RecN	Activities largely unknown, ATP binding	Z98268	Z95117
RecQ	DNA helicase		U00016
Sbcd	ATP-dependent dsDNA exonuclease	Z77137	
SSB (ssDNA-binding protein)	ssDNA binding	Z80775	L39923
RuvA	Holiday junction DNA helicase, complexes with RuvB	Z77724	U00011
RuvB	Holiday junction DNA helicase, complexes with RuvA	Z77724	U00011
RuvC	Holiday junction resolvase	Z77724	U00011
DNA topoisomerase I (*topA*)	Nicks one strand of a duplex to relax superhelical DNA	Z95436	
Helicase II (*uvrD*)	DNA helicase	Z95120 Z95208	U00016
DNA ligase	DNA ligase	Z93777 Z83018 Z83866 Z74020 Z70283	
DNA polymerase I (*polA*)	DNA polymerase, 5′-3′ exonuclease, 3′-5′ exonuclease	Z95554	Z46257
DNA gyrase Subunits: (*gyrA*; *gyrB*)	DNA gyrase, topoisomerase II, catalyses negative supercoiling	Z80233	Z70722

less DNA at the site of insertion (Gimble & Thorner 1992; Perler *et al.* 1992) and it has been postulated that inteins are selfish genetic elements of no consequence to the host organism. However, the discovery that the *recA* gene of *M. leprae* also encodes an intein but one which is inserted at a different site and encodes a different protein to the *M. tuberculosis* complex intein is indicative that their presence has been independently selected for (Davis *et al.* 1994). Inteins are not present in *recA* genes from other mycobacteria, thus their presence in *recA* appears to be associated with an intracellular pathogenic existence. The *M. tuberculosis recA* (with intein) is capable of complementing *recA-M. smegmatis* with respect to HR and repair of DNA damage (Frischkorn *et al.* 1998). Whether the presence of the intein affects function in *M. tuberculosis* has yet to be answered, but an extra level of regulation for *recA* expression is potentially provided by the requirement for efficient splicing of the protein to gain enzyme activity (Colston & Davis 1994). Other aspects of regulation of RecA and the SOS response are generally similar to those observed in other bacteria. As in *E. coli,* expression of *M. tuberculosis* RecA is induced by DNA damage (Movahedzadeh *et al.* 1997b). In *E. coli,* DNA damage activates RecA which in turn triggers the autocatalytic cleavage of the repressor protein LexA (Little 1984). LexA binds specifically to regions, known as SOS boxes, located upstream of its own gene, *recA* and other SOS genes, down-regulating their expression. Cleaved LexA does not bind DNA efficiently, thus allowing transcription of repressed genes. LexA from *M. tuberculosis* has been cloned and expressed and shown to bind to putative mycobacterial SOS boxes located in the promoter regions of *M. tuberculosis lexA* (Movahedzadeh *et al.* 1997a) and *recA* (Movahedzadeh *et al.* 1997b). The mycobacterial SOS boxes are not like those of *E. coli* but are similar to the Cheo box sequence involved in the regulation of DNA damage-inducible genes in *Bacillus subtilis* (see also Chapter 4). Cheo boxes are also located in the promoter regions of other mycobacterial recombination/DNA repair genes, e.g. *ruvA, ruvB, ruvC* and *recF.* Upstream of the promoter region of the *recA* gene there also appears to be an activator sequence (Movahedzadeh *et al.* 1997b).

2.2.2 *RecBCD and RecF*

The RecBCD pathway is the major pathway for HR in wild-type *E. coli* (Fig. 1.1). Although the exact pathway of recombination is highly dependent on the DNA substrates and on the genetic background of the host, it has been shown that RecA, RecBCD,

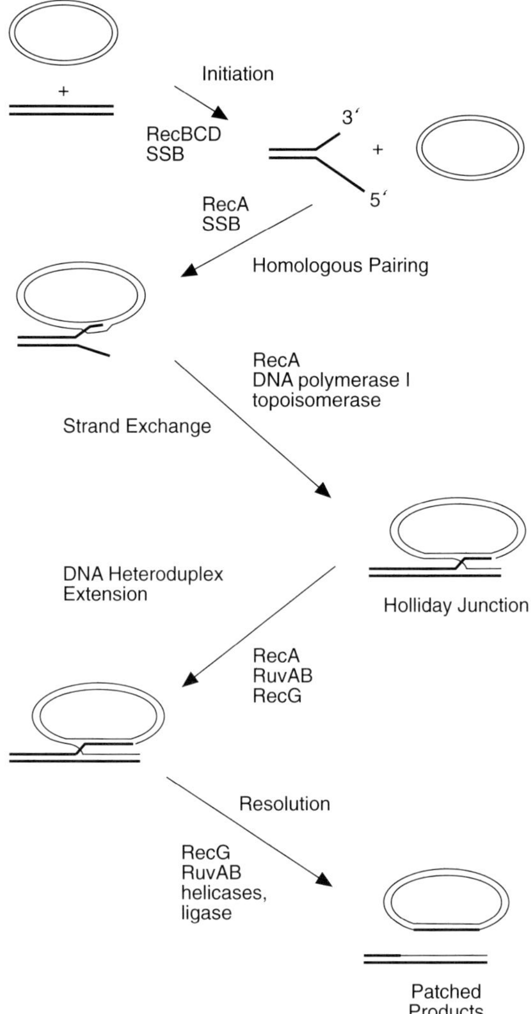

Fig. 1.1 Biochemical model for homologous recombination by RecBCD pathway in *Escherichia coli* (based on Kowalczykowski *et al.* 1994). All enzymes have homologues in *Mycobacterium tuberculosis* (see Table 1.1). Only the generation of a patched product is shown.

single-stranded DNA binding protein (SSB), DNA polymerase I, DNA gyrase, DNA ligase, and either RuvA, RuvB and RuvC, or RecG are all required for recombination (Kowalczykowski *et al.* 1994). Gene homologues for all of these proteins have been identified in *M. tuberculosis,* although with the exception of RecA none of the gene products has been studied to confirm biochemical activity. *recB, recC* and *recD* are the genes coding for the subunits of the RecBCD enzyme, exonuclease V. This enzyme is a DNA helicase with nuclease activity that can form ssDNA from dsDNA and it has been suggested that this is an important source of ssDNA substrate for the formation of the heteroduplex region by RecA (Smith *et al.* 1984). RecBCD has been shown to act preferentially on linear dsDNA rather than covalently closed or nicked circular DNA and has the ability to recognize a specific nucleotide sequence 5'-GCTGGTGG-3' known as a chi site and to cut the DNA strand at or near this (Taylor *et al.* 1985). The presence of chi sites greatly enhances the rate of recombination by the RecBCD pathway, but such sites have yet to be identified in mycobacteria. It seems reasonable to conclude that mycobacteria have a functional RecBCD pathway.

In *E. coli,* the RecF pathway acts predominantly on closed-circular substrates and is dependent on RecA but is entirely independent of RecBCD. For normal function it requires the gene products from *recF, recJ, recN, recO, recQ, recR, ruvA, ruvB, ruvC* and *recG* (Kowalczykowski *et al.* 1994). In mycobacteria, homologues of most of these genes have been identified. One exception to this is RecO which in *E. coli* works with RecF probably during the initial steps of recombination providing ssDNA suitable for RecA function. The exact function of RecO is not understood but it is essential to normal activity of the RecF pathway in *E. coli* (Kowalczykowski *et al.* 1994) and thus in mycobacteria, either an alternative enzyme is utilized or the *recO* homologue has simply been overlooked. A homologue of *recJ* has also yet to be identified in mycobacteria; however, *recN* is present and the gene product is known to have overlapping functions with RecJ. It seems likely that mycobacteria do have a functional RecF pathway, particularly as in the absence of a RecE-like pathway

there is a need for mechanisms to deal with recombination events on covalently closed DNA molecules. As in *E. coli,* the *recF* of mycobacteria is likely to be regulated by LexA, as there is an SOS box in the promoter region of the *recF* operon which is identical to that upstream of the *M. tuberculosis lexA* coding sequence.

2.3 Illegitimate recombination

Illegitimate recombination (IR) is a type of recombination that occurs between DNA molecules sharing no homology or with very short regions of homology, typically four to10 nucleotides. It occurs widely in both prokaryotes and eukaryotes and is responsible for major genome rearrangements including deletions, duplications, translocations and insertions. The mechanisms thought to be involved in IR are varied and in *E. coli* the mechanisms of short-homology-dependent and short-homology-independent IR are quite distinct. Short-homology-independent IR is mediated by subunit exchange of DNA gyrase whereas short-homology-dependent IR involves the formation of double-strand DNA breaks and then processing, annealing and ligation of DNA ends (Shimizu *et al.* 1997).

3 Recombination as a genetic tool

HR is an invaluable aid to the molecular geneticist in both prokaryote and eukaryote systems, providing the means by which DNA can be accurately excised, replaced or inserted into the chromosome. Targeted disruption of genes is commonly achieved using HR in a variety of bacteria (Winans *et al.* 1985; Miller & Mekalanos 1988; Stibitz *et al.* 1989) and is a technique that would be of great value to mycobacterial researchers, particularly since the *M. tuberculosis* genome project alone has provided ≈ 4000 predicted genes which need functional and phenotypic characterization. In principle, a simple approach utilizes a plasmid vector carrying a single homologous region of DNA which is complementary to part of a specific gene and entirely internal of the open reading frame. Introduction of this into the cell and resultant Campbell-type recombination (Campbell 1962) would integrate the vector into the open reading frame of the target gene on the chromosome thus creating two truncated genes (Fig. 1.2a). A more reliable approach involves using a vector with two regions of homologous DNA flanking an antibiotic marker. Recombination between both homologous vector sequences and the chromosome would result in excision of any chromosomal sequence between the homologous regions and integration of the antibiotic marker: allelic exchange (Fig. 1.2b). Mutant phenotypes can be confirmed by reintroduction of the wild type gene on a plasmid.

Similar approaches have been used in other bacteria to introduce promoterless reporter genes such as *lacZ*, *phoA* and *gfp* into the chromosome (Mahan *et al.* 1993; Kalogeraki & Winans 1997). Such fusions between target genes and reporters may be either transcriptional or translational and can be used to measure levels of gene expression. If the reporter is integrated to produce a fusion with the target protein, gene regulation at translational level can also be examined (Kalogeraki & Winans 1997; see also Chapter 4).

In addition to characterizing alleles by HR-mediated gene disruption or replacement, the production of rationally attenuated *M. tuberculosis* candidate vaccines will depend largely on the ability to accurately delete identified virulence genes by HR. The potential for mycobacteria to act in immunotherapy strategies and as carriers for heterologous vaccines may also depend on recombination strategies to produce stable expression of the heterologous antigens/epitopes from defined chromosomal loci.

3.1 Achieving homologous recombination in mycobacteria

The application of HR to mycobacterial genetics has been difficult in practice. The first problem encountered is how to deliver the mutated allele to the chromosome. One solution is to use 'suicide' vectors that, for ease of construction, can replicate in *E. coli*, but cannot replicate in mycobacteria. Thus, following introduction into the mycobacterium, expression of plasmid-borne markers can only be achieved by integration into the chromosome. However, the effectiveness of the 'suicide' strategy is limited by transformation efficiency, a constraint that does not apply to replicating vectors. With replicating vectors there comes a need for positive selection for the recombination event combined with a means of eliminating the plasmid. In other bacteria the use of conditionally replicative vectors has been successful. In this strategy the plasmid is allowed to replicate in the bacterium until a shift to non-permissive conditions, e.g. a rise in temperature, after which the plasmid can no longer replicate and expression of plasmid markers indicates integration into the chromosome. The following sections describe the successes and failures of these and variations on these strategies in mycobacteria.

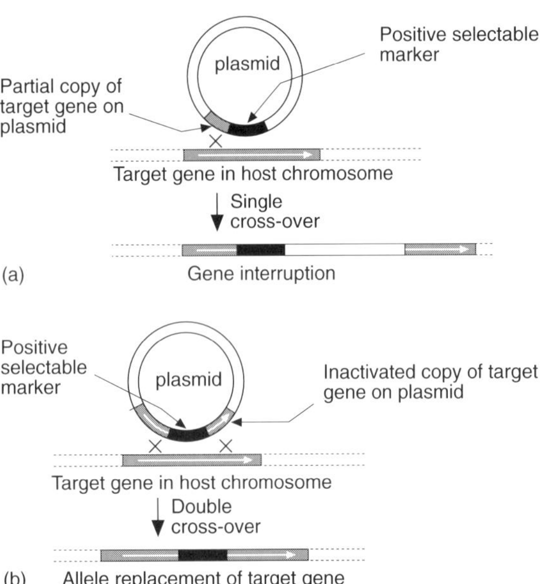

Fig. 1.2 (a) Campbell-type recombination, integrating a plasmid vector into a target gene via a single crossover. If the plasmid-borne homologous sequence is entirely internal of the target gene coding sequence, the resultant mutant will have undergone gene interruption and carry two truncated copies of the target gene. (b) Double crossover between a plasmid carrying two regions of homologous DNA results in allelic exchange/replacement.

3.1.1 *Mycobacterium smegmatis*

Suicide vectors

The first demonstration of HR in mycobacteria utilized the suicide vector strategy to knock out the orotidine monophosphate decarboxylase gene (*pyrF*) of *M. smegmatis* (Husson *et al.* 1990). The suicide vector consisted of pUC19 containing two contiguous regions of the *M. smegmatis pyrF* gene interrupted by the aminoglycoside phosphotransferase gene of Tn*903* (*aph*) which confers resistance to kanamycin (Kan^R) in *E. coli* and mycobacteria. The construct was introduced into *M. smegmatis* by electroporation and 10–500 transformants per microgram plasmid DNA were obtained. Southern blotting of the transformants revealed that ≈ 60% of the integrants had arisen by a single 'Campbell-type' crossover (Campbell 1962) resulting in a mero-diploid strain with one wild-type copy of *pyrF* plus an inactivated copy. The remaining 40% of transformants were derived from double crossover events with just the inactivated *pyrF* in the chromosome. Allelic exchange and creation of the null-mutant was confirmed by demonstration of uracil auxotrophy. Gene replacement was shown to occur most commonly by simultaneous double crossover but could also be derived at a frequency of 10^{-3} from single crossover transformants undergoing a second crossover to loop out the vector sequence.

Gene replacement of the *met* gene (Kalpana *et al.* 1991) and of a σ factor that regulates gene expression in response to certain environmental conditions (an extracytoplasmic function, or ECF, σ factor; Wu *et al.* 1997) of *M. smegmatis* have also been achieved with relative ease using a simple suicide delivery vector. However, other studies have indicated that recombination frequencies can vary considerably in different systems (Sander *et al.* 1995; Gordhan *et al.* 1996). Indeed, although plasmid integration occurs consistently at 10^{-3}–10^{-4}, the proportion of double crossover events appears to vary considerably from 0 to 40% and as a result various attempts have been made to increase the probability of isolating clones showing gene replacement. The incorporation of a counterselectable marker in the suicide plasmid provides a means by which single crossover plasmid integrants can be selected against, to leave only double crossover transformants that contain the mutated target gene but no other plasmid sequence (Fig. 1.3). This approach was adopted by Sander *et al.* (1995) who utilized the wild-type *rpsL* gene of *M. bovis* to confer dominant streptomycin sensitivity on *rpsL*⁻ streptomycin-resistant *M. smegmatis*. The *rpsL* gene was included in the suicide plasmid outside a mutated *pyrF* gene which contained the *aph* gene. Following transformation into *M. smegmatis*, double crossovers were selected for by a single-step double

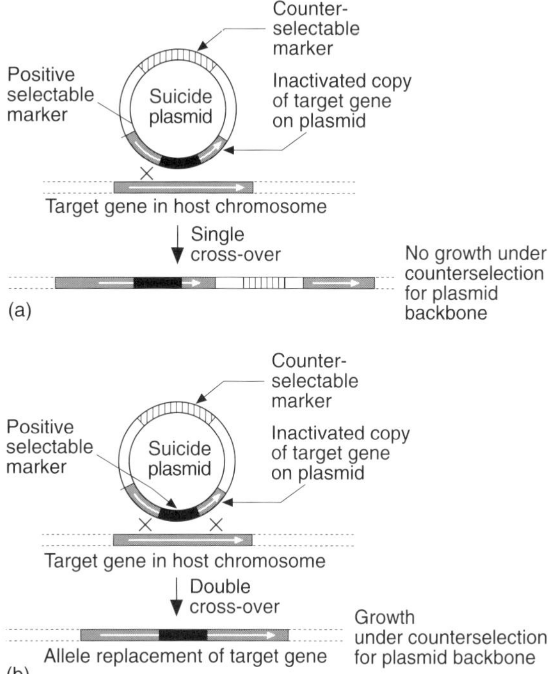

Fig. 1.3 Strategy to select for allele replacement utilizing a suicide vector containing a positive selectable marker flanked by two homologous DNA regions and with a counterselectable marker on the plasmid backbone. Simultaneous selection for both the positive and counterselectable markers prohibits growth of single-crossover mutants (a) but permits growth of double-crossover mutants (b).

selection on kanamycin and streptomycin. The Kan[R] Sm[R] transformants were shown to have undergone gene replacement of the *pyrF* gene as all were uracil auxotrophs and all were shown by Southern blotting to contain a single copy of the target gene. This strategy can be used with other markers. For example, the catalase–peroxidase gene (*katG*) of mycobacteria confers dominant isoniazid sensitivity on isoniazid-resistant mycobacteria that have a deleted *katG* (Norman *et al.* 1995). However, the drawback with such strategies is the need for host strains that carry antibiotic resistance markers.

The *sacB* gene encodes the *B. subtilis* secreted enzyme levansucrase (Gay *et al.* 1983) which catalyses the hydrolysis of sucrose and synthesis of levans (Dedonder 1996). In Gram-negative bacteria, *sacB* has been commonly used as a counterselectable marker (Ried & Collmer 1987; Hynes *et al.* 1989) as its expression in the presence of sucrose is lethal, although the biochemical basis for this lethality is not fully understood. Subsequently, the effectiveness of *sacB* as a counterselectable marker in Gram-positive bacteria including mycobacteria has been demonstrated (Jager *et al.* 1992; Pelicic *et al.* 1996a; see also Chapter 2). The major advantage of *sacB* is that it does not require an antibiotic-resistant host. When expressed on a plasmid in *M. smegmatis* and *M. bovis* bacille Calmette–Guérin (BCG) its selection efficiency on 10% sucrose was $\approx 5 \times 10^{-4}$ and $> 10^{-6}$, respectively (Pelicic *et al.* 1996a). When *sacB* was tested on a suicide plasmid containing the *aph* interrupted *pyrF* gene, all *M. smegmatis* transformants selected in a single-step double selection on kanamycin and sucrose were uracil auxotrophs demonstrating allele replacement (Pelicic *et al.* 1996b). It was also demonstrated that sequential selection of transformants on kanamycin and sucrose in a two-step procedure could be utilized to resolve gene replacement from Campbell-type integration of the plasmid and that this strategy could be used to create unmarked mutations in *M. smegmatis*. Unmarked mutations are often desirable as the incorporation of antibiotic markers in a mutant strain can restrict the choice of marker for further genetic manipulation and can restrict the use of the mutant

in, for example, vaccination studies. The counterselectable nature of the *pyrF* gene (Husson *et al.* 1990) itself has also been utilized to create strategies for integration of unmarked heterologous DNA into the chromosome of *M. smegmatis* (Knipfer *et al.* 1997). PyrF converts 5′-fluoro-orotic acid (5′-FOA) into a toxic nucleotide analogue that irreversibly inactivates thymidylate synthetase. A suicide plasmid containing the *aph* gene and also incorporating a cassette of heterologous DNA flanked by sequence homologous to the mycobacterial regions immediately upstream and downstream of the *pyrF* coding sequence, is introduced into *M. smegmatis*. Selection on kanamycin for single crossover integration of the plasmid is followed by selection on 5′-FOA for a second recombination event that loops out the vector sequence and the wild-type *pyrF* to leave the heterologous DNA accurately replacing the *pyrF* coding sequence (Knipfer *et al.* 1997).

Transformation with a linearized suicide plasmid or fragment of DNA can potentially increase the probability of detecting allele replacement because the linear DNA can be incorporated into the chromosome only by a double crossover and not a single (unless it recircularizes). In *E. coli* this strategy is inefficient unless the linear DNA contains chi sites (Dabert & Smith 1997) because the exonuclease activity of recBCD (exonuclease V) degrades foreign DNA with double-stranded ends (Kowalczykowski *et al.* 1994). In recBC and recD mutants transformation efficiency is restored (Symington *et al.* 1985; Russel *et al.* 1989). However, although *M. smegmatis* has an enzyme with exonuclease V-like properties (Winder & Barber 1973), and *M. tuberculosis* has genes encoding all three subunits of the enzyme (see Table 1.1), there was, in *M. smegmatis*, no reported difference in the rate of allelic exchange from circular compared with linear donor fragments (Kalpana *et al.* 1991; Gordhan *et al.* 1996).

To date HR leading to allele replacement in *M. smegmatis* has been achieved on a number of separate occasions using suicide vectors. Although only a limited number of alleles have been investigated, it can be estimated that the frequency of HR-mediated integration events using suicide plasmid vectors appears to be in the order of 10^{-4}–10^{-3}. The frequency of double crossover events leading to allele replacement appears more variable in different systems but is probably in the range of 10^{-6}–10^{-4}. How far these figures can be extrapolated to recombination at other loci remains to be seen.

Replicating vectors

Recently, a thermosensitive replicating vector has been used in conjunction with *sacB* to facilitate gene replacement of the *M. smegmatis* aminoglycoside 2′-*N*-acetyltransferase gene, *aac(2′)-Id* (Ainsa *et al.* 1997). A plasmid carrying a thermosensitive mycobacterial replicon, *sacB,* and an *aph* interrupted *aac(2′)-Id* was electroporated into *M. smegmatis* and transformants selected on kanamycin at 32°C. A single colony was grown in broth culture at 32°C and then plated out on kanamycin and 10% sucrose at the non-permissive temperature of 42°C. Several Kan[R], sucrose[R], thermoresistant colonies were obtained and one was demonstrated to have undergone gene replacement of *aac(2′)-Id* with the inactivated gene. The use of replicating vectors to deliver mutant alleles removes transformation efficiency as a limiting factor in the detection of recombination events. The problem with replicating vectors is separating the recombination phenotype from that of the plasmid and then eliminating the plasmid from the system to produce cured mutants. This study (Ainsa *et al.* 1997) demonstrates that by using a thermosensitive replicon and a counterselectable plasmid marker, plasmid-free gene replacement mutants can be selected for in a single step. Such a strategy will prove useful in isolating allelic exchange in systems where the frequency of double crossover is too low to be detected using a suicide delivery system.

3.1.2 *Mycobacterium tuberculosis complex*

Suicide vectors

In comparison to the fast-growing *M. smegmatis*, it has proved more difficult to harness HR for genetic techniques in slow-growing mycobacteria. The major

obstacles appear to be high levels of IR combined with low-efficiency transformation. Kalpana *et al.* (1991) constructed a suicide plasmid with an *aph*-interrupted *met* gene, mutations in which result in methionine auxotrophy. Introduction of the circular construct into *M. bovis* BCG and *M. tuberculosis* by electroporation resulted in little or no integration into the chromosome; however, the linearized plasmid integrated at a frequency of 10^{-5}–10^{-4} relative to the transformation frequency of a replicating vector. Southern blotting revealed that all the transformants had intact *met* genes and that the plasmid had integrated randomly in the chromosome, presumably via the linearized ends. To further demonstrate the illegitimate nature of this integration, linear DNA with no significant homology to mycobacterial sequences was shown to incorporate into the chromosomes of BCG and *M. tuberculosis* at a similar frequency; however, it did not integrate in *M. smegmatis*.

Following this discouraging study, a linearized suicide vector was used to introduce, by HR, a disrupted *uraA* gene into the chromosome of BCG (Aldovini *et al.* 1993). Although illegitimate recombination was again the dominant event, two out of 10 transformants analysed by Southern blot had integrated the whole plasmid into the chromosome at the *uraA* locus by a Campbell-type single crossover. This could only have occurred if the linearized plasmid had recircularized by illegitimate recombination, a common event in *E. coli* (Jasin & Schimmel 1984).

Successful allelic exchange using a suicide vector in slow-growing mycobacteria was first demonstrated by replacement of the *ureC* gene in *M. bovis* BCG (Reyrat *et al.* 1995). An *aph*-disrupted *ureC* was delivered on a linearized suicide plasmid and ≈ 4% of the Kan[R] transformants had undergone gene replacement. The frequencies of single crossover and illegitimate integration of the plasmid were not reported. The *ureC* null mutant provided an easily identifiable phenotype since techniques are available for simple, rapid selection of urease-negative colonies. However, it remains unresolved as to why the earlier attempts using similar strategies to replace the *uraA* and *met* genes had failed. Certainly some genes, or some regions of DNA, will be less recombinogenic than others and both *uraA* and *met* may have frequencies of recombination that preclude detection of gene replacement utilizing suicide delivery. Isolation of single-crossover events is much more likely than double crossovers but one study showed that by using a two-step strategy with an *aph*-disrupted *ureC* and the counterselectable marker, *sacB,* allelic exchange could be efficiently resolved from single-crossover-derived Kan[R] transformants (Pelicic *et al.* 1996c). This experiment utilized a closed circular plasmid which integrated at a frequency of 2×10^2 per microgram DNA and with no detected illegitimate recombination. Such high frequencies of integration of closed circular plasmids have not been observed in any other studies, suggesting that *ureC* is indeed highly recombinogenic. The study also demonstrates that the use of circular delivery vectors eliminates the overwhelming occurrence of illegitimate recombination.

The length of homologous sequence is also of critical importance in obtaining efficient HR. In both *B. subtilis* (Khasanov *et al.* 1992) and *E. coli* (Watt *et al.* 1996) the minimum length of homology appears to be about 70 bp and in *Campylobacter coli* $\approx$ 290 bp are required (Richardson & Park 1997). Above these values, the probability of HR events is generally found to be proportional to the length of homologous DNA taking part in the exchange. Balasubramanian *et al.* (1996) therefore used long linear homologous fragments of DNA (> 40 kb) excised from cosmid vectors as donors for gene replacement. Transformation of *M. tuberculosis* strains with both circular and linearized long DNA fragments, containing a copy of the leucine biosynthetic gene *leuD* inactivated by *aph* gene insertion, resulted in a relatively high recovery of plasmid integrants (about $1–3 \times 10^2$ per microgram DNA). However, only cells transformed with the linearized DNA gave rise to leucine auxotrophs, at a frequency of 6% of plasmid integrants and these were shown to have arisen as products of HR leading to allele exchange. The *leu*[+] Kan[R] transformants were found to be a mixture of spontaneous kanamycin-resistant mutants and illegitimate plasmid integrants. Transformation with smaller *leuD::aph* fragments resulted in no detectable HR.

Thus, the use of long substrates can enable the isolation of allelic exchange at apparently poorly recombinogenic loci. However, the manipulation of large regions of DNA and cosmids for the construction of recombination plasmids is relatively difficult. A compromise approach was adopted by Azad *et al.* (1996, 1997) whereby smaller homologous DNA fragments ($\approx$ 8 kb) were subcloned from cosmids into pUC19 and internal regions of $\approx$ 2 kb deleted and replaced with the hygromycin B phosphotransferase gene, *hyg*, from *Streptomyces hygroscopicus*. Using this strategy with a linearized plasmid, the mycocerosic acid synthase gene, *mas,* was replaced in *M. bovis* BCG in one of 38 hygromycin-resistant transformants with 8% single-crossover integration (Azad *et al.* 1996). No transformants were obtained with a circular plasmid. Using the same strategy but with the addition of *sacB* as a counterselectable marker, the gene cluster responsible for biosynthesis of phthiocerol and phenolphthiocerol, *pps*, was also disrupted by allelic exchange at a remarkable frequency of 12% of the transformants (Azad *et al.* 1997). Using shorter homologous DNA fragments also has the advantage that transformants can be rapidly screened for specific recombination by polymerase chain reaction (PCR) across the site of recombination from chromosomal regions into the marker gene or mutation. There are several more recent examples of successful gene replacement using linearized suicide vectors, e.g. an acyl-CoA synthase gene in BCG (Fitzmaurice & Kolattukudy 1998), the 16-kDa α-crystallin gene in *M. tuberculosis* (Yuan *et al.* 1998) and the antioxidant gene *noxR1* in *M. tuberculosis* and BCG (Stewart *et al.* unpublished data).

In summary, genes from bacteria of the *M. tuberculosis* complex can in many cases be mutagenized by delivery of a mutated allele from a suicide vector. In most experiments, linearized substrates are favoured for HR-mediated integration of plasmids and allelic exchange. It is notable that linear plasmid gives rise to a surprisingly high rate of single-crossover integration. To achieve this the linear plasmid must recircularize but the exact nature of this integration is not clear as linear plasmid can give rise to a higher frequency of single-crossover integration events than circular plasmid. The complicating factor of linear substrates is that in *M. tuberculosis* complex they give rise to high levels of illegitimate recombination which necessitates the need for extensive screening of transformants unless a counterselectable plasmid marker is used and single-event double crossover occurs at a high enough frequency to be selected for in a single-step strategy. It is also true that a number of genes have proved poorly recombinogenic and their replacement by suicide vector strategies has failed due to low frequency transformation. Recent advances in transformation procedures of *M. tuberculosis* complex species (Wards & Collins 1996) may help to abrogate this problem.

Replicating vectors

In an attempt to overcome the reliance of HR on transformation efficiency, Norman *et al.* (1995) reasoned that the use of a replicating vector would remove transformation as a limiting step and would also extend the time that plasmid DNA is available for recombination; the higher plasmid copy number would also increase the opportunity for recombination. The replicating vector consisted of a pUC18 backbone with the mycobacterial origin of replication from pAL5000 (Snapper *et al.* 1988) and an *aph*-interrupted target gene, *accBC*, which codes for a subunit of acyl-CoA carboxylase (Norman *et al.* 1994). To select against single crossovers but for allelic replacement in the absence of free plasmid, the counterselectable marker, *katG* (Zhang *et al.* 1992) was included on the plasmid and the host strain was a *katG*-deleted BCG. In transformant colonies with a translucent appearance, allelic exchange was detected by PCR but a cured stable mutant was not obtained. The problem appeared to be associated with stable persistence of the replicating plasmid, despite the counterselection provided by *katG*, and also by the possibly lethal consequences of loss of this gene.

The use of replicating vectors to deliver the mutated allele requires that the allelic replacement event can be selected for by a phenotype distinct from that conferred by free plasmid. Baulard *et al.* (1996)

constructed an *E. coli–Mycobacterium* shuttle vector containing a streptomycin resistance gene and a promoterless *aph* gene immediately downstream of a promoterless fragment of the BCG *hsp60* gene. The free plasmid confers streptomycin resistance (Sm^R) but not kanamycin resistance. However, Campbell-type integration of the plasmid at the chromosomal *hsp60* locus would enable expression of the *aph* gene from the *hsp60* promoter on the chromosome, thus providing a means of positive selection for the recombination event. Following electroporation of the plasmid into *M. smegmatis*, 10% of the Sm^R transformants were also Kan^R and were shown by PCR to have undergone Campbell-type integration at the *hsp60* locus. In BCG, no Sm^R transformants were Kan^R but following broth culture in medium supplemented with streptomycin and subsequent plating onto kanamycin-containing medium, several hundred Kan^R colonies were obtained and PCR and Southern analysis confirmed Campbell-type plasmid integration. This experiment demonstrated that the combination of replicating vectors for efficient delivery of recombination substrates, and positive selection for the recombination event itself provides an effective means for achieving HR and could be elaborated to produce gene replacement. However, it is only feasible using target genes which are expressed *in vitro* and it is likely to require the use of an efficient counterselectable plasmid marker to produce cured mutants.

A system which utilizes both a counterselectable plasmid marker and a conditionally replicative origin of replication has now been developed and shown to be an effective mechanism for gaining cured allelic exchange mutants in *M. tuberculosis* (Pelicic *et al.* 1997). The system utilizes a plasmid containing a thermosensitive origin of replication and the counterselectable marker *sacB*. The plasmid carried an *aph*-interrupted copy of the poorly recombinogenic *purC* gene and was introduced into *M. tuberculosis* by electroporation; transformants were grown on kanamycin at 32°C. Neither the thermosensitive origin nor the *sacB* marker was individually capable of effective counterselection for the plasmid. However, selection for Kan^R on 2% sucrose at the non-permissive temperature of 39°C yielded 100% plasmid-free allelic exchange mutants with the expected purine auxotrophic phenotype.

3.1.3 Other mycobacteria

High levels of illegitimate recombination are not a feature of all slow-growing mycobacteria. Efficient HR and gene replacement was demonstrated in *M. intracellulare* (Marklund *et al.* 1995), a slow-growing mycobacterium closely related to *M. avium*. A mycobacteriophage integrative vector carrying a gentamicin resistance gene, *aacC1*, was electroporated into *M. intracellulare*. A non-integrative suicide plasmid carrying an *aph*-interrupted *aacC1* was then introduced in circular and linear forms and single-crossover Gen[R] Kan[R] transformants were detected from both substrates. Further growth allowed for detection of a second crossover event leading to gene replacement of the *aacC1* gene and a Gen[s] Kan[R] phenotype. A similar two-step selection was also used to introduce an unmarked mutation into the *katG* gene of *M. intracellulare* (Marklund *et al.* 1998). Unlike the *M. tuberculosis* complex, *M. intracellulare* appears to be similar to *M. smegmatis* in that HR can occur relatively readily between the chromosome and both circular and linearized plasmid substrates, and there is no dominant illegitimate integration of linear substrates.

Allelic replacement has also been demonstrated in the animal and human pathogen, *M. marinum* (Ramakrishnan *et al.* 1997). A circular suicide plasmid carrying *sacB* and 9.2 kb of homologous sequence surrounding a Tn*3-aph* cassette was used to disrupt the photoene synthase gene, *crtB*, by HR. Following transformation of the bacteria, Campbell-type plasmid integration had occurred in all Kan[R], sucrose-sensitive colonies and at a variable frequency of 5×10^{-5}–7.5×10^{-3} relative to the transformation efficiency with a replicating plasmid. Further selection on 10% sucrose and kanamycin showed that a second crossover, looping out the *sacB*-containing sequence and effecting allelic exchange, occurred at a frequency of about 2×10^{-4}.

4 Conclusions

The above experiments have demonstrated that both fast-growing mycobacteria like *M. smegmatis* and slow growers such as the *M. tuberculosis* complex are capable of HR. As a result of the recent *M. tuberculosis* and *M. leprae* genome projects, it is possible to conclude that mycobacteria have pathways of recombination analogous to the RecF and RecBCD pathways of *E. coli*. However, there is still a great paucity of knowledge concerning recombination in mycobacteria. While experimental observations indicate that recombination in *M. smegmatis* has similar characteristics to that in *E. coli*, a number of important differences have been highlighted between the *M. tuberculosis* complex and *M. smegmatis*.

1 The frequency of HR in the *M. tuberculosis* complex is lower than in *M. smegmatis*.

2 In all except one experiment with *M. tuberculosis* complex, HR has occurred only between linear DNA substrates and the chromosome and not closed circular substrates and the chromosome. In *M. smegmatis* recombination between the chromosome and linear or circular substrates occurs at equal frequency.

3 In the *M. tuberculosis* complex there is high-frequency IR of linear DNA substrates into the chromosome and low-frequency IR of closed circular substrates. In *M. smegmatis* there is low-frequency illegitimate recombination of both linear and closed circular substrates.

To explain these differences one is instantly drawn to the most obvious difference between *M. tuberculosis* (and *M. leprae*) and other mycobacteria including *M. smegmatis*: the presence of an intein in RecA. The presence of inteins in *M. tuberculosis* and *M. leprae* RecA has been independently selected for and it has been proposed that its presence provides a level of regulation beneficial to an intracellular, pathogenic way of life (Davis *et al.* 1994). If this regulation results in reduced levels of activated RecA in the cell, LexA cleavage will be reduced leading to repression of *recA*, *lexA* and all LexA-regulated genes (SOS genes) including *recF* (Fig. 1.4). As both RecF and RecBCD pathways are dependent on RecA, the reduction in

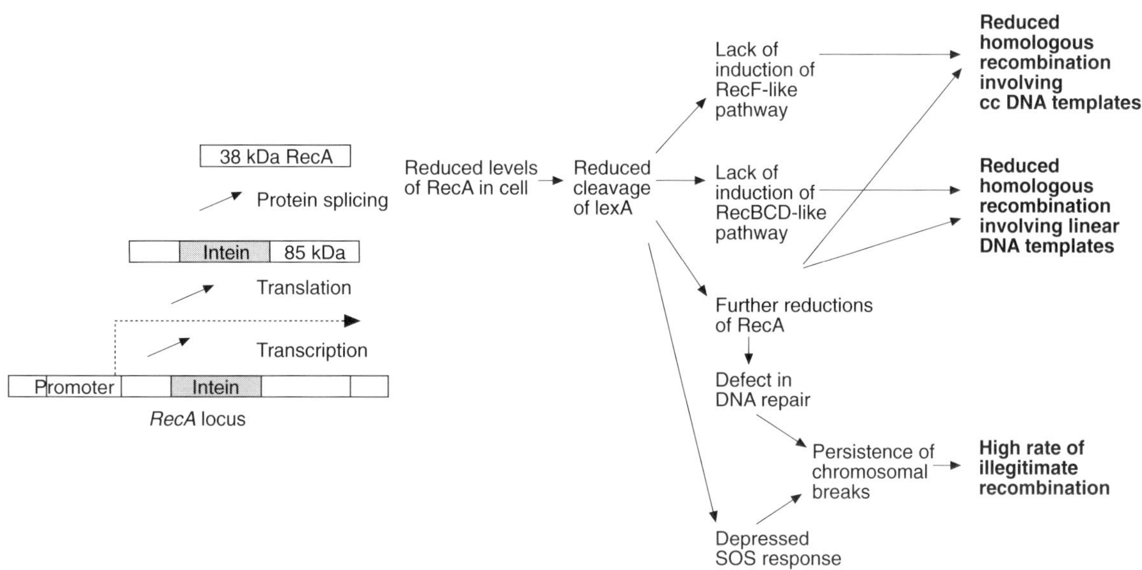

Fig. 1.4 Model to explain the deficiencies in homologous recombination but high level illegitimate recombination in *Mycobacterium tuberculosis* species complex.

RecA leads directly to repression of both pathways, thus explaining the low frequency of recombination observed in *M. tuberculosis* complex compared with *M. smegmatis* which has no RecA intein. The RecF pathway will be further repressed as the reduced cleavage of LexA will repress expression of *recF*, *ruvA*, *ruvB*, *ruvC* and probably other genes involved, via LexA binding to Cheo-like SOS boxes. As the RecF pathway in *E. coli* deals predominantly with recombination of closed circular substrates, this may explain the observed deficiency in this process by *M. tuberculosis* complex. Genes of the RecBCD pathway, e.g. *ruvA*, *ruvB*, *ruvC* and probably *ssb,* will also be repressed by reduced LexA cleavage. If RecBCD processes DNA with single-stranded ends as in *E. coli*, this would lead to poor processing of linear DNA substrates, which is true in *M. tuberculosis* complex although HR with linear substrates appears to be less repressed than with circular substrates. Against the above explanation is the fact that, in *recA- M. smegmatis*, the *M. tuberculosis recA* fully complements with respect to HR (Frischkorn *et al.* 1998). However, until a *recA- M. tuberculosis* strain is available to directly

compare the performance of RecA expressed with and without the intein in its native cell, it will not be possible to conclusively determine the effect of the intein on recombination.

It is not fully understood how illegitimate recombination occurs, but it has been suggested that it requires end-to-end joining of DNA strands. The high levels of IR observed in *M. tuberculosis* between linear DNA and the chromosome would thus require breaks to be present in chromosomal DNA. Chromosomal breaks arising from radiation damage and errors introduced during DNA metabolism would normally be repaired by systems involving RecA and RecF. Thus, decreased levels of RecA in the cell combined with depressed RecF activity may lead to persistence of chromosomal breaks whose ends may then act as targets for IR events (McFadden 1996). It has yet to be resolved whether the observed IR in mycobacteria is short-homology dependent or short-homology independent.

Despite problems with illegitimate recombination and low-frequency HR, the molecular geneticist is now equipped with techniques which make allelic exchange a realistic, if not yet routine, option. This should enable the characterization of genes involved in virulence and pathogenicity and provides a means by which the genomes of mycobacteria may be manipulated to create rationally attenuated vaccines. However, there is still a great paucity of knowledge concerning recombination and DNA repair in mycobacteria and its relevance in *M. leprae* and *M. tuberculosis* to an intracellular, pathogenic way of life.

5 References

Ainsa, J.A., Perez, E., Pelicic, V., Berthet, F.X., Gicquel, B. & Martin, C. (1997) Aminoglycoside 2′-N-acetyltransferase genes are universally present in mycobacteria: Characterization of the aac (2′) -lc gene from *Mycobacterium tuberculosis* and the aac (2′) -ld gene from *Mycobacterium smegmatis*. *Molecular Microbiology* **24**, 431–441.

Aldovini, A., Husson, R.N. & Young, R.A. (1993) The *uraA* locus and homologous recombination in *Mycobacterium bovis* BCG. *Journal of Bacteriology* **175**, 7282–7289.

Azad, A.K., Sirakova, T.D., Fernandes, N.D. & Kolattukudy, P.E. (1997) Gene knockout reveals a novel gene cluster for the synthesis of a class of cell wall lipids unique to pathogenic mycobacteria. *Journal of Biological Chemistry* **272**, 16741–16745.

Azad, A.K., Sirakova, T.D., Rogers, L.M. & Kolattukudy, P.E. (1996) Targeted replacement of the mycocerosic acid synthase gene in *Mycobacterium bovis* BCG produces a mutant that lacks mycosides. *Proceedings of the National Academy of Sciences of the USA* **93**, 4787–4792.

Balasubramanian, V., Pavelka, M.S., Bardarov, S.S. *et al.* (1996) Allelic exchange in *Mycobacterium tuberculosis* with long linear recombination substrates. *Journal of Bacteriology* **178**, 273–279.

Baulard, A., Kremer, L. & Locht, C. (1996) Efficient homologous recombination in fast-growing and slow-growing mycobacteria. *Journal of Bacteriology* **178**, 3091–3098.

Campbell, A. (1962) Episomes. *Advances in Genetics* **11**, 101–145.

Cole, S.T., Brosch, R., Parkhill, J. *et al.* (1998) Deciphering the biology of *Mycobacterium tuberculosis* from the complete genome sequence. *Nature* **393**, 537–544.

Colston, M.J. & Davis, E.O. (1994) The ins and outs of protein splicing elements. *Molecular Microbiology* **12**, 359–363.

Dabert, P. & Smith, G.R. (1997) Gene replacement with linear DNA fragments in wild-type *Escherichia coli*: Enhancement by chi sites. *Genetics* **145**, 877–889.

Davis, E.O., Jenner, P.J., Brooks, P.C., Colston, M.J. & Sedgwick, S.G. (1992) Protein splicing in the maturation of *M. tuberculosis* recA protein: a mechanism for tolerating a novel class of intervening sequence. *Cell* **71**, 201–210.

Davis, E.O., Sedgwick, S.G. & Colston, M.J. (1991) Novel structure of the recA locus of *Mycobacterium tuberculosis* implies processing of the gene product. *Journal of Bacteriology* **173**, 5653–5662.

Davis, E.O., Thangaraj, H.S., Brooks, P.C. & Colston, M.J. (1994) Evidence of selection for protein introns in the recAs of pathogenic mycobacteria. *EMBO Journal* **13**, 699–703.

Dedonder, R. (1996) Levansucrase from *Bacillus subtilis*. *Methods in Enzymology* **8**, 500–505.

Fitzmaurice, A.M. & Kolattukudy, P.E. (1998) An acyl-CoA synthase (acoas) gene adjacent to the mycocerosic acid synthase (mas) locus is necessary for mycocerosyl lipid synthesis in *Mycobacterium tuberculosis* var. bovis BCG. *Journal of Biological Chemistry* **273**, 8033–8039.

Frischkorn, K., Sander, P., Scholz, M., Teschner, K., Prammananan, T. & Bottger, E.C. (1998) Investigation of mycobacterial *recA* function: protein introns in the RecA of pathogenic mycobacteria do not affect competency for homologous recombination. *Molecular Microbiology* **29**, 1203–1214.

Gay, P., Lecoq, D., Steinmetz, M., Ferrari, E. & Hoch, J.A. (1983) Cloning structural gene sacB, which codes for exoenzyme levansucrase of *Bacillus subtilis*—expression of the gene in *Escherichia coli*. *Journal of Bacteriology* **153**, 1424–1431.

Gillen, J.R., Karu, A.E., Nagaishi, H. & Clark, A.J. (1977) Characterization of the deoxyribonuclease determined by lambda reverse as exonuclease VIII of *Escherichia coli*. *Journal of Molecular Biology* **113**, 27–41.

Gimble, F.S. & Thorner, J. (1992) Homing of a DNA endonuclease gene by meiotic gene conversion in *Saccharomyces cerevisiae*. *Nature* **357**, 301–306.

Gordhan, B.G., Andersen, S.J., De Meyer, A.R. & Mizrahi, V. (1996) Construction by homologous recombination and phenotypic characterization of a DNA polymerase domain polA mutant of *Mycobacterium smegmatis*. *Gene* **178**, 125–130.

Husson, R.N., James, B.E. & Young, R.A. (1990) Gene replacement and expression of foreign DNA in mycobacteria. *Journal of Bacteriology* **172**, 519–524.

Hynes, M.F., Quandt, J., O'Connell, M.P. & Puhler, A. (1989) Direct selection for curing and deletion of rhizobium plasmids using transposons carrying the *Bacillus subtilis* sacB gene. *Gene* **78**, 111–120.

Jager, W., Schafer, A., Puhler, A., Labes, G. & Wohlleben, W. (1992) Expression of the *Bacillus subtilis* sacB gene leads to sucrose sensitivity in the gram-positive bacterium *Corynebacterium glutamicum* but not in *Streptomyces lividans*. *Journal of Bacteriology* **174**, 5462–5465.

Jasin, M. & Schimmel, P. (1984) Deletion of an essential

gene in *Escherichia coli* by site-specific recombination with linear DNA fragments. *Journal of Bacteriology* **159**, 783–786.

Kalogeraki, V.S. & Winans, S.C. (1997) Suicide plasmids containing promoterless reporter genes can simultaneously disrupt and create fusions to target genes of diverse bacteria. *Gene* **188**, 69–75.

Kalpana, G.V., Bloom, B.R. & Jacobs, W.R. Jr (1991) Insertional mutagenesis and illegitimate recombination in mycobacteria. *Proceedings of the National Academy of Sciences of the USA* **88**, 5433–5437.

Khasanov, F.K., Zvingila, D.J., Zainullin, A.A., Prozorov, A.A. & Bashkirov, V.I. (1992) Homologous recombination between plasmid and chromosomal DNA in *Bacillus subtilis* requires approximately 70 bp of homology. *Molecular General Genetics* **234**, 494–497.

Knipfer, N., Seth, A. & Shrader, T.E. (1997) Unmarked gene integration into the chromosome of *Mycobacterium smegmatis* via precise replacement of the *pyrF* gene. *Plasmid* **37**, 129–140.

Kowalczykowski, S.C., Dixon, D.A., Eggleston, A.K., Lauder, S.D. & Rehrauer, W.M. (1994) Biochemistry of homologous recombination in *Escherichia coli*. *Microbiological Reviews* **58**, 401–465.

Kowalczykowski, S.C. & Eggleston, A.K. (1994) Homologous pairing and DNA strand-exchange proteins. *Annual Review of Biochemistry* **63**, 991–1043.

Little, J.W. (1984) Autodigestion of lexA and phage-lambda repressors. *Proceedings of the National Academy of Sciences of the USA* **81**, 1375–1379.

Mahan, M.J., Slauch, J.M. & Mekalanos, J.J. (1993) Selection of bacterial virulence genes that are specifically induced in host tissues. *Science* **259**, 686–688.

Marklund, B.I., Mahenthiralingam, E. & Stokes, R.W. (1998) Site-directed mutagenesis and virulence assessment of the *katG* gene of *Mycobacterium intracellulare*. *Molecular Microbiology* **29**, 999–1008.

Marklund, B.I., Speert, D.P. & Stokes, R.W. (1995) Gene replacement through homologous recombination in *Mycobacterium intracellulare*. *Journal of Bacteriology* **177**, 6100–6105.

McFadden, J. (1996) Recombination in mycobacteria. *Molecular Microbiology* **21**, 205–211.

Miller, V.L. & Mekalanos, J.J. (1988) A novel suicide vector and its use in construction ofsertion mutations—osmoregulation of outer-membrane proteins and virulence determinants in *Vibrio cholerae* requires *toxR*. *Journal of Bacteriology* **170**, 2575–2583.

Mizrahi, V. & Andersen, S.J. (1998) DNA repair in *Mycobacterium tuberculosis*. What have we learnt from the genome sequence? *Molecular Microbiology* **29**, 1331–1339.

Movahedzadeh, F., Colston, M.J. & Davis, E.O. (1997a) Characterization of *Mycobacterium tuberculosis* LexA: Recognition of a Cheo (*Bacillus*-type SOS) box. *Microbiology* **143**, 929–936.

Movahedzadeh, F., Colston, M.J. & Davis, E.O. (1997b) Determination of DNA sequences required for regulated *Mycobacterium tuberculosis* RecA expression in response to DNA-damaging agents suggests that two modes of regulation exist. *Journal of Bacteriology* **179**, 3509–3518.

Nair, S. & Steyn, L.M. (1991) Cloning and expression in *Escherichia coli* of a *recA* homologue from *Mycobacterium tuberculosis*. *Journal of General Microbiology* **137**, 2409–2414.

Norman, E., Dellagostin, O.A., McFadden, J. & Dale, J.W. (1995) Gene replacement by homologous recombination in *Mycobacterium bovis* BCG. *Molecular Microbiology* **16**, 755–760.

Norman, E., Desmet, K.L., Stoker, N.G., Ratledge, C., Wheeler, P.R. & Dale, J.W. (1994) Lipid synthesis in mycobacteria—characterization of the biotin carboxyl carrier protein genes from *Mycobacterium leprae* and *Mycobacterium tuberculosis*. *Journal of Bacteriology* **176**, 2525–2531.

Pelicic, V., Jackson, M., Reyrat, J.M., Jacobs, W.R., Gicquel, B. & Guilhot, C. (1997) Efficient allelic exchange and transposon mutagenesis in *Mycobacterium tuberculosis*. *Proceedings of the National Academy of Sciences of the USA* **94**, 10955–10960.

Pelicic, V., Reyrat, J.-M. & Gicquel, B. (1996a) Expression of the *Bacillus subtillis sacB* gene confers sucrose sensitivity on mycobacteria. *Journal of Bacteriology* **178**, 1197–1199.

Pelicic, V., Reyrat, J.M. & Gicquel, B. (1996b) Generation of unmarked directed mutations in mycobacteria, using sucrose counter-selectable suicide vectors. *Molecular Microbiology* **20**, 919–925.

Pelicic, V., Reyrat, J.M. & Gicquel, B. (1996c) Positive selection of allelic exchange mutants in *Mycobacterium bovis* BCG. *FEMS Microbiology Letters* **144**, 161–166.

Perler, F.B., Comb, D.G., Jack, W.E. *et al.* (1992) Intervening sequences in an archaea DNA-polymerase gene. *Proceedings of the National Academy of Sciences of the USA* **89**, 5577–5581.

Ramakrishnan, L., Tran, H.T., Federspiel, N.A. & Falkow, S. (1997) A crtB homolog essential for photochromogenicity in *Mycobacterium marinum*: isolation, characterisation, and gene disruption via homologous recombination. *Journal of Bacteriology* **179**, 5662–5868.

Reyrat, J.-M., Berthet, F.-X. & Gicquel, B. (1995) The urease locus of *Mycobacterium tuberculosis* and its utilization for the demonstration of allelic exchange in *Mycobacterium bovis* bacillus Calmette–Guérin. *Proceedings of the National Academy of Sciences of the USA* **92**, 8768–8772.

Richardson, P.T. & Park, S.F. (1997) Integration of heterologous plasmid DNA into multiple sites on the genome of *Campylobacter coli* following natural transformation. *Journal of Bacteriology* **179**, 1809–1812.

Ried, J.L. & Collmer, A. (1987) An *npti-sacb-sacr* cartridge for constructing directed, unmarked mutations in gram-negative bacteria by marker exchange–eviction mutagenesis. *Gene* **57**, 239–246.

Russel, C.B., Thaler, D.S. & Dahlquist, F.W. (1989) Chromosomal transformation of *Escherichia coli recD* strains with linearized plasmids. *Journal of Bacteriology* **171**, 2609–2613.

Sander, P., Meier, A. & Bottger, E.C. (1995) *rpsl+*: a dominant selectable marker for gene replacement in mycobacteria. *Molecular Microbiology* **16**, 991–1000.

Shimizu, H., Yamaguchi, H., Ashizawa, Y., Kohno, Y., Asami, M., Kato, J. & Ikeda, H. (1997) Short-homology-independent illegitimate recombination in *Escherichia coli*: Distinct mechanism from short-homology-dependent illegitimate recombination. *Journal of Molecular Biology* **266**, 297–305.

Smith, G.R., Amundsen, S.K., Chaudhury, A.M. *et al.* (1984) Roles of *recBC* enzyme and chi sites in homologous recombination. *Cold Spring Harbor Symposia on Quantitative Biology* **49**, 485–495.

Snapper, S.B., Lugosi, L., Jekkel, A. *et al.* (1988) Lysogeny and transformation in mycobacteria: stable expression of foreign genes. *Proceedings of the National Academy of Sciences of the USA* **85**, 6987–6991.

Stibitz, S., Aaronson, W., Monack, D. & Falkow, S. (1989) Phase variation in *Bordetella pertussis* by frameshift mutation in a gene for a novel 2-component system. *Nature* **338**, 266–269.

Symington, l. S., Morrison, P. & Kolodner, R. (1985) Intramolecular recombination of linear DNA catalyzed by the *Escherichia coli* RecE recombination system. *Journal of Molecular Biology* **186**, 515–525.

Taylor, A.F., Schultz, D.W., Ponticelli, A.S. & Smith, G.R. (1985) RecBC enzyme nicking at Chi sites during DNA unwinding: location and orientation dependence of cutting. *Cell* **41**, 153–163.

Wards, B.J. & Collins, D.M. (1996) Electroporation at elevated temperatures substantially improves transformation efficiency of slow-growing mycobacteria. *FEMS Microbiological Letters* **145**, 101–105.

Watt, V.M., Ingles, C.J., Urdea, M.S. & Rutter, W.J. (1996) Homology requirements for recombination in *Escherichia coli*. *Proceedings of the National Academy of Sciences of the USA* **82**, 4768–4772.

Winans, S.C., Elledge, S.J., Krueger, J.H. & Walker, G.C. (1985) Site-directed insertion and deletion mutagenesis with cloned fragments in *Escherichia coli*. *Journal of Bacteriology* **161**, 1219–1221.

Winder, F.G. & Barber, D.S. (1973) Effects of hydroxyurea, nalidixic acid and zinc limitation on DNA polymerase and ATP-dependent deoxyribonuclease activities of *Mycobacterium smegmatis*. *Journal of General Microbiology* **76**, 189–196.

Wu, Q.-L., Kong, D., Lam, K. & Husson, R.N. (1997) A mycobacterial extracytoplasmic function sigma factor involved in survival following stress. *Journal of Bacteriology* **179**, 2922–2929.

Yuan, Y., Crane, D.D., Simpson, R.M. *et al.* (1998) The 16-kDa alpha-crystallin (Acr) protein of *Mycobacterium tuberculosis* is required for growth in macrophages. *Proceedings of the National Academy of Sciences of the USA* **95**, 9578–9583.

Zhang, Y., Heym, B., Allen, B., Young, D. & Cole, S. (1992) The catalase-peroxidase gene and isoniazid resistance of *Mycobacterium tuberculosis*. *Nature* **358**, 591–593.

Chapter 2 / Mobile genetic elements and plasmids: tools for genetic studies

CHRISTOPHE GUILHOT, MARY JACKSON & BRIGITTE GICQUEL

1 Introduction

Studies of plasmids and transposons have provided information about fundamental biological processes such as DNA replication, gene transfer, chromosome rearrangement and mutations. In addition, plasmids and transposons have also been used to develop an extensive array for genetic studies in bacteria.

With the re-emergence of mycobacterial diseases in the 1980s, the drive towards applying molecular genetics to the understanding of fundamental aspects of mycobacterial pathogens led to the use of plasmids and transposons as tools for genetic studies. Consequently, the efforts of many laboratories have been focused on the development of such tools. In this review, we present current knowledge of mycobacterial plasmids and transposons. Their biological features and their relationship with elements from other genera are described. Finally, some of the tools based on plasmids and transposons which have helped or are expected to help the mycobacterial research will be presented.

2 Plasmids in mycobacteria

2.1 Structure and occurrence

Plasmids are not commonly found in clinical isolates of the two major mycobacterial human pathogens: *Mycobacterium leprae* and *M. tuberculosis*. In contrast, they are widespread among various other mycobacteria including both fast and slow growing species (for review, see Falkinham & Crawford 1994). Some strains of the *M. fortuitum* complex carry up to five different coexisting plasmids (Labidi *et al.* 1984). The sizes of these naturally occurring plasmids are diverse (from 4.8 kb to 320 kb) and they can be divided into two main groups according to their structure: covalently closed circular molecules and linear molecules.

Most studies of the occurrence of plasmids have been performed in mycobacteria from either the *M. avium–intracellulare* complex (MAI) or the *M. fortuitum* complex. The plasmids observed were first shown to be circular DNA by electron microscopy (Labidi *et al.* 1984; Crawford & Falkinham 1990). Plasmids in MAI are either small (<30 kb) or very large (>150 kb). Two groups of small plasmids have been distinguished in MAI according to their hybridization with two circular DNA molecules previously isolated from *M. avium*: pLR7 and pVT2. Most of the small plasmids detected in *M. avium* isolates belong to one of these groups (for review, see Falkinham & Crawford (1994)). There is a considerable size variation among plasmids of the pLR7 group (from 15 kb to 30 kb) whereas the pVT2 group seems to be more conserved; only plasmids of 12.9, 13.5 or 15.3 kb have been observed (Jucker & Falkinham 1990). Plasmids related to pLR7 do not seem to be confined to MAI. Indeed two plasmids exhibiting more than 60% DNA sequence identity with pLR7 in the region important for replication have been isolated from *M. scrofulaceum* and *M. fortuitum* (Qin *et al.* 1994; Beggs *et al.* 1995; Gavigan *et al.* 1997). One of these plasmids, pJAZ38, is mobilizable and has been transferred by conjugation from *M. fortuitum* to *M. smegmatis* (Gavigan *et al.* 1997).

Another important mycobacterial plasmid is pAL5000. It was first isolated from *M. fortuitum* and is the smallest known mycobacterial plasmid (Labidi *et al.* 1984; Labidi *et al.* 1985) at 4837 bp. This plasmid is the best characterized mycobacterial plasmid and has been extensively used for the construction of genetic tools (see section 4.1).

2.1.2 *Linear plasmids*

Recently, Picardeau and Vincent (1997) identified linear plasmids in 11 strains belonging to the related species: *M. celatum*, *M. xenopi* and *M. branderi*. All 11 strains contained 1–2 linear plasmids of sizes from 20 kb to 320 kb. The linear structure of these DNA molecules was established using three different approaches: exonuclease degradation, topoisomerase relaxation and electrophoretic mobility. A total of 14 different plasmids corresponding to five different hybridization groups were detected in the 11 strains analysed. No strain carried two plasmids belonging to the same group suggesting an incompatibility mechanism (Picardeau & Vincent 1997). Recently, linear plasmids have also been detected in *M. avium*

(Picardeau 1997). A small linear plasmid of 25 kb, pCLP, has been isolated from *M. celatum* and studied in more details (Picardeau & Vincent 1998). Its extremities were cloned and sequenced. They revealed features of invertron-like structure such imperfect terminal inverted repeats of 45 bp in length and several palindromic sequences. They were shown to bind proteins protecting the telomeres against exonuclease activity. The sequences of these extremities also exhibit an unusual base composition with a high $G+C$ content in the first 60 bp followed by alternating regions of high and low $G+C$ content (Picardeau & Vincent 1997, 1998). All these features are shared by linear plasmids found in other actinomycetes such as *Streptomyces* and *Rhodococcus*.

2.2 Functions encoded by mycobacterial plasmids

Very few phenotypes have been associated with plasmid DNA in mycobacteria. This is in part due to the extreme stability of these plasmids in their normal host. Consequently, the plasmid-encoded functions are difficult to dissociate from the chromosomal functions. Another reason is our inability to perform genetic studies in many of the species in which plasmids have been detected. Therefore, some functions have been shown to correlate with the presence of a plasmid but have not been rigorously demonstrated to be plasmid encoded. For example, a restriction modification system has been described in *M. avium* as being associated with a plasmid. The restriction enzyme produced is an isoschizomer of *XhoI* (Crawford & Falkinham 1990). In *M. scrofulaceum*, mercury and copper resistance (Hg^R, Cu^R) are suspected to be plasmid encoded. Indeed a derivative of a Hg^R and Cu^R strain having lost a 170-kb plasmid became sensitive to mercury and copper (Meissner & Falkinham 1984; Erarddi *et al.* 1987). In a *M. chelonae* strain, the ability to degrade the organic compound morpholine (Mor$^+$) seems to be associated with a 27.7-kb plasmid since an 1.8-kb insertion in a single BamHI fragment carried by this plasmid was found in all tested Mor$^-$ strains (Waterhouse *et al.* 1991). Finally, Fry *et al.* (1986) showed a close correlation between the presence of plasmids in *M. avium* and *M. scrofulaceum* and the ability to grow at 43°C or without oleic acid.

2.3 Replication of mycobacterial plasmids

Little information is available on mycobacterial plasmid replication. Only the regions important for replication in pAL5000 and in three plasmids from the pLR7 family have been characterized.

2.3.1 *pAL5000*

Plasmid pAL5000 was first cloned by Labidi *et al.* (1985). It replicates in many mycobacteria including *M. smegmatis, M. tuberculosis* complex and *M. avium* (Gicquel-Sanzey *et al.* 1989; Beggs *et al.* 1995), its copy number seems to be below five per cell (Ranes *et al.* 1990). Its 4837 bp sequence was first published by Rauzier *et al.* (1988). Five open reading frames (ORF) were identified as probably encoding proteins (Fig. 2.1). Insertional mutagenesis and subcloning first demonstrated that a 2.6-kb fragment containing ORF1, ORF2 and ORF5 was sufficient for replication (Snapper *et al.* 1988; Ranes *et al.* 1990). This region was further narrowed by Stolt and Stoker (1996a) to a 1605-bp DNA fragment containing only ORF1 and 2 (the region between positions 4325 and 1093 in the sequence reported by Rauzier *et al.* 1988). Deletion

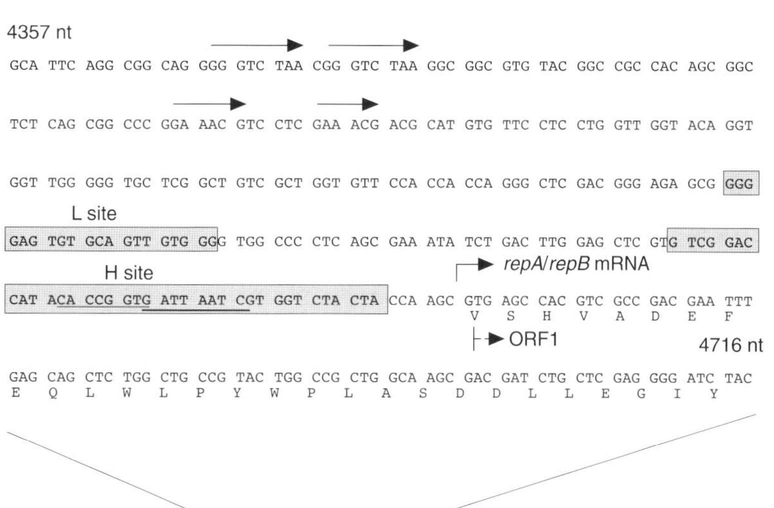

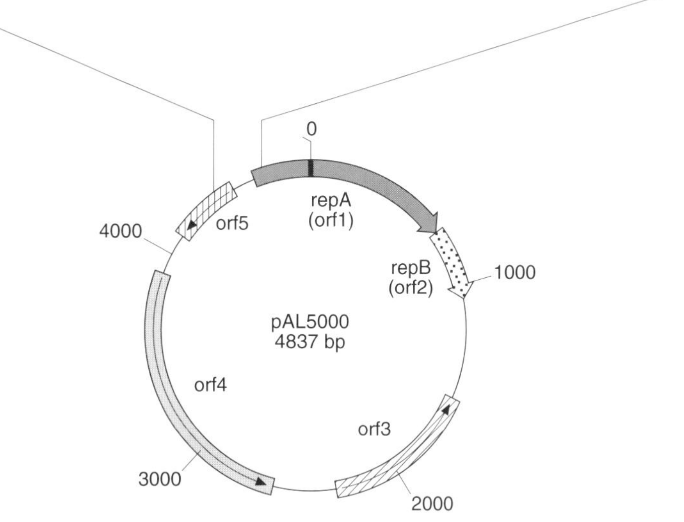

Fig. 2.1 Structural organization of the mycobacterial plasmid pAL5000. The sequence of the putative origin of replication region is shown. Arrows indicate direct repeats potentially involved in incompatibility and copy number. (The minimal region implicated in incompatibility and copy number extends from nt 4175 to nt 4620.) Shaded boxes show the two RepB binding sites. Within the H site, two palindromic sequences are underlined. The second is the putative 10 region of the promoter of the *repA/repB* genes. The start point of the *repA/repB* mRNA is indicated by a broken arrow.

analyses confirmed the involvement of these two ORFs in the replication of pAL5000: removing part of either ORF1 or ORF2 resulted in plasmids unable to replicate in *M. smegmatis* or *M. bovis* bacille Calmette–Guérin (BCG) (McAdam *et al.* 1995; Stolt & Stoker 1996a). However, one report suggested that ORF1 is not essential for replication in *M. fortuitum* (Villar & Benitez 1992). A possible explanation for this may be that some unidentified factors in *M. fortuitum* complement the lack of the ORF1 product. The two putative genes corresponding to ORF1 and ORF2 were renamed *repA* and *repB*, respectively (Stolt & Stoker 1996a).

Recent work by Stolt & Stoker (1996a,b) provided new insight into functions important for pAL5000 replication. They showed that RepA and RepB act *in trans* to allow the maintenance of a plasmid carrying a 760-bp fragment (the region between bases 3861 and 4620). This shows that this fragment contains the origin of replication. In view of the minimal region necessary for autonomous replication, the origin of replication can thus be mapped between positions 4325 and 4620. The *repA* and *repB* genes seem to overlap by one nucleotide (Rauzier *et al.* 1988). Stolt & Stoker (1996b) demonstrated that they are cotranscribed and that the mRNA starts at position 4632. The putative product of *repA* is similar to the replication proteins of the ColE2 plasmid family (Stolt & Stoker 1996a). The deduced translation product of *repB* shows a helix-turn-helix motif of DNA binding proteins (Rauzier *et al.* 1988).

Recombinant RepA and RepB proteins have been purified as fusions to the maltose binding protein of *Escherichia coli* (Stolt & Stoker 1996b). As predicted by Rauzier *et al.* (1988), RepB is able to bind DNA. In contrast, no DNA-binding activity was detected for RepA in spite of its similarities with other replication proteins known to interact with DNA. RepB binds two different sites in the *ori* region: site H (positions 4590–4623) binds with high affinity (binding constant, $K_d \approx 25$ nmol/l) and site L (positions 4534–4564) binds with a lower affinity (binding constant, $K_d \approx 300$ nmol/l). Site H is located only 9 bp upstream from the transcription start site and covers part of the promoter of the *repA* and *repB* genes suggesting that RepB contributes to the regulation of its own expression. It binds to both DNA strands in a region containing two palindromes (see Fig. 2.1). The L site is distant from the promoter (84 bases upstream from the *repA*/*repB* mRNA start), and RepB seems to bind this site more strongly on one than the other strand of the DNA. Stolt and Stoker (1996b) suggested that the L site is the origin of replication.

The region upstream from ORF1 contains several direct repeats, involved in copy number regulation and incompatibility: Stolt and Stoker (1996b) showed that doubling the number of these repeats in the same orientation decreases the plasmid copy number; and different plasmids carrying this region seem to be incompatible (Stolt & Stoker 1996a).

2.3.2 pLR7 family

Three plasmids belonging to the same family have been investigated recently: pLR7, pMSC262 and pJAZ38 (Qin *et al.* 1994; Beggs *et al.* 1995; Gavigan *et al.* 1997). They were isolated from three distantly related mycobacterial species, *M. avium*, *M. scrofulaceum* and *M. fortuitum*, respectively (Crawford & Bates 1984; Meissner & Falkinham III 1984; Gavigan *et al.* 1997). pLR7 and pMSC262 are ≈ 15 kb long and can replicate in many different mycobacteria. However, attempts to transform mc²155 and LR222, two hypertransformable mutants of *M. smegmatis* strain mc²6, with these two plasmids were unsuccessful (Qin *et al.* 1994; Beggs *et al.* 1995). In contrast, the plasmid pJAZ38, which is slightly larger (≈ 16.9 kb), was stably maintained in mc²155.

The regions essential for replication of each of these plasmids in different mycobacteria were cloned and sequenced. These regions exhibit more than 60% DNA sequence similarity between them. Putative ORFs encoding replication proteins (Rep) were identified. The potential Rep proteins of pLR7 and pJAZ38 are 53% identical and 66% similar. Surprisingly, they exhibited only 41% identity and 53% similarity with the putative Rep protein of pMSC262. This is unexpectedly low in view of the very high identity found at the nucleic acid level. This led Beggs *et al.* (1995) to look at the other ORFs in pMSC262. They identified another potential Rep protein encoded on the complementary strand, and which is more than 60% identical and 74% similar to the pLR7 Rep protein. It is shorter than its pLR7 and pJAZ38 counterparts (282 aa vs. 356 aa for pLR7 and 369 aa for pJAZ38) and the similarity stops after approximately the 180th amino acid. Downstream from this position, the sequence encoded in the +1 reading frame is very similar to the C-terminal parts of the pLR7 and pJAZ38 Rep proteins (Fig. 2.2). Possibly there was a mistake during the sequencing of pMSC262.

```
                 1                                                        50
pLR7             .......... .......... .......... ......VPYA GVPCWTGTQ.
pJAZ38           .......... .......... ..MPAPSEFV GLELDADAYA GVPCWSGGPA
pMSC262          MVCTRPWAAR LAAAEHEHRQ RSEAAAVAAI VLELGEAPYA GVPCWTGRLE
Consensus        ---------- ---------- ---------- --------YA GVPCW-G---

                 51                                                       100
pLR7             RWAQWTVPVA YDLRYDTDVR PHMGANQISR RALLRIAEAR ARYADYATGR
pJAZ38           HWAHVTVAVA YDVHYAM.VR PRMCNGGIAR TTLIVIAAAM AQTADWDTGR
pMSC262          RWARWTVAVA YDCRYDAQVR PVMPGNPISR RALLAIARAR ARYADHATGR
Consensus        -WA--TV-VA YD--Y---VR P-M----I-R --L--IA-A- A--AD--TGR

                 101                                                      150
pLR7             DCRPSNERLA TDTGYDVRTI QRASTVLRLL GVATEVLRGR QRTRIERLAS
pJAZ38           NCRPTNEQLE AATGFDERTI QGAHECLRLL GVATEVLRGR QRTYTERMPP
pMSC262          NCRPSNERLA ADTGYSVRTV QRADTVLRLL GVATEVLRGR QRTRVERLAS
Consensus        -CRP-NE-L- --TGY--RTI Q-A---LRLL GVATEVLRGR QRT--ERL--

                 151                                                      200
pLR7             WRVGDRGRGW ASVWALHDHR LLNRVIHKVQ SVLSPHPRSG PVRDQHVRQD
pJAZ38           WRMGDRHRGW PSLWALHGNP HIARVVHSLS PHLE...RSQ ATTKNSPLKR
pMSC262          WRVGDRGRGW ASVWALHDNP QLARFVQR.. ..LSPHPRSG PVRDQPSGKD
Consensus        WR-GDR-RGW -S-WALH--- ---R-V---- --L----RS- ----------

                 201                                                      250
pLR7             VVTTRNRRRT GAGNRGAARR ARPDGYGLAL AKTWRAHPQA PPWCHRHSPT
pJAZ38           LVTTQGGRKR PAPK..PARR RAPDEAGRRL ATRWRADRHA PPWVRTYAAD
pMSC262          VVTTGA
pMSC262          ......MALR AAGRGGAVRR RAPDGGGLAL ARAWRAAAHA PPWARRHTAG
                       ↑ putative frameshift in pMSC262

Consensus        -VTT------ -AG-----RR --PD--G--L A--WRA---A PPW--R----

                 251                                                      300
pLR7             AWAAILAAPA AAGWTPRDLN QLITDWLGV. GHWIPDTPHK PIGLLGAILA
pJAZ38           SWAAMLAAPA AAGWTPADLT ALVRDWLS.T GHWVPDVPAR PIALLGTMLA
pMSC262          AWAALLAGPA AYGWTPRDLN ALITDWAAVT GRRIPDHPHK PIGLLGAMLA
Consensus        AWAA-LA-PA A-GWTPRDL- -LI-DW---- G--IPD-P-- PI-LLGA-LA

                 301                                                      350
pLR7             WHGPEN.LAE RPAALDEARE AQARAANEQL RRAESATSHR AHLAGRAAAQ
pJAZ38           WHTSHNSLED RPAALDEARE AEELAAARRC VRDQF.RAHD EYATDRATAQ
pMSC262          WHGREH.LAE RPAALDEARE VAERAA.... ........HR ALH...AAQR
Consensus        WH-----L-E RPAALDEARE ----AA---- --------H- ------A---

                 351                                                    399
pLR7             AAQSGPGRAE AFAALAAARQ RSAQRRTAQA AAEQARIDAL IERARTPRR
pJAZ38           AALDGPGHAA ARQAVAEAVR RAACKRTVVV AAETAQFAAM VQTARSPR.
pMSC262          AAHGEHERA. ..APRAAPRZ .......... .......... .........
Consensus        AA------A- -----A---- ---------- ---------- ---------
```

Fig. 2.2 Alignment of the potential Rep proteins from pLR7 family plasmids. The Higgins and Sharp (1989) method (PILEUP; GCG, University of Wisconsin) was used for alignment. Amino acids conserved in the three sequences are shown in the consensus. The putative frameshift in the Rep protein from pMSC262 is indicated by an arrow.

The regions important for replication of these plasmids exhibit other common features. In the region located upstream from the potential *rep* genes, pLR7 and pMSC262 share more than 78% DNA sequence identity and contain numerous inverted repeats. In the same region, the sequence 5′-GTCT-CACGGTAGCGCATCC-3′ is almost perfectly conserved in the three plasmids (only a single difference in pJAZ38).

Clearly, these three plasmids are related. Interestingly, pMSC262 and pJAZ38 are compatible with pAL5000 derivatives in *M. bovis* BCG *and M. smegmatis*, respectively. This feature could be useful for developing genetic tools for experiments requiring more than one plasmid.

3 Transposable elements in mycobacteria

3.1 Occurrence, variety and distribution

The first mycobacterial insertion sequence (IS) was discovered in the mid-1980s, when molecular biology tools started to be used to study mycobacteria (McFadden *et al.* 1987). Since then, the combination of search for repetitive elements, hybridization with antibiotic resistance genes, use of transposon traps, serendipity and sequencing of whole genomes have led to isolation of more than 37 different mycobacterial IS. Most were initially identified because of their sequence similarities with other known mobile elements. Activity has only been demonstrated for IS*6100*, IS*6110*, IS*6120*, IS*900*, IS*1096* and IS*1110*. These elements are between 880 bp and 2260 bp long and found in the chromosome in 1 to more than 20 copies. A summary of the main features of the best characterized IS are given in Table 2.1.

Most of these IS exhibit both structural and sequence similarities with elements found in other bacteria. Three large families of elements include most of the characterized mycobacterial transposons: the IS*3*, IS*110* and IS*256* families. Although members of these families are found in many distantly related mycobacterial species, the host range for any particular element is very narrow, and most are found in only one species. This suggests a very limited hori-

Table 2.1 Main features of mycobacterial insertion sequences.

Mobile element	Length (bp)	IR (bp)	DR (bp)	Insertion site	Family	Strain Specificity
IS*6100*	880	14	8	Random	IS6	*M. fortuitum* FC1
IS*6110/986*	1361	28	3 or 4	Random	IS3	*M. tuberculosis* complex
IS*1141*	1587	22	ND	ND	IS3	*M. intracellulare*
IS*1137*	1364	32	3	ND	IS3	*M. smegmatis, M. chitae*
IS*900*	1451	None	None	CATGN(4–6)*CNCCTT	IS110	*M. paratuberculosis*
IS*901/IS902*	1472	None	None	CATN(7)*TTCCNTTC	IS110	*M. avium* RFLP type A/I
IS*1110*	1457	None	None	CATN(7)*TTCCNTTC	IS110	*M. avium*
IS*1081*	1324	—	8	Random	IS256	*M. tuberculosis* complex
IS*6120*	1486	24	9	Random	IS256	*M. smegmatis, M. aurum*
IS*1245*	1402	—	ND	ND	IS256	*M. avium*
IS*1311*	ND	—	ND	ND	IS256	*M. avium*
IS*1395*	1436	—	ND	ND	IS256	*M. xenopi*
IS*1407*	1431	—	9	ND	IS256	*M. celatum*
IS*1408*	1432	—	ND	ND	IS256	*M. branderi*
IS*1511*	ND	—	ND	ND	IS256	*M. gordonae*
IS*1512*	1429	15	ND	ND	IS256	*M. gordonae*
IS*1096*	2260	26	8	A + T rich site	Tn3926	*M. smegmatis*
ISmyco	968	17	4	ND	IS402	*M. tuberculosis*

ND, not done; RFLP, restriction fragment length polymorphism.

zontal transfer of these elements from one species to another.

The genome of *M. tuberculosis* H37Rv has now been sequenced (Cole *et al.* 1998). This allows a global view of the distribution of transposons and 'transposon-like' elements in its chromosome. They are scattered throughout the chromosome with the exception of a region of ≈ 800 kb, located between positions 4310 kb and 690 kb (numbering based on the sequence published by Cole *et al.* 1998), which seems to be devoid of insertion sequences. The explanation of this phenomenon is unclear. Recently, it was shown that in the *M. tuberculosis* strain W, a strain involved in a major multidrug resistant tuberculosis outbreak in New York city, one copy of IS*6110* is near the replication origin (between *dnaA* and *dnaN*). This suggests that there are permissive insertion sites within this region (Kurepina *et al.* 1997).

3.2 Members of the IS*3* family

Three elements of the IS*3* family have been initially isolated from mycobacteria: IS*6110* (also known as IS*986* and IS*987*), IS*1137*, IS*1141* (Thierry *et al.* 1990; Via & Falkinham 1993; Garcia *et al.* 1994). Their membership has been established on the basis of similarities in the terminal inverted repeats (IR), the presence of two ORFs (ORFA and B) in phase 0 and −1, respectively, and similarities in the amino acid sequences of ORFBs (similarities shared by retrovirus integrases) (Fayet *et al.* 1990; Chandler & Fayet 1993) (Fig. 2.3). In IS*911* and IS*150*, IS*3* family members isolated from *Shigella dysenteriae* and *E. coli*, respectively, the transposase is generated by a −1 translational frameshift at the end of ORFA and resulting in the fusion of ORFA and ORFB products. Possible frameshift windows are also present in IS*6110*, IS*1137* and IS*1141* (Fig. 2.3). Recently, the analysis of the H37Rv genome sequence revealed at least two other members of this family (Cole *et al.* 1998; see also Chapter 5).

IS*6110* has been the most extensively studied of the three mycobacterial elements. It is found in 0 to up to 20 copies in the chromosome of *M. tuberculosis* complex strains. This element is extensively used as a marker for epidemiological studies by restriction fragment length polymorphism (RFLP) (for review see Small & van Embden 1994 and Chapters 6 and 7). IS*6110* can transpose in *M. smegmatis* and generates cointegrates (Fomukong & Dale 1993), suggesting a replicative mechanism of transposition. Following its insertion a duplication of 3–4 bp occurs at the target site (a feature also shared by other members of the IS*3* family). The insertion sites of IS*6110* in *M. tuberculosis* do not display any consensus sequence (Mendiola *et al.* 1992; Fang & Forbes 1997). Nevertheless two *M. tuberculosis* chromosomal regions have been identified as being hot spots for the insertion of IS*6110*. The first is the DR region. This locus is composed of directly repeated sequences (DR) of 36 bp interspersed by non-repetitive segments of 36–41 bp in length. The number of DR ranges from 10 to 50 in different *M. tuberculosis* complex strains. Hermans *et al.* (1991) showed that 13 of 14 different isolates of *M. tuberculosis* carried at least one copy of IS*6110* in this region. The second hot spot is 267 bp in length and was named *ipl* (IS*6110* preferential *locus*). In a study of 84 isolates of *M. tuberculosis*, 74% were shown to harbour a copy of IS*6110* in at least six locations of the *ipl* region. These insertions were all in the same orientation (Fang & Forbes 1997). Surprisingly, this *ipl* region is located inside an IS-like element, IS*1547*, sharing similarities with members of the IS*110* family (see below).

3.3 Members of the IS*110* family

Four insertion sequences identified in mycobacteria belong to the IS*110* family: IS*900*, IS*901*, IS*902* and IS*1110* (Green *et al.* 1989; Kunze *et al.* 1991; Moss *et al.* 1992; Hernandez-Perez *et al.* 1994). IS*902* is almost identical to IS*901* (more than 98% DNA sequence identity, Moss *et al.* 1992) and they were isolated from the closely related species *M. paratuberculosis* and *M. avium*, respectively. This family of elements are 1451–1472 bp long and their putative transposases are very similar to that of IS*116* from *Streptomyces clavuligerus* and to a lesser extent to those of IS*110* and IS*117* from *Streptomyces* spp. (Kunze *et al.* 1991; Hernandez-Perez *et al.* 1994). These proteins contain

(a)

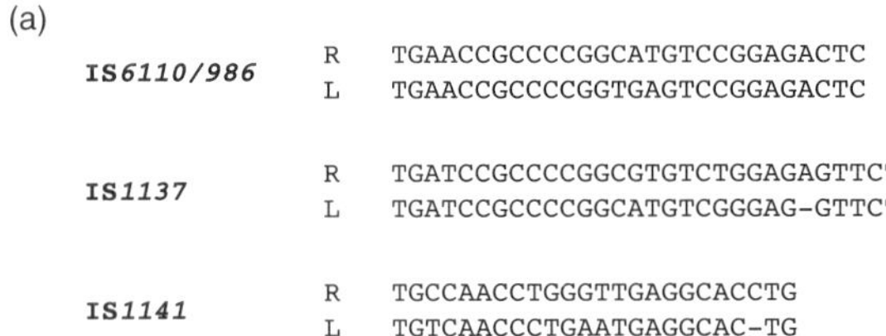

(b)

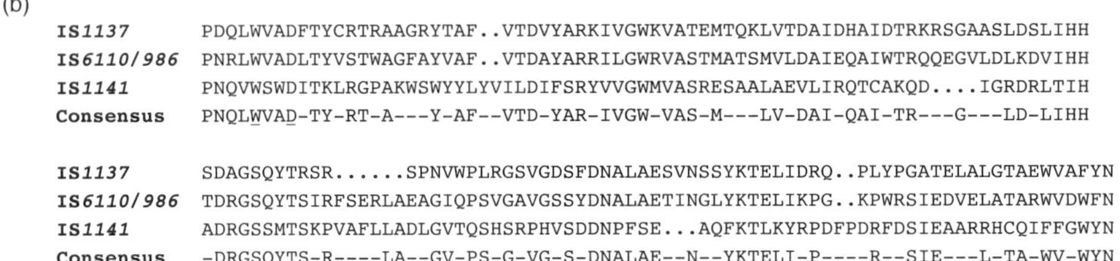

(c)

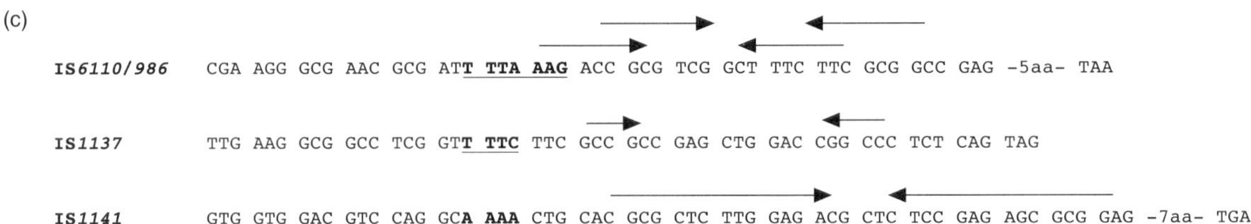

Fig. 2.3 Features indicating the membership of IS*6110*, IS*1137* and IS*1141* in the IS*3* family. (a) Sequences of the terminal inverted repeats of these three elements. The 5′-TG sequence is present in the three insertion sequences (ISs). (b) Partial alignment of the ORFB products of these ISs. The amino acids common to these sequences and retroviral integrases are underlined. The Higgins and Sharp (1989) method (PILEUP; GCG, University of Wisconsin) was used for alignment. Amino acids conserved in at least two sequences are shown in the consensus. (c) Potential frameshift windows. Bold underlined characters indicate the potential slippage sequence and arrows show inverted repeats that may form RNA secondary structure.

two highly conserved motifs also found in proteins responsible for the inversion of DNA in *Moraxella bovis* (Hernandez-Perez *et al.* 1994).

The features of IS*900*, IS*1110*, IS*901* and IS*116* are highly unusual for prokaryotic IS. They do not have terminal inverted repeats nor do they duplicate the target sequence during transposition. The cases of IS*110* and IS*117* are not as clear since they contain imperfect IRs. However, IS*117* does not generate target sequence duplication. The extremities of elements from this family and their junctions with the recipient molecules exhibit some similarities (Fig. 2.4). These ISs all have strong specificity of insertion and, in all the cases, identical sequences are found in the target site and in the transposon. Following insertion, these sequences are found flanking the rest of the transposon. A circular intermediate of transposition has been demonstrated for one element of this family, IS*117*. All these features suggest an unusual mechanism of transposition which does not fit the models of standard conservative or replicative mechanisms described by Shapiro (1979).

The specificity of insertion of IS*900* and IS*901* could explain the poor polymorphism observed in RFLP

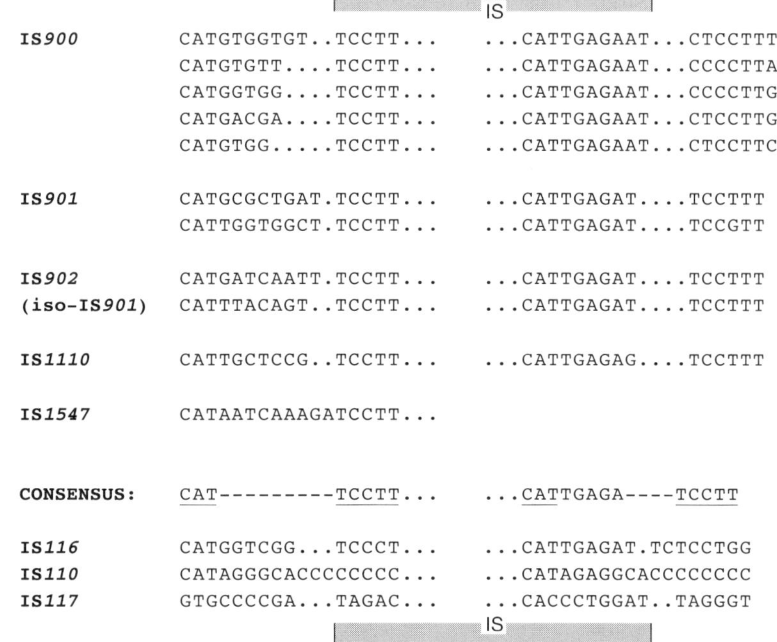

Fig. 2.4 Insertion sites and sequences of the extremities of IS*110* family members. The sequences are from Hernandez-Perez *et al.* (1994), Leskiw *et al.* (1990) and Fang & Forbes (1997). The extremities of IS*110* were not identified precisely and the figure shows one possible junction for this element. The underlined bases indicate the potential ribosome binding site and start codons found in the target sites and the extremities of these elements.

studies using these ISs as probes (see Chapter 8). However, IS*900* can be used to differentiate *M. avium* ssp. *paratuberculosis* (where it is present) from *M. avium* ssp. *avium* or *silvaticum* (where it is absent) (Kunze *et al.* 1992). Also, IS*901* has only been found in *M. avium* strains infecting birds and other animals but not in strains isolated from AIDS patients or the environment (Kunze *et al.* 1991).

The insertion sites of the closely related IS*116*, IS*900*, IS*901* and IS*1110* are very similar (Fig. 2.4). The consensus sequence presents a possible ribosome binding site (RBS) and a start codon on the complementary strand. The consequence is that the insertion orientates the transposase gene in the opposite direction from the potential expression signals present in the target sequence. Insertions are between an RBS and a start codon. Due to the right end sequence of these transposons, a new start codon is introduced 6–8 bp downstream from the RBS of the target site. This codon is the beginning of an ORF that spans the entire element (except in IS*1110* where a frameshift shortens this ORF). Upon transposition, the ORF can be expressed under the control of an external promoter. The chimeric translation signal in IS*900* (RBS from the target sequence and the start codon from the IS) has been shown to be functional by Murray *et al.* (1992). They fused this ORF to the *lacZ* gene of *E. coli* and showed that the RBS from the target sequence and a promoter located upstream from an IS*900* insertion site gave high level of β-galactosidase activity. The role of this ORF remains unknown but a regulatory function has been proposed (Hernandez-Perez *et al.* 1994). Indeed, the motility of IS*1110* in which this ORF is disrupted is higher than that of other elements of the family. Another interesting feature of these elements is that the left end sequence restores the disrupted translation signal by inserting a new RBS 4–8 bases upstream from the start codon of the target site. Thus, it is conceivable, but not established that the insertion preserves the expression of the downstream gene (Fig. 2.4).

Recently, other mycobacterial elements of this family were identified in *M. tuberculosis* (Fang &

Forbes 1997; Cole *et al.* 1998). These IS-like elements exhibit some of the features common to IS*900*, IS*901* and IS*1110* including a similar transposase and insertion site (Fig. 2.4).

3.4 Members of the IS*256* family

This family includes more than 25% of the known mycobacterial IS elements. Part of the reason for this apparent abundance is the active research done in many slow growing species to isolate markers suitable for typing strains and for epidemiological studies. Members of this family have been isolated from distantly related mycobacterial species including: *M. tuberculosis* complex, *M. smegmatis*, *M. avium*, *M. xenopi* and *M. gordonae* (Collins & Stephens 1991; Guilhot *et al.* 1992a; Guerrero *et al.* 1994; Picardeau *et al.* 1996; Picardeau *et al.* 1997). IS*6120* is clearly divergent from the other mycobacterial members of this family as assessed from similarities of the encoded putative transposases. The IS*6120* transposase exhibits less than 49% similarity and 28% identity with those of other members for which the similarity values are above 62% similar and 46% identical. These two IS*256* family subclasses also exhibit different structural features (Fig. 2.5).

IS*6120* is 1486 bp long with 24 bp imperfect terminal inverted repeats. Upon transposition, it generates a 9-bp duplication of the target site. There is no apparent specificity of insertion (Guilhot *et al.* 1992a). These features are common to IS*256*, IS*T2* and IS*Rm3*, all non-mycobacterial members of this family. Using IS*6120* as a probe, strong hybridization signals have been obtained with the genomic DNA of *M. aurum* L1, suggesting the presence of closely related elements in this species (C. Guilhot, unpublished data 1991).

Elements of the other subclasses present a different structure at their extremities. They have two sets of inverted repeats: IR1 and IR2. IR1 is 15 bp long. It was first believed that IR1 was the terminal sequence of these ISs but a closer look revealed a second set of inverted repeats, IR2, which may delimit these elements (Guerrero *et al.* 1994; Picardeau *et al.* 1997). IR2s are located 33–48 bp upstream from IR1 at the

left end and up to 5 bp downstream from IR1 at the right extremity. A 2-bp DR, 5′-TT-3′, flanks the mobile element if IR2 is indeed the terminus. However, several lines of evidence suggest that the element itself extends 6 bp upstream from IR2 on the left side and 6 bp downstream from IR2 on the right side. First, using these boundaries, DRs 8 or 9 bp long can be revealed, which are consistent in size with those of the IS*6120* subclass. These DRs do not display any consensus sequence. Second, these six extra base pairs flanking IR2 are highly conserved among the different elements of this subclass. Third, in *M. tuberculosis* H37Rv, the five copies of IS*1081* that have been sequenced (one of them appeared to be truncated) all have the same sequence for these 6 bp (Fig. 2.6) and three are flanked by 8 bp direct repeats. Therefore, we propose a new delimitation for these elements including the 6 bp flanking the IR2s. The new length of these ISs are presented in Table 2.1. We suggest that these elements generate a 8 or 9 bp DR of the target site following transposition. This unusual terminal organization is also found in IS*1164* from *Rhodococcus rhodochrous* (Komeda *et al.* 1996).

Phylogenetic analysis of the transposase genes in IS*1081* subclass elements, 16S rRNA, and superoxide dismutase suggest that the common ancestor of *M. bovis*, *M. xenopi*, *M. branderi* and *M. celatum* acquired an IS element of the IS*256* family before these species diverged. In contrast, IS*1511* and IS*1512* seem to have been transferred to *M. gordonae* at a different time than their homologues of the *M. bovis* and *M. xenopi* group (Picardeau *et al.* 1997).

3.5 Other elements

Several transposons or IS-like elements which do not belong to the families described above have been isolated from various mycobacteria. The main features of some of these elements are shown in Table 2.1. Two of them, IS*6100* and IS*1096*, are of special interest because they were used to construct transposon mutagenesis systems for mycobacteria (see section 4.2.2).

IS*6100* was first isolated from *M. fortuitum* FC1 as part of a composite transposon, Tn*610*, with two

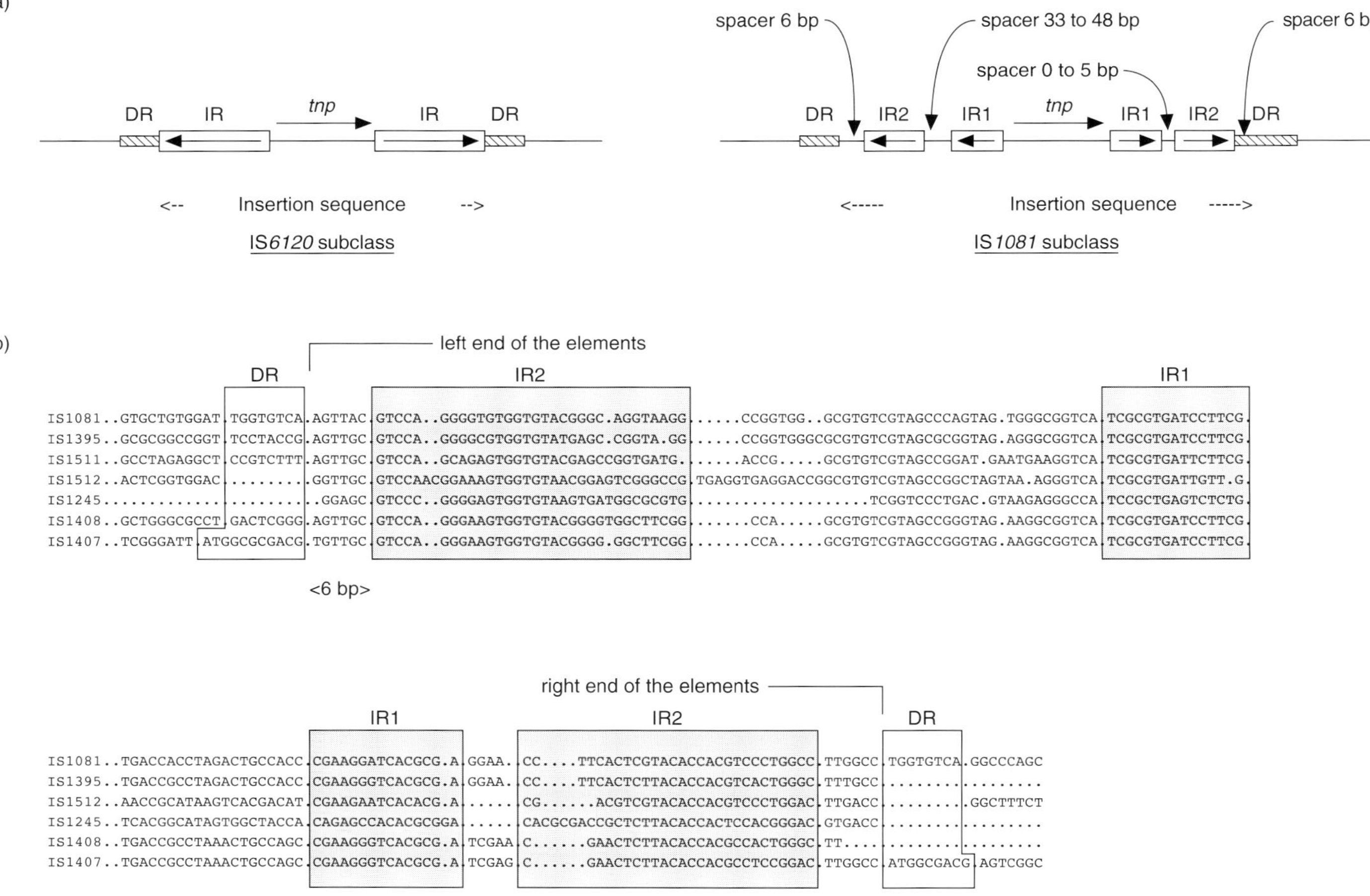

Fig. 2.5 Structural features of the elements of the IS256 family. (a) Structure of the two subclasses of elements. DR indicates direct repeats generated by the insertion of the element. IRs are the terminal inverted repeats. *tnp* indicates the putative transposase of these elements. (b) Alignment of the extremities of mycobacterial elements of the IS1081 subclass. The data are from Picardeau *et al.* (1997) and Picardeau (1997). In the case of IS1512, there are no DR, possibly because of a rearrangement which occurred after the insertion, of a composite transposon, or because of the formation of a cointegrate. In the case of IS1407, the DR are imperfect, there is an extra duplication of the dinucleotide GC at the left junction.

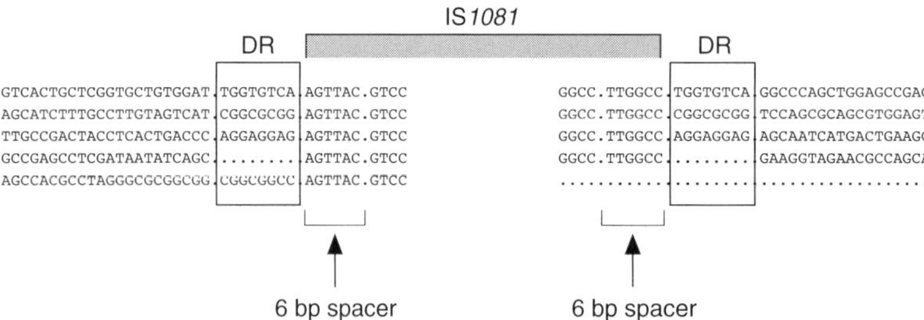

Fig. 2.6 Sequences of the five different insertion sites of IS1081 in the chromosome of H37Rv. These sequences are from the Genbank database. Three of the five insertion sites exhibit DRs 8 bp long. One shows no DR (possible explanations are given in the legend to Fig. 2.5). One copy of IS1081 seems to be truncated.

copies of IS6100 in opposite directions flanking a sulphonamide resistance gene (Martín *et al.* 1990). IS6100 exhibits similarities to elements of the IS6 family and like them transposes by a replicative mechanism. It appears to have no insertion specificity. Several copies of IS6100 were found in a plasmid harboured by a *Flavobacterium* species (Kato *et al.* 1994). A derivative of Tn610 carrying a kanamycin cassette was used to generate insertional mutant libraries of *M. smegmatis* (see section 4.2.2).

The second element which has been used as a mutagen is IS1096. It was isolated from *M. smegmatis* mc²155. The sequence of IS1096 revealed two large ORFs, ORFA and ORFR, exhibiting slight similarities with the transposase of Tn3926 and the resolvases of Tn1000 and Tn552, respectively (Cirillo *et al.* 1991). Insertion into ORFA blocked the transposition of this element. In contrast, mutation in ORFR did not decrease the transposition frequency or lead to the formation of a cointegrate (as expected if ORFR encodes a resolvase) (McAdam *et al.* 1995). These experiments suggested that ORFA encodes the transposase but that the putative product of ORFR is not involved in the resolution of a putative cointegrate intermediate. Sequencing of numerous insertion sites of IS1096 derivatives in *M. tuberculosis* complex and mapping these insertions in the chromosome of H37Rv suggested a clear preference for A+T rich sites but no real local or regional specificity of insertion (McAdam *et al.* 1995; Bardarov *et al.* 1997; Pelicic *et al.* 1997).

4 Genetic tools derived from mycobacterial plasmids and transposons: examples of utilization

Numerous derivatives of the naturally occurring plasmids, mobile genetic elements and bacteriophages have been constructed to obtain mycobacterial cloning vectors and mutagenesis systems.

4.1 Cloning vectors

4.1.1 *Extrachromosomal vectors*

Most of the extrachromosomal cloning vectors are derived from the *M. fortuitum* plasmid pAL5000 which replicates in both fast- and slow-growing mycobacteria and is present in three to five copies in the cytoplasm. Shuttle plasmids carrying the pAL5000 replicon and various *E. coli* replicons in addition to selectable markers have been constructed. These vectors allow both the convenient manipulation of DNA in *E. coli* and the expression of the cloned genes in mycobacteria. Such cloning vectors have been used for a variety of applications including the expression of homologous and heterologous genes, the study of mycobacterial promoter activities and the assessment of protein localization.

Derivatives of pAL5000 have, for example, been used to produce as cytoplasmic, secreted or membrane-associated proteins, a variety of antigens from different viral (Aldovini & Young 1991; Winter *et al.* 1995), parasitic (Abdelhak *et al.* 1995) and bacterial

(Stover *et al.* 1993) pathogens in recombinant BCG (rBCG). Several of these rBCG strains induce specific cellular and humoral immune responses in mice (Aldovini & Young 1991; Stover *et al.* 1991; Winter *et al.* 1995). Protective immunity has also been reported in the same animal (Stover *et al.* 1993; Abdelhak *et al.* 1995). pAL5000-derived expression vectors have also been used to produce mycobacterial proteins in mycobacterial hosts, to produce soluble, correctly processed recombinant proteins that better resemble the native product than would the same protein produced in *E. coli* (Harth *et al.* 1997). Previous studies showed, for example, that the 19 kDa of *M. tuberculosis* produced in *M. smegmatis* was glycosylated, unlike the recombinant 19-kDa antigen produced in *E. coli*, and that the *M. smegmatis* form of the protein was more potent in stimulating 19-kDa protein-reactive T-cell lines (Garbe *et al.* 1993). Similarly, unlike the MPT64 protein of *M. tuberculosis* purified from *M. smegmatis*, the mycobacterial protein purified from *E. coli* was unable to elicit delayed-type hypersensitivity (DTH) in *M. tuberculosis*-sensitized guinea pigs, or to induce substantial interferon-γ (IFN-γ) secretion by stimulated lymphocytes from *M. tuberculosis*-infected animals (Roche *et al.* 1996). Differences in the immunogenicity and ability to form multimeric complexes have also been reported for the *E. coli*- and *M. smegmatis*-purified forms of the *M. leprae* 35-kDa antigen (Triccas *et al.* 1996).

Reporter gene systems have been developed to quantify relative promoter strengths and study gene regulatory sequences by generating gene fusions to the β-galactosidase *lacZ* gene (Timm *et al.* 1994), the chloramphenicol acetyltransferase (CAT) gene (Das Gupta *et al.* 1993) or the *Pseudomonas* catechol 2,3-dioxygenase gene *XylE* (Curcic *et al.* 1994). Reporter systems based on quantification of the fluorescence of *Aequorea victoria* green fluorescent protein (GFP) (Dhandayuthapani *et al.* 1995; Kremer *et al.* 1995) or luciferase light emission (D. Portnoï, personal communication 1996) have been constructed to assess the activity of mycobacterial promoters in culture media, macrophages or *in vivo* in mice. Another application of these reporter systems is the identification in *M.*

smegmatis of mycobacterial DNA sequences encoding exported proteins by using *phoA* gene fusions (Lim *et al.* 1995). The direct selection of DNA sequences encoding exported proteins in mycobacteria from the *M. tuberculosis* complex, for example by using the Tn*phoA* methodology (Manoil & Beckwith 1985), unfortunately cannot be applied because of the lack of a solid medium compatible with both the growth of these species and the detection of a phoA activity.

An *E. coli*–mycobacteria shuttle cosmid (pYUB18) was produced using the pAL5000 replicon, an *E. coli* origin of replication and the *cos* site of bacteriophage λ and allows large DNA fragments to be cloned (Jacobs *et al.* 1991). This vector was used to construct the cosmid library of *M. tuberculosis* H37Rv, which was used to establish the physical map of the *M. tuberculosis* genome (Philipp *et al.* 1996). Finally, thermosensitive derivatives of the pAL5000 replicon have been constructed (Guilhot *et al.* 1992b): vectors carrying these thermosensitive replicons are conditionally replicative vectors, able to replicate at permissive temperature (30°C) but are efficiently lost by *M. smegmatis* when the temperature is shifted to 39°C. These vectors have proved to be useful tools for transposon mutagenesis in *M. smegmatis* (Guilhot *et al.* 1994). Unfortunately, these replicons are only partially thermosensitive in mycobacteria from the *M. tuberculosis* complex (Pelicic *et al.* 1997).

Other mycobacterial vectors have been derived from the *M. scrofulaceum* plasmid pMSC262 (Goto *et al.* 1991; Qin *et al.* 1994) which replicates in a variety of mycobacteria including *M. fortuitum*, *M. phlei*, *M. bovis* BCG and some *M. smegmatis* strains. The *M. avium* pLR7 plasmid (Beggs *et al.* 1995), which replicates in *M. bovis* BCG, *M. tuberculosis* and *M. avium* but not in *M. smegmatis*, has also been used for the development of mycobacterial genetic tools, as has the *M. fortuitum* pJAZ38 plasmid (Gavigan *et al.* 1997), which replicates in *M. smegmatis* and *M. fortuitum*. Both pMSC262 and pLR7 derivatives seem to be compatible with pAL5000-derived vectors in *M. bovis* BCG. So does a pJAZ38 based-plasmid in *M. smegmatis*. Lazraq *et al.* (1991) showed that the origin of mycobacteriophage D29 could be used in the construction of replicating vectors for mycobacteria. Some plasmids from Gram-negative and Gram-positive bacteria are also functional in *M. bovis* BCG and *M. smegmatis*: The *Corynebacterium* plasmid pNG2 (Radford & Hodgson 1991) is a broad-host-range plasmid which is able to replicate in *E. coli*, and has a high copy number in both *E. coli* and mycobacteria. The *E. coli* plasmid RSF1010 is also a broad host vector capable of replication in *M. bovis* BCG and *M. smegmatis* (Gormley & Davies 1991): it has the particularity of being transferable from *E. coli* to *M. smegmatis* by conjugation.

4.1.2 *Integrating vectors*

Vectors able to promote the site-specific integration of the entire plasmid DNA into the mycobacterial chromosome have been developed (see Chapter 3). Such vectors lack a mycobacterial replicon but carry a DNA fragment containing the attachment site (*attP*) and integrase gene (*int*) of a temperate bacteriophage. When introduced into mycobacteria, the expressed *int* gene catalyses the site-specific insertion of the entire vector into the chromosomal bacterial attachment site, *attB,* by recombination. This insertion is stable, as the excision of the integrating plasmid DNA requires a phage-encoded excisionase which is absent from the bacteria. Thus, a potentially valuable feature of integrating vectors is that, unlike extrachromosomal vectors, they can be efficiently maintained in mycobacteria without antibiotic selection. This is of particular interest for the construction of live recombinant BCG vaccines (Stover *et al.* 1991). Such vectors are also of great utility where stable expression of genes or gene fusions in single copy on the chromosome are preferable. This is, for example, the case when studying promoter activities. To date, integrative vectors have been constructed using the *attP–int* components from phages FRAT 1 (Haeseleer *et al.* 1993), L5 (Lee *et al.* 1991), Ms6 (Moniz-Pereira *et al.* 1995) and the *Streptomyces* plasmid pSAM2 (Martín *et al.* 1991).

4.2 Insertional mutagenesis systems

4.2.1 *General aspects*

Mutagenesis systems are important genetic tools for investigating protein functions and mycobacterial virulence mechanisms. However, the difficulty in creating mutants of almost all mycobacterial species greatly hampered such studies until recently. For studying mycobacterial virulence, alternative strategies were used based on the transfer of virulence determinants to non-pathogenic bacteria (Arruda *et al.* 1993; King *et al.* 1993) and to attenuated strains of mycobacteria from the *M. tuberculosis* complex using virulent *M. tuberculosis* or *M. bovis* chromosomal DNA (Pascopella *et al.* 1994; Collins *et al.* 1995). Although these studies led to the identification of genes associated with virulence activities (such as the entry into epithelial cells and haemolytic activity) or conferring growth advantages *in vivo*, only a minority of mycobacterial virulence genes have been identified. Similarly, cloning strategies and classical biochemical methods have been used to identify genes involved in specific mycobacterial physiological processes. Such studies have, for example, led to the characterization of gene clusters encoding enzymes involved in steps of cell-wall biogenesis (Yuan *et al.* 1995; Dubnau *et al.* 1997). However, these cloning strategies are limited to the cases where the gene to be characterized is absent from the recipient strain, and *E. coli* is not a good recipient strain as it might not be able to express all mycobacterial functions such as those involved in virulence or biosynthesis of mycobacterial-specific compounds. The use of random and site-specific mutagenesis systems should allow not only a better understanding of the physiology of mycobacteria and of the pathogenic mechanisms involved in the host–pathogen interactions, thus leading to the design of new antimycobacterial drugs, but also the construction of rationally attenuated strains of mycobacteria with vaccine potential.

Both transposon mutagenesis and allelic exchange are rare genetic events in mycobacteria (see Chapter 1). Thus, mutagenesis systems are highly dependent on efficient delivery vectors and markers to: (i) circumvent the relatively low transformation efficiencies in mycobacteria; and (ii) efficiently detect mutagenesis events. The following paragraphs describe some of the vectors and strategies that have been developed to overcome the major genetic obstacle to the progress in our understanding of mycobacteria: the lack of mutagenesis systems.

4.2.2 Transposon mutagenesis systems

A transposon mutagenesis system is a two-component system involving an insertion element and a delivery vector. The insertion element should have a detectable (preferably high) frequency of transposition, be absent from the target mycobacterial strain and transpose randomly into any part of the bacterial chromosome. Depending on the bacteria to mutagenize, several transposons seem to be good candidates for a transposon mutagenesis system. Efforts have mainly focused on *M. smegmatis* and *M. tuberculosis* complex strains. Three insertion sequences are suitable candidates for mutagenizing mycobacteria from the *M. tuberculosis* complex: IS*6100*, IS*1096* and IS*6120*. IS*6100* or IS*6110* could be used for *M. smegmatis*. All these elements may also be useful mutagens in species such as *M. avium* or *M. paratuberculosis*.

Regarding delivery vectors, suicide plasmids have been used to demonstrate transposition of IS*1096* (McAdam *et al.* 1995) in *M. bovis* BCG and of IS*6100*, IS*900* and IS*6110* in *M. smegmatis*. However, low electroporation efficiencies in addition to low frequencies of transposon hopping allowed no more than 100 mutants per experiment to be obtained in *M. bovis* BCG (McAdam *et al.* 1995) or *M. smegmatis* (Martín *et al.* 1990; England *et al.* 1991; Fomukong & Dale 1993).

An alternative to circumvent these low transformation efficiencies is to use conditionally replicative plasmids. Such transposon delivery plasmids are propagated in the strain to be mutagenized under permissive conditions, allowing transposition events to occur. Then, the application of non-permissive conditions, under which the vector is no longer able to

replicate, results in the selection of insertion events, i.e. transposition mutants. The result is independent of electroporation efficiencies because obtaining only one transformant containing the delivery vector under permissive conditions is enough to perform the subsequent transposition experiment. This strategy has been successfully used in *M. smegmatis* in which representative transposition mutant libraries were constructed (Guilhot *et al.* 1994). The transposon used was Tn*611*, a derivative of Tn*610* containing a kanamycin cassette, and the vector was a thermosensitive derivative of pAL5000, able to replicate at 30°C but not at 39°C in fast-growing mycobacteria (Guilhot *et al.* 1992b). More recently, derivatives of this thermosensitive vector have been used to generate representative *M. bovis* BCG and *M. tuberculosis* insertional mutant libraries (Pelicic *et al.* 1997). The pAL5000-derived thermosensitive origin of replication is only partially thermosensitive in mycobacteria of the *M. tuberculosis* complex. Consequently, a series of conditionally replicative vectors combining the counterselective properties of the *sacB* gene (Pelicic *et al.* 1996a) and of the thermosensitive replicon have been constructed. These vectors were used to deliver two different kanamycin cassette-containing derivatives of IS*1096*, Tn*5367* and Tn*5368* (McAdam *et al.* 1995) to *M. tuberculosis* and *M. bovis* BCG. After electroporation of the delivery vectors, transformants were selected on kanamycin (Km) plates and propagated in liquid medium at permissive temperature (32°C). The bacterial cultures were then plated on kanamycin–sucrose plates at non-permissive temperature (39°C). This selection step applied both sucrose and temperature counterselective pressures and resulted in the direct selection of insertion mutants, i.e. bacteria having integrated the IS*1096*::*Km* derivatives and lost the rest of the vector. More than 10⁶ mutants of *M. tuberculosis and M. bovis* BCG per experiment were obtained.

Another strategy to deliver a transposon efficiently to almost the entire population of mycobacterial cells, is to use conditionally replicative phage systems. Bardarov *et al.* (1997) have constructed conditionally replicating TM4 and D29 phasmid vectors (see Chapter 3). These chimeric molecules, which replicate in *E. coli* as plasmids and in mycobacteria as phages, can replicate in *M. smegmatis* at 30°C but not at 37°C in *M. smegmatis, M. phlei, M. bovis* BCG and *M. tuberculosis*. Using IS*1096* derivatives Tn*5367* and Tn*5368* as transposons, transposition was demonstrated in a *M. phlei* strain and in *M. tuberculosis* Erdman using the TM4-derived delivery phage system, and in *M. bovis* BCG using the D29-derived delivery phage system. In each case transposition was random and yielded thousands of mutants (Bardarov *et al.* 1997).

The feasibility of generating mutant libraries of *M. tuberculosis* should now provide a powerful boost to the identification of mycobacterial virulence genes. For instance, mutants affected in virulence genes could be selected *in vivo* using the tagged transposons insertional mutagenesis system developed by Hensel *et al.* (1995) in *Salmonella*. The isolation of avirulent mycobacterial mutants, for example auxotrophs or mutants affected in virulence genes, should provide new live vaccine candidates against tuberculosis and tools for the investigation of genes preferentially expressed *in vivo* (Mahan *et al.* 1993). The identification of virulence genes will also help in defining new drug targets in mycobacteria. Moreover, modified transposons bearing truncated reporter genes could be used to locate promoters and study their expression. Transposons carrying a truncated *phoA* gene (Tn*phoA*) could be used to identify new exported mycobacterial proteins (Manoil & Beckwith 1985) and to inactivate the gene(s) encoding them. Genes encoding homologous or heterologous protective antigens could also be stably introduced into the chromosome of avirulent mycobacteria using transposon-based tools.

4.2.3 Site-specific mutagenesis systems

Use of suicide vectors

One-step procedures. The classic allelic exchange protocol consists of transforming a mycobacterial strain with a suicide vector carrying a copy of the target gene inactivated by the insertion of a selection marker (generally an antibiotic resistant cassette)

then selecting transformants. If antibiotic markers are used, as the suicide vector has no mycobacterial replicon, antibiotic-resistant transformants can only result from the integration of all or part of the delivery plasmid into the bacterial chromosome. Although this approach to allelic exchange is feasible in the fast growing *M. smegmatis* (Husson *et al.* 1990), it is much more difficult in slow growing mycobacteria. The rarity of allelic exchange events and the relatively high frequencies of illegitimate recombination means that hundreds of antibiotic-resistant transformants may have to be screened to isolate one allelic exchange mutant (see Chapter 1). Allelic exchange was nevertheless demonstrated in *M. bovis* BCG (Reyrat *et al.* 1995; Azad *et al.* 1996) and *M. tuberculosis* (Balasubramanian *et al.* 1996) using linearized suicide vectors, although allelic exchange mutants comprised fewer than 6% of all transformants.

An alternative to tedious screening of transformants to isolate an allelic exchange mutant is to include a counterselectable marker in the delivery vector. After electroporation, the application of both antibiotic and counterselections results in the direct selection of transformants that have integrated the antibiotic resistance gene and lost the rest of the vector, i.e. transformants resulting from allelic exchange events. This strategy has been successfully used in *M. smegmatis* using the *rpsL* (Sander *et al.* 1995) and *sacB* genes (Pelicic *et al.* 1996b) and in *M. bovis* BCG using the *katG* and *sacB* genes (Norman *et al.* 1995; Azad *et al.* 1997). One drawback of the *rpsL* and *katG* counterselection-based systems is that they require the use of antibiotic-resistant strains of mycobacteria (streptomycin- and isoniazid-resistant strains, respectively). This is not the case with the counterselectable marker *sacB*.

Two-step procedures. The direct selection of allelic exchange mutants described above requires efficient transformation of the mycobacterial host. Unfortunately, even with adequate markers to facilitate the detection of allelic exchange events, the poor electroporation efficiencies obtained with slow-growing mycobacteria do not always allow the direct selection of mutants (Kalpana *et al.* 1991; Aldovini *et al.* 1993). Thus, an alternative strategy using the same suicide delivery vectors bearing a counterselectable marker but based on a two-step procedure was developed. In the first step, a transformant resulting from a single homologous recombination event is selected and propagated in liquid medium. In the second step, the grown culture is plated under counterselective conditions to select for the loss of the counterselectable marker, reasoning that the excision of this marker is the result of a second intrachromosomal crossing-over between the wild copy of the target gene and its disrupted copy. Thus, provided the antibiotic resistance selection is applied, the procedure leads to the selection of allelic exchange mutants. Using *sacB* as the counterselectable marker, this procedure permitted the isolation of *pyrF* mutants of *M. smegmatis* (Pelicic *et al.* 1996b) and of urease-deficient mutants of *M. bovis* BCG (Pelicic *et al.* 1996c). In addition, this type of procedure is compatible with the generation of unmarked mutations. This is of particular interest in vaccine applications where the presence of antibiotic markers is undesirable.

Use of replicative vectors

For mutagenesis of very poorly transformable strains of mycobacteria, it might be preferable to resort to conditionally replicative vectors. The thermosensitive *sacB* vector described above for transposon mutagenesis systems has also been used for allelic exchange experiments in *M. tuberculosis*. By replacing the transposon carried with the delivery vector by a kanamycin cassette-disrupted copy of the *M. tuberculosis purC* gene (Jackson *et al.* 1996) and applying the same procedure, purine auxotrophs of *M. tuberculosis* were constructed (Pelicic *et al.* 1997). As for transposition, the major advantages of this system are: (i) it is not dependent on transformation efficiencies; and (ii) the delivery vector replicates in the mycobacteria until the counterselections are applied, allowing time for allelic exchange events to occur and thus, presumably, increasing the number of mutants per experiment.

5 Concluding remarks

During the last 10 years, our knowledge of mobile elements and plasmids in mycobacteria has grown very rapidly. However, some basic aspects of these systems remain to be clarified: how circular and linear plasmids replicate, what are the functions they encode, how are they transferred, and what are the transposition mechanisms of mycobacterial transposons (especially those of the IS*256* and IS*110* families)? The numerous common structural features that mycobacterial plasmids and transposons share with their counterparts in phylogenetically related species, such as *Streptomyces* spp., should, however, allow us to benefit from the knowledge acquired in these other bacteria.

Many different tools have been developed from plasmids and transposons to facilitate investigations of the biology of mycobacteria. A collection of systems is now available, allowing many genetic manipulations to be performed: gene transfer, monitoring of gene expression both *in vitro* and *in vivo*, study of protein localization, random or site specific insertional mutagenesis and others. The availability of these tools, combined with the knowledge of the sequence of the entire genome of various mycobacteria, should greatly and rapidly increase our understanding of the biology of mycobacteria and shed a light on their pathogenic mechanisms during the interaction with the host.

6 Acknowledgements

M.J. is a recipient of a Fondation Mérieux fellowship. This work was supported by European Economic Community Biotech (BIO-CT92–0520) and Biomed (BMH1-CT94–1171) program grants, the National Institute of Health grant AI 35207 and Institut Pasteur.

7 References

Abdelhak, S., Louzir, H., Timm, J. *et al.* (1995) Recombinant BCG expressing the leishmania surface antigen Gp63 induces protective immunity against

Leishmania major infection in BALB/c mice. *Microbiology* **141**, 1585–1592.

Aldovini, A., Husson, R.N. & Young, R.A. (1993) The *uraA* locus and homologous recombination in *Mycobacterium bovis* BCG. *Journal of Bacteriology* **175**, 7282–7289.

Aldovini, A. & Young, R.A. (1991) Humoral and cell-mediated immune responses to live recombinant BCG-HIV vaccines. *Nature* **351**, 479–482.

Arruda, S., Bomfim, G., Knights, R., Huima-Byron, T. & Riley, L.W. (1993) Cloning of an *M. tuberculosis* DNA fragment associated with entry and survival inside cells. *Science* **261**, 1454–1457.

Azad, A.K., Sirakova, T.D., Rogers, L.M. & Kolattukudy, P.E. (1996) Targeted replacement of the mycocerosic acid synthase gene in *Mycobacterium bovis* BCG produces a mutant that lacks mycosides. *Proceedings of the National Academy of Sciences of the USA* **93**, 4787–4792.

Azad, A.K., Sirakova, T.D., Fernandes, N.D. & Kolattukudy, P.E. (1997) Gene knockout reveals a novel gene cluster for the synthesis of a class of cell wall lipids unique to pathogenic mycobacteria. *Journal of Biological Chemistry* **272**, 16741–16745.

Balasubramanian, V., Pavelka, M.S., Jr, Bardarov, S.S. *et al.* (1996) Allelic exchange in *Mycobacterium tuberculosis* with long linear recombination substrates. *Journal of Bacteriology* **178**, 273–279.

Bardarov, S., Kriakov, J., Carriere, C. *et al.* (1997) Conditionally replicating mycobacteriophages: a system for transposon delivery to *Mycobacterium tuberculosis*. *Proceedings of the National Academy of Sciences of the USA* **94**, 10961–10966.

Beggs, M.L., Crawford, J.T. & Eisenach, K.D. (1995) Isolation and sequencing of the replication region of *Mycobacterium avium* plasmid pLR7. *Journal of Bacteriology* **177**, 4836–4840.

Chandler, M. & Fayet, O. (1993) Translational frameshifting in the control of transposition in bacteria. *Molecular Microbiology* **7**, 497–503.

Cirillo, J.D., Barletta, R.G., Bloom, B.R. & Jacobs, W.R.J. (1991) A novel transposon trap for mycobacteria: isolation and characterization of IS*1096*. *Journal of Bacteriology* **173**, 7772–7780.

Cole, S.T., Brosch, R., Parkhill, J. *et al.* (1998) Deciphering the biology of *Mycobacterium tuberculosis* from the complete genome sequence. *Nature* **393**, 537–544.

Collins, D.M., Kawakami, R.P., Lisle, G.W.,D., Pascopella, L., Bloom, B.R. & Jacobs, W.R. (1995) Mutation of the principal s factor causes loss of virulence in a strain of the *Mycobacterium tuberculosis* complex. *Proceedings of the National Academy of Sciences of the USA* **92**, 8036–8040.

Collins, D.M. & Stephens, D.M. (1991) Identification of an insertion sequence, IS*1081*, in *Mycobacterium bovis*. *FEMS Microbiology Letters* **83**, 11–16.

Crawford, J. & Bates, J.H. (1984) Phage typing of mycobacteria. In: *The Mycobacteria: a Sourcebook* (eds G. P. Kubica & L. G. Wayne). New York: Marcel Dekker, pp. 123–132.

Crawford, J.T. & Falkinham J.O., III (1990) Plasmids of the *Mycobacterium avium* complex. In: *Molecular Biology of the Mycobacteria* (ed. J. Mcfadden). London: Harcourt Brace Jovanovich Publishers, pp. 97–119.

Curcic, R., Dhandayuthapani, S. & Deretic, V. (1994) Gene expression in mycobacteria: transcriptional fusions based on *xylE* and analysis of the promoter region of the response regulator *mtrA* from *Mycobacterium tuberculosis*. *Molecular Microbiology* **13**, 1057–1064.

Das Gupta, S.K., Bashyam, M.D. & Tyagi, A.K. (1993) Cloning and assessment of mycobacterial promoters by using a plasmid shuttle vector. *Journal of Bacteriology* **175**, 5186–5192.

Dhandayuthapani, S., Via, L.E., Thomas, C.A., Horowitz, P.M., Deretic, D. & Deretic, V. (1995) Green fluorescent protein as a marker for gene expression and cell biology of mycobacterial interactions with macrophages. *Molecular Microbiology* **17**, 901–912.

Dubnau, E., Lanéelle, M.A., Soares, S. *et al.* (1997) *Mycobacterium bovis* BCG genes involved in the biosynthesis of cyclopropyl keto- and hydroxy-mycolic acids. *Molecular Microbiology* **23**, 313–322.

England, P.M., Wall, S. & McFadden, J. (1991) IS*900*-promoted stable integration of foreign gene into mycobacteria. *Molecular Microbiology* **5**, 2047–2052.

Erarddi, F.X., Failla, M.L. & Falkinham, J.O., III (1987) Plasmid-encoded copper resistance and precipitation by *Mycobacterium scrofulaceum*. *Applied Environmental Microbiology* **53**, 1951–1954.

Falkinham, J.O., III & Crawford, J.T. (1994) Plasmids. In: *Tuberculosis: Pathogenesis, Protection, and Control* (ed. B. Bloom). Washington: American Society for Microbiology, pp. 185–198.

Fang, Z.G. & Forbes, K.J. (1997) A preferential locus for IS*6110* insertions in *M. tuberculosis* complex strains. *Journal of Clinical Microbiology* **35**, 479–481.

Fayet, O., Ramond, P., Polard, P., Prère, M.F. & Chandler, M. (1990) Functional similarities between retroviruses and the IS*3* family of bacterial insertion sequences? *Molecular Microbiology* **4**, 1771–1777.

Fomukong, N.G. & Dale, J.W. (1993) Transpositional activity of IS*986* in *Mycobacterium smegmatis*. *Gene* **130**, 99–105.

Fry, K.L., Meissner, P.S. & Falkinham, J.O., III (1986) Epidemiology of infection by nontuberculous mycobacteria. VI. Identification and use of epidemiologic markers for studies of *M. avium*, *M. intracelluare* and *M Scrofulaceum*. *American Review of Respiratory Diseases* **134**, 39–43.

Garbe, T., Harris, D., Vordermeier, M., Lathigra, R., Ivanyi, J. & Young, D. (1993) Expression of the *Mycobacterium tuberculosis* 19-kilodalton antigen in *Mycobacterium smegmatis*: immunological analysis and evidence of glycosylation. *Infection and Immunity* **61**, 260–267.

Garcia, M.J., Guilhot, C., Lathigra, R. *et al.* (1994) Insertion sequence IS1137, a new IS3 family element from *Mycobacterium smegmatis*. *Microbiology* **140**, 2821–2828.

Gavigan, J.-A., Ainsa, J.A., Pérez, E., Otal, I. & Martin, C. (1997) Isolation by genetic labeling of a new mycobacterial plasmid, pJAZ38, from *Mycobacterium fortuitum*. *Journal of Bacteriology* **179**, 4115–4122.

Gicquel-Sanzey, B., Moniz-Pereira, J., Gheorghiu, M. & Rauzier, J. (1989) Structure of pAL5000, a plasmid from *M. fortuitum* and its utilization in transformation of mycobacteria. *Acta Leprologica* **7** (Suppl. 1), 208–211.

Gormley, E.P. & Davies, J. (1991) Transfer of plasmid RSF1010 by conjugation from *Escherichia coli* to *Streptomyces lividans* and *Mycobacterium smegmatis*. *Journal of Bacteriology* **173**, 6705–6708.

Goto, Y., Taniguchi, H., Udou, T., Mizuguchi, Y. & Tokunaga, T. (1991) Development of a new host vector system in mycobacteria. *FEMS Microbiology Letters* **83**, 277–282.

Green, E.P., Tizard, M.L.V., Moss, M.T. *et al.* (1989) Sequence and characteristics of IS900, an insertion element identified in a human Crohn's disease isolate of *Mycobacterium paratuberculosis*. *Nucleic Acids Research* **17**, 9063–9073.

Guerrero, C., Bernasconi, C., Burki, D., Bodmer, T. & Telenti, A. (1994) IS1245: a novel insertion element from *Mycobacterium avium* is a specific marker for analysis of strain relatedness. *Journal of Clinical Microbiology* **33** (2), 304–307.

Guilhot, C., Gicquel, B., Davies, J. & Martin, C. (1992a) Isolation and analysis of IS6120, a new insertion sequence from *Mycobacterium smegmatis*. *Molecular Microbiology* **6**, 107–113.

Guilhot, C., Gicquel, B. & Martin, C. (1992b) Temperature-sensitive mutants of the *Mycobacterium* plasmid pAL5000. *FEMS Microbiological Letters* **98**, 181–186.

Guilhot, C., Otal, I., Rompaey, I.V., Martín, C. & Gicquel, B. (1994) Efficient transposition in mycobacteria: construction of *Mycobacterium smegmatis* insertional mutant libraries. *Journal of Bacteriology* **176**, 535–539.

Haeseleer, F., Pollet, J.F., Haumont, M., Bollen, A. & Jacobs, P. (1993) Stable integration and expression of the *Plasmodium falciparum* circumsporozoite protein coding sequence in mycobacteria. *Molecular Biochemistry and Parasitology* **57**, 117–126.

Harth, G., Lee, B.-Y. & Horwitz, M.A. (1997) High-level heterologous expression and secretion in rapidly growing nonpathogenic mycobacteria of four major *Mycobacterium tuberculosis* extracellular proteins considered to be leading vaccine candidates and drug targets. *Infection and Immunity* **65**, 2321–2328.

Hensel, M., Shea, J.E., Gleeson, C., Jones, M.D., Dalton, E. & Holden, D.W. (1995) Simultaneous identification of bacterial virulence genes by negative selection. *Science* **269**, 400–403.

Hermans, P.W.M., van Soolingen, D., Bik, E.M., Haas, P.E.W., Dale, J.W. & van Embden, J.D.A. (1991) Insertion element IS987 from *Mycobacterium bovis* BCG is located in a hot-spot integration region for insertion elements in *Mycobacterium tuberculosis* complex strains. *Infection and Immunity* **59**, 2695–2705.

Hernandez-Perez, M., Fomukong, N.G., Hellyer, T., Brown, I.N. & Dale, J.W. (1994) Characterization of IS1:110, a highly mobile genetic element from *Mycobacterium avium*. *Molecular Microbiology* **12**, 717–724.

Higgins, D.G. & Sharp, P.M. (1989) Fast and sensitive multiple sequence alignments on a microcomputer. *Computer Applied Bioscience* **5**, 151–153.

Husson, R.N., James, B.E. & Young, R.A. (1990) Gene replacement and expression of foreign DNA in mycobacteria. *Journal of Bacteriology* **172**, 519–524.

Jackson, M., Berthet, F.X., Otal, I. *et al.* (1996) The *Mycobacterium tuberculosis* purine biosynthetic pathway: isolation and characterization of the *purC* and *purL* genes. *Microbiology* **142**, 2439–2447.

Jacobs, W.R.J., Kalpana, G.V., Cirillo, J.D. *et al.* (1991) Genetic systems for mycobacteria. In: *Methods in Enzymology* (ed. J. H. Miller). New York: Horcourt Brace, pp. 537–555.

Jucker, M.T. & Falkinham, J.O., III (1990) Epidemiology of infection by nontuberculous mycobacteria. IX. Evidence for two DNA homology groups among small plasmids in *M. avium, M. intracellulare* and *M. scrofulaceum*. *American Review of Respiratory Diseases* **142**, 858–862.

Kalpana, G.V., Boom, B.R. & Jacobs, W.R.J. (1991) Insertional mutagenesis and illegitimate recombination in mycobacteria. *Proceedings of the National Academy of Sciences of the USA* **88**, 5433–5437.

Kato, K., Ohtsuki, K., Mitsuda, H., Yomo, T., Negoro, S. & Urabe, I. (1994) Insertion sequence IS6100 on plasmid pOAD2, which degrades nylon oligomers. *Journal of Bacteriology* **176**, 1197–1200.

King, C.H., Mundayoor, S., Crawford, J.T. & Shinnick, T.M. (1993) Expression of contact-dependent cytolytic activity by *Mycobacterium tuberculosis* and isolation of the genomic locus that encodes the activity. *Infection and Immunity* **61**, 2708–2712.

Komeda, H., Koayashi, M. & Shimizu, S. (1996) Characterization of the gene cluster of high-molecular-mass nitrile hydratase (H-NHase) induced by its reaction product in *Rhodococcus rhodochrous* J1. *Proceedings of the National Academy of Sciences of the USA* **93**, 4267–4272.

Kremer, L., Baulard, A., Estaquier, J., Poulain-Godefroy, O. & Locht, C. (1995) Green fluorescent protein as a new expression marker in mycobacteria. *Molecular Microbiology* **17**, 913–922.

Kunze, Z., Portaels, F. & McFadden, J.J. (1992) Biologically distinct subtypes of *Mycobacterium avium* differ in possession of insertion sequence IS*901 Journal of Clinical Microbiology* **30** (9), 2366–2372.

Kunze, Z.M., Wall, S., Appelberg, R., Silva, M.T., Portaels, F. & McFadden, J.J. (1991) IS*901*, a new member of a widespread class of atypical insertion sequences, is associated with pathogenicity in *Mycobacterium avium*. *Molecular Microbiology* **5**, 2265–2272.

Kurepina, N., Bifani, P.J., Connell, N. *et al.* (1997) Population analysis of *Mycobacterium tuberculosis* based on IS*6110* insertion site mapping. Thirty-Second US–Japan Cooperative Medical Science Program, Tuberculosis–Leprosy Research Conference, 68–72.

Labidi, A., Dauguet, C., Goh, K.S. & David, H.L. (1984) Plasmid profiles of *Mycobacterium fortuitum* complex isolates. *Current Microbiology* **11**, 235–240.

Labidi, A., David, H.L. & Roulland-Dussoix, D. (1985) Restriction endonuclease mapping and cloning of *Mycobacterium fortuitum* var. *fortuitum* plasmid pAL5000. *Annals of the Institute Pasteur/Microbiology* **136**, 209–215.

Lazraq, R., Houssaini-Iraqui, M., Clavel-Sérès, S. & David, H.L. (1991) Cloning and expression of the origin of replication of mycobacteriophage D29 in *Mycobacterium smegmatis. FEMS Microbiological Letters* **80**, 117–120.

Lee, M.H., Pascopella, L., Jabos, W.R.J. & Hatfull, G.F. (1991) Site-specific integration of mycobacteriophage L5: integration-proficient vectors for *Mycobacterium smegmatis, Mycobacterium tuberculosis,* and bacille Calmette–Guérin. *Proceedings of the National Academy of Sciences of the USA* **88**, 3111–3115.

Leskiw, B.K., Mevarech, M., Barritt, L.S. *et al.* (1990) Discovery of an insertion sequence, IS*116*, from *Streptomyces clavuligerus* and its relatedness to other transposable elements from actinomycetes. *Journal of General Microbiology* **136**, 1251–1258.

Lim, E.M., Rauzier, J., Timm, J. *et al.* (1995) Identification of *Mycobacterium tuberculosis* DNA sequences encoding exported proteins by using *phoA* gene fusions. *Journal of Bacteriology* **177**, 59–65.

Mahan, M.J., Slauch, J.M. & Mekalanos, J.J. (1993) Selection of bacterial virulence genes that are specifically induced in host tissues. *Science* **259**, 686–688.

Manoil, C. & Beckwith, J. (1985) TnphoA: a transposon probe for protein export signals. *Proceedings of the National Academy of Sciences of the USA* **82**, 8129–8133.

Martín, C., Mazodier, P., Mendiola, M.V. *et al.* (1991) Site-specific integration of *Streptomyces* plasmid pSAM2 in *Mycobacterium smegmatis. Molecular Microbiology* **5**, 2499–2502.

Martín, C., Timm, J., Rauzier, J., Gómez-Lus, R., Davies, J. & Gicquel, B. (1990) Transposition of an antibiotic resistance element in mycobacteria. *Nature* **345**, 739–743.

McAdam, R.A., Weisbrod, T.R., Martin, J. *et al.* (1995) In

vivo growth characteristics of leucine and methionine auxotrophic mutants of *Mycobacterium bovis* BCG generated by transposon mutagenesis. *Infection and Immunity* **63**, 1004–1012.

McFadden, J.J., Butcher, P.D., Thompson, J., Chiodini, R.J. & Hermon-Taylor, J. (1987) The use of DNA probes identifying restriction-fragment-length polymorphisms to examine the *Mycobacterium avium* complex. *Molecular Microbiology* **1**, 283–291.

Meissner, P.S. & Falkinham, J.O., III (1984) Plasmid-encoded mercuric reductase in *Mycobacterium scrofulaceum. Journal of Bacteriology* **157**, 669–672.

Mendiola, M.V., Martin, C., Otal, I. & Gicquel, B. (1992) Analysis of the regions responsible for IS*6110* RFLPs in a single *Mycobacterium tuberculosis* strain. *Research in Microbiology* **143**, 767–772.

Moniz-Pereira, J., Anes, E., Vieira, A., Garcia, M. & Pimentel, M. (1995) Construction of stable recombinant mycobacteria strains using a mycobacteriophage Ms6 based integrative vector. The Second Delivery System Meeting, Paris, France.

Moss, M.T., Malik, Z.P., Tizard, M.L.V., Green, E.P., Sanderson, J.D. & Hermon-Taylor, J. (1992) IS*902*, an insertion element of the chronic-enteritis-causing *Mycobacterium avium* subsp. *Silvaticum Journal of General Microbiology* **138**, 139–145.

Murray, A., Winter, N., Lagranderie, M. *et al.* (1992) Expression of *Escherichia coli* β-galactosidase in *Mycobacterium bovis* BCG using an expression system isolated from *Mycobacterium paratuberculosis* which induced humoral and cellular immune responses. *Molecular Microbiology* **6**, 3331–3342.

Norman, E., Dellagostin, O.A., McFadden, J. & Dale, J.W. (1995) Gene replacement by homologous recombination in *Mycobacterium bovis* BCG. *Molecular Microbiology* **16**, 755–760.

Pascopella, L., Collins, F.M., Martin, J.M. *et al.* (1994) Use of *vivo* complementation in *Mycobacterium tuberculosis* to identify a genomic fragment associated with virulence. *Infection and Immunity* **62**, 1313–1319.

Pelicic, V., Jackson, M., Reyrat, J.M., Jacobs, W.R., Gicquel, B. & Guilhot, C. (1997) Efficient allelic exchange and transposon mutagenesis in *Mycobacterium tuberculosis. Proceedings of the National Academy of Sciences of the USA* **94**, 10955–10960.

Pelicic, V., Reyrat, J.M. & Gicquel, B. (1996a) Expression of the *Bacillus subtilis sacB* gene confers sucrose sensitivity on mycobacteria. *Journal of Bacteriology* **178**, 1197–1199.

Pelicic, V., Reyrat, J.-M. & Gicquel, B. (1996b) Generation of unmarked directed mutations in mycobacteria, using sucrose counterselectable suicide vectors. *Molecular Microbiology* **20**, 919–925.

Pelicic, V., Reyrat, J.M. & Gicquel, B. (1996c) Positive

selection of allelic exchange mutants in *Mycobacterium bovis* BCG. *FEMS Microbiological Letters* **144**, 161–166.

Philipp, W.J., Poulet, S., Eiglmeier, K. *et al.* (1996) An integrated map of the genome of the tubercle bacillus, *Mycobacterium tuberculosis* H37Rv, and comparison with *Mycobacterium leprae*. *Proceedings of the National Academy of Sciences of the USA* **93**, 3132–3137.

Picardeau, M. (1997) *Etude genotypique des mycobacteries atypiques*. PhD Dissertation, Université Paris VI.

Picardeau, M., Bull, T. & Vincent, V. (1997) Identification and characterization of IS-like elements in *Mycobacterium gordonae*. *FEMS Microbiological Letters* **154**, 95–102.

Picardeau, M., Varnerot, A., Rauzier, J., Gicquel, B. & Vincent, V. (1996) *Mycobacterium xenopi* IS*1395*, a novel insertion sequence expanding the IS*256* family. *Microbiology* **142**, 2453–2461.

Picardeau, M. & Vincent, V. (1997) Characterization of large linear plasmids in mycobacteria. *Journal of Bacteriology* **179**, 2753–2756.

Picardeau, M. & Vincent, V. (1998) Mycobacterial linear plasmids have an invertron-like structure related to other linear replicons in actinomycetes. *Microbiology* **144**, 1981–1988.

Qin, M., Taniguchi, H. & Mizuguchi, Y. (1994) Analysis of the replication region of a mycobacterial plasmid, pMSC262. *Journal of Bacteriology* **176**, 419–425.

Radford, A.J. & Hodgson, A.L.M. (1991) Construction and characterization of a *Mycobacterium–Escherichia coli* shuttle vector. *Plasmid* **25**, 149–153.

Ranes, M.G., Rauzier, J., Lagranderie, M., Gheorghiu, M. & Gicquel, B. (1990) Functional analysis of pAL5000, a plasmid from *Mycobacterium fortuitum*: construction of a 'mini' *Mycobacterium/Escherichia coli* shuttle vector. *Journal of Bacteriology* **172**, 2793–2797.

Rauzier, J., Moniz-Pereira, J. & Gicquel-Sanzey, B. (1988) Complete nucleotide sequence of pAL5000, a plasmid from *Mycobacterium fortuitum*. *Gene* **71**, 315–321.

Reyrat, J.M., Berthet, F.X. & Gicquel, J.M. (1995) The urease locus of *M. tuberculosis* and its utilization for the demonstration of allelic exchange in *M. bovis* BCG. *Proceedings of the National Academy of Sciences of the USA* **92**, 8768–8772.

Roche, P.W., Winter, N., Triccas, J.A., Feng, C. & Britton, W.J. (1996) Expression of *Mycobacterium tuberculosis* MPT64 in recombinant *M. smegmatis*: purification, immunogenicity and application to skin tests for tuberculosis. *Clinical Experimental Immunology* **103**, 226–232.

Sander, P., Meier, A. & Böttger, E.C. (1995) *rpsL*+: a dominant selectable marker for gene replacement in mycobacteria. *Molecular Microbiology* **16**, 991–1000.

Shapiro, J.A. (1979) Molecular model for the transposition and replication of bacteriophage Mu and other transposable elements. *Proceedings of the National Academy of Sciences of the USA* **76**, 1933–1937.

Small, P.M. & van Embden, J.D.A. (1994) Molecular epidemiology of tuberculosis. In: *Tuberculosis: Pathogenesis, Protection, and Control* (ed. B. Bloom). Washington: American Society for Microbiology, pp. 569–582.

Snapper, S.B., Lugosi, L., Jekkel, A. *et al.* (1988) Lysogeny and transformation in mycobacteria: stable expression of foreign genes. *Proceedings of the National Academy of Sciences of the USA* **85**, 6987–6991.

Stolt, P. & Stoker, N.G. (1996a) Functional definition of regions necessary for replication and incompatibility in the *Mycobacterium fortuitum* plasmid pAL5000. *Microbiology* **142**, 2795–2802.

Stolt, P. & Stoker, N.G. (1996b) Protein–DNA interactions in the *ori* region of the *Mycobacterium fortuitum* plasmid pAL5000. *Journal of Bacteriology* **178**, 6693–6700.

Stover, C.K., Bansal, G.P., Hanson, M.S. *et al.* (1993) Protective immunity elicited by recombinant bacille Calmette–Guérin (BCG) expressing outer surface protein A (OspA) lipoprotein: a candidate Lyme disease vaccine. *Journal of Experimental Medicine* **178**, 197–209.

Stover, C.K., Cruz, V.F.D.L., Fuerst, T.R. *et al.* (1991) New use of BCG for recombinant vaccines. *Nature* **351**, 456–460.

Thierry, D., Cave, M.D., Eisenach, K.D. *et al.* (1990) IS**6**:*110*, an IS-like element of *Mycobacterium tuberculosis* complex. *Nucleic Acids Research* **18**, 188.

Timm, J., Perilli, M.G., Duez, C. *et al.* (1994) Transcription and expression analysis, using *lac*Z and *pho*A gene fusions, of *Mycobacterium fortuitum* β-lactamase genes cloned from a natural isolate and a high-level β-lactamase producer. *Molecular Microbiology* **12**, 491–504.

Triccas, J.A., Roche, P.W., Winter, N. *et al.* (1996) A 35-kilodalton protein is a major target of the human immune response to *Mycobacterium leprae*. *Infection and Immunity* **64**, 5171–5177.

Via, L.E. & Falkinham, J.O., III (1993) GenBank, L10239.

Villar, C.A. & Benitez, J. (1992) Functional analysis of pAL5000 in *Mycobacterium fortuitum*. *Plasmid* **28**, 166–169.

Waterhouse, K.V., Swain, A. & Venables, W.A. (1991) Physical characterization of plasmids in a morpholine-degrading mycobacterium. *FEMS Microbiological Letters* **80**, 305–310.

Winter, N., Lagranderie, M., Gangloff, S., Leclerc, C., Gheorghiu, M. & Gicquel, B. (1995) Recombinant BCG strains expressing the SIV$_{mac251}$ *nef* gene induce proliferative and CTL responses against nef synthetic peptides in mice. *Vaccine* **13**, 471–478.

Yuan, Y., Lee, R.E., Besra, G.S., Belisle, J.T. & Barry, C.E., III (1995) Identification of a gene involved in the biosynthesis of cyclopropanated mycolic acids in *Mycobacterium tuberculosis*. *Proceedings of the National Academy of Sciences of the USA* **92**, 6630–6634.

Chapter 3 / Mycobacteriophages

GRAHAM F. HATFULL

1 Introduction

Mycobacteriophages are viruses of the mycobacteria. They are extremely prevalent in nature and over 250 individual types have been isolated, mostly from soil samples (reviewed in Redmond 1963; Barksdale & Kim 1977; Mizuguchi 1984; Hatfull & Jacobs 1994). A chief motivation for the past interest in mycobacteriophages and reason for their isolation is the peculiarities of host preferences that characterizes each individual phage type. The phages can therefore be used as tools for typing mycobacterial isolates in a rapid and reproducible fashion (Snider *et al.* 1984). In more recent times, molecular methodologies for typing mycobacterial strains have been developed which reveal the relationships between strains with considerably finer tuning than can be accomplished with phages (see Chapters 6–8). However, these molecular methods require a degree of technological sophistication and expense that may not be universally available and phage typing still provides a useful tool in such environments.

In the past 10 years, mycobacteriophages have assumed a dominant role in the development of mycobacterial genetics. This should not be surprising since viruses have frequently come to the assistance of important biological systems with naive genetics that are in need of development. With the mycobacteria, the uptake of mycobacteriophage TM4 DNA was the first demonstration of efficient introduction of DNA into these organisms, albeit via the preparation of mycobacterial spheroplasts (Jacobs *et al.* 1987). Moreover, since mycobacteriophages have relatively small genomes compared to their hosts, recombinant phages could be constructed and used to make some of the first stable recombinant mycobacteria expressing foreign genes (Snapper *et al.* 1988). As more detailed studies on mycobacteriophages have progressed they have provided insights into mycobacterial gene structure and expression and provided new tools for mycobacterial genetics.

Mycobacteriophages are attractive objects for dissection because of their relatively small genome sizes (50–150 kb) and their intimate relationships with their mycobacterial hosts. By themselves, the phages are inert, represent no biological hazard and can easily be isolated in large numbers, facilitating both physical analysis and the isolation of mutants, even those that arise at low frequencies. However, when only a single particle of a virulent mycobacteriophage encounters its bacterial host it is able to completely reprogramme the host metabolism and direct it towards the replication of viral DNA, synthesis of virion proteins and formation of phage progeny. Ultimately the bacterium is lysed and viral particles released to find new host cells and repeat the process. Typically, the phage contains only those genes needed for virion structure and assembly and those required for directing the host machinery towards phage gene expression and DNA replication. Other core metabolic functions are provided by the host. Temperate phages have a greater intimacy with their hosts, establishing stable lysogens in which the phage DNA in inserted into the host genome and phage gene expression is silenced. Bacteriophages of all types continue to provide insights into novel mechanisms for a broad spectrum of biological processes.

What aspects of mycobacterial genetics and physiology are most amenable to understanding with mycobacteriophages? An obvious starting point—although still very poorly understood—is what molecules on the surfaces of mycobacteria are required for phage infections? These surface structures contribute substantially to the host range of the phages, since without the required receptors infection does not occur. Many mycobacteriophages have broad host ranges among mycobacterial species reflecting common receptor types; others have quite restricted host ranges and infect only limited species; for example, DS6A infects only members of the *Mycobacterium tuberculosis* complex (Redmond & Carter 1960). Following initial adsorption of the phage particle to the cell, phage DNA must be injected through the cell wall and the cell membrane, into the cytoplasm of the cell; mycobacteriophages are thus useful tools for probing these structures.

Once the DNA is in the cytoplasm, phage genes must be expressed. Virtually all phages use the host transcription machinery for initial gene expression and frequently rely on the host for all of their transcription. Phages therefore contain promoters which are recognized by the host and which are usually very active. Temperate phages must carefully regulate this transcription in order to form stable lysogens and may therefore provide additional insights into how mycobacterial transcription can be regulated. Once transcribed, phage genes must be translated using the host protein biosynthetic machinery and thus the genetic code, translation initiation and termination signals, and codon biases of the phage typically reflect those of the host. Phage DNA replication must also occur, although in many cases the phage encodes genes that provide additional replication functions or which modify the host replicative apparatus to direct it towards the virus. Phages may provide particular insights into mycobacterial DNA replication since they have presumably evolved to productively infect mycobacterial cells regardless of the state of chromosomal DNA replication. In the slow-growing mycobacteria in particular, only a relatively small proportion of the cells may be active in DNA replication. Finally, mycobacteriophages should illuminate other specialized host functions, such as bacterial proteins required for phage integration and chaperones needed for the assembly of viral particles.

Mycobacteriophages are also a valuable resource for enhancing mycobacterial genetic systems. While many such applications can be envisaged, they certainly include the use of phage-encoded functions to develop plasmid vectors or selectable markers and the use of phages to efficiently and specifically deliver genes to mycobacteria. For example, phages can be used to deliver reporter genes, transposons, antisense genes or genes encoding foreign antigens. This delivery can be accomplished without special treatment of the cells and without the need for selection of a subset of cells that have received the DNA.

In this chapter, I will discuss some of the better-characterized mycobacteriophages—particularly L5, D29, TM4 and I3—in some detail, and discuss ways in which these phages can be used to enhance our understanding of mycobacterial genetics.

2 Mycobacteriophage L5

Mycobacteriophage L5 is perhaps the most well-understood of all the mycobacteriophages. It was first isolated in Japan by Doke (1960) from a lysogenic strain of *Mycobacterium smegmatis* and forms turbid plaques on lawns of *M. smegmatis* strains derived from ATCC607 (including mc²155). It is almost identical to phage L1, which was isolated at the same time; the only apparent difference between the two is that L1 is temperature sensitive for plaque formation. The restriction maps for L1 and L5 are identical for all enzymes that have been tested. L5 represents a prototype for what appears to be a family of closely related mycobacteriophages that includes D29 (see below), probably FRAT1 and an additional member isolated in France (Hatfull & Jacobs 1994).

L5 is a temperate phage and stable *M. smegmatis* lysogens can be isolated from the turbid area of infected cells; these lysogens contain an integrated L5 prophage inserted at a single chromosomal locus (Snapper *et al.* 1988; Lee *et al.* 1991). They behave as typical phage lysogens, in that they are immune to superinfection by L5 and related phages, and release phage particles into the culture supernatant. However, L5 lysogens are not readily inducible by ultraviolet irradiation. While many mycobacteriophages may be competent to form pseudolysogens, only L5 and its relatives have been demonstrated to have the properties of true temperate phages.

The host range of L5 has in the past been somewhat ill-defined, although it is now clear that it efficiently infects both fast-growing strains such as *M. smegmatis* and slow-growing strains such as bacille Calmette–Guérin (BCG) (Fullner & Hatfull 1997). The reason for the past ambiguity lies in the specific requirements for L5 adsorption, which are different for *M. smegmatis* and BCG. Moreover, these conditions are different from those needed for the closely related phage, D29, to infect both fast- and slow-growing mycobacteria (Fullner & Hatfull 1997). L5 does not appear to infect bacterial species other than mycobacteria (Hatfull & Jacobs 1994).

2.1 Virion structure and assembly

L5 has an unremarkable morphology, with an icosahedral head—which contains the DNA—attached to a long flexible tail, features common to hundreds or thousands of other phages that have been described.

At the tip of the tail there appears to be a small spike which presumably is involved in making specific contacts with the host bacterium; no side-tail fibers have been identified. Like other phages, the head and tail structures appear to be assembled independently, and both tails and head-like structures can be isolated from infected cells (Hatfull & Sarkis 1993).

Visualization of virion proteins by SDS-gel electrophoresis reveals about 10 readily identifiable bands, and these can be assigned as head or tail components by comparison with the purified substructures (Hatfull & Sarkis 1993). One additional protein is present in the head-like particles but is absent from virions and presumably is required only for head assembly. A particularly interesting feature of the L5 structure is that the major head protein subunits appear to be covalently crosslinked to each other in a manner described in detail for the unrelated coliphage, HK97 (Popa *et al.* 1991); as a consequence, even in the presence of SDS, most of the head protein subunits are not dissociated and fail to enter the stacking gel. A small proportion of head subunits do enter the separating gel but migrate as (presumably) incompletely crosslinked pentameric and hexameric assemblies (180 and 210 kDa, respectively). N-terminal amino acid sequencing of the putative hexamer and pentamer produces a single sequence that corresponds to the product of gene *17* whose predicted molecular weight is only 35 kDa (Hatfull & Sarkis 1993). N-terminal sequencing of other virion proteins has been useful in determining which genes encode them.

2.2 Genome organization

L5 particles contain a linear genome of 52 297 bp whose entire DNA sequence has been determined. The ends of the viral DNA contain 9-base complementary 3′-extended single-stranded cohesive termini (*cos*) that can pair to form a circular genome following infection (Oyaski & Hatfull 1992). The attachment site used for phage integration (*attP*) is close to the centre of the genome. The region between the leftmost end and *attP* is referred to as the left arm, and the region between *attP* and the right end as the right arm (Fig. 3.1). A total of 93 genes have been identified, including three that encode tRNAs and 90 putative protein-encoding genes, although a substantial proportion of these are unusually small. However, there are few non-coding regions, with the notable exception of about 1 kb at the extreme right end of the genome (Fig. 3.1). The standard genetic code is used by L5 although as a consequence of the high $G + C$ content (63.2%) there is a predominance of G and C in the third position of codons. Translation initiation occurs at AUG, GUG and UUG codons, with approximately equal usage of AUG and GUG. At least five genes initiate with UUG (Hatfull & Sarkis 1993).

The left and right arms of L5 form separate functional and transcriptional domains. All of the genes in the left arm (with the exception of the integrase gene immediately to the left of *attP*) are transcribed in the rightwards direction and encode virion structure and assembly functions. The cluster of three tRNA genes is also in the left arm (Fig. 3.1), but their specific role in L5 biology is not known. In contrast, all of the genes in the right arm (with the exception of *34.1*, immediately to the right of *attP*) are transcribed leftwards and many are involved in DNA (or nucleotide) metabolism and regulation. As discussed further below, the two arms are also transcribed at different times during lytic growth, such that the right-arm genes are expressed early, and the left-arm genes are expressed late.

Elucidation of L5 gene functions is complicated by the lack of sequence similarity to other phages (or bacteria) at either the nucleotide or protein level. However, as the sequence databases grow, progressively more alignments are emerging and currently about 10% of the putative L5-encoded proteins have sequence similarity to proteins from other sources (Hatfull & Sarkis 1993; Ford *et al.* 1998a). In a few cases, these include phage-encoded proteins (e.g. L5 gp14 which is related to a structural protein of lactococcal phage r1t) but several are related to bacterial proteins such as DNA polymerase (gp44) and ribonucleotide reductase (gp50). Some L5-encoded proteins (e.g. gp36) are related to bacterial proteins of *M. tuberculosis* which are probably components of resident

prophages (Hendrix *et al.* 1999; see also Chapter 5). We anticipate the number of matches will continue to grow as the databases enlarge.

The cluster of three tRNA genes located ≈ 4 kb from the left end represent something of a puzzle. Examination of the proposed tRNA secondary structures suggests that they are charged with asparagine, tryptophan and glutamine, respectively, but have anticodons which pair with reasonably common codons (although there is only a single tryptophan codon). It thus seems unlikely that they are used to compensate for rare codons within phage genes. It is also plausible that they simply provide a boost of protein synthesis during phage production and represent the surviving members of what was once a larger tRNA cluster. However, the necessity of the tRNA genes for phage growth in any host has yet to be demonstrated, and even if they are required there are numerous functions outside of protein synthesis in which they could be involved.

A notable departure of the L5 genome organization from that of λ and its relatives is the unidirectional nature of transcription in the right arm. Genetic studies have shown that the L5 repressor is encoded by gene *71*, which is located in an approximately colinear position to the *cI* repressor gene in λ. One consequence of this is that transcription of the upstream genes (*72–89*) during lytic growth will also result in synthesis of repressor, even though its action promotes a shutting down of lytic gene expression. In λ this does not occur since the genes immediately adjacent to *cI* are transcribed in the opposite direction. Thus, this feature of L5 genome organization raises significant questions as to which molecular events govern the lytic–lysogenic decision.

A second puzzle arising from genome organization concerns the control of phage integration and excision. In particular, the integrase gene is located immediately to the left of the *attP* site, such that the binding sites for integrase protein lie at the extreme 5′ end of the gene. Integrase must be expressed in two particular circumstances during phage propagation, one is following phage infection in order to form stable lysogens, and the other is during prophage excision. The second scenario is complicated by the

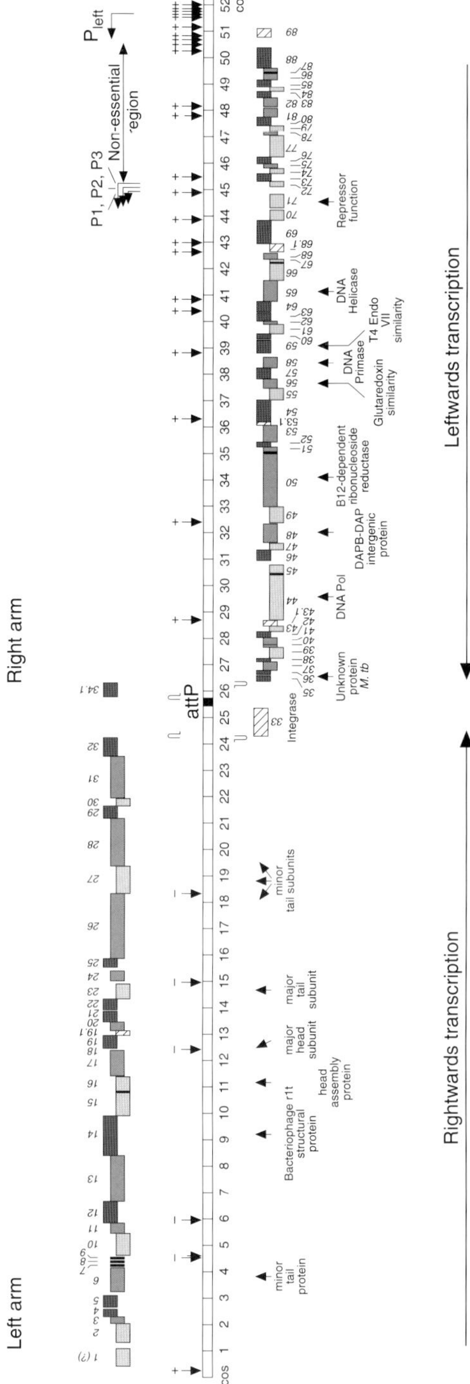

Fig. 3.1 Organization of the L5 genome. The 52 297-bp mycobacteriophage L5 genome is shown as a horizontal bar with coordinates marked at 1-kb intervals. Putative genes are shown as boxes, with those above the bar transcribed in the rightwards direction and those below in the leftwards direction. Putative gene functions are indicated. The locations of the *cos* ends, the *attP* site and promoters for repressor synthesis (P1, P2 and P3) and early lytic transcription (P$_{\text{left}}$) are shown. The positions of repressor-binding sites are indicated by vertical arrows above the genome with the orientation shown as – or +.

fact that when integrated, bacterial DNA extends to the right of the attachment site (*attR*) and is unlikely to contain promoters for integrase expression.

Thus, the idea that a promoter for integrase expression lies between the *int* gene and the positions of strand exchange at *attP* becomes attractive and preliminary experiments suggest a possible transcription start site between *int* and the leftmost integrase binding site of *attP* (M. Levin and G. F. Hatfull, unpublished observations). This in turn raises the question as to how integrase expression is regulated in a lysogen: there are no repressor binding sites in this region and perhaps integrase autoregulates its own synthesis, i.e. it acts as a transcription repressor as well as a recombinase. If this is the case, then the lysogenic attachment sites (*attL* and *attR*) are presumably occupied by Int and the only reason that excision does not occur is through tight down-regulation of a phage-encoded excisionase protein. A putative promoter overlapping the *attP* site could also account for synthesis of integrase following infections, since the promoter would in effect be derepressed until sufficient integrase was synthesized. While these musings are clearly speculative, they illustrate the perplexing consequences that arise from apparently subtle features of genome organization. They also help to formulate models for gene regulation that can readily be tested.

2.3 Gene expression and its regulation

Overview of expression patterns

Expression of L5 genes during lytic growth can be readily monitored by using thermo-inducible mutants of L5 that form stable lysogens at 30°C but initiate lytic growth when shifted to 42°C (Donnelly-Wu *et al.* 1993). Short metabolic labelling of the proteins synthesized at various times during lytic growth is very revealing and shows that there are two temporally separated phases of expression: early lytic growth that begins within five minutes of induction and continues for another 20–25 min, and late expression which begins about 20 min after induction and continues until cell lysis about 2 h later (Hatfull & Sarkis 1993). In general, the early genes

seem to be those in the right arm, and the late genes those in the left arm. The most prominent of these late proteins is the 35-kDa primary product of gene *17*, the major head subunit, and chasing of the labelled proteins with cold methionine results in loss of the 35-kDa band and the enhancement of higher-molecular-weight, crosslinked forms of the protein (Hatfull & Sarkis 1993). All phage gene expression utilizes the host RNA polymerase and is sensitive to rifampicin at all times during lytic growth (Hatfull & Sarkis 1993).

A second notable feature of these experiments is the reduction in expression of host genes within only a few minutes of induction. Presumably one or more phage-encoded early genes are responsible for specifically inhibiting host gene expression without influencing phage gene expression. However, we know little about the mechanism of this inhibition, which could occur by destruction of host DNA, inhibition of mRNA synthesis—at transcription initiation or mRNA stability—or regulation of host protein biosynthesis. Regulation of mRNA stability is quite an attractive possibility, since the phage genes could be expressed via as few as two long operons (one early, one late), and the 5′ end of at least the early mRNA (initiated at the P_{left} promoter; see below) is unusually stable (Hatfull & Sarkis 1993; Nesbit *et al.* 1995). Thus, a phage-encoded exoribonuclease specific for the 5′ ends of mRNAs (but to which the early transcript was not susceptible) could account for the observed patterns of protein synthesis.

Early lytic expression

Transcription of the early lytic genes initiates at a promoter, P_{left}, located at the right end of the genome (position 51 672) and proceeds leftwards (Fig. 3.1). It appears that P_{left} may be the only promoter required for expression of the early lytic genes, although it is hard to rule out the presence of others located in the right arm. The P_{left} promoter is inactive or very quiet in L5 lysogens but is rapidly turned on following induction of lytic growth (Nesbit *et al.* 1995), and significant protection of S1 probes continues well into the late lytic period. However, substantial protection of the same probe is also seen after incubation in rifampicin, suggesting that at least the 5′ ends of these P_{left}-initiated transcripts are found at late times due to prolonged mRNA stability rather than continued transcription initiation. The P_{left} promoter is recognized by *M. smegmatis* RNA polymerase *in vitro* and the promoter is highly active in *M. smegmatis* in the absence of all other phage genes.

The L5 phage repressor

What regulates the activity of P_{left}? Genetic studies have shown that L5 gene *71* has features of typical phage repressor genes, in that it is sufficient to confer superinfection immunity and is the location of several mutations giving rise to a clear plaque (i.e. non-temperate) phenotype. The amino acid sequence of the 183 amino acid protein contains a putative helix-turn-helix motif close to its N-terminus suggesting it binds DNA (Donnelly-Wu *et al.* 1993). DNA binding was confirmed by overexpression and purification of gp71 from *Escherichia coli* and demonstrating that it binds specifically to P_{left} DNA (Brown *et al.* 1997). There is a single binding site at the P_{left} promoter that overlaps the putative −35 hexamer required for RNA polymerase recognition such that gp71 binding could modulate transcription initiation. Curiously, there are 30 related sites scattered throughout the L5 genome—mostly in short intergenic regions—and 24 of these are substrates for binding of gp71 *in vitro*. Comparison of the sequences of these sites reveals a fairly tight 13-bp consensus (5′-GGTGGc/aTGTCAAG) which lacks obvious symmetry (Brown *et al.* 1997). Because of this sequence asymmetry, each of the sites can be assigned a specific orientation in the L5 genome (Fig. 3.1), revealing a remarkable correlation between site orientation and the direction of transcription.

It is unlikely that each of these binding sites acts as an operator site to control transcription initiation since several sites are in the left arm which is not transcribed until late in lytic growth. It seems rather that they play a role in regulating transcription elongation since placing a single site between the BCG hsp60 promoter and the firefly luciferase (*FFlux*)

reporter gene results in a polar effect on downstream gene expression in a gp71- and site-orientation-dependent manner (Brown *et al.* 1997). It is proposed that these sites are named 'stoperator' sites. The mechanism for the regulation is not known, although it presumably is not a physical roadblock to the transcription apparatus because of the strong dependence on site orientation and the relatively poor affinity of gp71 for these sites ($K_d = \approx 10\,nmol/l$). A possible biological explanation for these sites may be the need for L5 to silence phage gene expression during lysogeny, because even low levels of expression of genes that are antagonistic to bacterial growth (of which there must be many) would place lysogens at a selective disadvantage to non-lysogens. This need for prophage silencing must be common to all temperate phages, but different strategies can be employed to accomplish it. λ and its related phages contain numerous transcription terminators of both the ρ-factor-dependent and ρ-factor-independent types that could serve this function and which do not interfere with lytic growth due to the action of antitermination systems. It is plausible that L5 does not possess similar antitermination mechanisms.

Late lytic expression

Little is known about the location, number or control of late lytic promoters. Efforts to locate mRNA 5′ ends have produced ambiguous results, with the only signals found during late lytic growth being located within the 3′ end of gene *1*. Moreover, the signals are spread across a reasonably large region and differ in intensity from experiment to experiment, suggesting that they could arise from either processing or degradation of the mRNA rather than *bona fide* initiation events. While we anticipated that a late promoter would be located upstream of gene *1*, we note that the codon usage within gene *1* is poor and it may not be a functional gene.

There is also little known about how late gene expression is regulated. One attractive model is that late transcription requires a transcriptional activator that is encoded in the early transcriptional unit within the right arm. Recently, we have observed that prophage excision does not occur until the transition from early to late lytic expression (see below); since there is evidence that the excisionase gene is close to *attP*, the timing of excision could simply result from the time required for transcription complexes starting at P_{left} to travel 25 kb and reach the excisionase gene. The timing of late gene transcription could be regulated similarly if a putative activator gene were located at the end of the early operon.

2.4 Site-specific integration

Formation of L5 lysogens is accompanied by integration of the phage genome into the host chromosome by site-specific recombination. In its general respects, L5 integration is similar to that of other temperate bacteriophages, where recombination between a specific phage attachment site (*attP*) and a bacterial attachment site (*attB*) gives rise to an integrated prophage joined to the host chromosome by left (*attL*) and right (*attR*) attachment junction sites. The reaction is catalysed by the phage-encoded integrase protein, the product of gene *33* which is located immediately to the left of *attP* (Fig. 3.1).

Comparison of the *attP*, *attB*, *attL* and *attR* sequences reveals a 43-bp common core sequence within which strand exchange occurs (Lee *et al.* 1991). In *attB*, this common core overlaps the 3′ end of a tRNA^{gly} gene such that a complete tRNA^{gly} gene is reconstituted at *attL* following integration. Reconstitution of the tRNA gene appears to be essential for cell viability indicting that there is no redundancy of this tRNA function in *M. smegmatis* (Peña *et al.* 1997). Strand exchange occurs at one end of this common core and integrase cleaves seven bases apart with 5′ extensions. This seven-base overlap region corresponds to the anticodon loop of the tRNA^{gly} gene (Peña *et al.* 1996).

L5 integrase catalyses efficient integrative recombination *in vitro* (Lee & Hatfull 1993). Although the reaction conditions are fairly simple, requiring *attP* and *attB* DNAs, Int and a simple buffer, little recombination is observed unless an extract of *M. smegmatis* is also included indicating the requirement for a host protein factor (Lee & Hatfull 1993). This mycobacterial integration host factor (mIHF) has been purified

and characterized, and shown to be a novel 104 amino acid, heat-stable protein that is not obviously related at the sequence level to the HU/IHF family or any other family of proteins (Pedulla *et al.* 1996). The role of mIHF is curious since, unlike the well-studied IHF of *E. coli*, it does not bind with any sequence preference to *attP* DNA. However, it alters the manner in which Int binds to *attP*, and when both proteins are present, an intasome complex is formed that contains both Int and mIHF proteins (Pedulla *et al.* 1996).

A simple model for the role of mIHF in integration is that it facilitates the formation of Int-mediated protein bridges via looping of the DNA (Fig. 3.2). DNase I footprinting shows that in the absence of the host factor Int binds to several regions in *attP* DNA; one overlaps the end of the common core where strand exchange occurs (the core-type sites) and others that flank the core to the left and the right (the arm-type sites; Peña *et al.* 1997). Seven individual arm-type sites have been identified (P1–P7), three to the left and four to the right of the core (Fig. 3.2). With the exception of the P3 site, all of the others are present as pairs of directly repeated sites (i.e. P1 and P2, P4 and P5, P6 and P7). However, the P6 and P7 pair as well as P3 are dispensable for integrative recombination and the minimal active *attP* site is about 250 bp from P1 to P5 (Peña *et al.* 1997). mIHF appears to play a particularly important role in promoting the formation of an intramolecular bridge via simultaneous binding of Int to the core and the P4/P5 sites (Fig. 3.2). The P1/P2 pair of sites, while required for integration, are not needed for the formation of this intasome (C. E. A. Peña & G. F. Hatfull, unpublished observations 1998). Nevertheless, this intasome appears to be a true recombinational intermediate, since adding *attB* DNA leads to the formation of recombinant products. We also observe that upon addition of *attB*, a synaptic complex is formed in which *attB* is also present (Fig. 3.2). The P1/P2 pair of sites is required for the formation of this synaptosome and we suspect that they participate in forming Int-mediated intermolecular protein bridges between *attP* and *attB* (Fig. 3.2).

Prophage excision occurs during induction of L5 lysogens but does not seem to occur until the onset of

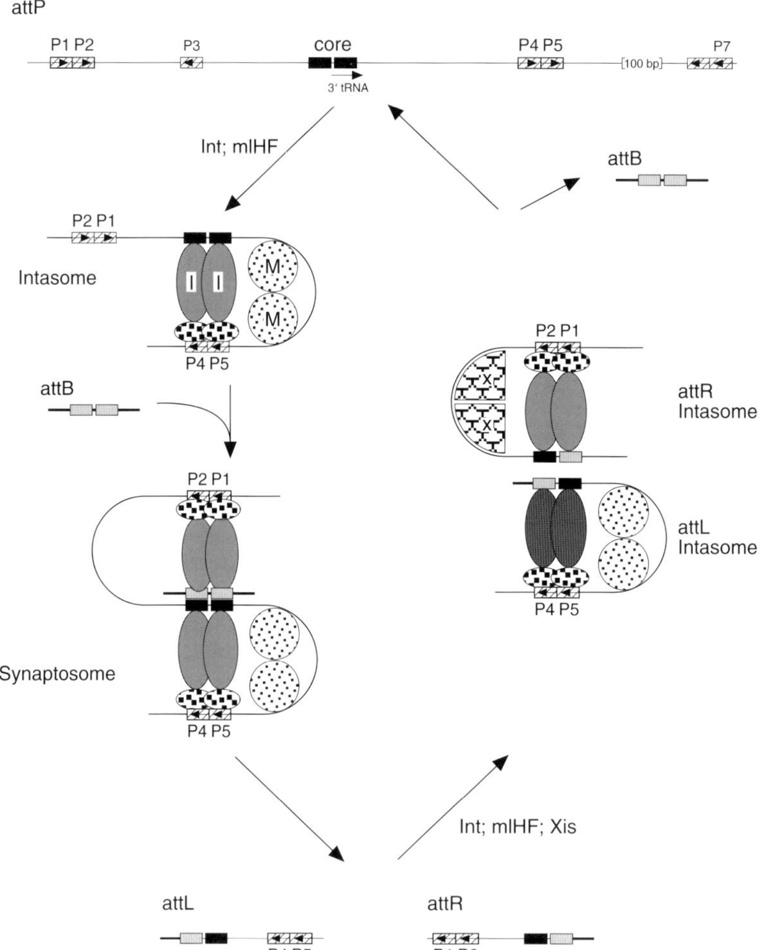

Fig. 3.2 Integration and excision of L5. The figure shows a schematic representation of pathways for integration and excision of L5. The phage attachment site (attP) is shown at the top with the core-type (core) and arm-type (P1–P7) integrase binding sites; P1, P2, P4 and P5 are the only arm-type sites needed for integration. The phage-encoded integrase (Int; I) protein and the host-encoded mycobacterial Integration Host Factor (mIHF; M) bind to *attP* DNA and form an intasome in which Int forms protein bridges between the P4/P5 arm-type sites and the core and mIHF stabilizes a sharp DNA bend. The P1/P2 arm-type sites are not occupied by Int. A synaptosome is formed by association of *attB* DNA *via* additional protein bridges between two more protomers of Int bound to the P1/P2 arm-type sites in *attP* and the core-type sites in *attB* DNA. Strand exchange occurs within this synaptic complex to produce the *attL* and *attR* attachment junctions. The intasome and synaptosome structures are supported by experimental evidence. We postulate that excision could occur by the formation of an *attL* intasome which is not significantly different to the *attP* intasome and an attR intasome, in which excisionase (Xis; X) binds between the core and P2 and promotes the formation of protein bridges between P1/P2 and the core. The *attL* and *attR* intasomes may interact via protein–protein interactions within which strand exchange occurs to yield the *attP* and *attB* products of excisive recombination. This model for excision is speculative.

late lytic growth (C. E. A. Peña & G. F. Hatfull, unpublished observations 1998). We assume that L5 encodes an excisionase protein that is required for excision, although the gene encoding this activity has yet to be identified. However, integration-proficient vectors (see below) that contain a 5-kb fragment of L5 surrounding *attP* form recombinants that are less stable than those constructed with a shorter segment of L5 DNA (Lee *et al.* 1991). Presumably, the larger segment contains the putative excisionase gene, but we do not yet know which of several open reading frames is responsible. The recent observation that *attR* DNA (containing the core and P1/P2 arm-type sites; Fig. 3.2) is released as free DNA following integrative recombination—and thus is unable to form an intasome complex with Int and mIHF—suggests the possibility that excisionase could possibly stimulate excision (i.e. recombination between *attL* and *attR*) by promoting the formation of *attR* intasomes (Fig. 3.2; C. E. A. Peña & G. F. Hatfull, unpublished observations 1998).

L5 integration occurs efficiently in both fast- and slow-growing mycobacteria (Lee *et al.* 1991). In part, this results from the presence of the conserved tRNA^{gly} gene that contains the necessary sequences for *attB* function. The 43-bp common core actually differs between *M. smegmatis* and BCG at one position, although it does not influence integration and occurs within the short variable loop of the tRNA, such that tRNA function is also not altered (Lee *et al.* 1991). However, the tRNA^{gly} expressed from an L5 lysogen of BCG is structurally distinct from that made in non-lysogenic BCG. *M. tuberculosis* and BCG also contain mIHF which appears to be highly conserved, even though the flanking sequences are not (Pedulla *et al.* 1996).

2.5 Molecular determinants of L5 infection

Rather little is known about how L5 recognizes its mycobacterial hosts and infects its DNA. L5-resistant mutants can be readily isolated but have not been characterized in detail (Donnelly-Wu *et al.* 1993). However, injection of L5 DNA is influenced by a gene

(*mpr*) that confers resistance when it is overexpressed (Barsom & Hatfull 1996). The mechanism whereby Mpr-overexpressing strains are resistant to L5 is not clear, although it is specific to L5 and its close relative D29, and does not significantly alter infection of *M. smegmatis* by other mycobacteriophages. Preliminary observations suggest that Mpr is localized to the mycobacterial membrane, and—perhaps not surprisingly—substantial overexpression leads to cell death (Barsom & Hatfull 1996). The entire *mpr* gene is not needed for this lethality and a segment encoding the first 60 amino acids is sufficient. Suppressor mutants that survive high-level *mpr* expression can be readily isolated but most of these are still phage resistant indicating that the secondary mutations are in other genes. These also have a variety of altered colony morphologies suggesting that tolerance of *mpr* overexpression could result from changes in membrane-associated secretory systems (Barsom & Hatfull 1996). These secondary mutations have yet to be mapped.

3 Mycobacteriophage D29

Mycobacteriophage D29 was isolated from soil by Froman in 1954 and infects a broad range of mycobacterial species including *M. smegmatis* and *M. tuberculosis* (Froman *et al.* 1954; Russell *et al.* 1963); it has also been shown to adsorb to *M. leprae* cells (David *et al.* 1984). It is not a temperate phage and kills a large proportion of cells that it encounters. Previous studies demonstrated that D29 efficiently adsorbs to *M. smegmatis* and initiates DNA replication soon after infection, with progeny phage appearing about 90 min after infection (Sellers *et al.* 1962; David *et al.* 1980).

The similarity of the morphologies of L5 and D29, the close correlation of L5 and D29 virion proteins, as well as the observation that D29 is subject to L5 superinfection immunity (i.e. it does not infect an L5 lysogen) suggest that L5 and D29 are closely related bacteriophages. Although the L5 and D29 genomes have quite distinct restriction maps (Lazraq *et al.* 1989; Oyaski & Hatfull 1992) they strongly cross-hybridize to each other (Donnelly-Wu *et al.* 1993). It is therefore not surprising that DNA sequencing of the D29 genome reveals it to be a very close relative of L5. Thus, while D29 has been frequently referred to as a virulent phage, it can be more accurately described as a clear-plaque variant of a temperate parent.

3.1 Genome organization

The DNA sequence of the D29 genome shows it be 49 136 bp in length with a *G* + C content of 63.3% (Ford *et al.* 1998); the genetic organization is shown in Fig. 3.3. Perhaps the most striking feature of the D29 genome is its similarity to that of L5. The two phage genomes are closely related at the DNA sequence level, although this cannot simply be stated as a percentage similarity due to the substantial numbers of insertions and deletions that are required to align them. However, in the left arm, the genomes are essentially colinear and the DNA sequences have about 80% similarity. The right arms have more discontinuities, with regions of high similarity interspersed with regions of low similarity (Ford *et al.* 1998).

The close relationship of L5 and D29 enables the assignment of some small genes that were not initially recognized in L5. For example, genes *19.1*, *43.1*, *53.1*, *68.1* and *89* have good codon usage in both phages and have been included in the revised L5 map shown in Fig. 3.1. The comparison also suggests that L5 gene *34* may have been misassigned, since the reading frame is not conserved in D29; however, another reading frame, designated *34.1*, is present in both, and a majority of sequence differences lie in the codon third position.

Most of the predicted protein sequences of D29 are related to L5 proteins and produce similar database search results. However, there are three examples of D29 putative protein sequences that identify significant similarities, but that do not have a counterpart in L5. One of these is the product of D29 gene *10* (gp10), that matches HI1415, a gene of unknown function identified by sequencing the *Haemophilus influenzae* genome (Fleischmann *et al.* 1995). Examination of the relationship between L5 gp10

and D29 gp10 shows that the D29 protein is about 200 amino acids longer than L5 gp10, as a result of additional sequences internal to the protein. It is this additional segment that matches HI1415, which itself is about 200 residues in length. The other two D29 matches identified are *36.1*, which matches deoxycytidinylate deaminases and *59.2* which matches non-haem haloperoxidases (Ford *et al.* 1998).

D29 contains the same three tRNA genes present in L5, but encodes two additional tRNAs, predicted to be tRNA^{glu} and tRNA^{tyr}. The intergenic regions are shorter in D29 than in L5, and the group of five D29 tRNA genes occupy only a slightly larger region than the three L5 genes. It is tempting to speculate that the L5 organization was derived from a precursor state that included a larger set of tRNA genes.

3.2 Lysogenization

A notable difference between the L5 and D29 genomes is a large (3.6 kb) deletion in the right arm of D29 that removes part of gene *71* through to part of gene *82*. The loss of a large segment of D29 gene *71* presumably accounts for the inability of D29 to form stable lysogens, although its genome organization and overall similarity to L5 indicates that it most likely originated from a temperate parental bacteriophage. Presumably, this was a relatively recent event, since D29 is capable of lysogenizing *M. smegmatis* provided that the L5 gp71 repressor is present (Ford *et al.* 1998). In light of this observation, it is perhaps not surprising that other features of L5 lysogeny are also present in D29. In particular, a sequence related to the L5 P_{left} promoter can be identified, and—although there are differences in the −10 sequence—a putative repressor binding site overlaps the −35 motif (Ford *et al.* 1998). The genome also contains many additional putative repressor binding sites which conform to a similar 13-bp consensus sequence to that in L5. Most are located in analogous positions to the L5 stoperator sites and are in the same orientation relative to the direction of transcription. Finally, there is evidence that D29 also has an active integration apparatus.

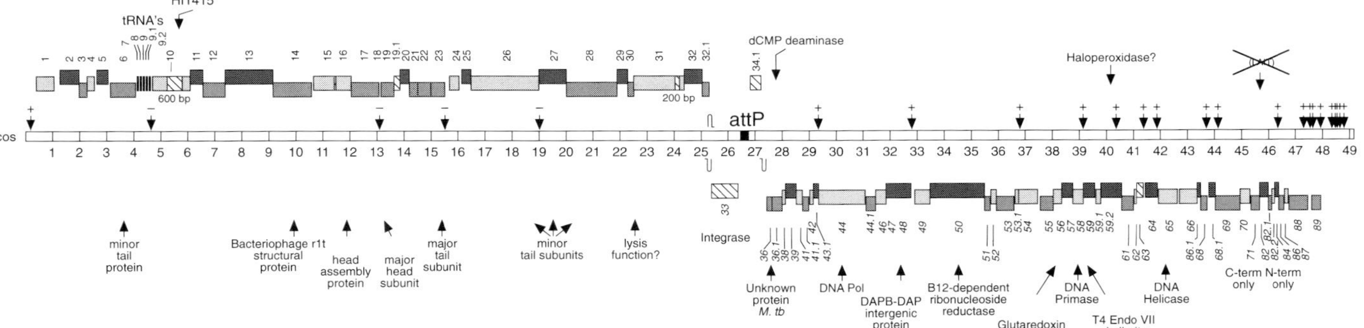

Fig. 3.3 Genome organization of mycobacteriophage D29. The 49 136-bp genome of mycobacteriophage D29 is shown as a horizontal bar with coordinates marked at 1-kb intervals. Putative genes are shown as boxes, with those above the bar transcribed in the rightwards direction and those below in the leftwards direction. Substantial portions of the D29 genome map are similar to that of L5 shown in Fig. 3.1, although there is a 3.4-kb deletion at the right end that removes part of gene *71*, part of gene *82* and the intervening DNA.

3.3 Integration

A region at the centre of the D29 genome corresponds to the *attP* site (Peña *et al.* 1997; Ford *et al.* 1998). In particular, the common core is present, although only 42 bp are identical to *attB* DNA; the non-identical base is at one extreme end, immediately to the left of the leftmost cut site (Peña *et al.* 1996). Comparison of the surrounding sequences shows that they are only about 70% identical to L5 (with inclusion of a few small gaps for optimal alignment) but having related sites in positions of the L5 arm-type sites (Peña *et al.* 1996). The consensus of the D29 arm-type site is related but not identical to the L5 arm-type consensus sequence. It is plausible that the D29 and L5 integrases exercise some degree of specificity in their binding to these arm sites.

A region of D29 containing *attP* and *int* (*33*) was shown previously to confer the ability to transform *M. smegmatis* when present on plasmid vectors, and was proposed to be an origin of replication (David *et al.* 1992). However, this seems unlikely and these vectors presumably transform via integration. This is confirmed by more detailed studies showing that recombinant plasmids containing D29 *attP* and *int* transform via integration into a specific site in the *M. smegmatis* genome (Ribeiro *et al.* 1997). Furthermore, formation of D29 lysogens in the presence of L5 repressor is accompanied by insertion of the D29 genome into the L5 (and D29) *attB* site (Ford *et al.* 1998).

3.4 Evolutionary origins

The evolution of bacteriophages is complex largely because of the expansive opportunities for lateral exchange of genes. In addition, they replicate extremely rapidly providing great opportunities for selection of new variant forms. Temperate phages may also act as donors and receivers of genetic information when present as resident prophages in infected cells. Comparison of D29 and L5 begins to shed some light on how these phages evolve.

A notable feature of the right arms of the D29 and L5 genomes is their apparent mosaicism, whereby segments of high similarity are punctuated by regions that are apparently unrelated. Typically, the junctions between these are at gene boundaries and there are numerous cases throughout the right arms. This scenario reflects a similar relationship to that seen in the family of λ-like phages of *E. coli* (Hendrix *et al.* 1999). While the presence of microhomologies derived from the sequence determinants for translation initiation and perhaps termination could promote recombination, we think it is more likely that it is only the rare illegitimate events at these boundaries that give rise to new combinations of genes or functional cassettes.

The observation that D29 contains a DNA segment that is homologous with an *H. influenzae* gene (HI1415) is remarkable. Closer examination of the *H. influenzae* genome surrounding HI1415 suggests that it is part of a defective prophage located within the *trp* operon. At least six genes in this putative prophage are related to known phage sequences, including the *H. influenzae* HP1 tail-fibre gene, the terminase from *Bacillus subtilis* phage SF6, the *rha* gene from *Salmonella* phage P22, the antirepressor gene of *E. coli* phage 933W and an integrase gene, in addition to the D29 gene *10* homologue. This amazing assortment of genes reflects the breadth to which phages genes can communicate with each other and the folly of using host-based classification methods for describing phage evolution.

4 Mycobacteriophage TM4

4.1 Introduction

Mycobacteriophage TM4 has been employed for a number of useful applications in mycobacterial genetics, although less is known about its overall biology than that of L5 and D29. It was isolated following induction of a putative lysogen of the *M. avium* complex (Timme & Brennan 1984) but infects both fast- and slow-growing mycobacteria. However, even though it efficiently infects *M. tuberculosis*, it infects BCG with a greatly reduced efficiency of plating, although host range mutants of TM4 that do infect BCG can be readily isolated (Jacobs *et al.* 1993).

Infection of *M. smegmatis* or BCG by TM4 produces only faintly turbid plaques indicating that it is unlikely that TM4 forms stable lysogens at high frequency, at least in these strains. Morphologically, TM4 is similar to L5 and D29, with an icosahedral head and a long, flexible, non-contractile tail (Timme & Brennan 1984).

4.2 Genome structure

The TM4 genome is ≈ 50 kb in length and contains cohesive ends, forming ladders in pulsed-field gel electrophoresis. The DNA is not evidently similar to any of the other well-characterized mycobacteriophages as judged by DNA hybridization or restriction digestion. However, the ease with which TM4 shuttle phasmids can be constructed (see below) suggests that there may be substantial portions of the TM4 genome that are nonessential for growth.

Recently, the complete sequence of the 52 797-bp genome of TM4 has been determined (Ford *et al.* 1998b) and a map of the genome organization established (Fig. 3.4). A large proportion of the genome contains protein-coding genes and there are few non-coding regions; in contrast to L5 and D29, all of the reading frames are transcribed in the same direction. In accord with its non-temperate behaviour, TM4 does not appear to encode an integrase protein and no attachment site has been located. Nevertheless, a region from the left end of the genome to ≈ 23 kb is remarkably similar in organization to the left arms of L5 and D29. Several reading frames in this region encode proteins with sequence similarities to L5 or D29 proteins, including gene *5* (similar to L5/D29 gene *14*), gene *16* (similar to L5/D29 gene *25*), gene *25* (similar to gene L5/D29 gene *32*) and gene *26* (similar to D29 gene *32.1*). N-terminal sequencing of TM4 virion proteins indicates that gene *14* encodes the major tail subunit (which is analogous but not homolous to L5/D29 gene *23*) and gene *21* encodes a minor tail protein (Fig. 3.4). A particular curiosity is that TM4 genes *15* and *16* are arranged like λ genes *G* and *T* and L5/D29 genes *24* and *25*, such that they are expected to be expressed as TM4 gp15 and gp15 to gp16 via a trans-

lational frameshift at the 3′ end of gene *15*. The flanking genes (TM4 *14* and *17*) have analogous functions to those in L5 (L5/D29 *23* and *26*, encoding the major tail and tail-length determinant proteins, respectively) but the sequence similarity is restricted to TM4 gene *16* and L5 gene *25*. Several other reading frames in this region encode proteins with similarity to those encoded by the temperate lactococcal phage r1t. It thus seems likely that TM4 genes *1–26* constitute an operon involved in virion structure and assembly with a global organization similar to that in mycobacteriophages L5 and D29 as well as phage λ.

If TM4 genes *1–26* are required for virion assembly, the remainder of the TM4 genes (*27–92*) may be involved in DNA metabolism and regulatory events. However, with the exception of gene *88* that encodes a homologue of L5 gene *78*, this region is quite different to the right arms of L5 and D29 and database searching has not identified a DNA polymerase (or RNA polymerase) or other activities involved in nucleotide or DNA metabolism. However, several open reading frames—including genes *29*, *55*, *67* and *70*—are related to putative genes in cryptic prophages of *M. tuberculosis* and *67* and *70* may be similar to glutaredoxins and primasome proteins, respectively. Perhaps the most enticing similarity is between the product of TM4 gene *49* and the *Streptomyces* WhiB protein which is transcriptional regulator required for sporulation. However, the specific role for gene *49* in TM4 growth is not known.

Rather little is known about TM4 gene expression, although preliminary observations suggest that expression is temporally divided into early and late patterns. The position of promoters is not known, although presumably the host RNA polymerase is used, since there are few open reading frames long enough to encode a viral RNA polymerase (although this cannot be excluded). It is plausible that gene *26* is the last of the late genes (as is its homologue, *32.1* in D29) and there is a short non-coding region of about 150 bp that could contain regulatory signals for genes *27–92*). The genes *27–39* are closely linked and the next intergenic gap does not appear until between genes *39* and *40*, and then between genes *40* and *41*.

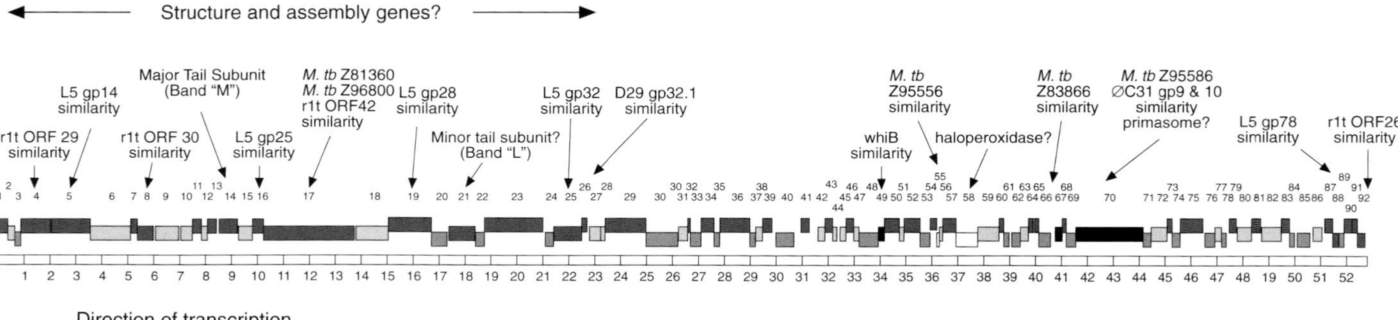

Fig. 3.4 Genome organization of mycobacteriophage TM4. The 52 797-bp genome of mycobacteriophage TM4 is shown as a horizontal bar with coordinates marked at 1-kb intervals. Putative protein-coding genes are shown as boxes above the bar; all of the TM4 genes identified are transcribed in the rightwards direction. Genes 1–26 probably encode virion structure and assembly proteins and there is sequence similarity between some of the putative products of these genes and L5 genes as indicated; TM4 gene *26* is similar to D29 *32.1* which is the last gene in the late operon. Other sequence similarities are noted including those corresponding to *Mycobacterium tuberculosis* predicted gene products.

This latter region does contain a sequence from which the transcribed RNA could fold into a secondary structure reminiscent of an RNase III recognition site, although it is not obvious what role it might play in TM4 development.

The elucidation of the TM4 genome map should now greatly facilitate efforts to understand its biology. Since numerous shuttle phasmids have been constructed that contain cosmids inserted at various places in the phage genome, identification of the points of insertion should reveal which parts are non-essential. In addition, through defined S1 probes, use of specific primers, and particular probes, it should be possible to map TM4 transcripts and elucidate the details of gene expression and regulation. In addition, host-range mutations can be located to identify TM4 proteins involved in host recognition.

5 Mycobacteriophage I3

Mycobacteriophage I3 was first described in 1971 by Sundaraj and Ramakrishnan as a generalized transducing phage of the mycobacteria. It is morphologically different to L5, D29 and TM4 and has a large hexagonal head and a contractile tail. Analysis of virion proteins reveals two (34 kDa and 70 kDa) that are highly abundant and probably constitute the major subunits of the head and tail; the gene encoding the larger of the two has been cloned and the sequence reported (Ramesh & Gopinathan 1994). However, the complete I3 genome sequence has yet to be determined and the overall genome organization is not known.

The I3 genome is circularly permuted and packaged by a 'headful' mechanism. A peculiar aspect of the I3 genome is the presence of short single-stranded gaps an average of 10 nucleotides in length. These short single-stranded regions are found at random throughout I3 genomes with 13–14 gaps in each DNA molecule. They do not appear to have any adverse effect on infectivity of I3 phage particles or the ability of I3 DNA to transfect *M. smegmatis* (Reddy & Gopinathan 1986). However, if the genome is denatured then the DNA characteristically breaks down to a heterogeneous-sized range of fragments that are all substantially shorter than the length of the genome. The role of these gaps in I3 is not known.

Phage I3 is the only general transducing phage of the mycobacteria to be described. The phage appears to transduce a variety of markers in *M. smegmatis* including auxotrophic and antibiotic resistance genes, although the frequency of transduction is fairly low (Raj & Ramakrishnan 1971; Saroja & Gopinathan 1973). The transfer of these markers is inhibited by I3 antiserum but not by DNase supporting the idea that these markers are transferred by transducing particles of I3.

6 Phage-derived genetic tools

6.1 Introduction

The above discussion of specific mycobacteriophages highlights the novel biological features that can be discovered in this fascinating group of viruses. However, there are numerous utilities for mycobacteriophages that can be envisaged, ranging from the direct application of mycobacteriophages to phage therapy and phage typing to the extraction of well-defined phage functions for specific purposes. The use of mycobacteriophages for phage typing has been quite highly developed and provides a relatively cheap and rapid means of classifying mycobacterial isolates. While not as quick or informative as some genomic approaches its cost-effectiveness makes it useful in particular circumstances.

As individual mycobacteriophages are dissected at the molecular level, a large panoply of phage-derived tools can be imagined. Phage L5 should be a particularly useful source since it is a temperate phage and contains characteristic integration and regulatory systems that are expected to be of considerable utility (Hatfull 1994). While several of these specific phage utilities will be described here, it is clear that there are many other potential applications that have yet to be explored.

6.2 Mycobacterial vector systems

Several different types of mycobacteriophage vector systems have been described. One intriguing class includes shuttle phasmids, which are chimeric molecules that replicate in *E. coli* as large circular plasmids, but grow as phages in mycobacteria (Jacobs *et al.* 1987). These take advantage of the fact that many mycobacteriophage genomes are of similar size to phage λ (45–50 kb) such that virtually the entire mycobacteriophage genome can be inserted into a λ cosmid vector and packaged into λ particles. These can then be recovered in *E. coli* and further be manipulated using *E. coli* rather mycobacteria. Recombinants that maintain all of the required functions for growth can be recovered as plaque-forming units following electroporation of mycobacteria with shuttle phasmid DNA (Hatfull & Jacobs 1994). These shuttle phasmids are particularly useful because the cosmid moiety can be altered to include genetic elements such as reporter genes or transposons that can be efficiently delivered to mycobacterial cells.

A second group of vectors are integration-proficient plasmids, which cannot replicate in mycobacteria but contain the integration apparatus (*attP* and *int*) of a temperate mycobacteriophage. The first of these to be described were derived from phage L5 (Lee *et al.* 1991) and transform with high efficiency to produce stable recombinants in which the plasmid DNA is integrated at the phage attachment site, *attB*. Similar vectors have been constructed from mycobacteriophages FRAT1 (Haeseleer *et al.* 1993), Ms6 (Anes *et al.* 1992) and the integrative *Streptomyces* plasmid pSAM1 (Martin *et al.* 1991; Seone *et al.* 1997). These vectors are particularly useful when it is necessary to construct recombinants that are genetically stable in the absence of selection, such as recombinant vaccine strains (Stover *et al.* 1991), or using *in vivo* selection approaches (Pascopella *et al.* 1994). The high frequency of transformation is also important since it enables the generation of highly representative recombinant libraries with integrated DNA's. Finally, since the presence of genes on extrachromosomal vectors can result in significant phenotypic effects, integration-proficient vectors provide a relatively simple method for analysing gene function and expression without such complications (Barsom & Hatfull 1996).

There are several specific circumstances where it is desirable to construct mycobacterial recombinants that do not carry antibiotic resistance genes, especially in recombinants designed for use as live human vaccines. An alternative selectable marker has been developed from the repressor gene (*71*) of L5, which confers superinfection immunity (Donnelly-Wu *et al.* 1993). Plasmid vectors containing L5 gene *71* can be introduced into mycobacteria by electroporation, followed by direct selection of solid medium supplemented with mycobacteriophages such as D29. This selection can be applied to both fast- and slow-growing mycobacteria and in the contexts of extrachromosomal and integrative plasmid vectors (Donnelly-Wu *et al.* 1993).

Mycobacteriophages harbour strong transcriptional promoters that can be utilized for the expression of foreign genes. However, the only well-characterized lytic promoter of mycobacteriophages is the P_{left} promoter of L5 which is active in early lytic growth (Nesbit *et al.* 1995). This is a strong promoter and gives high levels of expression of reporter genes in both fast-and slow-growing mycobacteria (Brown *et al.* 1997). The P_{left} promoter is also under the control of the phage repressor protein (gp71) which tightly down-regulates it. While further vector development is necessary, this should constitute a useful system for regulated gene expression of foreign genes in mycobacteria.

6.3 Transposon-delivery vehicles

The ability of bacteriophages to inject at least one genome into every cell present in a bacterial culture makes them ideally suited to the delivery of transposons, which generally move at relatively low frequencies (Bardarov *et al.* 1997). Moreover, since the transposition events can be selected directly, without significant bacterial growth, large numbers of independent transposition events can be generated.

The construction of transposon-delivery vehicles is facilitated by the isolation of conditionally replicating phage mutants. Temperature-sensitive derivatives of mycobacteriophages TM4 and D29 have been described that grow at 30°C but are defective in replication at 37°C (Bardarov *et al.* 1997). Shuttle phasmid derivatives of these were generated that carry the mycobacterial transposon, Tn*5378* (a kanamycin-resistant derivative of IS*1096*) or mini-Tn*10* (kan). Infection of fast- or slow-growing mycobacteria at 37°C with these phages results in the production of kanamycin-resistant derivatives in which the transposon has hopped from the phage onto the mycobacterial chromosome. Hundreds or thousands of derivatives can be isolated, although not all combinations of phages and hosts work equally well. Specifically, a TM4 derivative efficiently delivers Tn*5367* to *Mycobacterium phlei* and *M. tuberculosis* whereas D29 efficiently delivers Tn*5367* to BCG. The insertions occur at random with little or no target preference and apparently within all sectors of the *M. tuberculosis* genome (Bardarov *et al.* 1997; see also Chapter 2).

6.4 Luciferase reporter phages

Mycobacteriophages are also effective tools for the introduction of reporter genes. The idea of constructing recombinant phages carrying the firefly luciferase gene (*FFlux*) is attractive since mycobacteria do not normally emit light, and thus the presence of mycobacterial cells can be determined by infecting with the recombinant phage and detecting the emission of photons (Jacobs *et al.* 1993). Since the phages can potentially infect all of the mycobacterial cells present, and relatively few photons can be readily measured, this system should be able to detect mycobacteria with considerable sensitivity. By utilizing mycobacteriophages with different but well-defined host specificities it should be feasible to elucidate the nature of the infected host, similarly to the manner of phage typing. Most importantly though, the emission of light requires the expression of the luciferase gene, the presence of intracellular adenosine triphosphate (ATP), and the ability of the cells to take up the substrate, luciferin (Jacobs *et al.* 1993); dead cells do not make light. Thus, these phages have the potential to empirically determine drug susceptibility profiles by assessing whether the presence of an antibiotic influences the pattern of light production. The achievement of a system that

enables the rapid and inexpensive determination of drug susceptibility profiles of *M. tuberculosis* clinical isolates would meet an important clinical need.

The feasibility of using luciferase reporter phages has been demonstrated by the construction of TM4 shuttle plasmids that carrying the *FFlux* gene. Infection of *M. tuberculosis* cultures with this phage results in substantial light production which is sensitive to the presence of antibiotics (Jacobs *et al.* 1993). Moreover, light production can be used to effectively determine the drug-susceptibility profiles of cultured *M. tuberculosis* strains. While TM4 has a rather broad host range, the signals from *M. tuberculosis* can be effectively distinguished from those of other mycobacteria by the use of *p*-nitro-α-acetylamino-β-hydroxypropiophenoate (NAP) that selectively inhibits *M. tuberculosis* (Riska *et al.* 1997).

A potential limitation of the use of TM4-derived luciferase reporter phages is that phage-mediated lysis of the host cells results in loss of further light production. A set of additional TM4 luciferase shuttle phasmids were constructed and the one giving the highest luciferase expression was identified; most of the cosmid insertions were at the end of the right arm of the TM4 genome suggesting that this region contains non-essential functions; conditionally replicating phages were then isolated that would not promote lysis of the infected cells (Carrière *et al.* 1997). This results in a substantial improvement in the sensitivity of detection and the determination of the drug susceptibilities of as few as 120 BCG cells can be determined within 24 h (Carrière *et al.* 1997).

Luciferase reporter phages have also been constructed from phages L5 and D29 (Sarkis *et al.* 1995; Pearson *et al.* 1996). L5 luciferase phages were constructed by recombination resulting in a defined insertion of the reporter gene into the tRNA cluster close to the left end of the genome. A consequence of the manipulation was the fortuitous generation of a transcriptional promoter at the junction of phage and luciferase sequences resulting in high levels of luciferase expression in lysogens of *M. smegmatis* (Sarkis *et al.* 1995). Thus, phages with the greatest propensity to lysogenize produce the largest luciferase signals and those that either cannot lysogenize or do so at a reduced frequency produce substantially lower signals. Since *M. smegmatis* has a reasonably rapid growth rate, subsequent growth of the infected cells permits the identification of extremely small numbers of cells (< 10) in a fairly short period of time (Sarkis *et al.* 1995). The L5 derivative that produces the greatest signals from *M. smegmatis* is also a host-range mutant that does not infect slow-growing mycobacteria and these therefore have somewhat limited utility for the analysis of *M. tuberculosis* (Fullner & Hatfull 1997). Phage D29 luciferase phages effectively infect both *M. smegmatis* and BCG but do not have the lysogenic properties of L5 (Pearson *et al.* 1996).

7 Perspectives

Mycobacteriophages have clearly played important roles in understanding the molecular biology of the mycobacteria, shedding new light on mycobacterial genetics and fueling new developments in the control of mycobacterial diseases. What can mycobacteriophages tell us about mycobacterial virulence?

It is becoming ever apparent that bacteriophages play a central role in bacterial pathogenesis. In particular, phage-associated genes have been implemented in the virulence of *E. coli*, *Vibrio cholerae*, *Shigella dysenteriae*, *Corynebacterium diphtheriae*, as well as other Gram-negative and Gram-positive bacteria. As more complete bacterial genome sequences are determined, it is becoming clear that resident prophages or shipwrecked cryptic prophages constitute a significant portion of the genome. It also is apparent that these prophages contribute to the overall properties of the bacteria through expression of phage-encoded or phage-associated genes. Moreover, many of these genes can be transmitted horizontally from cell to cell producing new pathogenic varieties of bacteria.

It is thus fair to ask whether phage-encoded or phage-associated genes contribute to the pathogenicity of *M. tuberculosis*. At the time of writing the genomic sequence of one strain of *M. tuberculosis* has been determined and a second is close to completion. From the sequences reported so far, at least three cryptic prophages or phage remnants can be identified (see Chapter 5). The investigation of mycobacteriophage biology greatly facilitates the identification of these elements since they contain homologues of genes found in L5, D29 and TM4, and the characterization of additional mycobacteriophage genomes will further simplify this analysis. It is thus reasonable to assume that mycobacteriophages will persist as central players in understanding both the molecular biology and virulence of the mycobacteria.

8 References

Anes, E., Portugal, I. & Moniz-Pereira, J. (1992) Insertion into the *Mycobacterium smegmatis* genome of the *aph* gene through lysogenization with the temperate mycobacteriophage Ms6. *FEMS Microbiological Letters* **74**, 21–25.

Bardarov, S., Kriakov, J., Carriere, C. *et al.* (1997) Conditionally replicating mycobacteriophages: a system for transposon delivery to *Mycobacterium tuberculosis*. *Proceedings of the National Academy of Sciences of the USA* **94**, 10961–10966.

Barksdale, L. & Kim, K.-S. (1977) *Mycobacterium*. *Bacteriological Reviews* **41**, 217–372.

Barsom, E.K. & Hatfull, G.F. (1996) Characterization of a *Mycobacterium smegmatis* gene that confers resistance to phages L5 and D29 when overexpressed. *Molecular Microbiology* **21**, 159–170.

Brown, K.L., Sarkis, G.J., Wadsorth, C. & Hatfull, G.F. (1997) Transcriptional silencing by the mycobacteriophage L5 repressor. *EMBO Journal* **16**, 5914–5921.

Carrière, C., Riska, P.F., Zimhony, O. *et al.* (1997) Conditionally replicating luciferase reporter phages: Improved sensitivity for rapid detection and assessment of drug susceptibilities of *Mycobacterium tuberculosis*. *Journal of Clinical Microbiology* **35**, 3232–3239.

David, H.L., Clavel, S. & Clement, F. (1980) Absorption and growth of the bacteriophage D29 in selected mycobacteria. *Annals of Virology (Institut Pasteur)* **131**, 167–184.

David, H., Clement, F., Clavel-Seres, S. & Rastogi, N. (1984) Abortive infection of *Mycobacterium leprae* by the mycobacteriophage D29. *International Journal of Leprosy* **52**, 515–523.

David, M., Lubinsky-Mink, S., Ben-Zvi, A., Ulitzer, S., Kuhn, J. & Suissa, M. (1992) A stable *Escherichia coli-Mycobacterium smegmatis* plasmid shuttle vector containing the mycobacteriophage D29 origin. *Plasmid* **28**, 267–271.

Doke, S. (1960) Studies on mycobacteriophages and

lysogenc mycobacteria. *Journal of Kunamoto Medical Society* **34**, 1360–1371.

Donnelly-Wu, M., Jacobs, W.R. & Hatfull, G.F. (1993) Superinfection immunity of mycobacteriophage L5: applications for genetic transformation of mycobacteria. *Molecular Microbiology* **7**, 407–417.

Fleischmann, R.D., Adams, M.D., White, O. *et al.* (1995) Whole-genome random sequencing and assembly of *Haemophilus influenzae* Rd. *Science* **269**, 496–512.

Ford, M.E., Sarkis, G.J., Belanger, A.E., Hendrix, R.W. & Hatfull, G.F. (1998a) Genome structure of mycobacteriophage D29: Implications for phage evolution. *Journal of Molecular Biology* **279**, 143–164.

Ford, M.E., Stenstrom, C., Hendrix, R.W. & Hatfull, G.F. (1998b) Mycobacteriophage TM4: genome structure and gene expression. *Tubercle and Lung Disease* **79**, 63–73.

Froman, S., Will, D.W. & Bogen, E. (1954) Bacteriophage active against virulent *Mycobacterium tuberculosis* I. Isolation and activity. *American Journal of Public Health* **44**, 1326–1333.

Fullner, K.J. & Hatfull, G.F. (1997) Mycobacteriophage L5 infection of *Mycobacterium bovis* BCG: Implications for phage genetics in the slow-growing mycobacteria. *Molecular Microbiology* **26**, 755–766.

Haeseleer, F., Pollet, J.-F., Haumont, M., Bollen, A. & Jacobs, P. (1993) Stable integration and expression of the *Plasmodium falciparum* circumsporozoite protein coding sequence in mycobacteria. *Molecular Biochemistry and Parasitology* **57**, 117–126.

Hatfull, G.F. (1994) Mycobacteriophage L5: a toolbox for tuberculosis. *ASM News* **60**, 255–260.

Hatfull, G.F. & Jacobs, W.R. Jr (1994) Mycobacteriophages: Cornerstones of Mycobacterial Research. In: *Tuberculosis: Pathogenesis, Protection and Control* (ed. B. R. Bloom). Washington DC: American Society for Microbiology, pp. 165–183.

Hatfull, G.F. & Sarkis, G.J. (1993) DNA sequence, structure and gene expression of mycobacteriophage L5: a phage system for mycobacterial genetics. *Molecular Microbiology* **7**, 395–405.

Hendrix, R.W., Smith, M.C.M., Burns, R.N., Ford, M.E. & Hatfull, G.F. (1999) Evolutionary relationships among diverse bacteriophages and prophages: All the world's a phage. *Proceedings of the National Academy of Sciences of the USA* **96**, 2192–2197.

Jacobs, W.R. Jr, Tuckman, M. & Bloom, B.R. (1987) Introduction of foreign DNA into mycobacteria using a shuttle phasmid. *Nature* **327**, 532–535.

Jacobs, W.R., Jr, Barletta, R., Udani, R. *et al.*(1993) Rapid assessment of drug susceptibilies of *Mycobacterium tuberculosis* by means of luciferase reporter phages. *Science* **260**, 819–822.

Lazraq, R., Moniz-Pereira, J., Clavel-Séres, S., Clément, F. & David, H.L. (1989) Restriction map of

Mycobacteriophage D29 and its deletion mutant F5. *Acta Leprologica* **7**, 234–238.

Lee, M.H., Pascopella, L., Jacobs, W.R. & Hatfull, G.F. (1991) Site-specific integration of mycobacteriophage L5: integration-proficient vectors for *Mycobacterium smegmatis*, *Mycobacterium tuberculosis*, and bacille Calmette–Guérin. *Proceedings of the National Academy of Sciences of the USA* **88**, 3111–3115.

Lee, M.H. & Hatfull, G.F. (1993) Mycobacteriophage L5 integrase-mediated site-specific recombination *in vitro*. *Journal of Bacteriology*. **175**, 6838–6841.

Martin, C., Mazodier, P., Mediola, M.V. *et al.* (1991) Site-specific integration of the *Streptomyces* plasmid pSAM2 in *Mycobacterium smegmatis*. *Molecular Microbiology* **5**, 2499–2502.

Mizuguchi, Y. (1984) Mycobacteriophages. In: *The Mycobacteria: a Sourcebook,* (eds G. P. Kubica & L. G. Wayne). New York: Marcel Dekker, pp. 641–662.

Nesbit, C.E., Levin, M.E., Donnelly-Wu, M.K. & Hatfull, G.F. (1995) Transcriptional regulation of repressor synthesis in mycobacteriophage L5. *Molecular Microbiology* **17**, 1045–1056.

Oyaski, M. & Hatfull, G.F. (1992) The cohesive ends of mycobacteriophage L5 DNA. *Nucleic Acids Research* **20**, 3251.

Pascopella, L., Collins, F.M., Martin, J.M. *et al.* (1994) Use of *vivo* complementation in *Mycobacterium tuberculosis* to identify a genomic fragment associated with virulence. *Infection and Immunity* **62**, 1313–1319.

Pearson, R.E., Jurgensen, S., Sarkis, G.J., Hatfull, G.F. & Jacobs, W.R. Jr (1996) Construction of D29 shuttle phasmids and luciferase reporter phages for detection of mycobacteria. *Gene* **183**, 129–136.

Pedulla, M.L., Lee, M.H., Lever, D.C. & Hatfull, G.F. (1996) A novel host factor for integration of mycobacteriophage L5. *Proceedings of the National Academy of Sciences of the USA* **93**, 15411–15416.

Peña, C.E.A., Stoner, J.E. & Hatfull, G.F. (1996) Positions of strand exchange in mycobacteriophage L5 integration and characterization of the *attB* site. *Journal of Bacteriology* **178**, 5533–5536.

Peña, C., Lee, M.H., Pedulla, M.L. & Hatfull, G.F. (1997) Characterization of the mycobacteriophage L5 attachment site, *attP*. *Journal of Molecular Biology* **266**, 76–92.

Popa, M.P., McKelvey, T.A., Hempel, J. & Hendrix, R.W. (1991) Bacteriophage HK97 structure: Wholesale covalent cross-linking between the major head shell subunits. *Journal of Virology* **65**, 3227–3237.

Raj, C.V.S. & Ramakrishnan, T. (1971) Genetic studies in mycobacteria: Isolation of auxotrophs and mycobacteriophages for *Mycobacterium smegmatis* and their use in transduction. *Journal of the Indian Institute of Science* **53**, 126–140.

Ramesh, G.R. & Gopinathan, K.P. (1994) Structural proteins of mycobacteriophage I3, Cloning, expression and sequence analysis of a gene encoding a 70-kDa structural protein. *Gene* **143**, 95–100.

Reddy, A.B. & Gopinathan, K.P. (1986) Presence of random single-stranded gaps in mycobacteriophage I3 DNA. *Gene* **44**, 227–234.

Redmond, W.B. (1963) Bacteriophages of mycobacteria: a review. *Advances in Tuberculosis Review* **12**, 191–229.

Redmond, W.B. & Carter, J.C. (1960) A bacteriophage specific to *Mycobacterium tuberculosis* varieties *hominis* and *bovis*. *American Review of Respiratory Disease* **82**, 781–786.

Ribeiro, G., Viveiros, M., David, H.L. & Costa, J.V. (1997) Mycobacteriophage D29: contains an integration system similar to that of the temperate mycobacteriophage L5. *Microbiology* **143**, 2701–2708.

Riska, P.F., Jacobs, W.R. Jr, Bloom, B.R., McKitrick, J. & Chan, J. (1997) Specific identification of *Mycobacterium tuberculosis* with luciferase reporter mycobacteriophage: Use of *p*-nitro-α-acetylamino-β-hydroxy-propiophenoate. *Journal of Clinical Microbiology* **35**, 3225–3231.

Russell, R.L., Jann, G.J. & Froman, S. (1963) Lysogeny in the mycobacteria. II. Some phage–host relationships of lysogenic mycobacteria. *American Review of Respiratory Diseases* **88**, 528–538.

Sarkis, G.J., Jacobs, W.J. Jr & Hatfull, G.F. (1995) L5 luciferase reporter mycobacteriophages: a sensitive tool for the detection and assay of live mycobacteria. *Molecular Microbiology* **15**, 1055–1067.

Saroja, D. & Gopinathan, K.P. (1973) Transduction of isoniazid susceptibility-resistance and streptomycin resistance in mycobacteria. *Antimicrobial Agents and Chemotherapy* **4**, 643–645.

Sellers, M.I., Baxter, W.L. & Runnals, H.R. (1962) Growth characteristics of mycobacteriophages D28 and D29. *Canadian Journal of Microbiology* **8**, 389–399.

Seone, A., Navas, J. & Lobo, J.M. (1997) Targets for pSAM2 integrase-mediated site-specific integration in the *Mycobacterium smegmatis* chromosome. *Microbiology* **143**, 3375–3380.

Snapper, S.B., Lugosi, L., Jekkel, A. *et al.* (1988) Lysogeny and transformation in mycobacteria: stable expression of foreign genes. *Proceedings of the National Academy of Sciences of the USA* **85**, 6987–6991.

Snider, D.E., Jones, W.D. Jr & Good, R.C. (1984) The usefulness of phage typing *Mycobacterium tuberculosis* isolates. *American Review of Respiratory Diseases* **130**, 1095–1099.

Stover, C.K., de la Cruz, V.F., Fuerst, T.R. *et al.*(1991) New use of BCG for recombinant vaccines. *Nature* **351**, 456–460.

Sundaraj, C.V. & Ramakrishnan, T. (1971) Transduction in *Mycobacterium smegmatis*. *Nature* **228**, 280–281.

Timme, T.L. & Brennan, P.J. (1984) Induction of bacteriophage from members of the *Mycobacterium avium*, *Mycobacterium intracellulare*, *Mycobacterium scrofulaceum* serocomplex. *Journal of General Microbiology* **130**, 2059–2066.

Chapter 4 / Gene expression and regulation

JULIANO TIMM, MANUEL GOMEZ & ISSAR SMITH

1 Introduction

Most scientists involved in mycobacterial research are ultimately interested in pathogenic members of this genus and how to prevent or cure the diseases they cause. To acquire information that can be used to develop new strategies for diagnosis, vaccines and antimycobacterial agents, it is essential to study the genetics and physiology of mycobacterial pathogens. In order to survive and grow in the host, pathogenic mycobacteria must circumvent the hostile environment of the phagocytic cells and the granuloma. Therefore, it is important to understand how these bacteria respond to this stressful environment and control the expression of genes that encode factors essential for the disease process.

The major mechanism regulating gene expression in prokaryotes is transcriptional control, and it is reasonable to predict that this is also true for mycobacteria. Although studies of mycobacterial RNA polymerases were first performed 20 years ago, the identification of mycobacterial promoters, σ factors and other transcriptional regulators has only recently begun.

This chapter will focus on mycobacterial gene expression and regulation. The first section summarizes methods for studying gene expression and the second and third sections review our knowledge of the basic transcriptional machinery and translational signals, respectively. Finally, the last section discusses mycobacterial responses to several types of environmental change and some properties of mycobacterial pathogens that are relevant to their intracellular lifestyle.

2 Methods for studying mycobacterial gene expression

This section describes techniques for monitoring gene expression in mycobacteria. All methods discussed here have been first developed for other organisms and later adapted, with only slight modifications, to mycobacteria. As a result, they have been extensively discussed in methodological handbooks, e.g. those for *Bacillus* (Harwood & Cutting 1990), *Escherichia coli* (Miller 1972) or *Streptomyces* (Hopwood *et al.* 1985). Detailed protocols can also be found in general laboratory manuals (Sambrook *et al.* 1989). We will therefore only briefly review these techniques and concentrate on providing examples of their applications in mycobacteria.

We will also highlight methods for isolating mycobacterial genes expressed preferentially during infection. Some of these techniques, such as *in vivo* expression technology (IVET) or differential fluorescence induction (DFI), are relatively new but have already allowed the identification of virulence factors in other systems (Heithoff *et al.* 1997b).

Techniques for studying gene expression fall in two categories: 'genetic methods', the study of gene fusions, and 'biochemical methods' for the detection of transcription (mRNA) or translation (protein) products.

2.1 Genetic methods

Gene fusion is a powerful tool for the study of gene expression. Usually the gene of interest is fused to a 'reporter gene' encoding an enzyme whose activity is assayed by a colorimetric or fluorimetric method. A selectable phenotype can also be linked to the reporter gene allowing the selection for genes which are turned on and off in different conditions.

Fusions are constructed by creating chimeric DNAs between the gene of interest and a reporter gene. If the reporter gene is placed under the control of both transcriptional and translational signals of the target gene and a hybrid protein is produced, the construct is defined as a translational or gene fusion. When an intact reporter gene containing its own translational start signals is placed downstream from the promoter and the transcriptional start point of the target gene, the construct is called a transcriptional or operon fusion.

There are basically two ways to construct gene fusions. One takes advantage of the properties of transposable genetic elements. Such an element can be modified to harbour a reporter gene without affecting its transposition functions. This transposon-based fusion generator is then delivered to the bacteria and gene fusions are created by random insertion of the element throughout the chromosome, producing a library of gene fusions. Although efficient transposon delivery systems are now available for

mycobacteria (see Chapter 2), engineered transposons containing reporter genes have not yet been described.

The second method of fusion construction is based on recombinant DNA. The fusion is created *in vitro* using a plasmid or phage that carries the reporter gene. Table 4.1 lists several mycobacteria–*E. coli* shuttle plasmids designed for the creation of fusions to a variety of reporter genes. One should take two precautions when using plasmid vectors for fusion construction. A common problem is that readthrough from plasmid promoters interferes with the monitoring of target promoter activity. This problem is overcome in most cases through the incorporation of transcription terminators upstream of the reporter gene. Another possible problem is that some promoters respond differently when placed in a multicopy plasmid. This has been observed with the regulatory sequences of the heat-shock gene *hsp60* from *Mycobacterium bovis* bacille Calmette–Guérin (BCG). An *hsp60–lacZ* fusion produced β-galactosidase constitutively when placed on a plasmid, but when integrated in the BCG chromosome, showed an increase in β-galactosidase levels in response to stress with heat, acid and peroxide (Stover *et al.* 1991). The reasons for these discrepancies between extrachromosomal and integrated fusions are not completely understood, but have been also observed in other organisms (Zuber *et al.* 1987). These results prompted us and others to design vectors which allow integration of gene fusions in mycobacterial chromosomes through the integrative functions of the mycobacterial phage L5 (see Chapter 3).

In the past 6 years several reporter genes have been successfully used in mycobacteria. We will briefly discuss the advantages and limitations of each reporter system and provide examples of some of their specific applications.

2.1.1 *'Classical reporter genes'*

lacZ

This was the first reporter gene used in mycobacteria. The *lacZ* gene product, *E. coli* β-galactosidase, is prob-

Table 4.1 Mycobacterial vectors for the construction of fusions to reporter genes.

Vector	Type[a]	Marker[b]	Reporter gene and fusion type	Reference[c]	Comment
pJEM12–13–14	rep	kan	*lacZ*, translational	1	Allow the construction of translational fusions to *lacZ* in all reading frames and contain a transcription terminator.
pJEM15	rep	kan	*lacZ*, transcriptional	1	Derivative of pJEMs containing a *cII-lacZ* fusion and a synthetic ribosome binding site
pSM128	int	str	*lacZ*, transcriptional	2	Derivative of pJEM15 containing the L5 *int* cassette
pSD7	rep	kan	*cat*, transcriptional	3	Contains transcription and translation terminators
pRCX3	rep	kan	*xylE*, transcriptional	4	Does not contain a transcription terminator
pJEM11	rep	kan	*phoA*, translational	5	Contains a truncated *phoA* and a transcription terminator
pMH66	int	hyg	*FFlux*, transcriptional	6	Contains a transcription terminator, *lacZ* and *FFlux* (see text)
pFPV27	rep	kan	*mgfp*, transcriptional	7	Contains a mutant *gfp* gene isolated by Cormack *et al.* (1996) (see text), but not a transcription terminator

[a] rep, replicative; int, integrative.
[b] kan, kanamycin; str, streptomycin; hyg, hygromycin.
[c] 1, Timm *et al.* (1994b); 2, J. Timm, unpublished; 3, Das Gupta *et al.* (1993); 4, Curcic *et al.* (1994); 5, Lim *et al.* (1995); 6, Mdluli *et al.* (1996); 7, Barker *et al.* (1998).

ably the most widely used reporter of gene expression in prokaryotes, the main reason for this being that fusions to *lacZ* are easy to analyse. β-galactosidase activity can be readily detected in bacterial colonies by adding the compound 5-bromo-4-chloro-3-indolyl-β-ᴅ-galactoside (X-gal) to the medium; clones expressing *lacZ* exhibit a blue colour, the intensity of which provides an estimation of the levels of *lacZ* expression. β-galactosidase activity can also be quantitatively assayed in bacterial extracts through an inexpensive colorimetric assay which employs the compound *o*-nitrophenyl-β-ᴅ-galactoside (ONPG) or, by using the fluorescent substrates fluorescein-di-β-ᴅ-galactoside (FDG) or 4-methylumbelliferyl-β-ᴅ-galactopyranoside (MUG), when the levels are low or when the number of bacteria is reduced as is the case in infected macrophages. In addition, it is possible to construct both operon and gene fusions to *lacZ* and, consequently, to study gene regulation at the post-transcriptional level.

In mycobacteria, *lacZ* fusions were useful in studying the regulation of several genes *in vitro*, e.g. *M. fortuitum blaF* (Timm *et al.* 1994b) and *M. tuberculosis purCL* (Jackson *et al.* 1996). Libraries of *lacZ* fusions containing *M. tuberculosis* and mycobacterial phage DNA have also been constructed and screened for new regulatory elements (Barletta *et al.* 1992; Timm *et al.* 1994a; Jain *et al.* 1997). *lacZ* fusions were also employed to study gene regulation *in vivo*, i.e. fusions expressed by mycobacteria growing within host cells. Using FDG, which can cross the cell envelopes of mycobacteria and mammalian cells, it was possible to directly monitor the activity of various promoters in internalized BCG cells without macrophage lysis (Dellagostin *et al.* 1995).

The cat *gene*

This constitutes an interesting alternative to *lacZ* since it provides a selectable phenotype. The *cat* gene product, chloramphenicol acetyltransferase (CAT), inactivates chloramphenicol, an antibiotic that is quite stable and can be used with slow and rapidly growing mycobacteria, as both exhibit very low levels of natural resistance. This property was used by Tyagi and collaborators to develop a 'promoter trap' for mycobacteria. This system is based on a plasmid (pSD7) containing a promoterless *cat* preceded by a transcription terminator and cloning sites. Genomic libraries can be constructed in this vector and screened for promoters which support growth of *M. smegmatis* in the presence of different concentrations of chloramphenicol (Das Gupta *et al.* 1993; Jain *et al.* 1997).

CAT catalyses the acetylation of chloramphenicol using acetyl-CoA, which liberates a free CoA sulf-hydryl group. Although spectrophotometric methods for assaying CAT have been described (Kleanthous & Shaw 1984), the most sensitive and widely used is based on the detection of ^{14}C-labelled chloramphenicol on thin-layer chromatograms (Gorman *et al.* 1982). CAT activity is reported to be low when it is fused to another polypeptide, therefore gene fusions to *cat* are usually transcriptional.

xylE

Another alternative to *lacZ* is the *xylE* gene from *Pseudomonas*. The product of *xylE*, catechol 2,3-dioxygenase (CDO), converts catechol into a compound with a yellow colour which permits the screening for colonies harboring active transcriptional fusions (Konyecsni & Deretic 1988). CDO activity can also be quantified in mycobacterial extracts and whole cells by a simple colorimetric assay. *xylE* was used to analyse the expression of the response regulator gene *mtrA* from *M. tuberculosis* (Curcic *et al.* 1994).

phoA

The *phoA* gene product, *E. coli* alkaline phosphatase, functions as a reporter of protein export since, usually, translational fusions to a truncated *phoA* produce proteins with alkaline phosphatase activity only when the target gene encodes an exported protein containing a signal sequence (Hoffman & Wright 1985). This property permitted the development of a genetic method for identifying genes encoding exported proteins which has recently been adapted to mycobacteria. It is based on a plasmid (pJEM11) containing a truncated *phoA* placed downstream of a transcription terminator and cloning sites. Genomic libraries can be constructed in this vector and screened in *M. smegmatis* for clones with alkaline phosphatase activity, which exhibit a blue colour in the presence of the chromogenic substrate 5-bromo-4-chloro-3-indolyl phosphate (X-P). This approach permitted the isolation of previously uncharacterized putative exported proteins from *M. tuberculosis* (Lim *et al.* 1995). *phoA* has also been employed in studies where the target gene encoded a known exported protein, e.g. the *M. fortuitum* β-lactamase gene (Timm *et al.* 1994b) and the *M. tuberculosis* 85A gene (Kremer *et al.* 1995a).

The PhoA assay is done with sonicated extracts and *p*-nitrophenylphosphate as the substrate (Brockman & Heppel 1968).

IVET systems

Recently, a special genetic reporter system, IVET, was described, which uses the host (cell cultures or an animal) as 'selective medium' to enrich for gene fusions that are preferentially expressed *in vivo*. IVET is basically a promoter trap in which the reporter gene encodes a protein required for pathogenesis (e.g. an enzyme of a biosynthetic pathway indispensable *in vivo*) or one that confers antibiotic resistance (e.g. CAT). Usually, IVET vectors contain a second reporter gene which can be monitored *in vitro* and allows counterscreening for fusions expressed in these conditions. For example, Heithoff *et al.* (1997a) employed a IVET vector containing a tandem *cat–lacZ* to isolate several *Salmonella in vivo* induced genes, i.e. fusions conferring a Lac phenotype to bacteria under laboratory conditions, but which were resistant to chloramphenicol when inside macrophages or when injected into mice. A similar system has now been developed for mycobacteria, which uses *inhA* as a reporter (Dubnau *et al.* 1997). This gene confers isoniazid (INH) resistance to *M. tuberculosis* when expressed on a multicopy plasmid. Thus, conditionally expressed promoters can be identified by selection for INH resistance under *in vivo* conditions (growth within macrophages or animals treated with INH), followed by screening for INH sensitivity under laboratory conditions.

2.1.2 Bioluminescent reporters: luciferases

Bioluminescence refers to the process of visible light emission in living organisms mediated by an enzyme catalyst. These enzymes are generically called luciferases and their substrates are often named luciferins. The major advantage of luciferases as reporters is the sensitivity of luminescent assays which require usually a few seconds and allow the detection of these enzymes in the range of picograms. Moreover, since the reaction products are transient and the half-lives of luciferases are short, this system allows monitoring of gene activity over time (Meighen 1993; Bronstein *et al.* 1994). By combining a charge-coupled device (CCD) camera and a microscope, it is also possible to image light emission at the single-cell level (Pettersson *et al.* 1996).

Luciferases from bacteria (Lux) as well as from fireflies (Luc or FFlux) have been used as reporters. Since the light-emitting reactions catalysed by one or the other type of luciferases are quite distinct, the potential applications vary (Hill *et al.* 1993). Lux catalyses the oxidation of fatty aldehydes. As the genes for the biosynthesis of these compounds have been cloned, they can be supplied to the bacteria *in trans*, avoiding therefore the exogenous delivery of any substrate. Contag *et al.* (1995) showed that *Salmonella* containing *lux* fusions and the fatty aldehyde biosynthetic genes *in trans* could be tracked in living animals and tissue culture cells. On the other hand, FFlux uses firefly luciferin as a substrate in a reaction that consumes adenosine triphosphate (ATP). They may function therefore as reporters of intracellular ATP levels, and consequently, of cell viability (Stanley & Williams 1969).

Although *M. smegmatis* strains expressing the *lux*

genes from *Vibrio harveyi* were used initially as a tool for assaying antimicrobial agents (Andrew & Roberts 1993), *FFlux* is now the reporter of choice for this type of studies. Both mycobacterial phages (Carriere *et al.* 1997) and plasmids (Hickey *et al.* 1996) harbouring *FFlux* have been described.

Recently, *FFlux* has also been used as a reporter in studies of mycobacterial promoters. An integrative plasmid (pMH66) has been described containing a *lacZ* gene under the control of a constitutive promoter and a promoterless *FFlux* preceded by a polylinker (Mdluli *et al.* 1996). In this system, *lacZ* functions as an internal control and promoter activities are expressed by luciferase/β-galactosidase ratios.

Luciferase *in vitro* assays require a luminometer, an apparatus for rapid injection of luciferin and record of light emission. Several luminometers and assay kits are now available, and a survey was published recently (Stanley 1997).

Green fluorescent proteins

Green fluorescent proteins (GFP) are proteins found in marine invertebrates that contain a chromophore and emit green fluorescence when excited with light of the appropriate wavelength. Since this process does not require any exogenously added cofactor, the production of GFP can be detected in living cells without invasive treatments, by means of spectrofluorometry, fluorescence microscopy or fluorescence-based flow cytometry. *gfp* is now the reporter of choice for gene expression monitoring and as a tag in studying protein localization in a variety of organisms (Cubitt *et al.* 1995). Recently, fluorescence-activated cell sorters (FACS)-optimized mutants of GFP were obtained that fluoresce more intensely in bacterial systems (Cormack *et al.* 1996). The existence of these mutants allowed the development of a method named DFI which uses FACS to separate clones expressing *gfp* fusions from a heterogeneous bacterial population (Valdivia & Falkow 1996). Pools of bacteria transformed with promoter libraries constructed in *gfp*-based vectors can be readily screened for clones showing modulated expression under different conditions. Using this method, Valvidia and Falkow were able to identify

Salmonella genes that are induced by acid stress (Valdivia & Falkow 1996) or that are preferentially expressed within macrophages (Valdivia & Falkow 1997). Now this technique has been adapted to *M. marinum* (Barker *et al*. 1998).

The versatility of *gfp* as a reporter for mycobacteria has also been shown by several other groups. *gfp* fusions to previously characterized mycobacterial promoters, such as BCG p*hsp60*, were used to show that GFP fluorescence could be monitored in mycobacterial suspensions using spectrofluorometry and flow cytometry, and also in mycobacteria growing inside macrophages through epifluorescence or laser scanning confocal microscopy. Moreover, *gfp*-tagged mycobacteria could also be detected in infected tissues of mice and in preparations of macrophage vesicles (Dhandayuthapani *et al*. 1995; Kremer *et al*. 1995b; Parker & Bermudez 1997).

2.2 Biochemical methods

Despite the power of gene-fusion techniques, there are instances where use of biochemical methods may be preferable because of their rapidity. For example, primer extension analysis can be used to study transcription of several genes in the same strain. Similarly, two-dimensional protein gels permit the direct visualization of the entire protein content of a bacterial population. Another advantage of biochemical methods is that they directly measure the transcription and translation products from genes found in their normal chromosomal context; the stability and activity of the products are not affected, as they often are when fused to a reporter enzyme.

2.2.1 *Methods based on RNA analysis*

Methods for RNA isolation from mycobacteria have been described by many groups. In general, protocols developed for Gram-positive bacteria, which contain cell walls that are not readily susceptible to lysis, also work well for mycobacteria. They are based on either enzymatic lysis (Bashyam & Tyagi 1994), sonication, or physical disruption with beads (Cheung *et al*. 1994; Mangan *et al*. 1997). It is important to note that soni-

cation should be avoided when dealing with pathogenic mycobacteria. And since bacterial RNA species usually have short lives, enzymatic incubations are not recommended. We therefore suggest the use of glass or silica/ceramic beads to break the mycobacteria cell wall.

RNA analysis through techniques like northern blot, primer extension, S1 mapping and RNase protection assay have been successfully achieved in mycobacteria. All these techniques have been extensively discussed elsewhere (Sambrook *et al*. 1989). Here we will concentrate our attention on methods for detection of differential gene expression whose starting material is RNA.

cDNA subtractive hybridization was originally developed by Duguid *et al*. (1988) to identify genes whose transcripts were modulated in scrapie-infected tissue. Since then it has been successfully applied to a variety of biological problems in many systems, including prokaryotes, e.g. sporulation in *Bacillus subtilis* (Mathiopoulos & Sonenshein 1989). In mycobacteria this technique has been used to isolate a *M. avium* macrophage-induced gene (Plum & Clark-Curtiss 1994), and *M. tuberculosis* genes which are differentially expressed in the virulent strain H37Rv compared to the attenuated strain H37Ra (Kinger & Tyagi 1993). In any case the principle is the same, involving first isolation of RNA from two cell populations. In the study conducted by Plum and Clark-Curtiss (1994), one population was composed of *M. avium* growing in liquid broth and the other was composed of *M. avium* recovered from macrophages. The RNA is then converted to cDNA, followed by hybridization between the two cDNA libraries, removal of the hybridized species, which represent mRNAs from genes comparatively expressed in both populations and ribosomal RNAs, and finally recovery of the cDNAs that are unique to one pool. These cDNAs can then be cloned directly or used to probe genomic libraries.

The technique of differential display (DD) reverse transcriptase polymerase chain reaction (RT-PCR) is based on the use of random primers to PCR-amplify cDNA derived from RNA isolated from two cell populations (as above). The PCR amplification is done in the presence of radiolabelled nucleotides and the amplified fragments are resolved on a DNA sequencing gel, producing fingerprints specific for each population (Liang & Pardee 1992). Bands which are unique to or more abundant in one population can then be isolated, cloned and sequenced. Since in prokaryotes poly(A)-tailed mRNAs are rare, cDNA synthesis is also performed in the presence of random primers, typically hexamers (Abu Kwaik & Pederson 1996), which increases the chance of artifacts. Despite this problem, DD RT-PCR has been successfully used to identify a macrophage-induced *Legionella pneumophila* locus (Abu Kwaik & Pederson 1996) and genes from uropathogenic *E. coli* which are induced upon P-pilus-mediated attachment to human red blood cells (Zhang & Normark 1996). To our knowledge this technique has not yet been used to isolate mycobacterial genes preferentially expressed within the macrophage, but has been employed to identify macrophage genes which are regulated during infection with various mycobacteria (Ragno *et al.* 1997).

A new method based on RT-PCR has been recently developed in our laboratory (Manganelli *et al.* 1998). In this method, the amount of cDNA produced (which is proportional to the amount of mRNA present in the RNA preparation) is measured by a real-time quantitative PCR assay which uses a family of novel fluorescent probes called molecular beacons. Molecular beacons are hairpin-shaped oligonucleotide probes that undergo a conformational change when they bind to their targets, resulting in the restoration of fluorescence of an internally quenched fluorophore (Tyagi & Kramer 1996). When present in a PCR reaction where their target is the amplification product, molecular beacons emit fluorescence whose intensity is a direct measure of the amplicon concentration. This system presents many advantages over 'traditional' RT-PCR: it is very sensitive and allows quantitative determinations to be made over a wide range of target concentrations. Moreover, as these assays are carried out in sealed tubes, the risk of contaminating untested samples is eliminated. This system can be used to monitor changes in gene expression in response to various conditions by normalizing the levels of mRNAs observed in different RNA preparations to those of an internal standard, i.e. a gene whose mRNA levels remain constant during exposures to the varying conditions.

2.2.2 Methods based on protein analysis

Early studies on molecular biology of mycobacteria, which aimed to identify prominent mycobacterial antigens potentially useful for the development of new vaccines and immunodiagnostic tests, relied mainly on protein analysis by immunotechniques like western blot and crossed immunoelectrophoresis, and sodium dodecyl sulphate (SDS)/polyacrylamide gel electrophoresis (PAGE).

More recently, one-dimensional SDS PAGE and western blotting were used to monitor the response of mycobacteria to stresses (e.g. heat shock) and to investigate mycobacterial proteins released during growth. For example, it was possible to detect an increase in the synthesis of several *M. tuberculosis* or BCG proteins following heat shock by metabolic labelling with ^{35}S-methionine and ^{14}C-amino acids (Patel *et al.* 1991).

It was, however, the introduction of two-dimensional gel electrophoresis (O'Farrell 1975) that enabled researchers to better define the response of mycobacteria to different stresses. For example, Sherman *et al.* (1995) analysed the response of various mycobacterial species to hydrogen peroxide (H_2O_2) and showed that in the slow-growing species *M. avium* and *M. bovis* BCG this response was impaired, i.e. resulted in the induction of very few proteins, in comparison to that observed in *M. smegmatis*. This method was also used to compare the proteins synthesized by *M. tuberculosis* growing within human macrophages with those made in response to various stress conditions (heat shock, low pH, H_2O_2) during extracellular growth (Lee & Horwitz 1995). The results showed that several proteins were induced and many others repressed depending upon the stimulus and that some macrophage-induced proteins were absent during extracellular growth. Similarly, investigators sought to characterize proteins which were preferentially synthesized in *M. tuberculosis* during late exponential and stationary-phase growth *in vitro* (Yuan *et al.* 1996). At least seven proteins were shown to be maximally synthesized 1–2 weeks following the end of log-phase growth, and one of these, an α-crystallin homologue (16-kDa protein), appeared to constitute a predominant stationary-phase protein (see section 5.5).

3 Transcriptional apparatus

In this section we review what is known about the basic transcription machinery in mycobacteria, giving special emphasis to mycobacterial σ factors and mycobacterial promoters.

3.1 Bacterial RNA polymerases

Most of what is known about RNA polymerase (RNAP) and the process of transcription in prokaryotes is based on studies performed with *E. coli* (Burgess 1976; Record *et al.* 1996). However, the basic structure of RNAP seems to be conserved in most bacteria (Burgess 1976). The prokaryotic RNAP is an oligomeric enzyme composed basically of five subunits. Two copies of the α and one copy of each of the β and β′ subunits form a stable complex, the core RNAP (E). The core RNAP interacts transiently with the σ subunit (Helmann & Chamberlin 1988; Gross *et al.* 1992) to form the RNAP holoenzyme (Eσ). Although the core RNAP is able to initiate transcription at non-specific sites, the σ subunit gives the RNAP the ability to bind to certain promoter sequences and initiate transcription at specific sites. Promoter recognition is determined by the contact between σ and DNA sequences that, for *E. coli* σ70, are centred −10 and −35 bp upstream of the transcription start point (TSP). The α subunit (36 kDa, in *E. coli*) is involved in the initial binding of RNAP to some promoters by either: (i) interacting through the C-terminal domain (CTD) with sequences upstream of the −35 region (Ross *et al.* 1993); or (ii) interacting through the CTD with class I transcriptional activators (Ishihama 1993). The β and β′ subunits (150 and 155 kDa, respectively, in *E. coli*) interact directly with the DNA template and the RNA,

at least during elongation, and form the channel and grooves through which DNA and RNA presumably move during transcription. The active site of the RNAP is located in the β protein (Kashlev *et al.* 1990), which is the subunit that binds rifampicin (Stetter & Zillig 1974).

The affinity of the σ subunit for a specific promoter is the main determinant for the efficiency of the transcription initiation. Therefore, σ factors are major players in regulating transcription. All bacteria have a primary σ factor, such as *E. coli* σ70 and *Streptomyces coelicolor* σhrdB, which are essential for growth and direct most RNA synthesis in exponentially growing cells. In addition, bacteria may produce several alternative σ factors that can replace the primary σ factor and change RNAP promoter specificity. Alternative σ factors are not required for growth under normal conditions although there are exceptions (De Las Penas *et al.* 1997; Keseler & Kaiser 1997). The ability to induce the synthesis, or to activate, alternative σ factors is a mechanism to alter the sets of genes expressed under different circumstances. Alternative σ factors have been implicated in various processes such as expression of virulence determinants (Finlay & Falkow 1997), stationary-phase gene expression and general stress response (Loewen & Hengge-Aronis 1994; Hecker *et al.* 1996), or sporulation (Haldenwang 1995).

Two families of non-related σ factors have been identified: the σ70 and the σ54 family (Merrick *et al.* 1987). All primary and most of the alternative σ factors belong to the σ70 family. The comparison between the amino acid sequences of members of the σ70 family allowed the definition of several regions and subregions within the proteins, conserved to different extents, that are associated with distinct functions (Helmann & Chamberlin 1988; Lonetto *et al.* 1992). Among them, the usually acidic subregion 1.1, present only in primary σ factors and *E. coli* σS, is involved in restricting the DNA binding properties of free σ factors (Dombroski *et al.* 1993). Subregions 2.4 and 4.2 are implicated in recognition of the −10 and −35 regions, respectively.

On the basis of sequence similarity, Lonetto and collaborators defined three groups of σ factors within the σ70 family (Lonetto *et al.* 1992, 1994). Group 1 comprises the primary σ factors, which are highly similar (at least 51% identical), especially in subregions 2.3, 2.4 and 4.2. Group 2 includes alternative σ factors that, although similar to primary σ factors, are dispensable for growth, like *E. coli* σ^S and *S. coelicolor* σ^{hrdA}, σ^{hrdC} and σ^{hrdD}. Their similarity with members of group 1, especially in the DNA binding regions, suggests that they may recognize similar promoters. Group 3 includes the other alternative σ factors, which are quite divergent from members of groups 1 and 2. Group 3 σ factors can be further classified in subgroups of alternative σ factors that are related both by sequence similarity and, often, by function, like those involved in heat-shock response, sporulation, flagellar synthesis or extracytoplasmic function (ECF). The divergence between the different subgroups reflects different promoter specificities and specialized functions in which they are involved.

3.2 Mycobacterial transcription and RNA polymerases

Low transcription rates have been suggested as one of the factors determining mycobacterial slow growth (Woodley *et al.* 1972; Wheeler & Ratledge 1994; see also Chapter 11). Since the RNA/DNA ratio is 10 times lower in *M. tuberculosis* than in *E. coli*, while the generation time (during which the cellular RNA content doubles) is at least 20-fold longer, it was calculated that the net rate of RNA synthesis per DNA template is 200-fold lower in *M. tuberculosis* than in *E. coli* (Harshey & Ramakrishnan 1977). The *in vivo* elongation rate for total *M. tuberculosis* RNA was found to be 10 times lower than in *E. coli* (Harshey & Ramakrishnan 1977) and *in vitro* experiments with purified RNAP from *M. smegmatis*, using bacteriophage T7 DNA as template, indicated RNA elongation rate and specific activity about two- and threefold lower, respectively, than those for *E. coli* RNAP (Chamberlin *et al.* 1979; Wiggs *et al.* 1979). Therefore, mycobacterial RNAP has intrinsically low activities. However, this explains only in part the reduced transcriptional rate in *M. tuberculosis*. Harshey and Ramakrishnan speculated that the number of RNAP molecules engaged in RNA synthesis at a specific time and the frequency of transcription initiation must be lower in *M. tuberculosis* than in *E. coli* (Harshey & Ramakrishnan 1977).

There is no information concerning the number of RNAP molecules per cell in mycobacteria. However, the number and size of its components was established by SDS-PAGE analysis of purified RNAP from *M. smegmatis* (Wiggs *et al.* 1979). The results indicated a canonical α$_2$ββ′σ structure with α subunits of 43–48 kDa and β and β′ subunits of 120–150 kDa. The same structure was observed for *M. tuberculosis* RNAP (I. Smith, unpublished observation 1997) and has been confirmed by the cloning and sequencing of the genes encoding *M. tuberculosis* RNAP subunits. In addition, some of the genes encoding the RNAP α, β and β′ subunits (encoded by *rpoA*, *rpoB* and *rpoC*, respectively) from *M. smegmatis* and *M. leprae* have been identified. The *rpoA* gene from *M. tuberculosis* is part of the cluster of ribosomal protein (r-protein) genes *rpsM–rpsK–rpsD–rpoA–rplQ*, which is similar to the *E. coli* α operon (Keener & Nomura 1996). *M. leprae rpoB* and *rpoC* genes (Honore *et al.* 1993) are located downstream of *rplL*, as in other bacteria in which the β operon has the structure *rplJ–rplL–rpoB–rpoC*. However, a gene that is not present in the *E. coli* β operon, *mkl* (supposedly encoding an ABC transporter), is located between *rplL* and *rpoB* in *M. leprae*. In *M. tuberculosis*, the *rpoB–rpoC* cluster is not preceded by *mkl* or *rplL* (Miller *et al.* 1994). *M. smegmatis rpoB* has also been cloned but the sequence available only includes 194 bp upstream and 50 bp downstream of the coding region (Hetherington *et al.* 1995). It is not possible to predict whether the different location of *rpoB–rpoC* in *M. tuberculosis*, relative to *E. coli* and *M. leprae*, or the presence of *mkl* in *M. leprae*, have some effect in *rpoB–rpoC* regulation. The different organization of the loci encoding for r-proteins and RNAP subunits may reflect the modular evolution of the loci encoding the transcriptional and translational machinery (Honore *et al.* 1993). In this sense, it is relevant that in *E. coli rpoB–rpoC* and *rpoA* expression is not coupled with the expression of the upstream r-protein genes as result of mechanisms like attenuation and mRNA processing (Keener & Nomura 1996). The promoters used to transcribe *rpoA*, *rpoB* or *rpoC* in mycobacteria have not been identified yet, although sequences resembling the −10 and −35 consensus sequences for *E. coli* σ[70] promoters were found upstream of *M. tuberculosis rpoB* and *rpoC* (Miller *et al.* 1994).

3.3 Mycobacterial σ factors

The *M. tuberculosis* genome encodes up to 13 σ factors, which have been named as σ[A] to σ[M] (Gomez *et al.* 1997a; Cole *et al.* 1998). Some of them, as well as some of their homologues in *M. smegmatis*, have been partially characterized. σ[A] and σ[B] are similar to primary σ factors from other bacteria. σ[F] is similar to σ factors that regulate gene expression during stationary phase and sporulation in some Gram-positive bacteria. The other 10 σ factors belong to the ECF subfamily. No genes encoding σ factors of the σ[54] family have been identified in mycobacteria so far.

σ[A] *and* σ[B]

The apparent molecular weights are 65 and 40 kDa, respectively. These were identified as σ factors present in highly purified preparations of *M. smegmatis* RNAP, since they were recognized by a monoclonal antibody specific for an epitope conserved in most members of the σ[70] family. The corresponding genes, *sigA* and *sigB* (named originally *mysA* and *mysB*, respectively) were cloned and shown to be separated only by 3 kb (Predich *et al.* 1995). The same organization was found in *M. tuberculosis* and *M. leprae* (Doukhan *et al.* 1995). Immediately downstream of *sigB* a gene, *ideR*, was identified as the mycobacterial homologue of the *Corynebacterium diphtheriae dtxR* (Dussurget *et al.* 1996). A similar cluster containing genes homologous to the mycobacterial *sigA–sigB–ideR* genes seems to be present in *Brevibacterium lactofermentum* (Oguiza *et al.* 1995, 1996) and probably exists in *C. diphtheriae*, since an incomplete open reading frame (ORF), similar to *sigB*, is present upstream of *dtxR* (Doukhan *et al.* 1995).

M. smegmatis, *M. tuberculosis* and *M. leprae* σ[A] proteins are 99–100% identical in regions 1.2, 2, 3 and 4 (Doukhan *et al.* 1995). When comparing the entire protein sequences, the identity drops to 83–85% mainly because of differences in region 1.1, which

has different lengths and sequences in each of the three species. Because of the high similarity of σ[A] with *S. coelicolor* σ[hrdB] (about 87% identity in regions 1.2, 2, 3 and 4) and the presence of a long region 1.1, *sigA* was suggested to encode the primary σ factor in mycobacteria (Predich *et al.* 1995). σ[A] was later found to be essential in *M. smegmatis*, since inactivation of *sigA* by allelic replacement is only possible in a merodiploid strain that contains an extra copy of the gene (Gomez *et al.* 1998). The *sigA* promoter was mapped in *M. smegmatis* and *M. bovis* BCG (Gomez *et al.* 1998) and found to match the consensus sequence proposed by Bashyam *et al.* (1996). The same regulatory sequence is present upstream of *M. tuberculosis* and *M. leprae sigA* genes. The essential character of *sigA* in *M. smegmatis*, the abundance of σ[A] in *M. smegmatis* RNAP purified from exponentially growing cells, the similarity of σ[A] with σ[hrdB] and the conservation of the *sigA* regulatory and coding regions in *M. smegmatis*, *M. tuberculosis*, *M. bovis and M. leprae* strongly suggest that *sigA* codes for the primary σ factor in mycobacteria. The amino acid sequence of region 2.4 in σ[A] is almost identical to that of *S. coelicolor* σ[hrdB] and *E. coli* σ[70], suggesting that they recognize very similar −10 hexamers. However, the differences between subregions 4.2 of the mycobacterial σ[A] or *S. coelicolor* σ[hrdB] (which are almost identical to each other) and that of *E. coli* σ[70] are more numerous and may reflect different −35 hexamer specificities. These differences could explain why a majority of mycobacterial promoters function poorly in *E. coli* (Bashyam *et al.* 1996).

It was found that a point mutation in *M. bovis sigA* (also known as *rpoV*), producing a single amino acid change in the C-terminal part of the protein (R522 to H), in region 4.2, was responsible for the attenuated phenotype of the *M. bovis* strain ATCC 35721 (Collins *et al.* 1995). This finding made *sigA* one of the few mycobacterial genes that fulfil Koch's molecular postulates defining a virulence gene (Falkow 1988). The attenuated phenotype of *M. bovis* ATCC 35721 could be explained by the production of a mutant form of σ[A] with altered promoter specificity (Collins 1996). However, the R522H mutation is located in a region that has been involved in the interaction between σ factors and class II activators (Baldus *et al.* 1995). Therefore, it is possible that the attenuated phenotype of strain ATCC35721 is the result of the defective expression of a set of genes dependent on a transcriptional activator. No evidence of specific differences in *M. bovis* gene expression as consequence of the *sigA* R522H mutation has been published so far. However, a pair of isogenic *M. smegmatis* strains differing only in the presence of a mutation homologous to R522H in *sigA* was constructed. The *M. smegmatis* mutant strain showed altered sensitivity to several antibiotics, which was interpreted as result of altered cell-wall permeability (J. Timm and I. Smith, unpublished results 1998).

M. smegmatis, *M. tuberculosis* and *M. leprae* σ^B proteins are 91–95% identical (Doukhan *et al.* 1995). Unlike σ^A, σ^B has a short subregion 1.1 (25–29 amino acids) that is highly conserved in the three species. σ^B shows most similarity to the mycobacterial σ^A and to *S. coelicolor* σ^{hrdB} and σ^{hrdA} (64, 62 and 53% identity, respectively, in regions 1.2, 2, 3 and 4). The phylogenetic analysis of 73 σ factors of groups I and II placed the mycobacterial σ^A and σ^B, *S. coelicolor* σ^{hrdB} and σ^{hrdA} and *B. lactofermentum* σ^A and σ^B in a cluster clearly separated from the one including those members of group II involved in stationary-phase gene expression (*E. coli* σ^S and homologues in other species) (Gruber & Bryant 1997), suggesting that σ^B is not the mycobacterial homologue of σ^S. The function of σ^{hrdA} is unknown as the inactivation of h*dA* or its combined inactivation with h*dC* and h*dD* does not produce obvious phenotypic consequences in *S. coelicolor* (Buttner & Lewis 1992). *B. lactofermentum sigA* and *sigB* have not been inactivated yet but transcription of both genes reportedly decreases in stationary phase (Oguiza *et al.* 1997).

The comparison of subregions 2.4 and 4.2 of mycobacterial σ^A with the same regions of σ^B indicates 80% and 40% identity, respectively, although the amino acid residues implicated so far in direct contact with DNA (Lonetto *et al.* 1992; Moran Jr. 1993) are identical in both σ factors (Predich *et al.* 1995). It is not possible to predict if the differences present in subregions 2.4 and 4.2 cause different promoter specificities.

M. smegmatis sigB has been inactivated by allelic replacement. The *sigB* mutant does not show any obvious growth impairment but it is more sensitive to a variety of reagents (e.g. the respiration inhibitor sodium azide, redox cycling drugs like plumbagine or reducing agents like dithiothreitol) and more resistant to INH (Smith *et al.* 1997). The phenotype of strains in which *sigB* or *ideR* has been inactivated is different, e.g. siderophore production is still regulated by iron in the *sigB* defective strain, indicating the absence of a polar effect of the *sigB* mutation on *ideR*. *sigB* is transcribed during exponential growth from a promoter similar to that of *sigA*, which is conserved in *M. smegmatis*, *M. tuberculosis* and *M. leprae* (M. Gomez and I. Smith, unpublished results 1998). In *M. tuberculosis*, *sigB* transcription increases after heat shock, SDS treatment or shift to low aeration conditions (Manganelli *et al.* 1999).

σ^F

sigF was identified in *M. tuberculosis* by PCR amplification using degenerate primers based on sequences conserved in subregions 2.1 and 2.3 of the σ^{70} family of σ factors (DeMaio *et al.* 1996). Southern blot experiments indicated the presence of sequences similar to *sigF* in *M. bovis* BCG, *M. avium*, *M. triviale* and *M. gordonae*, but not in *M. smegmatis* or *M. abscessus*. A gene with similarity to *sigF* was also sequenced from *M. leprae*. *M. tuberculosis sigF* encodes a 261 amino acid protein that is similar to *S. coelicolor* σ^F and *B. subtilis* σ^B and σ^F (41, 30 and 26% identity, respectively). *S. coelicolor* and *B. subtilis* σ^F are required during sporulation while *B. subtilis* σ^B controls a large regulon induced during stationary phase or in response to general stress. In *B. subtilis*, both σ^F and σ^B activities are controlled, in part, by a 'switching partner' mechanism, in which an anti-σ factor can bind either to the σ factor, blocking its activity, or to an anti-anti-σ factor, that releases the σ subunit from the σ-anti-σ complex. The anti-anti-σ and the anti-σ factor are encoded by genes located upstream of both *sigF* and *sigB*. A gene, *usfX*, encoding a protein with weak similarity to the anti-σ factors that control σ^B and σ^F activity in *B. subtilis*, was found immediately upstream of the mycobacterial *sigF*. Upstream of *usfX* another ORF, *usfY*, encodes a protein with no significant similarity with the *B. subtilis* σ^B or σ^F anti-anti-σ factors (DeMaio *et al.* 1997). *sigF* transcription was measured in *M. bovis* BCG by RNA protection assay and found to be induced after cold shock, during stationary phase, nitrogen depletion and, to a lesser extent, after oxidative or microaerophilic stress and alcohol shock. No expression was detected during logarithmic phase (DeMaio *et al.* 1996). Overproduction of σ^F prevented growth of *M. bovis* BCG, but not of *M. smegmatis*, this being interpreted as indication that σ^F could be involved in the expression of *M. tuberculosis*-complex genes that are required for, or induced during, growth arrest in response to starvation or stress (DeMaio *et al.* 1997). The similarity of the mycobacterial σ^F with σ factors involved in post-exponential gene expression, sporulation or general stress response raises interesting possibilities about its function, especially since *sigF* is apparently present only in slow-growing pathogenic mycobacteria and it is assumed that general and specific stress responses are survival mechanisms in many pathogens. Finally, the ability of *M. tuberculosis* cells to enter what is considered a dormant state during latent infection could be the result of mechanisms triggering some type of postexponential differentiation in which σ^F could be involved, as discussed later in this chapter.

σ^E and the other mycobacterial members of the extracytoplasmic function subfamily

The ECF subfamily (Lonetto *et al.* 1994) includes σ factors from different organisms that control a variety of functions in response to extracytoplasmic stimuli. In *E. coli*, σ^E activity is induced by the presence of misfolded proteins in the cell wall and controls the so-called periplasmic heat-shock response (Missiakas & Raina 1997). *degP*, encoding a serine protease that localizes in the periplasmic space where its function is to degrade abnormally folded proteins, is one of the σ^E dependent genes. σ^E had been considered an alternative, dispensable σ factor in *E. coli*, required for survival at high temperatures. However, it has been shown that σ^E is essential at all temperatures (De Las Penas *et al.* 1997). Another *E. coli* σ factor that belongs

to the ECF subfamily, FecI, is a regulator of ferric citrate transport genes (Angerer *et al.* 1995). The genes immediately downstream of *E. coli sigE* and *fecI*, which are *rseA* and *fecR*, respectively, code for regulators of the corresponding σ factors (Ochs *et al.* 1995; Rouviere *et al.* 1995).

A gene encoding an ECF σ factor, *sigE*, was originally identified as result of the *M. leprae* genome sequencing project and then cloned from *M. tuberculosis*, *M. avium* and *M. smegmatis* (Wu *et al.* 1997). Southern blot analysis indicated the presence of the same gene in *M. bovis and M. fortuitum*. *M. tuberculosis sigE* codes for a 257 amino acid protein, σ^E, which is 82–92% identical to the homologues in other mycobacteria. A truncated form of this protein (including regions 1.2–4) was overproduced in *E. coli* as a His-tagged fusion protein, purified and its activity demonstrated in reconstitution experiments. In these experiments σ^E conferred specific RNA polymerizing activity to *M. smegmatis* core RNAP, using the *M. tuberculosis* and *M. smegmatis sigE* promoter regions as templates. Therefore, σ^E could contribute to the expression of its own gene. Several transcription start points (TSPs) for *M. smegmatis* and *M. bovis* BCG *sigE* were identified by *in vivo* primer extension, suggesting that *sigE* may be transcribed from several promoters subjected to complex regulation. Immediately downstream of *sigE*, an ORF, *orf2*, was hypothesized to encode a negative regulator of σ^E. However, no similarity with previously identified proteins was found. A third ORF encoding a protein highly similar to DegP was identified immediately downstream of *orf2*, resembling the situation of *Pseudomonas aeruginosa* in which a gene homologous to *degP* is also located downstream of *sigE* (*algU*). A *sigE*-defective *M. smegmatis* strain failed to induce the protective response induced by sublethal concentrations of H$_2$O$_2$. Also, the defect of σ^E increased the sensitivity of *M. smegmatis* to SDS, acid pH and high temperature (Wu *et al.* 1997). In agreement with the deduced role of σ^E in stress survival, RT-PCR studies on the *M. tuberculosis sigE* transcript indicated that its expression is induced by SDS treatment and heat shock (Manganelli *et al.* 1998).

Among the other ECF σ factor genes identified in the *M. tuberculosis* genome, at least six have been shown to be expressed: *sigC, sigD, sigG, sigH, sigI* and *sigM* (Manganelli *et al.* 1998). *sigH* and *sigM* are induced by heat shock, while *sigI* is induced by moderate cold shock (Manganelli *et al.* 1998). The *M. smegmatis* gene *sloA*, homologue of *M. tuberculosis* *sigM*, has also been shown to be induced by heat shock (Salazar *et al.* 1997).

3.4 Mycobacterial promoters

The high degree of conservation of the main σ factor in different bacterial species, specially in region 2.4, and to a lesser extent in region 4.2, suggests similar promoter specificity. Accordingly, the consensus sequence for promoters recognized by the main σ factor in *E. coli* and *B. subtilis* is highly conserved (Harley & Reynolds 1987; Helmann 1995).

The analysis of a compilation of 69 mycobacterial promoters identified by primer extension or S1 mapping, from several mycobacterial species, indicated that most of them are similar to those recognized by *E. coli* σ70 (M. Gomez and I. Smith, unpublished observations 1998). For the purpose of deducing consensus sequences, promoters were classified in three groups: group A (47 out of 69), containing promoters with −10 and −35 hexamer sequences similar to the *E. coli* σ70 consensus; group B (16 out of 69), containing promoters in which −10, but no obvious −35 hexamers, similar to the *E. coli* σ70 consensus were observed; and group C (six out of 69) that had no obvious −10 or −35 sequences similar to the *E. coli* σ70 consensus. Examples of promoters belonging to each of the groups are given below. A complete list of the promoters, with their sequences, is available from the authors upon request.

The *G*+C content of the 69 mycobacterial promoters included in groups A, B and C increased in the order group A<group B<group C (Table 4.2).

Table 4.2 Consensus for mycobacterial promoters.

Group		No[a]	−35[b]	−10[b]	G + C%[c]
	E. coli	112	$T_{82}T_{83}G_{78}A_{64}C_{53}a_{44}$	$t_{38}g_{38}N_{99}T_{82}A_{90}T_{52}A_{59}a_{49}T_{89}$	
	B. subtilis	236	$T_{88}T_{84}G_{78}A_{64}C_{52}A_{54}$	$T_{60}G_{52}N_{99}T_{94}A_{96}T_{60}A_{76}A_{74}T_{94}$	33
	C. glutamicum	33	$t_{48}t_{48}G_{72}C_{51}c_{45}A_{51}$	$N\ G_{54}g_{42}T_{78}A_{72}N_{99}A_{57}a_{42}T_{84}$	67
	Streptomyces	29	$T_{86}T_{90}G_{99}A_{69}C_{66}R_{62}$	$T_{59}A_{86}g_{41}R_{69}R_{72}T_{99}$	57
A	*M. smegmatis*	18	$T_{78}T_{72}G_{72}a_{39}C_{56}W_{66}$	$t_{44}K_{72}T_{50}T_{94}A_{89}T_{61}a_{44}a_{39}T_{78}$	47
	M. tuberculosis	19	$T_{68}t_{42}G_{74}A_{63}C_{68}W_{58}$	$C_{42}G_{52}N_{99}T_{84}A_{84}N_{99}a_{37}a_{42}T_{84}$	54
	All	47	$T_{75}T_{62}G_{72}A_{53}C_{57}a_{40}$	$Y_{66}g_{45}N_{99}T_{83}A_{83}t_{42}a_{44}a_{40}T_{81}$	50
B	*M. smegmatis*	7	nc	$g_{42}G_{57}N_{99}T_{85}A_{85}N_{99}A_{57}c_{42}T_{74}$	51
	M. tuberculosis	8	nc	$a_{37}G_{75}G_{75}T_{75}A_{75}G_{62}G_{62}C_{75}T_{99}$	62
	All	16	nc	$g_{31}G_{62}G_{56}T_{82}A_{81}g_{43}R_{93}C_{56}T_{88}$	55
C	All	6	nc	nc	62
D	*M. paratuberculosis*	6	$T_{99}G_{83}M_{50}C_{66}G_{99}T_{50}$	$C_{87}g_{37}G_{62}c_{37}C_{50}S_{99}$	64

The consensus sequences for groups A, B and C were deduced from a compilation of sixty-nine mycobacterial promoters identified by primer extension or S1 mapping (M. Gomez and I. Smith, unpublished). The collection of *M. paratuberculosis* promoters obtained by Bannantine *et al.* (1997) was treated as a different group and the consensus sequence is shown separately. The consensus sequences for *E. coli* σ70 (Hawley and McClure, 1983), *B. subtilis* σA (Helmann, 1995), *C. glutamicum* (Patek *et al.*, 1996) and streptomycete promoters (Strohl, 1992) are shown for comparison.
[a] Number of sequences used in the analysis.
[b] The numbers in the consensus sequences indicate the percentage of conservation. N = any nucleotide; K = G or T; Y = C or T; R = A or G; W = A or T; M = C or A; S = G or C. nc = not conserved.
[c] G + C content of the promoter region (from −100 to +1 for *B. subtilis* promoters; from −50 to +1 for *C. glutamicum* and streptomycete promoters; from −42 to +1, for mycobacterial promoters).

Transcription start point

The initiating nucleotide at the TSP of the 69 sequences was frequently G (48%), followed by A (30%), similarly to that reported for *Corynebacterium glutamicum* promoters (Patek *et al.* 1996). *E. coli* RNAP has also preference for purine starts, with the difference that A occurs more frequently (Hawley & McClure 1983). The deduced consensus sequence for the TSP (−1 to +2) was $C_{42}R_{78}N$ (where R is a purine, N is any nucleotide and the numbers indicate the percentage of conservation).

Distance between transcription start point and the −10 hexamer

The distance between the TSP and the proposed −10 hexamers in groups A and B promoters was most frequently 7 (44%) or 6 bp (28%), ranging from 4 to 8 bp, in a distribution similar to that described for *B. subtilis* σ^A or *E. coli* σ^{70} promoters (Harley & Reynolds 1987; Helmann 1995).

Distance between the −10 and the −35 hexamers

The distance between the −10 and −35 hexamers was 16–18 bp in 59% of group A promoters but many of them had longer spacers. This distribution is different from that of *E. coli* (Harley & Reynolds 1987) or *B. subtilis* (Helmann 1995), in which 90% of the promoters recognized by the primary σ factor have spacers of 16–18 bp or 17 bp, respectively.

Consensus sequences for groups A and B promoters

The consensus sequences for promoters belonging to groups A and B are shown in Table 4.2. A similar consensus sequence has been deduced from another study (Mulder *et al.* 1997). The conservation of the −10 and −35 hexamers in group A, and the −10 hexamer in group B, and their similarity to the *E. coli* σ^{70} consensus suggests that they are recognized by the homologous primary σ factor in mycobacteria, σ^A. The functionality of mycobacterial −10 and −35 hexamers resembling the *E. coli* σ^{70} consensus has

been demonstrated, by obtaining mutations that alter promoter activity, in some cases, for example, *M. fortuitum blaF* (Timm *et al.* 1994b) and *M. smegmatis rpsL* (Kenney & Churchward 1996). The similarity to the *E. coli* σ^{70} consensus was limited, however, in several respects. First, the observed degree of conservation, specially for the −35 hexamer, was lower than that for *E. coli* σ^{70} or *B. subtilis* σ^A promoters (Table 4.2), as has been described for *C. glutamicum* (Patek *et al.* 1996) and streptomycetes (Strohl 1992) and also pointed out in a previous compilation of 14 *M. smegmatis* and 10 *M. tuberculosis* promoters (Bashyam *et al.* 1996). Second, the average length of the spacer between −10 and −35 hexamers in group A promoters was larger, as mentioned above. Third, the consensus sequence for groups A and B promoters is richer in $G+C$ than that of the *E. coli* σ^{70} consensus, specially for the *M. tuberculosis* promoters. The special composition of the latter sequences, observed also in a previously cited compilation of mycobacterial promoters, could be linked to the lower strength of *M. tuberculosis* promoters relative to those of *M. smegmatis* (Das Gupta *et al.* 1993; Bashyam *et al.* 1996).

These three differences could be responsible for the fact that mycobacterial genes are usually better expressed in streptomycetes than in *E. coli* (Kieser *et al.* 1986) and could also be related to the low transcription rates associated with mycobacterial slow growth.

The lower degree of conservation found in mycobacterial, *C. glutamicum* and streptomycete promoters, specially for the −35 hexamers, could be the result of having multiple σ factors with slightly different promoter specificities. In addition, it has been suggested that low $G+C$ content in the chromosome of an organism may impose a selective pressure resulting in the evolution of RNAP with higher specificity for more conserved $A+T$-rich promoters (Morrison & Jaurin 1990). Therefore, actinomycetes and other organisms with high $G+C$ content in the chromosome could be characterized by having RNAPs with less promoter specificity than that of organisms with lower $G+C$ content.

Another possible effect of the high $G+C$ content of the mycobacterial chromosome is that negative supercoiling could be important to unwind some promoters during the process of strand opening. Supercoiling of the DNA template was found to be required for *in vitro* transcription of the *M. bovis* BCG *hsp60* and other promoters by *M. smegmatis* RNAP (Levin & Hatfull 1993), although this is not an absolute requirement (Predich *et al.* 1995).

Group B could represent a subtype of σ^A dependent promoters in which the –35 region is not conserved. In some cases, the absence of conserved –35 hexamers could be compensated by the presence of an extended –10 motif. Some *E. coli* promoters are characterized by having no obvious –35 hexamer, but having an extension of the –10 region with consensus sequence TNTGN immediately upstream of the –10 hexamer. This motif has been shown, through mutation studies, to be important for promoter activity (Record *et al.* 1996). A similar motif has been identified in other bacteria including *B. subtilis* (TNTGN) (Helmann 1995) and corynebacteria (GG) (Patek *et al.* 1996). Accordingly, the occurrence of one or two Gs in the two positions immediately upstream of the –10 hexamer was higher in group B than in group A of mycobacterial promoters (Table 4.2). The functionality of an extended –10 motif has been demonstrated by mutational studies for the *M. smegmatis rpsL* gene (Kenney & Churchward 1996) and for the *M. smegmatis* S16 and *M. tuberculosis* T125 promoters (Bashyam & Tyagi 1998).

Some of the promoters included in groups A and B are known to be regulated, for example, *M. smegmatis fxbA*, *M. tuberculosis recA* and *lexA*, *M. leprae* 18-kDa gene (homologous to *M. tuberculosis acg*), BCG *hsp60* tsA, as discussed later in this chapter, and the mycobacteriophage L5 promoters P1, P2, P3 and Pleft (see Chapter 3).

Group C promoters

Group C includes promoters that do not resemble the *E. coli* σ^{70} consensus. The special features of these promoters could be associated with low levels of expression or the existence of recognition sites for regulators, such as specialized σ factors or transcription factors. In fact, probably all the promoters in group C are regulated. Among them, the *M. smegmatis sigE* promoter P2, and the *M. bovis* BCG *sigE* promoters P1 and P2, could be autoregulated, as discussed previously. Other group C promoters are BCG *hsp60* tsB, the second heat-shock inducible promoter located upstream of *M. bovis* BCG *hsp60* (Stover *et al.* 1991), and the *groEL1* promoter (Kong *et al.* 1993). *M. bovis* BCG *hsp60* tsB seems to have special characteristics since its not recognized by *M. smegmatis* RNAP *in vivo* or *in vitro* (Levin & Hatfull 1993).

Mycobacterium paratuberculosis *promoters*

A collection of *M. paratuberculosis* promoters was obtained by screening, in *M. smegmatis*, a library of DNA fragments from *M. paratuberculosis* cloned in a promoter probe vector. The analysis of the regions upstream of the TSP revealed conserved sequences centred at –10 and –35 positions with consensus sequence markedly different to the mycobacterial or the *E. coli* σ^{70} consensus (group D in Table 4.2) (Bannantine *et al.* 1997). The $G+C$ content of the *M. paratuberculosis* promoters was also higher and the initiating nucleotide was frequently T. The differences found between *M. paratuberculosis* promoters and the mycobacterial promoter consensus could be related to the extremely slow growth rate of *M. paratuberculosis*, assuming that the *M. paratuberculosis* promoter collection is representative and that the primary σ factor in this species is similar to that of other mycobacteria.

3.5 Transcriptional regulators

The *M. tuberculosis* genome codes for more than 100 putative transcriptional regulators (Cole *et al.* 1998). Transcription regulators modulate the interaction between RNAP and promoters. Repressors bind to sequences known as operators that usually overlap with the promoter or the TSP, generally blocking transcription initiation. Activators facilitate the interaction of the RNAP with the promoter by binding to upstream sequences and interacting with the α-CTD or the σ factor (Gralla & Collado-Vides 1996). Repressors and activators can be regulated at the level of expression, by covalent modification (phosphorylation) or by binding to other cofactors or regulatory proteins. Transcription factors fall into different families of proteins that share similarity at the amino acid level, the position of the DNA binding site within the protein and, often, the mechanism regulating their activity. Examples of transcription regulator families are the response regulators (RR), XylS/AraC, LysR and Fnr/CRP families (Ninfa 1996).

RR are a widespread family of transcription factors that become active when phosphorylated by a histidine kinase that acts as sensor protein (Parkinson 1993). These two-component systems mediate the response to various environmental signals and have also been involved in the control of virulence genes in pathogens (Miller *et al.* 1989). Eleven complete two-component systems have been identified in the *M. tuberculosis* genome (Cole *et al.* 1998). Three of them have been partially characterized: MtrA/MtrB, RegX3/SenX3 and TcrA.

The *M. tuberculosis* gene *mtrA* codes for a protein similar to *P. aeruginosa* PhoB, which is involved in the response to phosphate starvation, and *S. coelicolor* A3(2) AsfQ1. Downstream of *mtrA*, a gene (*mtrB*) coding for a protein similar to known histidine kinases in other two-component systems was found. The functional similarity of MtrA with other RR was proven by demonstrating that MtrA can be phosphorylated by CheA, a typical histidine protein kinase. A possible role of MtrA–MtrB in virulence was speculated as the expression of a *mtrA–gfp* transcriptional fusion was induced during intracellular growth of *M. bovis* BCG in macrophages (Via *et al.* 1996).

M. tuberculosis RegX3 (Wren *et al.* 1992; Supply *et al.* 1997) is highly similar to *M. tuberculosis* MtrA and to *E. coli* and *B. subtilis* PhoB. The RegX3 homologue in *M. leprae* contains frameshifts suggesting that is a pseudogene. Another gene, *senX3*, encoding a protein similar to *M. tuberculosis* MtrB and to the sensor protein from *E. coli* and *B. subtilis* PhoR was found upstream of *regX3*. SenX3 was predicted to have two transmembrane domains, which is a feature characteristic of other sensor proteins like *E. coli* PhoB or

Salmonella PhoP. The expression of *lacZ* transcriptional fusions constructed with different fragments of the *senX3–regX3* region indicated that both genes are part of a polycistronic operon. The *senX3–regX3* intercistronic region includes several copies of repetitive elements called mycobacterial interspersed repetitive units (MIRUs) of unknown function. Interestingly, MIRUs are also present in the *mtrA–mtrB* intercistronic region (Supply *et al.* 1997).

M. tuberculosis TcrA is similar to the RRs PhoP and OmpR from *E. coli* and *Salmonella*. The *tcrA* gene was detected by Southern hybridization in mycobacterial species belonging to the *M. tuberculosis* complex, but not in *M. smegmatis*, *M. marinum* or *M. avium*. An *M. tuberculosis tcrA* defective strain was constructed by allelic replacement. This strain exhibited normal growth *in vitro* but two-dimension gel electrophoresis analysis of total extracts revealed altered levels of several proteins, suggesting a global regulatory function for TcrA (Buchmeier *et al.* 1997).

A gene encoding a protein, V38K, belonging to the AraC/XylS family of transcription factors (Gallegos *et al.* 1997) was isolated in *M. tuberculosis* H37Rv (Das Gupta *et al.* 1993). The similarity of V38K with VirF and VirFy, which regulate the expression of genes required for invasion in *Shigella* and *Yersinia*, respectively, and to several regulators of adhesin production in enterotoxigenic bacteria, indicated a possible role as regulator of virulence genes in *M. tuberculosis* (Das Gupta *et al.* 1993). No further characterization of this protein has been reported. However, nucleotides –58 to –9 relative to the first base in the start codon proposed for V38K are identical to the *M. tuberculosis* promoter sequence T150 characterized by Bashyam *et al.* (1996). A search against the *M. tuberculosis* genomic sequence indicated that the T150 sequence is unique. Therefore, the gene encoding V38K is transcribed in *M. tuberculosis* from a promoter resembling the *E. coli* σ^{70} consensus.

Several transcription factors are probably involved in *M. smegmatis* acetamidase regulation. Transcriptional fusions to *cat* were used to demonstrate that induction of acetamidase is at the level of transcription and that a 1.5-kb region, immediately upstream of the acetamidase gene, contains elements involved in repressing transcription in the absence of acetamide (Parish *et al.* 1997). This region contains three ORFs that, with the acetamidase gene, could constitute an operon. The first two ORFs, plus another one transcribed divergently from the putative operon, encode proteins similar to known positive and negative regulators from other organisms (Parish *et al.* 1997). The inducible acetamidase promoter has been used to conditionally express a *hisD* antisense message to cause His auxotrophy in *M. smegmatis* (Parish & Stoker 1997).

Other transcription regulators that have been studied in mycobacteria are the iron-dependent regulator IdeR, the oxidative stress response regulator OxyR and the SOS response regulator LexA. Their role in mycobacterial physiology is discussed in the final section of this chapter. Genes encoding proteins similar to known transcription factors from other organisms such as Fur, involved in iron uptake and oxidative stress regulation in enteric bacteria, HspR, involved in heat-shock response in Gram-positive bacteria, or Fnr/CRP, a family of transcription factors involved in regulating anaerobic gene expression and catabolite repression in many bacteria, have also been identified in the *M. tuberculosis* genome; possible functions of these proteins are discussed later.

4 Translational signals and codon usage

Although translational machinery is not covered in this chapter, we briefly discuss in this section what is known about ribosomal binding sites (RBS), start codons and patterns of codon usage in mycobacteria. Consideration of these three features is useful when analysing DNA sequences in search of coding regions. In addition, they contribute to determining the level of expression of a particular gene.

Initial binding of prokaryotic mRNAs to ribosomes is mediated by the interaction of the 3′ end of the 16s rRNA with a complementary sequence about seven nucleotides long, known as the RBS or Shine–Dalgarno box, located four to seven nucleotides upstream of the start codon in the mRNA

(Draper 1996). The analysis of the few compilations of mycobacterial putative RBS that have been published indicate similarity to the prokaryotic consensus in size, sequence and position (Dale & Patki 1990; Hatfull & Sarkis 1993; Honore *et al.* 1993; Bannantine *et al.* 1997). However, the limited number of RBS compared makes it difficult to estimate what level of complementarity between RBS and anti-RBS is characteristic of mycobacteria. For example, poor complementarity to the anti-RBS sequence was observed in a compilation of 12 *M. leprae* genes that were expected to be expressed at high levels (Honore *et al.* 1993). Extensive pairing between RBS and 16S rRNA may be not required for translation in mycobacteria, as has been suggested for streptomycetes (Strohl 1992). In addition, some mycobacterial genes have been identified in which the TSP coincides with the first nucleotide of the start codon: *M. fortuitum blaF* (Timm *et al.* 1994b), *M. tuberculosis purC* (Jackson *et al.* 1996) and *M. leprae oxyR* (Dhandayuthapani *et al.* 1997). The same was observed for the *FFlux* reporter gene carried in the recombinant L5 mycobacteriophage phGS18 (Sarkis *et al.* 1995) and 11 other actinomycete genes (Strohl 1992). The relative abundance of genes in which transcription initiates at the translational start codon could indicate that the presence of RBS is not an absolute requirement for translation in actinomycetes and that alternative mechanisms for the interaction between mRNA and ribosomes exist. Nevertheless, regions containing anti-RBS sequences in the 16s rRNA from all mycobacterial species studied, including *M. smegmatis*, *M. marinum*, *M. tuberculosis*, *M. paratuberculosis*, *M. avium* and *M. leprae*, are identical to each other and closely related to those from streptomycetes and *E. coli* (Suzuki *et al.* 1988; Liesack *et al.* 1990; Kempsell *et al.* 1992; Ji *et al.* 1994a,b). For coding region prediction purposes, mycobacterial RBS should be similar to a segment of the sequence 5′AGAAAGGAGG3′, which is the reverse-complementary sequence of the conserved mycobacterial anti-RBS sequence.

In *E. coli*, AUG is the preferred start codon, followed by GUG. The analysis of the *M. tuberculosis* genome suggests that AUG is also preferred (Cole *et al.* 1998).

However, the frequency of GUG initiation codons is higher than in *E. coli*, reflecting the higher *G+C* content of the chromosome. The rare use of TTG as start codon has been suggested for the *dnaA* gene of *M. leprae*, *M. smegmatis* and *M. tuberculosis* (Salazar *et al.* 1996) and established for five genes of the L5 mycobacteriophage (Hatfull & Sarkis 1993).

Compilations of codon usage in mycobacteria have been published for genes from *M. leprae* (Dale & Patki 1990; Honore *et al.* 1993), *M. tuberculosis* complex (Dale & Patki 1990; Andersson & Sharp 1996), *M. paratuberculosis* (Bannantine *et al.* 1997), L5 mycobacteriophage (Hatfull & Sarkis 1993) and mycobacterial species in general (Wada *et al.* 1992). The observed bias for codons with G or C in the third position has been interpreted as result of the high G+C content of the mycobacterial DNA. The percentage of codons ending with G or C was 73% in a compilation of codon usage for the 12 *M. leprae* genes mentioned above, while for *M. tuberculosis*-complex genes this percentage was 83% (Andersson & Sharp 1996). The stronger bias in *M. tuberculosis* vs. *M. leprae* correlates with the different G+C content of their chromosomes (65% vs. 57%). The bias is also evident in the lower frequency of codons containing A in any position (Honore *et al.* 1993). Three classes of genes have been established in *E. coli* based on different patterns of codon usage. Class II genes are highly expressed during exponential growth in rich medium and have a strongly biased pattern of codon usage that reflects the relative abundance of tRNA species in those growth conditions. Class I contains genes involved in most metabolic processes and their pattern of codon usage is much less biased. Class III contains genes probably exchanged by horizontal gene transfer, which are characterized by an even lower bias (Henaut & Danchin 1996). When the pattern of codon usage for *M. tuberculosis*-complex genes expected to be expressed at either high or low level was compared with the pattern of codon usage for homologous genes from *E. coli* or *B. subtilis*, it was concluded that the correspondence between level of expression and preferential use of certain codons also exists in *M. tuberculosis* complex. The patterns were less heterogeneous than in *E. coli*, though, probably

reflecting the absence of selective pressure associated with fast growth (Andersson & Sharp 1996).

5 Gene regulation and physiology

Previous sections of this chapter presented methods for studying mycobacterial gene expression and described the basic transcriptional machinery that performs this essential function. In this section, we discuss regulated systems of gene expression that have been studied in mycobacteria. Since our knowledge in many of these areas is still rudimentary, the discussions will include relevant comparative data obtained in other organisms.

5.1 Heat shock

The heat-shock response is probably the most highly studied system in which several genes are regulated coordinately by a specific stimulus. In all living organisms, ranging from human beings to bacteria, including mycobacteria, highly conserved polypeptides, collectively known as heat-shock proteins, e.g. DnaK, GroES and GroEL, show increased synthesis when cells are exposed to elevated temperatures. Other stresses like exposure to ethanol and heavy metals can also elicit a similar response. Some of these proteins function as chaperones under non-stress conditions to correctly fold newly synthesized polypeptides or to aid in the translocation of proteins. The increased production of chaperones and specific proteases like DegP (see below) after heat shock and related stresses helps eliminate, by refolding and proteolysis, the denatured proteins.

DnaK and GroEL are immunodominant mycobacterial antigens as specific antibodies to these proteins are observed during mycobacterial diseases of humans and mice. These and other observations suggest that their synthesis may be elevated upon bacterial infection of the mammalian host (Shinnick *et al.* 1988; Young *et al.* 1988). As is the case in streptomycetes (Mazodier *et al.* 1991; Duchene *et al.* 1994; De Leon *et al.* 1997), there are two *groEL* genes in mycobacterial species, one of which is linked to *groES* (Rinke de Wit *et al.* 1992). This latter arrangement, *groESL,* is found in all bacteria. In addition, *M. tuberculosis* contains two *dnaJ* genes (see below).

While the nature of the heat-shock response is highly conserved, it is regulated by distinct mechanisms in different life forms. In eukaryotes, the phosphorylation of a conserved transcription factor present before stress permits the activation of heat-shock genes (Sorger 1990). In some cases, e.g. *Drosophila*, activation is at the level of transcription elongation, not initiation (Morimoto *et al.* 1992). In *E. coli and* other Gram-negative bacteria, heat shock causes a rise in the levels of two σ factors, σ^{htpR} and σ^E. The latter is a member of the ECF family of alternative σ factors, described previously in this chapter. It functions during extreme heat shock as part of an RNAP to transcribe *htpR*, encoding σ^{htpR}, and other heat-shock genes (Bukau 1993). The mechanisms modulating the levels of the heat-shock σ factors are quite complex, involving control of gene transcription, as well as mRNA and protein stability. *Streptomyces, Bacillus* and related spore formers do not have heat-shock σ factors of the σ^{htpR} class. However, a group of heat-shock genes in *B. subtilis* is transcribed by RNAP containing an alternative σ factor, σ^B, unrelated to the ECF family, that is also involved in several stress responses (Hecker *et al.* 1996).

Measurement of RNA levels and analysis of transcriptional fusions to *lacZ* indicate that regulation of the mycobacterial heat-shock response is mainly at the level of gene transcription, as occurs in other bacteria (Patel *et al.* 1991; Stover *et al.* 1991). Examination of the *M. tuberculosis* genome does not reveal any ORFs with homology to σ^{htpR}. Homologues to *B. subtilis* σ^B exist in actinomycetes (see below), but there are no data suggesting their involvement in the mycobacterial or streptomycete heat-shock response. However, there is at least one functional *sigE* gene of the ECF class, encoding σ^E, in streptomycetes (Jones *et al.* 1997), and both *B. subtilis* (Sorokin *et al.* 1997) and mycobacterial species (Salazar *et al.* 1997; Cole *et al.* 1998) possess multiple ECF σ factor genes. The role of the σ^E in streptomycete heat shock is not known, but one of the mycobacterial ECF σ factors, σ^E (Wu *et al.* 1997), and one of the *B. subtilis* homologues, known as σ^X

(Huang *et al.* 1997), are involved in heat-shock response since disruptions of these genes result in lower survival following heat shock. In agreement with its suggested function, transcription of *M. tuberculosis sigE* is induced after heat shock (Manganelli *et al.* 1998). *M. tuberculosis sigE* is closely linked to a homologue of *E. coli degP,* which is induced after extreme heat shock and is transcribed by RNAP containing σ^E in *E. coli* (Pallen & Wren 1997). Heat-inducible transcription has also been described for two additional *M. tuberculosis* ECF σ factors, σ^H and σ^M (Manganelli *et al.* 1998), and one *M. smegmatis* ECF σ factor, encoded by *sloA* (Salazar *et al.* 1997).

While alternative σ factors do play a role in heat-shock response in *Actinomyces* and *Bacillus*, much evidence indicates that these organisms use the major housekeeping RNAP to transcribe many of the genes encoding heat-shock proteins. Negative control plays an important role in heat-shock regulation in these and several other bacterial species. They possess a conserved sequence known as CIRCE (controlling inverted repeat of chaperone expression) which contains an inverted repeat and a 9-bp spacer (TTAGCACTC-N9-GAGTGCTAA) at or near the TSP of certain heat-shock genes (Hecker *et al.* 1996). In *B. subtilis*, a regulator, HcrA, encoded by the first gene of the *hcrA–dnaK–grpE–dnaJ* operon, binds to CIRCE sequences and represses heat-shock genes (Yuan & Wong 1995; Hecker *et al.* 1996). ORFs quite similar to HcrA are found in several bacteria, including the genera *Clostridium, Staphylococcus, Lactococcus* and *Mycobacterium*. There is no direct evidence indicating that these ORFs, if translated, function as the *B. subtilis* counterpart, but their sequence similarity to HcrA and in some cases the linkage of the ORFs to *dnaK* suggest that heat-shock genes in these organisms may be regulated as in *B. subtilis*. In *M. tuberculosis*, the *hcrA* homologue is linked to the second *dnaJ* locus and not to the *dnaK* operon. While the existence of *hcrA* has not yet been demonstrated in streptomycetes, the existence of CIRCE elements in their *groESL1 and groE2* heat-shock gene promoters suggest its presence. These actinomycetes possess an operon containing *dnaK, grpE* and *dnaJ* that presents multiple inverted repeats, different from CIRCE, in its promoter region. The last gene in this operon encodes a protein, HspR, that binds to these inverted repeats (Bucca *et al.* 1995). *hspR* is induced after heat shock along with the other genes in the *dnaK* operon and its gene product functions as an autogenous repressor of this operon. HspR does not repress the streptomycete *groE* genes, however, suggesting a system that requires double negative control (Bucca *et al.* 1997). After heat shock, *hcrA* repression is lifted by a currently unknown mechanism and ultimately the *dnaK–grpE–dnaJ–hspR* operon is induced; the HspR produced may then serve to down-regulate the heat-shock response. Analysis of the *M. tuberculosis* genomic sequence also shows the presence of a *dnaK–grpE–dnaJ–hspR* operon, complete with inverted repeats in the promoter region that show high sequence similarity to the streptomycete *dnaK* operon (in addition to the above-mentioned mycobacterial operon that has *hcA* and the *dnaJ2* gene). It is not known why actinomycetes need two repressors for the regulation of heat-shock genes. However, the mycobacterial *hcrA–dnaJ2* operon does not have sequences resembling CIRCE or the HspR binding palindrome. Thus, the mycobacterial HcrA may be constitutively produced.

5.2 Iron regulation

As an essential cofactor of enzymes with important cellular functions such as DNA and amino acid synthesis, respiration and detoxification, iron is required by all living organisms. Low levels of this element also constitute a regulatory signal that triggers the expression of certain bacterial virulence genes, e.g. those encoding some toxins and haemolysins, etc. Since the concentration of free iron in the environment and serum (10^{-18} mol/l) is too low for all life forms (except lactobacilli), they have developed systems to solubilize and assimilate environmental iron that is ordinarily found in the insoluble ferric hydroxide form. Many microorganisms secrete siderophores, i.e. low-molecular-weight compounds that bind with high affinity and solubilize Fe^{3+} and then are transported into the cell by membrane-

bound receptors. Other organisms have receptors that can directly bind to and internalize host iron-containing complexes like transferrin or lactoferrin. The mycobacterial iron uptake components performing these functions are described in Chapter 14.

Since high levels of iron can be toxic to living organisms, its uptake into cells is tightly regulated. This element can catalyse the formation of highly reactive oxygen species (ROS) from the H_2O_2 and superoxides that are produced during normal aerobic respiration. Eukaryotes regulate iron uptake by post-transcriptional mechanisms (Qian & Tang 1995) while bacteria negatively regulate this process at the level of transcription. In many Gram-negative and some Gram-positive bacteria the repressor is Fur. When this protein forms a complex with its corepressor, ferrous iron, it is able to repress iron uptake genes by binding to operator sites in their promoter regions. When iron levels are low, deferrated Fur cannot bind to these genes, allowing the synthesis of the uptake machinery. This leads to an increase in intracellular iron so that active Fur–Fe complexes can form and again repress the uptake genes. DtxR, an analogous repressor of the iron-uptake machinery has been described in *Corynebacterium diphtheriae* (Tao *et al.* 1994). In mycobacteria, IdeR, a homologue of DtxR, has been isolated and shown to negatively regulate the synthesis of exochelins and mycobactins, the two types of mycobacterial siderophores (Dussurget *et al.* 1996). At present only one gene involved in iron acquisition, *fxbA*, has been described in mycobacteria. *fxbA* encodes a putative formyltransferase that is necessary for exochelin biosynthesis in *M. smegmatis and* is negatively regulated by high iron levels (Fiss *et al.* 1994). The promoter region of this gene contains a sequence resembling the consensus DtxR operator site found in corynebacterial iron-regulated genes: T(A/T)AGGTTAG(G/C)CTAACCT(T/A)A.   When complexed to ferrous iron, IdeR binds to the *fxbA* iron box, *in vitro*, and represses this gene *in vivo* (Dussurget *et al.* 1999b). A gene encoding an IdeR/DtxR homologue has also been found in streptomycetes (Gunter-Seeboth & Schupp 1995) and one iron-regulated gene, *desA*, with a DtxR-like iron box sequence in its promoter has been described (Gunter *et al.* 1993). On the basis of these sequence similarities, but not on direct experimental evidence, it is likely that streptomycete iron uptake is regulated in a manner similar to that of coryneforms and mycobacteria.

As predicted, inactivation of the *M. smegmatis* *ideR* causes deregulation of siderophore synthesis. However, deregulation is partial, indicating that IdeR does not account for all adaptive responses to iron availability and suggests that additional mechanisms control the mycobacterial iron uptake machinery. A similar diversity of iron regulation systems in one bacterium has been described in *Pseudomonas* (Venturi *et al.* 1995) and in *B. subtilis* where three functional Fur proteins, YqkL, YqfV and YgaG, have been found (Bsat *et al.* 1998). Recent genetic and genome analysis has identified two ORFs in *M. tuberculosis* showing significant similarity to the Fur protein found in Gram-negative bacteria (Pagan-Ramos *et al.* 1997; Smith *et al.* 1997) and another ORF (SirR) that is related to IdeR (Cole *et al.* 1998). It is not known whether these genes, *furA*, *furB* and *sirR*, are transcribed *in vivo*, and whether their gene products function in iron regulation. Significantly, one of the ORFs, *furA*, is found directly upstream of the *katG* gene in all mycobacteria and the two genes are cotranscribed (Pagan-Ramos *et al.* 1997). Interestingly, one of the three *B. subtilis fur* homologues (*ygaG* or *perR*) is found immediately downstream a peroxidase gene and 10 kb from *katA*, encoding the major catalase in this organism (Bsat *et al.* 1998). A role for the mycobacterial FurA in the regulation of *katG* expression has not been shown yet, but as described below, a relationship between iron regulation and oxidative stress has already been demonstrated in *M. smegmatis*.

The importance of siderophores in mycobacterial pathogenesis has been questioned (Lambrecht & Collins 1993; Wheeler & Ratledge 1994). There is also little information about the actual iron levels in the mycobacterium-containing phagosome, i.e. if iron is limiting in this structure, as it is in those enveloping certain intracellular pathogens like *Legionella* and *Salmonella* (Byrd & Horwitz 1989; Garcia-del Portillo *et al.* 1992). Definitive answer to these questions must await the isolation of *M. tuberculosis* mutants that cannot make siderophores so that bacterial survival can be tested in macrophages.

5.3 Oxidative stress response

As discussed above, aerobically respiring cells produce superoxides and H_2O_2 that must be inactivated before they give rise to more dangerous ROS. In addition, cells can be threatened by these compounds arising from the environment. The most striking example is the respiratory burst that is produced by neutrophils when they phagocytose bacteria that have bound to Fc receptors. Superoxide dismutases (SODs) and catalases serve to detoxify the ROS and, in many bacteria, these enzymes are induced after exposure of cells to these compounds. The oxidative stress response has been most widely studied in enteric bacteria. In these organisms, superoxide stress causes activation of the positive regulators SoxR and SoxS that are necessary for the expression of *sodA*, encoding a superoxide dismutase and several other genes encoding stress response enzymes. H_2O_2 stress activates OxyR, which then becomes a positive regulator of several genes including those encoding KatG, a catalase peroxidase and AhpCF, subunits of the alkyl hydroxyperoxidase (Ahr) (Demple 1991). Many bacteria also possess another catalase, KatE, which is not induced by external H_2O_2 as is KatG, but is more abundant in stationary-phase cultures. In enteric bacteria, the gene encoding KatE is transcribed by RNAP containing the stationary-phase σ factor, σ^S (Hengge-Aronis 1993). In *B. subtilis*, *katE* is transcribed during stationary growth by σ^B containing RNAP (Engelmann *et al.* 1995).

In mycobacteria, little is known about the regulation of the genes encoding SodA, KatG and related enzymes like Ahr. Certain mycobacteria like *M. smegmatis* also contain a functional KatE and levels of the enzyme increase in the stationary phase (Dubnau *et al.* 1996; Dussurget *et al.* 1999b). However, genes encoding proteins homologous to SoxR, SoxS, σ^S or the *B. subtilis* σ^B have not been described in these bacteria. Significantly, a null mutation in the gene encoding *M. smegmatis* σ^E (Wu *et al.* 1997) causes enhanced sensitivity to H_2O_2. Null mutations in the *B. subtilis sigX*, encoding σ^X, also cause enhanced sensitivity to H_2O_2, through an unknown mechanism (Huang *et al.* 1997). Although an *oxyR* homologue is found in *M. leprae*, closely linked to the *ahpCF* operon as is the case in enteric bacteria (Dhandayuthapani *et al.* 1997), its function has not been assessed *in vivo*. Significantly, there is an *oxyR* homologue in *M. tuberculosis*, also linked to the *ahpCF* operon, but its gene product is not functional as the putative ORF has many deletions and termination codons (Deretic *et al.* 1995; Sherman *et al.* 1995).

The role of the host oxidative burst and the mycobacterial oxidative stress response in virulence has been a controversial subject, as studies on the survival of *M. tuberculosis* in guinea-pig macrophages gave contradictory results regarding the role of ROS in the killing of *M. tuberculosis* in these cells (Jackett *et al.* 1978; O'Brien & Andrew 1991; O'Brien *et al.* 1991). It has also been reported that entry of *M. tuberculosis* into macrophages does not induce an oxidative burst, as these bacteria use complement receptors, not those for Fc (Schlesinger *et al.* 1990). However, recent results have shown that *M. bovis katG* mutants show attenuated virulence in guinea pigs (Wilson *et al.* 1995). This is consistent with studies in other pathogens in which mutant bacteria lacking SodA, SodB or KatG were less virulent (Franzon *et al.* 1990; Tsolis *et al.* 1995). The necessity for a robust *M. tuberculosis* oxidative stress response for survival in the infected human host is indirectly supported by other studies in which alveolar macrophages from patients with active tuberculosis had a three- to four-fold greater oxidative burst (production of H_2O_2) than similar cells isolated from healthy subjects (Kuo *et al.* 1996).

There is also a relationship between iron levels, Fur and the regulation of the genes encoding enzymes of the bacterial oxidative stress response. The *E. coli* Fur is a repressor of the *sodA* gene encoding the magnesium-dependent SOD (Niederhoffer *et al.* 1990) and mutations in *fur* cause an iron overload leading to oxidative stress and DNA damage (Touati *et al.* 1995). *Pseudomonas fur* mutants also show lower total catalase and SOD activities (Hassett *et al.* 1996). On the other hand, mutations in the *S. typhimurium fur* gene allow mutants to survive better in macrophages, a phenomenon presumably caused by higher levels of SodA (Tsolis *et al.* 1995). Also, a *C. diphtheriae* gene encoding the Ahr subunit AhpC is negatively regulated by iron (Tai & Zhu 1995). Ahrs are involved in response to oxidative stress caused by H_2O_2 and organic peroxides (Bsat *et al.* 1996; Dhandayuthapani *et al.* 1996). The connection between iron regulation and oxidative stress response is also important in mycobacteria as null mutations in the *M. smegmatis ideR* cause increased sensitivity to superoxides and H_2O_2, a phenotype caused by lowered levels of SodA and catalase: peroxidase activity (KatG) observed in the mutants (Dussurget *et al.* 1996). The mechanism by which IdeR controls SodA and KatG is not known, but the levels of these proteins and their mRNAs, as measured by immunoassays and RNA determinations, are lower in *ideR* mutants. This suggests a positive regulatory role for IdeR at the level of *sodA* and *katG* gene expression (Dussurget *et al.* 1999a).

5.4 SOS response

Bacterial cells respond to DNA-damaging agents in a characteristic manner so that several genes involved in DNA synthesis and repair, cell division and recombination become induced. In *E. coli* and most other Gram-negative bacteria, these genes are largely under the control of a repressor, LexA. In unstressed conditions, this protein binds to an operator sequence (CTGT-N8-ACAG) called the SOS box in the promoter region of SOS-induced genes, causing repression. After DNA damage, the RecA protein becomes activated and facilitates the autocatalytic cleavage of LexA. The reduced ability of the cleaved protein to bind to its operator sites lifts the repression of the DNA damage-inducible genes. The genes for LexA and RecA also have SOS boxes, are repressed by LexA and are induced by DNA damage via the same mechanism (Walker 1996). *B. subtilis* has a LexA homologue, but its binding site on target genes (GAAC-N4-GTTC), called the Cheo box, is different from the Gram-negative counterpart. The *M. tuberculosis lexA* and *recA* promoter regions also contain Cheo boxes and the *M. tuberculosis* LexA binds to these regions (Durbach *et al.* 1997; Movahedzadeh *et al.* 1997). The *recA* gene is induced after exposure of *M. tuberculosis* to DNA-damaging agents, and though the effect is small, two- to threefold, this seems to occurs through the inactivation of LexA as in other bacteria.

5.5 Growth and stationary-phase regulation

Little is known about the mechanisms that control growth in mycobacterial species. Studies in other bacteria, largely *E. coli*, have indicated a relationship between growth rate and ribosomal RNA (rRNA) levels. The multiple (seven) *rrn* cistrons encoding rRNA have tandem promoters p1 and p2, of which only the former is growth regulated. Amino acid starvation, via the stringent response, induces the synthesis of the nucleotides pppGpp and ppGpp that repress transcription from the stronger p1 promoter (Keener & Nomura 1996). A 'discriminator' sequence, GCGCCNCC, found between the −10 and the TSP of the p1 promoters, but not p2 promoters is essential for growth and stringent regulation (Keener & Nomura 1996). It has been reported that the rate of rRNA synthesis in *M. tuberculosis* and levels of rRNA are ≈ 10% of these parameters in *E. coli* (see section 3.2; see also Chapter 11 for additional discussion of mycobacterial growth rates and dormancy). The slow growth rates of mycobacteria have been ascribed to their low numbers of *rrn* genes, e.g. one copy in the slow-growing pathogens like *M. tuberculosis* and *M. leprae* and two copies in the faster-growing saprophytes like *M. smegmatis.* This compares with six or more copies in even faster-growing *Streptomyces*, *Bacillus* and enteric bacteria.

This overly simple idea was questioned since the growth rates of pathogenic and non-pathogenic bacteria can vary widely and are independent of the number of *rrn* genes, i.e. some fast-growing mycobacteria like *M. chelonae and M. abscessus* have only one *rrn* gene (Domenech *et al.* 1994). This indicates that other factors are implicated in growth-rate control. Recent experiments have cast more doubt on this 'quantitative *rrn* mode' of mycobacterial growth

control as elimination of one of the two *M. smegmatis rrn* genes has no effect on growth rate (Sander *et al.* 1996). Sequencing of the *rrn* genes of *M. tuberculosis*, *M. leprae* and other slow-growing mycobacteria shows that they are highly conserved, including their upstream regulatory regions (Gonzalez-y-Merchand *et al.* 1997). One of the two *M. smegmatis* genes, *rrnA* is also homologous to the single copy found in these pathogens. *In vivo* mapping studies have localized several TSPs for the *rrnA* genes of both species, while only one promoter has been detected, *in vitro*, as well as *in vivo* for the *M. smegmatis rrnB* promoter (Predich *et al.* 1995; Gonzalez-y-Merchand *et al.* 1997). The multiple *rrnA* promoters show different levels of activity, i.e. *M. smegmatis rrnA* p2 is more active than p1 or pCL1, and the *M. tuberculosis rrnA* pCL1 is more active than p1, p2 being deleted in this species. It is not presently known whether growth in different conditions will cause differential regulation of these promoters. Growth-rate studies have not been carried out since mycobacterial species do not show easily alterable generation times. The stringent response has not been studied in mycobacteria, and the discriminator sequence found in *E. coli rrn* p1 promoters is not found in any of the mycobacterial *rrn* promoters. However, ORFs highly similar to *spoT* and *relA*, coding for enzymes of the ppGpp synthesis, can be detected in the *M. tuberculosis* genome sequencing projects, but their functions remain unknown. Streptomycetes do regulate ribosomal synthesis after amino acid starvation by the formation of ppGpp, as demonstrated by the relaxed phenotype observed upon the inactivation of the *S. coelicolor relA* gene.

The stationary phase of the growth cycle is an important area of study in bacteria. Most economically and medically important secondary metabolites like antibiotics and secreted enzymes are produced at this stage in *Bacillus* and *Streptomyces.* In addition, our understanding of cellular differentiation has been greatly enhanced by elegant studies on sporulation performed in these genera. Other bacteria do not go through an elaborate morphological differentiation like sporulation when nutrients or other essential components like oxygen become limiting, but major alterations in cell structure and metabolism can occur and this has been studied most extensively in *E. coli*. It is useful to briefly review what happens to these cells during the transition to the stationary (non-growing) phase since there may be parallels in mycobacteria. Upon the cessation of growth, *E. coli* cells undergo changes in cell shape, cell envelope structure and general metabolism (Huisman *et al.* 1996). They become smaller, rounder and accumulate specific lipids and carbohydrates while others diminish. The peptidoglycan layer of the cell wall shows changes in composition and increases in thickness, and unsaturated membrane lipids become highly cyclopropanated. Many cytoplasmic changes also occur, including the synthesis of new proteins including enzymes like the catalase KatE, and those involved in trehalose production and the cyclopropanation of unsaturated membrane lipids. Many of the genetic determinants of these proteins are transcribed by an RNAP containing σ^s. Higher levels of this σ factor, resulting from increased *rpoS* transcription, translation and σ^s stabilization, are observed in early stationary phase and ppGpp is necessary for the increased *rpoS* expression. These multiple changes give resistance to various environmental insults, e.g. antibiotic treatment, osmotic shock and oxidative stress, etc.

Mutations preventing formation of products appearing in stationary phase frequently cause inability to survive in the non-growing state. Interestingly, the cyclopropanation of unsaturated cell-membrane lipids observed during *E. coli* stationary phase is analogous to the situation in mycobacterial pathogens that have cyclopropanated mycolic acids in place of the unsaturated species found in non-pathogens like *M. smegmatis* (Brennan & Nikaido 1995). It has been suggested that mycolic acid modifications confer resistance to ROS found in the macrophage environment that could crosslink and destroy mycolic acids with unsaturated double bonds (Yuan *et al.* 1995).

It is postulated that the stationary phase or related quiescent states (variously known as dormancy or persistence) may also be important for *M. tuberculosis* virulence. After an initial infection, in which the mammalian host can arrest the disease by the formation of granulomas, these bacteria can persist in a latent form, with the absence of symptoms in the host. In humans, this 'disease-free state' can last for many years. Upon a suitable stimulus, frequently a challenge to the immune system, growth of pre-existing *M. tuberculosis* resumes and the clinical manifestations of the disease return.

Viable tubercle bacilli can be cultured from lung lesions in a sputum-negative host years after infection. There have also been reports of morphological variants of *M. tuberculosis* recovered, in a dormant state, from patients with long-term tuberculosis or from infected animals, starting with the finding of Much's granules over 90 years ago (Stanford 1987) to more recent reports (Khomenko 1987). These aberrant cells have been described as being smaller, possessing a rounded shape, with thickened and electrodense cell walls and loss of acid-fast staining. Many studies have been performed to understand the amazing ability of *M. tuberculosis* to persist in a 'dormant' form in an infected host. Among these experiments, initiated in systematic fashion over 40 years ago are those comparing various biochemical parameters in virulent *M. tuberculosis* grown *in vitro* with those in bacteria recovered from infected animals. The results indicated changes in lipid composition, pathogenicity and immunogenicity, as well as a shift towards anaerobic metabolism in bacteria grown *in vivo* (Segal & Bloch 1956, 1957). Other experiments have focused on the biochemistry and physiology of *M. tuberculosis* before, during and after the cessation of growth caused by a gradual decrease in oxygen availability (Wayne 1994). In these studies, the cells in stationary phase retained viability for long periods of time and showed lower levels of oxidative stress response enzymes, like SOD and catalase, and increased amounts of the glyoxalate pathway enzymes, isocitrate lyase and glycine dehydrogenase. In addition, higher levels of nitrate reductase are observed (Wayne & Hayes 1998). This suggests a transition to anaerobic metabolism, reminiscent of the change in physiology noted in *M. tuberculosis* grown in animals, discussed above. Not unexpectedly, a homologue of Fnr can be found in the *M. tuberculosis* genomic databases. This protein, extensively studied in enteric bacteria, is a positive regulator of anaerobic respiratory genes like those encoding nitrate and nitrite reductases and is a repressor of aerobic respiratory genes, e.g. *sodA*. (Lynch & Lin 1996). Stationary cultures of *M. tuberculosis* are also sensitive to metronidazole, while growing cultures are resistant. This drug is only useful against anaerobic bacteria as it must be reduced to be effective (Wayne & Sramek 1994). Conversely, other antibiotics like INH have no effect on the cells in the dormant phase, while they are bactericidal for exponentially growing *M. tuberculosis* and other mycobacteria. At the same time, as cells are kept in the low-oxygen-induced resting state, there are also morphological changes. Two weeks after initiation of dormancy, *M. tuberculosis* cells develop a highly thickened cell wall that, on the basis of staining properties, has a higher lipid content (Cunningham & Spreadbury 1997). This is suggestive of a resistant spore- or cyst-like coat. Little is currently known about proteins that may play a role in the dormant stage, but a new protein, URB-1, was observed in stationary cultures of *M. tuberculosis* (Wayne 1994). Recent work has shown that this protein is a known antigen of *M. tuberculosis,* the 16-kDa protein and that it is specifically induced during the transition from high to low (about 2%) oxygen tension (Yuan *et al.* 1996). The 16-kDa protein is homologous to α-crystallin proteins of other organisms that can act as molecular chaperones and it also can protect proteins against heat denaturation. Its gene, *acg*, is only found in members of the *M. tuberculosis* complex and a related protein is found in *M. leprae*. Reporter fusion studies have shown that the expression of the *M. tuberculosis acg* is regulated at the transcriptional level and deletion analysis has localized a region between 160 and 130 bp upstream from the TSP that, when removed, allows oxygen-independent expression of the gene. This strongly suggests that regulation is negative and that a repressor, normally binding to the promoter region, is removed when oxygen levels drop (Yuan *et al.* 1998).

Other reporter studies have shown that the *M. leprae acg* homologue, when introduced into BCG, is specifically induced in macrophages (Dellagostin *et al.* 1995).

One can question the relevance of the *in vitro* dormancy models to the actual latency observed in hosts infected with *M. tuberculosis*. However, there are many observations that support the significance of these studies. The responses to gradual oxygen depletion, discussed above, i.e. enhanced viability and the biochemical and morphological changes, are not observed in similarly treated *M. smegmatis.* These responses in *M. tuberculosis,* except for α-crystallin protein induction, all require gradual and not rapid oxygen depletion, a situation the bacteria would be expected to face in lung granulomas. The thickening of the cell walls is also reminiscent of the microscopic appearance of the so-called Much granules, described above. Probably the strongest evidence comes from the virulence phenotype of a *M. tuberculosis acg* mutant. These cells show lower survival in human macrophages and are less virulent for mice (Yuan *et al.* 1998; C. Barry III, personal communication). Reporter gene experiments have also indicated that the *M. tuberculosis acg,* like its *M. leprae* counterpart is induced in macrophages (Yuan *et al.* 1998).

Other studies have focused on the genetics of dormancy, with the idea of finding the mycobacterial homologues of genes from streptomycetes and *B. subtilis* that are known to regulate late growth phenomena like the entry into stationary phase and sporulation. As described before (section 3.3), a gene homologous to the *sigF* genes from streptomycetes and *B. subtilis* and to the *B. subtilis sigB* has been found (DeMaio *et al.* 1996). While it is tempting to speculate that σF is involved in α-crystallin gene transcription, the different environmental conditions that induce their genetic determinants are not identical. Since a null mutation in the *sigF* gene has not yet been obtained, its role in mycobacterial physiology and virulence is not known. A family of genes in mycobacterial species showing homology to the *whiB* sporulation gene of streptomycetes has also been isolated, but its role in mycobacterial physiology is also not known at the present time (Gomez *et al.* 1997b).

6 Conclusions and future prospects

Much work has been performed in the area of mycobacterial gene expression in recent years, as this chapter demonstrates. Tools for studying the regulation of gene expression have been developed. Using these methods, the responses to various environmental insults like oxidative stress and iron starvation are being actively studied. However, we expect that there will be a dramatic increase of knowledge in the genetics of mycobacterial physiology and virulence in the immediate future. The complete sequence of the *M. tuberculosis* H37Rv genome is now available, soon to be followed by that of the clinical *M. tuberculosis* strain CSU no.93. This information is already making possible the rapid identification of genes isolated by various methods and the comparison of mycobacterial sequences with well-characterized genes from other bacteria. It will soon be possible to measure the expression of all *M. tuberculosis* ORFs, whose sequences are derived from these genome databases, under a variety of environmental conditions and during infection, using the rapidly developing methods of functional genomic analysis. These techniques, in which a complete genome or thousands of genes are arrayed and then probed for differential gene expression in many environmental conditions, have been used in bacteria (Saizieu *et al.* 1998), yeast (DeRisi *et al.* 1997) and in normal and cancerous human cells (Schena *et al.* 1996; Zhang *et al.* 1997). The next step will be the inactivation of genes, identified by functional genomics, using the effective new methods of gene disruption (see Chapter 1) to elucidate their functions. Hopefully, the next major review of mycobacterial gene expression will reflect this 'new era' of mycobacterial research.

7 Acknowledgements

We gratefully acknowledge our many colleagues who contributed to this chapter with helpful discussions and unpublished data. Work in our laboratory is supported by NIH grants GM 19693 and GM 32651.

8 References

Abu Kwaik, Y. & Pederson, L.L. (1996) The use of differential display-PCR to isolate and characterize a *Legionella pneumophila* locus induced during the intracellular infection of macrophages. *Molecular Microbiology* **21**, 543–556.

Andersson, G.E. & Sharp, P.M. (1996) Codon usage in the *Mycobacterium tuberculosis* complex. *Microbiology* **142**, 915–925.

Andrew, P.W. & Roberts, I.S. (1993) Construction of a bioluminescent mycobacterium and its use for assay of antimycobacterial agents. *Journal of Clinical Microbiology* **31**, 2251–2254.

Angerer, A., Enz, S., Ochs, M. & Braun, V. (1995) Transcriptional regulation of ferric citrate transport in *Escherichia coli* K-12. FecI belongs to a new subfamily of sigma 70-type factors that respond to extracytoplasmic stimuli. *Molecular Microbiology* **18**, 163–174.

Baldus, J.M., Buckner, C.M. & Moran, C.P. Jr (1995) Evidence that the transcriptional activator Spo0A interacts with two sigma factors in *Bacillus subtilis*. *Molecular Microbiology* **17**, 281–290.

Bannantine, J.P., Barletta, R.G., Thoen, C.O. & Andrews, R.E. Jr (1997) Identification of *Mycobacterium paratuberculosis* gene expression signals. *Microbiology* **143**, 921–928.

Barker, L.P., Brooks, D.M. & Small, P.L. (1998) The identification of *Mycobacterium marinum* genes differentially expressed in macrophage phagosomes using promoter fusions to green fluorescent protein. *Molecular Microbiology* **29**, 1167–1170.

Barletta, R.G., Kim, D.D., Snapper, S.B., Bloom, B.R. & Jacobs, W., Jr (1992) Identification of expression signals of the mycobacteriophages Bxb1, L1 and TM4 using the *Escherichia–Mycobacterium* shuttle plasmids pYUB75 and pYUB76 designed to create translational fusions to the *lacZ* gene. *Journal of General Microbiology* **138**, 23–30.

Bashyam, M.D., Kaushal, D., Dasgupta, S.K. & Tyagi, A.K. (1996) A study of mycobacterial transcriptional apparatus: identification of novel features in promoter elements. *Journal of Bacteriology* **178**, 4847–4853.

Bashyam, M.D. & Tyagi, A. (1994) An efficient and high-yielding method for isolation of RNA from mycobacteria. *Biotechniques* **17**, 834–836.

Bashyam, M.D. & Tyagi, A. (1998) Identification and analysis of 'extended -10' promoters from mycobacteria. *Journal of Bacteriology* **180**, 2568–2573.

Brennan, P.J. & Nikaido, H. (1995) The envelope of mycobacteria. *Annual Review of Biochemistry* **64**, 29–63.

Brockman, R.W. & Heppel, L.A. (1968) On the localization of alkaline phosphatase and cyclic phosphodiesterase in *Escherichia coli*. *Biochemistry* **7**, 2554–2562.

Bronstein, I., Fortin, J., Stanley, P.E., Stewart, G.S. & Kricka, L.J. (1994) Chemiluminescent and bioluminescent reporter gene assays. *Analytical Biochemistry* **219**, 169–181.

Bsat, N., Chen, L. & Helmann, J. (1996) Mutation of the *Bacillus subtilis* alkyl hydroperoxide reductase (*ahpCF*) operon reveals compensatory interactions among hydrogen peroxide stress genes. *Journal of Bacteriology* **178**, 6579–6586.

Bsat, N., Herbig, A., Casillas-Martinez, L., Setlow, P. & Helmann, J. (1998) *Bacillus subtilis* contains multiple Fur homologues: identification of the iron uptake (Fur) and peroxide regulon (PerR) repressors. *Molecular Microbiology* **29**, 189–198.

Bucca, G., Ferina, G., Puglia, A.M. & Smith, C.P. (1995) The *dnaK* operon of *Streptomyces coelicolor* encodes a novel heat-shock protein which binds to the promoter region of the operon. *Molecular Microbiology* **17**, 663–674.

Bucca, G., Hindle, Z. & Smith, C.P. (1997) Regulation of the *dnaK* operon of *Streptomyces coelicolor* A3 (2): is governed by HspR, an autoregulatory repressor protein. *Journal of Bacteriology* **179**, 5999–6004.

Buchmeier, N.A., Wu, L. & Guiney, D.G. (1997). Analysis of a two-component response regulator gene (*tcrA*) uniquely present in the *Mycobacterium tuberculosis* complex. In: *ASM Conference on Tuberculosis: Past, Present and Future*, Copper Mountain, Colorado, Washington, DC: ASM Press, p. 16.

Bukau, B. (1993) Regulation of the *Escherichia coli* heat-shock response. *Molecular Microbiology* **9**, 671–680.

Burgess, R.R. (1976) Purification and physical properties of *E. coli* RNA polymerase. In: *RNA Polymerase* (eds R. Losick & M. Camberlin). Cold Spring Harbor: Cold Spring Harbor Laboratory Press, pp. 69–100.

Buttner, M.J. & Lewis, C.G. (1992) Construction and characterization of *Streptomyces coelicolor* A3 (2): mutants that are multiply deficient in the nonessential *hrd*-encoded RNA polymerase sigma factors. *Journal of Bacteriology* **174**, 5165–5167.

Byrd, T.F. & Horwitz, M.A. (1989) Interferon gamma-activated human monocytes downregulate transferrin receptors and inhibit the intracellular multiplication of *Legionella pneumophila* by limiting the availability of iron. *Journal of Clinical Investigations* **83**, 1457–1465.

Carriere, C., Riska, P.F., Zimhony, O. *et al.*(1997) Conditionally replicating luciferase reporter phages: improved susceptibility for rapid detection and assessment of drug susceptibility of *Mycobacterium tuberculosis*. *Journal of Clinical Microbiology* **35**, 3232–3239.

Chamberlin, M.J., Nierman, W.C., Wiggs, J. & Neff, N. (1979) A quantitative assay for bacterial RNA polymerases. *Journal of Biological Chemistry* **254**, 10061–10069.

Cheung, A.L., Eberhardt, K.J. & Fischetti, V.A. (1994) A method to isolate RNA from gram-positive bacteria and mycobacteria. *Analytical Biochemistry* **222**, 511–514.

Cole, S.T., Brosh, R., Parkhill, J. *et al.*(1998) Deciphering the biology of *Mycobacterium tuberculosis* from the complete genome sequence. *Nature* **393**, 537–544.

Collins, D.M. (1996) In search of tuberculosis virulence genes. *Trends in Microbiology* **4**, 426–430.

Collins, D.M., de Kawakami, R.P., Lisle, G.W., Pascopella, L., Bloom, B.R. & Jacobs, W.R. Jr (1995) Mutation of the principal sigma factor causes loss of virulence in a strain of the *Mycobacterium tuberculosis* complex. *Proceedings of the National Academy of Sciences of the USA* **92**, 8036–8040.

Contag, C.H., Contag, P.R., Mullins, J.I., Spilman, S.D., Stevenson, D.K. & Benaron, D.A. (1995) Photonic detection of bacterial pathogens in living hosts. *Molecular Microbiology* **18**, 593–603.

Cormack, B.P., Valdivia, R.H. & Falkow, S. (1996) FACS-optimized mutants of the green fluorescent protein (GFP). *Gene* **173**, 33–38.

Cubitt, A.B., Heim, R., Adams, S.R., Boyd, A.E., Gross, L.A. & Tsien, R.Y. (1995) Understanding, improving and using green fluorescent proteins. *Trends in Biochemical Sciences* **20**, 448–455.

Cunningham, A.F. & Spreadbury, C.L. (1997). Mycobacterial dormancy: the role of low oxygen tension in the development of cell wall thickening and in the upregulation of the 16-kDa alpha-crystallin homolog. In: *ASM Conference on Tuberculosis: Past, Present and Future*, Copper Mountain, Colorado. Washington, DC: ASM Press p. 35.

Curcic, R., Dhandayuthapani, S. & Deretic, V. (1994) Gene expression in mycobacteria: transcriptional fusions based on *xylE* and analysis of the promoter region of the response regulator *mtrA* from *Mycobacterium tuberculosis*. *Molecular Microbiology* **13**, 1057–1064.

Dale, J.W. & Patki, A. (1990) Mycobacterial gene expression and regulation. In: *Molecular Biology of the Mycobacteria* (ed. J. McFadden). San Diego: Academic Press, pp. 173–198.

Das Gupta, S.K., Bashyam, M.D. & Tyagi, A.K. (1993) Cloning and assessment of mycobacterial promoters by using a plasmid shuttle vector. *Journal of Bacteriology* **175**, 5186–5192.

De Las Penas, A., Connolly, L. & Gross, C.A. (1997) SigmaE is an essential sigma factor in *Escherichia coli*. *Journal of Bacteriology* **179**, 6862–6864.

De Leon, P., Marco, S., Isiegas, C., Marina, A., Carrascosa, J.L. & Mellado, R.P. (1997) *Streptomyces lividans groES, groEL1* and *groEL2* genes. *Microbiology* **143**, 3563–3571.

Dellagostin, O.A., Esposito, G., Eales, L.J., Dale, J.W. & McFadden, J. (1995) Activity of mycobacterial promoters during intracellular and extracellular growth. *Microbiology* **141**, 1785–1792.

DeMaio, J., Zhang, Y., Ko, C. & Bishai, W.R. (1997) *Mycobacterium tuberculosis sigF* is part of a gene cluster with similarities to the *Bacillus subtilis sigF* and *sigB* operons. *Tubercule and Lung Disease* **78**, 3–12.

DeMaio, J., Zhang, Y., Ko, C., Young, D.B. & Bishai, W.R. (1996) A stationary-phase stress-response sigma factor from *Mycobacterium tuberculosis. Proceedings of the National Academy of Sciences of the USA* **93**, 2790–2794.

Demple, B. (1991) Regulation of bacterial oxidative stress genes. *Annual Review of Genetics* **25**, 315–337.

Deretic, V., Philipp, W., Dhandayuthapani, S. *et al.*(1995) *Mycobacterium tuberculosis* is a natural mutant with an inactivated oxidative-stress regulatory gene: implications for sensitivity to isoniazid. *Molecular Microbiology* **17**, 889–900.

DeRisi, J.L., Iyer, V.R. & Brown, P.O. (1997) Exploring the metabolic and genetic control of gene expression on a genomic scale. *Science* **278**, 680–686.

Dhandayuthapani, S., Mudd, M. & Deretic, V. (1997) Interactions of OxyR with the promoter region of the *oxyR* and *ahpC* genes from *Mycobacterium leprae* and *Mycobacterium tuberculosis. Journal of Bacteriology* **179**, 2401–2409.

Dhandayuthapani, S., Via, L.E., Thomas, C.A., Horowitz, P.M., Deretic, D. & Deretic, V. (1995) Green fluorescent protein as a marker for gene expression and cell biology of mycobacterial interactions with macrophages. *Molecular Microbiology* **17**, 901–912.

Dhandayuthapani, S., Zhang, Y., Mudd, M.H. & Deretic, V. (1996) Oxidative stress response and its role in sensitivity to isoniazid in mycobacteria: characterization and inducibility of *ahpC* by peroxides in *Mycobacterium smegmatis* and lack of expression in *M. aurum* and *M. tuberculosis. Journal of Bacteriology* **178**, 3641–3649.

Dombroski, A.J., Walter, W.A. & Gross, C.A. (1993) Aminoterminal amino acids modulate sigma-factor DNA-binding activity. *Genes and Development* **7**, 2446–2455.

Domenech, P., Menendez, M.C. & Garcia, M.J. (1994) Restriction fragment length polymorphisms of 16S rRNA genes in the differentiation of fast-growing mycobacterial species. *FEMS Microbiological Letters* **116**, 19–24.

Doukhan, L., Predich, M., Nair, G. *et al.* (1995) Genomic organization of the mycobacterial sigma gene cluster. *Gene* **165**, 67–70.

Draper, D.E. (1996) Translational initiation. In: *Escherichia Coli and Salmonella Cellular and Molecular Biology* (ed. F. C. Neidhardt). Washington, DC: ASM Press, pp. 902–907.

Dubnau, E., Soares, S., Huang, T.J. & Jacobs, W.R. Jr (1996) Overproduction of mycobacterial ribosomal protein S13 induces catalase/peroxidase activity and hypersensitivity to isoniazid in *Mycobacterium smegmatis. Gene* **170**, 17–22.

Dubnau, E., Soares, S. & Smith, I. (1997). Promoters of *Mycobacterium tuberculosis* expressed in human macrophages. In: *First Latin American Symposium on Tuberculosis*, la Havana, Cuba, pp. 28–29.

Duchene, A.M., Kieser, H.M., Hopwood, D.A., Thompson, C.J. & Mazodier, P. (1994) Characterization of two *groEL* genes in *Streptomyces coelicolor* A3(2). *Gene* **144**, 97–101.

Duguid, J.R., Rohwer, R.G. & Seed, B. (1988) Isolation of cDNAs of scrapie-modulated RNAs by subtractive hybridization of a cDNA library. *Proceedings of the National Academy of Sciences of the USA* **85**, 5738–5742.

Durbach, S.I., Andersen, S.J. & Mizrahi, V. (1997) SOS inducing in mycobacteria: analysis of the DNA-binding activity of LexA-like repressor and its role in DNA damage induction of the *recA* gene from *Mycobacterium smegmatis. Molecular Microbiology* **26**, 643–653.

Dussurget, O., Rodriguez, G.M. & Smith, I. (1996) An *ideR* mutant of *Mycobacterium smegmatis* has a derepressed siderophore production and an altered oxidative-stress response. *Molecular Microbiology* **22**, 535–544.

Dussurget, O., Rodriguez, G.M. & Smith, I. (1999a) Protective role of the mycobacterial IdeR against oxygen species and isoniazid toxicity. *Tubercle Lung Disease* **79**, 99–106.

Dussurget, O., Timm, J., Gomez, M. *et al.* (1999b) Transcriptional control of the iron-responsive *fxbA* gene by the mycobacterial regulator IdeR. *Journal of Bacteriology* **181**.

Engelmann, S., Lindner, C. & Hecker, M. (1995) Cloning, nucleotide sequence, and regulation of *katE* encoding a sigma B-dependent catalase in *Bacillus subtilis. Journal of Bacteriology* **177**, 5598–5605.

Falkow, S. (1988) Molecular Koch's postulates applied to microbial pathogenicity. *Review of Infection and Disease* **10**, S274–S276.

Finlay, B.B. & Falkow, S. (1997) Common themes in microbial pathogenicity revisited. *Microbiological Molecular Biological Review* **61**, 136–169.

Fiss, E.H., Yu, S. & Jacobs W.R. Jr (1994) Identification of genes involved in the sequestration of iron in mycobacteria: the ferric exochelin biosynthetic and uptake pathways. *Molecular Microbiology* **14**, 557–569.

Franzon, V.L., Arondel, J. & Sansonetti, P.J. (1990) Contribution of superoxide dismutase and catalase activities to *Shigella flexneri* pathogenesis. *Infection and Immunity* **58**, 529–535.

Gallegos, M.T., Schleif, R., Bairoch, A., Hofmann, K. & Ramos, J.L. (1997) Arac/XylS family of transcriptional regulators. *Microbiological Molecular Biological Review* **61**, 393–410.

Garcia-del Portillo, F., Foster, J.W., Maguire, M.E. & Finlay, B.B. (1992) Characterization of the micro-environment of *Salmonella typhimurium*-containing vacuoles within MDCK epithelial cells. *Molecular Microbiology* **6**, 3289–3297.

Gomez, J., Chen, J.M. & Bishai, W. (1997a) Sigma factors

of *Mycobacterium tuberculosis*. *Tubercle Lung Disease* **78**, 175–183.

Gomez, M., Doukhan, L., Nair, G. & Smith, I. (1998) *sigA* is an essential gene in *Mycobacterium smegmatis*. *Molecular Microbiology* **29**, 617–628.

Gomez, J., Greenberg, B., Chater, K. & Bishai, W. (1997b) Identification of the *whm* family of genes in *M. tb. & M. leprae*: relationship to transcription factors required for sporulation in related bacteria. In: *32nd U S -Japan Cooperative Medical Science Program Tuberculosis-Leprosy Research Conference*, Cleveland, Ohio, pp. 12–16.

Gonzalez-y-Merchand, J.A., Garcia, M.J., Gonzalez-Rico, S., Colston, M.J. & Cox, R.A. (1997) Strategies used by pathogenic and nonpathogenic mycobacteria to synthesize rRNA. *Journal of Bacteriology* **179**, 6949–6958.

Gorman, C.M., Moffat, L.F. & Howard, B.H. (1982) Recombinant genomes which express chloramphenicol acetyltransferase in mammalian cells. *Molecular Cell Biology* **2**, 1044–1051.

Gralla, J.D. & Collado-Vides, J. (1996) Organization and function of transcription regulatory elements. In: *Escherichia Coli and Salmonella Cellular and Molecular Biology* (ed. F. C. Neidhardt). Washington, DC: ASM Press, pp. 1232–1245.

Gross, C.A., Lonetto, M. & Losick, R. (1992) Bacterial sigma factors. In: *Transcriptional Regulation* (ed. S. L. McKnight & K. R. Yamamoto). Cold Spring Harbor: Cold Spring Harbor Laboratory Press, pp. 129–176.

Gruber, T.M. & Bryant, D.A. (1997) Molecular systematic studies of eubacteria, using sigma70-type sigma factors of group 1 and group 2. *Journal of Bacteriology* **179**, 1734–1747.

Gunter, K., Toupet, C. & Schupp, T. (1993) Characterization of an iron-regulated promoter involved in desferrioxamine B synthesis in *Streptomyces pilosus*: Repressor-binding site and homology to the diphtheria toxin gene promoter. *Journal of Bacteriology* **175**, 3295–3302.

Gunter-Seeboth, K. & Schupp, T. (1995) Cloning and sequence analysis of the *Corynebacterium diphtheriae dtxR* homologue from *Streptomyces lividans* and *Streptomyces pilosus* encoding a putative iron repressor protein. *Gene* **166**, 117–119.

Haldenwang, W.G. (1995) The sigma factors of *Bacillus subtilis*. *Microbiological Reviews* **59**, 1–30.

Harley, C.B. & Reynolds, R.P. (1987) Analysis of *E. coli* promoter sequences. *Nucleic Acids Research* **15**, 2343–2361.

Harshey, R.M. & Ramakrishnan, T. (1977) Rate of ribonucleic acid chain growth in *Mycobacterium tuberculosis* H37Rv. *Journal of Bacteriology* **129**, 616–622.

Harwood, C.R. & Cutting, S.M. (1990) *Molecular Biological Methods for Bacillus*. John Wiley and Sons, New York.

Hassett, D.J., Sokol, P.A., Howell, M.L. *et al.*(1996) Ferric uptake regulator (Fur) mutants of *Pseudomonas aeruginosa* demonstrate defective siderophore-mediated iron uptake, altered aerobic growth, and decreased superoxide dismutase and catalase activities. *Journal of Bacteriology* **178**, 3996–4003.

Hatfull, G.F. & Sarkis, G.J. (1993) DNA sequence, structure and gene expression of mycobacteriophage L5: a phage system for mycobacterial genetics. *Molecular Microbiology* **7**, 395–405.

Hawley, D.K. & McClure, W.R. (1983) Compilation and analysis of *Escherichia coli* promoter DNA sequences. *Nucleic Acids Research* **11**, 2237–2255.

Hecker, M., Schumann, W. & Volker, U. (1996) Heat-shock and general stress response in *Bacillus subtilis*. *Molecular Microbiology* **19**, 417–428.

Heithoff, D.M., Conner, C.P., Hanna, P.C., Julio, S.M., Hentschel, U. & Mahan, M.J. (1997a) Bacterial infection as assessed by *in vivo* gene expression. *Proceedings of the National Academy of Sciences of the USA* **94**, 934–939.

Heithoff, D.M., Conner, C.P. & Mahan, M.J. (1997b) Dissecting the biology of a pathogen during infection. *Trends in Microbiology* **5**, 509–513.

Helmann, J.D. (1995) Compilation and analysis of *Bacillus subtilis* sigma A-dependent promoter sequences: evidence for extended contact between RNA polymerase and upstream promoter DNA. *Nucleic Acids Research* **23**, 2351–2360.

Helmann, J.D. & Chamberlin, M.J. (1988) Structure and function of bacterial sigma factors. *Annual Review of Biochemistry* **57**, 839–872.

Henaut, A. & Danchin, A. (1996) Analysis and predictions from *Escherichia coli* sequences, or *E. coli in silico*. In: *Escherichia Coli and Salmonella Cellular and Molecular Biology* (eds F. C. Neidhardt *et al.*). Washington, DC: ASM Press, pp. 2047–2066.

Hengge-Aronis, R. (1993) Survival of hunger and stress: the role of *rpoS* in early stationary phase gene regulation in *E. coli*. *Cell* **72**, 165–168.

Hetherington, S.V., Watson, A.S. & Patrick, C.C. (1995) Sequence and analysis of the *rpoB* gene of *Mycobacterium smegmatis*. *Antimicrobial Agents and Chemotherapy* **39**, 2164–2166.

Hickey, M.J., Arain, T.M., Shawar, R.M. *et al.*(1996) Luciferase *in vivo* expression technology: use of recombinant mycobacterial reporter strains to evaluate antimycobacterial activity in mice. *Antimicrobial Agents and Chemotherapy* **40**, 400–407.

Hill, P.J., Stewart, G.S. & Stanley, P.E. (1993) Bioluminescence and chemiluminescence literature. Luciferase reporter genes—*lux* and *luc*. Part 2. *Journal of Bioluminescence and Chemiluminescence* **8**, 267–291.

Hoffman, C.S. & Wright, A. (1985) Fusions of secreted proteins to alkaline phosphatase: an approach for studying protein secretion. *Proceedings of the National Academy of Sciences of the USA* **82**, 5107–5111.

Honore, N., Bergh, S., Chanteau, S. *et al.*(1993) Nucleotide sequence of the first cosmid from the *Mycobacterium leprae* genome project: structure and function of the Rif-Str regions. *Molecular Microbiology* **7**, 207–214.

Hopwood, D.A., Bibb, M.J., Chater, K.F. *et al.* (1985) *Genetic Manipulation of Streptomyces: a Laboratory Manual.* Norwich: The John Innes Foundation.

Huang, X., Decatur, A., Sorokin, A. & Helmann, J.D. (1997) The *Bacillus subtilis* sigma (X) protein is an extracytoplasmic function sigma factor contributing to survival at high temperature. *Journal of Bacteriology* **179**, 2915–2921.

Huisman, G.W., Siegele, D.A., Zambrano, M.M. & Kolter, R. (1996) Morphological and physiological changes during stationary phase. In: *Escherichia coli and Salmonella typhimurium. Cellular and Molecular Biology* (ed. F. C. Neidhardt). Washington, DC: American Society for Microbiology, pp. 1672–1682.

Ishihama, A. (1993) Protein–protein communication within the transcription apparatus. *Journal of Bacteriology* **175**, 2483–2489.

Jackett, P.S., Aber, V.R. & Lowrie, D.B. (1978) Virulence and resistance to superoxide, low pH and hydrogen peroxide among strains of *Mycobacterium tuberculosis*. *Journal of General Microbiology* **104**, 37–45.

Jackson, M., Berthet, F.X., Otal, I. *et al.* (1996) The *Mycobacterium tuberculosis* purine biosynthetic pathway: isolation and characterization of the *purC* and *purL* genes. *Microbiology* **142**, 2439–2447.

Jain, S., Kaushal, D., DasGupta, S.K. & Tyagi, A.K. (1997) Construction of shuttle vectors for genetic manipulation and molecular analysis of mycobacteria. *Gene* **190**, 37–44.

Ji, Y.E., Colston, M.J. & Cox, R.A. (1994a) Nucleotide sequence and secondary structures of precursor 16S rRNA of slow-growing mycobacteria. *Microbiology* **140**, 123–132.

Ji, Y.E., Colston, M.J. & Cox, R.A. (1994b) The ribosomal RNA (*rrn*) operons of fast-growing mycobacteria: primary and secondary structures and their relation to *rrn* operons of pathogenic slow-growers. *Microbiology* **140**, 2829–2840.

Jones, G.H., Paget, M.S., Chamberlin, L. & Buttner, M.J. (1997) Sigma-E is required for the production of the antibiotic actinomycin in *Streptomyces antibioticus*. *Molecular Microbiology* **23**, 169–178.

Kashlev, M., Lee, J., Zalenskaya, K., Nikiforov, V. & Goldfarb, A. (1990) Blocking of the initiation-to-elongation transition by a transdominant RNA polymerase mutation. *Science* **248**, 1006–1009.

Keener, J. & Nomura, M. (1996) Regulation of ribosome synthesis. In: *Escherichia Coli and Salmonella Cellular and Molecular Biology* (eds F. C. Neidhardt, R. C., J. L. III Ingraham, K. B. Low, B. Magasanik, M. Schaechter & H. E. Umbarger). Washington, DC: American Society for Microbiology, pp. 1417–1431.

Kempsell, K.E., Ji, Y., Estrada-G.I.C.E., Colston, M.J. & Cox, R.A. (1992) The nucleotide sequence of the promoter, 16S rRNA and spacer region of the ribosomal RNA operon of *Mycobacterium tuberculosis* and comparison with *Mycobacterium leprae* precursor rRNA. *Journal of General Microbiology* **138**, 1717–1727.

Kenney, T.J. & Churchward, G. (1996) Genetic analysis of the *Mycobacterium smegmatis rpsL* promoter. *Journal of Bacteriology* **178**, 3564–3571.

Keseler, I.M. & Kaiser, D. (1997) Sigma54, a vital protein for *Myxococcus xanthus*. *Proceedings of the National Academy of Sciences of the USA* **94**, 1979–1984.

Khomenko, A.G. (1987) The variability of *Mycobacterium tuberculosis* in patients with cavitary pulmonary tuberculosis in the course of chemotherapy. *Tubercle* **68**, 243–253.

Kieser, T., Moss, M.T., Dale, J.W. & Hopwood, D.A. (1986) Cloning and expression of *Mycobacterium bovis* BCG DNA in 'Streptomyces lividans'. *Journal of Bacteriology* **168**, 72–80.

Kinger, A.K. & Tyagi, J.S. (1993) Identification and cloning of genes differentially expressed in the virulent strain of *Mycobacterium tuberculosis*. *Gene* **131**, 113–117.

Kleanthous, C. & Shaw, W.V. (1984) Analysis of the mechanism of chloramphenicol acetyltransferase by steady-state kinetics. Evidence for a ternary-complex mechanism. *Biochemical Journal* **223**, 211–220.

Kong, T.H., Coates, A.R., Butcher, P.D., Hickman, C.J. & Shinnick, T.M. (1993) *Mycobacterium tuberculosis* expresses two chaperonin-60 homologs. *Proceedings of the National Academy of Sciences of the USA* **90**, 2608–2612.

Konyecsni, W.M. & Deretic, V. (1988) Broad-host-range plasmid and M13 bacteriophage-derived vectors for promoter analysis in *Escherichia coli* and *Pseudomonas aeruginosa*. *Gene* **74**, 375–386.

Kremer, L., Baulard, A., Estaquier, J., Content, J., Capron, A. & Locht, C. (1995a) Analysis of the *Mycobacterium tuberculosis* 85A antigen promoter region. *Journal of Bacteriology* **177**, 642–653.

Kremer, L., Baulard, A., Estaquier, J., Poulain-Godefroy, O. & Locht, C. (1995b) Green fluorescent protein as a new expression marker in mycobacteria. *Molecular Microbiology* **17**, 913–922.

Kuo, H.P., Ho, T.C., Wang, C.H., Yu, C.T. & Lin, H.C. (1996) Increased production of hydrogen peroxide and expression of CD11b/CD18 on alveolar macrophages in patients with active pulmonary tuberculosis. *Tubercle and Lung Disease* **77**, 468–475.

Lambrecht, R.S. & Collins, M.T. (1993) Inability to detect mycobactin in mycobacteria-infected tissues suggests an alternative iron acquisition mechanism by mycobacteria *in vivo*. *Microbial Pathogenesis* **14**, 229–238.

Lee, B.-Y. & Horwitz, M.A. (1995) Identification of macrophage and stress-induced proteins of *Mycobacterium tuberculosis*. *Journal of Clinical Investigation* **96**, 245–249.

Levin, M.E. & Hatfull, G.F. (1993) *Mycobacterium smegmatis*

RNA polymerase: DNA supercoiling, action of rifampicin and mechanism of rifampicin resistance. *Molecular Microbiology* **8**, 277–285.

Liang, P. & Pardee, A.B. (1992) Differential display of eukaryotic messenger RNA by means of the polymerase chain reaction. *Science* **257**, 967–971.

Liesack, W., Pitulle, C., Sela, S. & Stackebrandt, E. (1990) Nucleotide sequence of the 16S rRNA from *Mycobacterium leprae*. *Nucleic Acids Research* **18**, 5558.

Lim, E.M., Rauzier, J., Timm, J. *et al.* (1995) Identification of *Mycobacterium tuberculosis* DNA sequences encoding exported proteins by using *phoA* gene fusions. *Journal of Bacteriology* **177**, 59–65.

Loewen, P.C. & Hengge-Aronis, R. (1994) The role of the sigma factor sigma S (KatF) in bacterial global regulation. *Annual Review of Microbiology* **48**, 53–80.

Lonetto, M.A., Brown, K.L., Rudd, K.E. & Buttner, M.J. (1994) Analysis of the *Streptomyces coelicolor sigE* gene reveals the existence of a subfamily of eubacterial RNA polymerase σ factors involved in the regulation of extracytoplasmic functions. *Proceedings of the National Academy of Sciences of the USA* **91**, 7573–7577.

Lonetto, M., Gribskov, M. & Gross, C.A. (1992) The σ^{70} family: sequence conservation and evolutionary relationships. *Journal of Bacteriology* **174**, 3843–3849.

Lynch, A.S. & Lin, E.C.C. (1996) Responses to molecular oxygen. In: *Escherichia Coli and Salmonella Cellular and Molecular Biology* (ed. F. C. Neidhardt). Washington, DC: ASM Press, pp. 1526–1538.

Mangan, J.A., Sole, K.M., Mitchison, D.A. & Butcher, P.D. (1997) An effective method of RNA extraction from bacteria refractory to disruption, including mycobacteria. *Nucleic Acids Research* **25**, 675–676.

Manganelli, R., Dubnau, E., Tyagi, S., Kramer, F.R. & Smith, I. (1999) Differential expression of 10 sigma factor genes in *Mycobacterium tuberculosis*. *Molecular Microbiology* **31**, 715–724.

Mathiopoulos, C. & Sonenshein, A.L. (1989) Identification of *Bacillus subtilis* genes expressed early during sporulation. *Molecular Microbiology* **3**, 1071–1081.

Mazodier, P., Guglielmi, G., Davies, J. & Thompson, C.J. (1991) Characterization of the *groEL*-like genes in *Streptomyces albus*. *Journal of Bacteriology* **173**, 7382–7386.

Mdluli, K., Sherman, D., Hickey, M. *et al.* (1996) Biochemical and genetic data suggest that InhA is not the primary target for activated isoniazid in *Mycobacterium tuberculosis*. *Journal of Infection and Disease* **174**, 1085–1090.

Meighen, E.A. (1993) Bacterial bioluminescence: organization, regulation, and application of the *lux* genes. *FASEB Journal* **7**, 1016–1022.

Merrick, M., Gibbins, J. & Toukdarian, A. (1987) The nucleotide sequence of the sigma factor gene *ntrA* (*rpoN*) of *Azotobacter vinelandii*: analysis of conserved sequences in NtrA proteins. *Molecular General Genetics* **210**, 323–330.

Miller, J.F. (1972) *Experiments in Molecular Genetics*. Cold Spring Harbor: Cold Spring Harbor Laboratory Press.

Miller, L.P., Crawford, J.T. & Shinnick, T.M. (1994) The *rpoB* gene of *Mycobacterium tuberculosis*. *Antimicrobial Agents and Chemotherapy* **38**, 805–811.

Miller, J.F., Mekalanos, J.J. & Falkow, S. (1989) Coordinate regulation and sensory transduction in the control of bacterial virulence. *Science* **243**, 916–922.

Missiakas, D. & Raina, S. (1997) Protein misfolding in the cell envelope of *Escherichia coli*: new signaling pathways. *Trends in Biochemistry and Science* **22**, 59–63.

Moran, C.P. Jr (1993) RNA polymerase and transcription factors. In: *Bacillus Subtilis and Other Gram-Positive Bacteria* (eds A. L. Sonensheim, J. A. Hoch & R. Losick). Washington, DC: ASM Press, pp. 653–668.

Morimoto, R.I., Sarge, K.D. & Abravaya, K. (1992) Transcriptional regulation of heat shock genes. A paradigm for inducible genomic responses. *Journal of Biological Chemistry* **267**, 21987–21990.

Morrison, D.A. & Jaurin, B. (1990) *Streptococcus pneumoniae* possesses canonical *Escherichia coli* (sigma 70) promoters. *Molecular Microbiology* **4**, 1143–1152.

Movahedzadeh, F., Colston, M.J. & Davis, E.O. (1997) Determination of DNA sequences required for regulated *Mycobacterium tuberculosis* RecA expression in response to DNA-damaging agents suggests that two modes of regulation exist. *Journal of Bacteriology* **179**, 3509–3518.

Mulder, M.A., Zappe, H. & Steyn, L.A. (1997) Mycobacterial promoters. *Tubercle and Lung Disease* **78**, 211–223.

Niederhoffer, E.C., Naranjo, C.M., Bradley, K.L. & Fee, J.A. (1990) Control of *Escherichia coli* superoxide dismutase (*sodA* and *sodB*) genes by the ferric iron uptake regulation (*fur*) locus. *Journal of Bacteriology* **172**, 1930–1938.

Ninfa, A.J. (1996) Regulation of gene transcription by extracellular stimuli. In: *Escherichia Coli and Salmonella Cellular and Molecular Biology* (ed. F. C. Neidhardt). Washington, DC: ASM Press, pp. 1246–1262.

O'Brien, S. & Andrew, P.W. (1991) Guinea-pig alveolar macrophage killing of *Mycobacterium tuberculosis, in vitro*, does not require hydrogen peroxide or hydroxyl radical. *Microbial Pathogenesis* **11**, 229–236.

O'Brien, S., Jackett, P.S., Lowrie, D.B. & Andrew, P.W. (1991) Guinea-pig macrophages kill *Mycobacterium tuberculosis*, but killing is independent of susceptibility to hydrogen peroxide ot triggering of the respiratory burst. *Microbial Pathogenesis* **10**, 199–207.

O'Farrell, P.H. (1975) High-resolution two-dimensional electrophoresis of proteins. *Journal of Biological Chemistry* **250**, 4007–4021.

Ochs, M., Veitinger, S., Kim, I., Welz, D., Angerer, A. & Braun, V. (1995) Regulation of citrate-dependent iron transport of *Escherichia coli*: *fecR* is required for transcription activation by FecI. *Molecular Microbiology* **15**, 119–132.

Oguiza, J.A., Marcos, A.T., Malumbres, M. & Martin, J.F. (1996) Multiple sigma factor genes in *Brevibacterium lactofermentum*: characterization of *sigA* and *SigB*. *Journal of Bacteriology* **178**, 550–553.

Oguiza, J.A., Marcos, A.T. & Martin, J.F. (1997) Transcriptional analysis of the *sigA* and *sigB* genes of *Brevibacterium lactofermentum*. *FEMS Microbiological Letters* **153**, 111–117.

Oguiza, J.A., Tao, X., Marcos, A.T., Martin, J.F. & Murphy, J.R. (1995) Molecular cloning, DNA sequence analysis, and characterization of the *Corynebacterium diphtheriae* *dtxR* homolog from *Brevibacterium lactofermentum*. *Journal of Bacteriology* **177**, 465–467.

Pagan-Ramos, E., Song, J., Via, L.E. & Deretic, V. (1997). Loss of *oxyR* in *Mycobacterium tuberculosis*: implications for host–pathogen interactions and sensitivity to INH. In: *ASM Conference on Tuberculosis: Past, Present and Future*, Copper Mountain, Colorado, p. 13.

Pallen, M.J. & Wren, B.W. (1997) The HtrA family of serine proteases. *Molecular Microbiology* **26**, 209–221.

Parish, T., Mahenthiralingam, E., Draper, P., Davis, E.O. & Colston, M.J. (1997) Regulation of the inducible acetamidase gene of *Mycobacterium smegmatis*. *Microbiology* **143**, 2267–2276.

Parish, T. & Stoker, N.G. (1997) Development and use of a conditional antisense mutagenesis system in mycobacteria. *FEMS Microbiological Letters* **154**, 151–157.

Parker, A.E. & Bermudez, L.E. (1997) Expression of the green fluorescent protein (GFP) in *Mycobacterium avium* as a tool to study the interaction between Mycobacteria and host cells. *Microbial Pathogenesis* **22**, 193–198.

Parkinson, J.S. (1993) Signal transduction schemes of bacteria. *Cell* **73**, 857–871.

Patek, M., Eikmanns, B.J., Patek, J. & Sahm, H. (1996) Promoters from *Corynebacterium glutamicum*: cloning, molecular analysis and search for a consensus motif. *Microbiology* **142**, 1297–1309.

Patel, B.K., Banerjee, D.K. & Butcher, P.D. (1991) Characterization of the heat shock response in *Mycobacterium bovis* BCG. *Journal of Bacteriology* **173**, 7982–7987.

Pettersson, J., Nordfelth, R., Dubinina, E. *et al.* (1996) Modulation of virulence factor expression by pathogen target cell contact. *Science* **273**, 1231–1233.

Plum, G. & Clark-Curtiss, J.E. (1994) Induction of *Mycobacterium avium* gene expression following phagocytosis by human macrophages. *Infection and Immunity* **62**, 476–483.

Predich, M., Doukhan, L., Nair, G. & Smith, I. (1995) Characterization of RNA polymerase and two σ factor genes from *Mycobacterium smegmatis*. *Molecular Microbiology* **15**, 355–366.

Qian, Z.M. & Tang, P.L. (1995) Mechanisms of iron uptake by mammalian cells. *Biochimica Biophysica Acta* **1269**, 205–214.

Ragno, S., Estrada, I., Butler, R. & Colston, M.J. (1997) Regulation of macrophage gene expression following invasion by *Mycobacterium tuberculosis*. *Immunology Letters* **57**, 143–146.

Record, M.T., Jr, Reznicoff, W.S., Craig, M.L., McQuade, K.L. & Schlax, P.J. (1996) *Escherichia coli* RNA polymerase (Eσ70), promoters, and the kinetics of the steps of transcription initiation. In: *Escherichia Coli and Salmonella Cellular and Molecular Biology* (ed. F. C. Neidhardt). Washington, DC: ASM Press, pp. 792–821.

Rinke de Wit, T.F., Bekelie, S., Osland, A. *et al.* (1992) Mycobacteria contain two *groEL* genes: the second *Mycobacterium leprae groEL* gene is arranged in an operon with *groES*. *Molecular Microbiology* **6**, 1995–2007.

Ross, W., Gosink, K.K., Salomon, J. *et al.* (1993) A third recognition element in bacterial promoters: DNA binding by the alpha subunit of RNA polymerase. *Science* **262**, 1407–1413.

Rouviere, P.E., De Las Penas, A., Mecsas, J., Lu, C.Z., Rudd, K.E. & Gross, C.A. (1995) *rpoE*, the gene encoding the second heat-shock sigma factor, sigma E in *Escherichia coli*. *EMBO Journal* **14**, 1032–1042.

Saizieu, A., Certa, U., Warrington, J., Gray, C., Keck, W. & Mous, J. (1998) Bacterial transcript imaging by hybridization of total RNA to oligonucleotide arrays. *Nature Biotechnology* **16**, 45–48.

Salazar, L., Arraiz, N., Cassart, Y., Gomez, M. & Takiff, H.E. (1997) Expression of an *M. smegmatis* ECF sigma factor increases both at high temperature and during stationary phase. In: *ASM Conference on Tuberculosis: Past, Present and Future*, Copper Mountain, Colorado, p. 23.

Salazar, L., Fsihi, H., de Rossi, E. *et al.* (1996) Organization of the origins of replication of the chromosomes *of Mycobacterium smegmatis*, *Mycobacterium leprae* and *Mycobacterium tuberculosis* and isolation of a functional origin from *M. smegmatis*. *Molecular Microbiology* **20**, 283–293.

Sambrook, J., Fritsch, E.K. & Maniatis, T. (1989) *Molecular Cloning: a Laboratory Manual*. Cold Spring Harbor: Cold Spring Harbor Laboratory Press.

Sander, P., Prammananan, T. & Bottger, E. (1996) Introducing mutations into a chromosomal rRNA gene using a genetically modified eubacterial host with a single rRNA operon. *Molecular Microbiology* **22**, 841–848.

Sarkis, G.J., Jacobs, W.R. Jr & Hatfull, G.F. (1995) L5 luciferase reporter mycobacteriophages: a sensitive tool for the detection and assay of live mycobacteria. *Molecular Microbiology* **15**, 1055–1067.

Schena, M., Shalon, D., Heller, R., Chai, A., Brown, P.O. & Davis, R.W. (1996) Parallel human genome analysis: microarray-based expression monitoring of 1000 genes. *Proceedings of the National Academy of Sciences of the USA* **93**, 10614–10619.

Schlesinger, L.S., Bellinger-Kawahara, C.G., Payne, N.R. & Horwitz, M.A. (1990) Phagocytosis of *Mycobacterium tuberculosis* is mediated by human monocyte complement receptors and complement component C3. *Journal of Immunology* **144**, 2771–2780.

Segal, W. & Bloch, H. (1956) Biochemical differentiation of *Mycobacterium tuberculosis* grown *in vivo* and *in vitro*. *Journal of Bacteriology* **72**, 132–141.

Segal, W. & Bloch, H. (1957) Pathogenic and immunogenic differentiation of *Mycobacterium tuberculosis* grown *in vivo* and *in vitro*. *American Review of Tubercular and Pulmonary Disease* **75**, 495–500.

Sherman, D.R., Sabo, P.J., Hickey, M.J. *et al.*(1995) Disparate responses to oxidative stress in saprophytic and pathogenic mycobacteria. *Proceedings of the National Academy of Sciences of the USA* **92**, 6625–6629.

Shinnick, T.M., Vodkin, M.H. & Williams, J.C. (1988) The *Mycobacterium tuberculosis* 65-kilodalton antigen is a heat shock protein which corresponds to common antigen and to the *Escherichia coli* GroEL protein. *Infection and Immunity* **56**, 446–451.

Smith, I., Dussurget, O., Timm, J., (1997) Sigma factors, iron regulation and oxidative stress. In: *ASM Conference on Tuberculosis: Past, Present and Future*, Copper Mountain, Colorado, Washington DC: ASM Press, p. 25.

Sorger, P.K. (1990) Yeast heat shock factor contains separable transient and sustained response transcriptional activators. *Cell* **62**, 793–805.

Sorokin, A., Bolotin, A., Purnelle, B. *et al.* (1997) Sequence of the *Bacillus subtilis* genome region in the vicinity of the lev operon reveals two new extracytoplasmic function RNA polymerase sigma factors SigV and SigZ. *Microbiology* **143**, 2939–2943.

Stanford, J.L. (1987) Much's granules revisited. *Tubercle* **68**, 241–242.

Stanley, P.E. (1997) Commercially available luminometers and imaging devices for low-light level measurements and kits and reagents utilizing bioluminescence or chemiluminescence: survey update 5. *Journal of Bioluminescence and Chemiluminescence* **12**, 61–78.

Stanley, P.E. & Williams, S.G. (1969) Use of the liquid scintillation spectrometer for determining adenosine triphosphate by the luciferase enzyme. *Analytical Biochemistry* **29**, 381–392.

Stetter, K.O. & Zillig, W. (1974) Transcription in Lactobacillaceae. DNA-dependent RNA polymerase from *Lactobacillus curvatus*. *European Journal of Biochemistry* **48**, 527–540.

Stover, C.K., de la Cruz, V.F., Fuerst, T.R. *et al.* (1991) New use of BCG for recombinant vaccines. *Nature* **351**, 456–460.

Strohl, W.R. (1992) Compilation and analysis of DNA sequences associated with apparent streptomycete promoters. *Nucleic Acids Research* **20**, 961–974.

Supply, P., Magdalena, J., Himpens, S. & Locht, C. (1997) Identification of novel intergenic repetitive units in a mycobacterial two-component system operon. *Molecular Microbiology* **26**, 991–1003.

Suzuki, Y., Nagata, A., Ono, Y. & Yamada, T. (1988) Complete nucleotide sequence of the 16S rRNA gene of *Mycobacterium bovis* BCG. *Journal of Bacteriology* **170**, 2886–2889.

Tai, S.S. & Zhu, Y.Y. (1995) Cloning of a *Corynebacterium diphtheriae* iron-repressible gene that shares sequence homology with the AhpC subunit of the alkyl hydroperoxide reductase of *Salmonella typhimurium*. *Journal of Bacteriology* **177**, 3512–3517.

Tao, X., Schiering, N., Zeng, H.Y., Ringe, D. & Murphy, J.R. (1994) Iron, DtxR, and the regulation of diphtheria toxin expression. *Molecular Microbiology* **14**, 191–197.

Timm, J., Lim, E.M. & Gicquel, B. (1994a) *Escherichia coli*-mycobacteria shuttle vectors for operon and gene fusions to *lacZ*: the pJEM series. *Journal of Bacteriology* **176**, 6749–6753.

Timm, J., Perilli, M.G., Duez, C. *et al.* (1994b) Transcription and expression analysis, using *lacZ* and *phoA* gene fusions, of *Mycobacterium fortuitum* beta-lactamase genes cloned from a natural isolate and a high-level beta-lactamase producer. *Molecular Microbiology* **12**, 491–504.

Touati, D., Jacques, M., Tardat, B., Bouchard, L. & Despied, S. (1995) Lethal oxidative damage and mutagenesis are generated by iron in Δ*fur* mutants of *Escherichia coli*: protective role of superoxide dismutase. *Journal of Bacteriology* **177**, 2305–2314.

Tsolis, R.M., Baumler, A.J. & Heffron, F. (1995) Role of *Salmonella typhimurium* Mn-superoxide dismutase (SodA) in protection against early killing by J774 macrophages. *Infection and Immunity* **63**, 1739–1744.

Tyagi, S. & Kramer, F.R. (1996) Molecular beacons: probes that fluoresce upon hybridization. *Nature Biotechnology* **14**, 303–308.

Valdivia, R.H. & Falkow, S. (1996) Bacterial genetics by flow cytometry: rapid isolation of *Salmonella typhimurium* acid-inducible promoters by differential fluorescence induction. *Molecular Microbiology* **22**, 367–378.

Valdivia, R.H. & Falkow, S. (1997) Fluorescence-based isolation of bacterial genes expressed within host cells. *Science* **277**, 2007–2011.

Venturi, V., Weisbeek, P. & Koster, M. (1995) Gene regulation of siderophore-mediated iron acquisition in

Pseudomonas: not only the Fur repressor. *Molecular Microbiology* **17**, 603–610.

Via, L.E., Curcic, R., Mudd, M.H., Dhandayuthapani, S., Ulmer, R.J. & Deretic, V. (1996) Elements of signal transduction in *Mycobacterium tuberculosis*: *in vitro* phosphorylation and *in vivo* expression of the response regulator MtrA. *Journal of Bacteriology* **178**, 3314–3321.

Wada, K., Wada, Y., Ishibashi, F., Gojobori, T. & Ikemura, T. (1992) Codon usage tabulated from the GenBank genetic sequence data. *Nucleic Acids Research* **20**, 2111–2118.

Walker, G.C. (1996) The SOS response of *Escherichia coli*. In: *Escherichia Coli and Salmonella Cellular and Molecular Biology* (ed. F. C. Neidhardt). Washington, DC: American Society for Microbiology, 1400–1416.

Wayne, L.G. (1994) Dormancy of *Mycobacterium tuberculosis* and latency of disease. *European Journal of Clinical Microbiological Infection and Disease* **13**, 908–914.

Wayne, L.G. & Hayes, L.G. (1998) Nitrate reduction as a marker for hypoxic shiftdown of *Mycobacterium tuberculosis*. *Tubercule Lung Disease* **79**, 127–132.

Wayne, L.G. & Sramek, H.A. (1994) Metronidazole is bactericidal to dormant cells of *Mycobacterium tuberculosis*. *Antimicrobial Agents and Chemotherapy* **38**, 2054–2058.

Wheeler, P.R. & Ratledge, C. (1994) Metabolism of *Mycobacterium tuberculosis*. In: *Tuberculosis: Pathogenesis, Protection, and Control* (ed. B. R. Bloom). Washington, DC: American Society for Microbiology Press, pp. 353–385.

Wiggs, J.L., Bush, J.W. & Chamberlin, M.J. (1979) Utilization of promoter and terminator sites on bacteriophage T7 DNA by RNA polymerases from a variety of bacterial orders. *Cell* **16**, 97–109.

Wilson, T.M., de Lisle, G.W. & Collins, D.M. (1995) Effect of *hA* and *katG* on isoniazid resistance and virulence of *Mycobacterium bovis*. *Molecular Microbiology* **15**, 1009–1015.

Woodley, C.L., Kilburn, J.O., David, H.L. & Silcox, V.A. (1972) Susceptibility of mycobacteria to rifampin. *Antimicrobial Agents and Chemotherapy* **2**, 245–249.

Wren, B.W., Colby, S.M., Cubberley, R.R. & Pallen, M.J. (1992) Degenerate PCR primers for the amplification of fragments from genes encoding response regulators from a range of pathogenic bacteria. *FEMS Microbiological Letters* **78**, 287–291.

Wu, Q.L., Kong, D., Lam, K. & Husson, R.N. (1997) A mycobacterial extracytoplasmic function sigma factor involved in survival following stress. *Journal of Bacteriology* **179**, 2922–2929.

Young, D., Lathigra, R., Hendrix, R., Sweetser, D. & Young, R.A. (1988) Stress proteins are immune targets in leprosy and tuberculosis. *Proceedings of the National Academy of Sciences of the USA* **85**, 4267–4270.

Yuan, Y., Crane, D.C. & Barry, C.E. III (1996) Stationary phase-associated protein expression in *Mycobacterium tuberculosis*: function of the mycobacterial alpha-crystallin homolog. *Journal of Bacteriology* **178**, 4484–4492.

Yuan, Y., Crane, D.C., Simpson, R.M. *et al.*(1998). The 16-kDa α-crystallin (Acr) protein of *Mycobacterium tuberculosis* is required for growth in macrophages. *Proceedings of the National Academy of Sciences of the USA* **95**, 9578–9583.

Yuan, Y., Lee, R.E., Besra, G.S., Belisle, J.T. & Barry,C.E. III (1995) Identification of a gene involved in the biosynthesis of cyclopropanated mycolic acids in *Mycobacterium tuberculosis*. *Proceedings of the National Academy of Sciences of the USA* **92**, 6630–6634.

Yuan, G. & Wong, S.L. (1995) Isolation and characterization of *Bacillus subtilis groE* regulatory mutants: evidence for *orf39* in the *dnaK* operon as a repressor gene in regulating the expression of both *groE* and *dnaK*. *Journal of Bacteriology* **177**, 6462–6468.

Zhang, J.P. & Normark, S. (1996) Induction of gene expression in *Escherichia coli* after pilus-mediated adherence. *Science* **273**, 1234–1236.

Zhang, L., Zhou, W., Velculescu, V.E. *et al.*(1997) Gene expression profiles in normal and cancer cells. *Science* **276**, 1268–1272.

Zuber, P., Healy, J.M. & Losick, R. (1987) Effects of plasmid propagation of a sporulation promoter on promoter utilization and sporulation in *Bacillus subtilis*. *Journal of Bacteriology* **169**, 461–469.

Chapter 5 / Genomics of *Mycobacterium tuberculosis* and *Mycobacterium leprae*

STEPHEN V. GORDON, KARIN EIGLMEIER, ROLAND BROSCH, THIERRY GARNIER, NADINE HONORÉ, BART G. BARRELL & STEWART T. COLE

1 Introduction

The elucidation of the entire genome sequences of *Mycobacterium tuberculosis* and *Mycobacterium leprae* will herald a new era in mycobacteriology, pushing a once neglected genus firmly to the forefront of bacterial research. The goal of deciphering the complete genetic complement of these infamous pathogens was a particularly attractive one, especially when one takes into account the considerable difficulties inherent in their manipulation. For example, both pathogens have to be handled under biohazard-containment facilities, and while *M. tuberculosis* has a slow growth rate in *in vitro* culture, *M. leprae* cannot be cultivated at all on artificial media and has an extremely long generation time of 2 weeks or more in experimentally infected animals. With these difficulties in mind, a clone-based approach for sequencing the genome was decided on as this approach would fulfil several criteria.

1 The characterization of random clones from the cosmid library would result in the establishment of a physical map in a 'bottom-up' manner. This would allow first estimates of genome size to be made, an especially important point in the case of *M. leprae* where so far no chromosomal DNA of a quality suitable for pulsed-field gel electrophoresis (PFGE) has been isolated and for which the construction of a physical map from a 'top-down' method was therefore not feasible.

2 Due to the size of the cloned chromosomal DNA fragments the clones harbour complete operons and adjacent genes. This will be valuable when analysing genes with related functions or in studying some of the unusually large operons present in the genome (30–40 kb).

3 The arrays of overlapping cosmids represent a renewable source of chromosomal DNA from every region of the genome. This results in a modularity of the clones which can be distributed and conveniently handled in a non-pathogenic, easily cultivable host bacterium, thus removing the need for tedious preparations of mycobacterial chromosomal DNA.

4 Selected clones represent ideal starting material for systematic sequencing. They can be sequenced to completion and thus give access to extended, thoroughly analysed and annotated regions of the chromosome at an early stage of the project, long before the genome is completely sequenced. Problems arising in regions where the data assembly is difficult due to the presence of repetitive elements can be resolved more easily when working with longer stretches of cloned DNA. Once the sequence of a clone is finished, it becomes a 'ready-to-use' module.

Deciphering the entire genome sequence of *M. tuberculosis* H37Rv represents an historic milestone in our fight against this pathogen (Cole *et al.* 1998). At the time of writing, the genome sequencing projects of *M. leprae* and *M. tuberculosis* CSU 93 are also close to completion. In fact *M. leprae* was the first mycobacterium to be subjected to systematic DNA sequencing, with the project being undertaken in a clone-by-clone approach with the participation of several laboratories (Honoré *et al.* 1993; Fsihi *et al.* 1996a; Smith *et al.* 1997) and latterly with the Sanger Centre, Hinxton, UK. This was followed by *M. tuberculosis* H37Rv and then by the *M. tuberculosis* strain CSU 93, a clinical isolate being sequenced by a whole-genome-shotgun approach at The Institute for Genomic Research (TIGR, Rockville, MD, USA). Scientists working in the mycobacterial field will therefore have at their disposal the complete sequence of two strains of *M. tuberculosis*, as well as a set of overlapping clones representing the entire genome of the H37Rv strain. Furthermore, the genome of a fourth mycobacterium, *Mycobacterium avium*, is proposed to be sequenced by TIGR and has an expected completion date of the year 2000. This explosion of information will place the mycobacteria among the best described bacterial genera, a reversal of the situation that existed a mere 5–10 years ago. With all this information in our hands, it will be possible to compare the genome sequences of the different mycobacterial species with each other, as well as with those from other sequenced microbes such as *Escherichia coli, Helicobacter pylori, Haemophilus influenzae, Bacillus subtilis*, etc. This should in turn reveal genes specific to the mycobacteria, allowing us unprecedented insight into the fundamental biology of the mycobacteria. Moreover, it will become possible to explore the evolution of the mycobacteria at the genetic level and to verify the existing phylogenetic tree.

2 The sequencing of the *M. leprae* genome

2.1 Library construction and establishment of the physical map

The first decision was to determine what strain to sequence. To our knowledge no *M. leprae* strain had ever been cloned, so working with clinical isolates was the generally accepted solution. Hence, for the construction of the *M. leprae* cosmid library that formed the basis of the sequencing project, chromosomal bacterial DNA was isolated from armadillo-derived *M. leprae*, originally isolated from a patient from Tamil Nadu (Eiglmeier *et al.* 1993). This DNA was blunt-ended and used to produce a cosmid library in the vector Lorist6 (Gibson *et al.* 1987; Eiglmeier *et al.* 1993). About 1000 independent clones, corresponding theoretically to 10 genome equivalents assuming a typical bacterial chromosome size of 3–4 Mb, were subjected to modified fingerprint analysis for characterization (Coulson *et al.* 1986; Sulston *et al.* 1988; Eiglmeier *et al.* 1993). Using this method, overlapping cosmids were identified and assembled into contiguous stretches of genomic DNA (contigs), establishing the first version of the physical map. By hybridization, either using complete cosmids or suitable end-fragments of cosmids as probes, the map was refined and completed, but the number of contigs could not be reduced to less than four. Our understanding of the basic organization of the genome was further improved by positioning on the map all of the *M. leprae* genes and loci that had been cloned up to then. This approach permitted the confirmation of the contig topologies and led to the assumption that the quasi-totality of the *M. leprae* genome was represented by the four contigs. It was concluded at that point that the missing regions of the chromosome were probably small and by summing up the contigs (1200 kb, 800 kb, 400 kb and 380 kb) the genome was estimated to be about 2.8 Mb in size. Because most bacterial chromosomes characterized so far are circular, it was presumed that this was also the case for *M. leprae*. However, only the complete genome sequence will confirm this hypothesis as no data from PFGE for *M. leprae* are available.

The physical map obtained by this 'bottom-up' approach permitted us to outline the first characteristics of the *M. leprae* genome and to select cosmids which would be part of the sequencing project. The first set of clones subjected to complete sequencing was composed of single, non-overlapping cosmids from distinct and well-separated regions of the chromosome which represented the backbone of the sequencing project. In the next step, their DNA sequences served as landmarks and were compared with sequences obtained from the extremities of cosmids belonging to the corresponding regions of the chromosome. Clones whose end-sequences showed the smallest overlap were then subjected to complete sequencing, guaranteeing a minimal redundancy of sequence data. The closure of the remaining gaps between the contigs (which may contain sequences unclonable in cosmids), should be possible by long-range polymerase chain reaction (PCR) and subsequent sequencing of the resulting products. An overview of the *M. leprae* genome map, displaying the four contigs into which the cosmids are arranged, is shown in Fig. 5.1. Shaded and boxed cosmids indicate those clones that are already sequenced (as of April 1999). Unboxed cosmids are mapped but not sequenced. The expected completion date for the project is October 1999.

2.2 Insights from the *M. leprae* genome sequence

Initial observations into the biology of the leprosy bacillus could be drawn even at an early stage of the project, when the physical and genetic map was being constructed. For example, the single ribosomal RNA operon was positioned on the map and, in contrast to most other bacteria so far analysed, this operon is not located close to the origin of replication *oriC* (Cole & Saint Girons 1994). A similar situation was found in *M. tuberculosis* H37Rv, where the *rrn* locus and the origin of replication are separated by ≈ 1.6 Mb (Philipp *et al.* 1996b). This arrangement may

be a reason why these species show such slow growth rates, although of course other factors may also be involved. Furthermore, the positions of 29 copies of the *M. leprae* specific family of repetitive elements RLEP (Woods & Cole 1990) were determined by hybridization, with the different copies being classified into four families. It became evident from this mapping that the distribution of these elements in the chromosome is not completely random, with several copies located close to each other.

The publication of the first set of sequenced *M. leprae* cosmids permitted further insights into the organization of the chromosome, but due to the unfinished status of the project, conclusions about the presence or absence of certain genes and metabolic pathways as well as comparisons with the minimal gene set proposed by Mushegian and Koonin (1996) are premature. Nevertheless, some insights can be gleaned from the available data.

As would be expected, the chromosome of *M. leprae* shows regions of high similarity to those of *M. tuberculosis*. This similarity is most striking in the *oriC* region, where, as well as the highly conserved gene order, many open reading frames (ORFs) of unknown function are conserved between the two species (Fsihi *et al.* 1996a; Salazar *et al.* 1996). For example, the *M. leprae* cosmid B1770 shows extensive similarity to the *M. tuberculosis* cosmid MTCY10H4, with the genes for *recF*, *gyrB*, *gyrA*, *pabS*, *pknB*, *pknA*, *pbpA*, as well as tRNAs for isoleucine, alanine and leucine present on both cosmids at the same relative positions (Fsihi *et al.* 1996a). Further, an ORF of unknown function that lies directly downstream of the *M. leprae recF* gene has an identical homologue (97.9% identity in 187 amino acid overlap) in the same position in *M. tuberculosis*.

This conservation of genes and gene order in the *oriC* region points to the possible lethal effect of any mutations in this area, hence constraining the capacity of this locus to diverge between the species. On the other hand, in regions of the chromosome where this selective pressure is lacking, it is likely that the *M. leprae* chromosome will have accumulated extensive mutations relative to *M. tuberculosis*. This is most

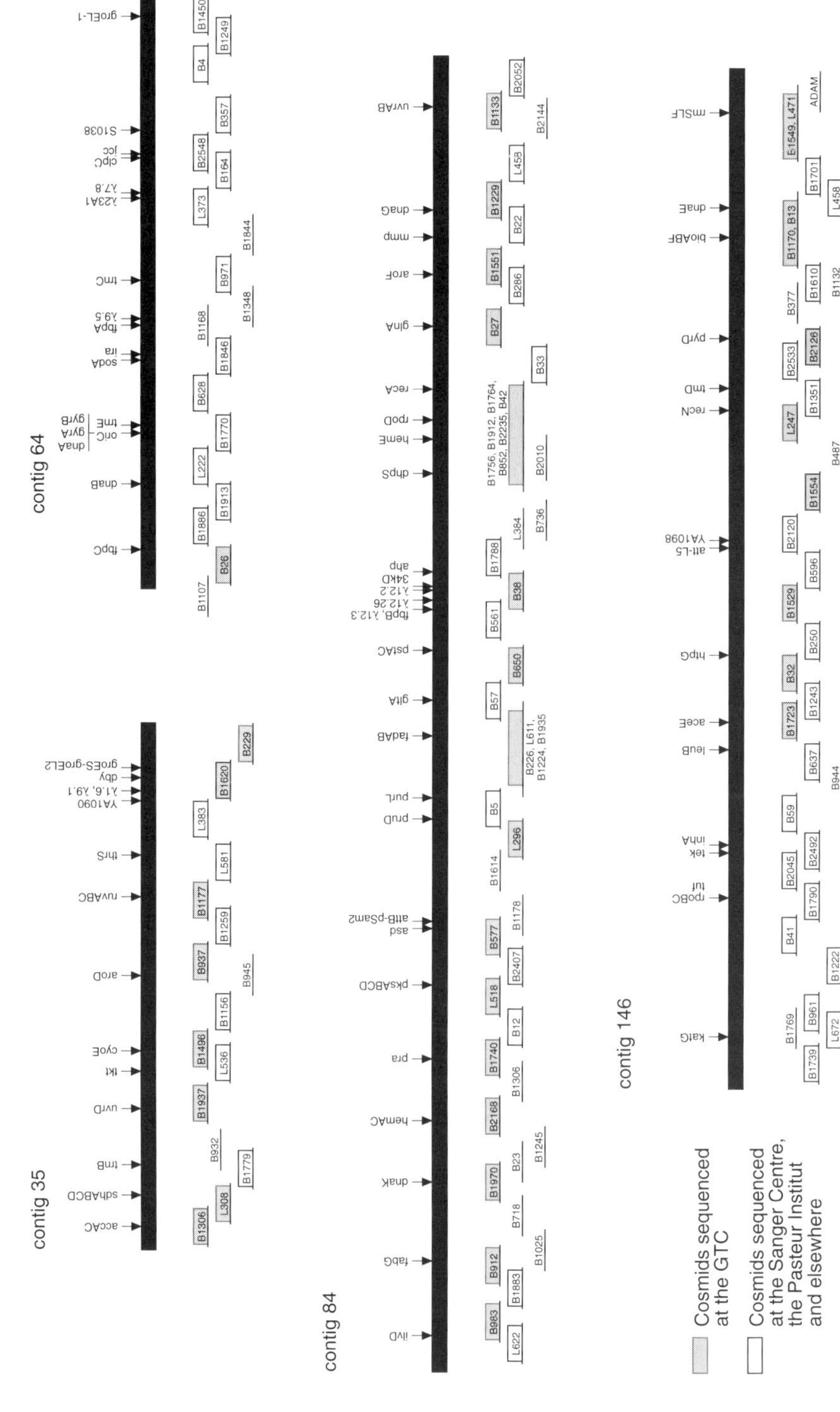

Fig. 5.1 Genomic organization of *Mycobacterium leprae*. The four contigs of the *M. leprae* genome are represented as thick black lines. The positions of various genes that were mapped to the contigs are shown over the lines. The cosmids that make up the contigs are shown under the line, with shaded cosmids representing those sequenced at Genome Therapeutics Corporation (GTC) and the boxed cosmids indicating those sequenced at the Sanger Centre and elsewhere. Unboxed cosmids have been mapped but not sequenced.

strikingly seen in the region between the conserved ribosomal operons *str*, which contains the genes *rpsL–rpsG–fusA–tuf*, and S10, containing 11 ribosomal protein genes from *rpsJ* to *rpsQ* (see Plate 1, between pp. 102 and 103). This region in *M. tuberculosis* encompasses ~14.4 kb and contains 16 ORFs, including genes with similarity to an L-lactate dehydrogenase (*lldD*, 34.5% identity over a 380 amino acid overlap) and coenzyme pyrrolo-quinoline-quinone synthesis protein E (*pqqE*, 23.9% identity over 377 amino acids). However, this same region in *M. leprae* spans ≈ 11.9 kb and contains just a single ORF coding for a protein of 220 amino acids of unknown function. A similar situation is seen upstream of the *str* operon, between the conserved *rpoC* gene and the start of the *str* operon, a region that in *M. tuberculosis* spans ≈ 14.2 kb and contains 13 ORFs, yet in *M. leprae* is ≈ 9 kb and contains just a single ORF that codes for an endonuclease IV purinase, an ORF that is conserved in the same position in *M. tuberculosis*. This loss of coding capacity in *M. leprae* undoubtedly reflects the degenerate evolution of the *M. leprae* progenitor strain, which as it became more dependent on an obligate intracellular lifestyle accumulated mutations in genes that were no longer essential. This is further reflected by the multiple pseudogenes scattered throughout the *M. leprae* genome, vestiges of once functional genes rendered inoperative by mutation. Sections of the *M. leprae* chromosome can contain extensive stretches of pseudogenes, with, for example, the cosmid MLCB2052 possessing a region of over 16 kb where virtually all the ORFs are pseudogenes.

A surprising example of mutational inactivation of a gene was found with the *katG* gene of *M. leprae* (Eiglmeier *et al.* 1997; Nakata *et al.* 1997). The sequence of this region revealed that the *katG* gene is a pseudogene, containing a mutated start codon (GTA rather than GTG) and two deletions of 41 and 40 codons, respectively, as well as multiple frameshifts. Hence the use of isoniazid in drug regimens for the treatment of leprosy is inappropriate as the drug can never be activated by the bacillus. Furthermore, work with *M. bovis* has shown that the catalase–peroxidase is a virulence factor, as mutants lacking KatG activity were less virulent (Wilson *et al.* 1995). This mirrors the work of Middlebrook who showed that loss of catalase activity in tubercle bacilli was associated with loss of virulence (Middlebrook 1954). The role of catalase is most likely to protect the bacillus from the reactive oxygen species that the macrophage produces in response to phagocytosis of bacteria. Hence, it is quite unexpected that *M. leprae*, a pathogenic intracellular mycobacterium, should be able to dispense with one of its major defences against oxidative stress. This indicates that either *M. leprae* has some other mechanism to withstand reactive oxygen, or that it fails to trigger a respiratory burst.

In terms of repeated DNA in the genome of *M. leprae*, a new family of small repetitive elements, REP1, were identified which are scattered throughout the genome (Smith *et al.* 1997). The cosmid L383 furthermore contains a transposase pseudogene fragment with homology to IS*1081*, with a putative intact transposase gene just downstream of it that shows homology to the IS*3411* element from *E. coli*, yet lacks any obvious inverted or direct repeats. Whether this putative IS element is active or not is unclear, but it is striking that in an organism where so many pseudogenes have arisen there are so few cases of IS-element-mediated disruption of ORFs. Whether this points merely to an absence of active, or promiscuous, IS elements in the *M. leprae* genome remains to be seen.

Curiously, at least three ORFs (*gyrA*, *recA*, *xheA*) seem to encode proteins that can undergo an autocatalytic protein-splicing mechanism, removing a protein intron (intein) from a precursor molecule. The inteins are believed to correspond to homing endonucleases and two of them, present in GyrA and RecA, respectively, were subjected to detailed analysis (Davis *et al.* 1994; Fsihi *et al.* 1996b). Surprisingly, the intein contained in the GyrA protein was found to be distributed in some strains of *Mycobacterium flavescens*, *Mycobacterium gordonae* and *Mycobacterium kansasii*. It is interesting, taking into account the current rarity of intein containing precursor proteins in the eubacterial world, that three have been found in *M. leprae*.

3 The sequencing of the *M. tuberculosis* H37Rv genome

3.1 Library and map construction

As with *M. leprae*, a fundamental resource for the analysis of the *M. tuberculosis* genome was the availability of cosmid libraries. Initially, three discrete cosmid cloning vectors were used, namely pYUB18, pYUB328 and pYUB412, each with particular characteristics that would be useful for downstream experiments. The shuttle vector pYUB18 (Jacobs *et al.* 1991) facilitates the investigation of gene products and the dissection of the pathogenicity of *M. tuberculosis* as the clones can be easily introduced into surrogate hosts, such as the fast-growing *M. smegmatis*, where faithful gene expression can be expected. Clones based on the *E. coli* cosmid vector pYUB328 (Balasubramanian *et al.* 1996) will certainly be useful in allelic exchange experiments, as the DNA insert is flanked by restriction sites for *Pac*I, a rare cutting restriction endonuclease with no recognition sites in *M. tuberculosis* H37Rv. The third vector, the integration-proficient pYUB412 (Pavelka & Jacobs 1996), is an *E. coli–Mycobacterium* shuttle vector lacking a mycobacterial origin of replication but carrying the mycobacteriophage L5 attachment site (*attP*) and integrase gene (*int*) allowing integration at the attachment site (*attB*) present in certain mycobacterial genomes.

Contigs were established by characterization of individual clones from the three libraries by fingerprinting and hybridization with either suitable probes or entire cosmids in a chromosome walking approach. The resulting map consisted of 16 contigs and displayed the positions of more than a hundred known genes and loci. In contrast to *M. leprae*, chromosomal DNA suitable for PFGE can be prepared from *M. tuberculosis*, thus allowing a physical map to be established and an integrated map to be constructed from the combined data sets (Philipp *et al.* 1996b). The first step in the generation of the physical map of *M. tuberculosis* was the choice of appropriate restriction enzymes, with this being dictated by the relatively high $G+C$ content (66%) of the

M. tuberculosis chromosome. The only enzymes that gave a manageable number of restriction fragments were enzymes recognizing sites containing exclusively A and T residues: *Asn*I (recognizing ATTAAT) and *Dra*I (TTTAAA) which generated 47 and 35 fragments, respectively. Sixteen of the identified *Dra*I restriction sites were associated with IS*6110* as this insertion element contains a unique *Dra*I site. To establish the order of the restriction fragments, linking clones spanning rare-cutter sites were identified in the libraries and used as probes in hybridization experiments, thus permitting unambiguous linking of most of the fragments. To complete the map, particular *Asn*I or *Dra*I macrorestriction fragments were used as probes in reciprocal cross-hybridization experiments. Finally, confirmation of the established genome map was obtained by two-dimensional PFGE of *M. tuberculosis* H37Rv DNA reciprocally digested with *Dra*I and *Asn*I, thus allowing all existing restriction sites to be accounted for. The results obtained from the different PFGE experiments indicated that the chromosome has a circular topology and is ≈ 4.4 Mb in size. Extrachromosomal elements could not be detected.

Although the cosmid clones shown on the integrated map (Philipp *et al.* 1996b) selected by this approach represented the basic template DNA source for the start of the H37Rv sequencing project, problems such as underrepresentation of certain regions of the chromosome in the cosmid libraries, unstable or chimeric inserts, and the relatively small insert size complicated the efforts to sequence the entire genome. These problems were overcome by shotgun sequencing 48 clones from a bacterial artificial chromosome (BAC) library containing H37Rv DNA inserts up to 104 kb, and by sequencing about 40 000 clones derived from a whole genome shotgun library. The BAC library of *M. tuberculosis* (Brosch *et al.* 1998) comprises about 5000 clones and was constructed using the vector pBeloBAC11 (Kim *et al.* 1996) which combines a simple phenotypic screen for recombinant clones with the stable propagation of large inserts (Shizuya *et al.* 1992). The BAC cloning system is based on the *E. coli* F-factor, whose replication is strictly controlled and thus ensures stable maintenance of large constructs (Willets & Skurray 1987). A central advantage of the BAC cloning system over cosmid vectors is that the F-plasmid is present in only one or a maximum of two copies per cell, reducing the potential for recombination between DNA fragments and, more importantly, avoiding the lethal overexpression of cloned bacterial genes. Comparison of the sequence data from the termini of several hundred BAC clones with the sequences obtained from the cosmids allowed a minimal overlapping BAC map to be established. Figure 5.2 shows a map of the *M. tuberculosis* chromosome with the location of the BAC and cosmid clones indicated. In the final stages of sequencing this canonical set of 68 BAC clones that covers essentially the complete genome provided templates for gap-filling as these clones carried regions of the genome which were underrepresented or missing from cosmid or plasmid libraries. Hence, through the combined efforts of fully sequencing selected cosmids and BACs, and the sequencing of whole genome shotgun clones, the sequence of the entire genome of H37Rv was determined.

3.2 The mycobacterial database MycDB

Comparisons between the published mycobacterial sequences and genomes would be tremendously facilitated by centralization of all the newly obtained information in one database and by unifying the presentation of the data as well as the gene nomenclature and the sequence annotation. This would permit interested researchers to obtain a more 'user-friendly' overview of the molecular microbiology of mycobacteria. A first step in this direction was made by Bergh and Cole (1994) with the construction of MycDB, an integrated mycobacterial database in which it is aimed to centralize all data linked to mycobacteria. It contains the sequences of mycobacterial genes, identified antigens, available antibodies, physical maps of *M. leprae* and *M. tuberculosis*, the annotated *M. tuberculosis* cosmids, and selected references extracted from MedLine. It can be accessed via the World Wide Web: http://www.

pasteur.fr/mycdb, or at the mirror site: http://kiev. physchem.kth.se/MycDB.html.

The MycDB database is based on the ACeDB software that was originally developed for the *Caenorhabditis elegans* genome project but since then has been adapted for several other projects (Durbin & Thierry-Mieg 1991). The ACeDB database manager can accept a number of different 'objects' which can be related to each other and these relationships displayed. Most of the currently employed terms and subjects in molecular microbiology can be entered into the database as an 'object' in a corresponding class (e.g. loci, gene, clone, sequence, antibody, antigen, etc.). This database manager has the advantage of flexibility and can be expanded any time new classes should arise as will undoubtedly be the case after the completion of the mycobacterial sequencing projects. Typically, information about an object of a certain class is represented in a tree structure in a text format window, covering all the known aspects of the object. However, some classes have special displays that show the data in a more graphic manner. By double-clicking on the displayed aspects, additional information can be obtained, leading in a stepwise fashion to other objects (Bergh & Cole 1998). More recently, a relational database has been developed that provides detailed information about the genome, genes, proteins and protein families of *M. tuberculosis*. TubercuList is easy to consult via the World Wide Web: http://www.pasteur.fr/Bio/TubercuList.

3.3 Surprises from the *M. tuberculosis* H37Rv sequence

3.3.1 *Lipid metabolism*

The complexity and range of lipids produced by *M. tuberculosis* suggested a sophisticated cellular apparatus for lipid metabolism, but the sheer number of genes dedicated to lipid metabolism was unexpected. The bacillus contains examples of every known lipid and polyketide biosynthetic system, with ≈ 250 enzymes involved in fatty acid metabolism compared with ≈ 50 in *E. coli* (Cole *et al.* 1998).

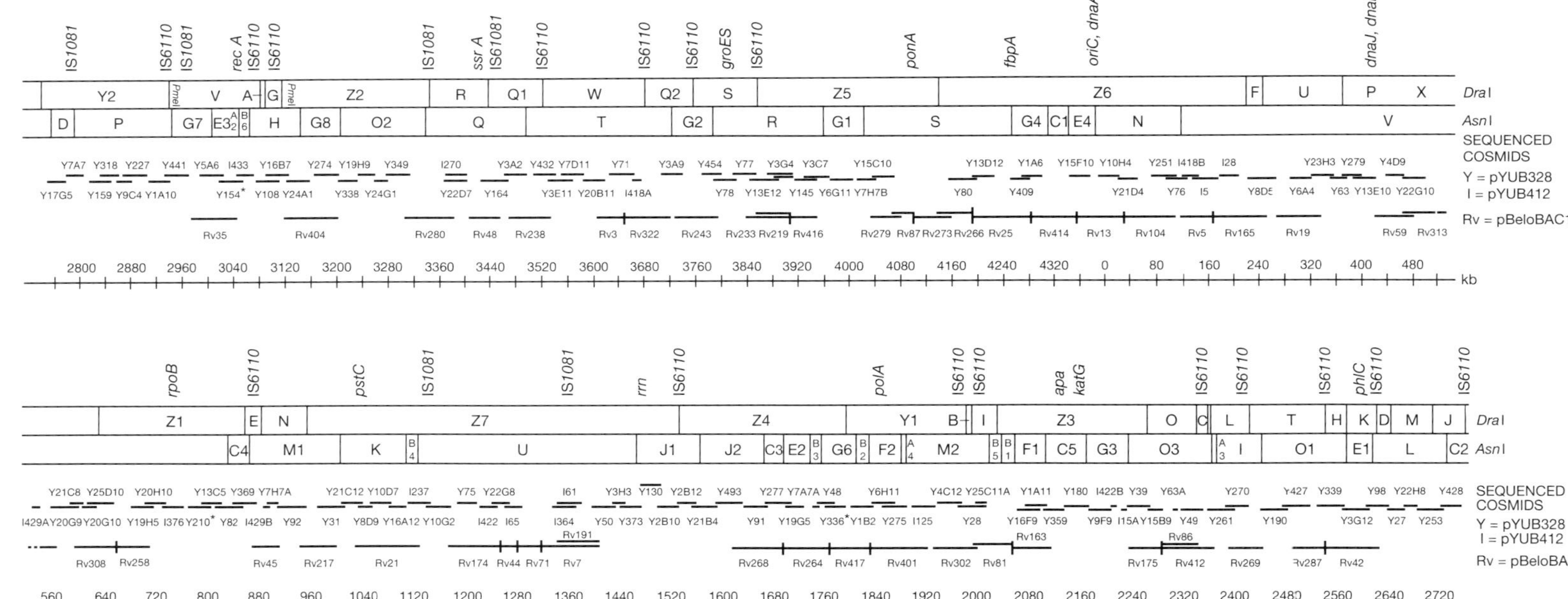

Fig. 5.2 The genomic organization of *M. tuberculosis* H37Rv. The figure shows the restriction fragment map of *M. tuberculosis* using the rare-cutting endonucleases *Dra*I and *Asn*I. The cosmid and BAC clones which were sequenced are shown under the restriction map, with the 'Y' designation indicating clones based on the pYUB328 shuttle vector, 'I' clones representing clones based on the integrative vector pYUB412, and 'Rv' indicating BAC clones constructed with the pBeloBAC11 vector. The scale of the map in kilobases is shown under the BAC clones, with '0' indicating the origin of replication.

The genome encodes over 100 enzymes for lipid degradation. In addition to the classic *fadA/fadB* β-oxidation system there exist 36 *fadD* alleles that code for acyl-CoA synthase, 36 *fadE* genes encoding acyl-CoA dehydrogenase, and 21 *echA* genes for enoyl-CoA hydratase/isomerase (Cole *et al.* 1998). The remarkable multiplicity of degradative enzymes suggests that *M. tuberculosis* has an unparalleled ability to exploit lipids as a carbon source. This may well have important implications for *in vivo* growth, where host-derived lipids could supply the main food source for the bacillus.

Biosynthesis of fatty acids is achieved through two systems, fatty acid synthase (FAS) I and FAS II. The FAS I system is a single, multifunctional polypeptide that occurs mainly in fungi and vertebrates, while FAS II is a multienzyme complex found in bacteria and plants. The mycobacteria are unusual in that they have been shown to encode both FAS I and FAS II activities (Barry *et al.* 1998). In *M. tuberculosis* the FAS I system is encoded by the *fas* gene and produces short-chain lipids that are then elongated to full length mycolic acids by the FAS-II complex. Among the components of FAS II are the enoyl reductase InhA and the β-ketoacyl synthase KasA, both targets for the potent antimycobacterial drug isoniazid (Banerjee *et al.* 1994; Mdluli *et al.* 1998). Unsaturation of mycolic acids appears to be catalysed by three aerobic terminal desaturases, encoded by *desA*1–3, that resemble ACP (acyl carrier protein)-ester utilizing plant desaturases. This suggests that unsaturation of the mermycolate chain may take place while the acyl group is still bound to the ACP.

The genome encodes polyketide synthase systems (PKS) of type I, e.g. mycocerosic acid synthase, and type II, e.g. the phenolphthiocerol system (Cole *et al.* 1998). In addition, *M. tuberculosis* possesses enzymes of the chalcone and stilbene PKS superfamily, a class that is phylogenetically distinct and shows no similarity at the amino acid level to other FAS or PKS systems (Hopwood 1997). As the chalcone and stilbene superfamily had previously only been shown in higher plants their discovery in *M. tuberculosis* was highly unexpected. Another unusual PKS in *M. tuberculosis* is encoded by *pks12*, whose putative

protein product has a predicted molecular weight of 432 kDa. This protein shows homology to the erythronolide synthase of *Saccharopolyspora erythraea*, although its product in *M. tuberculosis* is unknown. It is tempting to speculate, however, that the polyketides of *M. tuberculosis*, like many polyketides, have immunosuppressive activity.

3.3.2 New insertion sequences

During annotation of the sequence data, more than 50 loci were found with similarity to transposase enzymes. Among these were the 16 copies of IS*6110* (Thierry *et al.* 1990), six copies of IS*1081* (Collins & Stephens 1991), two copies of the IS-like element described by Mariani *et al.* (1993) and two copies of IS*1547* (EMBL accession number Y13470) (see also Chapter 2). However, many of the putative transposase enzymes were previously undescribed, leading to the identification of a whole new range of IS elements in the genome. Among these new transposases were members of the IS*3*, IS*21*, IS*30*, IS*100*, and IS*256* families, with some IS fitting no previous classification. These elements of uncertain lineage could, however, be grouped on the basis of their similarity to each other. Table 5.1 lists the IS elements uncovered to date and their relationships. In total, IS elements contribute over 70 kb of sequence to the genome of *M. tuberculosis* H37Rv. While a complete review of these new IS sequences is beyond the scope of this chapter, some of their salient features are outlined below.

The transposase enzymes coded for by IS*1532*, IS*1533*, and IS*1534* showed significant levels of homology at their C terminus to transposases of the IS*21* family. The elements possess large inverted repeats (48 bp, 54 bp and 49 bp, respectively) containing internal direct repeats, a feature believed to play some role in the binding of the transposase to the ends of the elements (Mahillon & Chandler 1998). They generated 4–5-bp direct repeats of the target DNA, a feature common to members of this family. The IS*1534* element may be defective in *M. tuberculosis* H37Rv as it contains a nonsense mutation that truncates the upstream ORF Rv3636.

Table 5.1 New insertion elements *M. tuberculosis* H37Rv.

Family	Element	Size	IR	DR	Genome position	Accession no.
IS3	IS*1540*	1162 bp	ND	ND	3847640-3848802	Z95389
	IS*1604*	1408 bp	ND	ND	3116814-3118222	Z81331
IS5	IS*1560*	1567 bp	25 bp	2 bp	3799983-3801550	AL009198
	IS*1560'*	1512 bp	ND	ND	1158919-1160431	Z92539
IS21	IS*1532*	2609 bp	48 bp	4 bp	3843787-3846396	Z77165
	IS*1533*	2212 bp	54 bp	5 bp	3288363-3290575	Z83858
	IS*1534*	2129 bp	49 bp	5 bp	4075657-4077786	Z95436
IS30	IS*1603*	1327 bp	32 bp	ND	3557290-3556817	AL021646
IS*110*	IS*1558*	1212 bp	13 bp	ND	2720642-2721854	Z81451
	IS*1558'*	803 bp	ND	ND	2439143-2439946	AL021957
	IS*1607*	1227 bp	ND	ND	2260441-2261668	Z74025
	IS*1608'*	1031 bp	ND	ND	3753327-3754360	AL009198
	IS*1608'*	1031 bp	ND	ND	4318338-4319369	Z83864
IS256	IS*1553*	1292 bp	13 bp	ND	4078503-4079795	Z95436
	IS*1554*	1435 bp	15 bp	ND	102546-1026891	Z95210
	IS*1552'*	844 bp	ND	ND	3849292-3850136	Z81331
IS*1535*	IS*1535*	2322 bp	17 bp	ND	1027039-1029361	Z95210
	IS*1536*	1391 bp	ND	ND	701363-702754	Z97182
	IS*1537*	1889 bp	ND	ND	4301540-4303429	Z97188
	IS*1538*	2055 bp	ND	ND	3333755-3335810	Z83018
	IS*1539*	2057 bp	ND	ND	3194136-3196193	Z74024
	IS*1602*	2052 bp	ND	ND	3100172-3102224	AL008967
	IS*1605'*	287 bp	ND	ND	921573-921860	AL022004
ISL3	IS*1557-1*	1513 bp	20 bp	ND	1468140-1469653	Z73419
	IS*1557-2*	1513 bp	28 bp	ND	4252851-4254364	AL022076
	IS*1557'*	516 bp	ND	ND	832350-832866	AL021958
	IS*1555'*	398 bp	ND	ND	3115741-3116139	Z81331
	IS*1561'*	1319 bp	ND	ND	3754292-3755611	AL009198
	IS*1606'*	330 bp	ND	ND	947309-947639	AL022004
Unknown	IS*1556*	1468 bp	ND	ND	2342940-2344408	Z73966

IR$_s$, inverted repeats; DRs, direct repeats; ND, not detected; elements that are truncated are marked with a prime (e.g. IS*1560'*).

A number of transposons related to IS256 from *Staphylococcus aureus* have been previously described in the mycobacteria, such as IS*1081* in the tubercle bacilli (Collins & Stephens 1991), and to this grouping we can now add IS*1552*, IS*1553* and IS*1554*. The single ORF present in these elements codes for a protein of 281 amino acids in IS*1552,* 409 amino acids in IS*1553*, and 439 amino acids in IS*1553*. The transposase encoded by IS*1552* is truncated since it is ≈120 amino acids shorter than expected for members of this group. In fact, it appears that IS*1552* has undergone horizontal transfer between *M. tuberculosis* and *Rhodococcus opacus*, as it shares 80% identity (91% similarity) over a 278 amino acid overlap with an IS element carried on the pHG201 plasmid of *R. opacus* (Grzeszik *et al.* 1997). Members of the genus *Rhodococcus*, like the mycobacteria, belong to the norcardiform taxonomic group, and contain mycolic

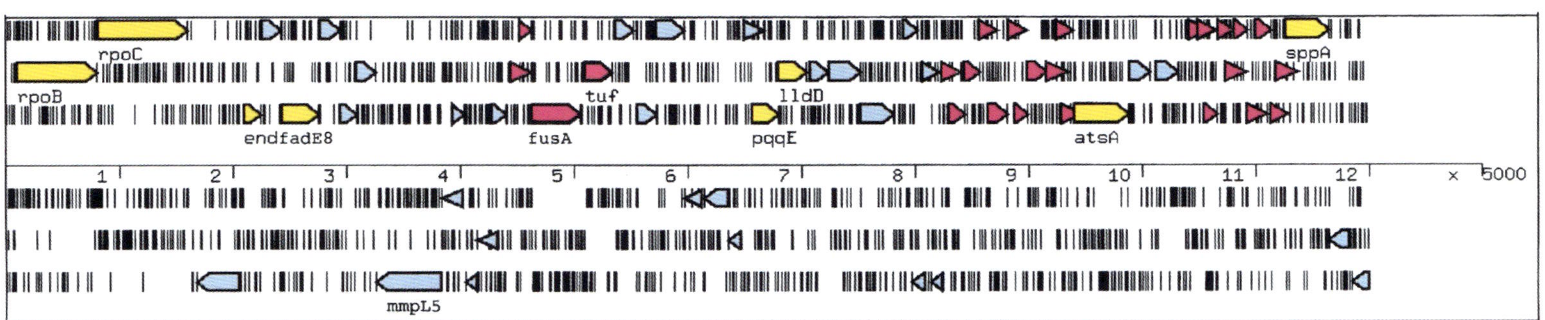

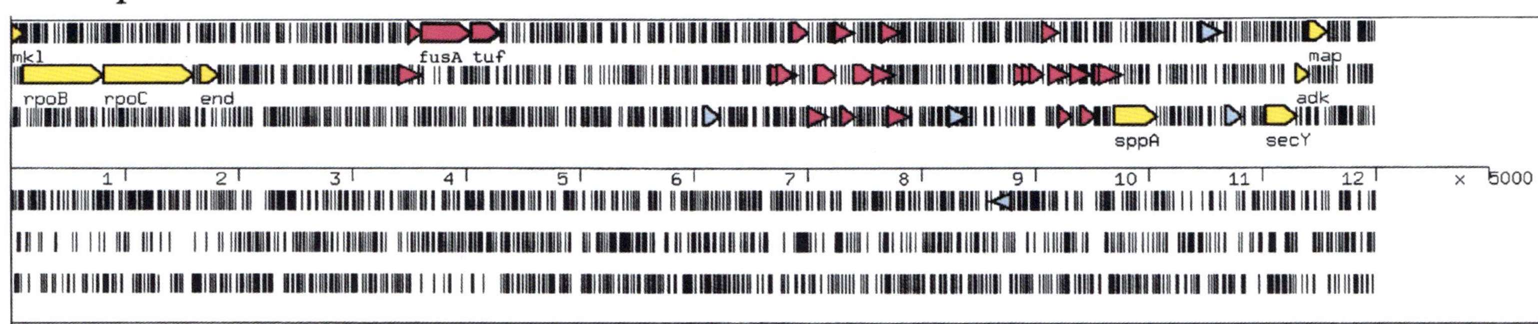

Plate 1 Local comparison of the *M. leprae* and *M. tuberculosis* genomes. A 60 kb region from both organisms is shown, with ORFs represented by rectangles or triangles, and with stop codons marked by short horizontal lines. The *str* and S10 ribosomal protein operons are shown in red, with genes involved in intermediary metabolism in yellow; all other ORFs are in blue. Gene names are shown for a selection of ORFs to serve as landmarks.

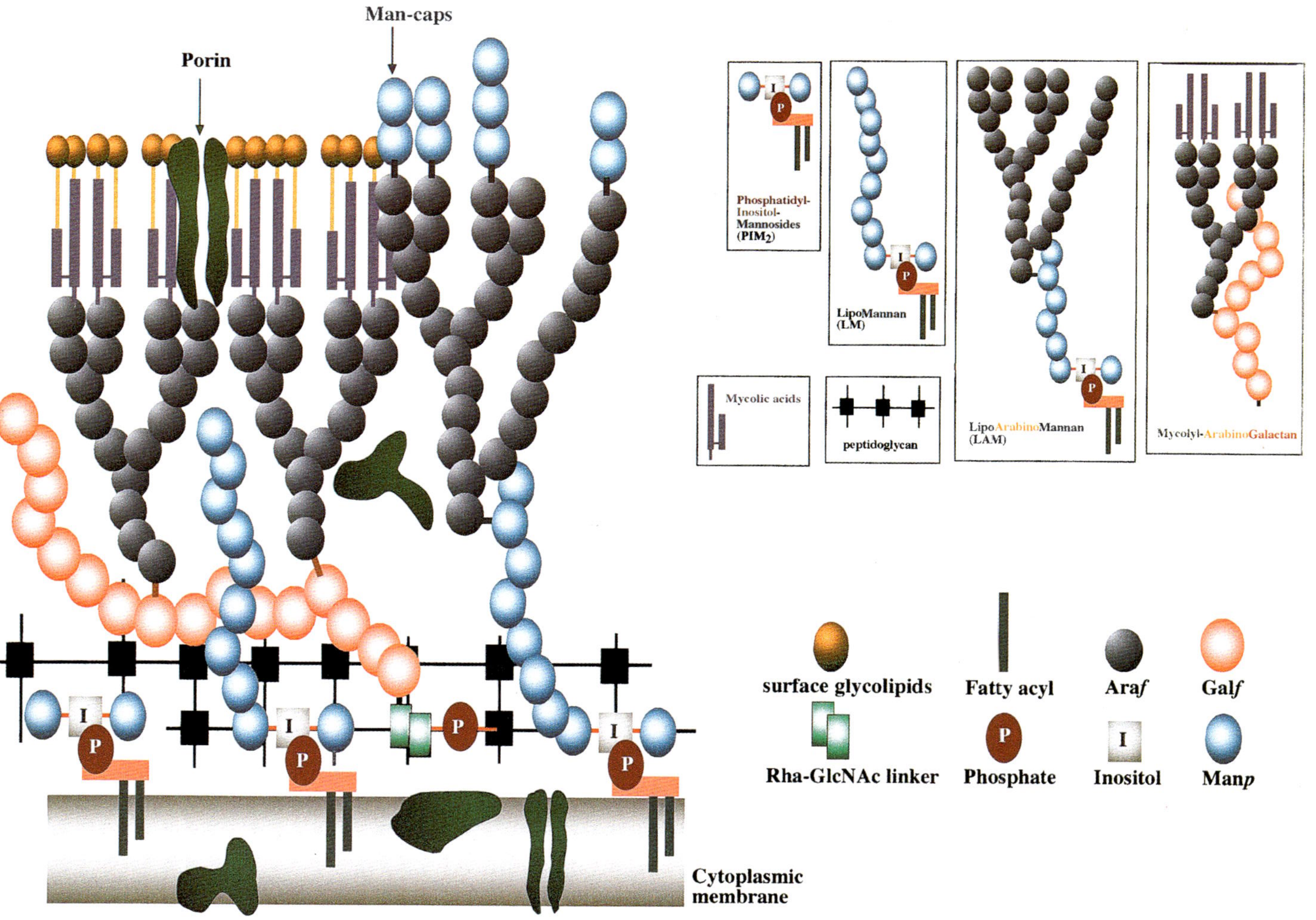

Plate 2 A model of the cell-wall core of *Mycobacterium* and associated lipids and lipoglycans.

acids, have a $G+C$ content of 63–73%, and form substrate mycelia. The relatedness of these two bacterial genera makes it highly tempting to speculate that IS*1552* was transferred from *Rhodococcus* into *M. tuberculosis*, perhaps while these genera shared an ecological niche.

A new IS family, designated the IS*1535* family, was also described (Gordon *et al*. 1999). These IS elements contain two ORFs coding for proteins of 193–195 and 247–550 amino acids, respectively. Significantly, the upstream proteins of IS*1536*, IS*1537*, IS*1538* and IS*1602* contain a resolvase motif, suggesting that this protein has resolvase activity. The proteins encoded by the downstream ORFs show weak homology with transposase enzymes and possess very basic pI values, in the range of 11.23–11.78, which may play a role in the binding of the protein to the negatively charged DNA substrate molecule.

The presence of this large collection of IS and repetitive elements in the genome of *M. tuberculosis* H37Rv begged the question as to whether they were also present in the genomes of other tubercle bacilli. Hence, a PCR analysis was initiated to search for these elements in the reference strains *M. bovis*, *M. bovis* bacille Calmette–Guérin (BCG) Pasteur, *M. africanum*, *M. microti* OV254, *M. tuberculosis* CSU 93 and 29 other clinical isolates. The results of this analysis (Gordon *et al*. 1999) revealed that while virtually all the IS elements were present in the same location in the strains tested, IS*1532* showed a relatively high degree of variation. This element was found to be absent from 10 of the 29 clinical *M. tuberculosis* isolates tested, as well as being absent from *M. tuberculosis* CSU 93, *M. bovis* and *M. bovis* BCG Pasteur. Therefore IS*1532* may be useful as a genetic marker for the differentiation of *M. tuberculosis* strains.

3.3.3 Prophage

The genome of *M. tuberculosis* H37Rv contains at least two prophages, phiRv1 and phiRv2 (present on cosmids MTCY336 and MTCY441, respectively), with a third putative prophage located at the region corresponding to MTCY16B7. The presence of proteins with homology to prohead protease enzymes, capsid proteins and phage integrase/excisionase enzymes in close proximity to each other were the necessary clues that allowed these prophage to be identified. Furthermore, phiRv1 and phiRv2 appear to be related to each other, with their respective prohead protease enzymes having 89.3% identity to each other over a 139 amino acid overlap, as well as two other ORFs of unknown function in the phage sharing over 80% identity. The possible prophage in MTCY16B7 is located upstream from the direct repeats (DR) region, a cluster of 36 bp direct repeats in the *M. tuberculosis* chromosome (Hermans *et al*. 1991). This putative phage encodes a protein with some homology to the bacteriophage P1 phd gene (35.9% identity in a 39 amino acid overlap), while the other ORFs are of unknown function. With the possible existence of a prophage so close to the DR region it is tempting to speculate that the DR region, and the IS*6110* contained therein, were originally introduced to the tubercle complex by this phage. Recombination between the repeats would then have lead to the variation that is seen at this locus (Hermans *et al*. 1991), with the IS*6110* element seeding the chromosome by migration out of the repeat region.

The sequence of phiRv1 is identical to that of the RD3 region described by Mahairas *et al*. (1996). This region was identified by genomic subtractive hybridization as being present in *M. bovis* but absent from *M. bovis* BCG Pasteur, hinting that loss of this region may have had some role in the attenuation of the BCG vaccine strain. However, as RD3 was also absent in 84% of clinical isolates of *M. tuberculosis* a link with virulence was discounted (Mahairas *et al*. 1996). The location of the insertional event is interesting, in that it is located in the biotin operon between the *bioB* and *bioD* genes. Hence, strains containing this prophage may be biotin auxotrophs due to disruption of the biotin operon. This is a possible explanation for the observation of Middlebrook that some clinical isolates of *M. tuberculosis* require biotin as a growth factor, with biotin being added as a supplement to mycobacterial growth media.

3.3.4 *The polymorphic GC-rich sequence and the major polymorphic tandem repeat*

The polymorphic GC-rich sequence (PGRS) and the major polymorphic tandem repeat (MPTR) were originally described as non-coding repetitive sequences in the genome of *M. tuberculosis* (Hermans *et al.* 1992; Ross *et al.* 1992; Poulet & Cole 1995). The PGRS was defined as multiple copies of the sequence CGGCGGCAA present at at least 26 discrete loci, while the MPTR consisted of the sequence GCCGGTGTTG, or its complement, arranged in tandem copies separated by 5 bp spacers. While both sequences have been exploited for their potential as epidemiological tools to differentiate *M. tuberculosis* strains (see Chapters 6 and 7), no serious attention was paid to their possible function in the mycobacteria. It is now clear, however, that the PGRS and MPTR are in fact genes encoding a large family of glycine, alanine and asparagine rich proteins. Multiple-sequence alignment of the PGRS and MPTR proteins further reveals that they belong to two broad families, designated the PE and the PPE families, respectively.

The PE family is so called after the presence of the motif proline–glutamic acid (PE) at positions 8 and 9 in a highly conserved N-terminal domain of ≈110 amino acids. This region is followed by a C-terminal domain that varies in size, sequence and repeat copy number, with a size range from 100 to ≈500 amino acids. This family contains 99 members, which can be subdivided into the PE and PE-PGRS subfamilies. The PE-PGRS family contains proteins with multiple repeats of a glycine–glycine–alanine or a glycine–glycine–asparagine motif, while the other subgrouping contains proteins with limited homology at their C terminus.

The PPE family resembles the PE family in that there exists a conserved N-terminal domain, here ≈180 amino acids, containing the motif proline–proline–glutamic acid (PPE) at positions 7–9, followed by a variable C-terminal region. This family contains 68 members, which break down into three subfamilies. The first of these families contains the MPTR sequences, which are characterized by repeats of the motif AsnXGlyXGlyXAsnXGly, while the second grouping contains the motif GlyXXSerValProXXTrp at position 350. The third grouping consists of proteins that contain the conserved 180 amino acid N-terminal domain but are unrelated at the C terminus.

Hence, with 167 members the PE and PPE protein families constitute one of the largest protein families in *M. tuberculosis*, which if they were all expressed would represent in the region of 4% of all protein species present. What role they play in the biology of the tubercle bacillus is not clear, yet some clues exist that may provide a starting point. For example, one member of the PGRS family is the 55-kDa fibronectin binding protein described by Abou-Zeid *et al.* (1991). This protein was shown to elicit an antibody response in eight out of 14 patients tested, indicating that either some patients fail to produce an immune response, or that this protein may not be produced in all strains of *M. tuberculosis*. Indeed, it is tempting to speculate that the whole PGRS family may have some role in the generation of antigenic variation by virtue of the variable C-terminal domain. Differential expression of the PGRS proteins, or perhaps strand slippage at the repeat sequences, could lead to the expression of a different repertoire of antigens, a strategy used successfully by pathogens such as *Haemophilus influenzae* and *Neisseria* species (Robertson & Meyer 1992). Further clues to the role of the PGRS can be inferred from the work by Musser and colleagues (Sreevatsan *et al.* 1997) who have shown that *M. tuberculosis* isolates from different geographical locations show a remarkable degree of homology at the genetic level. However, it is well known that probes based on the PGRS region can be used to differentiate *M. tuberculosis* clinical isolates (Ross *et al.* 1992; Strassle *et al.* 1997; S. Poulet and S.T. Cole, unpublished), so this region is obviously under some pressure to undergo rearrangement compared with other genes in the bacillus. With this is in mind, the function and role of the PGRS proteins certainly warrants detailed examination.

3.3.5 *Anaerobic genes*

Although the tubercle bacilli are classified as obligate aerobic organisms, it has been known for many years that they display some of the characteristics of facultative anaerobes when they grow *in vivo*. These features include a shift to the anaerobic mode of metabolism via the Embden–Meyerhof glycolytic pathway, a diminished oxidative response to glucose, glycerol as well as various glycolytic intermediates, and the stimulation of growth of primary clinical isolates by increased CO_2 levels (Segal 1984). With the genome sequence available to us, we can now probe the biochemical pathways of the bacillus to see whether the genes exist that could explain some of these physiological features.

One enzyme that may help to shed some light on the metabolism of *M. tuberculosis* in the host is fumarate reductase, an enzyme that plays a central role during the functioning of the tricarboxylic acid (TCA) cycle under anaerobic conditions. Under oxygen-limiting conditions, the genes for α-ketoglutarate dehydrogenase and succinate dehydrogenase become repressed, blocking the TCA cycle. Fumarate reductase allows the cycle to operate as a branched pathway, with the branch from oxaloacetate to succinate functioning as a reductive pathway with the final step from fumarate to succinate being catalysed by fumarate reductase. Hence the action of fumarate reductase may allow the TCA cycle to function in *M. tuberculosis* at the low oxygen tension prevailing in the host. In fact, genes suspected of being involved in adaptation to low oxygen partial pressure may be interesting candidates for inactivation with the aim of producing new vaccine strains.

Other genes for anaerobic metabolism in the genome are the genes for the formation of the nitrate reductase complex, which allows the utilization of nitrate as a terminal electron acceptor. However, as *M. bovis* strains lack nitrate reductase activity yet are fully virulent, it is unlikely that this enzyme would play some role in *in vivo* survival.

4 Comparative genomics

One of the burning questions in mycobacterial research is the basis for the attenuation of the vaccine strain *M. bovis* BCG. If we are to produce new vaccine strains for the treatment of tuberculosis we need to know the nature of the original mutation, or mutations, that attenuated the *M. bovis* progenitor strain. This knowledge would give us a foundation stone on which to base subsequent studies into new vaccine development. However, investigations into the genetic basis of virulence of the tubercle bacilli have always been fraught with difficulties, principally due to the lack of defined methods for the production of mycobacterial mutants altered in virulence. Thankfully, with the development of efficient methods for the generation of transposon mutants of *M. tuberculosis*, this should no longer be the case (Bardarov *et al.* 1997; Pelicic *et al.* 1997; see also Chapter 2). Furthermore, with the entire genome sequence of *M. tuberculosis* in our hands, we can now begin in earnest to explore the nature of mycobacterial virulence.

From the physical maps constructed for *M. tuberculosis* H37Rv and *M. bovis* BCG Pasteur it was possible to get an initial insight into zones of difference between the two genomes (Philipp *et al.* 1996a). Although the majority of *Asn*I restriction fragments were common between *M. bovis* BCG and *M. tuberculosis* (33 out of 47), significant variation in some fragment sizes was seen. This hinted at the presence of extra DNA in the *M. tuberculosis* H37Rv strain, although some of this variation could be attributed to the presence of repetitive elements, such as IS*6110* which exists in 16 copies in *M. tuberculosis* H37Rv but as a single copy in *M. bovis* BCG. *M. tuberculosis* H37Rv cosmids clones that mapped to these variable regions were hence used as probes against total *Eco*RI restriction digests of *M. bovis* and *M. bovis* BCG substrains Pasteur, Glaxo, Copenhagen, Moreau and Tokyo. With this strategy local comparisons between the genomes of *M. tuberculosis* H37Rv and *M. bovis* substrains could be made.

This analysis revealed the existence of various

genomic differences between *M. tuberculosis, M. bovis,* and *M. bovis* BCG substrains (Philipp *et al.* 1996a). The majority of these differences were the same as those identified by Stover and colleagues (Mahairas *et al.* 1996), namely the RD1, RD2 and RD3 regions that are absent from *M. bovis* BCG yet present in *M. bovis* and *M. tuberculosis.* Interestingly, however, a class of clones was found that identified polymorphisms between the various BCG strains tested. For example, the clones Y498 and Y414 identified regions that appeared to be conserved among *M. bovis, M. tuberculosis* but only some *M. bovis* BCG substrains. These genetic differences may be the source of the well known variation between the *M. bovis* BCG substrains that have been cloned throughout the world, in particular their varying ability to induce protective immune responses (Lagranderie *et al.* 1996).

The BAC library of *M. tuberculosis* presents an ideal tool with which to perform comparative genomics between the tubercle bacilli. As the library contains an average insert size of $\approx 70\,$kb the entire *M. tuberculosis* chromosome can be represented by less than 70 overlapping clones. Restriction digests of the clones can then be blotted to membranes and probed with radiolabelled total genomic DNA from, for example, *M. bovis* BCG. Restriction fragments that fail to produce a signal after hybridization must be absent from the probe DNA, hence identifying deleted regions. This analysis revealed that the region encompassed by the BAC clone Rv58 contained a deletion of 12.7 kb in *M. bovis* BCG Pasteur relative to *M. tuberculosis,* RD4, (Brosch *et al.* 1998). Further PCR experiments showed that the region was also deleted from the Danish, Glaxo, Copenhagen and Russian strains of BCG, as well as from *M. bovis.* Hence, it appears that the deletion is specific to bovine strains of the bacillus. The region contains 11 ORFs whose putative products show low similarity to proteins involved in polysaccharide biosynthesis (Brosch *et al.* 1998). Considering the possible role of mycobacterial polysaccharides in binding to host receptors (Ehlers & Daffé 1998), this region may alter the receptor specificity of *M. bovis* compared to *M. tuberculosis.*

5 Conclusions

In many ways, the elucidation of the complete genomic sequence of *M. tuberculosis* H37Rv can be seen as a new beginning for mycobacterial research. Since its original isolation in 1905 (Steenken *et al.* 1935) the H37Rv bacillus has provided the backbone for innumerable studies into the physiology, virulence, drug resistance and genetics of the tubercle bacilli. With the completion of the genome sequencing project we now have the possibility of looking back on this accumulated wealth of information and placing it in the context of the sequence data. As just one example, the contentious question as to whether *in vivo* growing *M. tuberculosis* can respire anaerobically was painstakingly investigated by analysing *in vivo*-grown bacilli for the presence of enzymes characteristic of anaerobic respiration (Segal 1984). With the sequence data, we are now able to identify genes whose products are known to be involved in anaerobic metabolism, validating earlier investigations as well as providing us with candidate genes whose expression may be vital for the *in vivo* survival of the bacilli.

The imminent completion of the genome sequence of *M. leprae* will provide a further dimension to our knowledge of the pathogenic mycobacteria. It will be possible to do a whole genome comparison with *M. tuberculosis,* identifying coding regions that have been lost in *M. leprae* due to deletions, frameshifts, etc., an analysis that could provide us with the reasons as to why *M. leprae* has resisted all attempts at *in vitro* culture. Such an analysis will also identify the genes in *M. leprae* that were maintained relative to *M. tuberculosis* and hence essential for *in vivo* growth, describing much about the intracellular milieu in which the bacillus persists. These types of *in silico* analyses will lend crucial insights into the physiology of *M. tuberculosis* and *M. leprae* without the need for 'hands-on' manipulation of the organism. Indeed, it is perhaps through the union of computer technology and biology that we may best accelerate our understanding of these pathogens.

6 Acknowledgements

Financial support from the Wellcome Trust, the Heiser Trust, the Association Française Raoul Follereau and the Groupement de Recherches et des Etudes des Genomes (GIP-GREG) is gratefully appreciated. S.V.G. received a Wellcome Trust travelling research fellowship.

7 References

Abou-Zeid, C., Garbe, T., Lathigra, R. *et al.* (1991) Genetic and immunological analysis of *Mycobacterium tuberculosis* fibronectin binding proteins. *Infection and Immunity* **59**, 2712–2718.

Balasubramanian, V., Pavelka, M.S., Jr, Bardarov, S.S. *et al.* (1996) Allelic exchange in *Mycobacterium tuberculosis* with long linear recombination substrates. *Journal of Bacteriology* **178**, 273–279.

Banerjee, A., Dubnau, E., Quemard, A. *et al.* (1994) *inh*A, a gene encoding a target for isoniazid and ethionamide in *Mycobacterium tuberculosis*. *Science* **63**, 227–230.

Bardarov, S., Kriakov, J., Carriere, C. *et al.* (1997) Conditionally replicating mycobacteriophages: a system for transposon delivery to *Mycobacterium tuberculosis*. *Proceedings of the National Academy of Sciences of the USA* **94**, 10961–10966.

Barry, C.E., III, Lee, R.E., Mdluli, K. *et al.* (1998) Mycolic acids: Structure, biosynthesis and physiological functions. *Progress in Lipid Research* **37**, 143–179.

Bergh, S. & Cole, S.T. (1994) MycDB: an integrated mycobacterial database. *Molecular Microbiology* **12**, 517–534.

Bergh, S. & Cole, S.T. (1998) Using MycDB on the World-Wide Web. In: *Mycobacteria protocols.* (eds T. Parish & N. G. Stoker). Totowa, NJ: Humana Press Inc., pp. 109–118.

Brosch, R., Gordon, S.V., Billault, A. *et al.* (1998) Use of a *Mycobacterium tuberculosis* H37Rv Bacterial Artificial Chromosome (BAC) library for genome mapping, sequencing and comparative genomics. *Infection and Immunity* **66**, 2221–2229.

Cole, S.T., Brosch, R., Parkhill, J. *et al.* (1998). Deciphering the biology of *Mycobacterium tuberculosis* from the complete genome sequence. *Nature* **393**, 537–544.

Cole, S.T. & Saint Girons, I. (1994) Bacterial genomics. *FEMS Microbiological Review* **14**, 139–160.

Collins, D.M. & Stephens, D.M. (1991) Identification of an insertion sequence, IS*1081*, in *Mycobacterium bovis*. *FEMS Microbiological Letters* **67**, 11–15.

Coulson, A., Sulston, J., Brenner, S. & Karn, J. (1986) Toward a physical map of the genome of the nematode *Caenorhabditis elegans*. *Proceedings of the National Academy of Sciences of the USA* **83**, 7821–7825.

Davis, E.O., Thangaraj, H.S., Brooks, P.C. & Colston, M.J. (1994) Evidence of selection for protein introns in the RecAs of pathogenic mycobacteria. *EMBO Journal* **13**, 699–703.

Durbin, R. & Thierry-Mieg, J. (1991) A *C. elegans* database. Documentation, code and data available from anonymous FTP servers at lirmm.lirmm.fr, cele.mrc-lmb.cam.ac.uk and ncbi.nlm.nih.gov.

Ehlers, M.R.W. & Daffé, M. (1998) Interactions between *Mycobacterium tuberculosis* and host cells: are mycobacterial sugars the key? *Trends in Microbiology* **6**, 328–335.

Eiglmeier, K., Fsihi, H., Heym, B. & Cole, S.T. (1997) On the catalase-peroxidase gene, *katG*, of *Mycobacterium leprae* and the implications for treatment of leprosy with isoniazid. *FEMS Microbiological Letters* **149**, 273–278.

Eiglmeier, K., Honoré, N., Woods, S.A., Caudron, B. & Cole, S.T. (1993) Use of an ordered cosmid library to deduce the genomic organization of *Mycobacterium leprae*. *Molecular Microbiology* **7**, 197–206.

Fsihi, H., De Rossi, E., Salazar, L. *et al.* (1996a) Gene arrangement and organization in a approximately 76 kb fragment encompassing the *oriC* region of the chromosome of *Mycobacterium leprae*. *Microbiology* **142**, 3147–3161.

Fsihi, H., Vincent, V. & Cole, S.T. (1996b) Homing events in the *gyrA* gene of some mycobacteria. *Proceedings of the National Academy of Sciences of the USA* **93**, 3410–3415.

Gibson, T.J., Rosenthal, A. & Waterston, R.H. (1987) Lorist6, a cosmid vector with *Bam*HI, *Not*I, *Sca*I and *Hind*III sites. *Gene* **53**, 283–286.

Gordon, S.V., Heym, B., Parkhill, J., Barrell, B. & Cole, S.T. (1999) New insertion sequences and a novel repeated sequence in the genome of *Mycobacterium tuberculosis* H37Rv. *Microbiology* **145**, 881–892.

Grzeszik, C., Lubbers, M., Reh, M. & Schlegel, H.G. (1997) Genes encoding the NAD-reducing hydrogenase of *Rhodococcus opacus* MR11. *Microbiology* **143**, 1271–1286.

Hermans, P.W., van Soolingen, D., Bik, E.M., da Hass, P.E., Dale, J.W. & van Embden, J.D. (1991) Insertion element IS987 from *Mycobacterium bovis* BCG is located in a hot-spot integration region for insertion elements in *Mycobacterium tuberculosis* complex strains. *Infection and Immunity* **59**, 2695–2705.

Hermans, P.W., van Soolingen, D. & van Embden, J.D. (1992) Characterisation of a major polymorphic tandem repeat in *Mycobacterium tuberculosis* and its potential use in the epidemiology of *Mycobacterium kansasii* and *Mycobacterium gordonae*. *Journal of Bacteriology* **174**, 4157–4165.

Honoré, N., Bergh, S., Chanteau, S. *et al.* (1993) Nucleotide sequence of the first cosmid from the *Mycobacterium leprae* genome project: structure and function of the Rif-Str regions. *Molecular Microbiology* **7**, 207–214.

Hopwood, D. (1997) Genetic contributions to understanding polyketide synthases. *Chemical Reviews* **97**, 2465–2497.

Jacobs, W.R., Jr, Pana, G.V., Cirillo, J.D. (1991) Genetic systems for mycobacteria. *Methods in Enzymology* **204**, 537–555.

Kim, U.J., Birren, B.W., Slepak, T., Mancino, V., Boysen, C. & Kang, H.L. (1996) Construction and characterization of a human bacterial artificial chromosome library. *Genomics* **34**, 213–218.

Lagranderie, M.R., Balazuc, A.M., Deriaud, E. & Leclerc, C.D. (1996) Comparison of immune responses of mice immunized with five different *Mycobacterium bovis* vaccine strains. *Infection and Immunity* **64**, 1–9.

Mahairas, G.G., Sabo, P.J., Hickey, M.J., Singh, D.C. & Stover, C.K. (1996) Molecular analysis of genetic differences between *Mycobacterium bovis* BCG and virulent *M. bovis*. *Journal of Bacteriology* **178**, 1274–1282.

Mahillon, J. & Chandler, M. (1998) Insertion sequences. *Microbiological Molecular Biological Review* **62**, 725–774.

Mariani, F., Piccolella, E., Colizzi, V., Rappuoli, R. & Gross, R. (1993) Characterization of an IS-like element from *Mycobacterium tuberculosis*. *Journal of General Microbiology* **139**, 1767–1772.

Mdluli, K., Slayden, R.A., Zhu, Y. *et al.*(1998) Inhibition of a *Mycobacterium tuberculosis*—Ketoacyl ACP synthase by isoniazid. *Science* **280**, 1607–1610.

Middlebrook, G. (1954) Isoniazid-resistance and catalase activity of tubercle bacilli. *American Review of Tuberculosis* **69**, 471–472.

Mushegian, A.R. & Koonin, E.V. (1996) A minimal gene set for cellular life derived by comparison of complete bacterial genomes. *Proceedings of the National Academy of Sciences of the USA* **93**, 10268–10273.

Nakata, N., Matsuoka, M., Kashiwabara, Y., Okada, N. & Sasakawa, C. (1997) Nucleotide sequence of the *Mycobacterium leprae katG* region. *Journal of Bacteriology* **179**, 3053–3057.

Pavelka, M.S. & Jacobs, W.R. (1996). Biosynthesis of diaminopimelate, the precursor of lysine and a component of peptidoglycan, is an essential function of *Mycobacterium smegmatis*. *Journal of Bacteriology* **178**, 6496–6507.

Pelicic, V., Jackson, M., Reyrat, J.-M., Jacobs, W.R., Gicquel, B. & Guilhot, C. (1997) Efficient allelic exchange and transposon mutagenesis in *Mycobacterium tuberculosis*. *Proceedings of the National Academy of Sciences of the USA* **94**, 10955–10960.

Philipp, W.J., Nair, S., Guglielmi, G., Lagranderie, M.,

Gicquel, B. & Cole, S.T. (1996a) Physical mapping of *Mycobacterium bovis* BCG Pasteur reveals differences from the genome map of *Mycobacterium tuberculosis* H37Rv and from *M. bovis*. *Microbiology* **142**, 3135–3145.

Philipp, W.J., Poulet, S., Eiglmeier, K. *et al.* (1996b) An integrated map of the genome of the tubercle bacillus, *Mycobacterium tuberculosis* H37Rv, and comparison with *Mycobacterium leprae*. *Proceedings of the National Academy of Sciences of the USA* **93**, 3132–3137.

Poulet, S. & Cole, S.T. (1995) Characterization of the highly abundant polymorphic GC-rich-repetitive sequence (PGRS) present in *Mycobacterium tuberculosis*. *Archives of Microbiology* **163**, 87–95.

Robertson, B.D. & Meyer, T.F. (1992) Genetic variation in pathogenic bacteria. *Trends in Genetics* **8**, 422–427.

Ross, B.C., Raios, K., Jackson, K. & Dwyer, B. (1992) Molecular cloning of a highly repeated DNA element from *Mycobacterium tuberculosis* and its use as an epidemiological tool. *Journal of Clinical Microbiology* **30**, 942–946.

Salazar, L., Fsihi, H., Da Rossi, E. *et al.* (1996) Organization of the origins of replication of the chromosomes of *Mycobacterium smegmatis, Mycobacterium leprae* and *Mycobacterium tuberculosis* and isolation of a functional origin from *M. smegmatis*. *Molecular Microbiology* **20**, 283–293.

Segal, W. (1984) Growth dynamics of *in vivo* and *in vitro* grown mycobacterial pathogens. In: *The Mycobacteria: a Sourcebook* (eds G. P. Kubica & L. G. Wayne). New York: Marcel Dekker, pp. 547–573.

Shizuya, H., Birren, B., Kim, U.J., Mancino, V. & Slepak, T. (1992) Cloning and stable maintenance of 300-kilobase-pair fragments of human DNA in *Escherichia coli* using an F-factor based vector. *Proceedings of the National Academy of Sciences of the USA* **89**, 8794–8797.

Smith, D.R., Richterich, P., Rubenfield, M. *et al.*(1997) Multiplex sequencing of 1.5 Mb of the *Mycobacterium leprae* genome. *Genome Research* **7**, 802–819.

Sreevatsan, S., Pan, X., Stockbauer, K.E. *et al.* (1997) Restricted structural genes polymorphism in the *Mycobacterium tuberculosis* complex indicates evolutionarily recent global dissemination. *Proceedings of the National Academy of Sciences of the USA* **94**, 9869–9874.

Steenken, W., Oatway, W.H. & Petroff, S.A. (1935) Biological studies of the tubercle bacillus. III. Dissociation and pathogenicity of the R and S variants of the human tubercle bacillus (H37). *Journal of Experimental Medicine* **60**, 515–540.

Strassle, A., Putnik, J., Weber, R., Fehr-Merhof, A., Wust, J. & Pfyffer, G.E. (1997) Molecular epidemiology of *Mycobacterium tuberculosis* strains isolated from patients in a human immunodeficiency virus cohort in Switzerland. *Journal of Clinical Microbiology* **35**, 374–378.

Sulston, J., Mallett, F., Staden, R., Durbin, R., Horsnell, T. &

Coulson, A. (1988) Software for genome mapping by fingerprinting techniques. *CABIOS* **4**, 125–132.

Thierry, D., Cave, M.D., Eisenach, K.D. *et al.* (1990) IS*6110*, an IS-like element of *Mycobacterium tuberculosis*. *Nucleic Acids Research* **18**, 188.

Willets, N. & Skurray, R. (1987). Structure and function of the F factor and mechanism of conjugation. In: *Escherichia coli and Salmonella typhimurium: Cellular and Molecular Biology* (eds F. C. Neidhardt *et al.*). Washington, DC: American Society of Microbiology, pp. 1110–1133.

Wilson, T.M., de Lisle, G.W. & Collins, D.M. (1995) Effect of *inhA* and *katG* on isoniazid resistance and virulence of *Mycobacterium bovis*. *Molecular Microbiology* **15**, 1009–1015.

Woods, S.A. & Cole, S.T. (1990) A family of dispersed repeats in *Mycobacterium leprae*. *Molecular Microbiology* **4**, 1745–1751.

Chapter 6 / Molecular epidemiology: human tuberculosis

PAUL VAN HELDEN

1 Introduction

Throughout recent history, epidemiology has made significant contributions to the study of health and disease, and has influenced health strategy and policy over many decades. Epidemiologists have alerted us to emerging epidemics and predicted the outcome of diseases. Traditional epidemiology can perhaps be simplistically seen as a descriptive science, since it often consists of collection of data (e.g. the incidence of disease in a given area) and the attempt to seek associations between incidence of disease and other factors, such as location of a common point serving as a source of infection. The data gathering is largely dependent on field work, and aims to associate the transmission of disease with contact between two individuals. Field work consisting of contact tracing is enormously difficult and dependent on the memories and willingness of individuals and communities to communicate with the field worker. Thus, traditional epidemiology is limited by the efficacy of the tools available and can be used only in a certain context.

In the case of infectious diseases, strain typing can be a very useful adjunct to clinical epidemiology. Strain typing may be phenotypic, where an expressed characteristic of the organism is examined, or may be genotypic. The recent marriage between molecular biology and epidemiology has provided a synergistic input into the study of disease and become known as molecular epidemiology. This technology should ideally form a part of all modern epidemiological studies. The typing of strains may be merely descriptive, such as is used in outbreak analysis, or may be applied in a broader context in terms of population genetics and disease dynamics. The techniques described below may therefore be applied in circumscribed settings (e.g. nosocomial transmission in a single ward) or be used in a wider, perhaps even global context, in order to model disease dynamics and monitor intervention strategies.

2 Genotype analysis, limitations, comparisons and assessment

2.1 General considerations

The use of any given genotyping system and its validity will be dependent on the stability of the genome of that organism. Prior to embarking on any genotype study, it is important that one have some insight into the degree of instability of the genome and factors that may promote instability. These factors would include recombination, transposition or any other genome changes. With respect to tuberculosis, it is perhaps fortunate for molecular epidemiologists that there is little or no evidence for panmixia. The population of *Mycobacterium tuberculosis* can thus be regarded as largely, if not entirely, clonal. So with respect to *M. tuberculosis*, particularly in this discussion, it is important to have a good understanding of clonality and therefore evolution and to define terms clearly. A strain or clone, as defined in this case, has a specific genotype and may occur in more than one patient. Similar strains, which are not genotypically identical, will be called clonal variants.

A number of different genotyping methods have been developed in the past few years. The more well-known ones will be outlined briefly below with some of their advantages and disadvantages. These techniques rely largely on the occurrence of repeated sequences in the genome of the organism. The number, location and variability of flanking sequences and restriction enzyme cutting sites with respect to these repeat sequences generate polymorphisms in the genome which can be used for typing, by various methods. Further information concerning various elements in the genome of *M. tuberculosis* can be obtained from Chapters 2 and 5. Similar techniques and technologies are available to type other mycobacteria and these are also discussed elsewhere in this book (Chapters 7 and 8).

2.2 Polymerase chain reaction-based polymorphic analysis

The advantages of this technology are that it is rapid, and can potentially be done using either purified DNA, a scrape from a colony (culture) or from unprocessed clinical material, e.g. a sputum sample. A specific polymorphic area of the genome may be used (Goyal *et al.* 1994) but the polymorphic information content of this methodology is limited and it is not suitable for large numbers of samples. Further polymerase chain reaction (PCR)-based applications include random amplification of polymorphic domains (RAPD). In this case a mix of non-specific primers is used to amplify unknown portions of the genome. The use of this technology tends to divide molecular biologists into two camps which are either supportive or quite opposed to this technology. Those who oppose this method in general, deride it for the lack of reproducibility and the fact that the domains amplified are unknown. In addition, the patterns generated give a series of fragments which can vary considerably in intensity, making it difficult to assign significance to the various members of each pattern. This is problematical for database entry (this will be discussed in detail later).

A third PCR-based method relies on the presence of an insertion element in the genome (known as IS*6110*, discussed later). A mixed-linker PCR method has been developed (Haas *et al.* 1993) which utilizes the sequence, location, and copy number of this element to produce a polymorphic pattern based on this insertion element. This technology has been used successfully, but has generally not been the most commonly used method for fingerprinting *M. tuberculosis* and to some extent may require a higher degree of expertise than is present in many laboratories. It may suffer from the same problems of fragment intensity and difficulty of assignment as seen for RAPD PCR. A major advantage of the PCR methodology is the possibility of genotyping a sample where no culture can be obtained, for example in archival material. In addition, very rapid genotyping is possible—this may be important in the case of suspected drug-resistant strain transmission. A future vision for PCR genotyping might be where a mixture of primer pairs is used to amplify a number of known highly polymorphic areas. A highly specific pattern containing a reasonable number of bands of defined intensity should be produced. The absence, presence or altered mobility of bands should serve as useful discriminators.

2.3 Spoligo typing

A specific domain (DR—direct repeat) within the genome of *M. tuberculosis* consists of a set of direct repeats; this locus is a preferential site for insertion of a copy of the IS*6110* element. The repeats consist of a 36-bp-conserved repeat, interspersed by a variable spacer sequence (35–41 bp). The sequence of the variants is known, and thus it is possible to synthesize oligonucleotides specific for each one of the spacers, and array them in a dot-blot type format. Amplification of the region followed by hybridization of the amplicons to the oligonucleotide array will therefore yield a very clear pattern which is relatively culture (strain) specific (Goyal *et al.* 1997). The absence or presence of a hybridization spot will indicate which repeats occur in any given sample and a variable intensity of spot will give an indication of a partial or complete sequence deletion. This system (known as spoligotyping) offers clear discriminatory power and has the advantage of not requiring very much interpretation, the results being generally unequivocal. Unfortunately, it is based entirely upon one locus in the genome and therefore is perhaps not as informative and powerful as many of the other techniques available. For example, it has recently been shown that strains with identical spoligotype may differ in IS*6110* pattern (Goyal *et al.* 1997; Kamerbeek *et al.* 1997). It could, however, be useful in low-incidence communities or when the study of an outbreak is urgent since it is relatively rapid.

2.4 Restriction fragment length polymorphism genotyping

Arguably the most extensively used technology for culture typing of *M. tuberculosis* depends on Southern blotting. A number of probes are in use and these will be discussed briefly.

IS6110

This IS*3*-like element, known also as IS*986* or IS*987*, occurs at various locations in the genome of *M. tuberculosis* and at variable copy number (Mazurek *et al.* 1991; Van Soolingen *et al.* 1993, 1994, 1995; Hermans *et al.* 1995). Since its sequence is apparently invariant (Dale *et al.* 1998), it is an ideal target sequence. The copy number of the element ranges from one to 25 copies per genome, although rare *M. tuberculosis* cultures have been reported with no inserts. Usually, the DNA obtained from a culture of

M. tuberculosis is digested with the restriction enzyme *Pvu*II (which cuts once within the element) generating a number of electrophoretically distinct fragments which can be detected by Southern blotting using a probe complementary to the 3′ region of IS*6110*. Since the element is sequence conserved, the fragments can easily be detected and there is usually little variation in intensity. This technology yields a very clear fingerprint pattern which is hypervariable and easy to interpret. Strains are recognized according to the IS*6110* copy number and electrophoretic mobility of this element plus associated flanking sequence. A polymorphic pattern may also be generated using a probe complementary to the 5′ region of the IS*6110* element. Using the two IS*6110* probes, it is thus possible to identify flanking sequence polymorphisms on both sides of the IS element (R.M. Warren *et al.* unpublished results).

Discriminatory power is a function of copy number, and it is widely thought that cultures with five copies require additional typing, while those with more than five copies yield reliable results (Braden *et al.* 1997; M. Richardson *et al.* unpublished results; Van Soolingen *et al.* 1993; Yang *et al.* 1996). This is not always true (Warren *et al.* 1996b; R.M. Warren unpublished results) and will be discussed later and under Recommendations.

The major polymorphic tandem repeat

The major polymorphic tandem repeat (MPTR) consists essentially of a 10-bp repeated sequence separated by 5 bp sequences. MPTR sequences occur in multiple copies in the genome of *M. tuberculosis*, but unfortunately there is little polymorphism in the MPTR-containing restriction enzyme fragments, although more polymorphism occurs in other species of mycobacteria (Hermans *et al.* 1992). This probe is therefore not commonly used as it has low powers of discrimination.

The polymorphic GC-rich sequence

The polymorphic GC-rich sequence (PGRS) occurs at ≈25–30 loci in the genome of *M. tuberculosis* (see

Chapter 5), where each one has a common 9-bp consensus repeat (GCCGCCGTT). The element may be repeated two to seven times at each locus. It shows a fairly high level of polymorphism and is therefore useful for typing (Ross *et al.* 1992; Van Soolingen *et al.* 1993; Yang *et al.* 1996; Braden *et al.* 1997). The restriction enzyme *Alu*1 is usually used but, since *Pvu*II sites are a subset of *Alu*I sites, it is possibly advisable to use a different enzyme in cases where a *Pvu*II digest using IS*6110* is also employed. A combination of these two probes will therefore yield additional information (this will be discussed later).

The direct repeat region

The direct repeat (DR) region of *M. tuberculosis* consists of the two elements usually flanking a preferred insert site of IS*6110* (Groenen *et al.* 1993; Van Soolingen *et al.* 1993). It is this repeat that is also used for the spoligotyping procedure described above. It shows limited polymorphism in *M. tuberculosis* and can thus be used for strain typing. *Alu*1 is the enzyme frequently used, although *Pvu*II is also useful, as it allows one to position the associated IS*6110* element if the same filter is re-used. Its disadvantage is that it occurs usually on either side of one IS*6110* element and therefore its location and low copy number confer limited discriminatory powers.

(GTG)₅ and related sequences

A number of synthetic oligonucleotide repeats have been tested for polymorphic power in the genome of *M. tuberculosis*. Thus far, the only one reported to provide a useful polymorphic multiple locus pattern is (GTG)₅ in combination with *Hin*f1 (Wiid *et al.* 1994). This probe has also been used to isolate specific sequences from *M. tuberculosis* and these probes (MTB484(1) and 2K4) in turn have been used as probes (Warren *et al.* 1996b). A pattern similar to that obtained for PGRS is obtained on analysis of the genome of *M. tuberculosis*, however, additional polymorphic and very strongly hybridizing fragments of higher molecular weight are obtained in addition to the weak hybridization of the other PGRS-containing

repeats (Warren *et al.* 1996b). This probe may therefore offer the advantage of detection of multiple, but weakly hybridizing loci as well as a highly polymorphic and clearly interpretable strong signal. The PGRS-like hybridization pattern in this case is reported to be clearer than that obtained using the PGRS probe itself.

Interpretation of results

The aim of strain typing is to obtain an informative pattern which can be compared to those of other cultures and matched when appropriate. Should one wish to compare the patterns of only a few cultures (e.g. 10–20) this is easily done. However, if one wishes to do a large number of culture comparisons, intersample variability makes pattern comparison by visual means impossible. It is therefore important that some additional analytical method be devised to enable pattern comparison and this requirement highlights the advantages and disadvantages of the various techniques. Clearly, a visual comparison of a large number of strains is impossible and the assistance of a computerized system is essential. Spoligotyping, for example, lends itself very well to computerization (e.g. spreadsheet), as the results are usually clear-cut.

In order to facilitate computer-based pattern matching, the experimental system used should yield a clear, unequivocal and reproducible pattern which can readily be scanned into a computer system and where the information may be manipulated by the operator. The optimal system will thus generate a well-separated and sharp multifragment pattern with bands of equal intensity. In this regard, the IS*6110*-based system is arguably the most suitable. A further and essential requirement is that a system of markers be incorporated so that comparisons of cultures examined at different times, and in distant locations is possible. A standardized protocol has been proposed (van Embden *et al.* 1993), which can be useful in examining not only local but national and global disease spread and dynamics. The running of external and internal standards which hybridize to other probes (Warren *et al.* 1996a,b) allows one to compensate for anomalous fragment migration during gel electrophoresis and Southern blotting. This allows the computer operator to assign a position to each fragment in a gel and load this into the database with maximum possible accuracy (Fig. 6.1).

However, despite computerization, there is a need for operator interpretation (the 'eye of the beholder'). The accuracy of band position and assignment is crucial, since the slightest misalignment or failure to enter the correct number and position of fragments in any pattern will imply that the accuracy of further pattern comparisons will be compromised. It is also essential that a degree of tolerance be built into the system with respect to fragment positions. This is, however, a source of error, since a high level of tolerance will compensate for relatively inaccurate assignment and will allow correct pattern matching, but will also lead to an overestimate in pattern matching if the patterns are non-identical, but related or similar (Fig. 6.1). When strains differ genotypically by minor electrophoretic band shift or by one band in a 23-fragment copy strain, too high a tolerance setting will give excessive clustering (see section 3.2) and yield incorrect results. Thus, the choice of method for pattern analysis must be carefully considered. In this regard, it is the author's opinion that the IS*6110*-based systems are the most user-friendly at present as an initial fingerprinting screening.

It is at this point that low copy number (IS*6110*) strains and the low copy number methods (e.g. DR), and others, such as PGRS, and $(GTG)_5$-based systems also demonstrate their weaknesses. The former have the problem of limited powers of discrimination and the latter systems generate hybridizing fragments that are not as easily distinguished or of such clear cut intensity as those found with IS*6110*. The problem inherent in these systems is that the assignment of patterns into a database is operator dependent and therefore subjective. Any inconsistency in operator assignment will therefore generate a database which will produce inaccurate results and conclusions.

2.5 Recommendations

Irrespective of the bacterial typing system used, pattern matching of cultures from different patients can be done. However, it is frequently the case that cultures appear to match with one probe, but show genotype differences using additional probes (Warren *et al.* 1996b; Yang *et al.* 1996; Berman *et al.* 1997; Goyal *et al.* 1997). This is particularly so for the low copy number IS*6110* cultures, illustrating the

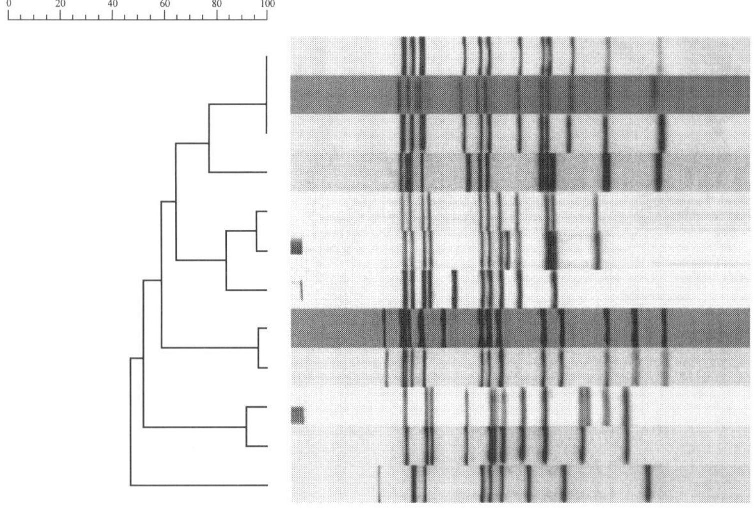

Fig. 6.1 Example of computerized pattern matching for cluster analysis of IS*6110* typed *Mycobacterium tuberculosis* strains from a community study. A portion of a large database is shown, with a similarity index indicated. Identical strains are shown by the horizontal lines at similarity index 100%. Note the minor position shift of some fragments which is acceptable for pattern matching at the tolerance limit set. A number of clonal variants can be seen.

importance of multiple probe analysis for enhancing molecular epidemiology.

The choice of analytical systems will also depend on the aim of the study. If the study comprises the examination of a limited number of cultures from what appears to be a small outbreak, then the use of one system may be adequate. However, if one wishes to examine a larger database of samples where clearly defined outbreaks are not evident, then multiprobe analysis is recommended. Should two forms of multiprobe analysis by computerization be required, the most suitable system is probably the double IS*6110* system (i.e. 5′ and 3′ probes). However, these probes will examine the same repeat elements at the same location in the genome and low copy number cultures will remain low copy, irrespective of whether one is using a 5′ or 3′ probe. The use of spoligotyping or a PGRS-like probe (Ross *et al.* 1992) or a (GTG)$_5$-based (MTB484(1)) (Warren *et al.* 1996b) probe offers additional powers of discrimination. Given that these probes are more problematical with respect to computerization, it is perhaps best to use the IS*6110* 3′ system in order to subgroup strains prior to doing secondary analysis. Once the initial grouping has been done, similar strains could be grouped on one electrophoretic system and a final visual analysis using additional probes is then possible. In all cases, markers are essential.

The use of the system chosen will essentially be dictated by the question asked, the scale of the study and the time scale addressed. However, if the study of tuberculosis is to assume global dimensions, a uniform system should be used. This problem has not yet been settled but possibly one of the most informative and user-friendly systems at this stage is an IS*6110* system followed by PGRS or MTB484(1).

3 Interpretation of results

3.1 Pattern matching

Once cultures of *M. tuberculosis* have been obtained from patients and subjected to genotyping followed by computerization and pattern matching, it is found that the cultures fall into two broad categories. One category consists of strains where any given strain has at least one other match of 100% identity, and the other broad category where an individual strain matches no other culture in the collection. Between these two extremes lie a large number of strains which bear some resemblance to one other in the culture collection which may vary from 99% similarity to no similarity (Fig. 6.1).

In order to establish the significance of identical cultures, a correlation with field work has been done in many cases. It has become evident from many studies that when two patients in a defined and relatively restricted area harbour the same strain, it can be fairly reasonably assumed that one patient has infected the other or they have been infected from a common source. In this case, it is also clear that the definition of identical must be carefully determined. In a small, low-incidence community, it is likely that cultures deemed identical using probes with even a fairly low discriminatory power represent a transmission chain. However, in a higher-incidence community the use of probes with higher discriminatory power is necessary. Field studies (contact tracing) in both high- and low-incidence communities have shown that 100% pattern matching using one system only often cannot be validated (Small *et al.* 1994; Warren *et al.* 1996a,b; Braden *et al.* 1997). However, the likelihood of contact and transmission increases substantially after confirmation of genotype match by multiple analytical systems. It has been possible in some studies to confirm contact in all cases within a small group of patients with identical strains.

The extent to which subcluster analysis should be done will essentially be dictated by the study objective, since it may theoretically be possible to define each culture as a different strain if the sequence of each genome was known. If pattern matching is less than 100%, non-identity is assumed. One should be aware that this does not guarantee absence of epidemiological relationship, since it has been shown in patients where reactivation has occurred, or in patients where tuberculosis persists over a long period, that cultures can be obtained which differ slightly from an original culture. It is clear in these cases that there is an epidemiological relationship and that changes in the genotype of the culture have occurred within one patient over a period of time.

This situation is probably relatively uncommon, but the quantitative effect on epidemiology is at this stage not known. Conversely, cultures taken from a large area or from geographically remote areas or over a long period of time may match, but will almost certainly not be epidemiologically linked (in terms of recent transmission (Braden *et al.* 1997)).

3.2 Cluster analysis

Cultures having at least one match within a community are assumed to represent cases where recent transmission is a cause of disease. Those strains which are unique are assumed to represent reactivation or latent disease. The proportion of recently transmitted cases can be calculated according to the formula [(number of clustered strains) – (number of clusters)]/total number of strains (Small *et al.* 1994). This formula has been invoked since it is thought that the index case for every cluster should be removed from the estimate, and gives an estimate of the extent of recently transmitted disease. However, in high-incidence communities it may be that a relatively large number of the same strain is in circulation, in which case it cannot necessarily be assumed that there is a single index case. It may therefore be argued that this formula is not always adequate.

It is generally accepted in clinical practice that when there is progression from infection to active disease within 2 years, the case is defined as recent transmission. It is thought therefore that the cluster estimate should perhaps be done within a 2-year window. However, the time period is contentious. A number of workers have reported that the extent of clustering rises sharply in the first few months of study and then levels off almost to a plateau with increasing time (Warren *et al.* 1999). This would suggest that there is incomplete sampling, resulting in underestimates of clusters and recent transmission. Given the natural history of the disease and the sector of the population in which the disease occurs in developed countries as well as the relatively low incidence of the disease, it is not surprising that the degree of clustering (representing recent transmission) is relatively low. However, it is curious that

independent studies in very different settings have produced similar estimates of transmission rate (Alland *et al.* 1994; Small & van Embden 1994; Small *et al.* 1994; Warren *et al.* 1996a), although it may vary considerably depending on the subgroup within a community (e.g. homeless people, institutionalized human immunodeficiency virus (HIV)-positive cases or prisons) (Koo *et al.* 1997; Moss *et al.* 1997). It is currently not understood why this is so and would suggest either that the experimental results are inadequate or that there is some aspect of the dynamics of the disease that is not yet understood.

If coverage in any study is less than 100%, there will be an underestimate of clustering. The results obtained from either random or non-random sampling, will also bias the estimate of clustering. Furthermore, the limited data available (given that tuberculosis is a long-term disease and that molecular epidemiology has not been in use for many years) suggests that the accumulative probability of a match over years does not increase linearly with time, but tends to plateau at a relatively low level (Warren *et al.* 1999). The process for the estimate of clustering thus needs revision: in addition to the adjustment required for index cases, adjustment for sampling, host migration, infection from outside the study community, new relapses, reinfection and current genome changes should possibly be taken into account. None of these problems has been adequately addressed as yet. In addition, a preanalysis period of at least 2 years may be required. This will allow for progression to active disease of new cases infected by primary index cases identified in the initial 2-year period and will increase cluster estimate accuracy.

4 Challenging dogmas

The study of tuberculosis by traditional epidemiological methods has produced much useful information, but has also been based on a number of assumptions. These have been made in spite of deficiencies in the tools available, e.g. casual contact is difficult to detect by field study. With the advent of molecular studies it is perhaps time that some of these hypotheses are checked and challenged if necessary. One of the assumptions is that the extent of recent transmission will be the determinant of incidence. Cluster analysis would suggest that this is not necessarily the case and that even in high incidence communities reactivation cases may dominate. A further presumption is that multiple case households are the result of transmission within that household. A study of 33 households in a high-incidence area would suggest that this is the case in no more than 50% of the households studied (Classen *et al.* 1999). However, where there is a close family relationship (e.g. mother–child) the transmission risk is much higher.

It has also long been thought that drug resistance largely reflects treatment failure and that the majority of cases of drug resistance are acquired resistance. Strain typing and analysis of genes involved in drug resistance and tuberculosis have recently shown that the majority of drug-resistant cases are transmitted and not acquired (Warren *et al.* 1997; A van Rie *et al.* 1999). These new findings illustrate how the advent of new tools can shed new light on the study of disease dynamics, which may have long-term implications for health. Complex traditional epidemiological studies can identify potential risk factors for tuberculosis (e.g. crowding and poor nutrition) and should be overlaid with studies on cluster analysis in order to ascertain more accurately the risk factors driving disease in any given community. It remains to be seen whether molecular epidemiology will revolutionize thinking concerning tuberculosis or confirm old beliefs.

5 Evolution, diversity, virulence and transposition

'You cannot go on being a good egg for ever. You must either hatch or rot.' (C. S. Lewis.)

The study of cultures obtained from different clusters may provide molecular biologists with clearly defined strains of different phenotype or challenge the notion of different phenotype (Victor *et al.* 1997). For example, it is possible that vertical transmission (e.g. from mother to offspring) may select for relatively benign organisms (low virulence), whereas

horizontal transmission (i.e. between unrelated hosts) may select for highly virulent organisms (Sorci *et al.* 1997). Phenotype (genotype) may also determine the pattern of disease (Cohen *et al.* 1995). Furthermore, it is possible that clonal expansion is indicative of success. Therefore, these studies could provide excellent source material for further studies on virulence, transmissibility or pathogenicity.

The number of individual strains of any given species of microorganism is quite astounding. This is particularly remarkable in *M. tuberculosis*, given that it is thought to be relatively recently evolved (perhaps only a few thousand years old) and that sequence analysis of individual genes or sections of the genome would suggest that the genome is conserved. There is a need for reconciliation between research results which suggest genome conservation in tuberculosis (sequence analysis) and the diversity of strains detected. At the same time, if we are to utilize different regions of the genome for studies in molecular epidemiology, it is important that we understand the rate of change of different genomic elements. It is essential that the probe uses target areas sufficiently stable to detect transmission, whereas it is also essential that this element evolves sufficiently that it provides us with adequate discriminatory powers.

Limited evidence suggests that different elements evolve at different rates, and that changes in one area of the genome may be independent of others. Many questions remain: for example, if one is utilizing a transposable element as a probe can one always assume that evolution proceeds by the acquisition of additional elements, or can there also be a decrease in copy number over time? We do not know whether there is only divergent evolution or whether convergent evolution is possible (via preferred insert sites). The latter could explain identical genotypes where there is no contact between patients. It is quite possible that most transposon inserts are stable but a random jump into an unstable area of the genome or into an active operon may create a hypermutable strain. In this case the progenitor may rapidly give rise to a large number of apparently unique strains in the community (J.W. Dale, personal communication 1997). This may be detectable if a number of probes are used to study the genome, although it is likely that recombinations in one area of the genome may well cause pattern variability which is then detected by other probes.

It is possible that the chronology of genome change could be established with the close collaboration of field workers and could be done in a well-defined area over a long period of time, because of the slow progression of this disease. A cluster analysis done in a community may also show the emergence of a new epidemic strain and thus illustrate evolution in action. It would advantageous to be able to pinpoint genome changes as recent or ancient in the context of epidemiology, but this is not yet possible. In a large database of strains (Fig. 6.2), where each strain is compared against the other, results suggest that there

has been substantial clonal expansion of some strains. It is likely that there is an epidemiological relationship within these families of strains, but they may reflect old chains of transmission. They may also represent virulent strains or those with some survival advantage (Bifani *et al.* 1996; Victor *et al.* 1997). It has not yet been possible to correlate this with traditional epidemiology and establish relationship by transmission, and therefore the significance and meaning of these families and whether they are indeed related and clonally expanded is as yet unconfirmed.

The rate of evolution of the species may be a function of opportunity, which will in turn be a function of variables such as the number of hosts experienced, the time spent resident in the host (active or latent disease; Wayne 1994), and migration of the host into

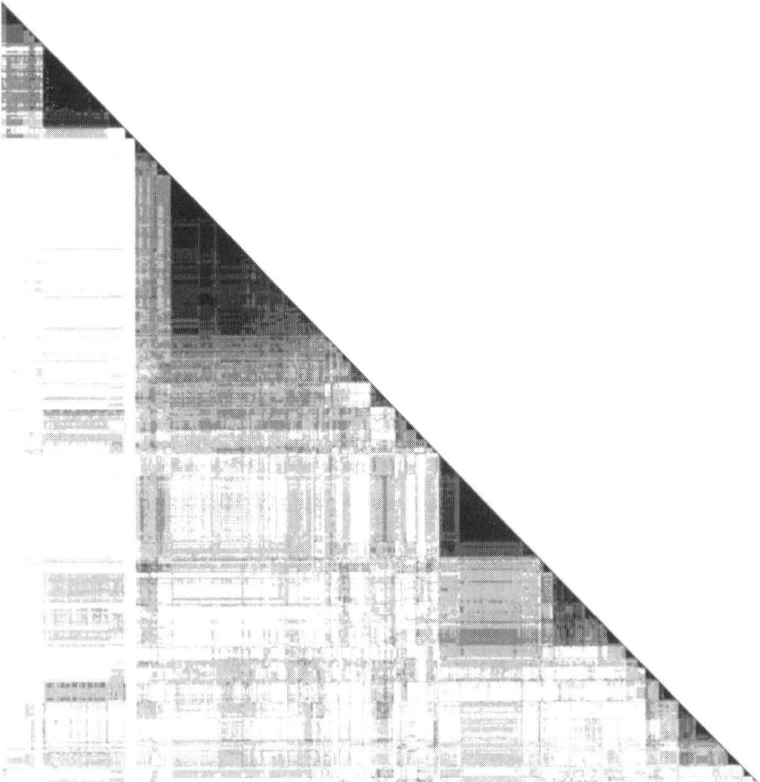

Fig. 6.2 Similarity matrix of a large database of strains, where each genotype is compared with every other. The intensity of the plot is indicative of the degree of similarity, with the darker shades and those close to the diagonal being most similar and representative of transmission chains (or clusters). The triangular regions may represent clonal expansion from progenitor strains.

a different population. The expansion of strains and clonal variants in a community can be represented as shown in Fig. 2.

It is possible that the colonization of a virgin population of hosts by *M. tuberculosis* could give rise to a different rate of evolution compared to colonization of a relatively resistant host population (where exposure to tuberculosis is extensive and extended). A comparison of *M. tuberculosis* cultures from relatively unexposed populations vs. those where tuberculosis has been resident for a long time may give us leads concerning the common ancestor or ancestral strains and the evolutionary process. The basic principle concerns the concept that clonal variants are likely to accumulate mutations as a function of their distance (in time) from a common ancestor (Austerlitz *et al.* 1997). The time will relate to the passage number (or host number). It could be that the strains surviving in low-incidence communities represent cultures closest to ancestral strains, or may also represent cultures which have adapted to a slow rate of transmission. Thus, low- and high-incidence areas as well as the location of disease in individual patients may represent unique niches for various members of the bacterial species and represent cultures which are phenotypically different. An abundance of mutations may have a quantitative effect on phenotype, but may not necessarily reduce fitness (Kashi *et al.* 1997).

It is interesting to note that there seems to be a limit to the number of loci of any given element in the genome of the organism. This may be due to spatial considerations, replicative burden, or other unknown factors. Thus, we see strains of *M. tuberculosis* with IS*6110* elements varying in copy number from 0 to 25. We may postulate that transposition would be expected to increase copy number, but it would seem that there must be a self-regulating limit.

One could also speculate that transposition will be a function of the copy number resident in any given strain. However, this seems not to be the case since there is no gradual increase in the proportion of strains containing an increasing copy number of IS*6110*. Presumably there is a selection process against transpositional mutations owing to the increasing likelihood of deleterious rearrangements

with increasing copy number. It has also been shown in other organisms that transposition of a transposable element may be independent of copy number (Vieira & Biemont 1997). Transposition or recombination events are essentially a natural experiment, since transposition into open reading frames (ORFs) may easily produce natural knock-out strains. Conversely, transposition into promotor or enhancer regions can effect phenotype quantitatively. These transpositional events may lead to either the disappearance of some strains or clonal expansion from successful genotypes (Fig. 6.2).

Until we can clearly define which areas of the genome are conserved, how genetic (and phenotypic) novelty is produced, and how 'simple' progenitors may give rise to more complex or different descendants, we cannot ascertain the effect of evolution on molecular epidemiology. The study of evolution of *M. tuberculosis* is likely to be of considerable importance for molecular epidemiology, since the effect of evolution on cluster analysis would result in an underestimate of recent transmission.

6 Dynamics of disease: implications for health policy

While cluster analysis can be extremely useful in an outbreak situation to identify point sources of the disease and routes of transmission, it can also be useful on a broader scale in studying the dynamics of disease in communities, within countries, as well as globally. The process of examining disease dynamics within a community and predicting the cause and outcome of an epidemic is essentially mathematically based (Blower *et al.* 1996; Brewer *et al.* 1996; De Leo & Dobson 1996; Garcia *et al.* 1997). These mathematical models offer a useful way to examine and predict population dynamics. However, the current problem with any model is the lack of hard data. With respect to tuberculosis modelling, it is an accurate estimate of the current, past and future transmission rates that is absent. Therefore, any methodology that assists with the estimation of transmission rate will be essential for the estimation of the reproductive rate of the organism and the accurate modelling of the epidemic. These theorems are useful for investigating control programmes and policies which will include protocols for therapy and vaccination. Any modelling and planning for a new control strategy can thus be knowledge based rather than based on assumptions, provided that molecular epidemiological techniques can give a proper quantitative estimate and contribute substantially to the understanding of disease dynamics.

New plans and policies with treatment modalities are needed to combat tuberculosis. There is currently a worldwide movement to utilize directly observed therapy (DOT) for tuberculosis control. This programme being advocated by the World Health Organization as a panacea has drawn much support and criticism. Proper quantitative assessment must be done in order to assess the effect of new control programmes such as DOTs. Molecular epidemiology is arguably our most quantitative tool and could provide accurate estimates of the effect of new therapies, thus assisting with planning and evaluation disease control. It can also not be assumed that results from one study area can be extrapolated to any other area, since geographical differences exist (Hermans *et al.* 1995; Warren *et al.* 1996a,b; Dobner *et al.* 1997). Programme modifications for local area requirements will almost certainly be required (Stead 1997).

7 Conclusion

Molecular epidemiology is the application and integration of molecular biology with traditional epidemiology to aid in the identification and monitoring of pathogens and dissemination of disease in communities and populations. These techniques make unequivocal identification of cultures possible. The spread of *M. tuberculosis* within small communities, within countries, across national borders and globally has been detected and traced. This information has been used in some cases to effectively locate index cases and prevent further spread of disease. Thus far, it has not provided significant input for policy and planning, although it is likely to do so in time. Not only does it make the dissection of past disease dynamics possible, but it makes the evaluation of his-

torical control policies possible, it allows the fairly rapid evaluation of current policies and thus provides input for improvements for future control. This can be done not only at local community level, but also internationally. Such information will enable both local authorities and international organizations to optimize efforts to control tuberculosis. Given the long-term nature of tuberculosis, this will become increasingly important as it is recognized that current control measures may not be optimal in all areas of the globe.

This branch of science has been developed only recently, and while we have learnt a great deal, many questions remain unanswered. As our interpretation of results improves, our data will become more reliable and provide accurate and assumption-free information to improve control of tuberculosis.

8 Acknowledgements

The MRC (SA), the University of Stellenbosch, Tygerberg Hospital, GlaxoWellcome Action TB Initiative and NIH grant RO1 A135265–03 for financial support. The European Union contract Biomed1-BMH1-CT93–1614 assisted with computer analysis. I would also like to thank my colleagues for many hours of stimulating discussion and exciting collaboration, particularly Drs R. Warren and J. Hauman, I. Wiid, W. Bourn, T. Victor, N. Beyers, A. van Rie, Prof. P. Donald, Ms. M. Richardson, S. Sampson, C. Classen. Finally, I would like to thank Dr Eileen van Helden for many forms of assistance not the least of which is editorial assistance in completing this chapter.

9 References

Alland, D., Kalkut, G.E., Moss, A.R. *et al.* (1994) Transmission of tuberculosis in New York City—an analysis by DNA fingerprinting and conventional epidemiological methods. *New England Journal of Medicine* **330**, 1710–1716.

Austerlitz, F., JungMuller, B., Godelle, B. & Gouyon, P.H. (1997) Evolution of coalescence times, genetic diversity and structure during colonization. *Theoretical Population Biology* **51**, 148–164.

Berman, W.J., Reves, R.R., Hawkes, A.P. *et al.* (1997) DNA fingerprinting with two probes decreases clustering of *Mycobacterium tuberculosis*. *American Journal of Respiratory and Critical Care Medicine* **155**, 1140–1146.

Bifani, P.J., Plikaytis, B.B., Kapur, V. *et al.* (1996) Origin and interstate spread of a New York city multidrug-resistant *Mycobacterium tuberculosis* clone family. *Journal of the American Medical Association* **275**, 452–457.

Blower, S.M., Small, P.M. & Hopewell, P.C. (1996) Control strategies for tuberculosis epidemics: new models for old problems. *Science* **273**, 497–500.

Braden, C.R., Templeton, G.L., Cave, M.D. *et al.* (1997) Interpretation of restriction fragment length polymorphism analysis of *Mycobacterium tuberculosis* isolates from a state with a large rural population. *Journal of Infectious Diseases* **175**, 1446–1452.

Brewer, T.F., Heymann, S.J., Colditz, G.A. *et al.* (1996) Evaluation of tuberculosis control policies using computer sumulation. *Journal of the American Medicine Association* **276**, 1888–1903.

Classen, C.N., Warren, R., Richardson, M. *et al.* (1999) Impact of social interactions in the community on the transmission of tuberculosis in a high incidence area. *Thorax* **54**, 136–140.

Cohen, J., Culotta, E. & Glanz, J. (1995) Differences in HIV strains may underlie disease patterns. *Science* **270**, 30–32.

Dale, J.W., Tang, T.H., Wall, S., Zainuddin, Z.F. & Plikaytis, B. (1998). Conservation of IS*6110* sequence in strains of *Mycobacterium tuberculosis* with single and multiple copies. *Tubercle and Lung Disease* **78**, 225–227.

De Leo, G.A. & Dobson, A.P. (1996) Allometry and simple epidemic models for microparasites. *Nature* **379**, 720–722.

Dobner, P., Bretzel, G., Rusch-Gerdes, S., Zeldmann, K., Rifai, M., Loscher, T. & Rinder, H. (1997) Geographic variation of the predictive values of genomic mutations associated with streptomycin resistance of *Mycobacterium tuberculosis*. *Molecular Cell Probes* **11**, 123–126.

van Embden, J.D.A., Cave, M.D., Crawford, J.T. *et al.* (1993) Strain identification of *M. tuberculosis* by DNA fingerprinting: recommendation for a standardized methodology. *Journal of Clinical Microbiology* **31**, 406–409.

Garcia, A., Maccario, J. & Richardson, S. (1997) Modelling the annual risk of tuberculosis infection. *International Journal of Epidemiology* **26**, 190–203.

Goyal, M., Saunders, N.A., van Embden, J.D.A., Young, D.B. & Shaw, R.J. (1997) Differentiation of *Mycobacterium tuberculosis* isolates by spoligotyping and IS*6110* restriction fragment length polymorphism. *Journal of Clinical Microbiology* **35**, 647–651.

Goyal, M., Young, D., Zhang, Y., Jenkins, P.A. & Shaw, R.J. (1994) PCR amplification of variable sequences upstream of *kat*G gene to subdivide strains of *Mycobacterium tuberculosis* complex. *Journal of Clinical Microbiology* **32**, 3070–3071.

Groenen, P.M., Bunschoten, A.E., van Soolingen, D. & van Embden, J.D.A. (1993) Nature of DNA polymorphism in the direct repeat cluster of *M. tuberculosis*, application for strain differentiation by a novel typing method. *Molecular Microbiology* **10**, 1057–1065.

Haas, W.H., Butler, W.R., Woodley, C.L. & Crawford, J.T. (1993) Mixed-linker polymerase chain reaction: a new method for rapid fingerprinting of isolates of the *Mycobacterium tuberculosis* complex. *Journal of Clinical Microbiology* **31**, 1293–1298.

Hermans, P.W.M., Messadi, F., Guebrexabher, H. *et al.* (1995) Analysis of the population structure of *Mycobacterium tuberculosis* in Ethiopia, Tunisia and the Netherlands: usefulness of DNA typing for global tuberculosis epidemiology. *Journal of Infectious Diseases* **171**, 1504–1513.

Hermans, P.W.M., van Soolingen, D. & van Embden, J.D.A. (1992) Characterization of a major polymorphic tandem repeat in *Mycobacterium tuberculosis* and its potential use in the epidemiology of *Mycobacterium kansasii* and *Mycobacterium gordonae*. *Journal of Bacteriology* **174**, 4157–4165.

Kamerbeek, J., Schouls, L., Kolk, A. *et al.* (1997) Simultaneous detection and strain differentiation of *Mycobacterium tuberculosis* for diagnosis of epidemiology. *Journal of Clinical Microbiology* **35**, 907–914.

Kashi, Y., King, D. & Soller, M. (1997) Simple sequence repeats as a source of quantitative genetic variation. *Trends in Genetics* **13**, 74–78.

Koo, D.T., Baron, R.C. & Rutherford, G.W. (1997) Transmission of *Mycobacterium tuberculosis* in a California State Prison, 1991. *American Journal of Public Health* **87**, 279–282.

Mazurek, G.H., Cave, M.D., Eisenach, K.D., Wallace, R.J. Jr, Bates, J.H. & Crawford, J.T. (1991) Chromosomal DNA fingerprint patterns produced with IS*6110* as strain-specific markers for epidemiologic study of tuberculosis. *Journal of Clinical Microbiology* **29**, 2030–2033.

Moss, A.R., Alland, D., Telzak, E. *et al.* (1997) A city-wide outbreak of multiple-drug-resistant strain of *Mycobacterium tuberculosis* in New York. *International Journal of Tubercle Lung Disease* **1**, 115–121.

Ross, B.C., Raios, K., Jackson, K. & Dwyer, B. (1992) Molecular cloning of a highly repeated DNA element from *M. tuberculosis* and its use as an epidemiological tool. *Journal of Clinical Microbiology* **30**, 942–946.

Small, P.M., Hopewell, P.C., Singh, S.P. *et al.* (1994) The epidemiology of tuberculosis in San Francisco. *New England Journal of Medicine* **330**, 1703–1708.

Small, P.M. & van Embden, J.D.A. (1994) Molecular epidemiology of tuberculosis. In: *Tuberculosis: Pathogenesis, Protection and Control* (ed. B. R. Bloom). Washington, DC: ASM Press, pp. 569–582.

Sorci, G., Moller, A.P. & Boulinier, T. (1997) Genetics of host–parasite interactions. *Ecology and Evolution* **12**, 196–200.

Stead, W.W. (1997) The origin and erratic global spread of tuberculosis. *Clinical Chest Medicine* **18** (1), 65–77.

Van Soolingen, D., de Haas, P.E.W., Herman, P.W.M., Groenen, P.M.A. & van Embden, J.D.A. (1993) Comparison of various repetitive DNA elements as genetic markers for strain differentiation and epidemiology of *Mycobacterium tuberculosis*. *Journal of Clinical Microbiology* **31**, 1987–1995.

Van Soolingen, D., de Haas, P.E.W., Hermans, P.W.M. & van Embden, J.D.A. (1994) DNA fingerprinting of *Mycobacterium tuberculosis*. *Methods in Enzymology* **235**, 196–205.

Van Soolingen, D., Qian, L.S., De Haas, P.E.W. *et al.* (1995) Predominance of single genotype of *Mycobacterium tuberculosis* in countries of East Asia. *Journal of Clinical Microbiology* **33**, 3234–3238.

Victor, T.C., Warren, R., Beyers, N. & van Helden, P.D. (1997) Transmission of multidrug-resistant strains of *Mycobacterium tuberculosis* in a high incidence community. *European Journal of Clinical Microbiological Infection and Disease* **16**, 548–549.

Vieira, C. & Biemont, C. (1997) Transposition rate of the 412 retrotransposable element is independent of copy number in natural populations of *Drosophila simulans*. *Molecular Biological Evolution* **14**, 185–188.

Warren, R., Hauman, J., Beyers, N. *et al.* (1996a) Unexpectedly high strain diversity of *Mycobacterium tuberculosis* in a high-incidence community. *South African Medicine Journal* **86**, 45–49.

Warren, R., Richardson, M., Sampson, S. *et al.* (1996b) Genotyping of *M. tuberculosis* with additional markers enhances accuracy in epidemiological studies. *Journal of Clinical Microbiology* **34**, 2219–2224.

Warren, R., Richardson, M., van der Spuy, G. *et al.* (1999) DNA fingerprinting and molecular epidemiology of tuberculosis: use and interpretation in an epidemic setting. *Electrophoresis* (in press).

Warren, R.M., Shah, S.S. & Alland, D. (1997) Multiple drug resistance: a world-wide threat. *Baillière's Clinical Infectious Diseases: Mycobacterial Diseases*, part 1 (ed. A. Malin & K. P. W. J. McAdam). Baillière Tindall, pp. 77–96.

Wayne, L.G. (1994) Dormancy of *Mycobacterium tuberculosis* and latency of disease. *European Journal of Clinical Microbiological Infection and Disease* **13**, 908–914.

Wiid, I.J.F., Werely, C., Beyers, N., Donald, P. & van Helden, P.D. (1994) Oligonucleotide (GTG)$_5$ as a marker for *Mycobacterium tuberculosis* strain identification. *Journal of Clinical Microbiology* **32**, 1318–1321.

Yang, Z.H., Chaves, F., Barnes, P.F. *et al.* (1996) Evaluation of method for secondary DNA typing of *Mycobacterium tuberculosis* with pTBN12 in epidemiologic study of tuberculosis. *Journal of Clinical Microbiology* **34**, 3044–3048.

Chapter 7 / Molecular epidemiology:
Mycobacterium bovis

DESMOND M. COLLINS

1 Introduction

Until 15 years ago, strain differentiation of mycobacterial species was difficult and in the case of *Mycobacterium bovis* virtually impossible. Methods such as phage susceptibility (Redmond *et al.* 1979), amino acid uptake, pyrolysis mass spectrometry, high-performance liquid chromatography and polyacrylamide gel electrophoresis of proteins provide very little discrimination (Olson *et al.* 1995). Only molecular techniques that utilize DNA features have enabled the production of good typing systems for *M. bovis* that can be used as epidemiological tools. Because these discriminating DNA methods are based on the detection of DNA differences, the terms DNA fingerprint and restriction fragment length polymorphism (RFLP) have become commonplace in typing schemes for *M. bovis*. This chapter will examine the history, current use and future prospects of these techniques as they apply to *M. bovis*. While the discussion will be very largely devoted to the techniques, it must be borne in mind that the major benefit from their use only occurs when they are used together with the results of classical epidemiological studies.

Tuberculosis caused by *M. bovis* is often referred to as bovine tuberculosis because it is an important infectious disease of cattle. This epithet is somewhat misleading as the organism infects a wide range of different mammalian hosts including humans in whom it can cause a disease that is indistinguishable from that due to *M. tuberculosis*. Although *M. bovis* usually causes less than 5% of the cases of human tuberculosis (O'Reilly & Daborn 1995), the tuberculosis problem is so huge that this zoonotic aspect of *M. bovis* combined with the importance of the global cattle industry makes *M. bovis* an organism of considerable importance.

M. tuberculosis and *M. bovis* are genetically very closely related (Imaeda 1985; Sreevatsan *et al.* 1997) and contain the same repetitive elements, but the historical development and current use of DNA fingerprinting methods for the two species display considerable differences for several major reasons. First, the only useful DNA fingerprinting method for *M. bovis* between 7 and 15 years ago was restriction endonuclease analysis (REA) (Collins & de Lisle 1984; 1985). Although this method works equally well for *M. tuberculosis* (Collins & de Lisle 1984) it has not been used in any significant way for that species. Second, the major method of DNA fingerprinting for *M. tuberculosis* is based on RFLPs of the insertion sequence IS*6110* which is present in five to 20 copies in most *M. tuberculosis* strains (see Chapters 2 and 6). In contrast, IS*6110* is only present in one copy in the majority of *M. bovis* strains and its use has played a relatively minor role in DNA fingerprinting of *M. bovis*. Third, most DNA fingerprinting of *M. bovis* has been on strains from animals where the epidemiological problems being addressed are somewhat different from those in human populations. Where the technique has been applied to strains of *M. bovis* from humans, the focus of the study has been to determine which animals are the source of infection or occasionally whether the spread of infection has been from animal to human or human to human. This is different from the situation with *M. tuberculosis* where the source of infection is always other humans.

2 Restriction endonuclease analysis

Bacterial REA or restriction fragment analysis as it is sometimes called was developed about 1980 (Marshall *et al.* 1981). In the basic method, DNA from a selected strain is digested with a suitable restriction endonuclease and the fragments produced are separated on the basis of size by electrophoresis for a few hours on 200 mm long submerged agarose gels. The pattern of fragments, or DNA fingerprint, is visualized by staining the gel in ethidium bromide and recording on film the fluorescence that occurs when the DNA fragments are exposed to ultraviolet light. The method has the potential, which has been realized in many cases, to produce a typing system for any bacterial species. In practice, the large number of fragments produced and difficulties in adequately separating them made it difficult to use this approach for typing *M. bovis*.

In the early 1980s we adapted the basic method by using a longer agarose gel, a longer running time, recirculating the buffer, and cooling the electrophoresis tank (Collins & de Lisle 1984; 1985; Collins *et al.* 1993). This enabled us to achieve good separation of the larger DNA fragments and to clearly observe a small number of fragment differences between different strains. The larger molecular size regions of two routine gel fingerprints of DNA from New Zealand strains of *M. bovis* are shown in Fig. 7.1. Most of the clearly observable differences between these strains are seen in the largest 10–15 fragments. In our typing system, a DNA sample is digested separately with each of the enzymes *Bst*EII, *Pvu*II and *Bcl*I. The type assigned is based on a combined result from all three restriction patterns. The most discriminating enzyme is *Bst*EII and use of the other two enzymes increases the number of types discriminated by about 90%. Over the last 15 years this method has been used to discriminate more than 1600 New Zealand strains of *M. bovis* into 165 different types. While strains of *M. bovis* are genetically so

Fig. 7.1 The larger fragment region of REA patterns after *Bst*EII digestion of DNA from *Mycobacterium bovis* isolated from 18 animals in New Zealand. There are 11 different patterns; the same patterns are shared by strains 3, 14,15; strains 4, 5; strains 8, 9 16; strains 11, 13, 18; other strain patterns are unique.

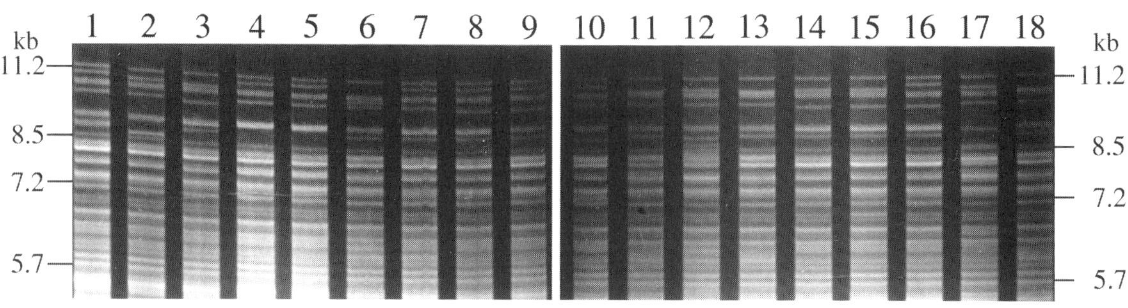

similar that there are few restriction fragment pattern differences between any two types, these differences appear very stable. Our knowledge of their stability is based on repeated *in vitro* culture of some strains and cumulative REA results over many years; in New Zealand, restriction types are geographically localized and there are many occasions on which *M. bovis* strains with the same restriction type have been isolated from wild animals in the same area over a 12–15-year period.

This method has been used by us in a number of epidemiological studies of bovine tuberculosis in New Zealand (Collins *et al.* 1986, 1988, 1993, 1994a; de Lisle *et al.* 1990, 1995) and other countries (Cousins *et al.* 1990; Collins *et al.* 1994b; Bolske *et al.* 1995) and was also used to characterize different strains of *M. bovis* bacille Calmette–Guérin (BCG) (Collins & de Lisle 1987). In our hands the technique has proved relatively robust but, while a few other groups have tried REA for *M. bovis* or *M. tuberculosis* (Imaeda 1985; Patel *et al.* 1986; Grange *et al.* 1990; Cousins *et al.* 1993), the technique has not gained general acceptance. This is partly because none of these groups employed the long gel format or in most cases the other modifications we espoused, and they were unable to achieve a satisfactory degree of fragment resolution. It is noteworthy that the group (Cousins *et al.* 1993) that observed most differences between strains of *M. bovis* also used procedures that were most similar to our own.

2.1 Pulsed-field gel electrophoresis

A method related to REA, pulsed-field gel electrophoresis (PFGE), which uses rare cutting restriction endonucleases and separates very large DNA fragments in the range 10–800 kb, was used successfully to discriminate different strains of *M. bovis* BCG (Zhang *et al.* 1995) into many different restriction types. Not surprisingly, PFGE is also capable of good discrimination between clinical *M. bovis* strains (Feizabadi *et al.* 1996; R. J. Wallace Jr, personal communication 1999), although not all studies have been successful (Olson *et al.* 1995). As with our modified REA technique, specialized equipment and technical experience are required for PFGE and this may limit its widespread use for *M. bovis*.

3 Repetitive elements used for typing

Repetitive DNA elements can be ideal features on which to base a DNA fingerprinting method. This is because the techniques of DNA isolation, restriction enzyme digestion, gel electrophoresis, Southern blotting onto nylon or nitrocellulose and probe hybridization that are required to produce DNA fingerprints are well established in all molecular biology laboratories. Provided copies of the repetitive element rarely shift from their chromosomal sites over the time frame of most epidemiological studies but are sufficiently mobile to shift over a longer time frame and have a varying copy number between about four and 30, it is relatively simple to use them as the basis of a good typing system.

3.1 IS*6110*

The insertion sequence IS*6110* (also known as IS*986* and IS*987*) largely fulfils these requirements for *M. tuberculosis* and has become the typing system of choice for that species (see Chapter 6). Unfortunately, in many studies, over 80% of *M. bovis* strains (more than 97% of strains in New Zealand) contain only a single copy of IS*6110* which is inserted at the same chromosomal site. Interestingly, this is the same site at which IS*6110* is inserted in the rare *M. tuberculosis* strains that only have a single copy of the element (Fomukong *et al.* 1994). Superficially, it would appear that IS*6110* would be of no use for typing the majority of *M. bovis* strains but in fact a moderate degree of strain differentiation can be achieved with this element (Fig. 7.2) due to the fact that the single copy of IS*6110* is inserted into a region of the chromosome called the direct repeat (DR) region which contains a large but variable number of 36-bp DNA sequences separated by 35–41-bp spacer regions (Hermans *et al.* 1991). This strain variability in the number of repeats can alter the size of the frag-

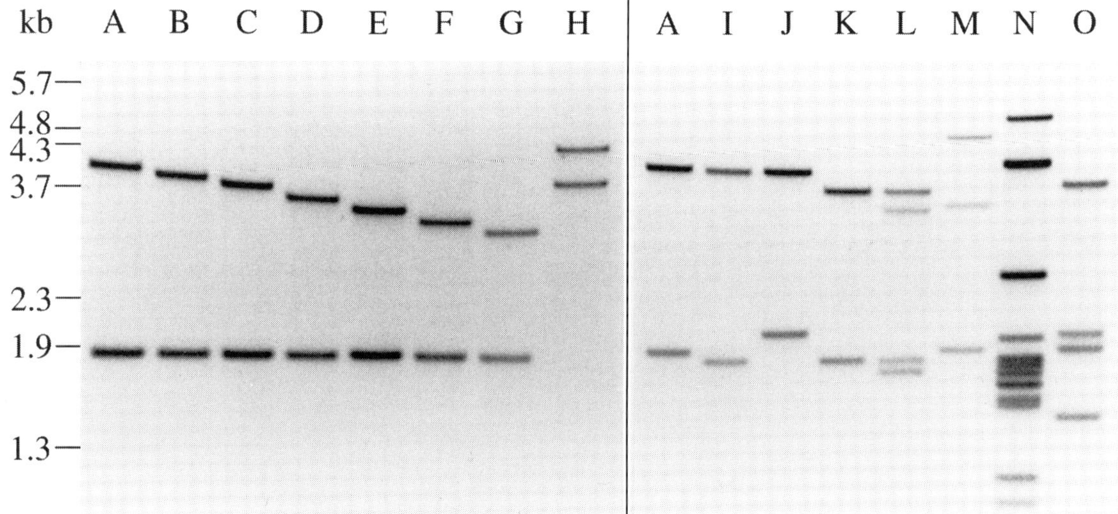

Fig. 7.2 Southern blot hybridization of IS*6110* probe to DNA from 15 strains of *M. bovis* isolated from animals in New Zealand. (From Collins *et al*. 1993, with permission.)

ments that hybridize to the IS*6110* probe because each fragment contains some of the DR elements in the flanking sequence as well as part of IS*6110*. There is a single site for the restriction enzyme *Pvu*II about one-third of the way through IS*6110* and when DNA from *M. bovis* is digested with this enzyme and the whole insertion sequence is used as a probe, two fragment lines are hybridized for each copy of IS*6110* in the chromosome. Use of the whole insertion sequence as a probe gives much better discrimination between *M. bovis* strains than does the standard method for *M. tuberculosis* which employs just one-half of IS*6110* as a probe (Collins *et al*. 1993; Skuce *et al*. 1994; Liebana *et al*. 1997). This is because the side of IS*6110* used as a probe for *M. tuberculosis* would detect only the lower fragment in Fig. 7.2, lanes A–K, and would not therefore discriminate between strains A–G. In a few countries, *M. bovis* strains with multiple copies of IS*6110* appear to be common (Rigouts *et al*. 1996; Liebana *et al*. 1997) and in these situations IS*6110* typing is more useful. However, the stability of RFLPs due to multiple copies of IS*6110* in *M. bovis* has been questioned (Rigouts

et al. 1996) and further work is required to establish their reliability.

3.2 Direct repeat region probe

The DR region itself is also used as a DNA probe for typing *M. bovis* and is reported to provide similar (Skuce *et al*. 1996) or better (Romano *et al*. 1996) discrimination between *M. bovis* strains than does IS*6110* typing. Because there is less than 4 kb of DR region (in the case of *M. bovis* BCG, 2.1 kb on one side of IS*6110* and 1.5 kb on the other) (Hermans *et al*. 1991) and the six-base cutting restriction enzyme *Pvu*II cuts rarely, more RFLPs are observed with the more frequent four-base cutting *Alu*I (Romano *et al*. 1996). This has recognition sites in some of the spacer sequences between the DR regions and has become the enzyme of choice for DR typing. An example of DR fingerprints for 14 *M. bovis* strains is given in Fig. 7.3.

It is important to realize, particularly for strains that have a single copy of IS*6110*, that the underlying genetic event that is being detected with the DR probe (such as the loss of several of the repeat units) is in some cases the same as that detected by the IS*6110* probe. In practice, it is difficult to quantify the relative typing abilities of the two probes. First, some

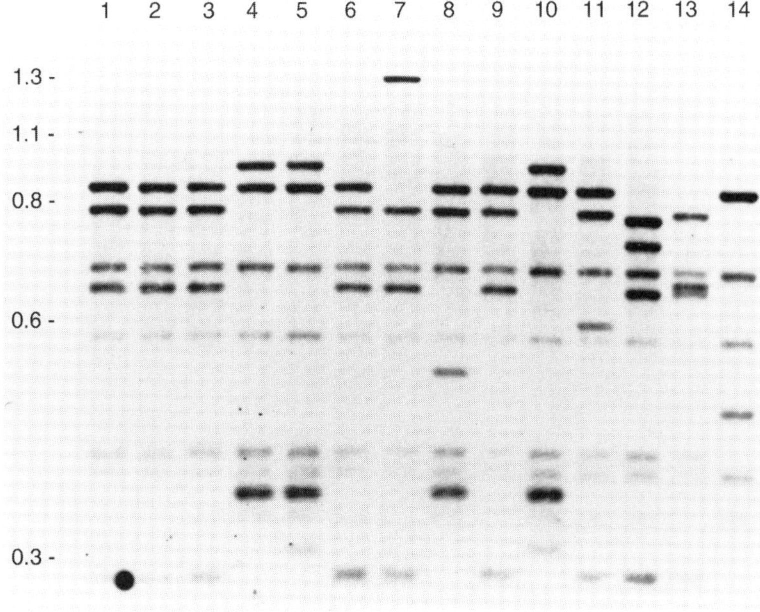

Fig. 7.3 Southern blot hybridization of DR probe to DNA from strains of *M. bovis* isolated from cattle (Lanes 1,2) and humans (Lanes 3–14) in The Netherlands. (From van Soolingen *et al.* 1994, with permission.)

groups use only half of IS*6110* as a probe and miss detecting any variation occurring in the flanking sequence associated with the other half (Romano *et al.* 1996). Second, while it is possible to observe IS*6110* RFLPs that are due to loss of one or two DR units (such as the differences in Fig. 7.2 in the upper fragments between lanes A and B, or between lanes F and G), these differences are too fine to be detected under the conditions used in some laboratories. It is probably best therefore to regard IS*6110* and DR typing as methods that have some degree of overlap but that together provide more discrimination than either method separately.

3.3 Polymorphic GC-rich sequence probe

The most discriminating repetitive element for typing *M. bovis* is the polymorphic GC-rich sequence (PGRS) discovered independently by Ross *et al.* (1992) and Doran *et al.* (1993). This consists of 26 or more clusters of 24-bp DRs widely spread through the chromosome (Poulet & Cole 1995). Unlike the IS*6110* sequence which has little variation between copy

sequences, the PGRS is a consensus sequence (Poulet & Cole 1995). Some PGRS copies differ more than others and some clusters contain more copies than others, and because of this there is more variation in the intensity of hybridized fragment lines than for IS*6110* fingerprints. In addition, because there are always many copies of PGRS, and some RFLPs only differ slightly, it is important to use electrophoresis and blotting techniques that give high resolution. An example of high-resolution PGRS fingerprints for DNA from 14 different strains is shown in Fig. 7.4. These patterns are for DNA from the same 14 strains as those in Fig. 7.3.

3.4 IS*1081*

The only other repetitive element that has been fully evaluated for its ability to type strains of *M. bovis* is the insertion element IS*1081*. This element is present in six copies in most *M. bovis* strains but shows very few RFLPs and is of little use for epidemiology (van Soolingen *et al.* 1992; Collins *et al.* 1993; Skuce *et al.* 1994) except for distinguishing *M. bovis* BCG

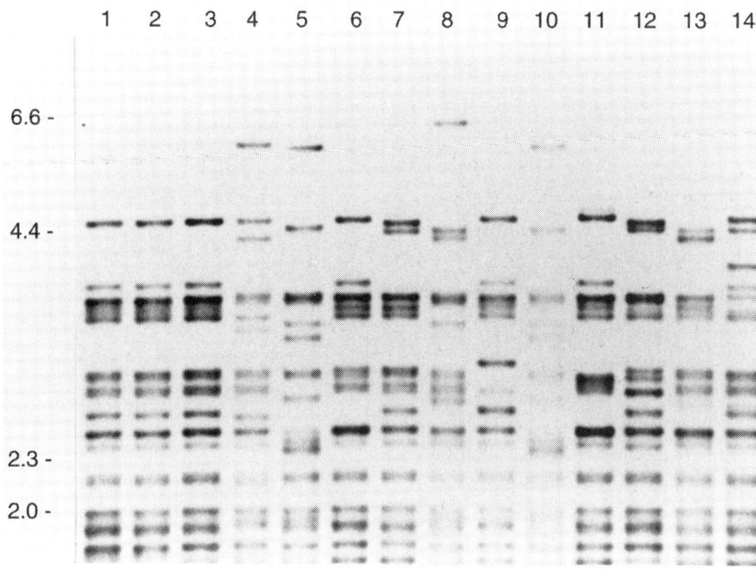

Fig. 7.4 Southern blot hybridization of polymorphic GC-rich sequence (PGRS) probe to DNA from the same strains as those in Fig. 7.3. (From van Soolingen *et al.* 1994, with permission.)

from most other *M. bovis* strains (van Soolingen *et al.* 1992).

4 Polymerase chain reaction-based typing methods

Typing methods based on either REA or hybridization with probes of repeated DNA elements require relatively large amounts of DNA that can only be obtained by secondary culture and this delays the typing of strains by some weeks. In contrast, PCR-based methods require little DNA and have the potential to be performed on DNA from primary culture or, as recently reported for *M. tuberculosis* (Kamerbeek *et al.* 1997), even directly from highly infected clinical samples. A number of PCR typing techniques have been developed for *M. bovis* or *M. tuberculosis* (Frothingham 1995; Beggs *et al.* 1996; Butcher *et al.* 1996; Glennon *et al.* 1997) but these have so far been applied to few *M. bovis* strains (Beggs *et al.* 1996; Glennon *et al.* 1997), provide no discrimination (Frothingham 1995) or rely on the presence of multiple copies of IS*6110* and would not therefore be expected to be useful for most *M. bovis* strains. An important factor that must be borne in mind in assessing any PCR-based method is that PCR is subject to artifactual problems that are often understated or unrecognized and it is often difficult to establish reproducible techniques (Tyler *et al.* 1997).

4.1 Spoligotyping

The only PCR approach reported to give reasonable discrimination of *M. bovis* strains is one based on measuring the presence or absence of each individual spacer sequence in the DR region. In the current version of the method, called spacer oligotyping (spoligotyping), the entire DR region is amplified by PCR and the products are used as a probe for hybridization to an array of 43 oligonucleotides each of which is specific to one spacer sequence in the DR region (Kamerbeek *et al.* 1997). Each spoligotype is a combination of the presence or absence of hybridization signal to each of the 43 oligonucleotides.

Whether fingerprinting with a DR probe discriminates only a subset of spoligotypes, or in addition also detects some types independently, depends on whether there is any variation between strains in the flanking sequences at each end of the DR region or whether individual spacer regions have mutations that cause or remove an *Alu*I site. In practice, DR typing distinguished between some *M. bovis* BCG daughter strains that spoligotyping did not distinguish (Goguet de la Salmoniere *et al.* 1997; Howard *et al.* 1997). However, since the immediate sources of the strains were not clear, and since BCG strains of the same daughter type from different sources are sometimes genetically different (Zhang *et al.* 1995), it is not possible to conclude from these studies which of the techniques is more discriminating. A very recent study of a large number of *M. bovis* strains found that DR typing and spoligotyping were equally discriminating (Cousins *et al.* 1998). In most cases, both methods discriminated strains into the same types but each of these methods distinguished between some strains that were not distinguished by the other method. Because of the substantial overlap of the two methods, they are treated as a single approach for comparison purposes below.

5 Comparison of techniques

Comparisons of various combinations of the four established typing systems for characterizing *M. bovis* strains (based on REA, IS*6110*, PGRS and DR/spoligotyping) have been made in a number of studies (Collins *et al.* 1993; Cousins *et al.* 1993; Skuce *et al.* 1994; van Soolingen *et al.* 1994; Collins & de Lisle 1995; Aranaz *et al.* 1996; Romano *et al.* 1996; Skuce *et al.* 1996). While different study designs and groups of strains were used, and in some cases optimal methods were not used for some typing systems, a number of conclusions can nevertheless be made.

Although most groups have chosen to use technically simpler methods, the best discrimination between *M. bovis* strains is achieved by REA. If the technique is performed optimally it provides better discrimination among most *M. bovis* strains (those with single copies of IS*6110*) than a combination of the three other established methods.

Apart from REA, none of the other established techniques gives sufficient discrimination to be used alone as a good general typing system.

When laboratory or economic resources are limited, typing methods based on any one of PGRS, IS*6110* or DR/spoligotyping by itself may provide sufficient information for some useful epidemiological conclusions to be made. However, it will be safer to draw conclusions from strains that have different fingerprints than from those that have the same fingerprint. This is because use of additional methods will frequently show that strains that appear identical by one method are in fact different.

If most of the strains in a particular study have a single copy of IS*6110*, then PGRS or DR/spoligotyping will provide more discrimination that IS*6110* typing.

In a few situations, most or a large number of the *M. bovis* strains being typed will contain multiple copies of IS*6110*. In these situations the advantages of using IS*6110* typing increase and the benefit from employing other additional methods is reduced.

Apart from the conclusions above, three further factors must be taken into account when establishing a system for typing *M. bovis*: cost, time and data comparison. While typing based on REA, IS*6110*, DR and PGRS share a common requirement for preparation of moderate amounts of DNA, it is more expensive and takes considerably more laboratory time to routinely perform several methods on every sample rather than one method. Affordability of the best typing systems will be a major issue for many laboratories who in practice may utilize a less discriminating typing system on grounds of cost. In addition, all these methods provide a result some considerable time after primary culture. This elapsed time from sampling to typing is not of concern for most applications but if a faster result is required then spoligotyping is currently the only established method available. Finally, the current literature on typing *M. bovis* strains does not reflect the difficulty of comparison of DNA fingerprints because few laboratories have yet reached the stage of typing many hundreds or thousands of *M. bovis* strains. As more strains are typed, the need to have a more discriminating typing system becomes more apparent and the adoption of such typing systems for large numbers of strains greatly increases the cost and time of fingerprint comparisons and ultimately necessitates the use of computer-aided analysis (see Chapter 6).

6 Epidemiological applications

Except in the case of New Zealand where REA typing of *M. bovis* strains has been performed for 15 years, studies in other countries are much more recent and so far have largely been designed with the twin aims of establishing a useful typing system as well as performing preliminary epidemiological studies. Hence, much more is known about the molecular epidemiology of *M. bovis* infection in New Zealand than in any other country and this is reflected in the discussion below. However, this disproportionate lack of knowledge from other countries is expected to change rapidly in the case of some countries where major typing projects are now in progress.

6.1 New Zealand

New Zealand has been unable to eliminate *M. bovis* infection from cattle and farmed deer in about one-third of the country despite over 25 years of intensive testing and culling. Classical epidemiological studies have established that this is due to the continual reinfection of domestic livestock by infected wildlife. While the Australian brushtail possum is the principal wildlife reservoir of infection, other species such as wild deer and feral ferrets may also have a role in the spread of infection. Our first epidemiological use of REA showed that *M. bovis* isolates in different parts of New Zealand had different REA types (Collins *et al.* 1986). Subsequently, we showed that possums, other wildlife species and farmed animals in the same area are often infected with the same REA types of *M. bovis* (Collins *et al.* 1988; de Lisle *et al.* 1995). These were important findings as they indicated that strains of *M. bovis* in New Zealand are not specific to particular animal hosts, and that there is transmission of infection between wildlife species and farmed animals in the same area. With this knowledge, REA was subsequently used to provide information on possible sources of infection by using selected *M. bovis* isolates (Collins *et al.* 1994a). Where there was a single REA type found among a cluster of animals in a previously disease-free area this indicated there had been a single source of infection. If that REA type had been seen before this result directed attention at a possible source to be investigated. If the type had not been seen before this information could be used to exclude possible sources. In many cases, REA has been able to

clearly indicate whether an infection in a farmed animal has come from infected local wildlife or from infected cattle or deer being transported onto the farm from another area. Such information is an important factor in determining the subsequent level of herd testing or wildlife control operations. REA has proved particularly useful in defining the extent and subsequent spread of infected wildlife in a number of areas including the MacKenzie Basin (de Lisle *et al.* 1995). In New Zealand, REA is now used on a routine basis as an integral part of the *M. bovis* control scheme.

6.2 Other countries

If DNA typing of *M. bovis* is to reach its full potential, other countries need to define their *M. bovis* problem using a broad approach similar to that used in New Zealand. It is only when they know the range of *M. bovis* types present in different species of animals and the geographical spread of those isolates that they will be in a sound position for *M. bovis* typing to contribute significantly to control of bovine tuberculosis within their countries. Preliminary indications are that some countries have similarities to New Zealand in both their strain types and the epidemiology of infection while in other countries the situation is very different.

In Ireland, preliminary studies have shown that badgers and cattle in the same areas are usually infected with the same *M. bovis* types (Collins *et al.* 1994b; Skuce *et al.* 1996). The same types also infect other wildlife and humans and over 90% of strains have a single copy of IS*6110*. Clearly, both Ireland and New Zealand each have a common pool of *M. bovis* strains that is being transmitted between different host species. Although, compared to the clear-cut role of possums in New Zealand, there is considerable disagreement on the importance of badgers in spreading *M. bovis* to cattle in Ireland, this uncertainty is likely to change as more detailed epidemiological studies which incorporate DNA typing of *M. bovis* are performed.

In Spain, the situation appears very different to that in Ireland and New Zealand. While the majority of cattle are infected with strains of *M. bovis* containing a single copy of IS*6110* (Gutierrez *et al.* 1995; Liebana *et al.* 1997), in one of these studies almost as many *M. bovis* strains from cattle contained multiple copies of IS*6110* (Liebana *et al.* 1997). Interestingly, *M. bovis* strains from goats always contained five to eight copies of IS*6110*. Future studies will be needed to determine whether these differences between goat and cattle strains represent host factors or geographical differences. However, the presence of so many strains containing multiple copies of IS*6110* is a striking feature of the Spanish situation.

In addition to New Zealand, Ireland and Spain, IS*6110* fingerprinting has also been used for typing cattle strains of *M. bovis* from the Netherlands (van Soolingen *et al.* 1994), Argentina (van Soolingen *et al.* 1994; Romano *et al.* 1996), Australia (Cousins *et al.* 1993), North America (Perumaalla *et al.* 1996) and Burundi (Rigouts *et al.* 1996). With the exception of Burundi, these strains were usually found to contain only a single copy of IS*6110*. Apart from New Zealand, relatively few strains from animals other than cattle have been typed but it appears that *M. bovis* strains with multiple copies of IS*6110* are common in some animal species (van Soolingen *et al.* 1994). These differences in strain type with respect to different animal species may reflect adaptation of *M. bovis* types to particular animal hosts or alternatively they may indicate that, in those countries where these differences occur, cattle and other animals rarely have opportunities to infect each other. Further studies will be required to resolve this question but meanwhile it would probably be advantageous in many countries to continue using IS*6110* as part of a typing scheme for *M. bovis*.

6.3 Confined outbreaks

As well as defining the overall nature of *M. bovis* infection in a region and being used on a regular basis to assist with disease control and eradication, DNA typing of *M. bovis* can play an important role in understanding confined outbreaks. In New Zealand, REA has been used to show that 12 domestic cats in one city were infected from a common source and that 26 other cats from various regions of the country were infected with *M. bovis* types which were similar or identical to those in other animal species in the same regions (de Lisle *et al.* 1990 and unpublished results 1998).

In Australia, REA, PGRS and IS*6110* typing showed that identical strains of *M. bovis* had infected captive seals (Cousins *et al.* 1990, 1993) and also an animal trainer who worked with these animals (Thompson *et al.* 1993). The strains had an unusual REA pattern and multiple copies of IS*6110*. The same or similar strains have also been found in wild seals in Australia (Cousins *et al.* 1993) and in a single cattle beast from a coastal New Zealand farm (Collins & de Lisle 1995). Clearly, while seal strains have some features that make them unusual, they remain infectious for humans and other animals. Seals infected with strains of *M. bovis* that contain multiple copies of IS*6110* have also been reported from Argentina (van Soolingen *et al.* 1994).

REA has also been used to conclusively show that *M. bovis* was independently transferred in infected deer from the UK to New Zealand and to Sweden (Bolske *et al.* 1995). In both cases the same strain was involved and in Sweden it caused a major outbreak of infection. Because these strains had six or seven copies of IS*6110* (Szewzyk *et al.* 1995) (see Fig. 7.2, lane N), IS*6110* typing could have been used equally well for this work.

6.4 Human isolates

Apart from the infected seal trainer described above, a number of studies which used a variety of typing methods have included some *M. bovis* strains from humans (Cousins *et al.* 1993; van Soolingen *et al.* 1994; Collins & de Lisle 1995; Szewzyk *et al.* 1995; Feizabadi *et al.* 1996; Romano *et al.* 1996; Skuce *et al.* 1996). In the largest of these studies, 20 strains from Argentina and 47 strains from the Netherlands were typed using RFLPs to IS*6110*, PGRS and DR (van Soolingen *et al.* 1994). Some of the PGRS and DR patterns for these strains are shown in Figs 7.3 and 7.4. In Argentina, humans are infected with the same or similar strains to those found in cattle. This is consistent with another study from Argentina (Romano *et al.* 1996) and also appears to be the case for a small number of human strains that have been typed from Australia (Cousins *et al.* 1993), New Zealand (D. M. Collins and G. W. de Lisle, unpublished results 1998) and Ireland (Skuce *et al.* 1996). This does not establish that cattle were the direct source of these infections as in all these countries similar or identical types have also been isolated from other animal species. In the Netherlands, there appeared to be two different groups of human isolates only one of which had similar types to those found in cattle (van Soolingen *et al.* 1994). Another interesting finding from the Netherlands was a group of five human strains of *M. bovis*, all with identical RFLPs with IS*6110*, PGRS and DR (van Soolingen *et al.* 1994). Since four of the patients lived with or near each other and had no other obvious source of infection this finding supports a long-held belief that human-to-human spread of *M. bovis* occurs.

7 Future outlook

In the last 10 years, DNA typing of *M. bovis* strains has progressed from being moderately used in New Zealand to being extensively used in New Zealand and moderately used in a small number of other countries. A range of different typing techniques, and often a combination of three techniques, is being employed because no one method has all the advantages of discrimination and ease of use that are evident for IS*6110* typing of *M. tuberculosis*. It is highly probable that a combination of different DNA typing methods for *M. bovis* will be used into the future and that there will be continued interest in investigating possible new techniques. Such techniques could be based on repetitive DNA sequences that have been identified during sequencing of the *M. tuberculosis* genome and that in some cases have already been found to discriminate between *M. bovis* strains (Frothingham & Meeker-O'Connell 1998). Alternatively, new bacterial DNA fingerprinting techniques (van Steenbergen *et al.* 1995; Janssen *et al.* 1996; Mazurek *et al.* 1996) that have yet to be applied to *M. bovis* strains may be found to give good discrimination.

Within a few years, a number of countries apart from New Zealand will have large databases of *M. bovis* types and will have needed to develop reliable ways of handling and comparing new DNA fingerprints to these databases. The difficulties of comparing large numbers of different types may cause a re-evaluation of which typing methods to use, as fingerprints from some methods are harder to compare than those from others. Ultimately, typing of very large numbers of strains can only be performed by computer-aided analysis. The author's group has implemented such analysis for REA typing by using a similar approach to that used for *M. tuberculosis* (Woelffer *et al.* 1996). While moderate success was achieved, the computer-aided system for REA is not as robust as that for IS*6110* because the patterns are much more complex and harder to resolve. Somewhat similar problems will probably also be encountered for computer-aided analysis of PGRS fingerprints.

Now that the value of typing *M. bovis* strains is becoming more widely accepted, its use in countries where the techniques have already been applied is expected to increase in the next few years and it will also be applied for the first time in other countries. This increased DNA fingerprinting of *M. bovis* strains will undoubtedly lead to a better understanding of the epidemiology of *M. bovis* infection and will contribute towards a better definition of the global bovine tuberculosis problem. There is now a wider appreciation than at any time in the last 30 years of the small but significant role of *M. bovis* in human tuberculosis. Currently, many laboratories culture suspect human tuberculosis samples under conditions that inhibit growth of *M. bovis*. Increased appreciation of the role of *M. bovis* should lead to more laboratories seeking to isolate *M. bovis* and it can be expected that in the next few years, DNA typing of these human strains will receive increased attention.

8 References

Aranaz, A., Liebana, E., Mateos, A. *et al.* (1996) Spacer oligonucleotide typing of *Mycobacterium bovis* strains from cattle and other animals: a tool for studying epidemiology of tuberculosis. *Journal of Clinical Microbiology* **34**, 2734–2740.

Beggs, M.L., Cave, M.D., Marlowe, C., Cloney, L., Duck, P. & Eisenach, K.D. (1996) Characterization of *Mycobacterium tuberculosis* complex direct repeat sequence for use in cycling probe reaction. *Journal of Clinical Microbiology* **34**, 2985–2989.

Bolske, G., Englund, L., Wahlstrom, H., de Lisle, G.W., Collins, D.M. & Croston, P.S. (1995) Bovine tuberculosis in Swedish deer farms: epidemiological investigations and tracing using restriction fragment analysis. *The Veterinary Record* **136**, 414–417.

Butcher, P.B., Hutchinson, N.A., Doran, T.J. & Dale, J.W. (1996) The application of molecular techniques to the diagnosis and epidemiology of mycobacterial diseases. *Journal of Applied Bacteriology* **81**, 53S–71S.

Collins, D.M. & de Lisle, G.W. (1984) DNA restriction endonuclease analysis of *Mycobacterium tuberculosis* and *Mycobacterium bovis* BCG. *Journal of General Microbiology* **130**, 1019–1021.

Collins, D.M. & de Lisle, G.W. (1985) DNA restriction endonuclease analysis of *Mycobacterium bovis* and other members of the tuberculosis complex. *Journal of Clinical Microbiology* **21**, 562–564.

Collins, D.M. & de Lisle, G.W. (1987) BCG identification by DNA restriction fragment patterns. *Journal of General Microbiology* **133**, 1431–1434.

Collins, D.M. & de Lisle, G.W. (1995) *Mycobacterium bovis* strains. In: *Tuberculosis in Wildlife and Domestic Animals* (eds F. Griffin & G. Lisle). Dunedin: University of Otago Press, pp. 202–204.

Collins, D.M., Erasmuson, S.K., Stephens, D.M., Yates, G.F. & de Lisle, G.W. (1993) DNA fingerprinting of *Mycobacterium bovis* strains by restriction fragment analysis and hybridisation with the insertion elements IS*1081* and IS*6110*. *Journal of Clinical Microbiology* **31**, 1143–1147.

Collins, D.M., Gabric, D.M. & de Lisle, G.W. (1988) Typing of *Mycobacterium bovis* isolates from cattle and other animals in the same locality. *New Zealand Veterinary Journal* **36**, 45–46.

Collins, D.M., de Lisle, G.W., Collins, J.D. & Costello, E. (1994b) DNA restriction fragment typing of *Mycobacterium bovis* isolates from cattle and badgers in Ireland. *The Veterinary Record* **134**, 681–682.

Collins, D.M., de Lisle, G.W. & Gabric, D.M. (1986) Geographic distribution of restriction types of *Mycobacterium bovis* isolates from brush-tailed possums (*Trichosurus vulpecula*) in New Zealand. *Journal of Hygiene* **96**, 431–438.

Collins, D.M., Radford, A.J., de Lisle, G.W. & Billman-Jacobe, H. (1994a) Diagnosis and epidemiology of bovine tuberculosis using molecular biological approaches. *Veterinary Microbiology* **40**, 83–94.

Cousins, D.V., Francis, B.R., Gow, B.L. *et al.* (1990) Tuberculosis in captive seals: bacteriological studies on an isolate belonging to the *Mycobacterium tuberculosis* complex. *Research in Veterinary Science* **48**, 196–200.

Cousins, D., Williams, S., Liebana, E. *et al.* (1998) Evaluation of four DNA typing techniques in epidemiological investigations of bovine tuberculosis. *Journal of Clinical Microbiology* **36**, 168–178.

Cousins, D.V., Williams, S.N., Ross, B.C. & Ellis, T.M. (1993) Use of a repetitive element isolated from *Mycobacterium tuberculosis* in hybridization studies with *Mycobacterium bovis*: a new tool for epidemiological studies of bovine tuberculosis. *Veterinary Microbiology* **37**, 1–17.

Doran, T.J., Hodgson, A.L.M., Davies, J.K. & Radford, A.J. (1993) Characterisation of a highly repeated DNA sequence from *Mycobacterium bovis*. *FEMS Microbiology Letters* **111**, 147–152.

Feizabadi, M.M., Robertson, I.D., Cousins, D.V. & Hampson, D.J. (1996) Genomic analysis of *Mycobacterium bovis* and other members of the *Mycobacterium tuberculosis* complex by isozyme analysis and pulse-field gel electrophoresis. *Journal of Clinical Microbiology* **34**, 1136–1142.

Fomukong, N.G., Tang, T.H., Al-Maamary, S. *et al.* (1994) Insertion sequence typing of *Mycobacterium tuberculosis*: characterisation of a widespread subtype with a single copy of IS*6110*. *Tubercle and Lung Disease* **75**, 435–440.

Frothingham, R. (1995) Differentiation of strains in *Mycobacterium tuberculosis* complex by DNA sequence polymorphisms, including rapid identification of *M. bovis* BCG. *Journal of Clinical Microbiology* **33**, 840–844.

Frothingham, R. & Meeker-O'Connell, W.A. (1998) Genetic diversity in the *Mycobacterium tuberculosis* complex based on variable numbers of tandem DNA repeats. *Microbiology* **144**, 1189–1196.

Glennon, M., Jager, B., Dowdall, D. *et al.* (1997) PCR-based fingerprinting of *Mycobacterium bovis* isolates. *Veterinary Microbiology* **54**, 235–245.

Goguet de la Salmoniere, Y., Li, H.M., Torrea, G., Bunschoten, A., van Embden, J. & Gicquel, B. (1997) Evaluation of spoligotyping in a study of the transmission of *Mycobacterium tuberculosis*. *Journal of Clinical Microbiology* **35**, 2210–2214.

Grange, J.M., Collins, J.D., O'Reilly, L.M., Costello, E. & Yates, M.D. (1990) Identification and characteristics of *Mycobacterium bovis* isolated from cattle, badgers and deer in the Republic of Ireland. *Irish Veterinary Journal* **43**, 33–35.

Gutierrez, M., Samper, S., Gacigan, J., Marin, J. & Martin, C. (1995) Differentiation by molecular typing of *Mycobacterium bovis* strains causing tuberculosis in cattle and goats. *Journal of Clinical Microbiology* **33**, 2953–2956.

Hermans, P.W.M., van Soolingen, D., Bik, E.M., de Haas, P.E.W., Dale, J.W. & van Embden, J.D.A. (1991) Insertion element IS*987* from *Mycobacterium bovis* BCG is located in a hot-spot integration region for insertion elements in

Mycobacterium tuberculosis complex strains. *Infection and Immunity* **59**, 2695–2705.

Howard, S.T., Laszlo, A. & Johnson, W.M. (1997) Genetic identification of *Mycobacterium bovis* BCG by restriction fragment length polymorphism analysis of the direct repeat region. *Journal of Clinical Microbiology* **35**, 965–968.

Imacda, T. (1985) Deoxyribonucleic acid relatedness among selected strains of *Mycobacterium tuberculosis, Mycobacterium bovis, Mycobacterium bovis* BCG, *Mycobacterium microti*, and *Mycobacterium africanum*. *International Journal of Systematic Bacteriology* **35**, 147–150.

Janssen, P., Coopman, R., Huys, G. *et al.* (1996) Evaluation of the DNA fingerprinting method AFLP as a new tool in bacterial taxonomy. *Microbiology* **142**, 1881–1893.

Kamerbeek, J., Schouls, L., Kolk, A. *et al.* (1997) Simultaneous detection and strain differentiation of *Mycobacterium tuberculosis* for diagnosis and epidemiology. *Journal of Clinical Microbiology* **35**, 907–914.

Liebana, E., Aranaz, A., Dominguez, L. *et al.* (1997) The insertion element IS*6110* is a useful tool for DNA fingerprinting of *Mycobacterium bovis* isolates from cattle and goats in Spain. *Veterinary Microbiology* **54**, 223–233.

de Lisle, G.W., Collins, D.M., Loveday, A.S., Young, W.A. & Julian, A.F. (1990) A report of tuberculosis in cats in New Zealand, and the examination of strains of *Mycobacterium bovis* by DNA restriction endonuclease analysis. *New Zealand Veterinary Journal* **38**, 10–13.

de Lisle, G.W., Yates, G.F., Collins, D.M., MacKenzie, R.W., Crews, K.B. & Walker, R. (1995) A study of bovine tuberculosis in domestic animals and wildlife in the MacKenzie Basin and surrounding areas using DNA fingerprinting. *New Zealand Veterinary Journal* **43**, 266–271.

Marshall, R.B., Wilton, B.E. & Robinson, A.J. (1981) Identification of leptospira serovars by restriction endonuclease analysis. *Journal of Medical Microbiology* **14**, 163–166.

Mazurek, G.H., Reddy, V., Marston, B.J., Haas, W.H. & Crawford, J.T. (1996) DNA fingerprinting by infrequent-restriction-site amplification. *Journal of Clinical Microbiology* **34**, 2386–2390.

O'Reilly, L.M. & Daborn, C.J. (1995) The epidemiology of *Mycobacterium bovis* infections in animals and man: a review. *Tubercle and Lung Disease* **76** (Suppl. 1), 1–46.

Olson, E.S., Forbes, K.J., Watt, B. & Pennington, T.H. (1995) Population genetics of *Mycobacterium tuberculosis* complex in Scotland analysed by pulsed-field gel electrophoresis. *Epidemiology and Infection* **114**, 153–160.

Patel, R., Kvach, J.T. & Mounts, P. (1986) Isolation and restriction endonuclease analysis of mycobacterial DNA. *Journal of General Microbiology* **132**, 541–551.

Perumaalla, V.S., Adams, L.G., Payeur, J.B. *et al.* (1996) Molecular epidemiology of *Mycobacterium bovis* in Texas and Mexico. *Journal of Clinical Microbiology* **34**, 2066–2071.

Poulet, S. & Cole, S.T. (1995) Characterization of the highly abundant polymorphic GC-rich-repetitive sequence (PGRS) present in *Mycobacterium tuberculosis*. *Archives of Microbiology* **163**, 87–95.

Redmond, W.B., Bates, J.H. & Engel, H.W.B. (1979) Methods for bacteriophage typing of mycobacteria. In: *Methods in Microbiology*, Vol. 13 (eds T. Bergan & J. R. Norris). New York: Academic Press, pp. 345–375.

Rigouts, L., Maregeya, B., Traore, H., Collart, J.P., Fissette, K. & Portaels, F. (1996) Use of DNA restriction fragment typing in the differentiation of *Mycobacterium tuberculosis* complex isolates from animals and humans in Burundi. *Tubercle and Lung Disease* **77**, 264–268.

Romano, M.I., Alito, A., Fisanotti, J.C. *et al.* (1996) Comparison of different genetic markers for molecular epidemiology of bovine tuberculosis. *Veterinary Microbiology* **50**, 59–71.

Ross, B.C., Raios, K., Jackson, K. & Dwyer, B. (1992) Molecular cloning of a highly repeated DNA element from *Mycobacterium tuberculosis* and its use as an epidemiological tool. *Journal of Clinical Microbiology* **30**, 942–946.

Skuce, R.A., Brittain, D., Hughes, M.S., Beck, L.-A. & Neill, S.D. (1994) Genomic fingerprinting of *Mycobacterium bovis* from cattle by restriction fragment length polymorphism analysis. *Journal of Clinical Microbiology* **32**, 2387–2392.

Skuce, R.A., Brittain, D., Hughes, M.S. & Neill, S.D. (1996) Differentiation of *Mycobacterium bovis* isolates from animals by DNA typing. *Journal of Clinical Microbiology* **34**, 2469–2474.

van Soolingen, D., de Haas, P.E.W., Haagsma, J. *et al.* (1994) Use of various genetic markers in differentiation of *Mycobacterium bovis* strains from animals and humans and for studying epidemiology of bovine tuberculosis. *Journal of Clinical Microbiology* **32**, 2425–2433.

van Soolingen, D., Hermans, P.W.M., de Haas, P.E.W. & van Embden, J.D.A. (1992) Insertion element IS*1081*-associated restriction fragment length polymorphisms in *Mycobacterium tuberculosis* complex species: a reliable tool for recognizing *Mycobacterium bovis* BCG. *Journal of Clinical Microbiology* **30**, 1772–1777.

Sreevatsan, S., Pan, X. & Musser, J.M. (1997) Restricted structural gene polymorphism in the *Mycobacterium tuberculosis* complex indicates evolutionarily recent global dissemination. *Proceedings of the National Academy of Sciences of the USA* **94**, 9869–9874.

van Steenbergen, T.J.M., Colloms, S.D., Hermans, P.W.M., de Graaff, J. & Plasterk, R.H.A. (1995) Genomic DNA fingerprinting by restriction fragment end labeling. *Proceedings of the National Academy of Sciences of the USA* **92**, 5572–5576.

Szewzyk, R., Svenson, S.B., Hoffner, S.E. *et al.* (1995) Molecular epidemiological studies of *Mycobacterium bovis* infections in humans and animals in Sweden. *Journal of Clinical Microbiology* **33**, 3183–3185.

Thompson, P.J., Cousins, D.V., Gow, B.L., Collins, D.M., Williamson, B.H. & Dagnia, H.T. (1993) Seals, seal trainers and mycobacterial infection. *American Review of Respiratory Diseases* **147**, 164–167.

Tyler, K.D., Wang, G., Tyler, S.D. & Johnson, W.M. (1997) Factors affecting reliability and reproducibility of amplification-based DNA fingerprinting of representative bacterial pathogens. *Journal of Clinical Microbiology* **35**, 339–346.

Woelffer, G.B., Bradford, W.Z., Paz, A. & Small, P.M. (1996) A computer-assisted molecular epidemiologic approach to confronting the reemergence of tuberculosis. *American Journal of Medical Science* **311**, 17–22.

Zhang, Y., Wallace, R.J. Jr & Mazurek, G.H. (1995) Genetic differences between BCG substrains. *Tubercle and Lung Disease* **76**, 43–50.

Chapter 8 / Molecular epidemiology: other mycobacteria

JOSEPH O. FALKINHAM, III

1 Introduction: applications of molecular epidemiology

The objective of molecular epidemiology is to use molecular markers (e.g. DNA, RNA or proteins) to determine whether different mycobacterial isolates of the same species and recovered from either the same or different sources are related. Specifically, it is to determine whether independently recovered isolates are members of the same clone or not. It is important that members of the same species be compared, otherwise a false conclusion of relatedness can be derived.

Molecular epidemiology, the use of molecular markers for strain typing, can be of immediate use to the patient and clinician or can be used to provide long-term, indirect benefits. In the short term, a source of infection can be identified, whether another patient, the hospital, or the environment that surrounds the patient. Also, pseudoinfections, through contamination of instruments or contamination in the laboratory, can be identified. If infection

is recurrent, typing can be used to determine if infection is due to reactivation of a dormant mycobacterium or due to a new infection. By recovering more than a single isolate from a patient or patient sample, it is possible to determine whether a patient is infected with more than a single clone. Such polyclonal infection is important, because the different clones from a single patient have been shown to possess different antibiotic susceptibilities (Von Reyn *et al.* 1995). Beyond those immediate uses, molecular epidemiology can be used to describe the population structure of mycobacteria. That can, in turn, lead to identification of subtypes and knowledge that selection of certain types is occurring in patients. Identification that certain types predominate in patients can be used to investigate the route of infection and the identification of virulence determinants.

2 Objectives of the chapter

The chapter is divided into three main sections. Section 3 provides the reader with a general background on the principles and use of molecular markers for the study of non-tuberculous mycobacteria. Examples are reviewed in section 4 which is divided by mycobacterial species. In that section, specific details are provided by the brief discussions of the approaches and results and inclusion of a comprehensive list of published references. The details are included to emphasize the general principles briefly stated in the first portion. The last section attempts to bring together many of the observations to provide a speculative, yet I hope, coherent picture of mycobacterial populations and the information provided by different molecular markers.

3 Principles and use of molecular markers

3.1 Criteria for selection of epidemiological markers

To be of use, methods for molecular epidemiology must permit each isolate to be typed (typeability) reproducibly and must discriminate between isolates (Maslow *et al.* 1993). Typeability refers to the ability of the method to identify the type of each strain. If a substantial number of strains cannot be typed, the utility of the method for epidemiological studies is limited. Because of the need to employ methods that will provide a type for every mycobacterial isolate, there has been a focus on DNA as a target, because all cells have DNA. Typing based on gene expression, such as biotyping, serotyping, phage typing and enzyme activities (e.g. multilocus enzyme electrophoresis), are all dependent upon finding conditions that ensure the equal expression of genes in all isolates.

Typing methods must be reproducible. Specifically, a method must yield the same type for an individual isolate in repeated experiments. Because some experimental steps are likely to have a strong influence on the results (e.g. impact of the DNA isolation method on the size of the fragments), standard protocols are required. In that manner, investigators are assured that the results are insensitive to operator and site differences and typing results from different laboratories can be compared directly. Reproducibility is also influenced by the 'robustness' of a particular typing method. A robust method would be one in which the same type can be obtained for an individual strain in spite of some differences in performing the method's individual steps (e.g. growth of isolates, method of DNA isolation).

It is critical that the patterns for an individual marker used for strain typing be shown to be relatively stable. The word 'relatively' was chosen on purpose. In order for the marker to be useful in typing there must be diversity in the population. However, there cannot be too much diversity. Thus, a marker that transposes too frequently may not be of use for typing, unless one wishes to examine clonal diversity (see discussion of IS*1110*). Stability of some widely used markers (i.e. insertion sequences) have not been analysed in great depth. Patterns for IS markers have been reported as 'stable' when changes are absent over several transfers. However, there has been no evidence of the power of the method to detect changed patterns. Thus, it is possible that changes occurring in 10% of the population would be missed.

Discrimination refers to the probability that two unrelated strains (i.e. not from the same clone) in a single population will be placed into different groups (e.g. types or clusters). The discriminatory power of individual typing methods can be calculated (i.e. discrimination index) and different methods can be compared on the basis of their discriminatory power (Hunter and Gaston 1988; Hunter 1990). The discrimination index is proportional to the number of different types in the population and the number of strains belonging to individual type groups (Hunter and Gaston 1988). It is logical that the discriminatory power be related to the number of types represented by the population. High numbers of types, or a combination of typing methods, provides more character states to distinguish between strains.

A molecular marker present in low copy number lacks discriminatory power. This is the case for single gene typing such as the *dnaJ* gene (Victor *et al.* 1996) or 32-kDa protein gene (Soini and Viljanen 1997) and ribosomal RNA (rRNA)-based typing, because mycobacteria have only one or two rRNA gene operons (Domenech *et al.* 1994). However, low copy number and weak discriminatory power do not mean that a marker cannot tell us something about a particular mycobacterial population. For example, typing data employing low copy number markers have provided insights into higher orders of population structure (e.g. geographical distribution) or selection of types in humans or animals (e.g. serotypes in *Mycobacterium avium*).

The effect of the distribution of strains within types also influences the discriminatory power of a typing method (Hunter & Gaston 1988; Hunter 1990). Reduction of discrimination index occurs when isolates are not distributed equally among types, but found predominantly in a few types. For example, although there are a substantial number of *M. avium* complex serotypes, 57% of *M. avium* complex isolates recovered from patients with acquired immune deficiency syndrome (AIDS) in the United States belonged to serotypes 8 and 4 (Yakrus and Good 1990). For DNA-based typing methods using specific genetic elements, the distribution of strains within types can be influenced by the existence of 'hotspots' for integration of the element. 'Hotspots' limit the number of different types within the population and thus reduce the heterogeneity and the discrimination index.

It is not essential that a marker used for strain typing be found exclusively in the species to be typed. For example, both the polymorphic GC-rich sequence (PGRS) (Ross *et al.* 1992b) and the major polymorphic tandem repeat (MPTR) (Hermans *et al.* 1992) originally discovered in *M. tuberculosis* (see Chapters 6 and 7) are present in other mycobacteria and can be used in typing, notably PGRS in *M. kansasii* (Ross *et al.* 1992a) and *M. ulcerans* (Jackson *et al.* 1995). However, the wide distribution of a marker can be a problem for closely related species. Specifically, isolates of different species can share the same profile. Further, cross-reactions between markers (e.g. IS*1081* and IS*1395*; Collins 1994; Picardeau *et al.* 1996) or between a marker and genomic DNA (Kent *et al.* 1995) have been noted. Thus, before isolates are typed, their species assignment must be correct.

High copy number of a particular target for one typing method may influence the discriminatory power of a second typing method. Large restriction fragment pulsed-field gel electrophoresis (LRF-PFGE) patterns may be influenced by the existence of high copy numbers of repeated sequences, such as insertion sequences (e.g. IS*1245* in *M. avium*) or repeated genetic elements (e.g. PGRS in *M. kansasii*). Whether the restriction endonuclease employed for generating the LRF-PFGE patterns cleaves, or does not cleave, the repeated sequence, the resulting number or size of LRF-PFGE bands will be influenced by the copy number of the element.

3.2 Epidemiological markers used for typing mycobacteria

A number of different molecular markers have been investigated, developed, and exploited for typing mycobacteria. Categories of markers and examples are listed in Table 8.1 and the markers used for individual *Mycobacterium* species, by category, are listed in Table 8.2. Within Table 8.1 are definitions of the methods and their abbreviations. The categories are provided because they either reflect differences in the molecular target whose heterogeneity is being utilized (e.g. single gene vs. repeated sequence), or the laboratory techniques are different (e.g. restriction endonuclease vs. polymerase chain reaction (PCR)). There may be another marker added to that list. It has been reported that *M. tuberculosis* DNA contains fragments able to hybridize to the enterobacterial repetitive intergenic consensus (ERIC) sequences (Sechi *et al.* 1998). ERIC probes may prove to be of use in typing other *Mycobacterium* species.

3.3 Technical issues and potential problems

In this section I would like to point out a number of technical issues that are of importance. For the most part they have to do with the inclusion of controls that will allow the investigator to determine whether factors, other than the isolate, are influencing the results. Because isolation of DNA, RNA or protein from cells requires that the cells be broken, most

Table 8.1 Categories of molecular markers used in typing mycobacteria.

Category	Examples	Description
Single-gene RFLP	*dnaJ*-RFLP 32-kDa protein	Variation in pattern of restriction fragments of a single gene
Repeat sequence RFLP	PGRS-RFLP IS*1245*-RFLP	Variation in pattern of restriction fragments of genomic DNA carrying a repeated sequence
Genomic RFLP	LRF-PFGE	Variation in pattern of large restriction fragments of total (genomic) DNA separated by pulsed-field gel electrophoresis
Plasmid typing	Plasmids	Presence or absence of plasmids representing different similarity groups or of restriction fragments of plasmids
Spacer typing	16S-23S rRNA	Variation in pattern of restriction fragments of region between two genes
PCR spacer typing	IS*1245*-IS*1311*	Variation in size and number of PCR fragments generated by amplification of sequences between repeated elements
Arbitrary primed PCR	AP-PCR (RAPD)	Variation in size and number of PCR products generated by PCR reactions using single primers
Multilocus enzyme electrophoresis	MEE	Variation in the electrophoretic mobility of individual enzymes
Serotype/chemotype	Serotype, chemotype	Presence of different antigenic surface molecules

RFLP, restriction fragment length polymorphism; PGRS, polymorphic GC-rich sequence; LRF-PFGE, large restriction fragment pulsed-field gel electrophoresis; PCR, polymerase chain reaction.

lysates contain DNases, RNases, and proteinases that are no longer separated from their substrates as they are in cells. Not only can these enzymes reduce the yield of the target, but the size of the molecules can be reduced to the extent that the lysate is unsuitable for use in fingerprinting. For example, a collection of *M. abscessus* isolates could not be typed by LRF-PFGE because of the presence of DNase activity in lysates (Wallace *et al.* 1993). It is not difficult to develop a simple method for detecting any DNase, RNase or proteinase activity and well worth the effort. Repeated freeze–thaw cycles of lysates used for fingerprinting can also lead to changes in the size of macromolecules and their suitability as substrates for reactions (e.g. restriction endonuclease digestion or PCR). Thus, we try to prepare fresh lysates and freeze aliquots to reduce the impact of repeated freezing and thawing. In addition to DNase activity, the presence of DNA methylation and methylase activity in lysates can influence the results of fingerprinting. For

example, differences in methylation of some *Pvu*II restriction sites in some populations of *M. tuberculosis* isolates altered the IS*6110*–restriction fragment length polymorphism (RFLP) pattern (van Soolingen *et al.* 1996).

Most typing methods involve gel electrophoresis. That, in turn, requires the use of molecular weight markers. Markers should not be restricted to purchased molecular-weight ladders, but should also include bands generated by a single reference strain of bacteria; hopefully of the same species and readily available to other investigators. In that fashion, gel-to-gel variation in mobility of any marker can be measured. That measurement provides day-by-day quality assurance data and is critical for determining whether two strains share the same band or not.

Patterns generating by any typing markers must be stable over repeated generations. Although many, if not most, of the papers reviewed here noted that patterns were stable (with the notable exception of

Table 8.2 Molecular markers employed for typing individual mycobacterial species.

Species	Category	Specific markers
M. kansasii	Single-gene RFLP	*hsp-65*, rRNA, *gyrA*
	Repeat sequence RFLP	PGRS, MPTR, IS*1652*
	Genomic RFLP	LRF-PFGE
	Spacer typing	16S-23S rRNA
	Arbitrary primed PCR	AP-PCR
M. avium complex	Single-gene RFLP	*dnaJ*, 32-kDa-protein gene, Plasmid
	Repeat sequence RFLP	IS*900*, IS*901*/IS*902*, IS*1245*, IS*1311*
	Genomic RFLP	LRF-PFGE
	Plasmid typing	Plasmid
	Spacer typing	16S-23S rRNA, IS*1245*-IS*1311*
	Arbitrary primed PCR	AP-PCR
	Multilocus enzyme	MEE
	Serotype	Serotype
M. xenopi	Repeat sequence RFLP	IS*1081*, IS*1395*
	Genomic RFLP	LRF-PFGE
M. haemophilum	Genomic RFLP	LRF-PFGE
M. malmoense	Single-gene RFLP	rRNA
	Arbitrary primed PCR	AP-PCR
	Chemotype	Chemotype
M, ulcerans	Single-gene RFLP	3′ 16S rRNA
	Repeat sequence RFLP	PGRS-RFLP
Rapidly growing	Single-gene RFLP	16S rRNA
	Genomic RFLP	LRF-PFGE
	Plasmid typing	Plasmids
	Multilocus enzyme	MEE

For abbreviation definitions, see Table 8.1.

IS*1110* patterns, Hernandez-Perez *et al.* 1994), the data did not demonstrate a great depth of analysis. Usually, investigators examined typing patterns in subcultures without measuring the number of generations between subcultures. Hence, the lower limit of detection of a change in a typing pattern was unknown. The alternative of comparing patterns of isolates recovered from the same environmental source, animal or human also lacks information concerning the level of detection of variant patterns.

Considering subcultures, one should not forget that the artificial, laboratory culture subjects microorganisms to selection. For example in *M. avium*, though the transparent colony type predominates among isolates upon primary culture, repeated subculture results in the appearance, through selection, of the more rapidly growing, opaque colony-forming variants. Thus, comparison of subcultures may simply represent the measurement of the effect of artificial laboratory culture on the marker's pattern. We have observed another, and equally troublesome, consequence of subculture and colonial variation in *M. intracellulare*. Transparent and opaque variants of a number of *M. intracellulare* strains had different plasmid DNA profiles (Erardi *et al.* 1985), due to the transposition of the insertion sequence, IS*1141*, from a plasmid to the chromosome (Via 1993). Thus, one must ensure that the colony morphology of isolates is the same, or another variable is introduced into a typing scheme.

Selection of types may occur. Some markers used in strain typing could be subject to selection that may not impact other markers. Thus, there could be a reduction of types among isolates using one typing method. For example, LRF-PFGE patterns may be more discriminatory than patterns displayed by multilocus enzyme electrophoresis (MEE), because some nucleotide changes in genes for enzymes analysed by MEE may be subject to selection. That is not necessarily the case for sites cleaved by restriction endonucleases used in LRF-PFGE analysis.

For typing methods involving proteins, enzymes (i.e. MEE) or the products of enzymatic action (i.e. serotyping), care must be taken to ensure that the presence or absence and amount of either is not subject to isolate-to-isolate variation, even when isolates are grown under the same conditions (witness the assumption that methylation of *Pvu*II sites was the same in all *M. tuberculosis* isolates; van Soolingen *et al.* 1996).

Finally, it should be understood that PCR primers for amplification or a probe for RFLP analysis may form hybrids with identical copies in the genome and related sequences. Thus, it is incorrect to assume, without further analysis, that all DNA sequences reacting with primers or probes prepared on the basis of sequence information of a specific DNA sequence (e.g. IS*1245*) contain identical copies of that marker.

3.4 Analysis of molecular epidemiological data

A variety of methods have been employed for presentation of typing data. First, the depth of the analysis (e.g. statistical power) must be justified (balanced) by the quality of the data. If the presentation is simply a side-by-side comparison of gel photographs or autoradiographs, then the eye can be used alone. The weakness of that approach is that often different gels are aligned and without actually measuring the mobilities of bands (in the gel photographs), it is impossible to get a feeling for the variation in mobility; especially if the gels do not share an isolate

in common. Molecular-weight markers are not sufficient for gel-to-gel comparisons. Coefficients for variation and that criteria used for defining strains of the same or different clones must be provided. Often this is manifested by calculation of a fragment length error tolerance. Commonly, a range of 3–5% has been reported. Further, the reader must know whether minor bands (e.g. weakly hybridizing or amplified) are included in any calculations.

If coefficients of variation and criteria for basing relatedness are provided (and justified), numerical characteristics describing the method of typing (e.g. the discrimination index (Hunter & Gaston 1988; Hunter 1990)) can be provided. The relatedness between isolates can be estimated by calculating the coefficient of similarity (CS) that equals $2\times$(number of shared bands)/total number of bands in both isolates (Dice 1945). Further, the range of values for placing one isolate in one group or another should be presented. When reviewing strain typing data, the relationship between number of bands, population size and probability that two isolates sharing the same profile are truly different should be calculated. Calculations leading to estimates of magnitude or percentage of base substitution (Cooper *et al.* 1989) have also been presented. Dendrograms have been used to visually display the relatedness of strains identified by sequence (De Smet *et al.* 1995; Soini *et al.* 1996), RFLP (Iinuma *et al.* 1997; Picardeau *et al.* 1997), and MEE (Wasem *et al.* 1991; Yakrus *et al.* 1992; Feizabadi *et al.* 1996). Typing data based on MEE includes average number of alleles per locus, mean genetic diversity and weighted genetic distance between electrophoretic types (Wasem *et al.* 1991; Yakrus *et al.* 1992).

Care must be taken in analysing the results of typing strains. There are few questions when strains share identical patterns and the marker displays a high degree of polymorphism. The same is true if strains share unrelated patterns. However, there have been attempts to recognize 'clusters' based upon the relatedness of strains. 'Clusters' have been defined in different papers based on different degrees of relatedness or degrees of band sharing. The range of values used to define 'clusters' has ranged from 67 to 80% of bands shared in common. Because of the demonstrated *in vitro* and *in vivo* stability of the markers used as targets for those studies, it is surprising that strains differing by one-third of the bands are described as 'related' or belonging to the same 'cluster'. However, those studies are quite valuable in those cases where other information identifies the strains as related based on other epidemiological data (e.g. common source of exposure). Specifically, that data will permit the calculation of the rate of divergence of mycobacterial strains and thus establish a true estimate of variance and hence a definition and guide for including strains within 'clusters'.

4 Applications

4.1 Molecular epidemiology of *Mycobacterium kansasii*

M. kansasii causes pulmonary infections in patients with underlying lung disease associated with smoking and chronic obstructive pulmonary disease (Lillo *et al.* 1990). In addition, *M. kansasii* is the second most common mycobacterium (after *M. avium*) causing disseminated infection in AIDS patients in the United States (Shafer & Sierra 1992; Witzig *et al.* 1995). In the absence of evidence of person-to-person transmission, an environmental source of infection has been proposed (Wolinsky 1979). Because of the repeated and almost exclusive isolation of *M. kansasii* from tap water (Engel and Berwald 1980; Wright *et al.* 1985) and its long-term survival in water (Joynson 1979), it has been proposed that one route of pulmonary *M. kansasii* infection is through aerosols (Collins & Yates 1984).

An outbreak of *M. kansasii* infection in one region of Australia prompted development of a DNA probe and application of RFLP methodology to investigate strains in the outbreak. The two methods for RFLP analysis were rRNA-RFLP and genomic RFLP using a whole chromosomal probe to highlight high-molecular-weight restriction fragments (Huang *et al.* 1991). The patient isolates and two of the seven water isolates of *M. kansasii* (all *M. kansasii*-AccuProbe reactive) shared identical rRNA and genomic RFLP

patterns (Huang *et al.* 1991). The two RFLP techniques offered only limited discrimination because only three bands were seen in rRNA-RFLP and eight to10 with genomic RFLP (Huang *et al.* 1991). The two water isolates, while sharing the same DNA fingerprints with the clinical isolates, lacked nitrate reductase activity. All the clinical isolates had nitrate reductase activity (Huang *et al.* 1991).

The PGRS, isolated from *M. tuberculosis* and shown to hybridize to the DNA of other mycobacteria, including *M. kansasii* (Ross *et al.* 1992b), was used to characterize a collection of *M. kansasii* strains recovered from throughout the world (Ross *et al.* 1992a). An *M. kansasii*-specific DNA probe, pMK1–9 (Huang *et al.* 1991) and the *M. kansasii*-AccuProbe were employed for identification. A total of 105 *M. kansasii* strains, identified by biochemical and cultural tests, were examined. Ninety-eight per cent of strains reacted with the AccuProbe and 85% reacted with pMK1–9 (Ross *et al.* 1992a). Eleven different PGRS-RFLP patterns were seen amongst the 105 strains. The strains that failed to hybridize with pMK1–9 had distinct PGRS-RFLP patterns and 19 of the 20 pMK1–9 probe-negative strains had a different 16S rRNA gene sequence (Ross *et al.* 1992a). The majority of the *M. kansasii*-probe non-reactive strains were of European origin (Ross *et al.* 1992a), suggesting a geographical distribution of subtypes of *M. kansasii*. A DNA sequence capable of hybridizing to all *M. kansasii* strains (p6123), including those that failed to react with either the AccuProbe or pMK1–9 probes, yet not to DNA of the other *Mycobacterium* species tested, was isolated (Yang *et al.* 1993b).

Another repetitive DNA sequence originating from *M. tuberculosis*, the MPTR, has been shown to be useful in typing *M. kansasii* and *M. gordonae* (Hermans *et al.* 1992). MPTR, which consists of 10-bp repeats separated by 5-bp non-repeated sequences, hybridizes to slow-growing mycobacteria that share susceptibility to rifampin (Hermans *et al.* 1992). Although the repeated sequence(s) in *M. kansasii* strains reacting with the MPTR probe are not identical to MPTR (i.e. hybridization was weaker with *M. kansasii* restriction fragments compared to *M. tuberculosis*), MPTR-RFLP analysis with *M. kansasii* offers a high degree of discrimination because 12–24 bands are illuminated with the probe (Hermans *et al.* 1992). Among 19 *M. kansasii* strains, nine different MPTR-RFLP patterns were observed. Generally, *M. kansasii* MPTR-RFLP types correlated with phage types, although representatives of one MPTR-RFLP type belonged to six different phage types (Hermans *et al.* 1992).

A repetitive DNA sequence, IS*1652*, cloned from a representative of the *M. kansasii* subgroup that failed to hybridize with pMK1–9 (Ross *et al.* 1992a), was absent in AccuProbe-and pMK1–9-reactive *M. kansasii* strains and other mycobacteria tested (Yang *et al.* 1993a). Between one and 10 copies of IS*1652* were found in different *M. kansasii* subgroup strains (Yang *et al.* 1993a). In spite of the fact that IS*1652* lacks a transposase, open reading frame (ORF), and inverted terminal repeats, IS*1652*-RFLP yielded different patterns for individual strains belonging to the same PGRS-RFLP group (Yang *et al.* 1993a).

PCR amplification of 16S–23S rRNA gene transcribed spacer led to the discovery of heterogeneity among a collection of *M. kansasii* strains (Abed *et al.* 1995). Three different patterns of PCR products were found with 22 *M. kansasii* strains (Abed *et al.* 1995). Unfortunately, no strains were examined that were in common with other studies of molecular epidemiology of *M. kansasii* and the sequences of the PCR products yielding different restriction fragment patterns were not disclosed. Consequently, the relationship of those three groups to the IS*1652*-containing *M. kansasii* types (Picardeau *et al.* 1997) or the *hsp-65*-RFLP types and their 16S-23S rRNA spacer sequences

(Alcaide *et al.* 1997) cannot be determined. Because the PCR products were not hybridized with an rRNA gene probe, it is impossible to determine whether all the PCR products represented amplified fragments of the 16S–23S rRNA gene spacer.

LRF-PFGE typing was performed on a collection of 84 clinical isolates of *M. kansasii* strains from Japan (Iinuma *et al.* 1997). Twenty-one and 16 different LRF-PFGE groups were identified among the strains using *Vsp*I and *Spe*I, respectively (Iinuma *et al.* 1997). Three of the LRF-PFGE types belonged to one of the *M. kansasii* subgroups defined by 16S rRNA sequence (Iinuma *et al.* 1997). Unfortunately, although the authors report using *Dra*I and *Xba*I for LRF-PFGE (in common with Picardeau *et al.* 1997), the results are not reported, preventing a comparison of groups.

Two thorough and comprehensive investigations of the molecular epidemiology of *M. kansasii* employing a variety of typing methods led to the identification of five different types (Alcaide *et al.* 1997; Picardeau *et al.* 1997). Clinical and water isolates of *M. kansasii* were analysed by DNA probes and MPTR-RFLP, IS*1652*-RFLP, RFLP analysis of a PCR product of the *hsp-65* gene, LRF-PFGE, amplified fragment length polymorphism (AFLP), and for the presence or absence of the *gyrA* intein. The five types defined by *hsp-65*-RFLP (Table 8.3) contained both clinical and water isolates (Picardeau *et al.* 1997). MPTR-RFLP and *hsp-65* RFLP yielded type-specific patterns, whereas PFGE and AFLP yielded polymorphic patterns among individuals belonging to a single *hsp-65*-RFLP type (Picardeau *et al.* 1997). The two types that carried IS*1652* were AccuProbe negative (see Table

Table 8.3 Epidemiological types of *Mycobacterium kansasii*. Data from Picardeau *et al.* (1997) and Alcaide *et al.* (1997).

hsp-65 RFLP type	p6123	AccuProbe	pMK1–9	IS*1652*	*gyrA* intein	LRF-PFGE
I	+	+	+	−	+	Types a–d
II	+	−	−	+	−	Types a–e
III	+	±	−	+	−	Types a–c
IV	+	−	?	−	+	One type
V	+	±	?	−	+	One type

For abbreviation definitions, see Table 8.1.

8.1), in agreement with Yang *et al.* (1993a), and IS*1652*-RFLP patterns were polymorphic in only one of those two (Picardeau *et al.* 1997). PCR-based amplification of the *M. kansasii*-species-specific probe (p6123) isolated and described by Yang *et al.* (1993b) was positive for all isolates in the study in which it was employed (Picardeau *et al.* 1997). Sequence heterogeneity of the analogue of the *mpb70* protein gene demonstrated by single-strand conformational polymorphism (SSCP) analysis (Woolford *et al.* 1997) also demonstrates the diversity of *M. kansasii*.

Clearly, development of these typing methods will contribute directly to describing the epidemiology of *M. kansasii*. First, typing will lead to the identification of environmental sources for *M. kansasii* infection. For example, *M. kansasii* isolates that failed to react with the AccuProbe were associated with disseminated disease in AIDS patients, whereas AccuProbe-positive, type I isolates were recovered from immunocompetent patients with pulmonary disease (Tortoli *et al.* 1994). Second, it could lead to the identification of genetic determinants for pathogenicity, by identifying *M. kansasii* types not (or rarely) associated with infection and hence, lacking pathogenicity determinants.

4.2 Molecular epidemiology of the *Mycobacterium avium* complex

M. avium and *M. intracellulare*, members of the *M. avium* complex (MAC), cause pulmonary infections in immunocompetent persons with predisposing lung conditions (e.g. pneumoconiosis and silicosis; Wolinsky 1979). In children, *M. avium* causes cervical lymphadenitis (Wolinsky 1995). In addition, *M. avium* has been shown to colonize cystic fibrosis patients (Kinney *et al.* 1989). Approximately 25–50% of US and European AIDS patients with profound immunodeficiency are infected with *M. avium* (Moore & Chaisson 1996). Although immunocompetent patients are infected with either *M. avium* or *M. intracellulare*, almost all AIDS patients are infected with *M. avium* (Drake *et al.* 1988; Guthertz *et al.* 1989). A wide range of *M. avium* serotypes have been recovered from non-AIDS patients and environmental samples, but a restricted a range is found among AIDS patient isolates. *M. avium* complex isolates of serotypes 1–6, 8–11, and 21 hybridized with a commercial *M. avium*-specific rRNA gene probe and those of serotypes 7 and 12–20 and 25 reacted with the *M. intracellulare*-specific probe (Saito *et al.* 1990). Among AIDS patients in the United States, serotypes 1, 4, and 8 predominate (Yakrus and Good 1990) whereas serotype 6 predominates amongst AIDS patient isolates in Sweden (Hoffner *et al.* 1990; Julander *et al.* 1996).

There is a wide diversity of genotypes within what is called the *M. avium* complex at the present time. Further, the taxonomy of this group is undoubtedly going to change. Three subspecies have been proposed; *M. avium* ssp. *avium*, *M. avium* ssp. *silvaticum*, and *M. avium* ssp. *paratuberculosis* (Thorel *et al.* 1990). The latter subspecies is the causative agent of Johne's disease in cattle and shares 100% DNA:DNA relatedness with subspecies *avium*. It can be distinguished from the other two subspecies by its absolute requirement for mycobactin and the presence of multiple copies of IS*900* (Thorel *et al.* 1990). As the evidence below will demonstrate, there is wide diversity of genotypes within the other two *M. avium* subspecies. Generally, it appears that many *M. avium* complex isolates recovered from animals, birds, and humans belong to either *M. avium* or *M. intracellulare* by cultural and biochemical tests or by hybridization (e.g. AccuProbe, Gen-Probe, San Diego, CA). However, the remaining isolates, including animal, bird and human isolates and many from environmental samples (Fonteyne *et al.* 1997), do not fall into those defined categories. Typing may assist the assignment of those isolates to distinct taxonomic groups.

PCR amplification of two sequences, DT1 and DT6, can be used for assignment of *M. avium* complex isolates to either *M. intracellulare* (DT1 positive) or *M. avium* (DT6 positive) (Devallois *et al.* 1996, 1997). For example, a number of *M. intracellulare* AccuProbe-negative isolates were shown positive for DT1 amplification (Devallois *et al.* 1997), thus clarifying their identification.

A wide variety of markers have been employed for epidemiological studies of members of the *M. avium* complex. These include single gene-RFLP (i.e.

32-kDa protein gene, *dnaJ* gene), insertion sequences (e.g. IS*900*, IS*901*/IS*902*, IS*1245*, and IS*1311*), spacer typing (e.g. IS*1245*-IS*1311* and 16S-23S ribosomal DNA internal transcribed spacer), LRF-PFGE, MEE, arbitrary primed PCR (AP-PCR), and plasmids. Because of the number of different molecular markers used for studies of *M. avium* and the fact that different markers have not been used with the same strains in the same study, the results with each marker will be considered separately.

4.2.1 Single-copy gene restriction fragment length polymorphism

Polymorphism of single, conserved genes in isolates of the *M. avium* complex has been reported. Although isolate-to-isolate variation was reported for both the 32-kDa protein (Soini *et al.* 1996) and *dnaJ* (Victor *et al.* 1996), there were too few bands to provide the discrimination required for epidemiological studies. However, RFLP analysis of single genes may prove useful in identifying relationships between groups within the *M. avium* complex. For example, in a collection of 22 *M. avium* complex isolates (i.e. reacted with the MAC AccuProbe and neither the *M. avium*, nor *M. intracellulare* probes), all had different sequences for the 32-kDa-protein gene (Soini *et al.* 1996). Sixteen of the 22 isolates could be grouped with *M. avium* and *M. intracellulare* reference strains and the remaining five fell into a single group; designated the *Mycobacterium avium*—*intracellulare* and *mycobacterium* species (MAIX) group (Soini *et al.* 1996). Members of that latter group were pigmented and lacked tellurite reductase activity (Soini *et al.* 1996). A wide range of heterogeneity of the *hsp65* gene was found among *M. intracellulare* DT1-positive isolates, demonstrating the diversity of that species (Devallois *et al.* 1997).

4.2.2 IS*900*

Although IS*900* is unique to *M. paratuberculosis*, the probe carrying the element (i.e. pMB22) also carries chromosomal DNA that flanks the site of insertion. Consequently, pMB22 can be used as a probe for RFLP analysis. PMB22-RFLP analysis yields multiple hybridizing bands with strains of *M. avium* (four to seven bands) and *M. intracellulare* (four to 12 bands) (De Smet *et al.* 1996). The majority of isolates of *M. avium* recovered from AIDS patients belong to a single type called A (Hampson *et al.* 1989; De Smet *et al.* 1996). That same RFLP pattern is found among *M. avium* isolates recovered from non-AIDS patients (De Smet *et al.* 1996). In addition, other *M. avium* isolates from AIDS patients demonstrated an 'A-related' profile, sharing two to four bands in common with the A-type (De Smet *et al.* 1996). Ten of 14 *M. intracellulare* isolates, all recovered from non-AIDS patients, each yielded a unique profile (De Smet *et al.* 1996). There was no correlation between pMB22-RFLP type and serotype (De Smet *et al.* 1996). Polyclonal *M. avium* infection has been demonstrated by differences in pMB22-RFLP patterns of isolates from a single patient (Visuvanathan *et al.* 1992).

M. avium isolates from European human immunodeficiency virus (HIV)-infected patients were predominantly the A-type (Hampson *et al.* 1989; Portaels *et al.* 1990; McFadden *et al.* 1992; Visuvanathan *et al.* 1992; De Smet *et al.* 1996). In contrast, *M. avium* isolates from HIV-infected African patients belonged to another type, H (Portaels *et al.* 1990; McFadden *et al.* 1992). Members of pMB22-RFLP type H were homogeneous, failed to grow at 45°C (i.e. not *M. avium*), reacted with the MAC AccuProbe, and failed to react with either the *M. avium*- or *M. intracellulare*-specific probes (Fonteyne *et al.* 1997). Based on the 16S rRNA gene sequence, the isolates were neither *M. avium* nor *M. intracellulare*, but represented a homogeneous group (Fonteyne *et al.* 1997).

Such RFLP-based typing data has been used to estimate population diversity of the species *M. avium* and *M. intracellulare*. Based on the observations that pMB22-RFLP profiles of 10 *M. intracellulare* strains were all unique and those of 17 *M. avium* strains belonged to a limited range of RFLP types, it was suggested '*M. intracellulare* is genetically more heterogeneous than *M. avium*' (De Smet *et al.* 1996). Although that may ultimately prove to be the case, the suggestion was not merited by the data. First, the population sizes examined were small. Second, *M.*

avium and *M. intracellulare* isolates from all possible sources were not included. Thus, all that can be said is that a limited spectrum of pMB22-RFLP types have been recovered from AIDS patients, confirming the data reporting the limited spectrum of serotypes in AIDS patients (Yakrus and Good 1990).

4.2.3 IS901 (IS902)

Using pMB22 as a probe, isolates of *M. avium*, particularly those recovered from animals, had band patterns that differed from those in most *M. avium* complex isolates (Kunze *et al.* 1991). Those different patterns were shown to be due to an IS*900*-related insertion sequence, designated IS*901* (Kunze *et al.* 1991). Independently, an identical insertion sequence designated IS*902* was isolated from a wood pigeon isolate of *M. avium* (Moss *et al.* 1992).

The presence or absence of IS*901* distinguishes two groups of *M. avium* isolates. In a study of 202 *M. avium* isolates, 48 of the 55 isolated from birds or other animals had IS*901* (IS*902*) (Kunze *et al.* 1991). Those were most likely representatives of serotypes 1, 2 or 3. In contrast, IS*901* (IS*902*) was absent in 127 *M. avium* isolates recovered from AIDS patients and 20 of environmental origin (Kunze *et al.* 1991). It is expected that those isolates would be members of serotypes 4, 6 or 8. Analysis of banding patterns of restriction endonuclease-digested DNA revealed that there were 10–12 copies of IS*902* in the genomes of wood pigeon strains of *M. avium* (Moss *et al.* 1992). Those findings were confirmed in a second study. Specifically, IS*901* (IS*902*) was absent in 81 *M. avium* isolates recovered from AIDS patients and in 36 of 40 *M. avium* isolated from non-AIDS patients (Kunze *et al.* 1992). IS*901* (IS*902*) was shown to be present in almost every *M. avium* strain of avian or other animal origin, with the exception of porcine strains (Kunze *et al.* 1992; Ahrens *et al.* 1995; Bono *et al.* 1995). In a third study, two of seven *M. avium* isolates from humans (HIV-infection status unknown) had IS*901* (IS*902*) and IS*901* was not found in any swine or environmental isolate (Nishimori *et al.* 1995).

Although three different IS*901* (IS*902*)-RFLP patterns were found among 10 different *M. avium* isolates (Moss *et al.* 1992), others reported that the IS*901* (IS*902*)-RFLP banding patterns in the *M. avium* isolates of bird and animal origin was quite uniform (Kunze *et al.* 1991; Kunze *et al.* 1992). The difference could be due to the fact that eight of the 10 strains included in the former study were recovered from wood pigeons (Moss *et al.* 1992). Although the limited heterogeneity of IS*901* (IS*902*)-RFLP patterns reduces its utility in strain typing, IS*901* (IS*902*)-RFLP typing may be used to indicate whether an *M. avium* isolate is of avian or animal origin (Kunze *et al.* 1991). It has been proposed that this subtype of the *M. avium* complex be assigned to the subspecies *M. avium* ssp. *silvaticum* (Thorel *et al.* 1990).

4.2.4 IS1110

The insertion sequence IS*1110* was identified in *M. avium* as a mobile genetic element capable of transposing to plasmid DNA and consequently increasing its size (Hernandez-Perez *et al.* 1994). Cultures arising from the growth of single colonies of an IS*1110*-containing strain were found to have three different patterns reflecting the transposition of IS*1110*. Between 25 and 50% of colonies showed evidence of transposition (Hernandez-Perez *et al.* 1994). Initially because of its lack of stability and narrow distribution (i.e. only four of 35 *M. avium* strains carried the element (Hernandez-Perez *et al.* 1994)), one would have concluded that IS*1110* would be unsuitable for typing strains of *M. avium*. However, recent studies using a full-length IS*1110* probe showed that IS*1110*-hybridizing sequences were present in 70% of 50 *M. avium* isolates and that patterns were stable (Hernandez-Perez *et al.* 1997). Because IS*1110*-hybridizing sequences were present in all types of *M. avium* isolates (e.g. AIDS and non-AIDS, IS*901*-hybridizing and IS*901* non-hybridizing) and the discrimination index for typing with IS*1110* was high (among isolates with IS*1110*-reacting sequences), typing with this insertion sequence may prove to be valuable; especially for other *M. avium* complex species and IS*1245*-negative isolates.

4.2.5 IS1245

IS*1245* was detected as an *M. avium* DNA fragment

that yielded a multi-banded RFLP pattern when used as a probe (Guerrero *et al*. 1995). IS*1245* is found only in *M. avium* ssp. *avium*, *M. avium* ssp. *paratuberculosis*, and *M. avium* ssp. *silvaticum* and absent from other mycobacteria; most notably *M. intracellulare* and *M. scrofulaceum* (Guerrero *et al*. 1995). *M. avium* isolates recovered from human patients had a high number of IS*1245* copies with a median number of 16–20 copies and a range of one to 27 (Guerrero *et al*. 1995; Picardeau and Vincent 1996). Those isolates would be expected to lack IS*901* and belong to serotypes 4, 6 or 8. Thirty-eight *M. avium* isolates fell into 27 types and IS*1245*-RFLP typing had a discrimination index of 0.97 in one study (Guerrero *et al*. 1995). Swine isolates of *M. avium* also displayed a high IS*1245* copy number (i.e. >8). Strains of avian and bovine origin had identical two-band patterns (Bono *et al*. 1995; Guerrero *et al*. 1995). Those avian and bovine strains would most likely contain IS*901* and belong to serotypes 1, 2 or 3.

4.2.6 IS*1311*

An IS*1245*-related mobile genetic element, IS*1311* has also been employed in typing strains of *M. avium* (Roiz *et al*. 1995). Nineteen different IS*1311*-RFLP patterns (i.e. two to 20 bands) were displayed among 75 *M. avium* strains (i.e. positive by *M. avium* probe, negative by *M. intracellulare* probe). IS*1311*-RFLP patterns from *M. avium* strains from AIDS patients were highly polymorphic (Roiz *et al*. 1995). Infection by more than a single *M. avium* strain was demonstrated by IS*1311*-RFLP typing in one patient (Roiz *et al*. 1995).

Significant information relevant to the epidemiology of *M. avium* has come from employment of *M. avium* IS*1311*-RFLP typing. First, clusters of *M. avium* strains with the same IS*1311*-RFLP type were recovered from groups of AIDS and non-HIV-infected patients at different times over the period 1989–92 (Roiz *et al*. 1995). That data suggest that the patients were infected from a common source, either another patient or environmental compartment (e.g. water). Second, *M. avium* strains of some IS*1311*-RFLP types, although recovered from non-HIV-infected individuals, were rarely recovered from AIDS patients (Roiz *et al*. 1995). These data are consistent with the hypoth-

esis that there is some degree of selection for or against certain *M. avium* IS*1311*-RFLP types in AIDS patients.

4.2.7 IS*1245*–IS*1311* 'spacer' typing

Using primers hybridizing to IS*1245* and IS*1311* to amplify the regions between these two high copy number insertion sequences, band patterns of the amplified products were used to discriminate between strains of *M. avium* (Picardeau & Vincent 1996). IS*1245*—IS*1311* 'spacer' typing offered the same level of discrimination as did IS*1245*-RFLP typing. IS*1245*—IS*1311* 'spacer' types of isolates recovered from single patients isolated over periods of 0.5–23 months remained the same, demonstrating the stability of the types. However, there were patients whose *M. avium* isolates had different profiles, indicating the presence of infection by more than a single *M. avium* clone (Picardeau & Vincent 1996).

4.2.8 16S–23S rDNA internal transcribed spacer

A wide heterogeneity exists in the sequence of the 16S–23S rDNA internal transcribed spacer (ITS) of members of the *M. avium* complex (Frothingham & Wilson 1993). Distinct sequences were designated 'sequevars'. Sixteen reference strains identified on the basis of cultural and biochemical tests and DNA probe reactivity as *M. avium* fell into four ITS sequevars (Mav-A to Mav-D) and 12 identified as *M. intracellulare* belonged to a single ITS sequevar (Min-A). Seven strains that either reacted with the *M. intracellulare* probe (one), both probes (two), or did not react with either probe, each had a unique ITS sequences; sequevars MAC-A through MAC-G (Frothingham & Wilson 1993).

M. avium complex isolates recovered from patients with disseminated or only pulmonary disease were characterized by ITS sequence (Frothingham & Wilson 1994). Isolates recovered from 13 HIV-infected patients with disseminated disease belonged to a narrow range of sequevars, either Mav-A (three patients) or Mav-B (10 patients). In contrast, isolates recovered from seven non-HIV-infected patients with only pulmonary disease belonged to a wide range of sequevars; MAC-A (four patients) and Min-A, MAC-A, and MAC-H (one patient each) (Frothingham & Wilson 1994).

Expansion of the number of ITS sequevars has come from study of *M. avium* and *M. intracellulare* strains from HIV-infected and non-HIV-infected humans and animals (De Smet *et al.* 1995). In addition to identifying new ITS sequevars, the data demonstrated that the Mav-B sequevar was found in 18 of 20 *M. avium* strains isolated from patients with AIDS and in all three *M. avium* strains recovered from children with cervical lymphadenitis (De Smet *et al.* 1995). In agreement with the data of Frothingham and Wilson (1994), *M. avium* complex strains isolated from non-HIV-infected patients belonged to *M. intracellulare* Min and *M. avium* complex MAC ITS sequevars (De Smet *et al.* 1995). Isolates of the *M. avium* complex recovered from cattle, pigs and pigeons belonged to sequevars Mav-A and Mav-B (De Smet *et al.* 1995).

4.2.9 LRF-PFGE

Members of the *M. avium* complex, including strains of *M. paratuberculosis*, displayed different LRF-PFGE patterns (Lévy-Frébault *et al.* 1989; Coffin *et al.* 1992). In fact, individual *M. paratuberculosis* strains did not fall into a single LRF-PFGE type, but displayed three different patterns (Lévy-Frébault *et al.* 1989; Coffin *et al.* 1992). Because cleavage with the enzymes *Ssp*I or *Dra*I yielded patterns with a substantial number of fragments (i.e. six to 24), the technique offers a high discriminatory power.

In a study of four reference strains and 35 isolates of the *M. avium* complex, 39 different LRF-PFGE patterns were reported using the enzymes *Xba*I or *Asn*I (Mazurek *et al.* 1993). Strains recovered from single patients over a 27 or 37-month period had the same PFGE-RFLP patterns (Mazurek *et al.* 1993). In another study, 38 different LRF-PFGE patterns were displayed by 121 *M. avium* isolates with the enzymes *Xba*I or *Ase*I (Burki *et al.* 1995). Although it might appear that the latter study did not show the same degree of discrimination as did the former, a substan-

tial proportion of the 121 *M. avium* strains fell into 'pseudo-outbreak patterns' due to laboratory contamination as a result of inadequate sterilization of a sampling needle (Burki *et al.* 1995).

Since those initial reports, LRF-PFGE-based RFLP analysis has been employed to demonstrate infection by more than a single *M. avium* complex strain in AIDS patients (Arbeit *et al.* 1993; Slutsky *et al.* 1994; Von Reyn *et al.* 1995) and showed that AIDS patients and waters to which they were exposed shared a single clone of *M. avium* (Von Reyn *et al.* 1994).

4.2.10 Multilocus enzyme electrophoresis

In contrast to the limited diversity of MEE types among members of the *M. tuberculosis* complex, isolates of the *M. avium* complex exhibit a great deal of polymorphism. In a collection of two *M. paratuberculosis* strains and 35 *M. avium* complex isolates representing a wide range of serotypes, all enzymes were polymorphic and a mean genetic diversity of 0.38 was obtained (Wasem *et al.* 1991). Six different electrophoretic types (ETs) were displayed by 10 *M. intracellulare* isolates, 17 ETs among 25 *M. avium* isolates, one ET for two *M. paratuberculosis* strains, and two widely different ETs for two *M. scrofulaceum* strains (Wasem *et al.* 1991). MEE patterns can also be used to distinguish those species from one another (Wasem *et al.* 1991).

MEE has been used to examine the diversity of *M. avium* isolates of serotypes 4 and 8 (Yakrus *et al.* 1992). The mean genetic diversity was 0.23 and only 10 of the 17 enzymes were polymorphic (Yakrus *et al.* 1992). The diversity would be expected to be lower because a narrower range of *M. avium* complex isolates was examined. The distribution of *M. avium* isolates among ETs was not uniform. One ET was found among 39% of the serotype 4 strains and another ET was displayed by 42% of serotype 8 strains, all recovered from patients with AIDS (Yakrus *et al.* 1992).

Among a collection of 18 reference strains and 72 human and 25 animal and bird *M. avium* complex isolates from Australia, 58 different ETs were identified and 15 of 17 enzymes were polymorphic (Feizabadi *et al.* 1996). The index of diversity was 0.28. No evidence of a relationship between ET type, serotype and geographical origin was found and the reference strains were distributed among different ETs (Feizabadi *et al.* 1996). Isolates from the same area and from different animals, namely pigs and chickens or pigs and humans, belonged to the same ET, suggesting a common source of infection (Feizabadi *et al.* 1996). Members from single MEE ETs were found with different LRF-PFGE patterns (Feizabadi *et al.* 1996). Two of six human patients harbored *M. avium* complex isolates belonging to different LRF-PFGE types.

4.2.11 Arbitrary primed polymerase chain reaction

Rapid discrimination of *M. avium* isolates has been demonstrated by AP-PCR (also called RAPD). In the first published study, three primers yielded PCR-amplified products whose patterns allowed discrimination between strains isolated from four different patients (Matsiota-Bernard *et al.* 1997). Each PCR reaction with a single primer resulted in between three and seven amplified products of different size. In one patient, two isolates yielded different AP-PCR patterns with all three primers (Matsiota-Bernard *et al.* 1997). The other three patients were infected with isolates of a single AP-PCR type.

Although AP-PCR appears to offer sufficient discrimination to distinguish between *M. avium*, *M. intracellulare* and *M. scrofulaceum*, and for epidemiological investigations, characterization of AP-PCR products demonstrated that only a limited portion of the genome was being examined. Using one AP-PCR product as a probe, it was found that a number of bands of different size produced by the same primer hybridized with the probe (L.E. Via, unpublished data 1999). This evidence suggests that a single primer initiated amplification from a single site and at a number of distant sites, producing a set of PCR fragments that shared sequence homology. In spite of this finding, AP-PCR offered the same degree of discrimination as IS*1245*-RFLP typing and therefore was useful in typing environmental and patient isolates of the *M. avium* complex (Via *et al.* in preparation).

4.2.12 Plasmids

Plasmids are present in a substantial proportion of *M. avium* complex isolates recovered from non-HIV-infected patients (Meissner & Falkinham 1986; Jensen *et al.* 1989; Hellyer *et al.* 1991), HIV-infected patients (Hellyer *et al.* 1991; Crawford & Bates 1986; Jucker & Falkinham 1990), swine (Masaki *et al.* 1989), and the environment (Meissner & Falkinham 1986). A higher frequency of *M. avium* complex isolates recovered from patients with AIDS have small plasmids (i.e. < 25 kb) compared to isolates from non-HIV-infected patients (Crawford & Bates 1986). Small *M. avium* plasmids fall into four different DNA:DNA similarity groups (Jucker & Falkinham 1990; Jucker 1991). Because individual plasmids of any of the four groups are found in *M. avium* complex isolates with any other, but not one of the same group, each represents a single incompatibility group (Jucker & Falkinham 1990; Jucker 1991).

Because a substantial proportion of *M. avium* complex isolates have plasmids and the plasmids are unusually stable (Crawford & Bates 1986; Jucker & Falkinham 1990), plasmids can be used as targets for typing strains. Using DNA probes specific for each of the four small plasmid incompatibility groups (Jucker & Falkinham 1990; Jucker 1991), dot-blot hybridization has been used to determine whether AIDS patients were infected by more than one clone of *M. avium*. Unique clones were defined as those having a different profile for the four plasmids. Among 30 *M. avium*-infected AIDS patients, 22 (73%) had isolates carrying plasmids and of those, 18 (82%) had more than a single *M. avium* isolate recovered from blood. Blood from nine of those 18 patients (50%) yielded two or more *M. avium* strains with different plasmid profiles (Eaton *et al.* in preparation). Although plasmid typing is still in its infancy, dot-blot hybridization can be easily performed and is a simple method of rapidly determining whether a patient is infected with one or more clones of the *M. avium* complex.

4.3 Molecular epidemiology of *Mycobacterium xenopi*

M. xenopi is a waterborne, opportunistic pathogen causing pulmonary infections (Costrini *et al.* 1981; Slosárek *et al.* 1993) and disseminated infections in liver transplant patients (McDiarmid *et al.* 1995) and in AIDS patients (Shafer & Sierra 1992; Jacoby *et al.* 1995). *M. xenopi* infections usually occur as outbreaks associated with the presence of *M. xenopi* in water to which patients were exposed (Costrini *et al.* 1981; Slosárek *et al.* 1993) and it has been proposed that infection occurs through the formation of *M. xenopi*-laden aerosols (Collins & Yates 1984). Some isolates identified initially as members of the *M. avium* complex were later shown to represent isolates of *M. xenopi* (Marx *et al.* 1995).

LRF-PFGE and IS-RFLP analyses have been shown to be of use in molecular epidemiological studies of *M. xenopi*. RFLP analysis has been performed using two related insertion sequences, IS*1081* and IS*1395*. IS*1081*, an insertion sequence originally identified in members of the *M. tuberculosis* complex, has been used for fingerprinting isolates of *M. xenopi* (Collins 1994). RFLP analysis of *M. xenopi* isolates demonstrated that a labelled IS*1081*-probe hybridized to between four and 13 individual restriction fragments (Collins 1994). A number of *M. xenopi* isolates recovered from different patients and regions shared the same IS*1081*-RFLP pattern (Collins 1994).

Using IS*1081* as a probe, a related insertion sequence, IS*1395*, was identified and cloned from a strain of *M. xenopi* (Picardeau *et al.* 1996). IS*1395* shares 86% sequence similarity with IS*1081*, suggesting that cross-hybridization was responsible for the reaction between IS*1081* and restriction fragments of genomic DNA of *M. xenopi* strains (Collins 1994). IS*1395* also shares 45% sequence similarity with IS*1245* and IS*1311* of the *M. avium* complex (Picardeau *et al.* 1996). The copy number of IS*1395* in *M. xenopi* strains was between three and 18 with a median at 14 (Picardeau *et al.* 1996). Although the high median copy number would suggest that the element offers a high degree of discrimination and

utility for epidemiological studies, the RFLP patterns demonstrated by 19 unrelated *M. xenopi* strains were relatively similar (Picardeau *et al.* 1996). Possibly, *M. xenopi* is a more genetically homologous species. Comparison of IS*1395*-RFLP and LRF-PFGE analyses demonstrated that the discrimination of both methods were similar, with the exception of those *M. xenopi* strains carrying few copies (i.e. three to five) of IS*1395* (Picardeau *et al.* 1996). That too, suggests genetic uniformity of *M. xenopi*. Approximately 25 bands were observed by LRF-PFGE and a number of bands were shared in common (Picardeau *et al.* 1996). Although IS*1395*-RFLP could distinguish strains with identical LRF-PFGE patterns, the latter technique may be of more widespread use in epidemiological studies because almost every *M. xenopi* strain yielded 20–25 bands by LRF-RFLP (Picardeau *et al.* 1996).

4.4 Molecular epidemiology of *Mycobacterium haemophilum*

A new species, *Mycobacterium haemophilum*, with a requirement for heme or complexed iron and a low-temperature optimum for growth was first described in 1978 as a causative agent of skin granulomas (Sompolinsky *et al.* 1978). *M. haemophilum* has also been shown to cause infections in healthy children (Dawson *et al.* 1981), renal transplant patients (Gouby *et al.* 1988), and bone marrow transplant patients (White *et al.* 1995) and is now recognized as an emerging pathogen of immunocompromised patients (Straus *et al.* 1994). Although attempts to recover *M. haemophilum* from possible water sources have been unsuccessful (Gouby *et al.* 1988), possibly because of its fastidious growth requirements, it is likely that water (in iron pipes?) is the source.

LRF-PFGE patterns of 19 *M. haemophilum* isolates have been analysed. Among the 19 isolates, four to seven bands were observed in *Xba*I digests of total DNA and six different types could be distinguished (Yakrus & Straus 1994). Evidence for six different types among just 19 isolates suggests genotypic heterogeneity that can be exploited in epidemiological studies. Identical patterns were observed for 12 isolates from five different hospitals in the New York City metropolitan area (Yakrus & Straus 1994). Although that evidence suggests a common source of infection, caution must be taken because of the rather limited discriminatory power of the method (i.e. four to seven bands per isolate).

4.5 Molecular epidemiology of *Mycobacterium malmoense*

As is the case with a number of other mycobacterial species, *M. malmoense* is a recently described species (Schröder & Juhlin 1977) of increasing significance for human health. *M. malmoense* has been reported as both a pulmonary pathogen (Zaugg *et al.* 1993) and the causative agent of disseminated disease in AIDS patients (Fakih *et al.* 1996).

Thin-layer chromatography of surface glycolipids has been used to type *M. malmoense* isolates (Katila *et al.* 1991). Thirty *M. malmoense* isolates representing five different glycolipid types were used to investigate whether rRNA-RFLP typing could be used for typing. Both 16S and 23S rRNA probes yielded three different patterns (Kauppinen *et al.* 1994b). Combining the data yielded five different rRNA-RFLP types, none of which was restricted to a single glycolipid type (Kauppinen *et al.* 1994b). Thus, a combination of both rRNA-RFLP and glycolipid typing would provide sufficient discrimination for epidemiological studies.

AP-PCR (or RAPD) typing has also been investigated for its utility for typing isolates of *M. malmoense*. Forty-five *M. malmoense* isolates were included in the study and two PCR primers selected on the number of bands generated (Kauppinen *et al.* 1994a). The patterns were reproducible and the two primers were reported to be able to distinguish differences between strains (Kauppinen *et al.* 1994a). There was no overlap between individual glycolipid, rRNA-RFLP, or AP-PCR types (Kauppinen *et al.* 1994a), providing further evidence that a combination of typing methods would be effective for epidemiological studies.

4.6 Molecular epidemiology of *Mycobacterium ulcerans*

M. ulcerans causes chronic, necrotizing ulcers on the skin of infected humans (Portaels *et al.* 1996). Increased numbers of cases of *M. ulcerans* infection have been reported in both Africa (Marston *et al.* 1995) and Australia (Johnson *et al.* 1996). Epidemiological studies have demonstrated an association between infection and close proximity to swamps or rivers (Marston *et al.* 1995). To date, *M. ulcerans* has not been isolated from the environment (Portaels *et al.* 1996), possibly because of its very slow growth; even for a mycobacterium! Three *M. ulcerans* types have been distinguished on the basis of the base sequence at the 3' end of the 16S rRNA (Portaels *et al.* 1996). Those types corresponded with the continent of origin of the isolates: type 1, Africa; type 2, Australia/South-East Asia; and type 3, America (Portaels *et al.* 1996). In addition, there were differences in the enzymatic characteristics of types (Portaels *et al.* 1996).

Australian and African isolates of *M. ulcerans* were typed using PGRS-RFLP (Jackson *et al.* 1995). Approximately 12–20 restriction fragments hybridized with the PGRS probe (i.e. pMB12) and 11 different RFLP patterns were recognized by the authors among 56 *M. ulcerans* isolates tested (Jackson *et al.* 1995). Multiple isolates from single patients yielded the same pattern and single isolates tested repeatedly had the same pattern (Jackson *et al.* 1995), indicating the stability of PGRS-RFLP patterns in *M. ulcerans*. The five American (Benin) isolates fell into two unique PGRS-RFLP types, the seven African (Zaire) isolates belonged to three unique PGRS-RFLP-types and the Australian patient iso-lates belonged to six different PGRS-RFLP types (Jackson *et al.* 1995), demonstrating a rather wide genetic diversity of *M. ulcerans* types. Isolates from patients who were thought to have been infected in North Queensland and Papua, New Guinea or Malaysia and Darwin displayed the same PGRS-RFLP type (Jackson *et al.* 1995), suggested a rather wide geographical distribution of types.

4.7 Molecular epidemiology of rapidly growing mycobacteria

There has been a growing recognition of the role of rapidly growing mycobacteria in nosocomial disease and pulmonary and disseminated infections (Wayne & Sramek 1992; Wallace 1994). Nosocomial disease has been associated with the presence of rapidly growing mycobacteria in supposedly sterile solutions (Ashford *et al.* 1997) and in water (Burns *et al.* 1991; Hector *et al.* 1992). Infections associated with cardiac surgery or mammaplasty are often caused by rapidly growing mycobacteria, primarily in the south-eastern United States (Wallace *et al.* 1989a,b). Progress and employment of molecular markers for rapidly growing mycobacteria has been hindered by the lack of clearly defined species. Recent recognition that the rapidly growing species causing the majority of infections, *M. abscessus, M. chelonae* and *M. fortuitum* (90%, Wallace 1994) were distinct species (Lévy-Frébault *et al.* 1986; Kusunoki & Ezaki 1992) has made application of a number of molecular techniques for epidemiological studies possible.

Plasmid typing, MEE, 16S rRNA gene RFLP, LRF-PFGE and AP-PCR have all been employed for epidemiological studies of rapidly growing mycobacteria. Although 16S rRNA gene RFLP was able to distinguish between rapidly growing *Mycobacterium* species (Domenech *et al.* 1994), it lacked the level of discrimination necessary for epidemiological studies (Yew *et al.* 1993). Interestingly, *M. abscessus* and *M. chelonae* have a single 16S rRNA gene copy, like slowly growing mycobacteria, and unlike other rapidly growing mycobacteria that have two (Domenech *et al.* 1994). The combination of plasmid typing and MEE provided sufficient discriminatory power to identify water as the source of an *M. fortuitum* strain leading to sternal and abdominal wounds associated with cardiac bypass-related surgery (Wallace *et al.* 1989a). Plasmid typing, alone, was sufficient to demonstrate the heterogeneity of *M. fortuitum* isolates associated with infections following augmentation mammaplasty (Wallace *et al.* 1989b). Identity of antibiotic-susceptibility and MEE patterns of *M. abscessus* isolates recovered from patients and from a solution added to blood culture medium, led to identification of the source of a pseudo-epidemic (Ashford *et al.* 1997).

LRF-PFGE analysis has been employed in epidemiological studies of nosocomial outbreaks of infections caused by *M. fortuitum, M. abscessus,* and *M. chelonae*. In a study of respiratory tract colonization, LRF-PFGE analysis and plasmid profiles were used to identify the drinking water system in a hospital as the source of *M. fortuitum* (Burns *et al.* 1991). In four nosocomial outbreaks of *M. fortuitum* infection, LRF-PFGE analysis demonstrated that outbreak isolates shared common fingerprints and led to the identification of water as the source of the outbreak strain (Hector *et al.* 1992). LRF-PFGE analysis patterns of *M. chelonae* isolates are polymorphic and stable and were used to demonstrate the identity of clustered *M. chelonae* isolates and pinpoint the environment as the source of infection (Wallace *et al.* 1993). Unfortunately, LRF-PFGE analysis of *M. abscessus* was not successful because DNase activity in lysates led to loss of genomic DNA in 13 of 30 (43%) isolates (Wallace *et al.* 1993). It is quite possible that DNase activity could also affect plasmid and 16S rRNA gene RFLP typing as well.

Because of the inability of LRF-PFGE to type all *M. abscessus* isolates, AP-PCR was investigated as a typing tool. AP-PCR analysis of 118 *M. abscessus* isolates, including strains with DNase activity used in the earlier LFR-PFGE study, demonstrated the utility of that fingerprinting technique (Zhang *et al.* 1997). AP-PCR distinguished the same clusters and isolates within clusters as did LRF-PFGE (Zhang *et al.* 1997). However, those authors also noted (as has this author) that AP-PCR patters are influenced strongly by the methodology and that purified DNA and multiple (yet single) primers are required (Zhang *et al.* 1997).

4.8 Mycobacteria in search of molecular markers

A number of *Mycobacterium* species are of significance to either human or animal health and, to date, lack demonstrated methods for molecular typing. The species include *M. marinum* (Wayne & Sramek 1992), *M. scrofulaceum* (Sanders *et al.* 1995), *M. simiae* (Valero *et al.* 1995), *M. szulgai* (Wayne & Sramek 1992) and the recently described *M. genavense* (Nadal *et al.* 1993). In contrast to those pathogens, the non-pathogen *M. gordonae* can be typed with either the PGRS (Ross *et al.* 1992b) or the MPTR (Hermans *et al.* 1992).

Typing methods for other pathogenic mycobacterial species appear to be at hand, although they are untested to date. Isolates of *M. szulgai* react with DNA probes for either the PGRS (Ross *et al.* 1992b) or the MPTR (Hermans *et al.* 1992). Thus, it may be possible to perform epidemiological studies using either of those two genetic elements as probes. Because isolates of *M. scrofulaceum* carry plasmids that hybridize with plasmids of the *M. avium* complex (Jucker & Falkinham 1990; Jucker 1991), plasmid-typing could be performed with that species.

The oligonucleotide $(GTG)_5$ (see Chapter 6) offers promise for typing non-tuberculous mycobacteria that lack a typing system (Cilliers *et al.* 1997). In a study of 90 isolates, representing most mycobacterial species, it was shown that $(GTG)_5$-RFLP patterns were stable and of sufficient heterogeneity to offer promise for epidemiological use (Cilliers *et al.* 1997). Because a limited number of isolates of a single species were included, no conclusions concerning the utility of this fingerprinting technique could be derived for a single species. Alternative methods for typing *Mycobacterium* species where no markers have been identified are ERIC-RFLP (Sechi *et al.* 1998), LRF-PFGE and AP-PCR. For both, typeability, reproducibility, and discriminatory power are high and neither requires the identification of a DNA probe.

5 What does this tell us about non-tuberculous mycobacterial populations?

The first and most important evidence from molecular epidemiology of non-tuberculous mycobacteria is their enormous diversity. Diversity within the *M. avium* complex alone (especially *M. intracellulare*) is significantly greater than that in the *M. tuberculosis* complex. At least five unique subtypes have been described in *M. kansasii* (Picardeau *et al.* 1997). Only in *M. xenopi* is there any indication of narrow genetic diversity (Picardeau *et al.* 1996). The wide genetic diversity of non-tuberculous mycobacterial species is consistent with their postulated environmental habitat. Environmental microorganisms are expected to be able to reproduce in a variety of niches. Diversity is of possibly great significance for the slow-growing mycobacteria whose reproduction rate puts them at a severe disadvantage in competition for limited resources in niches.

The existence of wide genotypic diversity indicates that mechanisms for generating genetic diversity are active in non-tuberculous mycobacteria. These include mutation, recombination and transposition. The heterogeneity of IS*1652*-RFLP patterns in *M. kansasii*, even though IS*1652* lacks a transposase ORF and inverted terminal repeats normally required for transposition (Yang *et al.* 1993a), demonstrates the existence of other genetic mechanisms (e.g. recombination) for generating genetic diversity in that species. Evidence that isolates of *M. avium*, *M. intracellulare* and *M. scrofulaceum* share identical plasmids (Jucker 1991; Jucker *et al.* in preparation) indicates that mechanisms for horizontal gene transmission exist.

Second, there is evidence for the selection of types; particularly within the *M. avium* complex. Although both *M. avium* and *M. intracellulare* are found in the natural environment and in non-immunocompromised individuals with pulmonary disease, *M. avium* predominates in AIDS patients (Drake *et al.* 1988; Guthertz *et al.* 1989). Further, within *M. avium* certain types (i.e. serotype 4, 6, or 8 and IS*901*-negative isolates) are found more frequently in US AIDS patients than other *M. avium* types. Predominance of a limited range of types is also seen in AIDS patients in Sweden; however, the predominating serotypes are different in the United States (Yakrus & Good 1990) and Sweden (Hoffner *et al.* 1990). In birds and wild animals other types predominate (i.e. serotypes 1, 2 or 3 and IS*901*-positive).

Geography appears to limit the range of diversity within the *M. avium* complex. A collection of 35 *M. avium* complex reference strains had a higher index of genetic diversity (i.e. 0.38; Wasem *et al.* 1991) than either a collection of *M. avium* complex isolates belonging to serotypes 4 and 8 (0.23; Yakrus *et al.* 1992) or a collection of 18 reference, 72 human, and 25 bird and animal *M. avium* complex isolates from Australia (i.e. 0.28; Feizabadi *et al.* 1996).

I like to envision single mycobacterial species as large circles. The diameter and area of the circle represent the total diversity of the species. The circle for the *M. tuberculosis* complex would be considerably smaller than the circle representing the *M. avium* complex. Within the large species circle are small circles representing the environmental compartments occupied by the corresponding *M. avium* complex genotypes. The genotypes define those compartments and are, perhaps, required to occupy those compartments. Humans, birds or animals are exposed when they enter one of those compartments. However, not all genotypes within a particular compartment are capable of infection (it is also understood that host factors, independently, influence susceptibility). One of the important and long-term goals of molecular epidemiology is to identify those compartments and the genotypes of the mycobacterial species. Further, it is a goal to identify those genotypes that are capable of infection.

But what molecular markers should we use and which one is the best? A proposed hierarchy for markers used in typing members of the *Mycobacterium avium* complex is presented in Table 8.4. The markers are loosely ranked on the basis of increasing discrimination index from the top (least discriminatory) to the bottom (most discriminatory). The point of the ranking is to illustrate that the question is not whether one marker is more or less discriminatory, but rather, what do the typing data tell us about mycobacterial populations. I have tried to categorize the different population levels within the *M. avium* complex from the most inclusive (i.e. *M. avium* complex) to the least inclusive (variants within clones). This is a hypothesis and as such, data are

Table 8.4 Hierarchy of markers for typing the *Mycobacterium avium* complex.

Marker category	Population discrimination level					
	Complex	Species	Subspecies	Geographic	Clone	Interclonal
16S rRNA sequence	+	+	+			
Single copy gene	?	+	+			
Insertion sequence						
IS*900*	−	+[a]	+	+	−	−
IS*901*/IS*902*	−	+	+[b]	?	+	− ?
IS*1245*	+	−	−	?	+	− ?
IS*1311*	+	−	−	?	+	− ?
IS*1110*	?	?	?	?	+	+
Multilocus enzyme	?	+	+	+	+	− ?
Spacer						
IS*1245*–IS*1311*	−	−	−	?	+	− ?
16S-23S ITS	?	?	+	?	+	+ ?
Plasmids	+ ?	−	−	−	+	− ?
AP-PCR	−	+ ?	+ ?	?	+	+ ?
LRF-PFGE	−	− ?	− ?	?	+	+ ?

For abbreviation definitions, see Table 8.1.

[a] Presence of IS*900* defines *M. paratuberculosis*.

[b] Presence of IS*901*/IS*902* defines *M. avium* subspecies *silvaticum*.

missing. My objective is to attempt to provide a framework for discussion of methods for characterizing mycobacterial populations. Two points stand out from this first formulation. First, some insertion sequences can provide knowledge that is useful at different population levels. Presence or absence of IS*900* and IS*901*/IS*902* defines species or subspecies. In addition, IS-RFLP analysis can define geographical and clonal types. Second, it is not known precisely to what extent the definition of types by some markers can be influenced by variation within a clone (i.e. interclonal variation). If a high probability of interclonal variation is expected, as it is for IS*1110*, that marker can be excluded from a study if the identification of clones is the objective. However, it is troubling that the magnitude of interclonal variation is not known with precision for a number of markers apparently useful for defining clones.

What is needed is a study of a sample of isolates of a single mycobacterial species (i.e. the whole circle) involving all markers. One such study for the *M. avium* complex is on the agenda of the International Working Group of Mycobacterial Taxonomy (IWGMT). Many of the studies reviewed in this chapter, while demonstrating the utility of a particular typing method, have been performed with a limited spectrum of isolates (i.e. just part of the circle). The initial and daunting task is identification and assembly of a collection of truly representative isolates. Hopefully, with this review and chapter completed, I can get started on that task.

6 Acknowledgements

I thank all my students and colleagues and the members of the International Working Group of Mycobacterial Taxonomy and the Mycobacteriology Division of the American Society for Microbiology for all their advice and counsel through the years.

Research in my laboratory has been supported by funds provided by the National Institute of Allergy and Infectious Disease, the Potts Foundation, the Heiser Program for Research in Leprosy, and the American Water Works Association Research Foundation.

7 References

Abed, Y., Bollet, C. & De Mico, P. (1995) Demonstration of *Mycobacterium kansasii* species heterogeneity by the amplification of the 16S–23S spacer region. *Journal of Medical Microbiology* **42**, 156–158.

Ahrens, P., Giese, S.B., Kalusen, J. & Inglis, N.F. (1995) Two markers, IS*901*–IS*902*: and p40, identified by PCR and by using monoclonal antibodies in *Mycobacterium avium* strains. *Journal of Clinical Microbiology* **33**, 1049–1053.

Alcaide, F., Richter, I., Bernasconi, C. *et al.* (1997) Heterogeneity and clonality among isolates of *Mycobacterium kansasii*: implications for epidemiological and pathogenicity studies. *Journal of Clinical Microbiology* **35**, 1959–1964.

Arbeit, R.D., Slutsky, A., Barber, T.W. *et al.* (1993) Genetic diversity among strains of *Mycobacterium avium* causing monoclonal and polyclonal bacteremia in patients with AIDS. *Journal of Infectious Disease* **167**, 1384–1390.

Ashford, D.A., Kellerman, S., Yakrus, M. *et al.* (1997) Pseudo-outbreak of septicemia due to rapidly growing mycobacteria associated with extrinsic contamination of culture. *Journal of Clinical Microbiology* **35** (Suppl.), 2040–2042.

Bono, M., Jemmi, T., Bernasconi, C., Burki, D., Telenti, A. & Bodmer, T. (1995) Genotypic characterization of *Mycobacterium avium* strains recovered from animals and their comparison to human strains. *Applied and Environmental Microbiology* **61**, 371–373.

Burki, D.R., Bernasconi, C., Bodmer, T. & Telenti, A. (1995) Evaluation of the relatedness of strains of *Mycobacterium avium* using pulsed-field gel electrophoresis. *European Journal of Clinical Microbiology and Infectious Disease* **14**, 212–217.

Burns, D.N., Wallace, R.J. Jr, Schultz, M.E. *et al.* (1991) Nosocomial outbreak of respiratory tract colonization with *Mycobacterium fortuitum*: demonstration of the usefulness of pulsed-field gel electrophoresis in an epidemiologic investigation. *American Review of Respiratory Diseases* **144**, 1153–1159.

Cilliers, F.J., Warren, R.M., Hauman, J.H., Wiid, I.J.D. & van Helden, P.D. (1997) Oligonucleotide (GTG)$_5$ as an epidemiological tool in the study of nontuberculous mycobacteria. *Journal of Clinical Microbiology* **35**, 1545–1549.

Coffin, J.W., Condon, C., Compston, C.A. *et al.* (1992) Use of restriction fragment polymorphisms resolved by pulsed-field gel electrophoresis for subspecies identification of mycobacteria in the *Mycobacterium avium* complex and for isolation of DNA probes. *Journal of Clinical Microbiology* **30**, 1829–1836.

Collins, D.M. (1994) DNA fingerprinting of *Mycobacterium xenopi* strains. *Letters in Applied Microbiology* **18**, 234–235.

Collins, C.H. & Yates, M.D. (1984) Infection and colonisation by *Mycobacterium kansasii* and *Mycobacterium xenopi*: aerosols as a possible source? *Journal of Infection* **8**, 178–179.

Cooper, G.L., Grange, J.M., McGregor, J.A. & McFadden, J.J. (1989) The potential use of DNA probes to identify and types strains within the *Mycobacterium tuberculosis* complex. *Letters in Applied Microbiology* **8**, 127–130.

Costrini, A.M., Mahler, D.A., Gross, W.M., Hawkins, J.E., Yesner, R. & D'Esopo, N.D. (1981) Clinical and roentgenographic features of nosocomial pulmonary disease due to *Mycobacterium xenopi*. *American Review of Respiratory Diseases* **123**, 104–109.

Crawford, J.T. & Bates, J.H. (1986) Analysis of plasmids in *Mycobacterium avium-intracellulare* isolates from persons with acquired immunodeficiency syndrome. *American Review of Respiratory Diseases* **134**, 659–661.

Dawson, D.J., Blacklock, Z.M. & Kane, D.W. (1981) *Mycobacterium haemophilum* causing lymphadenitis in an otherwise healthy child. *Medical Journal of Australia* **2**, 289–290.

De Smet, K.A.L., Brown, I.N., Yates, M. & Ivanyi, J. (1995) Ribosomal internal transcribed spacer sequences are identical among *Mycobacterium avium–intracellulare* complex isolates from AIDS patients, but vary among isolates from pulmonary disease patients. *Microbiology UK* **141**, 2739–2747.

De Smet, K.A.L., Hellyer, T.J., Khan, A.W., Brown, I.N. & Ivanyi, J. (1996) Genetic and serovar typing of clinical isolates of the *Mycobacterium avium–intracellulare* complex. *Tubercle and Lung Disease* **77**, 71–76.

Devallois, A., Picardeau, M., Goh, K.S., Sola, C., Vincent, V. & Rastogi, N. (1996) Comparative evaluation of PCR and commercial DNA probes for detection and identification to species level of *Mycobacterium avium* and *Mycobacterium intracellulare*. *Journal of Clinical Microbiology* **34**, 2756–2759.

Devallois, A., Picardeau, M., Paramasivan, C.N., Vincent, V. & Rastogi, N. (1997) Molecular characterization of *Mycobacterium avium* complex isolates giving discordant results in AccuProbe tests by PCR-restriction enzyme analysis, 16S rRNA gene sequencing, and DT1–DT6 PCR. *Journal of Clinical Microbiology* **35**, 2767–2772.

Dice, L.R. (1945) Measures of the amount of ecological association between species. *Ecology* **26**, 297–302.

Domenech, P., Menendez, M.C. & Garcia, M.J. (1994) Restriction fragment length polymorphism of 16S rRNA genes in the differentiation of fast-growing mycobacterial species. *FEMS Microbiology Letters* **116**, 19–24.

Drake, W.M., Herron, R.M. Jr, Hindler, J.A., Berlin, O.G.W. & Bruckner, D.A. (1988) DNA probe reactivity of *Mycobacterium avium* complex isolates from patients with and without AIDS. *Diagnostic Microbiology and Infectious Disease* **11**, 125–128.

Engel, H.W.B. & Berwald, L.G. (1980) The occurrence of *Mycobacterium kansasii* in tap water. *Tubercle* **61**, 21–26.

Erardi, F.X., Meissner, P.S. & Falkinham, J.O. III (1985) Simple method for obtaining mycobacterial clones with altered plasmid profiles. *Plasmid* **13**, 22.

Fakih, M., Chapalamadugu, S., Ricart, A., Corriere, N. & Amsterdam, D. (1996) *Mycobacterium malmoense* bacteremia in two AIDS patients. *Journal of Clinical Microbiology* **34**, 731–733.

Feizabadi, M.M., Robertson, I.D., Cousins, D.V. *et al.* (1996) Genetic characterization of *Mycobacterium avium* isolates recovered from humans and animals in Australia. *Epidemiology and Infection* **116**, 41–49.

Fonteyne, P.-A., Kunze, Z.M., De Beenhouwer, H. *et al.* (1997) Characterization of *Mycobacterium avium* complex related mycobacteria isolated from an African environment and patients with AIDS. *Tropical Medicine and International Health* **2**, 200–207.

Frothingham, R. & Wilson, K.H. (1993) Sequence-based differentiation of strains in the *Mycobacterium avium* complex. *Journal of Bacteriology* **175**, 2818–2825.

Frothingham, R. & Wilson, K.H. (1994) Molecular phylogeny of the *Mycobacterium avium* complex demonstrates clinically meaningful divisions. *Journal of Infectious Diseases* **169**, 305–312.

Gouby, A., Branger, B., Oules, R. & Ramuz, M. (1988) Two cases of *Mycobacterium haemophilum* infection in a renal-dialysis unit. *Journal of Medical Microbiology* **25**, 299–300.

Guerrero, C., Bernasconi, C., Burki, D., Bodmer, T. & Telenti, A. (1995) A novel insertion element from *Mycobacterium avium*, IS*1245*, is a specific target for analysis of strain relatedness. *Journal of Clinical Microbiology* **33**, 304–307.

Guthertz, L.S., Damsker, B., Bottone, E.J., Ford, E.G., Midura, T.F. & Janda, M. (1989) *Mycobacterium avium* and *M. intracellulare* infections in patients with and without AIDS. *Journal of Infectious Disease* **160**, 1037–1041.

Hampson, S.J., Portaels, F., Thompson, J., Green, E.P., Hermon-Taylor, J. & McFadden, J. (1989) DNA probes demonstrate a single highly conserved strain of *Mycobacterium avium* infecting AIDS patients. *Lancet* **i**, 65–68.

Hector, J.S.R., Pang, Y., Mazurek, G.H., Zhang, Y., Brown, B.A. & Wallace, R.J. Jr (1992) Large restriction fragment patterns of genomic *Mycobacterium fortuitum* DNA as strain-specific markers and their use in epidemiologic investigation of four nosocomial outbreaks. *Journal of Clinical Microbiology* **30**, 1250–1255.

Hellyer, T.J., Brown, I.N., Dale, J.W. & Easmon, C.S.F. (1991) Plasmid analysis of *Mycobacterium avium– intracellulare* (MAI) isolated in the United Kingdom from patients with and without AIDS. *Journal of Medical Microbiology* **34**, 225–231.

Hermans, P.W.M., van Soolingen, D. & van Embden, J.D.A. (1992) Characterization of a major polymorphic tandem repeat in *Mycobacterium tuberculosis* and its potential use in the epidemiology of *Mycobacterium kansasii* and *Mycobacterium gordonae*. *Journal of Bacteriology* **174**, 4157–4165.

Hernandez-Perez, M., Fomukong, N.G., Hellyer, T., Brown, I.N. & Dale, J.W. (1994) Characterization of IS*1110*, a highly mobile genetic element from *Mycobacterium avium*. *Molecular Microbiology* **12**, 717–724.

Hernandez-Perez, M., Kunze, Z.M., Brown, S., Yakrus, M.A., McFadden, J.J. & Dale, J.W. (1997) Strain variation in *Mycobacterium avium*: polymorphism of IS*1110*-related sequences. *International Journal of Infectious Diseases* **1**, 192–198.

Hoffner, S.E., Källenius, G., Petrini, B., Brennan, P.J. & Tsang, A.Y. (1990) Serovars of *Mycobacterium avium* complex isolated from patients in Sweden. *Journal of Clinical Microbiology* **28**, 1105–1107.

Huang, A.H., Ross, B.C. & Dwyer, B. (1991) Identification of *Mycobacterium kansasii* by DNA hybridization. *Journal of Clinical Microbiology* **29**, 2125–2129.

Hunter, P.R. (1990) Reproducibility and indices of discriminatory power of microbial typing methods. *Journal of Clinical Microbiology* **28**, 1903–1905.

Hunter, P.R. & Gaston, M.A. (1988) Numerical index of discriminatory ability of typing systems: an application of Simpson's index of diversity. *Journal of Clinical Microbiology* **26**, 2465–2466.

Iinuma, Y., Ichiyama, S., Hasegawa, Y., Shimokata, K., Kawahara, S. & Matsushima, T. (1997) Large-restriction-fragment analysis of *Mycobacterium kansasii* genomic DNA and its application in molecular typing. *Journal of Clinical Microbiology* **35**, 596–599.

Jackson, K., Edwards, R., Leslie, D.E. & Hayman, J. (1995) Molecular method for typing *Mycobacterium ulcerans*. *Journal of Clinical Microbiology* **33**, 2250–2253.

Jacoby, H.M., Jiva, T.M., Kaminski, D.A., Weymouth, L.A. & Portmore, A.C. (1995) *Mycobacterium xenopi* infection masquerading as pulmonary tuberculosis in two patients infected with the human immunodeficiency virus. *Clinical Infectious Diseases* **20**, 1399–1401.

Jensen, A.G., Bennedsen, J. & Rosdahl, V.T. (1989) Plasmid profiles of *Mycobacterium avium/intracellulare* isolated from patients with AIDS or cervical lymphadenitis and from environmental samples. *Scandinavian Journal of Infectious Diseases* **21**, 645–649.

Johnson, P.D.R., Veitch, M.G.K., Leslie, D.E., Flood, P.E. & Hayman, J.A. (1996) The emergence of *Mycobacterium ulcerans* infection near Melbourne. *Medical Journal of Australia* **164**, 76–78.

Joynson, D.H.M. (1979) Water: the natural habitat of *Mycobacterium kansasii*? *Tubercle* **60**, 77–81.

Jucker, M.T. (1991) Relationship of plasmids in *Mycobacterium avium*, *Mycobacterium intracellulare*. and *Mycobacterium scrofulaceum*. PhD Dissertation, Virginia Polytechnic Institute and State University, Blacksburg, Virginia.

Jucker, M.T. & Falkinham, J.O. III (1990) Epidemiology of infection by nontuberculous mycobacteria. IX. Evidence for two DNA homology groups among small plasmids in *Mycobacterium avium*, *Mycobacterium intracellulare*, and *Mycobacterium scrofulaceum*. *American Review of Respiratory Diseases* **142**, 858–862.

Julander, I., Hoffner, S., Petrini, B. & Ostlund, L. (1996) Multiple serovars of *Mycobacterium avium* complex in patients with AIDS. *Acta Pathologica, Microbiologica, Immunologia Scandinavia* **104**, 318–320.

Katila, M.-L., Brander, E., Jantzen, E., Huttunen, R. & Linkosalo, L. (1991) Chemotypes of *Mycobacterium malmoense* based on glycolipid profiles. *Journal of Clinical Microbiology* **29**, 355–358.

Kauppinen, J., Mäntyjärvi, R. & Katila, M.-L. (1994a) Random amplified polymorphic DNA genotyping of *Mycobacterium malmoense*. *Journal of Clinical Microbiology* **32**, 1827–1829.

Kauppinen, J., Pelkonen, J. & Katila, M.-L. (1994b) RFLP analysis of *Mycobacterium malmoense* strains using ribosomal RNA gene probes: an additional tool to examine intraspecies variation. *Journal of Microbiological Methods* **19**, 261–267.

Kent, L., McHugh, T.D., Billington, O., Dale, J.W. & Gillespie, S.H. (1995) Demonstration of homology between IS*6110* and *Mycobacterium tuberculosis* and DNAs of other *Mycobacterium* spp. *Journal of Clinical Microbiology* **33**, 2290–2293.

Kinney, J.S., Little, B.J., Yolken, R.H. & Rosenstein, B.J. (1989) *Mycobacterium avium* complex in a patient with cystic fibrosis: disease vs. colonization. *Pediatric Infectious Disease* **8**, 393–396.

Kunze, Z.M., Portaels, F. & McFadden, J.J. (1992) Biologically distinct subtypes of *Mycobacterium avium* differ in possession of insertion sequence IS*901*. *Journal of Clinical Microbiology* **30**, 2366–2372.

Kunze, Z.M., Wall, S., Appelberg, R., Silva, M.T., Portaels, F. & McFadden, J.J. (1991) IS*901*, a new member of a widespread class of atypical insertion sequences, is associated with pathogenicity in *Mycobacterium avium*. *Molecular Microbiology* **5**, 2265–2272.

Kusunoki, S., Ezaki, T. (1992) Proposal of *Mycobacterium peregrinum* sp. nov., nom. rev. & elevation of *Mycobacterium chelonae* subsp. *abscessus* (Kubica *et al.*) to species status: *Mycobacterium abscessus* comb. nov. *International Journal of Systematic Bacteriology* **42**, 240–245.

Lévy-Frébault, V., Grimont, F., Grimont, P.A.D. & David, H.L. (1986) Deoxyribonucleic acid relatedness of the

Mycobacterium fortuitum–Mycobacterium chelonae complex. *International Journal of Systematic Bacteriology* **36**, 458–460.

Lévy-Frébault, V.V., Thorel, M.-F., Varnerot, A. & Gicquel, B. (1989) DNA polymorphism in *Mycobacterium paratuberculosis*, 'wood pigeon mycobacteria', and related mycobacteria analyzed by field inversion gel electrophoresis. *Journal of Clinical Microbiology* **27**, 2823–2826.

Lillo, M., Orengo, S., Cernoch, P. & Harris, R.L. (1990) Pulmonary and disseminated infection due to *Mycobacterium kansasii*: a decade of experience. *Reviews of Infectious Diseases* **12**, 760–767.

Marston, B.J., Diallo, M.O., Horsburgh, C.R. Jr *et al.* (1995) Emergence of Buruli ulcer disease in the Dalao region of Cote d'Ivoire. *American Journal of Tropical Medicine and Hygiene* **52**, 219–224.

Marx, C.E., Fan, K., Morris, A.J., Wilson, M.L., Damiani, A. & Weinstein, M.P. (1995) Laboratory and clinical evaluation of *Mycobacterium xenopi* isolates. *Diagnostic Microbiology and Infectious Disease* **21**, 195–202.

Masaki, S., Konishi, T., Sugimori, G., Okamoto, A., Hayashi, Y. & Kuze, F. (1989) Plasmid profiles of *Mycobacterium avium* complex isolated from swine. *Microbiology and Immunology* **33**, 429–433.

Maslow, J.N., Mulligan, M.E. & Arbeit, R.D. (1993) Molecular epidemiology: application of contemporary techniques to the typing of microorganisms. *Clinical Infectious Diseases* **17**, 153–164.

Matsiota-Bernard, P., Waser, S., Tassios, P.T., Kyriakopoulos, A. & Legaris, N.J. (1997) Rapid discrimination of *Mycobacterium avium* strains from AIDS patients by random amplified polymorphic DNA analysis. *Journal of Clinical Microbiology* **35**, 1585–1588.

Mazurek, G.H., Hartman, S., Zhang, Y. *et al.* (1993) Large DNA restriction fragment polymorphism in the *Mycobacterium avium–M. intracellulare* complex: a potential epidemiologic tool. *Journal of Clinical Microbiology* **31**, 390–394.

McDiarmid, S.V., Blumberg, D.A., Remotti, H. *et al.* (1995) Mycobacterial infections after pediatric liver transplantation: a report of three cases and review of the literature. *Journal of Pediatric Gastroenterology and Nutrition* **20**, 425–431.

McFadden, J.J., Kunze, Z.M., Portaels, F., Labrousse, V. & Rastogi, N. (1992) Epidemiological and genetic markers, virulence factors and intracellular growth of *Mycobacterium avium* in AIDS. *Research Microbiology* **143**, 423–430.

Meissner, P.S. & Falkinham, J.O., III (1986) Plasmid DNA profiles as epidemiological markers for clinical and environmental isolates of *Mycobacterium avium*, *Mycobacterium intracellulare*, and *Mycobacterium scrofulaceum*. *Journal of Infectious Diseases* **153**, 325–331.

Moore, R.D. & Chaisson, R.E. (1996) Natural history of opportunistic disease in an HIV-infected urban clinical cohort. *Annals of Internal Medicine* **124**, 633–642.

Moss, M.T., Malik, Z.P., Tizard, M.L.V., Green, E.P., Sanderson, J.D. & Hermon-Taylor, J. (1992) IS*902*, an insertion element of the chronic-enteritis-causing *Mycobacterium avium* subsp. *silvaticum*. *Journal of General Microbiology* **138**, 139–145.

Nadal, D., Caduff, R., Kraft, R. *et al.* (1993) Invasive infection with *Mycobacterium genavense* in three children with the acquired immunodeficiency syndrome. *European Journal of Clinical Microbiology and Infectious Disease* **12**, 37–43,.

Nishimori, K., Eguchi, M., Nakaoka, Y., Onodera, Y., Ito, T. & Tanaka, K. (1995) Distribution of IS*901* in strains of *Mycobacterium avium* complex from swine by using IS*901*-detecting primers that discriminate between *M. avium* and *Mycobacterium intracellulare*. *Journal of Clinical Microbiology* **33**, 2102–2106.

Picardeau, M. & Vincent, V. (1996) Typing of *Mycobacterium avium* isolates by PCR. *Journal of Clinical Microbiology* **34**, 389–392.

Picardeau, M., Prod'hom, G., Raskine, L., LePennec, M.P. & Vincent, V. (1997) Genotypic characterization of five subspecies of *Mycobacterium kansasii*. *Journal of Clinical Microbiology* **35**, 25–32.

Picardeau, M., Varnerot, A., Rauzier, J., Gicquel, B. & Vincent, V. (1996) *Mycobacterium xenopi* IS*1395*, a novel insertion sequence expanding the IS256 family. *Microbiology* **142**, 2453–2461.

Portaels, F., Fonteyne, P.-A., de Beenhouwer, H., *et al.* (1996) Variability in 3' end of 16S rRNA sequence of *Mycobacterium ulcerans* is related to geographic origin of isolates. *Journal of Clinical Microbiology* **34**, 962–965.

Portaels, F., Kunze, Z., McFadden, J.J., Fonteyne, P.A. & Carpels, G. (1990) Mycobacterial diseases in AIDS patients from developing and developed countries. *Tegen Tuberculose* **2**, 67–71.

Roiz, M.P., Palenque, E., Guerrero, C. & Garcia, M.J. (1995) Use of restriction fragment length polymorphism as a genetic marker for typing *Mycobacterium avium* strains. *Journal of Clinical Microbiology* **33**, 1389–1391.

Ross, B.C., Jackson, K., Yang, M., Sievers, A. & Dwyer, B. (1992a) Identification of a genetically distinct subspecies of *Mycobacterium kansasii*. *Journal of Clinical Microbiology* **30**, 2930–2933.

Ross, B.C., Raios, K., Jackson, K. & Dwyer, B. (1992b) Molecular cloning of a highly repeated DNA element from *Mycobacterium tuberculosis* and its use as an epidemiological tool. *Journal of Clinical Microbiology* **30**, 942–946.

Saito, H., Tomioka, H., Sato, K., Tasaka, H. & Dawson, D.J. (1990) Identification of various serovar strains of *Mycobacterium avium* complex by using DNA probes specific for *M. avium* and *M. intracellulare*. *Journal of Clinical Microbiology* **28**, 1694–1697.

Sanders, J.W., Walsh, A.D., Snider, R.L. & Sahn, E.E. (1995) Disseminated *Mycobacterium scrofulaceum* infection: a potentially treatable complication of AIDS. *Clinical Infectious Diseases* **20**, 549–556.

Schröder, K.H. & Juhlin, I. (1977) *Mycobacterium malmoense* sp. nov. *International Journal of Systematic Bacteriology* **27**, 241–246.

Sechi, L.A., Zanetti, S., Dupré, I., Delogu, G. & Fadda, G. (1998) Enterobacterial repetitive intergenic consensus sequences as molecular targets for typing of *Mycobacterium tuberculosis* strains. *Journal of Clinical Microbiology* **36**, 128–132.

Shafer, R.W. & Sierra, M.F. (1992) *Mycobacterium xenopi, Mycobacterium fortuitum, Mycobacterium kansasii,* and other nontuberculous mycobacteria in an area of endemicity for AIDS. *Clinical Infectious Disease* **15**, 161–162.

Slosárek, M., Kubin, M. & Jaresová, M. (1993) Water-borne household infections due to *Mycobacterium xenopi*. *Central European Journal of Public Health* **1**, 78–80.

Slutsky, A.M., Arbeit, R.D., Barber, T.W. *et al.* (1994) Polyclonal infections due to *Mycobacterium avium* complex in patients with AIDS detected by pulsed-field gel electrophoresis of sequential clinical isolates. *Journal of Clinical Microbiology* **32**, 1773–1778.

Soini, H., Eerola, E. & Viljanen, M.K. (1996) Genetic diversity among *Mycobacterium avium* complex AccuProbe-positive isolates. *Journal of Clinical Microbiology* **34**, 55–57.

Soini, H. & Viljanen, M.K. (1997) Diversity of the 32-kilodalton protein gene may form a basis for species determination of potentially pathogenic mycobacterial species. *Journal of Clinical Microbiology* **35**, 769–773.

Sompolinsky, D., Lagziel, A., Naveh, D. & Yankilevitz, T. (1978) *Mycobacterium haemophilum* sp. nov., a new pathogen of humans. *International Journal of Systematic Bacteriology* **28**, 67–75.

Straus, W.L., Ostroff, S.M., Jernigan, D.B. *et al.* (1994) Clinical and epidemiologic characteristics of *Mycobacterium haemophilum*, an emerging pathogen in immunocompromised patients. *Annals of Internal Medicine* **120**, 118–125.

Thorel, M.F., Krichevsky, M. & Lévy-Frébault, V.V. (1990) Numerical taxonomy of mycobactin-dependent mycobacteria, emended description of *Mycobacterium avium*, and description of *Mycobacterium avium* subsp. *avium* subsp nov., *Mycobacterium avium* subsp. *paratuberculosis* subsp. nov. & *Mycobacterium avium* subsp *silvaticum* subsp. nov. *International Journal of Systematic Bacteriology* **40**, 254–260.

Tortoli, E., Simonetti, M.T., Lacchini, C., Penati, V. & Urbano, P. (1994) Tentative evidence of AIDS-associated biotype of *Mycobacterium kansasii*. *Journal of Clinical Microbiology* **32**, 1779–1782.

Valero, G., Peters, J., Jorgensen, J.H. & Graybill, J.R. (1995) Clinical isolates of *Mycobacterium simiae* in San Antonio, Texas. *American Journal of Critical Care Medicine* **152**, 1555–1557.

van Soolingen, D., de Haas, P.E.W., Blumenthal, R.M. *et al.* (1996) Host-mediated modification of *Pvu*II restriction in *Mycobacterium tuberculosis*. *Journal of Bacteriology* **178**, 78–84.

Via, L.E. (1993) Insertion sequence IS*1141*: discovery, characterization, and association with *Mycobacterium intracellulare* colonial variation. PhD Dissertation, Virginia Polytechnic Institute and State University, Blacksburg, Virginia.

Victor, T.C., Jordaan, A.M., van Schalkwyk, E.J., Coetzee, G.J. & van Helden, P.D. (1996) Strain-specific variation in the *dnaJ* gene of mycobacteria. *Journal of Medical Microbiology* **44**, 332–339.

Visuvanathan, S., Holton, J., Nye, P., Miller, R.F. & Moss, M.T. (1992) Typing by DNA probe of mycobacterial species isolated from patients with AIDS. *Journal of Infection* **25**, 259–265.

Von Reyn, C.F., Jacobs, N.J., Arbeit, R.D., Maslow, J.N. & Niemczyk, S. (1995) Polyclonal *Mycobacterium avium* infections in patients with AIDS: variations in antimicrobial susceptibilities of different strains of *M. avium* isolated from the same patient. *Journal of Clinical Microbiology* **33**, 1008–1010.

Von Reyn, C.F., Maslow, J.N., Barber, T.W., Falkinham, J.O., III & Arbeit, R.D. (1994) Persistent colonisation of potable water as a source of *Mycobacterium avium* infection in AIDS. *Lancet* **343**, 1137–1142.

Wallace, R.J. Jr (1994) Recent changes in taxonomy and disease manifestations of the rapidly growing mycobacteria. *European Journal of Microbiology and Infectious Disease* **13**, 953–960.

Wallace, R.J. Jr, Musser, J.M., Hull, S.I. *et al.* (1989a) Diversity and sources of rapidly growing mycobacteria associated with infections following cardiac surgery. *Journal of Infectious Diseases* **159**, 708–716.

Wallace, R.J. Jr, Steele, L.C., Labidi, A. & Silcox, V.A. (1989b) Heterogeneity among isolates of rapidly growing mycobacteria responsible for infections following augmentation mammaplasty despite case clustering in Texas and other southern coastal states. *Journal of Infectious Diseases* **160**, 281–288.

Wallace, R.J. Jr, Zhang, Y., Brown, B.A., Fraser, V., Mazurek, G.H. & Maloney, G.H. (1993) DNA large restriction fragment pattern of sporadic and epidemic nosocomial strains of *Mycobacterium chelonae* and *Mycobacterium abscessus*. *Journal of Clinical of Microbiology* **31**, 2697–2701.

Wasem, C.F., McCarthy, C.M. & Murray, L.W. (1991)

Multilocus enzyme electrophoresis analysis of the *Mycobacterium avium* complex and other mycobacteria. *Journal of Clinical Microbiology* **29**, 264–271.

Wayne, L.G. & Sramek, H.A. (1992) Agents of newly recognized or infrequently encountered mycobacterial diseases. *Clinical Microbiology Reviews* **5**, 1–25.

White, M.H., Papadopoulos, E.B., Small, T.N., Kiehn, T.E. & Armstrong, D. (1995) *Mycobacterium haemophilum* infections in bone marrow transplant recipients. *Transplantation* **60**, 957–960.

Witzig, R.S., Fazal, B.A., Mera, R.M. *et al.* (1995) Clinical manifestations and implication of coinfection with *Mycobacterium kansasii* and human immunodeficiency virus type 1. *Clinical Infectious Diseases* **21**, 77–85.

Wolinksy, E. (1979) Nontuberculous mycobacteria and associated diseases. *American Review of Respiratory Diseases* **119**, 107–159.

Wolinsky, E. (1995) Mycobacterial lymphadenitis in children: a prospective study of 105 nontuberculous cases with long-term follow-up. *Clinical Infectious Diseases* **20**, 954–963.

Woolford, A.J., Hewinson, R.G., Woodward, M. & Dale, J.W. (1997) Sequence heterogeneity of an *mpb70* gene analogue in *Mycobacterium kansasii*. *FEMS Microbiology Letters* **148**, 43–48.

Wright, E.P., Collins, C.H. & Yates, M.D. (1985) *Mycobacterium xenopi* and *Mycobacterium kansasii* in a hospital water supply. *Journal of Hospital Infection* **6**, 175–178.

Yakrus, M.A. & Good, R.C. (1990) Geographic distribution, frequency, and specimen source of *Mycobacterium avium* complex serotypes isolated from patients with acquired immunodeficiency syndrome. *Journal of Clinical Microbiology* **28**, 926–929.

Yakrus, M.A., Reeves, M.W. & Hunter, S.B. (1992) Characterization of isolates of *Mycobacterium avium* serotypes 4 and 8 from patients with AIDS by multilocus enzyme electrophoresis. *Journal of Clinical Microbiology* **30**, 1474–1478.

Yakrus, M.A. & Straus, W.L. (1994) DNA polymorphisms detected in *Mycobacterium haemophilum* by pulsed-field gel electrophoresis. *Journal of Clinical Microbiology* **32**, 1083–1084.

Yang, M., Ross, B.C. & Dwyer, B. (1993a) Identification of an insertion sequence-like element in a subspecies of *Mycobacterium kansasii*. *Journal of Clinical Microbiology* **31**, 2074–2079.

Yang, M., Ross, B.C. & Dwyer, B. (1993b) Isolation of a DNA probe for identification of *Mycobacterium kansasii*, including the genetic subgroup. *Journal of Clinical Microbiology* **31**, 2769–2772.

Yew, W., Wong, P., Woo, H., Yip, C., Chan, C. & Cheng, F. (1993) Characterization of *Mycobacterium fortuitum* isolates from sternotomy wounds by antimicrobial susceptibilities, plasmid profiles, and ribosomal ribonucleic acid gene restriction patterns. *Diagnostic Microbiology and Infectious Diseases* **17**, 111–117.

Zaugg, M., Salfinger, M., Opravil, M. & Lüthy, R. (1993) Extrapulmonary and disseminated infections due to *Mycobacterium malmoense*: case report and review. *Clinical Infectious Diseases* **16**, 540–549.

Zhang, Y., Rajagopalan, M., Brown, B.A. & Wallace, R.J. Jr (1997) Randomly amplified polymorphic DNA PCR for comparison of *Mycobacterium abscessus* strains from nosocomial outbreaks. *Journal of Clinical Microbiology* **35**, 3132–3139.

Chapter 9 / Molecular diagnostics

KATHLEEN D. EISENACH

1 Role of the laboratory in diagnosing mycobacterial diseases

The timely identification of persons infected with *Mycobacterium tuberculosis* and rapid laboratory confirmation of tuberculosis are two key factors for the treatment and prevention of the disease. With the increasing incidence of drug-resistant *M. tuberculosis* strains, early detection of drug resistance is an important task in the proper management of patients with tuberculosis. Other mycobacterial infections can cause significant morbidity and mortality, especially in immunocompromised hosts; thus, a rapid and specific diagnosis of these infections is important for the implementation of appropriate drug therapy.

A definitive diagnosis of tuberculosis and other mycobacterial infections requires identification of the causative organism in clinical specimens. Conventional procedures start with microscopic examination of smears for the presence of acid-fast bacilli, and continue with culture, followed by biochemical tests of the cultured organisms to identify the specific *Mycobacterium* species. The entire process often requires 4–6 weeks from the time of specimen collection, primarily because of the slow growth of mycobacteria. Determination of drug susceptibility of an isolate by culturing can add 3–6 weeks to this already long process. The radiometric BACTEC TB system (Becton Dickinson Microbiology Systems, Sparks, MD) and the new automated liquid culture systems shorten the time to detection and increase recovery rates. However, these culture systems require an average of 13–15 days to detect positive specimens. The BACTEC system also offers the NAP (p-nitro-α-acetylamino-β-hydroxypropiophenone) test for identification of *M. tuberculosis* complex isolates. Chromatographic methods for identification of cell-wall mycolic acids are used by some reference laboratories to provide a more rapid and definitive species identification. To expedite the detection of drug resistance, drug-containing media can be directly inoculated with the patient's smear-positive specimen. With the direct method, drug susceptibility results can be anticipated within 2–4 weeks of arrival of the specimen to the laboratory.

Novel molecular assays for diagnosis and drug susceptibility testing offer several potential advantages over the above methods including faster turnaround times, very sensitive and specific detection of nucleic acids, and minimal, or possibly no, prior culture. The need for new technologies for rapid diagnosis of tuberculosis is clear. Great enthusiasm aroused by molecular technologies has been evident in the field of mycobacterial research. The goals have been to develop reliable procedures that can detect and identify mycobacteria directly in clinical specimens, methods for testing antimycobacterial drug susceptibility, and methods for assessing bacillary loads in tuberculosis patients to determine the efficacy of chemotherapy. The significant advances that have been made in the last decade towards these goals are described in this chapter.

2 Amplification techniques for direct detection

The advent of nucleic acid probe technology offered promise of rapid, specific and direct microbial detection in clinical samples. However, laboratory experience demonstrated that if the number of target molecules in a clinical sample is low, the sensitivity of nucleic acid probes was unacceptably low. With the description of the polymerase chain reaction (PCR) for amplification of nucleic acids in 1987, researchers in the field quickly recognized the technology's potential to provide more sensitive tuberculosis diagnostics and possibly obviate the need for mycobacterial culture. By 1990, several PCR assays designed to amplify mycobacterial nucleotide sequences had been described. Subsequently other amplification technologies were developed and applied to the detection of mycobacteria.

2.1 The polymerase chain reaction

PCR is the most prominent gene amplification technology, being the most thoroughly investigated, widely adopted and extensively published method. PCR and other target amplification methods allow exponential multiplication of DNA or RNA sequences, beginning with as few as one copy and producing as many as one billion copies within a few hours. Amplified copies (PCR products or amplicons) can be detected and characterized with specific oligonucleotide probes by using a variety of formats. PCR involves exponential amplification using two oligonucleotide primers that anneal to opposite strands of the target DNA, and are extended with a thermostable DNA polymerase; the extended DNA product becomes the target for further amplification through multiple cycles of denaturation, annealing, and extension. Modifications of the basic procedure include the reverse transcriptase (RT) PCR, multiplex PCR, quantitative PCR and nested PCR.

When the target is RNA, the RNA is transcribed in an RT reaction and the cDNA products are then amplified in a traditional PCR, hence the designation RT-PCR. Ribosomal and messenger RNA sequences of mycobacteria are amplified in this manner. Applications of these assays are discussed later in the chapter.

Several independent amplifications carried out simultaneously in one tube with a mixture of primers is referred to as multiplex PCR. To establish a specific multiplex assay in which each template is amplified efficiently can be challenging since reaction conditions must be appropriate for each primer set. Multiplex PCR assays commonly consist of one primer set for amplification of the target sequence and a separate set for an internal control DNA sequence. Other multiplex formats include primers for multiple target sequences. For example, Kox *et al.* (1997) designed a multiplex PCR assay for co-amplification of the *M. tuberculosis* complex-specific IS*6110* and a highly conserved stretch of the 16S rDNA. PCR products of this multiplex assay were analysed in a reverse crossblot hybridization with species-specific probes and a *Mycobacterium*-specific probe. This multiplex PCR enabled identification of *M. tuberculosis* and the most important opportunistic mycobacteria in clinical specimens. An added advantage was the ability to detect simultaneous infections by more than one mycobacterial species.

Standard PCR amplification does not provide quantitative information regarding the absolute amount of nucleic acid in a sample. Quantification can be accomplished using several different PCR formats. The endpoint or limiting dilution method involves serially diluting the target sample and comparing the endpoint signal to a standard curve. More precise results can be obtained with a competitive PCR, although this type of assay is more difficult to establish. A control must be constructed so that it is amplified with the same primers, contains similar G+C content, and is of similar size as the target. Both templates must amplify with equal efficiency. A competitive PCR based on IS*6110 has* been developed by DesJardin *et al.* (1998) for the purpose of quantifying *M. tuberculosis* DNA in sputum samples. Endpoint dilution and competitive methods are labour intensive and require multiple reactions for each sample. To simplify this, DesJardin *et al.* (1998) developed another IS*6110* PCR using an automated, real-time PCR system. The basis for this assay is the ABI Prism 7700 Sequence Detection system (TaqMan; Applied Biosystems, Inc./Perkin Elmer, Foster City, CA) which uses a fluorogenic probe with the amount of fluorescence detected being proportional to the amount of accumulated PCR product. The amount of target DNA in a sample is interpolated from a standard curve that is generated with each run. Quantification of PCR products occurs real time during the exponential phase of amplification. Since no postamplification handling is necessary, this eliminates potential sources of carryover contamination and reduces handling time. Comparable results have been observed with the IS*6110* competitive and TaqMan PCRs.

In a nested PCR, a second round of amplification is performed, using the amplicon of the first reaction as a target and a pair of primers complementary to sequences within this amplicon. Nested PCR provides increased sensitivity, but this is achieved at the risk of cross-contamination, since the tubes containing amplicons have to be opened to add new reagents or transfer amplicons to a second reaction tube. Alternatively, a reaction can be run with two sets of primers in one tube, providing the primer pairs can be designed with different annealing temperatures. Nested PCR can increase the specificity of the reaction, since the internal primers anneal only if the amplicon has the corresponding expected sequence. Increased sensitivity has been achieved with nested PCR assays that target single-copy sequences, such as the 65-kDa and 38-kDa (Pab) genes (Hance *et al.* 1989; Miyazaki *et al.* 1993).

2.2 Targets for detection

A suitable target for amplification may be a single-copy gene in the mycobacterial genome or one that is present as a repeated sequence. The choice of target and design of primers within the gene target are equally important in terms of assay sensitivity and specificity. Both genus-specific and species-specific gene targets have been utilized. Some of the targets include the genes for insertion elements (IS*6110*, IS*1081*), immunodominant antigens (38-kDa antigen (Pab), 65-kDa protein, MPB70 (18 kDa), 85 protein complex (30/32 kDa), MPB64) and ribosomal sequences (16S rRNA, 23S rRNA). Predominant among these is the insertion sequence IS*6110*/IS*986* (McAdam *et al.* 1990; Thierry *et al.* 1990), which is typically present in multiple copies in *M. tuberculosis*. The high copy number of IS*6110* is thought to result in increased sensitivity, although given the scale of amplification involved in PCR, this is unlikely to be a significant factor. Within IS*6110*, the choice of primers can affect the PCR results. The primers of Eisenach *et al.* (1990) are widely used and demonstrate high sensitivity and specificity. There have been reports of false positives with the IS*6110* PCR which suggest some primers may lack specificity.

Since IS*6110* is a member of the widely distributed IS*3* family of insertion sequences (see Chapter 2), it is plausible that some primers will detect other IS*3*-like elements. A potentially more serious problem is the existence of strains that lack IS*6110*. However, there have been few reports of such strains and it is unlikely that false-negative PCR results are attributable to strains lacking this element.

Of the single-copy targets, the 65-kDa protein gene and the 16S rRNA gene have been frequently used. These highly conserved genes serve as *Mycobacterium*-specific targets. Careful design of primers and PCR conditions can provide an assay capable of detecting any mycobacterial species, with identification to species level provided by a second set of primers or by hybridization with species-specific probes (Brisson-Noël *et al.* 1989; Hance *et al.* 1989; Böddinghaus *et al.* 1990). Investigators have also resorted to the use of rRNA sequences as targets that can be amplified via RT-PCR. The advantages are that the 16S rRNA sequences are found in high copy numbers (≈2000 molecules/cell) with stretches of sequence that are highly variable among species and serve as targets for species-specific amplification, whereas other stretches are conserved and serve as a genus-specific target.

2.3 Technical aspects of amplification assays

Mycobacterial cell lysis methods

The main objectives of sample preparation are efficient release of mycobacterial nucleic acid and removal of any substances in the sample that may be inhibitory to the PCR; at the same time, it is desirable to avoid introducing chemicals that may themselves may be inhibitory. The challenge has been to develop a practical method combining these objectives with simplicity of operation. The methods that are suitable in the research laboratory have not proven suitable for the clinical setting. Lysis of mycobacteria can be difficult because of the thick lipid-rich cell wall, components of which can also contaminate the nucleic acid preparation. The methods used com-

monly involve a combination of physical disruption (boiling, sonication, glass bead beating, or cycles of freezing–thawing), chemical degradation (guanidinium salts, sodium hydroxide, sodium dodecyl sulphate (SDS), chelex agents), and enzymatic digestion (lysozyme, proteinase K). For simplicity, a crude lysate is frequently used in the amplification reaction; however, where possible it is preferable to use a purified sample in which the DNA is concentrated and interfering substances have been eliminated. The latter can be accomplished by the traditional method of phenol–chloroform extraction and ethanol precipitation. As a rapid and simple alternative, Eisenach *et al.* (1991) have used the GeneClean reagents (Bio101, La Jolla, CA) for purifying and concentrating DNA.

Controls and elimination of inhibitors

Heparin, haemoglobin, phenol, SDS, and other undefined substances in clinical specimens are potent inhibitors of *Taq* polymerase activity. Inhibition of *Taq* polymerase during the PCR can cause false-negative results, thus decreasing assay sensitivity. Inhibition rates have been reported as high as 23%. Inhibition occurs most often when crude lysates are used and can often be rectified by diluting the sample or purifying the DNA. Nested PCR formats, which enable dilution of the sample in the second round of PCR, have been applied with the explicit purpose of overcoming PCR inhibitors.

Although the precise nature of such inhibitors is not known, their presence may be monitored with control templates. Purified *M. tuberculosis* DNA may be spiked in duplicate test samples or back-spiked into PCR-negative samples, an endogenous gene may be co-amplified along with the target DNA, or genetically engineered or plasmid target DNA may be used as internal controls. The fastest and least expensive procedure is co-amplification with an internal control. Eisenach *et al.* (1991) were the first to describe adding an internal control which was amplified with the IS*6110* primers to the PCR reaction mixture. The control DNA was a plasmid containing the 3′ and 5′ ends of the 123-bp IS*6110* target and a large insert of DNA, resulting in a large 600-bp PCR product. The control product could be easily distinguished from the 123-bp sample product on ethidium bromide-stained gels.

Kolk *et al.* (1994) developed a novel strain of *M. smegmatis* with a modified IS*6110* sequence integrated into its genome. The efficacy of each step in the assay including the sample preparation method can be monitored by adding the modified *M. smegmatis* strain to the clinical sample. The most common control for cell lysis and DNA extraction is a standardized aliquot of a broth culture containing a known number of *M. tuberculosis* organisms which is included in each run.

Contamination

The risk of false-positive results due to the carryover of target DNA from a positive to a negative sample is a major concern in the clinical application of PCR diagnostics. Contamination is a severe problem in the context of the diagnosis of tuberculosis, whereby amplification of one to 100 template molecules is usually sought. Contaminating DNA may come from clinical specimens containing large numbers of *M. tuberculosis* organisms, from *M. tuberculosis* cultures used as cell lysis controls, or target DNA used as positive PCR controls. Most frequently, the problem arises from the accumulation of PCR amplicons in the laboratory. Amplification systems have been adapted to include use of dUTP and uracil DNA glycosylase as a strategy to eliminate amplicon carryover.

2.4 Commercial amplification tests

The commercial PCR test for the detection of *M. tuberculosis* complex is marketed by Roche Diagnostic Systems (Branchburg, NJ). The Roche Amplicor MTB amplifies a region of the 16S rDNA sequence that is genus specific and detects PCR products by hybridization with a *M. tuberculosis* complex-specific probe. Another version of this test, available outside the United States, employs additional species-specific probes that allow detection and identification of *M. avium* and *M. intracellulare* (F. Hoffmann-La Roche Ltd, Basel, Switzerland). The Amplicor MTB system can be automated with the Roche Cobas instrument which consists of a Thermocycler TC9600, a hybridization system based on magnetic particle separation, and a microwell plate reader. Results with the Cobas Amplicor MTB system are comparable to those of its manual version (Rajalahti *et al.* 1998).

The Gen-Probe Amplified *Mycobacterium tuberculosis* Direct (MTD) Test (Gen-Probe, Inc., San Diego, CA) is a transcription-mediated amplification (TMA) system. The basis of TMA is conversion of the target (in this case 16S rRNA) into cDNA by reverse transcriptase, using a primer containing an RNA polymerase promoter. The product can therefore be transcribed by RNA polymerase to produce large numbers of RNA transcripts, which become templates for reverse transcription and further transcription in a cyclic geometric amplification. RNA products are detected by a hybridization protection assay that uses an acridinium ester-labelled DNA probe complementary to the rRNA target. The MTD test can detect $<10^3$ copies of rRNA, equivalent to one bacillus, and <5 bacilli even in the presence of a high number of unrelated organisms, thus being *M. tuberculosis* complex specific (Jonas *et al.* 1993).

A modification of TMA uses the RNA self-sustained sequence replication reaction (termed 3SR) in which RNaseH degrades the RNA–DNA duplexes and allows conversion to dsDNA that has an RNA polymerase promoter site at each end (Compton 1991). Nucleic Acid Sequence-Based Amplification (NASBA) is a commercial development of the 3SR (Organon Teknika, Amsterdam, the Netherlands) (Van der Vliet *et al.* 1993).

Strand Displacement Amplification (SDA; BD Microbiology Systems) which was first described by Walker *et al.* (1992) is based upon the annealing to denatured target DNA of oligonucleotide primers which possess 5′ tails containing restriction sites for the enzyme *Hinc*II. An exonuclease-deficient DNA polymerase extends the 3′ end of the annealed primers and incorporates thiolated nucleotide dATPαs in the newly synthesized strand. Within the primer sequence, *Hinc*II nicks the unmodified strand of the hemiphosphorothiolated duplex providing a free 3′ end from which the polymerase can extend and displace the downstream DNA strand. Exponential amplification is achieved through continuous polymerization and displacement of both sense and antisense DNA templates. A prototype system, the BDProbeTec-SDA, has been evaluated in a few studies (Ichiyama *et al.* 1997). Newer SDA systems are under development which employ thermostable enzymes. Such assays offer the potential to enhanced specificity and are capable of achieving >10^9-fold amplification in as little as 15–20 min (Spargo *et al.* 1996). Recent developments include amplification of long targets up to 2 kb in size and linkage of SDA to an RT reaction. The use of SDA for quantitative detection of DNA and RNA is also being explored.

Increased sensitivity of mycobacterial detection can also be achieved with probe amplification technologies, one such method being the ligase chain reaction (LCR). LCR involves the joining, catalysed sequentially by polymerase and a thermostable ligase, of two oligonucleotide probes specific for adjacent sequences in the target DNA, once such segments are hybridized with the complementary sequences. The products of each cycle step serve as templates for the next cycle, resulting in exponential amplification. LCR is the basis of the Abbott LCx *M. tuberculosis* (MTB) assay (Abbott Laboratories, Abbott Park, IL). The LCx MTB assay employs four oligonucleotide probes, designed in pairs, that are complementary to the *M. tuberculosis* complex-specific gene encoding the protein antigen b (Pab). The paired probes are labelled with different haptens, one for capture and the other for detection, so that only joined products have both haptens and are detected in the microparticle enzyme immunoassay. LCR products are detected using the automated LCx Analyser, in which a sample of amplified product is automatically transferred to an incubation well and microparticles coated with anticapture hapten bind the amplification product as well as any unligated probes with capture hapten. Clinical evalua-

tions of the LCx MTB test have demonstrated sensitivities and specificities similar to PCR and other amplification methods (Ausina *et al.* 1997; Tortoli *et al.* 1997).

Signal amplification formats that have been applied to mycobacterial detection include the Q-Beta replicase amplification system (formerly Gene-Trak, Framingham, MA) and the branched-chain DNA Signal amplification assay (Chiron Corp., Emeryville, CA). Q-Beta replicase amplification is based on the use of 'detector' probes that are geometrically amplified by Q-Beta replicase (RNA-directed RNA polymerase) following hybridization to specific RNA targets (Lizardi *et al.* 1988). In this system, amplification occurs after sample matrix and unhybridized detector probe are removed from the reaction mixture. This is accomplished by using multiple rounds of reversible target capture on paramagnetic particles. The Q-Beta replicase assay, designed to target *M. tuberculosis* 23S rRNA, was found to be sensitive and specific for direct detection in sputum samples (Shah *et al.* 1995). The Galileo was developed as a prototype instrument (Vysis Inc., Downers Grove, IL) for automating the assay in a closed disposable cartridge, thereby simplifying the assay and preventing contamination of the assay from external sources (Smith *et al.* 1997). In the branched-DNA assay, signal amplification is achieved via hybridization of multiple alkaline phosphatase-labelled probes to branched-chain oligonucleotide probes with multiple binding sites, followed by incubation with a chemiluminescent substrate. The chemiluminescent output is directly proportional to the concentration of DNA target present in the specimen. Shen *et al.* (1994) described a branched-DNA assay, which uses multiple probes complementary to the IS*6110* sequence, for detection and semiquantification of *M. tuberculosis* in sputum specimens. When luminescent signals were compared to semiquantitative AFB readings and colony counts on solid media, a correlation was observed. Currently, there are no apparent plans to commercialize mycobacterial assays based on these two signal amplification methods.

2.5 Performance of in-house polymerase chain reaction assays and commercial amplification tests

Experience with in-house-developed PCR tests has demonstrated overall sensitivities and specificities in the range of 70–100%. This variability is not surprising given the fact that laboratories differ in terms of extraction procedures, target and primer sequences, sample input, PCR conditions and detection methods. For an extensive review on published results one should consult Richeldi *et al.* (1995), Herold *et al.* (1996), Sandin (1996) and Forbes (1997). In most studies, sensitivity and specificity have been calculated as a function of the culture technique, since this is the reference method and corresponding clinical information has often not been available. When discrepant results have been revised on the basis of a positive history for culture or the clinical diagnosis of tuberculosis, the specificity and positive predictive value of the PCR tests have increased. Studies in which sensitivities approach 100% were carried out with larger proportions of smear-positive specimens than commonly found in clinical populations. Separate analyses of smear-negative, culture-positive specimens have shown that the sensitivities are significantly lower than those of smear-positive, culture-positive specimens. The diagnostic yield is significantly increased if more than one specimen per patient is analysed. The commercial tests give results comparable to those obtained with in-house PCR assays. Generally, among the commercial tests the sensitivities and specificities have been equivalent. Occasionally, an in-house IS*6110* PCR has appeared to be more sensitive than the commercial tests, and the Gen-Probe MTD to be more sensitive than the Roche Amplicor MTB; however, these differences have not been statistically significant (Vuorinen *et al.* 1995; Dalovisio *et al.* 1996; Huang *et al.* 1996; Ichiyama *et al.* 1996; Piersimoni *et al.* 1997; Cohen *et al.* 1998).

Although several technical factors affect the performance of amplification tests on clinical samples, the key factor, as for microscopy and cultures, is the density of *M. tuberculosis* organisms in the specimen. A clear relationship between PCR performance and the number of *M. tuberculosis* organisms in sputum specimens has been found by Clarridge *et al.* (1993). Only 52% of the specimens with <50 cfu/mL were positive in the IS*6110* PCR. Of those with >100 cfu/mL, 98% were positive. In a detailed analysis of the sensitivity and specificity of the 16S rRNA amplification test (Gen-Probe MTD test), Jonas *et al.* (1993) found that, like DNA amplification, the sensitivity is dependent on the bacterial load in the specimen. The assay was positive on only 53% of those samples containing <100 cfu/mL. However, positivity was 100% on samples with greater than 1000 cfu/mL. These data indicate that the performance of the amplification tests may be insufficient to diagnose tuberculosis in patients with paucibacillary disease. To improve sensitivity, Gen-Probe has developed a second generation test MTD2 which uses a 10-fold increase in volume of pretreated specimen. In a recent comparison of the MTD1 and MTD2, Gamboa *et al.* (1998) observed increased sensitivity with both respiratory and non-respiratory specimens; however, the differences in sensitivities between the two methods was significant for only respiratory specimens.

2.6 Diagnosing extrapulmonary tuberculosis with amplification methods

The real value of PCR diagnosis is in situations where the clinical picture is less clear or where smear and culture are less reliable. It is extrapulmonary tuberculosis (meningitis, pleuritis, peritonitis, pericarditis, lymph-node tuberculosis, skin tuberculosis, etc.) for which a rapid and accurate laboratory diagnosis would be most beneficial. Limitations of smear and culture are due to the lower number of organisms normally present in these types of specimens. There have been many investigations of PCR amplification of cerebrospinal fluid for the diagnosis of tuberculous meningitis. These were limited by the small numbers of patients studied, lack of corresponding culture results, and inadequate clinical diagnoses. A wide range of sensitivities (32–100%) has been reported. More clinical studies are needed to evaluate the diagnostic yield of PCR in extrapulmonary tuberculosis.

2.7 Polymerase chain reaction for monitoring response to tuberculosis treatment

Sputum microscopy and culture have traditionally been used for monitoring treatment response in pulmonary tuberculosis. Both techniques have obvious limitations. PCR, which combines sensitivity with speed, has been investigated as a means for assessing bacterial clearance. Yuen *et al.* (1997) reported that persistence of *M. tuberculosis* DNA in sputum was associated with more underlying illness, high radiographic scores on the extent of involvement, previous drug treatment, high degree of sputum smear positivity, and multidrug resistance. In this study, 60% of the patients who were PCR positive after 6 months of treatment were clinical relapses whereas none of those who became PCR negative before 6 months relapsed. Similarly, Kennedy *et al.* (1994) observed that their relapsed cases had PCR positive sputa beyond 6 months suggesting that PCR may be useful for detecting relapses. In these studies, PCR conversion to negativity was seen at 1–2 months following smear and culture conversion, and it was proposed that PCR is a suitable method for assessing treatment response. Contrary to these observations, Hellyer *et al.* (1996) demonstrated the persistence of *M. tuberculosis* DNA in sputum >12 months after start of treatment and >6 months after conversion in some patients. DNA persistence was not associated with radiographic extent of disease or relapse. Although the differences observed in these studies cannot be readily explained, the presence of amplifiable DNA over such long periods in culture-negative patients is not surprising considering the exquisite sensitivity of PCR methods. It seems logical that the inability to distinguish live and dead organisms would preclude DNA amplification from use in therapeutic monitoring.

It has been suggested that quantitative rather than qualitative assessment of DNA levels might reflect bacterial load; such approach has been investigated by DesJardin *et al.* (1998). Competitive and real-time PCR assays for IS*6110* were used to quantify *M. tuberculosis* DNA in sputum samples serially collected during the course of therapy. As anticipated, the amount of DNA corresponded to the numbers of AFB on microscopy, however, neither the DNA level nor AFB count correlated with the number of cultivable bacilli after initiation of therapy. Thus, these tests were not considered appropriate markers of treatment efficacy. Similar data have been observed in mice, where quantitative estimates of DNA did not correspond to the numbers of bacilli cultured from the drug-treated animals (de Wit *et al.* 1995).

RNA is less stable than DNA and would appear to be a more suitable target for this purpose. Moore *et al.* (1996) used the Gen-Probe MTD to monitor rRNA in sputum from patients receiving therapy and observed a poor correlation between smear and culture results and the presence of rRNA. In contrast to DNA and rRNA, prokaryotic mRNA has a very short half-life and should be a reliable target for indicating the presence of viable organisms. DesJardin *et al.* (1996) have developed a quantitative RT-PCR which amplifies alpha antigen mRNA and shown this assay to be a reliable marker of bacterial viability (Hellyer *et al.* 1999). In a study of culture-positive patients receiving standard antituberculosis therapy, a precipitous drop in alpha antigen levels was observed in as little as four days after the start of therapy (DesJardin *et al.* 1999). The data suggest that ratios of DNA to mRNA levels may provide the most meaningful assessment of the efficacy of drug treatment. Further studies involving more patients are needed to determine whether this approach will be useful for identifying patients with a high risk of relapse.

3 Species identification of cultured mycobacteria

3.1 Nucleic acid probe identification

Nucleic acid probes for the identification of *M. tuberculosis* complex and *M. avium* complex were introduced by Gen-Probe in 1987. Use of these radioisotopic DNA probes for rapid identification of cultures at the species level was the first application of molecular biology techniques in the clinical mycobacteriology laboratory. By 1990, chemiluminescent probes (AccuProbes) were available, and probes for speciating *M. kansasii* and *M. gordonae* had been developed. The AccuProbes are single-stranded oligomers complementary to the rRNA of these particular species of mycobacteria. Lysis of the mycobacterial cells releases RNA, and the acridinium ester-labelled probe binds with the rRNA of the target organism to form a stable DNA–RNA hybrid. Detection of the hybrid is accomplished by a hybridization protection assay, which is the same method used to detect transcription products in the MTD test.

Initially, AccuProbes were developed for culture confirmation of organisms grown on solid mycobacterial media, however, they are now widely used to identify mycobacteria in liquid culture, e.g. BACTEC 12B medium (Evans *et al.* 1992; Telenti *et al.* 1994; Metchock & Diem 1995). The test requires about 10^6 organisms to produce clear-cut results. Overall sensitivity and specificity of the AccuProbes are close to 100%. Discrepant results are rare, but have been reported in a number of cases. Misidentification of *M. terrae*, *M. avium* complex, and *M. celatum* as *M. tuberculosis* complex has been reported (Martin *et al.* 1993; Stockman *et al.* 1993; Butler *et al.* 1994). Other minor limitations include the inability to distinguish members of the *M. tuberculosis* complex and with the MAC (*M. avium* complex) probe there is no distinction between *M. avium* and *M. intracellulare*.

With emphasis on the rapid detection of *M. tuberculosis*, probes combined with the BACTEC system offer the most easily available and reliable method for most clinical laboratories. DNA probes have been widely adopted in industrialized countries but are not used in developing countries because of cost. Alternative probe-based methods under research and development, such as the culture confirmation test based on the direct repeat locus and the cycling probe technology (ID Biomedical, Vancouver, Canada), may ultimately be more afford-

able and amenable to low technology settings (Beggs *et al.* 1996).

3.2 Polymerase chain reaction combined with restriction enzyme analysis

Amplification of a highly conserved gene combined with restriction enzyme analysis of PCR products has been applied to the identification of several commonly encountered mycobacterial species. Plikaytis *et al.* (1992) were the first to describe such an assay in which a portion of the highly conserved heat-shock protein 65 (*hsp65*) gene was amplified using primers common to all mycobacteria and the PCR product digested separately with two restriction enzymes. Telenti *et al.* (1993a) developed a similar method which differed in the 65-kDa primers and restriction enzymes used. A third method described by Vaneechoutte *et al.* (1993) was based on the 16S rDNA target. With these methods the restriction fragment patterns were distinctive for *M. tuberculosis*, *M. bovis*, *M. avium*, *M. intracellulare*, *M. kansasii* and *M. gordonae*; however, the patterns occasionally varied within a species. Members of the *M. tuberculosis* complex consistently displayed the same patterns and could not be differentiated on a species level. Strains of *M. avium* were tightly clustered, whereas *M. kansasii*, *M. intracellulare* and *M. gordonae* each showed greater variability within their clusters.

PCR-restriction enzyme pattern analysis can be performed on isolates from solid and BACTEC media. Lysis of mycobacteria is usually accomplished by mechanical means and crude lysates used in the PCR. Most patterns can be recognized visually, however, computer-assisted analysis facilitates pattern comparisons and storage of a large database. The method requires high-resolution gels and internal standards, since some fragments differ in size only by a few nucleotides. If the laboratory is also performing diagnostic PCR, one should consider using primers directed to an unrelated genomic region to the one targeted in the PCR-restriction assay. This is important for avoiding carryover of amplified products from the PCR-restriction assay to a sensitive diagnostic PCR.

3.3 Nucleic acid sequence determination

Direct sequencing of mycobacterial genes has become an increasingly important method for identifying mycobacterial species. This approach is also useful for the detection of growth-deficient mycobacterial species directly in clinical specimens and in taxonomic characterization of mycobacterial strains. Genes that have been examined include those for the 16S rRNA, *dnaJ*, superoxide dismutase, *hsp65*, and 32-kDa protein. In general, the gene chosen for PCR- and sequencing-based identification should be found in all relevant mycobacterial species and not in other bacteria, it should contain enough sequence diversity between different species to allow for easy identification, and there should be very little variation among the strains belonging to one species. The 16S rRNA gene, with its conserved and variable regions, has become the preferred target (Rogall *et al.* 1990a).

In the direct sequencing method described by Kirschner *et al.* (1993), preparation of nucleic acids was accomplished by simple mechanical disruption of the bacteria and a 1-kb fragment of 16S rRNA gene was PCR amplified. Because of the specificity of one of the primers, mycobacterial DNA was preferentially amplified, permitting the correct identification of mycobacteria in samples containing more than one organism. The other primer targeted a conserved region in *Escherichia coli* and was biotinylated to allow for the single-stranded solid-phase sequencing technique. A third primer was used in the sequencing reactions which provided the nucleic acid sequence of region A (Rogall *et al.* 1990a). Most mycobacterial species have a unique sequence in region A, therefore precise identification is possible by comparing the sequence of the unknown isolate with the known signature sequences (Rogall *et al.* 1990b). Although region A can be used for routine identification, the additional analysis of region B may be required for isolates that are indistinguishable on the basis of region A or for isolates that show unique sequences in region A, possibly indicating previously undescribed taxa.

16S rRNA sequence determination represents a highly accurate and rapid method for identifying mycobacteria. More advanced instrumentation, i.e. a rapid-ramping thermal cycler, an automated sequence detection system, and computer-assisted analysis would allow final identification to be completed within 1 day. This technique offers several advantages in terms of speed, accuracy and versatility; however, its use is restricted to reference or clinical research laboratories because of the cost and technical expertise required.

3.4 Methods for distinguishing species of the *Mycobacterium tuberculosis* complex

The similarity of *M. tuberculosis*, *M. bovis* and *M. africanum* in clinical presentation and treatment of these infections has resulted in the laboratory not fully identifying these species. However, it is important to differentiate members of the *M. tuberculosis* complex on the species level so that the incidence of *M. bovis* infections in humans and animals can be documented. Traditionally, strains of *M. bovis* and *M. tuberculosis* have been distinguished by several biochemical properties; however, these results are often not available for 6–8 weeks and not always reliable. With the two species being virtually identical on the genetic level it has been difficult to identify sequence diversity on which a molecular method could be based.

Early descriptions of the insertion element IS*6110* indicated that *M. tuberculosis* and *M. bovis* could be distinguished on the basis of IS*6110* copy number, with the *M. bovis* strains having one to two copies of IS*6110* and the *M. tuberculosis* strains having 10–15 copies. Plikaytis *et al.* (1991) described an IS*6110* PCR assay based on this assumption for differentiating the two species. Subsequent DNA fingerprint data demonstrated that this was an invalid approach since some *M. bovis* strains have high copy numbers of IS*6110* and vice versa. Del Portillo *et al.* (1996) proposed that PCR amplification of the *mtp40* gene could be used as a diagnostic tool for detecting *M. tuberculosis* infections and for differentiating them from *M. bovis* infections. The basis for this was that *mtp40* appeared to be present in only *M. tuberculosis* strains (Parra *et al.* 1991). An extensive evaluation of the *mtp40* PCR indicated that the *mtp40* gene is found in most, but not all, *M. tuberculosis* strains and is absent in most, but not all, *M. bovis* strains casting doubt on the reliability of this method (Weil *et al.* 1996).

Recently the nucleotide sequences for the *oryR* gene and the pyrazinamidase (*pncA*) gene were determined with point mutations being observed in the two species (Scorpio & Zhang 1996; Sreevatsan *et al.* 1996). Subsequently, molecular methods based on these differences were developed. One method uses PCR and single-stranded conformation polymorphism analysis to detect a single characteristic mutation in the *pncA* gene of *M. bovis* (Scorpio *et al.* 1997). Another amplifies a region of the *oryR* gene with the PCR products being subject to restriction analysis following digestion with *Alu*I (Sreevatsan *et al.* 1996). To provide a simpler method, De los Monteros *et al.* (1998) developed an allele-specific PCR method based on the *oryR* sequences which was shown to be reliable for distinguishing *M. bovis* from *M. tuberculosis.*

Another promising PCR method developed by M. Beggs (personal communication) takes advantage of sequence differences in the direct repeat (DR) locus. In this assay one primer was complementary to the region flanking the DR locus (a sequence which is conserved in all *M. tuberculosis* complex strains) and the other to a spacer sequence which appeared to be unique to *M. bovis* (Beggs *et al.* 1996). *M. bovis* strains consistently yielded a 580-bp product whereas most *M. tuberculosis* strains yielded no product. The occasional product observed with the *M. tuberculosis* strains could be easily distinguished on the basis of its larger size. The assay was developed in a multiplex format such that the primers for IS*6110* are included in the PCR, thus enabling the simultaneous confirmation of the presence of *M. tuberculosis* complex DNA. A large collection of *M. bovis* and *M. tuberculosis* strains from diverse host and geographical origin has been tested demonstrating the diagnostic utility of this assay. Advantages of this method are that the PCR assay is very simple to perform with cultured cells being

placed directly in the PCR reaction (no DNA isolation required) and the results are straightforward.

Bacille Calmette–Guérin (BCG), which is used as a vaccine against *M. tuberculosis*, a recombinant vehicle for multivalent vaccines, and as cancer immunotherapy, can cause disease in humans, especially those with cellular immunodeficiencies. Therefore, the ability to rapidly and specifically identify BCG can be clinically important. Several methods have been reported to differentiate BCG from other members of the *M. tuberculosis* complex. These include DNA fingerprinting methods with the DR and IS*1081* probes (van Soolingen *et al.* 1992; see also Chapter 6) and amplification of a specific region containing the major polymorphic tandem repeat followed by restriction enzyme analysis (Frothingham 1995). Recently, the RD1 region was found to be present in all virulent *M. bovis* and *M. tuberculosis* strains tested but deleted from all BCG strains tested (Mahairas *et al.* 1996). With this information, Talbot *et al.* (1997) developed a multiplex PCR to detect the RD1 deletion. The assay included two primers complementary to regions flanking the RD1 and one primer complementary to DNA within the RD1 sequence, with results based on the size of the PCR products. In an evaluation of a large, representative collection of BCG and other *M. tuberculosis* complex strains, the RD1 PCR gave consistent and easy to interpret results, thus demonstrating that it is a promising tool for the rapid and specific identification of BCG.

4 Molecular methods for drug susceptibility testing

4.1 Analysis of mutational hotspots in genes associated with drug resistance

In the last few years there has been considerable progress in our understanding of the mechanisms of action of antimycobacterial agents and the basis of resistance to these compounds (Musser 1995; Heym *et al.* 1996; see also Chapter 15). To date, there is information about 12 genes involved in resistance in *M. tuberculosis*. Of greatest interest is the basis of resistance to the two key drugs, isoniazid (INH) and rifampin (RMP), as resistance to these compounds is likely to influence patient care. This information has allowed the development of novel strategies for detecting resistance at the genotype level. These strategies have the potential to provide results more rapidly than the traditional methods that rely on growth or inhibition of growth in the presence of the individual drugs.

The majority of resistance to RMP involves missense mutations in a well-characterized region of the *rpoB* gene (encoding the subunit of the RNA polymerase) (Telenti *et al.* 1993b); thus, investigation of RMP resistance is relatively straightforward. In contrast, resistance to INH is associated with a variety of mutations affecting one or more genes such as those encoding catalase–peroxidase (*katG*) (Zhang *et al.* 1992), the enoly acyl carrier protein reductase involved in mycolic acid biosynthesis (*inhA*) (Banerjee *et al.* 1994), and the recently described alkyl-hydroperoxide reductase (*ahpC*), which is involved in cellular response to oxidative stress (Deretic *et al.* 1995; Wilson & Collins 1996). Investigation of INH resistance is more complex since the analysis of limited regions in all three genes is required.

New techniques undergoing clinical evaluation involve screening mutational hotspots in the genes encoding drug targets using PCR followed by analysis with either automated DNA sequencing, single-strand conformation polymorphism (PCR-SSCP) or solid-phase hybridization (Telenti & Persing 1996).

Sequencing remains the gold standard for the detection of mutations. Automated sequencing has been applied to the identification of mutations in genes involved in resistance to INH, RMP, streptomycin and fluoroquinolones (Kapur *et al.* 1995). However, a complete surveillance of all known mutation sites for some of these genes (e.g. *katG*) requires multiple reactions per isolate. The requirement of an automated sequencer and the technical demands of this multistep process limits this method to sophisticated reference laboratories.

The principle of PCR-SSCP is based on the fact that the two denatured strands of a PCR-amplified DNA molecule adopt stable intramolecular conformations;

changes can be easily recognized by their altered electrophoretic mobility compared to the wild-type pattern. This analysis can be performed on an automated sequencer to render results in 24 h, or manually using a modified silver-staining procedure or radioactive labelling during the amplification step. SSCP requires long electrophoresis steps under highly controlled conditions and technical expertise to ensure reproducibility. Furthermore, the electrophoresis patterns for some mutations can be very similar to wild-type patterns. As with automated sequencing, SSCP would be best suited for reference laboratories. Recently Telenti *et al.* (1997) set out to define the sensitivity and specificity of SSCP at the reference laboratory level. A blind assessment of the accuracy of targeted mutation analysis was conducted using selected regions of four genes (*katG*, *inhA*, *ahpC* and *rpoB*). PCR-SSCP successfully detected >96% of the RMP-resistant strains and 87% of the INH-resistant strains and was 100% specific.

These methods generally require large amounts of amplified product to achieve unambiguous results. Although this limitation does not apply to analysis of isolated colonies or BACTEC cultures, direct detection of drug resistance markers within a clinical specimen may be hindered by the presence of inhibitors or by small numbers of mycobacteria. To overcome these problems, Whelen *et al.* (1995) devised a single-tube heminested PCR that provided high sensitivity of detection of *M. tuberculosis* in sputum and sufficient product for subsequent analysis by sequencing or SSCP.

Hybridization of DNA to oligonucleotide probes is a well-established technique for detecting mutations. Successful hybridization under stringent conditions is dependent upon a perfect match between the target and a short probe. This is the basis of a commercial test designed to detect mutations within the 69-bp hypervariable region of the *rpoB* (Inno-LiPA Rif.TB, Innogenetics N.V., Zwijndrecht, Belgium). The line probe kit consists of a membrane strip onto which 10 oligonucleotide probes are immobilized; one specific for *M. tuberculosis* complex, five overlapping wild-type probes that encompass the entire hypervariable region, and four for specific *rpoB* mutations.

Biotinylated PCR products are hybridized with the probes, and hybrids are determined by an immuno-enzymatic procedure that results in a visual colour. Evaluations of the line probe assay with RMP-resistant strains have shown >90% concordance with phenotypic RMP susceptibility testing results (Cooksey *et al.* 1997; Telenti *et al.* 1997). In Cooksey's study, five resistant isolates (two with codon insertions and three which had no mutations in the 69-bp region) were identified as RMP sensitive by the line probe assay. Mutations in four isolates which demonstrated resistant subpopulations by phenotypic susceptibility testing were correctly identified. However, in Telenti's study the LiPA missed one isolate which contained a mixed population. De Beenhouwer *et al.* (1995) described using the LiPA for detecting RMP resistance directly in clinical specimens. Results from these studies indicate that the line probe assay may serve an important role as a rapid and convenient screen for rifampin resistance in *M. tuberculosis*.

Other novel strategies that have been described for detecting *rpoB* mutations include dideoxy finger-printing (Felmlee *et al.* 1995), heteroduplex formation analysis (Williams *et al.* 1994), RNA/RNA mismatch (Nash *et al.* 1997) and molecular beacon probes (Piatek *et al.* 1998). Dideoxy fingerprinting was recently used to enhance visibility of mobility shifts of *rpoB*-specific amplification products. This technique combines elements of dideoxy sequencing and SSCP, resulting in increased sensitivity for mutation detection. Heteroduplex formation involves mixing denatured PCR product from the test strain with product from a susceptible control strain. Hybridization results in formation of heteroduplex products which exhibit different electrophoretic mobility compared with homoduplex hybrids. The RNA/RNA mismatch assay involves transcription of single-stranded RNA from the test strain PCR products and complementary single-stranded RNA from PCR products from an RMP-susceptible strain. RNase cleaves the RNA/RNA duplex at any positions of base mismatch, and the RNase reactions are analysed by agarose gel electrophoresis. An advantage of this assay over SSCP and heteroduplex analysis is that a larger region of the gene can be screened giving the potential of detecting more mutations within a single assay. The mismatch assay is simple to perform and interpret and has the capability of detecting resistant subpopulations. It has also been used to detect macrolide resistance in *M. avium* (Nash & Inderlied 1996). Another novel approach to mutation analysis is the use of fluorogenic reporter molecules called molecular beacons (Tyagi & Kramer 1996) for allelic discrimination in a real time PCR assay. Molecular beacons are single-stranded probes that possess a stem-and-loop structure with the loop portion being complementary to the target sequence. A fluorescent moiety is attached to one end of the stem and a non-fluorescein, quenching moiety to the other. When the molecular beacon hybridizes with the complementary target, the probe–target hybrid being stronger and more stable that the stem hybrid, the fluorophore is separated from the quencher, permitting the fluorophore to fluoresce. The power of molecular beacons is their ability to hybridize only to target sequences that are perfectly complementary. Piatek *et al.* (1998) have successfully used molecular beacon analysis to detect a broad range of point mutations, as well as insertions and deletions, in the 81-bp region of the *M. tuberculosis rpoB*. Molecular beacons have also been designed to detect mutations in the three genes associated with INH resistance (D. Alland, personal communication).

Cleavase fragment length polymorphism (CFLP) is a new alternative to SSCP and sequencing for mutational screening. CFLP has been applied to the identification and positioning of *katG* mutations associated with INH resistance in *M. tuberculosis* (Brow *et al.* 1996). CFLP is based on the observation that denatured single strands of DNA can assume defined conformations, which can be detected and cleaved by structure-specific endonucleases such as Cleavase I. The cleavage patterns are characteristic of the sequence analysed so that each DNA has its own structural fingerprint. Point mutations change the structural conformation around the site of the mutation and are reflected as changes in the structural fingerprint. By detecting conformational changes with the use of enzymatic cleavage rather than electrophoretic mobility, sequence differences in much larger molecules are detectable. The Cleavase technology, under development by Third Wave Technologies (Madison, WI), has potential utility for differentiating drug-sensitive and drug-resistant strains and for distinguishing mycobacteria at the level of genus and species.

4.2 Metabolic assays for assessing viability in the presence of drugs

Methods for detecting specific genetic mutations have limited practical value owing to our incomplete understanding of all the mutations associated with the development of resistance. Molecular methods that provide a direct measurement of bacterial metabolism can potentially circumvent this problem. One such approach has been the application of the Gen-Probe MTD amplification assay to the BACTEC system (Kawa *et al.* 1989; Miyamoto *et al.* 1996; Martin-Casabona *et al.* 1997). Detection of rRNA in drug containing BACTEC vials can shorten the turn-around time for susceptibility results. However, the stability of the rRNA requires incubation in the presence of antimycobacterial agents for 3–5 days to obtain reliable discrimination of drug-sensitive and drug-resistant isolates. Cangelosi *et al.* (1996) recently described hybridization and RT-PCR assays for *M. tuberculosis* pre-16S rRNA. Results for RMP and ciprofloxacin were obtained within 24 and 48 h, respectively, of exposure to the drugs. This system was unable to detect any depletion of rRNA precursor molecules in the presence of INH or ethambutol. Quantitative analysis of mRNA as a marker of viability has been proposed by Hellyer *et al.* (1999) to be a useful method for rapid drug susceptibility testing. Using a quantitative RT-PCR assay, *M. tuberculosis* strains that were susceptible to INH and RMP showed marked reduction in alpha antigen mRNA expression within 24 h of exposure to these drugs. In contrast, alpha antigen mRNA levels in resistant strains were not reduced.

An innovative method for examining metabolic activity is the biological assay based on the luciferase reporter phage (LRP) technology (Jacobs *et al.* 1993; see also Chapter 3). The luciferase reporter phage is an ingenious tool for evaluating viability and involves infecting mycobacterial cells with a phage carrying the firefly luciferase gene. In the presence of adenosine triphosphate (ATP), found only in living organisms, luciferase produces light from its substrate luciferin. When mycobacteria are infected with the reporter phage and then treated with drugs, light is produced only by viable or drug-resistant mycobacteria. Recently, Riska *et al.* (1997) have shown that the LRP assay when coupled with BACTEC and the NAP compound can differentiate between *M. tuberculosis* complex and non-tuberculous mycobacteria and characterize drug susceptibility patterns within 24–48 h. Recent achievements indicate that the development of an even more efficient LRP protocol is possible. First, new reporter mycobacteriophages with higher degrees of sensitivity have been generated (Carrière *et al.* 1997). These phages can detect mycobacteria in BACTEC vials at growth indices as low as 10. Second, it has been possible to detect *M. tuberculosis* in processed AFB smear-positive sputum samples within 24–48 h, without the need for subculturing in the BACTEC system. Third, a new format for photographic detection of light output has been developed (Riska *et al.* 1999). The film-based LRP assay, when incorporated with the new mycobacteriophages that can induce prolonged light production in infected host cells, should further enhance the efficiency of this diagnostic technology.

4.3 The future of molecular diagnostics

Recent technological advances have enabled the clinical mycobacteriology laboratory to detect *M. tuberculosis* in clinical specimens and to screen for resistance to the commonly used antituberculosis drugs within 24–48 h. Many clinical laboratories are routinely using either a commercial amplification system or an in-house PCR assay to test acid-fast, smear-positive respiratory specimens for primary diagnosis. Unfortunately, these rapid diagnostic tests have not replaced acid-fast smears or mycobacterial cultures. Smear microscopy provides an index of the degree of contagiousness, facilitating informed decisions regarding public health measures. Mycobacterial cultures allow determination of complete drug susceptibility profiles, which are recommended for all patients to ensure optimal treatment. High sensitivity and specificity must be achieved with all sample types before amplification techniques can replace classic diagnostic methods. High specificity can be achieved if the laboratory staff is properly trained and complies with the stringent quality control requirements. Lack of sensitivity most likely results from the use of small sample volumes and irregular dispersion of the organisms in the paucibacillary samples. These shortcomings suggest the need for improved sample preparation methods and/or the performance of more than one test on each sample. Such issues continue to be addressed, and we can expect that the second- and third-generation tests will provide improved sensitivity and specificity. Among the most promising recent developments is the microarray technology (DNA chips) which, combined with DNA or RNA amplification, could provide rapid identification of a wide range of mycobacterial species and drug susceptibility results. The microbiology laboratory wish list for future amplification systems includes those that are speedy, automated, sensitive, specific, are not at risk of cross-contamination, affordable and amenable to quantification. Thanks to the efforts of numerous academic scientists and commercial firms, this potential is becoming a reality. Hopefully, the ensuing competition between assays will eventually result in decreased costs for this state-of-the-art revolutionary technology.

5 Acknowledgements

The author would like to thank Donald Cave, Tobin Hellyer, Lucy DesJardin, Marjorie Beggs, Vivian Jonas-Taggart, Bill Keating, David Alland and John Chan for helpful comments and providing information prior to publication.

6 References

Ausina, V., Gamboa, F., Gazapo, E. *et al.* (1997) Evaluation of the semiautomated Abbot LCx *Mycobacterium tuberculosis* assay for direct detection of *Mycobacterium tuberculosis* in respiratory specimens. *Journal of Clinical Microbiology* **35**, 1996–2002.

Banerjee, A., Dubnau, E., Quemard, A. *et al.* (1994) *inhA*, a gene encoding a target for isoniazid and ethionamide in *Mycobacterium tuberculosis*. *Science* **263**, 227–230.

Beggs, M.L., Cave, M.D., Marlowe, C., Cloney, L., Duck, P. & Eisenach, K.D. (1996) Characterization of *Mycobacterium tuberculosis* complex direct repeat sequence for use in cycling probe reaction. *Journal of Clinical Microbiology* **34**, 2985–2989.

Böddinghaus, B., Rogall, T., Flohr, T., Blöcker, H. & Böttger, E.C. (1990) Detection and identification of mycobacteria by amplification of rRNA. *Journal of Clinical Microbiology* **28**, 1751–1759.

Brisson-Noël, A., Gicquel, B., Lecossier, D., Lévy-Frébault, V., Nassif, X. & Hance, A.J. (1989) Rapid diagnosis of tuberculosis by amplification of mycobacterial DNA in clinical samples. *Lancet* **2**, 1069–1071.

Brow, M.A.D., Oldenburg, M.C., Lyamichev, V. *et al.* (1996) Differentiation of bacterial 16S rRNA genes and intergenic regions and *Mycobacterium tuberculosis katG* genes by structure-specific endonuclease cleavage. *Journal of Clinical Microbiology* **34**, 3129–3137.

Butler, W.R., O'Connor, S.P., Yakrus, M.A. & Gross, W.M. (1994) Cross-reactivity of genetic probe for detection of *Mycobacterium tuberculosis* with newly described species *Mycobacterium celatum*. *Journal of Clinical Microbiology* **32**, 536–538.

Cangelosi, G.A., Brabant, W.H., Britschgi, T.B. & Wallis, C.K. (1996) Detection of rifampin- and ciprofloxacin-resistant *Mycobacterium tuberculosis* by using species-specific assays for precursor rRNA. *Antimicrobial Agents and Chemotherapy* **40**, 1790–1795.

Carrière, C., Riska, P.F., Zimhony, O. *et al.* (1997). Conditional replicating luciferase reporter phages: Improved sensitivity for rapid detection and assessment of drug susceptibility of *Mycobacterium tuberculosis*. *Journal of Clinical Microbiology* **35**, 3232–3239.

Clarridge, J.E. III, Shawar, R.M., Shinnick, T.M. & Plikaytis, B.B. (1993). Large scale use of polymerase chain reaction for detection of *Mycobacterium tuberculosis* in a routine mycobacteriology laboratory. *Journal of Clinical Microbiology* **31**, 2041–2056.

Cohen, R.A., Muzaffar, S., Schwartz, D. *et al.* (1998) Diagnosis of pulmonary tuberculosis using PCR assays on sputum collected within 24 hours of hospital admission. *American Journal of Respiratory and Critical Care Medicine* **157**, 156–161.

Compton, J. (1991) Nucleic acid sequence-based amplification. *Nature (London)* **350**, 91–92.

Cooksey, R.C., Morlock, G.P., Glickman, S. & Crawford, J.T. (1997) Evaluation of a line probe assay kit for characterization of *rpoB* mutations in rifampin-resistant

Mycobacterium tuberculosis isolates from New York city. *Journal of Clinical Microbiology* **35**, 1281–1283.

Dalovisio, J.R., Montenegro-James, S., Kemmerly, S.A. *et al.* (1996) Comparison of the amplified *Mycobacterium tuberculosis* (MTB) direct test, amplicor MTB PCR and IS*6110*-PCR for detection of MTB in respiratory specimens. *Clinical Infectious Diseases* **23**, 1099–1106.

De los Monteros, L.E.E., Galán, J.C., Gutiérrez, M. (1998) Allele-specific PCR method based on *pncA* and *oxyR* sequences for distinguishing *Mycobacterium bovis* from *Mycobacterium tuberculosis*: Intraspecific *M. bovis pncA* sequence polymorphism. *Journal of Clinical Microbiology* **36**, 239–242.

De Beenhouwer, H., Lhiang, Z., Jannes, G. *et al.* (1995) Rapid detection of rifampicin resistance in sputum and biopsy specimens from tuberculosis patients by PCR and line probe assay. *Tubercle and Lung Disease* **76**, 425–430.

Del Portillo, P., Thomas, M.C., Martínez, E. *et al.* (1996) Multiprimer PCR system for differential identification of mycobacteria in clinical samples. *Journal of Clinical Microbiology* **34**, 324–328.

Deretic, V., Philipp, W., Dhandayuthapani, S., Mudd, M.H. *et al.* (1995) *Mycobacterium tuberculosis* is a natural mutant with an inactivated oxidative-stress regulatory gene: implications for sensitivity to isoniazid. *Molecular Microbiology* **17**, 889–900.

DesJardin, L.E., Chen, Y., Perkins, M.D., Teixeira, L., Cave, M.D. & Eisenach, K.D. (1998) Comparative use of the ABI 7700 (Taq Man) and competitive PCR for quantification of IS*6110* DNA in sputum during the treatment of tuberculosis. *Journal of Clinical Microbiology* **36**, 1964–1968.

DesJardin, L.E., Perkins, M.D., Teixeira, L., Cave, M.D. & Eisenach, K.D. (1996) Alkaline decontamination of sputum specimens adversely effects the stability of mycobacterial mRNA. *Journal of Clinical Microbiology* **34**, 2435–2439.

DesJardin, L.E., Perkins, M.D., Wolski, K. *et al.* (1999) Measurement of sputum *Mycobacterium tuberculosis* mRNA as a surrogate for response to chemotherapy. *American Journal of Respiratory Critical Care Medicine* **159** (in press).

de Wit, D., Wootton, M., Dhillon, J. & Mitchison, D.A. (1995) The bacterial DNA content of mouse organs in the Cornell model of dormant tuberculosis. *Tubercle and Lung Disease* **76**, 555–562.

Eisenach, K.D., Cave, M.D., Bates, J.H. & Crawford, J.T. (1990) Polymerase chain reaction amplification of a repetitive DNA sequence specific for *Mycobacterium tuberculosis*. *Journal of Infectious Diseases* **161**, 977–981.

Eisenach, K.D., Sifford, M.D., Cave, M.D., Bates, J.H. & Crawford, J.T. (1991) Detection of *Mycobacterium tuberculosis* in sputum samples using a polymerase chain

reaction. *American Review of Respiratory Diseases* **144**, 1160–1163.

Evans, K.D., Nakasone, A.S., Sutherland, P.A., De la Maza, L.M. & Peterson, E.M. (1992) Identification of *Mycobacterium tuberculosis* and *Mycobacterium avium-intracellulare* directly from primary BACTEC cultures by using acridinium-ester-labeled DNA probes. *Journal of Clinical Microbiology* **31**, 2427–2431.

Felmlee, T.A., Liu, Q., Whelen, A.C., Williams, D., Sommer, S.S. & Persing, D.H. (1995). Genotypic detection of *Mycobacterium tuberculosis* rifampin resistance: comparison of single-strand conformation polymorphism and dideoxy fingerprinting. *Journal of Clinical Microbiology* **33**, 1617–1623.

Forbes, B.A. (1997) Critical assessment of gene amplification approaches on the diagnosis of tuberculosis. *Immunological Investigations* **26**, 105–116.

Frothingham, R. (1995) Differentiation of strains in *Mycobacterium tuberculosis* complex by DNA sequence polymorphisms, including rapid identification of *M. bovis* BCG. *Journal of Clinical Microbiology* **33**, 840–844.

Gamboa, F., Fernandez, G., Padilla, E. *et al.* (1998) Comparative evaluation ofitial and new, Version of the Gen-Probe amplified *Mycobacterium tuberculosis* direct test for direct detection of *Mycobacterium tuberculosis* in respiratory and nonrespiratory specimens. *Journal of Clinical Microbiology* **36**, 684–689.

Hance, A.J., Grandchamp, B., Levy-Frebault, V. *et al.* (1989) Detection and identification of mycobacteria by amplification of mycobacterial DNA. *Molecular Microbiology* **3**, 843–849.

Hellyer, T.J., DesJardin, L.E., Hehman, G.L., Cave, M.D. & Eisenach, K.D. (1999) Quantitative analysis of mRNA as a marker for viability of *Mycobacterium tuberculosis*. *Journal of Clinical Microbiology* **37**, 290–295.

Hellyer, T.J., Fletcher, T.W., Bates, J.H. *et al.* (1996) Strand displacement amplification and the polymerase chain reaction for monitoring response to treatment in patients with pulmonary tuberculosis. *Journal of Infectious Diseases* **173**, 934–941.

Herold, C.D., Fitzgerald, R.L. & Herold, D.A. (1996) Current techniques in mycobacterial detection and speciation. *Critical Review in Clinical Laboratory Sciences* **33**, 83–138.

Heym, B., Philipp, W. & Cole, S.T. (1996) Mechanisms of drug resistance in *Mycobacterium tuberculosis*. *Current Topics in Microbiology and Immunology* **215**, 49–69.

Huang, T., Liu, Y., Lin, H.H., Huang, W.K. & Cheng, D.L. (1996) Comparison of the Roche Amplicor Mycobacterium assay and Digene Sharp signal system with in-house PCR and culture for detection of *Mycobacterium tuberculosis* in respiratory specimens. *Journal of Clinical Microbiology* **34**, 3092–3096.

Ichiyama, S., Iinuma, Y., Tawada, Y. *et al.* (1996) Evaluation of Gen-Probe amplified *Mycobacterium tuberculosis* direct test and Roche PCR-Microwell plate hybridization method (amplicor mycobacterium) for direct detection of mycobacteria. *Journal of Clinical Microbiology* **34**, 130–133.

Ichiyama, S., Ito, Y., Sugiura, F.(1997) Diagnostic value of the strand displacement amplification method compared to those of Roche amplicor PCR and culture for detecting mycobacteria in sputum samples. *Journal of Clinical Microbiology* **35**, 3082–3085.

Jacobs, W.R., Jr, Barletta, R.G., Udani, R. *et al.* (1993) Rapid assessment of drug susceptibilities of *Mycobacterium tuberculosis* by means of luciferase reporter phages. *Science* **260**, 819–822.

Jonas, V., Alden, M.J., Curry, J.I. *et al.* (1993) Detection and identification of *Mycobacterium tuberculosis* directly from sputum sediments by amplification of rRNA. *Journal of Clinical Microbiology* **31**, 2410–2416.

Kapur, V., Li, L.-L., Hamrick, M.R. *et al.* (1995) Rapid *Mycobacterium* species assignment and unambiguous identification of mutations associated with antimicrobial resistance in *Mycobacterium tuberculosis* by automated DNA sequencing. *Archives of Pathology and Laboratory Medicine* **119**, 66–73.

Kawa, D.E., Pennell, D.R., Kubista, L.N. & Schell, R.F. (1989) Development of a rapid method for determining the susceptibility of *Mycobacterium tuberculosis* to isoniazid using the Gen-Probe DNA hybridization system. *Journal of Clinical Microbiology* **33**, 1000–1005.

Kennedy, N., Gillespie, S.H., Saruni, A.O. *et al.* (1994) Polymerase chain reaction for assessing treatment response in patients with pulmonary tuberculosis. *Journal of Infections Diseases* **170**, 713–716.

Kirschner, P., Springer, B., Vogel, U. *et al.* (1993) Genotypic identification of mycobacteria by nucleic acid sequence determination: Report of a 2-year experience in a clinical laboratory. *Journal of Clinical Microbiology* **31**, 2882–2889.

Kolk, A.H.J., De Noordhoek, G.T., Leeuw, O., Kuijper, S. & van Embden, J.D.A. (1994). *Mycobacterium smegmatis* strain for detection of *Mycobacterium tuberculosis* by PCR used as internal control for inhibition of amplification and for quantification of bacteria. *Journal of Clinical Microbiology* **32**, 1354–1356.

Kox, L.F., Jansen, H.M., Kuijper, S. & Kolk, A.H.J. (1997) Multiplex PCR assay for immediate identification of the infecting species in patients with mycobacterial disease. *Journal of Clinical Microbiology* **35**, 1492–1498.

Lizardi, P.M., Guerra, C.E., Lomeli, H., Tussieluna, I. & Kramer, F.R. (1988) Exponential amplification of recombinant-RNA hybridization probes. *Biotechnology* **6**, 1197–1202.

Mahairas, G.G., Sabo, P.J., Hickey, M.J., Singh, D.C. & Stover, C.K. (1996) Molecular analysis of genetic differences between *Mycobacterium bovis* BCG and virulent *M. bovis*. *Journal of Bacteriology* **178**, 1274–1282.

Martin, C., Levy-Frebault, V., Cattier, B., Legras, A. & Goudeau, A. (1993) False positive results of *Mycobacterium tuberculosis* complex DNA probe hybridization with a *Mycobacterium terrae* isolate. *European Journal of Clinical Microbiology and Infectious Diseases* **12**, 309–310.

McAdam, R.A., Hermans, P.W.M., Van Soolingen, D. *et al.* (1990) Characterization of a *Mycobacterium tuberculosis* insertion sequence belonging to the IS3 family. *Molecular Microbiology* **4**, 1607–1613.

Martin-Casabona, N., Xairó Mimó, D., González, T., Rosello, J. & Arcalis, L. (1997) Rapid method for testing susceptibility of *Mycobacterium tuberculosis* by using DNA probes. *Journal of Clinical Microbiology* **35**, 2521–2525.

Metchock, B. & Diem, L. (1995) Algorithm for the use of nucleic acid probes for identifying *Mycobacterium tuberculosis* from BACTEC 12B bottles. *Journal of Clinical Microbiology* **33**, 1934–1937.

Miyamoto, J., Koga, H., Kohno, S., Tashiro, T. & Hara, K. (1996) New drug susceptibility test for *Mycobacterium tuberculosis* using the hybridization protection assay. *Journal of Clinical Microbiology* **34**, 1323–1326.

Miyazaki, Y., Koga, H., Kohno, S. & Kaku, M. (1993) Nested polymerase chain reaction for detection of *Mycobacterium tuberculosis* in clinical samples. *Journal of Clinical Microbiology* **31**, 2228–2232.

Moore, D.F., Curry, J.I., Knott, C.A. & Jonas, V. (1996) Amplification of rRNA for assessment of treatment response of pulmonary tuberculosis patients during antimicrobial therapy. *Journal of Clinical Microbiology* **34**, 1745–1749.

Musser, J.M. (1995) Antimicrobial agent resistance in Mycobacteria: Molecular genetic insights. *Clinical Microbiology Reviews* **8**, 496–514.

Nash, K.A. & Inderlied, C.B. (1996) Rapid detection of mutations associated with macrolide resistance in *Mycobacterium avium* complex. *Antimicrobial Agents and Chemotherapy* **40**, 1748–1750.

Nash, K.A., Gaytan, A. & Inderlied, C.B. (1997) Detection of rifampin resistance in *Mycobacterium tuberculosis* by use of a rapid, simple, and specific RNA/RNA mismatch assay. *Journal of Infectious Diseases* **176**, 533–536.

Parra, C.A., Londono, L.P., Del Portillo, P. & Patarroyo, M.E. (1991) Isolation, characterization, and molecular cloning of a specific *Mycobacterium tuberculosis* antigen: identification of a species-specific sequence. *Infection and Immunology* **59**, 3411–3417.

Piatek, A.S., Tyagi, S., Pol, A.C. *et al.* (1998) Molecular beacon sequence analysis for detecting drug resistance in *Mycobacterium tuberculosis*. *Nature Biotechnology* **16**, 359–363.

Piersimoni, C., Callegaro, A., Nista, D. *et al.* (1997) Comparative evaluation of two commercial amplification assays for direct detection of *Mycobacterium tuberculosis* complex in respiratory specimens. *Journal of Clinical Microbiology* **35**, 193–196.

Plikaytis, B.B., Eisenach, K.D., Crawford, J.T. & Shinnick, T.M. (1991) Differentiation of *Mycobacterium tuberculosis* and *Mycobacterium bovis* BCG by a polymerase chain reaction assay. *Molecular and Cellular Probes* **5**, 215–219.

Plikaytis, B.B., Plikaytis, B.D., Yakrus, M.A. *et al.* (1992) Differentiation of slowly growing *Mycobacterium* species, including *Mycobacterium tuberculosis*, by gene amplification and restriction fragment length polymorphism analysis. *Journal of Clinical Microbiology* **30**, 1815–1822.

Rajalahti, I., Vuorinen, P., Nieminen, M.M. & Miettinen, A. (1998) Detection of *Mycobacterium tuberculosis* complex in sputum specimens by the automated Roche Cobas Amplicor Mycobacterium Tuberculosis test. *Journal of Clinical Microbiology* **36**, 975–978.

Richeldi, L., Barnini, S. & Saltini, C. (1995) Molecular diagnosis of tuberculosis. *European Respiratory Journal* **20**, 689s–700s.

Riska, P.F., Jacobs, W.R. Jr, Bloom, B.R., McKitrick, J. & Chan, J. (1997) Specific identification of *Mycobacterium tuberculosis* with the luciferase reporter mycobacteriophage: Use of *p*-nitro-α-acetylamino-β-hydroxy-propiophenone. *Journal of Clinical Microbiology* **35**, 3225–3231.

Riska, P.F., Su, Y., Bardarou, S. *et al.* (1999) Rapid film-based determination of antibiotic susceptibilities of *Mycobacterium tuberculosis* strains by using a luciferase reporter phage and the Bronx box. *Journal of Clinical Microbiology* **37**, 1144–1149.

Rogall, T., Flohr, T. & Bottger, E.C. (1990b) Differentiation of mycobacterium species by direct sequencing of amplified DNA. *Journal of General Microbiology* **136**, 1915–1920.

Rogall, T., Wolters, J., Flohr, T. & Bottger, E.C. (1990a) Towards a phylogeny and definition of species at the molecular level within the genus *Mycobacterium*. *International Journal of Systematic Bacteriology* **40**, 323–330.

Sandin, R.L. (1996) Polymerase chain reaction and other amplification techniques in mycobacteriology. *Clinics in Laboratory Medicine* **16**, 617–640.

Scorpio, A. & Zhang, Y. (1996) Mutations in *pncA*, a gene encoding pyrazinamidase/nicotinamidase, cause resistance to the antituberculous drug pyrazinamide in tubercle bacillus. *Nature Medicine* **2**, 662–667.

Scorpio, A., Collins, D., Whipple, D., Cave, D., Bates, J. & Zhang, Y. (1997) Rapid differentiation of bovine and human tubercle bacilli based on a characteristic mutation in the bovine pyrazinamidase gene. *Journal of Clinical Microbiology* **35**, 106–110.

Shah, J.S., Liu, J., Buxton, D. *et al.* (1995) Q-Beta replicase-amplified assay for detection of *Mycobacterium tuberculosis* directly from clinical specimens. *Journal of Clinical Microbiology* **33**, 1435–1441.

Shen, L.-P., Kern, D., Zanki, S. *et al.* (1994) Detection and semiquantitation of *Mycobacterium tuberculosis* in sputum specimens directly using a signal amplification branched-DNA assay. American Society of Microbiology Meeting in Las Vegas, NV, May 1994.

Smith, J.H., Buxton, D., Cahill, P. *et al.* (1997) Detection of *Mycobacterium tuberculosis* directly from sputum by using a prototype automated Q-beta replicase assay. *Journal of Clinical Microbiology* **35**, 1477–1483.

Spargo, C.A., Fraiser, M.S., VanCleve, M. *et al.* (1996) Detection of *M. tuberculosis* DNA using thermophilic strand displacement amplification. *Molecular and Cellular Probes* **10**, 247–256.

van Soolingen, D., Hermans, P.W.M., De Haas, P.E.W. & van Embden, J.D.A. (1992). (1992) Insertion element IS*1081*-associated restriction fragment length polymorphisms in *Mycobacterium tuberculosis* complex species: a reliable tool for recognizing *Mycobacterium bovis* BCG. *Journal of Clinical Microbiology* **30**, 1772–1777.

Sreevatsan, S., Escalante, P., Pan, X. *et al.* (1996) Identification of a polymorphic nucleotide in oxyR specific for mycobacterium. *Journal of Clinical Microbiology* **34**, 2007–2010.

Stockman, L., Springer, B., Bottger, E.C. & Roberts, G.D. (1993) *Mycobacterium tuberculosis* nucleic acid probes for rapid diagnosis. *Lancet* **341**, 1486.

Talbot, E.A., Williams, D.L. & Frothingham, R. (1997) PCR identification of *Mycobacterium bovis* BCG. *Journal of Clinical Microbiology* **35**, 566–569.

Telenti, M., de Quiros, J.F.B., Alvarez, M., Rionda, M.J.S. & Mendoza, M.C. (1994) The diagnostic usefulness of a DNA probe for *Mycobacterium tuberculosis* complex (Gen-Probe) in BACTEC cultures versus other diagnostic methods. *Infection* **22**, 18–23.

Telenti, A., Honoré, N., Bernasconi, C. *et al.* (1997) Genotypic assessment of isoniazid and rifampin resistance in *Mycobacterium tuberculosis*: a blind study at reference laboratory level. *Journal of Clinical Microbiology* **35**, 719–723.

Telenti, A., Imboden, P., Marchesi, F. *et al.* (1993b) Detection of rifampicin-resistance mutations in *Mycobacterium tuberculosis*. *Lancet* **341**, 647–650.

Telenti, A., Marchesi, F., Balz, M., Bally, F., Böttger, E.C. & Bodmer, T. (1993a) Rapid identification of mycobacteria to the species level by polymerase chain reaction and restriction enzyme analysis. *Journal of Clinical Microbiology* **31**, 175–178.

Telenti, A. & Persing, D.H. (1996) Novel strategies for the detection of drug resistance in *Mycobacterium tuberculosis*. *Research in Microbiology* **147**, 73–79.

Thierry, D., Brisson-Noël, A., Vincent-Lévy-Frébault, V., Nguyen, S., Guesdon, J.-L. & Gicquel, B. (1990) Characterization of a *Mycobacterium tuberculosis* insertion sequence, IS*6110*, and its application in diagnosis. *Journal of Clinical Microbiology* **28**, 2668–2673.

Tortoli, E., Lavinia, F. & Simonetti, M.T. (1997) Evaluation of a commercial ligase chain reaction kit (Abbott LCx) for direct detection of *Mycobacterium tuberculosis* in pulmonary and extrapulmonary specimens. *Journal of Clinical Microbiology* **35**, 2424–2426.

Tyagi, S. & Kramer, F.R. (1996) Molecular beacons: probes that fluoresce upon hybridization. *Nature Biotechnology* **14**, 303–308.

Van der Vliet, G.M.E., Schukkink, R.A.F., Van Gemen, B., Schepers, P. & Klatser, P.R. (1993) Nucleic acid sequence-based amplification (NASBA) for the identification of mycobacteria. *Journal of General Microbiology* **139**, 2423–2429.

Vaneechoutte, M., De Beenhouwer, H., Claeys, G. *et al.* (1993) Identification of *Mycobacterium* species by using amplified ribosomal DNA restriction analysis. *Journal of Clinical Microbiology* **31**, 2061–2065.

Vuorinen, P., Miettinen, A., Vuento, R. & Hällström, O. (1995) Direct detection of *Mycobacterium tuberculosis* complex in respiratory specimens by Gen-Probe amplified *Mycobacterium tuberculosis* direct test and Roche amplicor *Mycobacterium tuberculosis* test. *Journal of Clinical Microbiology* **33**, 1856–1859.

Walker, G.T., Fraiser, M.S., Schram, J.L., Little, M.C., Nadeau, J.G. & Malinowski, D.P. (1992) Strand displacement amplification—an isothermal, *in vitro* DNA amplification technique. *Nucleic Acids Research* **20**, 1691–1696.

Weil, A., Plikaytis, B.B., Butler, W.R., Woodley, C.L. & Shinnick, T.M. (1996) The *mtp40* gene is not present in all strains of *Mycobacterium tuberculosis*. *Journal of Clinical Microbiology* **34**, 2309–2311.

Whelen, A.C., Felmlee, T.A., Hunt, J.M. *et al.* (1995) Direct genotypic detection of *Mycobacterium tuberculosis* rifampin resistance in clinical specimens by using single-tube heminested PCR. *Journal of Clinical Microbiology* **33**, 556–561.

Williams, D.L., Waguespack, C., Eisenach, K. *et al.* (1994) Characterization of rifampin resistance in pathogenic mycobacteria. *Antimicrobial Agents and Chemotherapy* **38**, 2380–2386.

Wilson, T.M. & Collins, D.M. (1996) *ahpC*, a gene involved in isoniazid resistance of the *Mycobacterium tuberculosis* complex. *Molecular Microbiology* **19**, 1025–1034.

Yuen, K.Y., Yam, W.C., Wong, L.P. & Seto, W.H. (1997) Comparison of two automated DNA amplification systems with a manual one-tube nested PCR assay for diagnosis of pulmonary tuberculosis. *Journal of Clinical Microbiology* **35**, 1385–1389.

Zhang, Y., Heym, B., Allen, D., Young, D. & Cole, S. (1992) The catalase-peroxidase gene and isoniazid resistance of *Mycobacterium tuberculosis*. *Nature* **358**, 501–593.

Chapter 10 / Immunodiagnosis of mycobacterial infection

GRAHAM H. BOTHAMLEY, DAVID CATTY, RICHARD CLIFTON-HADLEY, FRANK GRIFFIN, GLYN HEWINSON & JOHN POLLOCK

1 Introduction

In spite of many attempts to develop and introduce sensitive, specific and simple immunoassay methods for the diagnosis of mycobacterial disease in humans and other animals, this has not proved to be easy, and acid-fast staining of specimens, the culture and identification of the mycobacterial species remain the gold standard. Yet these tests are labour intensive, insensitive and slow. Being dependent upon the finding of bacilli, the approach may miss many cases where the bacillary counts are low or absent in samples; there is, for instance, no established reliable method for confirming extrapulmonary tuberculosis, including tuberculous meningitis in humans. There is some reliance on delayed hypersensitivity testing in humans and this is still almost the only recourse in cattle. This test is notorious for cross-reactivity. For these reasons there is a good case for developing new immunodiagnostic approaches to the problem of rapid, simple and reliable detection of infection. These would be immensely valuable in screening patients or animals to assist in early diagnosis. The high morbidity associated with disseminated and meningeal disease makes an early diagnosis and an early start to treatment particularly important. Infection need not give rise to disease, but reliable evidence of infection would permit early, preventive treatment. To be efficient, immunological tests must have an extremely high negative predictive value (as few false-negative results as possible), whereas false-positive reactions have economic rather than disease-related consequences.

It has long been known that there is much antigenic cross-reactivity between mycobacterial species. This has hampered the development of immunodiagnostic tests. However, steady advances in defining mycobacterial antigens over the last few years, assisted both by modern molecular and genetic approaches and the advent of monoclonal antibodies, has allowed considerable progress. In consequence several new immunodiagnostic tests have been advocated and these are reviewed in this chapter. Comparison is made between these and the delayed hypersensitivity reaction to tuberculin which remains the only immunodiagnostic test yet to be widely used.

2 Diagnosis of tuberculosis in humans and other species

2.1 Delayed hypersensitivity reactions

Tuberculin skin testing

The tuberculin skin test measures the delayed hypersensitivity reaction to a mixture of secreted and autolysed proteins from *Mycobacterium tuberculosis*. Tuberculin is prepared from autoclaved and filtered liquid cultures of *M. tuberculosis* after 8 weeks' growth. The protein content is precipitated from the filtrate with either trichloroacetic acid or ammonium sulphate (purified protein derivative—PPD). Seibert prepared a large batch of tuberculin in 1939 (PPD-S), which remains the international standard. The activity of all subsequent preparations has been defined such that 5 tuberculin units (5 TU) in 0.1 mL elicits an induration of the same size ±20% as 0.1 mg of PPD-S; new preparations contain a detergent (0.05% Tween 80) to prevent adsorption. The diagnostic test employs 5 TU, as epidemiological studies have shown that this amount gives the best discrimination between patients and controls (Comstock *et al.* 1981).

The tuberculin test can be used either as a diagnostic test to increase the suspicion of active tuberculosis or as a screening test to indicate exposure to tubercle bacilli in otherwise asymptomatic individuals (Bothamley 1997). When used as a diagnostic test, tuberculin is injected into the skin (intradermally and not subcutaneously) and the response measured as the diameter of induration (not erythema) at 48 or 72 h. The edge of induration is assessed with a ballpoint pen and the diameter measured in the long axis of the forearm. A positive test is:

1 > 5 mm of induration for those in contact with tuberculosis, those with concurrent human immunodeficiency virus (HIV) infection and patients with clinical or radiographic features of tuberculosis, and

2 > 10 mm of induration in individuals thought to have tuberculosis and who come from an area where tuberculosis is common, with social reasons for exposure to tuberculosis, who are residents of long-term care facilities including prisons, with medical conditions which increase the risk of tuberculosis and who are intravenous drug users, and

3 > 15 mm in those who have no known risk factor for tuberculosis or who come from an area of the world where non-specific tuberculin reactions are common due to exposure to non-tuberculous mycobacterial disease.

The sensitivity (the percentage of individuals with tuberculosis correctly identified by a positive tuberculin reaction) and specificity (the number of individuals without tuberculosis who are negative) of the tuberculin reaction can vary quite widely, depending on the diameter of induration defined as positive and the population examined (Bothamley 1995; Rose *et al.* 1995). In general, the tuberculin skin test shows a greater specificity in higher latitudes and in populations where tuberculosis is rare. In the tropics the specificity may fall to 28%, while in the elderly the sensitivity may be poor (Evans *et al.* 1996). A false-positive reaction can be due to exposure to other mycobacteria and this is used to explain the geographical variability of tuberculin reactivity (Comstock *et al.* 1981). Tuberculin reactivity usually develops after bacille Calmette–Guérin (BCG) vaccination, but the diameter of induration is smaller than with active tuberculosis (Sepulveda *et al.* 1990). False-negative reactions can occur in patients with concurrent infections (especially measles and HIV infection), or immunosuppression caused by drugs, malnutrition, stress and lymphoreticular disease. More importantly, tuberculin responses may be suppressed in patients with active tuberculosis (Howard *et al.* 1970), although reactivity often returns when the patient has received effective treatment. Thus, tuberculin testing must be interpreted in the light of the clinical picture and even then has a poor predictive value.

Tuberculin reactivity can be used as a screening test for exposure to tuberculosis, because very few apparently healthy subjects will not give a positive result if they have been exposed to tuberculosis. In order to test large numbers of subjects during short periods, a multipuncture head coated with heat-sterilized culture filtrate (the Tine test), or passed through a concentrated solution of PPD (20 mg/mL—Heaf test;

Heaf 1951), may be used. Both techniques allow rapid testing of subjects without the technical problems of an intradermal injection and can be read up to 1 week later. The reaction can be read at 48–72 h as millimetres of induration or a week later by a grading system (Joint Tuberculosis Committee of the British Thoracic Society 1994). The significance of the reaction is affected by previous BCG vaccination, such that Grade 2 (where the six papules merge to form a ring of induration) is positive in those without a BCG vaccination, whereas in those who have received BCG vaccination a positive reaction begins at Grade 3 (the indurations combine to form a weal 10 mm in diameter).

Purified antigens

The cross-reactivity of mycobacterial antigens has been the chief failing of tuberculin as a diagnostic test reagent. Individual components have been purified and used as skin test reagents. Tuberculin activity was thought to reside in a single 9.7-kDa peptide (Kuwabara 1975), but this suggestion was not confirmed (Toida *et al.* 1985). More recently, Antigen 85B/Antigen 6/MPT59, Antigen 85A/P32/MPT44 and P38/Antigen 5 (Daniel *et al.* 1982; Kadival *et al.* 1987; Wiker and Harboe 1992; Wilcke *et al.* 1996; Wilkinson *et al.* 1997) have been examined for their specificity in tuberculosis. However, even individual proteins of *M. tuberculosis* are similar to proteins from other environmental mycobacterial species and as each protein contains a number of T-cell epitopes which might stimulate delayed hypersensitivity, no single protein has been identified so far which gives a species-restricted response. One peptide, synthesized from the known sequence of the P38 antigen (amino acids 350–369), showed species specificity in guinea pigs and a delayed hypersensitivity response in a PPD+ human subject (Vordermeier *et al.* 1992). This approach shows promise in identifying T-cell epitopes which are unique to the *M. tuberculosis* complex and which could then be used as a diagnostic reagent, but the value of this peptide has not been confirmed.

The bacille Calmette–Guérin test

Patients with tuberculosis who have an enhanced T-cell immunity to tuberculosis antigens and who are inadvertently given the BCG vaccine develop an accelerated reaction to the vaccine. In areas of the world where tuberculosis is common, BCG has been advocated as a diagnostic test, especially in children where vaccination is especially valuable and in whom tuberculosis is difficult to diagnose (Göçmen *et al.* 1994). A papule > 5 mm in diameter appears within 24–48 h, rather than the more usual 2–3 weeks, and a pustule develops in 3–5 days in children who receive BCG vaccination and have tuberculosis. Even in malnourished children with tuberculosis, the reaction to BCG vaccination is significantly earlier than in those without tuberculosis.

2.2 Serology of human infection

2.2.1 Antibody tests

There have been three main approaches to antibody-based serological tests for human tuberculosis.

1 Defining the antibody-binding specificity pattern associated with active infection and thereby defining also the bacterial antigens which are serologically active and possibly disease specific. This is usually achieved by separating mycobacterial antigens on a polyacrylamide gel by electrophoresis and then using sera from patients with tuberculosis to immunostain a western blot of the gel.

2 Using antibodies of tuberculosis sera which have been preabsorbed using non-tuberculous mycobacterial antigens. The more infection-specific residual antibodies are then used to affinity-purify active antigens which can then act as the specific target for test serum antibodies in a solid-phase assay.

3 Applying monoclonal antibodies which identify antibody binding sites which are restricted to the *M. tuberculosis* complex. These are then used either to purify the antigen or to identify recombinant bacterial clones expressing the antigen, or to detect and measure human patients' antibody to the same epitope in a competition immunoassay.

Infection-related or specific antigens

P38 (the 38-kDa antigen of *M. tuberculosis*, antigen 5, antigen 78) was identified by all three of the above approaches (Daniel & Andersen 1978; Coates *et al.* 1981; Wiker *et al.* 1988). It has attracted the most interest as an infection-specific antigen, and it is in consequence the most widely evaluated in serum immunoassays (reviewed by Bothamley 1995). Independent groups have consistently demonstrated a sensitivity of 70% or more with a specificity of 95% when testing sera from proven tuberculosis patients and controls. The evaluation of this antigen is more advanced than that of other purified proteins and has been used in a number of controlled clinical trials, for instance in patients with suspected pulmonary tuberculosis (Bothamley 1995; Wilkinson *et al.* 1997) and as a screening test, where comparison was made with the sputum smear test (Daniel *et al.* 1986; Steele & Daniel 1991). The advantages of using the 38-kDa antigen include:

1 most anti-38-kDa serum antibody in infected people is directed towards the *M. tuberculosis*-restricted epitopes of the molecule;

2 BCG-vaccinated controls and patients with other respiratory illnesses, including non-tuberculous mycobacterial disease, do not form antibody to this antigen;

3 the antigen is available as a recombinant protein in large quantity which allows assay standardization;

4 near-patient testing has already proved feasible with prepared kits (Cole *et al.* 1996).

The 16-kDa antigen of *M. tuberculosis*, a homologue of the heat-shock protein α-crystallin (Verbon *et al.* 1992), appears to be preferentially expressed during a semidormant phase of growth (Yuan *et al.* 1996). Antibody to this antigen has been found in contacts of patients with tuberculosis, in primary tuberculosis in children, in tuberculosis meningitis and in patients treated for tuberculosis who fail to respond to treatment and are subsequently shown to have a positive sputum culture (Jackett *et al.* 1988; Chandramuki *et al.* 1989; Bothamley *et al.* 1992a,b, 1988). This suggests that the antibody response to this antigen may reflect the change in the protein's expression associated with the transition from a quiescent to an active state of the tubercle bacillus.

The 16-kDa antigen and other purified protein antigens (for instance members of the antigen 85 complex of secreted antigens) have been tested with sera from proven tuberculosis patients and healthy controls, but as yet do not form the basis of a commercially available test (Bothamley 1995). The sensitivity and specificity of tests with these reagents remains to be confirmed in controlled trials including patients with suspected tuberculosis.

Lipoarabinomannan (LAM), the most powerfully antigenic lipopolysaccharide of *M. tuberculosis*, has also been used as a diagnostic reagent and forms the basis of at least one commercial test. The sensitivity seems to be lower than with the protein antigens. No results have been published of properly controlled clinical trials of antibodies to LAM, examining sera from patients with suspected pulmonary tuberculosis.

Antigen 60 is a complex mixture of heat-stable antigens which are apparently associated with the ribosomal fraction of *M. tuberculosis*. The sensitivity and specificity of antibody assays using this reagent is reported to be high (Cocito 1991).

Monoclonal antibodies

A purified antigen has a number of potential antibody binding sites (epitopes). Not all these will be tuberculosis specific, and those that are not may well bind antibodies present in sera which have been induced to epitopes of cross-reactive antigens of other microbes, especially harmless mycobacteria. A tuberculosis patient's antibody, binding to a disease-specific epitope, will be masked by the abundance of cross-reactive binding events. This probably explains why serological tests for tuberculosis are frequently positive in subjects found to be without the disease (Grange 1972). The problem could only be addressed with the advent of monoclonal antibodies (Mabs) with restricted species and molecular specificity. These were progressively defined in a series of World Health Organization (WHO)-sponsored workshops (summarized by Khanolkar-Young *et al.* 1992). Tuberculosis-specific Mabs then began to be used in a new form of (competition) immunoassay to detect and measure antibody in patients' sera, the binding of which to the target epitope could be inhibited by a Mab of overlapping specificity (Hewitt *et al.* 1982). This method allows serum to be tested at higher concentrations than is possible in a standard enzyme-linked immunosorbent assay (ELISA) technique and provides focus to disease-specific epitope antibody responses. As a result, a competition modification of the original ELISA test has given a good diagnostic sensitivity in all forms of extrapulmonary and also in smear-negative pulmonary tuberculosis (Wilkins & Ivanyi 1990; Wilkins *et al.* 1991). The same test is valuable in distinguishing all forms of pulmonary tuberculosis, whether smear-positive or -negative, from other respiratory illnesses, and in identifying patients with reactivation of previously self-healed tuberculosis. Positive results correlated well with subjects for whom a positive decision was made to give chemoprophylaxis because of contact with cases of infectious tuberculosis (Bothamley & Rudd 1994).

Monoclonal antibodies have also been used to affinity-purify target antigens from crude mycobacterial extracts and culture fluids, and to identify bacterial clones expressing recombinant mycobacterial antigens. This has allowed rapid progress in the study of the structure and properties of antigens of likely diagnostic importance, especially in the context of epitope mapping. Patients may vary in their ability to 'see' and respond to different tuberculosis-specific epitopes of different antigens. For efficient antibody tests, epitopes have to be found which are recognized adequately by all patients. This may depend upon the finding of both T-cell and B-cell epitopes to which good responses are made in all people. This demands a high degree of promiscuity in human leucocyte antigen (HLA) haplotype presentation of key peptide epitopes and the common recognition of the peptide epitope motifs by T cells.

Conclusions for antibody-based tests

The role of a serological test in the control of tuberculosis is now becoming clearer. They may have a role in:

1 screening for infectious tuberculosis;
2 identifying patients with extrapulmonary and smear-negative pulmonary tuberculosis; and
3 detecting failure to complete treatment for tuberculosis by the use of 'early' antigens such as the 16-kDa antigen.

The number of antigens still to be evaluated, however, is large. Progress will come when more antibody tests to discrete antigens are applied in the proper clinical setting. So far the 38-kDa antigen is the best reagent for an antibody test, as shown in well-controlled trials. Although antibodies to the 16-kDa antigen have not so far been a good basis for a diagnostic test, the molecule may be very important to an understanding of the pathogenesis of different forms of tuberculosis.

2.2.2 Antigen-detection assays for human tuberculosis

It is logical that a good laboratory immunotest for tuberculosis could depend upon the detection of specific epitopes of bacillary antigens released during active infection. Ideally such a test would be able to give a measure of infection load and monitor the efficacy of treatment.

Mabs and recombinant DNA methods began to focus on mycobacteria in the 1980s, and this led to the probing, cloning, expression and preparation of a large number of recombinant mycobacterial antigens for which sequence data, epitope maps and antigenic cross-reactivity data are accumulating at a steadily increasing pace. Amongst the antigens described is a collection known to be secreted or released during growth which are of special interest for antigen diagnosis. Secreted proteins may be important in the infection process and in host immunity (see Chapter

18). Since pathogenic mycobacterial species invade and replicate within host macrophages, the proteins secreted within the host cell may be presented to the host immune system early in infection or after reactivation. Several of them, including the 38-kDa (*Escherichia coli* PhoS protein homologue)/antigen 5 and the 30/31-kDa antigen (antigen 6, MPT 59, antigen 85 complex) are strongly immunogenic and antibodies are regularly found to these in tuberculosis patients. Assuming that these antigens (or others) appear in the circulation and/or in sputum or other body fluids, then they might be detected by a solid-phase antigen-capture sandwich ELISA test using Mabs or polyclonal antibodies carrying a high degree of specificity to the infection.

Initial attempts to detect antigen in patients' sera gave disappointing results but the position has been improved by concentrating antigen in immune complexes with polyethylene glycol, a strategy that is frequently used with sera for detecting antigens of larger parasites. *M. tuberculosis* antigen 5 was detected in serum immune complexes of 80% of patients with pulmonary tuberculosis (Radhakrishnan *et al.* 1992). The level of immune complexes and antigen decreased with the duration of antituberculous chemotherapy, so the test could be used to assess the clinical response to therapy. In the author's (D.C.) laboratory a similar approach has been applied (Catty 1993). Results were sufficiently promising to run a recently completed 5-year trial of over 2000 patients, of whom about one-third had confirmed tuberculosis (pulmonary and extrapulmonary) and another third had a suspicion of tuberculosis but negative culture results. Only some of this second group responded to antituberculous chemotherapy. Using a number of different Mabs to *M. tuberculosis*-secreted antigens, and a polyclonal antibody collectively to the same antigens, the test had positive predictive values ranging from 74 to 90% according to the capture antibody used and whether antigen-negative patients of the second group, earmarked because they subsequently responded to chemotherapy, were included (unpublished data). The 65-kDa, the 30/31-kDa and the 38-kDa antigens could all be detected in dissociated serum immune complexes. This offers some promise that a panel of mixed capture Mabs might achieve better results.

Tuberculous meningitis (TBM) is notoriously difficult to diagnose. The few bacilli that may be present in cerebrospinal fluid (CSF) make both the smear test and culture poorly sensitive. Mycobacterial antigen detection in CSF is one approach to the problem. Krambovitis *et al.* (1984) used latex particles coated with rabbit antibodies to *M. tuberculosis* membrane antigen in an agglutination assay with CSF. All tested TBM cases were positive with a specificity of 99%. Chandramuki *et al.* (1985) coated sheep red cells with a Mab against LAM in a CSF haemagglutination test which had marginally less sensitivity and false positive results in some disease control patients with pyogenic meningitis.

Conclusion

Antigen detection in both pulmonary and extrapulmonary tuberculosis is a potentially useful additional immunodiagnostic approach that lends itself to a number of simple assay formats. The systems, however, need extensive field testing with a range of antibody specificities, and trial specimens should be used concurrently in antibody and antigen tests as it is likely that a combination of the two assays will provide the best results.

2.3 Diagnosis of bovine tuberculosis

Bovine tuberculosis is a disease of cattle caused by infection with *M. bovis*, an organism which can infect many other species, including humans (Neill *et al.* 1994a). The disease continues to cause problems as a zoonotic health risk and as a barrier to agricultural trade in several countries (Daborn & Grange 1993; Caffrey 1994). To combat these problems, which in several regions are compounded by the presence of wildlife reservoirs of infection, many countries have implemented programmes designed to eradicate the disease. Detection of diseased animals is generally based on the measurement of immune responses.

Tuberculin skin testing

As with human tuberculosis control, the tuberculin skin test remains the most widely used test in the diagnosis of bovine tuberculosis. However, the same problems of sensitivity and specificity remain and the high false-positive rating incurs considerable expense where infected herds are slaughtered (Francis *et al.* 1978).

Skin testing in cattle has followed several formats and protocols depending on the country and situation (Francis *et al.* 1978; Monaghan *et al.* 1994). The antigens currently used are tuberculins, or PPDs, and responses are generally measured as increases in skin thickness at 72 h after intradermal injection. In several countries, the protocol involves measuring responses to PPD prepared from *M. bovis* (PPDB) (the single intradermal test). However, in regions where cattle are exposed to environmental mycobacteria, this test would result in a large number of false positives (Anon 1942; Leslie *et al.* 1975). This outcome can be explained as PPDs are complex mixtures of mycobacterial products and it has been shown that some of these components are common to several mycobacterial species (Jackett *et al.* 1988; Chan *et al.* 1990) and other organisms (Chaparas *et al.* 1970; Thorns & Morris 1986). Indeed it has been shown that cattle which are experimentally infected with *M. bovis* develop T-cell responses to PPD from *M. avium* (PPDA) as well as to PPDB (Pollock *et al.* 1994). The cross-reactivity of the PPD antigens has led to the development of a comparative intradermal test, where delayed-type hypersensitivity (DTH) responses to PPDB are compared with responses to PPDA injected at a different skin site. Under most circumstances, using that test, an animal is classified as positive if the response to PPDB is greater than the response to PPDA and meets several criteria. However, the test still lacks absolute sensitivity and specificity (Pritchard 1988; Monaghan *et al.* 1994). For this reason, research into alternative tests and diagnostic antigens is ongoing.

Several attempts have been made to develop serological tests for bovine tuberculosis. However, these have usually lacked adequate specificity, perhaps due to the use of complex antigens such as *M. bovis* soni-

cate or PPD (Auer 1987; Ritacco *et al.* 1987; Hanna *et al.* 1989) or have poor sensitivity when defined antigens such as MPB70 have been employed (Fifis *et al.* 1989; Harboe *et al.* 1990). It has become accepted that, as in other mycobacterial diseases, cattle with tuberculosis exhibit a spectrum of immune responses in which cell-mediated immunity (CMI) dominates in early disease with antibody developing in the later stages of infection (Ritacco *et al.* 1991). This may explain the limited potential of serological tests in countries with established eradication policies and the widespread usage of DTH responses, as measured by skin testing, for the detection of bovine tuberculosis.

Other tests

Because CMI responses are dominant in early bovine tuberculosis, several studies have investigated alternatives to DTH skin testing. Initially, *in vitro* responses to mycobacterial antigens were looked at using the lymphocyte proliferation assay (Outteridge & Lepper 1973; Thoen *et al.* 1980). However, probably because the test antigen was still PPD, there were no benefits of improved sensitivity and specificity and the tests were unwieldy. The detection of cytokines produced by T cells in response to mycobacterial antigen offered an alternative and a bioassay showed the potential of measuring the release of interferon-γ (IFN-γ) in T-cell responses (Wood *et al.* 1990). This was later modified to an ELISA system for IFN-γ measurement (Wood *et al.* 1991). This has proved useful in Australia and New Zealand. Field trials in some other countries have revealed specificity problems (Neill *et al.* 1994b; Monaghan *et al.* 1997). This may relate to the continued use of PPD as the test antigen. Documentation of animals which are positive in the IFN-γ test but negative by skin testing, and vice versa, indicates that disease diagnosis based on a single test parameter may never be perfect.

The search for specific antigens

Lack of specificity of PPD reagents has prompted research to identify antigens for improved diagnosis (Fifis *et al.* 1991). Whilst many discrete mycobacterial antigens have been described (Khanolkar-Young *et al.* 1992) and studied in human tuberculosis, few have been studied in bovine disease. Fifis *et al.* (1994) looked at a panel of purified *M. bovis* antigens for recognition by T cells from cases of bovine tuberculosis and concluded that the dominant antigen depended on the stage of infection. One immunodominant antigen of *M. bovis* which has received considerable attention is MPB70 (Nagai *et al.* 1981). This induces both T- and B-cell responses in cattle (Fifis *et al.* 1994) and has *M. bovis*-specific epitopes (Wood *et al.* 1988), but there are cross-reactions with *Nocardia* spp. (Harboe & Nagai 1984). The epitopic structure of MPB70 and other *M. bovis* antigens in bovine immune responses has been studied in the search for specific diagnostic tools (Billman-Jacobe *et al.* 1990; Radford *et al.* 1990; Pollock *et al.* 1994, 1995), but the potential of these reagents remains to be confirmed. Some studies have focused on secreted mycobacterial antigens which should be early targets of the immune response and thus good reagents for early diagnosis (Andersen *et al.* 1991). Screening of secreted proteins in mycobacterial short-term culture filtrates has identified ESAT-6 as a dominant target for IFN-γ-producing T cells in early bovine tuberculosis (Pollock & Andersen 1997a). This preferential recognition in CMI along with the fact that this antigen is not expressed by non-pathogenic, environmental mycobacteria, means that ESAT-6 has great potential for improved diagnosis of bovine tuberculosis (Pollock & Andersen 1997b).

Conclusion

The precise definition of the specificity of bovine anti-tuberculosis responses is required for the logical development of improved immunodiagnostic tests. Whilst progress has been made on several fronts, further understanding of the interactions between defined antigens and bovine T cells is still needed to support further efforts for disease eradication.

2.4 Diagnosis of tuberculosis in badgers

The control of bovine tuberculosis remains a significant problem in the UK, especially in the south-west of England where the rising incidence of the disease with extension into areas with no recent history of infection is attributed to a reservoir of *M. bovis* infection in badgers (Muirhead *et al.* 1974; Gallagher & Nelson 1979; Stuart & Wilesmith 1988; Report 1997). *M. bovis* infection of badgers was first described in Switzerland (Bouvier *et al.* 1957) and was not recorded in the UK until 1971 when a tuberculous badger was discovered on a farm that had recently suffered an outbreak of tuberculosis in cattle. Since then it has been demonstrated that naturally infected badgers may transmit disease to cattle (Little *et al.* 1982; Cheeseman *et al.* 1988). However, over a quarter of a century since the discovery of *M. bovis* infection in badgers, it remains difficult to use non-lethal methods to determine whether individual badgers are infected with *M. bovis*.

Diagnosis by culture

Clinical diagnosis of *M. bovis* infection in badgers by sampling faeces, urine and tracheal aspirates followed by standard culture procedures, has been shown to be an insensitive method of detecting infection (Pritchard *et al.* 1987; Newell *et al.* 1997). Recent evidence from a longitudinal study of natural *M. bovis* infection in badgers has shown that they can excrete *M. bovis* intermittently over a number of years (Newell *et al.* 1997). In addition, culturing *M. bovis* from clinical samples is an impractical method of diagnosis for control purposes due to the time required for bacterial growth in culture. An immunodiagnostic assay is therefore the method of choice.

Tuberculin skin testing

At first it was believed that badgers did not have the normal mammalian pathways of immunity (Morris *et al.* 1978) because many of the badgers investigated were at the end-stage of disease where there is a high bacterial load but very little cellular immune response due to anergy. In these early studies Morris *et al.* (1978) and Little *et al.* (1982) found that delayed hypersensitivity skin responses and *in vitro* lymphocyte transformation responses (LT) to PPD of bovine tubercle bacilli were invariably negative whether badgers were uninfected, naturally infected or experimentally infected. In later studies, Higgins (1985) found that PPD stimulated a small increase in skin thickness with a timing consistent with a DTH response in naturally infected badgers. Neither erythema, palpable oedema nor induration were observed in this study but a distinct histological reaction was present. Subsequently, Pritchard *et al.* (1986) found the skin test to have low sensitivity and specificity in a naturally infected population and to be of little practical value.

Serological assays

Attention therefore turned to the development of a serological assay. This was viewed as the only practical method of diagnosing tuberculosis in badgers (Dunnet *et al.* 1986) despite the fact that serological tests using PPD, including the complementation fixation test (Little *et al.* 1982) and ELISA (Morris *et al.* 1979) had also been shown to be unreliable and Higgins and Gatrill (1984) had reported that antibody responses to a variety of antigens were low in comparison with those of rabbits.

Early attempts to develop an ELISA for detecting *M. bovis* infection in badgers in the field were unsuccessful due to the lack of specificity using crude mycobacterial extracts which contained many cross-reactive antigens (Morris *et al.* 1978). To overcome this, a murine monoclonal anti-badger IgG antibody was produced and used to investigate the specificity of the antibody response of badgers infected with *M. bovis* (Goodger *et al.* 1994a). The Mab was directed against badger IgG heavy chain and reacted with badger and dog IgGs but not with cat, rabbit, mouse, guinea pig, bovine or ferret IgGs. This antibody-detection system functioned well in both ELISA and western blot and was used to identify a 25-kDa serodominant antigen of *M. bovis* which was conserved in all field strains of *M. bovis* tested. This was also the first antigen to be recognized by sera from experimentally infected badgers and seroconversion occurred ~ 32 weeks after infection. The antigen was partially purified from sonicated *M. bovis* bacilli using water precipitation and ion exchange chromatogra-

phy and its purification was monitored with a mouse monoclonal antibody, MBS43, which was specific for the 25-kDa antigen. The gene encoding this antigen, MPB83, has been cloned and sequenced (Hewinson *et al.* 1996; Matsuo *et al.* 1996).

The partially purified antigen was used to develop an indirect ELISA system for the assay of badger sera for specific antibodies (Goodger *et al.* 1994b). A presumed negative badger population was used to calculate the assay's threshold of seropositivity and, using this value, its sensitivity (37%) and specificity (98%) were determined in a second population of known culture status (Goodger *et al.* 1994b). The ELISA system has been adapted for field use and can be carried out in a mobile laboratory (Nolan & Goodger 1993). Evaluation of the ELISA assay in the individual animal in the field has revealed a specificity of 94% and a sensitivity of 41% while evaluation of the assay as a screening test in social groups of animals gives a specificity of 84% and a sensitivity of 73% (Clifton-Hadley *et al.* 1995). The test is currently being used in an effort to provide strategies for the control of bovine tuberculosis. One strategy currently on trial is based on the identification of those setts containing serologically positive animals with subsequent removal of all animals from those setts (Rees & Meldrum 1995).

Studies on the kinetics of serum antibody responses to MPB83 between 1982 and 1995 in a badger social group naturally infected with *M. bovis* have revealed that a number of cubs develop transient seropositivity within the first 6–8 months of life but then remain culture-negative for up to 5 years (Newell *et al.* 1997). Recent analysis of field data is consistent with these observations and suggests that the specificity of the ELISA at the population level decreases from 84% when adults only are included to 79.5% when cubs are also included, suggesting that there may be a disproportionate number of cubs which are culture-negative but ELISA-positive (Newell *et al.* 1997). The source or role of these antibodies in cubs is currently unknown. However, it is evident that ELISA-positivity in cubs is not necessarily an indicator of disease associated with concurrent or imminent excretion of organisms.

Other tests

Although measurement of T-cell responses could provide a more sensitive diagnostic test for detecting tuberculosis infection in badgers, this type of assay at present requires at least 24 h for completion and therefore could be difficult to implement in field programmes aimed at identifying and removing infected animals. Nevertheless, both lymphocyte transformation tests and IFN-γ assays for badgers are being developed.

Early attempts at developing lymphocyte transformation (LT) tests using PPD were unsuccessful (Morris *et al.* 1978; Little *et al.* 1982). This may have been due to early problems with methodology. Later experiments using blood from experimentally infected badgers established better conditions for an LT assay (Mahmood *et al.* 1987; Stuart *et al.* 1988). This assay required medium containing 10% autologous serum and whole live BCG Glaxo organisms as stimulating antigen. The authors reported that soluble extracts of mycobacteria (e.g. PPD) were usually toxic to the badger lymphocytes and postulated that this might explain the previously published failure to demonstrate lymphocyte responses in badgers using PPD as the antigen. Further explanations might be the effect of preservative in antigen preparations or an inhibitory effect of some serum samples in the culture. However, recent work in our laboratory using blood from naturally infected badgers at various stages of *M. bovis* infection has established that LT assays are not dependent on the source of sera and that antigen-specific lymphocyte proliferation for badgers using PPD is possible.

The gene encoding badger IFN-γ has been cloned and will be described elsewhere. The gene sequence shows striking homology with that of dog IFN-γ. Work is underway to produce an ELISA system for IFN-γ measurement in badgers and it is expected that the assay will be available within the next few years.

Conclusion

At present a serological assay based on the detection of the presence of IgG antibodies against the *M. bovis* antigen MPB83 is used for the diagnosis of tuberculosis in badgers. However, this assay lacks sensitivity in the individual animal. Assays to detect T-cell responses are being developed which could improve the sensitivity of diagnosis but these may prove difficult to implement in field programmes aimed at removing infected animals.

2.5 Diagnosis of tuberculosis in *Cervidae*

The intradermal skin test

Tuberculosis caused by *M. bovis* has been diagnosed histologically at necropsy in wild deer in the United States (Towar *et al.* 1965) and Europe (Wilson & Harrington 1976) and was first diagnosed in a herd of domesticated New Zealand deer in 1979 (Beatson *et al.* 1984). Attempts to use herd-based diagnostic intradermal tests for deer showed that the conventional bovine intradermal skin test protocol needed to be modified. Deer have thin skins which require extreme care to ensure that the intradermal inoculation is applied with precision. Failure to produce an intradermal bleb significantly reduces the sensitivity of the test (Anon 1995). The optimal skin site is the mid cervical neck region. Care must be taken to shave the hair evenly to within 1–2 mm of the skin so that the test can be applied and read accurately. Any visible, measurable or palpable reaction is considered positive, when the test is read at 72 h postinjection. Under these conditions the single mid-cervical skin test (MCST), using 0.1 mg of *M. bovis* tuberculin (PPDB), as a 0.1-mL inoculum, has a sensitivity of around 80% (Griffin & Buchan 1994).

The comparative cervical skin test

A complication with tuberculosis diagnosis in deer is the high level of non-specific sensitization which results from exposure to saprophytic mycobacteria (Clifton-Hadley & Wilesmith 1991). When the single intradermal test is used, major problems occur due to the poor specificity and low predictive value of the test (Griffin & Cross 1989). To circumvent these problems, a comparative cervical skin test (CCT) has been developed. The CCT uses *M. bovis* tuberculin (PPDB, 0.1 mg) and *M. avium* tuberculin (PPDA, 0.05 mg) at separate skin test sites in the neck. While the CCT produces increased levels of specificity, its sensitivity is significantly lower when used under field conditions (Kollias *et al.* 1982; Beatson *et al.* 1984). Whereas CCT used under controlled conditions in experimentally infected deer has a sensitivity of 91% (Corrin *et al.* 1993), its use under field conditions has produced sensitivity values ranging from 31% (Griffith 1989) to 64% (Griffin & Cross 1989). It is accepted that CCT has lower sensitivity than MCST because of the technical difficulties of applying and reading CCT in the field. A positive reaction is indicated by an increase in skin thickness at the PPDB site which is at least 2 mm, and greater than that found at the PPDA site. Concerns surrounding the sensitivity of CCT in deer are such that the current New Zealand standards (Anon 1995) for tuberculosis testing in deer preclude use of CCT in herds without a detailed history or those with a recent history of tuberculosis.

Apart from the questionable sensitivity of CCT, the use of MCST as a primary herd test causes significant suppression of CCT reactivity for up to 90 days (Corrin *et al.* 1993). This requires an unacceptably long interval between skin tests to confirm the status of MCST+ animals, using CCT as an ancillary test.

Laboratory tests and combined tests

Ancillary laboratory-based tuberculosis tests have been developed for use with herd skin testing programmes for farmed deer. A composite blood test for tuberculosis (BTB), which uses lymphocyte transformation (LT) and an antibody ELISA, has been developed (Griffin *et al.* 1994). These measure relative responses to PPDB vs. PPDA with mononuclear cells or serum from deer, to distinguish between non-specific background reactivity and specific responses to *M. bovis*. An attempt to develop more specific tests using antigens such as MBP70 gave more specificity but at the cost of a significant loss of sensitivity (Griffin *et al.* 1991). Detailed necropsy, histological and microbiological studies have been carried out on more than 200 animals with tuberculosis. Only those animals which had *M. bovis* isolated by culture from necropsy specimens were used to determine assay sensitivity. A similar number of animals from tuberculosis-free herds was used to determine test specificity. The ancillary serial BTB test is used to clarify the status of MCST+ animals whilst the ELISA test alone is used in parallel with MCST to identify MCST– tuberculous animals which fail to react to the skin test.

Considerable advantage can be gained by combining cell-based tests (LT or MCST) with an antibody (ELISA) test, as the composite results can improve diagnostic sensitivity without a major reduction in specificity. In one group of 102 tuberculous deer the sensitivity for LT alone was 90% and for ELISA 85%. When combining the tests and interpreting the results in parallel, if either or both tests are positive a BTB+ result was obtained with an overall sensitivity of 96% (Griffin *et al.* 1994). Thus, combining laboratory tests increases diagnostic sensitivity. There was a similar increase in test sensitivity when the results of MCST (82% alone) and ELISA (85% alone) were combined; interpreted in parallel the composite result was 95% (Griffin *et al.* 1994).

When using composite tests it is important to select assays which analyse different pathways (homoral vs. cellular) of immune response, so that they can target different populations of infected animals. There is emerging evidence that whilst the skin test is efficient in targeting acutely infected animals, it is far less sensitive in diagnosing those with chronic infection, a group that can be readily detected with the antibody test.

Specificity values for LT and ELISA were 98% and 100%, respectively, for a group of 200 animals from nine non-infected deer herds. In this group the composite specificity of BTB was 98%. Specificity values obtained on more than 50 000 deer over a 10-year period is 98.5, indicating that the test can accurately identify non-infected, skin test($\pm$) deer. Combining results of cell tests and an antibody test gains diagnostic sensitivity without a significant drop in specificity. It is important, however, to establish what influence the intradermal injection of tuberculin has on subsequent laboratory blood tests. Samples taken prior to skin test and for the succeeding weeks show that MCST causes significant suppression in LT for 3–14 days postinjection of tuberculin. Antibody levels in the ELISA test also increases dramatically in tuberculosis-infected animals from 4 to 28 days post-skin test. This effect is not seen in non-tuberculous deer (Griffin *et al.* 1994). Tests prior to and at 14 days post-MCST show that the sensitivity of ELISA is significantly enhanced by MCST, increasing from 46% before MCST to 85% in the weeks after skin testing. The dramatically suppressive impact of MCST on LT in the period immediately following skin testing, plus the enhanced antibody response in infected animals, has led to the recommendation that blood tests should be carried out between 14 and 28 days post-MCST injection.

In our opinion, there is a strong practical issue as to whether wildlife immunoassays for tuberculosis infection in animals should maximize sensitivity or specificity, and the bias should be towards maximizing sensitivity unless public debate defines that any wastage of animals is not justified.

3 Diagnosis of leprosy

The diagnosis of leprosy is essentially clinical. Immunodiagnostic tools have been developed, but their positive reactions in contacts of leprosy patients has given them a role in understanding the natural history of the disease, rather than any diagnostic merit.

3.1 Leprosin and lepromin

Delayed hypersensitivity to extracts of *M. leprae* is used in indeterminate leprosy to indicate the polarity of disease, whether tuberculoid (few bacilli with predominantly neural complications) or lepromatous (many bacilli with infiltration of the skin especially in cooler areas of the body), and thereby to assist in determining the prognosis for an individual patient. Lepromin is prepared from *M. leprae* bacilli which have multiplied within the nine-banded armadillo and then been extracted from the tissues, purified by differential centrifugation or chloroform extraction

and standardized at $4–16 \times 10^6$ bacilli/mL inoculum. Leprosin is prepared as a sonicated and filtered extract of leprosy bacilli which has been sterilized by irradiation. Two types of delayed hypersensitivity are described, the Fernandez and the Mitsuda reaction.

The Fernandez reaction occurs at 48–96 h after intradermal injection and indicates a tuberculoid polarity (in the $\approx 50\%$ of those with tuberculoid leprosy who respond at all) or close contact with lepromatous leprosy in an otherwise asymptomatic individual. This reaction occurs when using leprosin and is variable with lepromin, according to the amount of soluble material in the latter preparation.

The Mitsuda reaction is the more classical measure of delayed hypersensitivity to lepromin and appears as an area of induration 7–10 days after intradermal injection. The late reaction is thought to be due to the slow release of soluble antigens from the leprosy bacilli. A positive reaction (5 mm) occurs in patients with tuberculoid leprosy, but is characteristically absent in lepromatous leprosy. In areas where leprosy is endemic, up to 70% of the healthy population may give a positive Mitsuda reaction.

3.2 Antibody measurement and its clinical value

Antibody levels in leprosy correlate well with bacterial index and are therefore usually high in patients with lepromatous disease. Two specific reagents are available, phenolic glycolipid and its derivatives and a 35-kDa antigen of *M. leprae*. The former has an epitope which can be mimicked by a synthetic trisaccharide conjugated to bovine serum albumin (Cho *et al.* 1984). IgM antibody binding is the diagnostic measure in this test. A competition assay using the monoclonal antibody ML04 measured antibody indirectly and gave a similar sensitivity and specificity to the phenolic glycolipid (Mwatha *et al.* 1988) and appeared to have a role in monitoring the effectiveness of chemotherapy as antibody levels correlated well with the bacterial index (Roche *et al.* 1993). The 35-kDa protein antigen was identified as a recombinant clone using the monoclonal antibody ML04; when manufactured in *E. coli* the material was not serologically active, but when prepared from recombinant *M. smegmatis*, the material proved effective in the measurement of both antibody binding in lepromatous and IFN-γ production in tuberculoid leprosy (Triccas *et al.* 1996).

4 Diagnosis of non-tuberculous mycobacterial infections

4.1 Non-tuberculous mycobacteria and acquired immunodeficiency syndrome

The group with the highest incidence of non-tuberculous mycobacterial infections are represented by patients who are immunocompromised. A large number of immunodiagnostic tests have been evaluated in patients with acquired immune deficiency syndrome (AIDS), but all have a low sensitivity, and molecular techniques or mycobacterial culture afford the best diagnostic yield.

4.2 Mycobacterial sensitin

An extract of *M. avium* was used extensively in analysing the false-positive tuberculin rates in the south-eastern United States in the 1950s and was used as evidence that environmental mycobacteria might be responsible for these reactions (Edwards *et al.* 1969). Examination of a number of different mycobacterial preparations showed again the broad cross-reactivity of these antigenic mixtures (Magnusson 1986). An excellent study by Huebner *et al.* (1992) which corrected many of the shortcomings of previous studies confirmed the suspicion that these reagents have no place in the clinical diagnosis of non-tuberculous mycobacterial disease even in the immunocompetent host.

5 Conclusions

With the present state of development of immunodiagnostic tests for mycobacterial diseases, at least for human tuberculosis, the finding of mycobacteria by acid-fast staining or culture must remain the diagnostic method of choice for some time to come. Herd testing of cattle and deer for 'bovine' tuberculosis relies heavily on skin testing, although newer cell and antibody laboratory immunodiagnostic tests are beginning to have a role. In the future we can expect such laboratory aids to have a valuable role also in screening for human mycobacterial infections, across the spectrum of disease, and especially in the diagnosis of forms of tuberculosis where the staining and culture of bacilli is negative or where these gold standards cannot be applied. The simultaneous use of cellular and serological tests represents the most helpful advance to date in the diagnosis of human and animal disease. In serological tests, few reagents have been evaluated adequately in the field and the reported sensitivity and specificity of antibody tests towards some antigens may overestimate their true clinical value. The specific detection of antigen in the circulation or other fluids of people infected with *M. tuberculosis* is another immunoassay approach. This might find its value as a single test or one combined with an antibody test. It may offer particular attractions in that it may be specific for active disease. More reagent evaluation in field conditions is needed for this approach also. Serological tests have an additional role in assessing prognosis and the success of treatment in both tuberculosis and leprosy, which is especially valuable with the rise of multidrug-resistant organisms and the problems of adherence to long courses (many months) of treatment.

Improvements in immunological approaches to mycobacterial disease diagnosis will undoubtedly come as more is understood about the complex immune responses associated with these infections. The need to find good immunodiagnostic tests has undoubtedly been a major stimulus in the past for unravelling much of what we currently know about the host response when the body is invaded by mycobacteria. Nevertheless progress has been extremely slow considering how long tuberculin and the skin test has been available. It is an irony that it is only with the recent concern about the re-emergence of tuberculosis as a global epidemic that there has been any substantial growth of research targeted to this disease. This is now, however, occurring across many fronts and out of this may come new targets and new reagents, as well as new approaches, for diagnosis. Better and simpler diagnostic methods are certainly on some priority lists and we can expect to see a more rapid advance in immunodiagnostic methods in the future.

6 References

Andersen, P., Askgaard, D., Ljunhquist, L., Benzon, M.W. & Heron, I. (1991) T-cell proliferative responses to antigens secreted by *Mycobacterium tuberculosis. Infection and Immunity* **59**, 1558–1563.

Anon (1942) The tuberculin test. *Veterinary Record* **54**, 191–192.

Anon (1995) *Quality Standards for Deer Tuberculosis Testing.* Wellington, New Zealand: New Zealand Veterinary Association.

Auer, L.A. (1987) Assessment of an enzyme linked immunosorbent assay for the detection of cattle infected with *Mycobacterium bovis. Australian Veterinary Journal* **64**, 172–176.

Beatson, N.S., Hutton, J.B. & de Lisle, G.W. (1984) Tuberculosis-test and slaughter. *Proceedings of the New Zealand Veterinary Association Deer Branch* **1**, 18–27.

Billman-Jacobe, H., Radford, A.J., Rothel, J.S. & Wood, P.R. (1990) Mapping of the T and B cell epitopes of the *Mycobacterium bovis* protein, MPB70. *Immunology and Cell Biology* **68**, 359–365.

Bothamley, G.H. (1995) Serological diagnosis of tuberculosis. *European Respiratory Journal* **8** (Suppl. 20), 676–688s.

Bothamley, G.H. (1997) Immunological procedures for diagnosis of tuberculosis and leprosy. In: *Manual of Clinical Laboratory Immunology*, 5th edn (eds N. R. Rose, E. Conway de Macario, J. D. Folds, H. C. Lane & R. M. Nakamura). Washington, DC: American Society for Microbiology Press, pp. 534–542.

Bothamley, G.H., Beck, J.S., Potts, R.C., Grange, J.M., Kardjito, T. & Ivanyi, J. (1992a) Specificity of antibodies and tuberculin response after occupational exposure to tuberculosis. *Journal of Infectious Diseases* **166**, 152–156.

Bothamley, G.H. & Rudd, R.M. (1994) Clinical evaluation of a serological assay using a monoclonal antibody (TB72) to the 38 kDa antigen of *Mycobacterium tuberculosis. European Respiratory Journal* **7**, 240–246.

Bothamley, G.H., Rudd, R., Festenstein, F. & Ivanyi, J. (1992b) Clinical value of the measurement of *Mycobacterium tuberculosis*-specific antibody in pulmonary tuberculosis. *Thorax* **47**, 270–275.

Bothamley, G., Udani, P., Rudd, R., Festenstein, F. & Ivanyi, J. (1988) Humoral response to defined epitopes of tubercle bacilli in adult pulmonary and child

tuberculosis. *European Journal of Clinical Microbiology* **7**, 639–645.

Bouvier, G., Burgisser, H. & Schneider, P.A. (1957) Observations sur les maladies du gibier, des oiseaux et des poissons. *Schweiz. Arch. Für Tierheilkunde* **99**, 461–477.

Caffrey, J.P. (1994) Status of bovine tuberculosis eradication programmes in Europe. *Veterinary Microbiology* **40**, 1–4.

Catty, D. (1993) Monoclonal antibodies in diagnosis. In: *Molecular and Antibody Probes in Diagnosis* (eds M. R. Walker & R. Rapley). Chichester: John Wiley & Sons, pp. 229–252.

Chan, S.L., Reggardia, Z., Daniel, T.M., Girling, D.J. & Mitchison, D.A. (1990) Serodiagnosis of tuberculosis using an ELISA with antigen 5 and glycolipid antigens. *American Review of Respiratory Diseases* **142**, 385–390.

Chandramuki, A., Allen, P.R.J., Keen, M. & Ivanyi, J. (1985) Detection of mycobacterial antigens and antibodies in the CSF of patients with tuberculous meningitis. *Journal of Medical Microbiology* **20**, 239–247.

Chandramuki, A., Bothamley, G.H., Brennan, P.J. & Ivanyi, J. (1989) Levels of antibody to defined antigens of *Mycobacterium tuberculosis* in tuberculous meningitis. *Journal of Clinical Microbiology* **27**, 821–825.

Chaparas, S.D., Maloney, C.J. & Hendrik, S.R. (1970) Specificity of tuberculins and antigens from various species of mycobacteria. *American Review of Respiratory Diseases* **101**, 74.

Cheeseman, C.L., Wilesmith, J.W., Stuart, F.A. & Mallison, P.J. (1988) Dynamics of tuberculosis in a naturally infected badger population. *Mammalian Review* **18**, 61–72.

Cho, S.N., Fujiwara, T., Hunter, S.W., Rea, T.H., Gelber, R.H. & Brennan, P.J. (1984) Use of an artificial antigen containing the 3,6-di-O-methyl-D-glucopyranosyl epitope for the serodiagnosis of leprosy. *Journal of Infectious Diseases* **150**, 311–322.

Clifton-Hadley, R.S., Sayers, A.R. & Stock, M.P. (1995) Evaluation of an ELISA for *Mycobacterium bovis* infection in badgers (*Meles meles*). *Veterinary Record* **137**, 555–558.

Clifton-Hadley, R.S. & Wilesmith, R.W. (1991) Tuberculosis in deer; a review. *Veterinary Record* **129**, 5–12.

Coates, A.R.M., Hewitt, J., Allen, B.W., Ivanyi, J. & Mitchison, D.A. (1981) Antigenic diversity of *Mycobacterium* tuberculosis and *Mycobacterium bovis* detected by means of monoclonal antibodies. *Lancet* **ii**, 167–169.

Cocito, C.G. (1991) Properties of the mycobacterial antigen complex A60 and its application to the diagnosis and prognosis of tuberculosis. *Chest* **100**, 1687–1693.

Cole, R.A., Lu, H.M., Shi, Y.Z., Wang, J., De Hua, T. & Zhou, A.T. (1996) Clinical evaluation of a rapid immunochromatographic assay based on the 38 kDa antigen of *Mycobacterium tuberculosis* in patients with pulmonary tuberculosis in China. *Tubercle and Lung Disease* **77**, 363–368.

Comstock, G.W., Daniel, T.M., Snider, D.E., Edwards, P.Q., Hopewell, P.C. & Vandiviere, H.M. (1981) The tuberculin skin test. *American Review of Respiratory Diseases.* **124**, 356–363.

Corrin, K.C., Carter, C.E., Kissling, R.C. & de Lisle, G.W. (1993) An evaluation of the comparative tuberculin skin test for detecting tuberculosis in deer. *New Zealand Veterinary Journal* **41**, 12–20.

Daborn, C.J. & Grange, J.M. (1993) HIV/AIDS and its implications for the control of animal tuberculosis. *British Veterinary Journal* **149**, 405–417.

Daniel, T.M. & Andersen, P.A. (1978) The isolation by immunosorbent affinity chromatography and physiochemical characterisation of *Mycobacterium tuberculosis* antigen 5. *American Review of Respiratory Diseases* **117**, 533–539.

Daniel, T.M., Balestrino, E.A., Balestrino, O.C. *et al.* (1982) The tuberculin specificity in humans of *Mycobacterium tuberculosis* antigen 5. *American Review of Respiratory Diseases* **126**, 600–606.

Daniel, T.M., de Murillo, G.L., Sawyer, J.A. *et al.* (1986) Field evaluation of enzyme-linked immunosorbent assay (ELISA) of antibody to *Mycobacterium tuberculosis* antigen 5. *American Review of Respiratory Diseases* **134**, 662–665.

Dunnet, G.M., Jones, D.M. & McInerney, J.P. (1986) *Badgers and Bovine Tuberculosis—Review of Policy*. London: HMSO.

Edwards, L., Aquaviva, F., Livesay, V., Cross, F. & Palmer, C. (1969) An atlas of sensitivity to tuberculin, PPD-B, and histoplasmin in the United States. *American Review of Respiratory Diseases* **99**, 3–18.

Evans, D.J., Barker, R.J. & Geddes, D.M. (1996) Tuberculin skin tests. *Lancet* **348**, 1512–1513.

Fifis, T., Corner, L.A., Rothel, J.S. & Wood, P.R. (1994) Cellular and humoral immune responses in cattle to *Mycobacterium bovis* antigens. *Scandinavian Journal of Immunology* **39**, 267–274.

Fifis, T., Costopoulos, C., Radford, A.J., Bacic, A. & Wood, P.R. (1991) Purification and characterisation of major antigens from a *Mycobacterium bovis* culture filtrate. *Infection and Immunity* **59**, 800–807.

Fifis, T., Plackett, P., Corner, L.A. & Wood, P.R. (1989) Purification of a major *Mycobacterium bovis* antigen for the diagnosis of bovine tuberculosis. *Scandinavian Journal of Immunology* **29**, 91–101.

Francis, J., Seiter, R.J., Wilkie, I.W., O'Boyle, D., Lumsden, M.J. & Frost, A.J. (1978) The sensitivity and specificity of various tuberculin tests using bovine PPD and other tuberculins. *Veterinary Research* **103**, 420–435.

Gallagher, J. & Nelson, J. (1979) Cause of ill health and natural death in badgers in Gloucestershire. *Veterinary Record* **105**, 546–551.

Göçmen, A., Kiper, N., Ertan, Ü., Kalayci, Ö. & Özçlik, U. (1994) Is the BCG test of diagnostic value in tuberculosis? *Tubercle and Lung Disease* **75**, 54–57.

Goodger, J., Nolan, A., Russell, W.P. *et al.* (1994b) Serodiagnosis of *Mycobacterium bovis* infection in badgers: development of an indirect ELISA using a 25 kDa antigen. *Veterinary Record* **135**, 82–85.

Goodger, J., Russell, W.P., Nolan, A. & Newell, D.G. (1994a) Production and characterisation of a monoclonal badger anti-immunoglobulin G and its use in defining the specificity of *Mycobacterium bovis* infections in badgers by western blot. *Veterinary Immunology and Immunopathology* **40**, 243–252.

Grange, J.M. (1972) The humoral response in tuberculosis: its nature, biological role and diagnostic usefulness. *Advances in Tuberculosis Research* **21**, 1–78.

Griffin, J.F.T. & Buchan, G.S. (1994) Aetiology, pathogenesis and diagnosis of *Mycobacterium bovis* infection in deer. *Veterinary Microbiology* **40**, 193–205.

Griffin, J.F.T. & Cross, J.P. (1989) Diagnosis of tuberculosis in New Zealand farmed deer: an evaluation of intradermal skin testing and laboratory techniques. *Irish Veterinary Journal* **42**, 102–107.

Griffin, J.F.T., Cross, J.P., Chinn, D.N., Rodgers, C.R. & Buchan, G.S. (1994) Diagnosis of tuberculosis due to *Mycobacterium bovis* in New Zealand red deer (*Cervus elaphus*) using a composite blood test and antibody assays. *New Zealand Veterinary Journal* **42**, 173–179.

Griffin, J.F.T., Nagai, S. & Buchan, G.S. (1991) Tuberculosis in domesticated red deer: comparison of purified protein and the specific protein MBP70 for *in vitro* diagnosis. *Research in Veterinary Science* **50**, 279–285.

Griffith, L.M. (1989) Experiences with skin tests in the field. *Publications of the Veterinary Deer Society* **3**, 19–29.

Hanna, J., Neill, S.D. & O'Brien, J.J. (1989) Use of PPD and phosphatide antigens in an ELISA to detect the serological response in experimental bovine tuberculosis. *Research in Veterinary Science* **47**, 43–47.

Harboe, M. & Nagai, S. (1984) MPB70, a unique antigen of *Mycobacterium bovis* BCG. *American Review of Respiratory Diseases* **129**, 444–452.

Harboe, M., Wiker, H.G., Duncan, S.R. *et al.* (1990) Protein G based enzyme immunosorbent assay for anti-MPB70 antibody in bovine tuberculosis. *Journal of Clinical Microbiology* **28**, 913–921.

Heaf, F. (1951) The multiple puncture tuberculin test. *Lancet* **ii**, 151–153.

Hewinson, R.G., Michell, S. Ll, Russell, W.P., McAdam, R.A. & Jacobs, W.R., Jr (1996) Molecular characterization of MPT83: a seroreactive antigen of *Mycobacterium tuberculosis* with homology to MPT70. *Scandinavian Journal of Immunology* **43**, 490–499.

Hewitt, J., Coates, A.R.M., Mitchison, D.A. & Ivanyi, J. (1982) The use of murine monoclonal antibodies in the serodiagnosis of tuberculosis. *Journal of Immunological Methods* **55**, 205–211.

Higgins, D.A. (1985) The skin inflammatory response of the badger. *British Journal of Experimental Pathology* **66**, 643–653.

Higgins, D.A. & Gatrill, A.J. (1984) A comparison of the antibody responses of badgers (*Meles meles*) and rabbits (*Oryctolagus cuniculus*) to some common antigens. *International Archives of Allergy and Applied Immunology* **75**, 219–226.

Howard, W.L., Klopfenstein, M.D., Steininger, W.J. & Woodruff, C.E. (1970) The loss of tuberculin reactivity in certain patients with active pulmonary tuberculosis. *Chest* **57** (6), 530–534.

Huebner, R.E., Schein, M.F., Cauthen, G.M. *et al.* (1992) Evaluation of the clinical usefulness of mycobacterial skin test antigens in adults with pulmonary mycobacterioses. *American Review of Respiratory Diseases* **145**, 1160–1166.

Jackett, P.S., Bothamley, G.H., Btara, H.V., Mistry, A., Young, D.B. & Ivanyi, J. (1988) Immunodominance and specificity of mycobacterial antigens and their epitopes in the serology of tuberculosis. *Journal of Clinical Microbiology* **26**, 2313–2318.

Joint Tuberculosis Committee of the British Thoracic Society. (1994) Control and prevention of tuberculosis in the United Kingdom: Code of Practice 1994. *Thorax* **49**, 1193–1200.

Kadival, G.V., Chaparas, S.D. & Hussong, D. (1987) Characterisation of serologic and cell-mediated immunity of a 38 kDa antigen isolated from *Mycobacterium tuberculosis*. *Journal of Immunology* **139** (7), 2447–2451.

Khanolkar-Young, S., Kolk, A.H.J., Andersen, A.B. *et al.* (1992) Results of the third immunology of leprosy/immunology of tuberculosis anti-mycobacterial monoclonal antibody workshop. *Infection and Immunity* **60**, 3925–3927.

Kollias, G.V., Thoen, C.O. & Fowler, M.E. (1982) Evaluation of comparative cervical tuberculin skin testing in cervids naturally exposed to mycobacteria. *Journal of the American Veterinary Medicine Association* **181**, 1257–1262.

Krambovitis, E., McIllmurray, M.B., Lock, P.E., Hendrickse, W. & Holzel, H. (1984) Rapid diagnosis of tuberculous meningitis by latex particle agglutination. *Lancet* **ii**, 1229–1231.

Kuwabara, S. (1975) Purification and properties of tuberculin-active protein from *Mycobacterium tuberculosis*. *Journal of Biological Chemistry* **250** (7), 2556–2562.

Leslie, I.W., Herbert, C.N., Burn, K.J., McClancy, B.N. & Donnelly, W.J.C. (1975) Comparison of the specificity of human and bovine tuberculin PPD for testing cattle. *Veterinary Record* **96**, 332–341.

Little, T.W.A., Naylor, P.F. & Wilesmith, J.W. (1982) Laboratory study of *Mycobacterium bovis* infection in badgers and calves. *Veterinary Record* **111**, 550–557.

Magnusson, M. (1986) Tuberculins, other mycobacterial sensitins and 'new tuberculins'. *European Journal of Respiratory Disease* **69**, 129–134.

Mahmood, K.H., Rook, G.A.W., Stanford, J.L., Stuart, F.A. & Pritchard, D.G. (1987) The immunological consequences of challenge with bovine tubercle bacilli in badgers (*Meles meles*). *Epidemiology and Infection* **98**, 155–163.

Matsuo, T., Matsuo, H., Ohara, N. *et al.* (1996) Cloning and sequencing of an MPB70 homologue corresponding to MPB83 from *Mycobacterium bovis* BCG. *Scandinavian Journal of Immunology* **43**, 483–489.

Monaghan, M.L., Doherty, M.L., Collins, J.D., Kazda, J.F. & Quinn, P.J. (1994) The tuberculin test. *Veterinary Microbiology* **40**, 111–124.

Monaghan, M., Quinn, P.J., Kelly, A.P. *et al.* (1997) A pilot trial to evaluate the γ-interferon assay for the detection of *Mycobacterium bovis*-infected cattle under Irish conditions. *Irish Veterinary Journal* **50**, 229–232.

Morris, J.A., Stevens, A.E., Little, T.W.A. & Stuart, P. (1978) Lymphocyte unresponsiveness to PPD tuberculin in badgers infected with *Mycobacterium bovis*. *Research in Veterinary Science* **25**, 390–392.

Morris, J.A., Stevens, A.E., Stuart, P. & Little, T.W.A. (1979) A pilot study to assess the usefulness of ELISA in detecting tuberculosis in badgers. *Veterinary Record* **104**, 14.

Muirhead, R.H., Gallagher, J. & Burn, K.J. (1974) Tuberculosis in wild badgers in Gloucester: epidemiology. *Veterinary Record* **95**, 552–555.

Mwatha, J., Moreno, C., Sengupta, U., Sinha, S. & Ivanyi, J. (1988) A comparative evaluation of serological assays for lepromatous leprosy. *Leprosy Review* **59**, 195–199.

Nagai, S., Matsumoto, J. & Nagasuga, T. (1981) Specific skin-reactive protein from culture filtrate of *Mycobacterium bovis* BCG. *Infection and Immunity* **31**, 1152–1160.

Neill, S.D., Cassidy, J., Hanna, J. *et al.* (1994b) Detection of *Mycobacterium bovis* infection in skin test-negative cattle with an assay for bovine interferon-gamma. *Veterinary Record* **135**, 134–135.

Neill, S.D., Pollock, J.M., Bryson, D.B. & Hanna, J. (1994a) Pathogenesis of *Mycobacterium bovis* infection in cattle. *Veterinary Microbiology* **40**, 41–52.

Newell, D.G., Clifton-Hadley, R.S. & Cheeseman, C.L. (1997) The kinetics of serum antibody responses to natural infections with *Mycobacterium bovis* in one badger social group. *Epidemiology and Infection* **118**, 173–180.

Nolan, A. & Goodger, J. (1993) Diagnosis of tuberculosis in live badgers. In: *The Badger* (ed. Hayden, T.J). Proceedings of a Seminar held on 6–7 March 1991. Dublin: Royal Irish Academy, pp. 128–130.

Outteridge, P.M. & Lepper, A.W.D. (1973) The detection of tuberculin-sensitive lymphocytes from bovine blood by uptake of radio-labelled nucleotides. *Research in Veterinary Science.* **14**, 296–305.

Pollock, J.M. & Andersen, P. (1997a) Predominant recognition of the ESAT-6 protein in the first phase of infection with *Mycobacterium bovis* in cattle. *Infection and Immunity* **65**, 2587–2592.

Pollock, J.M. & Andersen, P. (1997b) The potential of the ESAT-6 antigen secreted by virulent mycobacteria for specific diagnosis of tuberculosis. *Journal of Infectious Diseases* **175**, 1251–1254.

Pollock, J.M., Douglas, A.J., Mackie, D.P. & Neill, S.D. (1994) Identification of bovine T-cell epitopes for three *Mycobacterium bovis* antigens: MPB70, 19,000 MW and MPB57. *Immunology* **82**, 9–15.

Pollock, J.M., Douglas, A.J., Mackie, D.P. & Neill, S.D. (1995) Peptide mapping of bovine T-cell epitopes for the 38 kDa tuberculosis antigen. *Scandinavian Journal of Immunology* **41**, 85–93.

Pritchard, D.G. (1988) A century of bovine tuberculosis 1888–1988: conquest and controversy. *Journal of Comparative Pathology* **99**, 357–399.

Pritchard, D.G., Stuart, F.A., Brewer, J.I. & Mahmood, K.H. (1987) Experimental infection of badgers (*Meles meles*) with *Mycobacterium bovis*. *Epidemiology and Infection* **98**, 145–154.

Pritchard, D.G., Stuart, F.A., Wilesmith, J.W. *et al.* (1986) Tuberculosis in East Sussex. III. Comparison of post mortem and clinical methods for the diagnosis of tuberculosis in badgers. *Journal of Hygiene, Cambridge* **97**, 27–36.

Radford, A.J., Wood, P.R., Billman-Jacobe, H., Geyson, H.M., Mason, T.J. & Tribbick, G. (1990) Epitope mapping of the *Mycobacterium bovis* secretory protein MPB70 using overlapping peptide analysis. *Journal of General Microbiology* **136**, 265–272.

Radhakrishnan, V.V., Mathai, A. & Sundaram, P. (1992) Diagnostic significance of circulating immune complexes in patients with pulmonary tuberculosis. *Journal of Medical Microbiology* **36**, 128–131.

Rees, W.H.G. & Meldrum, K.C. (1995) Regional and county status reports; Great Britain. In: *Mycobacterium bovis Infection in Animals and Humans* (ed. C. O. Thoen & J. H. Steele). Iowa: Iowa State University Press, pp. 250–256.

Report (1997) Bovine tuberculosis in badgers. *Twentieth Report by the Ministry of Agriculture, Fisheries and Food.* London: MAFF Publications.

Ritacco, V., de Kantor, I.N., Barrera, L. *et al.* (1987) Assessment of the sensitivity and specificity of enzyme-linked immunosorbent assay (ELISA) for the detection of mycobacterial antibodies in bovine tuberculosis. *Journal of Veterinary Medicine Series B* **34**, 119–125.

Ritacco, V., Lopez, B., Barrera, L., Errico, F., Nader, A. & de Kantor, I.N. (1991) Reciprocal cellular and humoral immune responses in bovine tuberculosis. *Research in Veterinary Science* **50**, 365–367.

Roche, P.W., Britton, W.J., Faibus, S.S., Naupane, K.D. & Theuvenet, W.J. (1993) Serological monitoring of the response to chemotherapy in leprosy patients. *International Journal of Leprosy* **61**, 35–43.

Rose, D.N., Schechter, C.B. & Adler J.J. (1995) Interpretation of the tuberculin skin test. *Journal of General Internal Medicine* **10**, 635–642.

Sepulveda, R.L., Ferrer, X., Latrach, C. & Sorensen, R.U. (1990) The influence of Calmette–Guérin bacillus immunisation on the booster effect of tuberculin testing in healthy young adults. *American Review of Respiratory Diseases* **142**, 24–28.

Steele, B.A. & Daniel, T.M. (1991) Evaluation of the potential role of serodiagnosis of tuberculosis in a clinic in Bolivia by decision analysis. *American Review of Respiratory Diseases* **143**, 713–716.

Stuart, F.A., Mahmood, K.H., Stanford, J.L. & Pritchard, D.G. (1988) Development of diagnostic tests for, and vaccination against, tuberculosis in badgers. *Mammalian Review* **18**, 74–75.

Stuart, F.A. & Wilesmith, J.W. (1988) Tuberculosis in badgers: a review. *Revue Scientifique et Technique de L'Office International Des Epizooties* **7** (4), 929–935.

Thoen, C.O., Jarnagin, J.L., Muscoplat, C.C., Cram, L.S., Johnson, D.W. & Harrington, R. (1980) Potential use of lymphocyte blastogenic responses in diagnosis of bovine tuberculosis. *Comparative of Immunology, Microbiology and Infectious Diseases* **3**, 355–361.

Thorns, C.J. & Morris, J.A. (1986) Shared epitopes between mycobacteria and other micro-organisms. *Research in Veterinary Science* **41**, 275–276.

Toida, I., Yamamoto, S., Takuma, S., Susuzuki, T. & Hwara, M. (1985) Lack of tuberculin reactivity of synthetic peptides. *Infection and Immunity* **50**, 614–619.

Towar, D.R., Scott, R.M. & Goyings, L.S. (1965) Tuberculosis in a captive deer herd. *American Journal of Veterinary Medicine* **26**, 339–346.

Triccas. J.A., Roche, P.W., Winter, N., Feng, C.G., Butler, C.R. & Britton, W.J. (1996) A 35-kilodalton protein is a major target of the human immune response to *Mycobacterium leprae*. *Infection and Immunity* **64**, 5171–5177.

Verbon, A., Hartskeerl, R.A., Schuitema, A., Kolk, A.H.J., Young, D.B. & Lathigra, R. (1992) The 14 kDa antigen of *Mycobacterium tuberculosis* is related to the α-crystallin family of low molecular weight heat shock proteins. *Journal of Bacteriology* **174**, 1352–1358.

Vordermeier, H.M., Harris, D.P., Mehrotra, P.K. *et al.* (1992) *Mycobacterium tuberculosis*-complex specific T-cell stimulation and delayed type hypersensitivity reactions induced by a peptide from the 38-kilodalton protein. *Scandinavian Journal of Immunology* **35**, 711–718.

Wiker, H.G. & Harboe, M. (1992) The antigen 85 complex: a major secretion product of *Mycobacterium tuberculosis*. *Microbiological Reviews* **56** (4), 648–661.

Wiker, H.G., Harboe, M., Bennedsen, J. & Closs, O. (1988) The antigens of *Mycobacterium tuberculosis*, H37Rv, studied by crossed immunoelectrophoresis: comparison with a reference system for *Mycobacterium bovis*, BCG. *Scandinavian Journal of Immunology* **27**, 223–229.

Wilcke, J.T.R., Jenson, B.N., Rava, P., Andersen, A.B. & Haslov, K. (1996) Clinical evaluation of MPT64 and MPT59, two proteins secreted from *Mycobacterium tuberculosis*, for skin test reagents. *Tubercle and Lung Disease* **77**, 250–256.

Wilkins, E.G.L. & Ivanyi, J. (1990) Potential value of serology in the diagnosis of extrapulmonary tuberculosis. *Lancet* **336**, 641–644.

Wilkins, E.G.L., Bothamley, G.H. & Jackett, P.S. (1991) A rapid simple ELISA to measure antibody to individual epitopes in the serodiagnosis of tuberculosis. *European Journal of Clinical Microbiology and Infectious Diseases* **10**, 559–563.

Wilkinson, R.J., Hasløv, K., Rappuoli, R. *et al.* (1997) Evaluation of the recombinant 38-kilodalton antigen of *Mycobacterium tuberculosis* as a potential immunodiagnostic reagent. *Journal of Clinical Microbiology* **35** (3), 553–557.

Wilson, P. & Harrington, R. (1976) A case of bovine tuberculosis in fallow deer. *Veterinary Record* **98**, 74.

Wood, P.R., Corner, L.A. & Plackett, P. (1990) Development of a simple, rapid *in vitro* cellular assay for bovine tuberculosis based on the production of interferon-γ. *Research in Veterinary Science* **49**, 46–49.

Wood, P.R., Corner, L.A., Rothel, J.S. *et al.*(1991) Field comparisons of an interferon-γ assay and the intradermal tuberculin test for the diagnosis of bovine tuberculosis. *Australian Veterinary Journal* **68**, 286–290.

Wood, P.R., Ripper, J., Radford, A.J. *et al.* (1988) Production and characterisation of monoclonal antibodies specific for *Mycobacterium bovis*. *Journal of General Microbiology* **134**, 2599–2604.

Yuan, Y., Crane, D.D. & Barry, C.E. (1996) Stationary phase-associated protein expression in *Mycobacterium tuberculosis*: Function of the alpha-crystallin homolog. *Journal of Bacteriology* **178**, 4484–4492.

Chapter 11 / Mycobacterial growth and dormancy

M. JOSEPH COLSTON & ROBERT A. COX

1 Introduction

1.1 Mycobacterial growth rates

Mycobacteria vary enormously in the rate at which they grow. Although they are conveniently divided into two groups based on growth rates—fast- and slow-growers—even this division hides a wealth of diversity. Some species, most notably *Mycobacterium leprae*, have not been grown at all *in vitro*. It seems likely that there are many other species, which are not pathogenic or are opportunistic pathogens, that have not been identified because they cannot be grown; *M. genavense* was, for example, identified only recently using molecular techniques and could not be grown in culture (Bottger *et al.* 1992; Coyle *et al.* 1992). Other species, such as *M. ulcerans* and *M. paratuberculosis* only produce visible growth after a month or more in culture (Wayne & Kubica 1986). More typically, the slow-growing mycobacteria produce visible colonies on solid medium within 10–28 days; most isolates of *M. tuberculosis* produce visible growth after about 15 days, although this will vary depending on the state of the inoculum. Even

fast-growing species have optimal growth rates which are slower than the optimal growth rates of many other bacteria. However, it is important to realize that when we talk of growth rates we are usually referring to *optimal* growth; all bacteria *can* grow slowly under appropriate conditions. However, slow-growing mycobacteria are unable to grow rapidly, even under optimal conditions.

1.2 Mycobacterial dormancy

In addition to having slow optimal growth rates, mycobacteria are also able to undergo prolonged periods of dormancy. Most bacteria can survive for periods of time without dividing (Kaprelyants *et al.* 1993; Koch 1997), but the phenomenon has particular significance in the case of mycobacterial pathogens because of the clinical ramifications. For example, there are many reports of the clinical manifestations of leprosy developing many years, sometimes decades, after exposure to the organism. Similarly, clinical tuberculosis is often a reactivation of an old, subclinical infection which has occurred many years previously. In addition to having clinical consequences for the individual patient, this ability to survive in a dormant state for many years enables the pathogens to remain within small population groups, making eradication of the disease extremely difficult (Smith & Moss 1994).

The ability to enter into prolonged dormancy is not restricted to the period immediately following infection. Treatment with antimycobacterial drugs will kill the vast majority of the mycobacterial population within a few days or weeks, but may leave a population of 'persisters' (Waters *et al.* 1974), which are genetically drug sensitive, but which survive drug treatment by entering into a 'dormant' state. One of the greatest challenges for the development of novel intervention strategies is the development of agents which will eradicate these dormant or persisting bacteria.

In this chapter we have summarized our current understanding of slow growth, cell maintenance and dormancy of mycobacteria. It should be noted, however, that our detailed knowledge is limited, and experimental data relating to these phenomena are scarce. Thus, in establishing our models we have had to rely on limited amounts of hard information, much of it reported many years ago, and to make several assumptions drawn from work carried out with non-mycobacterial species. However, we believe that the general principles which we have used are applicable to understanding mycobacterial growth, and the hypotheses and conclusions that we draw are accessible to experimental verification.

1.3 Definition of slow growth

Traditionally mycobacteria are classified as either fast-growing or slow-growing according to whether colonies appear on a solid medium within 5 days (fast-growers) or longer than 5 days (slow-growers). To facilitate comparison with other bacteria in this chapter we arbitrarily define slow growth as 0.2 doublings or fewer per hour (generation time >5 h). This definition includes pathogenic mycobacteria growing in a favourable liquid medium maintained at optimum conditions. For example, *M. tuberculosis* growing optimally has a generation time of 14–15 h,

equivalent to ≈0.07 doublings per hour (Wayne & Kubica 1986). Fast-growers are defined as bacteria with optimal growth rates (μ) of 0.2–1.0 doublings per hour.

Even fast-growing mycobacteria grow more slowly than many other bacteria; thus, for purposes of comparison with, for example, bacteria such as *Escherichia coli* and *Bacillus subtilis* we define a third category which we will refer to as ultra-fast-growers. Ultra-fast-growers, under optimum conditions, proliferate with several doublings per hour ($\mu > 1\,h^{-1}$) and genome replication spans more than one cell-division cycle (Cooper & Helmstetter 1968; Helmstetter & Cooper 1968). Such bacteria give rise to new-born cells which have more than one genome equivalent per cell. In contrast, new-born cells of fast- and slow-growing bacteria, including mycobacteria (Hiriyanna & Ramakrishnan 1986), are thought to have a single genome which is replicated within the cell-division cycle. When growth conditions are less favourable ($\mu < 1\,h^{-1}$) ultra-fast-growers resemble fast-growers and slow-growers, with genome replication being started and completed within the cell-division cycle. This is shown schematically in Fig. 11.1.

2 The bacterial cell cycle

Before considering the factors that are likely to influence the growth rate of mycobacteria it is helpful to understand current knowledge of the cell-division cycle as it relates to other eubacteria. Investigation of the growth of bacteria has led to an understanding of the principal features of cell proliferation (for review see Cooper 1991a), in which ribosome and protein biosynthesis are key elements. For example, the growth rate of *E. coli* is related to features of macromolecular synthesis by eqn 1 (Bremer 1975).

$$\mu = (60/\ln 2)\{[\psi_s\,\alpha_p\,\beta_p\,\beta_r \in_{rRNA} \in_{pep}]/$$
$$[L_{rrn}\,L_{rpo}/(1-f_t)]\}^{0.5} \tag{1}$$

where μ is the growth rate (doublings/h); ψ_s is the fraction of RNA polymerase (Rpo) synthesizing rRNA and tRNA; α_p is the fraction of total protein that is Rpo; β_p is the fraction of active Rpo; β_r is the fraction of active ribosomes; $\in_{rRNA}$ is the elongation rate

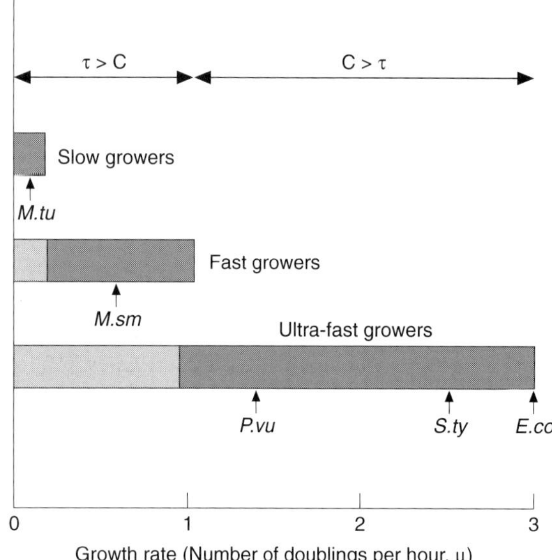

Fig. 11.1 Definition of slow-, fast- and ultra-fast-growers. A bacterium is classified by its optimum growth rate. Each category is defined by the range of maximum growth rates shown by the dark shaded area. The light shaded area indicates that all bacteria are capable of growing at less than optimal rates when conditions are suboptimal; that is, all bacteria are capable of slow growth. When the growth is less than one doubling per hour the genome is usually replicated within the cell-division cycle; in other words, the generation time, τ, exceeds the period, C, required for DNA replication ($\tau > C$). Each new-born cell has a single genome. Usually, when the growth rate exceeds one doubling per hour the replication of the genome takes place over more than one cell division cycle ($C > \tau$) so each new-born cell then has more than one genome equivalent. These properties characterize ultra-fast growth. *M.tu, M. tuberculosis* (Harshey & Ramakrishnan 1977); *M.sm, M. smegmatis* (G. Ellard personal communication 1997); *P.vu, Proteus vulgaris* (Schaechter *et al.* 1962); *S.ty, Salmonella typhimurium* (Schaechter *et al.* 1962); *E.co, Escherichia coli* (Koch, 1979).

of stable RNA (rRNA and tRNA); ϵ_{pep} is the peptide chain elongation rate; L_{rrn} is the size (nucleotides) of pre-rRNA; L_{rpo} is the size (aa) of core RNA polymerase; and f_t is the fraction of stable RNA that is tRNA.

Essentially this equation describes the rate of rRNA and protein synthesis. The terms ψ_s, α_p, β_p and ϵ_{rRNA}

describe the transcription of rRNA operons, and β_r and ϵ_{pep} describe features of protein synthesis. It reflects the knowledge that in balanced growth, cell proliferation is an exponential process (Maaløe & Kjeldgaard 1966) and emphasizes the importance of protein biosynthesis. The equation also reveals the central role of ribosomes in cell growth by focusing on the role of rRNA synthesis and the translational activity of ribosomes.

2.1 Cell-wall synthesis, DNA replication and synthesis of cytoplasmic components

The bacterial cell-division cycle is conveniently considered in three steps, namely, cell-wall synthesis, DNA replication, and synthesis of cytoplasmic components.

Cell-wall synthesis

The high surface area to volume ratio of rod-like bacteria ensures efficient diffusion of nutrients into the cell and excretion of material from the cell to the environment. Cell shape is maintained by means of a cell wall that is required to withstand turgor pressures of up to 50 atmospheres. The growth of the cell wall of rod-like bacteria has been considered in detail (for review see Cooper 1991b). The synthesis of the peptidoglycan component of the cell wall is considered to take place by a continuous process of growth based on localized cleavage of covalent bonds and the enlargement of the cell wall during the repair process. The growth of the cell wall is thought to be close to exponential.

The initiation of septum formation in the cell wall at the site of cell division is a critical event in cell proliferation. The protein FtsZ is important to septum formation (for review see Erickson 1997; Lutkenhaus & Addinall 1997). Protein FtsZ is tubulin-like, has guanosine triphosphatase (GTPase) activity and an ability to polymerize to form the Z-ring which is a cytoskeletal element essential to septum formation. Protein SulA (gene product of *sfiA*), which is a component of the SOS response (Huisman & D'Ari 1981) is rapidly induced following DNA cleavage and appears to block cell division by interacting with protein FtsZ (Lutkenhaus 1983; Bi & Lutkenhaus 1990; Dai *et al.* 1994; Huang *et al.* 1996), indicating a potential link between DNA replication and septum formation. The period between the initiation of chromosome replication and the appearance of the first signs of septum formation is designated 'U', and 'T' is the time between the appearance of the first signs of septum formation and cell separation (Fig. 11.2a).

DNA replication

Studies of several bacteria have revealed that DNA replication usually starts from the origin of replication and proceeds bidirectionally to the termination site. The replication may be represented as shown in Fig. 11.2. Three stages B, C and D are indicated in the figure. Stage B, the time between cell division and the initiation of DNA replication, has no defined functional significance but indicates the possibility that DNA replication may not always immediately follow cell division. Stage C represents the period of time during which the genome is replicated. Stage D (between the end of DNA synthesis and cell division) is required to allow the two genomes to separate and to move to sites appropriate to cell division. A theory was developed by Helmstetter and Cooper (1968) defining the amount of DNA per cell in a culture in balanced growth in terms of C, D and (see eqn 2).

$$G = (\tau/C\ln 2)\,[2^{(C+D)/\tau} - 2^{D/\tau}] \tag{2}$$

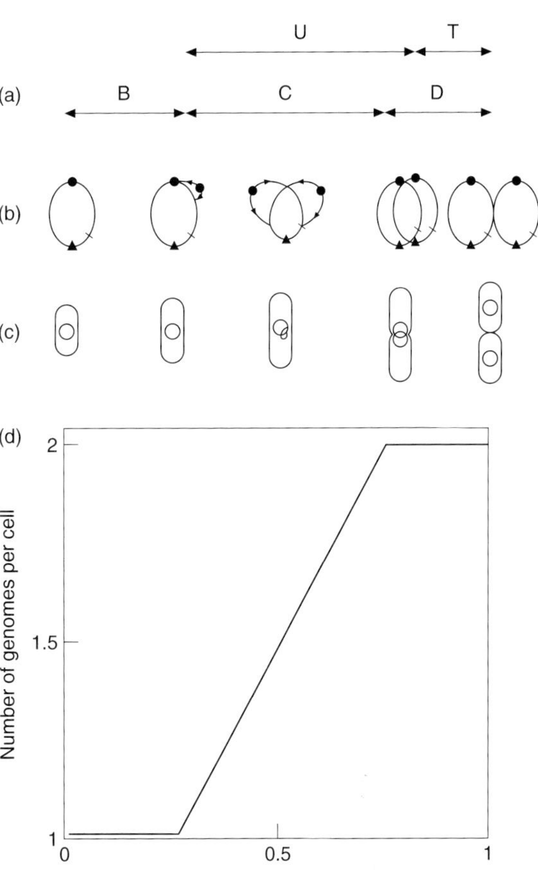

Fig. 11.2 The cell-division cycle is illustrated with reference to an idealized view of *M. tuberculosis* with a 24-h cycle. (a) Definitions of terms used in the description of the cell-division cycle. U is the period between the initiation of chromosome replication and the appearance of the first signs of septum formation; T is the period between the first signs of septum formation and cell separation; B, is the period between cell division and the initiation of DNA replication; C, is the period of DNA replication; D, is the period between the end of DNA replication and completion of septum formation. (b) Illustration of the DNA replication cycle. From left to right: the genome prior to the initiation of replication; ●, origin of replication; \, location of the *rrn* operon; ▲, termination of replication. The start of DNA replication illustrating the bidirectional nature of the event. Illustration of the point at which the *rrn* operon is duplicated. Completion of DNA replication without separation of the chromosomes. Separation of the chromosomes and movement to sites appropriate for cell separation. (c) Illustration of cell growth, chromosome replication and septum formation. (d) Kinetics of DNA replication illustrating the B, C and D periods and the linear nature of DNA replication. The time scale shown in (d) applies also to (a), (b) and (c).

where G is the average number of chromosome equivalents of DNA per cell where a chromosome equivalent is the mass of a single non-replicatory chromosome; C, D and the generation time, are usually expressed in minutes for fast-growers and hours for slow-growers.

This equation also applies to the special case of DNA replication when τ is less than C, that is when a round of replication is initiated before the previous round is completed, and hence the cell contains multiple chromosome equivalents (Cooper & Helmstetter 1968; Helmstetter & Cooper 1968). This mechanism is important to organisms such as *E. coli* which are capable of ultra-fast growth but not to fast-growing or slow-growing mycobacteria, where one round of chromosome replication is completed before the next starts (Hiriyanna & Ramakrishnan 1986).

The parameters B, C and D of DNA replication have been obtained for *E. coli* B/rA and *E. coli* B/rK with generation times ranging from 22.5 min to 1020 min (for review see Helmstetter 1987). The C period reflects the overall elongation rate ($2\epsilon_{DNA}$) of the bidirectional fork, i.e. twice the rate at which a replication fork moves in one direction around the chromosome. The average elongation rate (ϵ_{DNA}) per fork is $(g/2C)$ bp/min; g is the size (bp) of the genome. The rate at which the genome replicates ($2\epsilon_{\Delta NA}$) is given by the ratio g/C bp/min when μ is less than 1 doubling per hour. The elongation rate (ϵ_{DNA}) measured in base pairs incorporated per second and growth rate (μ) are linked by eqn 3 when μ is less than 1.

$$(\epsilon_{DNA})^3 = 27500\mu \tag{3}$$

A maximum value for ϵ_{DNA} is attained in fast-growing cultures, where the growth rate, μ, is greater than 1.

It is thought that the processes leading to chromosome replication and cell separation follow independent but interactive pathways. Although the two processes may proceed in parallel, the timing of events are linked, so the time between initiation of DNA replication and cell division (C+D) is equal to the time between the first signs of cell-wall septum formation and cell division (U+T), as shown in eqn 4.

$$C + D = U + T \tag{4}$$

The possibility of a biosynthetic link between DNA replication and the initiation of septum formation is inherent in the definition of U and T, and is implied by the interaction between FtsZ, the protein involved in septum formation, and SulA, which is induced following DNA cleavage. A linkage between DNA replication and cell division has been demonstrated in *M. smegmatis* (Klann *et al.* 1998).

The synthesis of cytoplasmic components

The synthesis of rRNA is regarded as the rate-determining step of ribosome synthesis, and the correlation between ribosome synthesis and protein synthesis has been long established (see for example Schaechter *et al.* 1958). The number of ribosomes per cell is known to increase exponentially during balanced growth from ρ_0 ribosomes in a new-born cell to $2\rho_0$ ribosomes in a cell that is about to divide. The number of ribosomes, ρ_a, in a cell aged a is given by eqn 5 where a is a fraction of τ, the generation or doubling time; i.e. $a=0$ for a new-born cell, and $a=1$ for a cell immediately prior to division.

$$\rho_a = \rho_0\, e^{a\ln 2} = \rho_0 2^a \tag{5}$$

The role of RNA polymerase (Rpo) in rRNA synthesis is made clear in eqn 1. The average number of ribosomes per cell (ρ) is related to the growth rate μ (the number of doublings per hour) by eqn 6.

$$\rho^{0.5} = 20 + 96\mu \tag{6}$$

The average number (N_R) of Rpo molecules per cell is given by eqn 7. The mass of protein (P fg) per average cell is given by eqn 8.

$$N_R^{0.5} = 17 + 37\mu \tag{7}$$

$$P^{0.5} = 6.4 + 6\mu \tag{8}$$

Equations 6–8 represent empirical relationships derived from the extensive experimental data of Bremer and Dennis (1987) for *E. coli* with growth rates in the range 0.6–2.5 doublings per hour.

2.2 Derivation of a model for slow growth

If we apply eqns 3, 6, 7 and 8 to calculate the properties of *E. coli* growing slowly, i.e. with growth rates of

Table 11.1 Estimated properties of 'average' *Escherichia coli* cells with generation times (τ) of 5–1000 h.

μ (h^{-1})	τ (h)	Mass of protein (fg)[a]	Protein/RNA[b]	No. of ribosomes[c] (ρ)	No. of Rpo cores[d] (N_R)	DNA synthesis $2\varepsilon_{DNA}$[e] (kbp/min)	C (mins)[f]	C/τ
0.2	5	58	12	1540	595	37.4	124	0.42
0.1	10	49	19	880	430	23.5	197	0.33
0.05	20	45	23	615	355	14.8	314	0.27
0.01	100	42	29	440	300	5.1	910	0.15
0.001	1000	41	31	400	290	1.1	4220	0.07

[a] Derived from equation 8.
[b] The RNA content (fg) was derived from equation 6 on the basis of the assumption that one ribosome has 2.75×10^{-18} g RNA, and that rRNA comprises 83% of total RNA.
[c] Derived from equation 6.
[d] Derived from equation 7.
[e] Derived from equation 3.
[f] Derived from the equation C = genome size (kbp)/$2\varepsilon_{DNA}$.

0.001–0.2 h^{-1}, we can predict that the contents of an average cell include:

1 several hundred ribosomes when the growth rate is slow ($\mu = 0.001$ h^{-1}) to more than 1000 ribosomes at faster growth rates ($\mu = 0.2$ h^{-1});

2 several hundred RNA polymerase units (α_2, β and β' subunits) of which fewer than 10 are engaged in rRNA synthesis. It is also predicted that the period of DNA synthesis (C) increases as the growth rate decreases. However, it appears from the ratio C/τ, that as the growth rate decreases and the generation time increases, DNA synthesis occupies a smaller fraction of the generation time.

In the following section we have taken the properties of slow-growing *E. coli* (i.e. with growth rates in the 'slow-growing mycobacterial range' of 0.001–0.2 divisions per hour) as deduced from these equations (see Table 11.1), and compared them with the observed properties of slow-growing mycobacteria, which optimally grow with a growth rate of ≈ 0.07 divisions per hour, in order to see if the mathematical models (and hence the assumptions on which they are based) hold true for slow-growing mycobacteria.

3 Proliferation of mycobacteria

In the above section we have derived a model of slow growth based on *E. coli* growing in suboptimal conditions. This has enabled us to draw conclusions about the composition of slow-growing bacteria and identify the effect of various growth-limiting factors on growth rates. In this section we will review what (limited) information is available on the content of slow-growing mycobacteria. If the mathematical model shown in eqn 1 is appropriate to describe the growth of *M. tuberculosis*, it should be possible to identify which factor(s) are important in contributing to the difference in growth rate between it and *E. coli*. In addition to knowledge about the chemical composition of *M. tuberculosis*, we will also need to obtain information about the synthesis of *M. tuberculosis* DNA, the rate of peptide chain elongation, the rate of RNA synthesis and other parameters which will enable us to test the fit of eqn 1.

3.1 The chemical composition of *Mycobacterium tuberculosis*

Before reviewing our knowledge of the mycobacterial cell cycle, and applying the model of slow growth outlined above, we will first consider what is known about the chemical composition of mycobacteria. Most of the available data come from Winder and Rooney (1970), who reported a comprehensive study of the chemical composition of *M. bovis* bacille Calmette–Guérin (BCG), a member of the *M. tuberculosis* family. The mycobacterium was grown in two different media. In each case the generation time was ≈24 h. By making a number of simple calculations based on the definitions of the units used, and the data generated by Winder and Rooney we arrive at the results presented in Table 11.2, in which we have calculated the number of cells per millilitre, as well as the amounts of DNA, RNA, protein, carbohydrate and lipid for an 'average' cell. This leads us to deduce a protein content of *M. bovis* BCG which is approximately four times the amount expected based on the calculations for slow-growing *E. coli* (Table 11.1).

The information given in Table 11.2 is based on the overall properties of the culture and provides average properties of the cell. The 'average' cell does not correspond to a clearly defined physiological state. It would be more helpful to have insight into the composition of a new-born daughter cell immediately following division, especially since the inherited resources profoundly influence a cell's development. A particular example is the ratio (r) of the mass of RNA (m_{RNA}) to the mass of DNA (m_{DNA}) which reflects the average number (ρ) of ribosomes per cell and the average number (G) of genomes per cell. The parameters r and G are related, as shown in eqn 9, through the fraction (f_r) of RNA that is rRNA, the mass (m_r) of the RNA moiety of the ribosome and the mass (g) of the genome.

$$\rho/G = f_r\, g\, r/m_r \tag{9}$$

The average quantities ρ and G are also related to ρ_0 and G_0, the values found in a new-born daughter

Table 11.2 Chemical composition of *M. bovis* BCG (Glaxo) during balanced growth[a].

Medium (growth rate, h⁻¹)	Age of culture[c] (days)	No. of 'average' cells/ml (×10⁻⁸)	DNA[d] No. of genomes	Mbp	fg	RNA[d] Mb	fg	Protein[e] fg	amino acids × 10⁻⁸	Protein (fg)/RNA (fg)	Carbohydrate (fmoles glucose equivalents)[f]	Lipid (fg)
A*	2	2.7				25.6	14.1	167	9.35	12	0.3	76
0.042	3	5.7	1.45 ± 0.25	6.38	6.9	23.5	12.9	143	8.01	11	0.3	76
	4	10.9				23.0	12.7	158	8.85	12	0.5	92
B†	5	2.1	1.45 ± 0.25	6.38	6.9	18.2	10.0	120	6.72	12	0.4	90
0.042	7	7.7				16.1	8.9	115	6.16	13	0.4	108

[a] Data of Winder and Rooney (1970) for growth at 37°C.

[b] The 'average' cell has a composition such as total protein/total number of cells. The reference point is the number of genomes (1.45 ± 0.25) per 'average' cell calculated by means of equation 2 with $T = 24$ h, $C = 10.33$ h and $D = 6$ h; the range indicated reflects uncertainty in the length of the D period based on the extreme values $D = 1$ h and $D = 13$ h. The size of the genome was taken as 4.4 Mbp (Philipp *et al.*, 1996).

[c] Inoculated cultures were left to stand for 4 days and were then shaken for the period shown.

[d] The amounts of RNA and DNA were originally expressed as μg atoms P; 1 μg atom of DNA P is equivalent to the DNA P of 4.7×10^{10} 'average' cells; Mb, megabases; Mbp, megabase pairs.

[e] Amounts of protein were originally expressed as g N; 85% of total N was shown to be protein. Protein g N was converted to g protein by the factor 1 g N ≡ 6.23 g protein. This factor was computed for a protein having the composition of total *E. coli* protein. The data given in Arnstein and Cox (1992) were used. The yield of cells was reported as the amount (mg) of insoluble N. The number of cells was calculated using the amount of protein per average cell.

[f] Glucose was used as the standard for estimations of carbohydrate.

* A: the medium contained per litre, glycerol (75.5 g), ammonium ions (approx. 0.5 g), asparagine (4 g), pancreatic casein hydrolysate (3 g).

† B: the medium was the same as A, except that the pancreatic casein hydrolysate was omitted.

cell. The synthesis of RNA and the replication of DNA proceed at different rates, as discussed above. The synthesis of RNA is exponential (see eqn 5). However, ρ, ρ_0, G and G_0 are related by eqn 10, where E and F are constants ($\rho = E_0$; $G = FG_0$)

$$\rho/G = E\,\rho_0/F\,G_0 \tag{10}$$

In order to estimate E and F, it is necessary to identify a mathematical model for the cell cycle. The simplest (canonical) model, which has been widely used in studies of microbial physiology (for review see Koch 1987), is based on four concepts. First, cellular components such as RNA and protein increase exponentially. Second, a cell divides when it achieves a critical size and age. Third, each cell in the population is characterized by the same critical size and age at division. Fourth, each cell that divides generates two equivalent daughter cells. Although this model is evidently an oversimplification it is capable of providing valuable insights into bacterial proliferation.

According to the canonical model the fraction, φ_a, of the cell population within an incremental age range is given by eqn 11

$$\varphi_a = 2\ln2\,e^{-a\ln2} \tag{11}$$

The number of ribosomes (ρ_a) in this fraction is $\rho_a\,\varphi_a$. The number of ribosomes per average cell (ρ) is given by eqn 12.

$$\rho = 2\ln2 \int_{a=0}^{a=1} \rho_0 e^{-a\ln2}\,\mathrm{d}a \tag{12}$$

It can be shown, by means of eqn 12 that $\rho = 2\ln2 \cdot \rho_0$. In other words, in eqn 10, $E = 1.3863$.

The constant F in eqn 10 may be evaluated, G is obtained by means of eqn 2 and the new-born cell has a single genome ($G = 1$). Hence numerically $F = G$. However, it is known that C is ≈ 620 min for *M. tuberculosis* ($\mu = 0.042\,\mathrm{h}^{-1}$) (Hiriyanna & Ramakrishnan 1986) so that a range of values may be deduced for F corresponding to extreme values of D. This procedure leads to the estimation, $F = 1.45 \pm 0.25$, for values of $C + D$ ranging from 11.33 to 23.00 h. Hence, $E \approx F$ so that $r \approx \rho_0/G_0$. This information is used in discussions of the mycobacterial cell-division cycle which follow. The average value corresponding to $D \approx 6$ h is judged to be most likely (see Table 11.2) because it is probable that the B period is significant. The composition of a new-born daughter cell (Table 11.3) was calculated on this basis. The information given in the table is based on the assumption that data for *M. tuberculosis* H37Rv are also valid for *M. bovis* BCG.

The RNA/DNA ratios were measured for several other species of mycobacteria, namely *M. smegmatis*, *M. microti* and *M. leprae* (G. A. Ellard, unpublished data 1997). These ratios and the number of ribosomes per new-born cell (ρ_0) deduced from these are presented in Table 11.4; data for *M. bovis* BCG are included for comparison. The data reveal that the slow-growing mycobacteria have a higher ribosomal content than is expected on the basis of the properties derived for a model slow-grower (see Table 11.1).

3.2 The mycobacterial cell-division cycle

There are few reported studies that are directly relevant to the mycobacterial cell-division cycle. Nevertheless, the available information provides insight into the reasons for mycobacterial slow growth. Bacterial growth reflects three principal

Table 11.3 Composition of a new-born cell of *M. bovis* BCG during balanced growth.

Medium	Growth rate	Number of genomes (G_0)	Number of ribosomes (ρ_0)	Mass of protein (fg)	Carbohydrate (fmoles glucose equivalents)	Lipid (fg)
A*	0.042	1	2900	146	0.20	56
B*	0.042	1	2100	107	0.25	68

* See Footnotes to Table 11.2 for composition of Medium A and Medium B.

Table 11.4 Ribosome content of new-born cells of several mycobacterial species during balanced growth.[a]

Species	Medium	μ (h^{-1})	τ (h)	Genome size (bp $\times 10^{-6}$)	MassRNA/mass DNA (r)[d]	No. of ribosomes (ρ_0)
M. smegmatis	L	0.440	2.25	5.7	6.5[b]	12 800
M. microti	TG	0.052	19.0	4.4	2.6[b]	4 000
M. bovis BCG	A	0.042	24.0	4.4	1.9[c]	2 900[e]
M. bovis BCG	B	0.042	24.0	4.4	1.3[c]	2 100
M. leprae	iv	0.0037	270.0	2.8	3.0[b]	3 000[f]

L, Lemco (nutrient) broth; TG, Tween–glutamate medium (Davidson *et al.*, 1982); A, see medium A, legend to Table 11.2; B, see medium, B legend to Table 11.2; iv, grown *in vivo* in armadillos (Levy, 1976).

[a] Please note that in late stationary phase values of ρ_0 were found to be substantially lower than in balanced growth. For example, the following values were deduced fom $m_{RNA}:m_{DNA}$ ratios (r); *M. bovis* BCG $\rho_0 \approx 1000$ ribosomes (Winder & Rooney, 1970); *M. microti* $\rho_0 \approx 2000$ ribosomes (G. A. Ellard, personal communication 1997); *M. smegmatis* $\rho_0 \approx 4000$ ribosomes (G. A. Ellard, personal communication 1997).

[b] Dr. G. A. Ellard, personal communication 1997. Values of r were obtained by methods based on those described by Ellard (1991).

[c] Winder & Rooney (1970).

[d] Values of ρ/G were calculated by means of equation 9 using values $f_r = 0.83$ and $\Sigma\, m_r = 4.8\,$kb. Values of ρ_0 were obtained by means of equation 10 where $E = 1.4$, and $F = 1.45 \pm 0.25$ (see text).

[e] The RNA content was reported as 20 fg/cell for *M. bovis* BCG ($\tau \approx 16\,$h) by isolating total RNA from an estimated number of cells (Mangan *et al.*, 1997) corresponding to $\rho_0 \approx 4300$ ribosomes.

[f] $\rho \approx 4000$ ribosomes was estimated from the yield of RNA obtained from an estimated number of cells (Estrada-G. *et al.*, 1988) corresponding to $\rho_0 \approx 2850$ ribosomes.

processes, namely, cell-wall synthesis, chromosome replication and synthesis of cytoplasmic components. The corresponding mycobacterial processes are discussed below with reference to members of the *M. tuberculosis* family.

3.2.1 Cell-wall synthesis

The mycobacterial cell wall is more complex than the walls of *E. coli* and *B. subtilis* (for review see Daffé & Draper 1998), and few studies relate cell-wall synthesis to the cell cycle. In the absence of evidence to the contrary we expect that mycobacterial cell-wall synthesis will proceed at a rate that is close to exponential. We also expect septum formation leading to cell separation to be linked to the initiation of chromosome replication as described in eqn 4.

One important difference between the cell wall of mycobacteria and that of other bacteria is the presence of a highly hydrophobic outer layer of complex lipids (mycolic acids) which, it is believed, forms a permeability barrier similar to the outer membrane of Gram-negative bacteria (Minnikin 1982). While it has been thought that this barrier could limit the ingress of hydrophilic nutrients into the mycobacterial cell and hence have an effect on growth rates, the demonstration that mycobacteria in general (Jarlier & Nikaido 1994; Trias & Benz 1994; Brennan & Nikaido 1995) and *M. tuberculosis* in particular (Senaratne *et al.* 1998) possess porin molecules which form aqueous channels, probably through this outer hydrophobic layer, suggests that these organisms have developed specialized mechanisms for the acquisition of hydrophilic nutrients.

3.2.2 Mycobacterial DNA synthesis

DNA replication in mycobacteria has not been extensively studied. The origins and termini of the *M. tuberculosis* and *M. leprae* genomes have been identified (Fsihi *et al.* 1996; Philipp *et al.* 1996) and an origin was isolated from *M. smegmatis* (Salazar *et al.* 1996; Qin *et al.* 1997). The time taken for the bidirectional replication fork to move from the origin to terminus

Table 11.5 Features of mycobacterial DNA synthesis.[a]

Parameter		E. coli B	M. sm (SN2)	M. tu (H37Rv)
Growth rate (doublings/h)		0.730	0.330	0.042
Generation time		82 ± 4 mins	180 ± 20 mins	24 ± 2 h
Genome size (kbp)		$4\,640$[b]	$5\,700$[c]	4400[d]
C period (min)		55	105	620
C/τ		0.67	0.58	0.43
Elongation rate	observed[e]	42750	27150	3500
(nucleotides/min)	calculated[f]	44300	26100	6600

M. sm, M. smegmatis; M. tu, M. tuberculosis.
[a] Hiriyanna & Ramakrishnan (1986).
[b] Genebank accession number U00096.
[c] Baess (1984).
[d] Philipp *et al.* (1996).
[e] 2ε = genome size (bp)/C period (s).
[f] calculated by means of equation 3.

(the C period) has been measured for both *M. tuberculosis* H37Rv and *M. smegmatis* SN2, using *E. coli* as the reference species to validate the experimental procedures (Hiriyanna & Ramakrishnan 1986). In these studies it was determined that the rate for *M. tuberculosis* was about 11 times slower than for *M. smegmatis*, and 13–18 times slower than for *E. coli*. The observed elongation rate ($\in_{DNA}$) for *M. tuberculosis* was 3200 nucleotides per minute which differs from the value calculated by means of eqn 3 (Table 11.5) by not more than twofold. Thus, the results determined by Hiriyanna and Ramakrishnan confirm the expectation inherent in eqn 3 that the period of DNA synthesis, C, increases as the growth rate decreases but takes place over a smaller fraction (C/τ) of the generation time in the slow-growing *M. tuberculosis* than in fast-growing bacteria.

The period of DNA synthesis (C) is more readily measured than the period between the end of DNA synthesis and cell separation (D). As yet no data are available for the D period of a mycobacterium. A nominal value of 0.25 is indicated in Fig. 11.1.

3.2.3 *Synthesis of mycobacterial cytoplasmic components*

A cell's capacity for protein biosynthesis is an impor-

tant factor in cell proliferation, as indicated in eqn 1. The principles of protein biosynthesis are thought to apply to bacteria in general. For example, we anticipate that regulatory mechanisms will maintain the fraction of active ribosomes (β_r) at ≈ 0.80 for a very wide range of species including mycobacteria. The pool of inactive ribosomes comprises newly synthesized subunits and subunits newly released from mRNA. Similarly, we anticipate that the ratio rRNA : tRNA : mRNA is very similar for all species and that the values found for *E. coli* also apply to mycobacteria. However, the number of ribosomes etc. per cell and rate of formation of peptide bonds will depend on the conditions of growth. It could be said that the rate at which a cell grows reflects the extent to which the cells resources are allocated to protein biosynthesis.

The rate of peptide bond formation

The importance of protein biosynthesis is expressed quantitatively in eqn 13 where the amount of protein P, is expressed as the number of amino acid residues and $\in_{pep}$ denotes the number of peptide bonds formed per minute (see Bremer & Dennis 1987).

$$\mu = (60/\ln2)\,(\rho/P)\,\beta_r \in_{pep} \tag{13}$$

Both ρ and P can be calculated for *M. bovis* (see Table

11.2), leaving $\in_{pep}$ as the only unknown. The values of $\in_{pep}$ obtained for *M. bovis* BCG growing in medium A (which contained, in g/L, glycerol, 7.5; NH_4^+, approx. 0.5; asparagine, 0.4; and pancreatic casein hydrolysate, 3.0) or medium B (which was the same as medium A except that pancreatic casein hydrolysate was omitted) were identical; namely 2.21 peptide bonds per second. Thus, although the absolute values of ρ and P differ according to the growth medium, the ratio (ρ/P) has the same value irrespective of whether cells were grown in medium A and medium B. It can be said that the efficiency of protein biosynthesis is the same for *M. bovis* grown in medium A as in medium B.

The rate of precursor rRNA synthesis

Current knowledge of mycobacterial proteins involved in RNA synthesis extends to Rpo (Honore *et al.* 1993) and σ factors (Doukhan *et al.* 1995; Predich *et al.* 1995). σ factors are needed for the initiation of transcription and play a role in promoter recognition. They dissociate from Rpo once the initiation phase is completed. Particular σ factors recognize particular classes of promoters. The results reveal substantial homologies between the mycobacterial proteins and their *E. coli* counterparts. It is interesting however, that while they are almost identical to the principal σ factor of *E. coli* in the region responsible for the binding to the 6 bp of the -10 box, they differ much more in the region involved in the binding to the 6 bp of the -35 box, located at positions -10 and -35, respectively, upstream from the transcription start point.

The rate of synthesis of precursor-rRNA (pre-rRNA) was measured (Harshey & Ramakrishnan 1977) for *M. tuberculosis* with a generation time of 10 h. Usually *M. tuberculosis* growing in optimum conditions has a generation time of 14–15 h (Wayne & Kubica 1986). Thus, the observed rate of 7.6 min for pre-rRNA, which was obtained for vigorously growing *M. tuberculosis* provides a guide for the *M. tuberculosis* family in general. However, by analogy with *E. coli*, the rate of pre-rRNA synthesis is expected to be largely independent of growth rate (Bremer & Dennis 1987).

The size of the pre-rRNA of *M. tuberculosis* is 5550 bp and includes the leader and spacer-1 region separating the 16S rRNA and 23S rRNA genes (Kempsell *et al.* 1992), the spacer-2 region separating the 23S and 5S genes and a trailer region (Ji *et al.* 1994b). The time required for pre-rRNA synthesis in strain H37Rv ($\mu=0.1\,h^{-1}$) was found to be 7.6 min (Harshey & Ramakrishnan 1977). Hence the average rate of rRNA chain elongation is 12.2 nucleotides per second.

Coupled transcription and translation

The rate of mRNA chain elongation is believed to be about half that of rRNA chain elongation, i.e. ≈ 6 nucleotides, or 2 codons per second for *M. tuberculosis* (Harshey & Ramakrishnan 1977). In bacteria transcription and translation are coupled, i.e. the growing mRNA chain is translated as it is being synthesized. Accordingly, a correlation exists between the rate of mRNA synthesis and the rate of polypeptide synthesis. For example, for transcription and translation to be coupled the rate of polypeptide synthesis must at least equal the rate of codon synthesis. This appears to be the case since the rate of polypeptide synthesis was estimated to be 2.2 amino acids which correlates well with the above mentioned rate of ≈ 2 codons per second. The corresponding rates (Bremer & Dennis 1987) observed for *E. coli* ($\mu=1.0\,h^{-1}$) are as follows: the mRNA elongation rate was found to be 50 nucleotides (16.7 codons) per second compared with a polypeptide chain elongation rate of 16 amino residues per second. Thus, in both *E. coli* ($\mu=1.0\,h^{-1}$) and *M. tuberculosis* ($\mu=0.042\,h^{-1}$) the mRNA elongation rate (codons per second) is approximately equal to the polypeptide chain elongation rate (amino acids per second), in accord with the coupled transcription/translation model.

Transcription of rRNA operons

Synthesis of rRNA is a key determinant in controlling the number of ribosomes per cell and hence in controlling the cell's capacity to synthesize protein. Many eubacteria have multiple rRNA (*rrn*) operons; *E. coli* for example has seven, noncontiguous *rrn* operons distributed asymmetrically around the origin of replication (Ellwood & Nomura 1982). The advantages of having multiple *rrn* operons are not fully known; however, while optimal growth rates can be obtained with only five operons provided the organisms are grown on complex medium (Condon *et al.* 1993), all seven operons are required for rapid adaptation to changes in nutrient availability and temperature (Condon *et al.* 1995).

In contrast to *E. coli*, mycobacteria have either a single, or just two *rrn* operons. Members of the *M. tuberculosis* family have a single operon (Bercovier *et al.* 1986; Suzuki *et al.* 1988; Kempsell *et al.* 1992) which is closely related to the single *rrn* operon of other slow growers. These operons are designated *rrnA$_s$*, the subscript 's' denoting slow growth (Ji *et al.* 1994a). Most rapid growers have a second *rrn* operon (Bercovier *et al.* 1986), designated *rrnB$_f$* ('f' denoting fast growth) (Gonzalez-y-Merchand *et al.* 1996). Thus, one mechanism by which growth rates and ribosome synthesis are linked in mycobacteria is by increasing the number of *rrn* operons per genome. Given that *E. coli* requires five *rrn* operons to grow maximally, and seven to be capable of rapid adaptation to environmental changes, we would anticipate that *M. tuberculosis* with a single *rrn* operon would be incapable of rapid growth and adaptation.

Some rapid-growing mycobacteria also have a single *rrn* operon (Domenech *et al.* 1994; Gonzalez-y-Merchand *et al.* 1997). In these species (e.g. *M. abscessus* and *M. chelonae*) more rapid growth from a single *rrn* operon appears to have been achieved by the acquisition of additional promoters by a process of duplication of sequence motifs (Gonzalez-y-Merchand *et al.* 1997). Thus while the *rrn*A operon of *M. tuberculosis* has two rather weak promoters, the single *rrn*A operon of *M. abscessus* has five promoters, the additional three of which are much more powerful. The *rrn*A$_f$ operon of *M. smegmatis* also has a very powerful additional promoter, and this species also has an additional operon, *rrn*B$_f$. Thus, compared with *M. tuberculosis*, the fast-growers have an increased potential for synthesis of rRNA, by a combination of increased gene dosage and by increased promoter activity (Fig. 11.3).

The durations of the *C* and *D* periods influence pre-rRNA synthesis in the following way. Suppose that a particular *rrn* operon is located at a position within the chromosome so that it is replicated at time $B+xC$ during the cell-division cycle. The contribution of the operons to pre-rRNA synthesis is as follows. The new-born cell has a single copy of this operon which is available for transcription for time τ. The operon will be replicated at time $B+xC$ and the new copy will function for the remainder $((1-x)\,C+D)$ of the cell-division cycle. Thus, the position of the operon within the genome determines the length of time for which the newly replicated operon is available for pre-rRNA synthesis. Thus, the 'average' *M. tuberculo-*

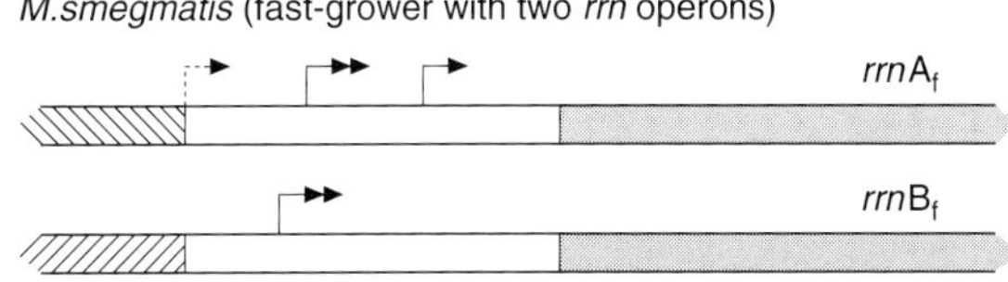

Fig. 11.3 Different strategies used by fast- and slow-growing mycobacteria to synthesize rRNA. *Mycobacterium tuberculosis* has a single *rrn* operon with one weak (   ) and one moderate (   ) putative promoter. *M.chelonae* has a single *rrn* operon with one weak, one moderate, and three strong (   ) putative promoters. *M. smegmatis* has two *rrn* operons. One operon, *rrn*A$_f$, has one weak, one moderate and one strong putative promoter; and the other, *rrn*B$_f$, has a single strong putative promoter. For details see Gonzalez-y-Merchand *et al.* (1997).

sis cell will have more than one *rrn* operon. The number, N_{rrn}, may be evaluated from eqn 14;

$$N_{rrn} = 2^{[(1-x)\,C+D]/\tau} \qquad (14)$$

for discussion see Bremer and Churchward (1977). The *rrn* operon of *M. tuberculosis* is located so that x is approximately two-thirds of the distance from the origin to the terminus (Philipp *et al.* 1996). Hence, N_{rrn}, was calculated to be equal to 1.35 ± 0.2 operons per 'average' cell, by substituting extreme values of D ($D=1$ h and $D=12$ h) in eqn 14.

Initiation of pre-rRNA synthesis

A key feature of pre-rRNA synthesis is the rate of initiation of transcription of *rrn* operons. Within a population of cells rRNA synthesis proceeds exponentially. However, for simplicity we consider the steady-state synthesis of pre-rRNA by a single operon and its replicate. The number of pre-rRNA transcripts synthesized during the life time of the cell is π_0, which leads to the synthesis of ρ_0 ribosomes; i.e. numerically, π_0 is approximately equal to ρ_0. Pre-rRNA synthesis will proceed from time 0 until time $(B+xC)$h at a particular linear rate with an average rate, i, initiations/min. After the operon is replicated transcription will proceed at twice the initial rate (see eqn 15) where, C and D are measured in hours.

$$\pi_0 = 60i\{\tau + (1-x)\,C+D\} \qquad (15)$$

A rate, i, initiations/min implies that the interval between initiations is $(1/i)$ min; during this period each functional polymerase will have moved $60\epsilon_{rRNA}/i$ bp. Hence, η, the average number of Rpo molecules per operon of L_{rrn} nucleotides is given by eqn 16.

$$\eta = iL_{rrn}/60\epsilon_{rRNA} \qquad (16)$$

The number of Rpos engaged in pre-rRNA synthesis is the product of the number of Rpos/operon (η) and the number (N_{rrn}) of operons/average cell. The fraction of total protein actively involved in pre-rRNA (N_{rrn} synthesis) ($\psi_s\,\alpha_p\,\beta_p$ in eqn 1) is readily calculated by taking into account the number of amino acids per Rpo (aa/Rpo) and total proteins (P amino acids/average cell), as shown in eqn 17.

$$\psi_s\,\alpha_p\,\beta_p = N_{rrn}.\,\eta.\,(aa/Rpo)/P \qquad (17)$$

After Rpo has formed an initiation complex the promoter is not available for further initiation events until the enzyme has moved away. The movement of the Rpo away from the promoter is termed 'promoter clearance'. The maximum possible rate of initiation (i_{lim}) is achieved when the enzyme binds to the promoter and immediately moves away. In this case, promoter clearance is the rate-determining step. The enzyme binds to the promoter forming an initiation complex which occupies I bp, i.e. the bound Rpo is required to move I bp before the promoter is cleared. The polymerase moves along the operon at an average rate of ϵ_{RNA} bp/s; hence, the time taken for promoter clearance is (I/ϵ_{RNA}). The number of initiations/min (i_{lim}) is given by eqn 18.

$$i_{lim} = 60(\epsilon_{RNA}/I) \qquad (18)$$

The ratio i/i_{lim} provides a measure of the allocation of the cell's resources to pre-rRNA synthesis.

Features of the transcription of *rrn* operons of members of the *M. tuberculosis* family (*M. microti* and *M. bovis* BCG) and of *M. leprae* are summarized in Table 11.6. The data include the number of initiations/min (see eqn 15), the ratio i/i_{lim} (see eqn 18), and the number of Rpo units/operon (see eqn 16). For comparison, data are included for the fast-grower *M. smegmatis*.

The fast-grower *M. smegmatis* has two *rrn* operons. One of these, *rrn*A$_f$, is a homologue of the single (*rrn*A$_s$) operon of *M. tuberculosis* and other slow-growers (Gonzalez-y-Merchand *et al.* 1996). At high growth rates (e.g. $\tau=2.5$ h) the second operon (*rrn* B$_f$) makes a similar contribution to pre-rRNA synthesis as *rrn*A$_f$, as judged by primer extension studies of the pre-rRNA fraction (Gonzalez-y-Merchand *et al.* 1998). Thus, the task of synthesizing ≈ 12800 pre-rRNA transcripts (see Table 11.4) appears to be equally divided between *rrn*A$_f$ and *rrn*B$_f$ operons; that is, each operon is required to provide ≈ 6400 transcripts in 2.25 h. The rate of initiation of transcription which is needed to achieve this target was estimated by means of eqn 15. The calculation was made on the basis of two assumptions. First, the loca-

Table 11.6 Summary of features of pre-rRNA synthesis[a].

Species	μ	τ (h)	No. of initiations/min/operon (i)	i/i_{lim}	No. of Rpo units/operon (η)	ε_{RNA} (nucleotides/s)
Slow-growers						
M. microti	0.052	19.0	2.60	0.230	17.0	12.2
M. bovis BCG[b]	0.042	24.0	1.60	0.140	11.0	12.2
M. bovis BCG[c]	0.042	24.0	1.12	0.100	7.4	12.2
M. leprae[d]	0.0037	270.0	0.10	0.009	1.0	12.2
Fast-grower						
M. smegmatis[e]	0.44	2.25	26.00	≤ 1.00	74.0	≥ 28.6

[a] The number of initiations per min (i) was calculated by means of equation 15 using values of π_0 equivalent to ρ_0 in Table 11.4. The ratio i/i_{lim} was calculated by means of equation 18. The number of Rpo units/operon (η) was calculated using equation 16.
[b] Grown in medium A (see Table 11.2).
[c] Grown in medium B (see Table 11.2).
[d] No data are available for Rpo of *M. leprae*; an elongation rate identical with that of Rpo of *M. tuberculosis* was used.
[e] For discussion, see text. The number of Rpo units per *rrn* operon is maximum because $\varepsilon_{rRNA} = 28.6$ nucleotides/s is the minimum value that ensures promoter clearance at the rate required for 26 initiations/min/operon and $I = 65$ bp (see eqn 18).

tion of the rrnA$_f$ operon within the genome of *M. smegmatis* is similar to the location of the rrnA$_s$ of *M. tuberculosis*, in other words, $x = 0.67$ in eqn 15. Second, the characteristic features of DNA replication are in the same proportion of the generation time as reported for *M. smegmatis* growing with a generation time of 3 h (see Table 11.5), i.e. $C = 1.32$ h. To maximize pre-rRNA synthesis the values $B = 0$ and $D = 0.93$ h were assigned, the calculation reveals that 26 initiations of transcription per minute are needed to synthesize 6400 ribosomes. The minimum rate (see eqn 18) for rRNA chain elongation required to achieve promoter clearance is 28.6 nucleotides per second; at this elongation rate the rrnA$_f$ operon would be loaded with ≈ 74 Rpo units. Thus, in order to synthesize the required number of ribosomes the *M. smegmatis* Rpo has a minimum elongation rate (28.6 nucleotides per second) that is more than twice the rate (12.2 nucleotides per second) found for *M. tuberculosis*. It is not known whether the Rpo enzymes of *M. tuberculosis* and *M. smegmatis* are significantly different or whether the substrates (rrnA$_s$ and rrnA$_f$ operons) are packed differently in the chromosome and so differ in their accessibilities to Rpo (Robinow & Kellenberger 1994).

3.3 Quantitative analysis of the *Mycobacterium tuberculosis* cell-division cycle

The growth of *E. coli* is successfully described by equations such as eqn 1 and others (Bremer & Dennis 1987) which identify growth-limiting parameters and predict their effect on growth rate:

$$\mu = (60/\ln 2)\{[\psi_s\, \alpha_p\, \beta_p\, \beta_r \in_{rRNA} \in_{pep}]/$$
$$[L_{rrn}\, L_{rpo}/(1 - f_t)]\}^{0.5} \tag{1}$$

In order to apply eqn 1 to *M. tuberculosis* we needed to derive information about:
1 the fraction of total protein that is Rpo (α_p);
2 the fraction of Rpo that is synthesizing rRNA and tRNA (ψ_s);
3 the fraction of active Rpo (β_p);
4 the fraction of active ribosomes (β_r);
5 the elongation rate of stable RNA, i.e. rRNA and tRNA ($\in_{rRNA}$);

6 the peptide chain elongation rate ($\in_{pep}$);

7 the size, in nucleotides, of pre-rRNA;

8 the size, in amino acids, of core Rpo (L_{rpo}); and

9 the fraction of stable RNA that is tRNA (f_t).

In the previous section we were able to derive or estimate values for each of these factors for *M. tuberculosis* and so we are now able to test whether the parameters specified in eqn 1 are sufficient to account for the characteristic features of mycobacterial growth, exemplified by *M. tuberculosis*.

The values of the appropriate parameters listed in Table 11.7 are amalgamated from results obtained for *M. tuberculosis* H37Rv and *M. bovis* BCG. In view of the close relationship between the two species this procedure is reasonable. Two sets of data are shown in the Table 11.7; one set for cells grown in medium A and one set for medium B (Table 11.2). The results show that values of generation times and growth rates calculated by means of eqn 2 agree within ±5% of the observed values.

The observation that cells grown in medium A and medium B have the same growth rate but differ in chemical composition is explained by the cell regulation of ribosome synthesis. For example, essentially

Table 11.7 Parameters used in computing the growth rates of *M. tuberculosis* and *E. coli* by means of equation 1.

Quantity	Symbol	*M. tu*		*E. co*	*E. co/M. tu*
		med A	med B		
No. of nucleotides/pre-rRNA	L_{rrn}	5550		6000	1.08[a]
No. of amino acids/Rpo core	L_{rpo}	2428		3407	1.40[a]
Fraction of stable RNA that is rRNA	(1-fs)	0.83		0.83	1.00[a]
Fraction of ribosomes engaged in protein synthesis	β_r	0.80		0.80	1.00[a]
Elongation rate of polypeptide (aa residues/s)	ε_{pep}	2.2		16.0	7.27[a]
Elongation rate of pre-rRNA (nucleotides/s)	ε_{rRNA}	12.2		85.0	6.97[a]
No. of *rrn* operons/genome		1		7.0	7.00
No. of *rrn* operons in average cell [A]	N_{rrn}	1.35		14.00	10.37[a]
No. of Rpo copies/operon [B]		11.0	7.4	13.4	1.22
No. of Rpo copies needed for steady state pre-rRNA synthesis [A.B]		16.0	10.7	188.0	11.75
Mass of protein/average cell (fg)	P	156	118	156	1.00
Fraction of total protein that is Rpo engaged in pre-rRNA synthesis [(A.B)/P]	$\alpha_p \cdot \beta_p \cdot \psi_s$	4.60×10^{-5}	4.21×10^{-5}	7.13×10^{-4}	16.93[a]
Calculated growth rate (h⁻¹)	μ	0.040	0.042	0.952	
Observed growth rate (h⁻¹)	μ	0.042	0.042	1.000	

[a] Ratio of values of a parameter specified in equation 1 based on *M. tuberculosis* grown in medium A.

M. tu, *M. tuberculosis*; *E. co*, *Escherichia coli*; med A, medium A (see Table 11.2); med B, medium B (see Table 11.2).

the same fraction of total protein is directly involved in pre-rRNA synthesis (see Table 11.7). This agreement between the observed and calculated values of growth rates suggests that the parameters specified in eqn 1 are sufficient to account for the slow growth of *M. tuberculosis*.

A quantitative explanation of the different growth characteristics of *M. tuberculosis* and *E. coli* is suggested by a comparison of *M. tuberculosis* having a growth rate of 0.042 doublings per hour ($\mu = 0.042\,h^{-1}$ and the generation time, τ, is 24 h)) with *E. coli* with a growth rate of 1 doubling per hour ($\mu = 1\,h^{-1}$, and $\tau = 60\,min$). Both species have similar protein contents (≈ 150 fg per cell).

Parameters for the two species are compared in Table 11.7. It is evident that the difference in the growth rates of *E. coli* and *M. tuberculosis* is *not* attributable to a single factor but to a combination of three. *E. coli* is favoured by a 7-fold faster rate of rRNA elongation (ϵ_{rRNA}), a 7.5-fold faster rate of peptide chain elongation (ϵ_{pep}), and a 15.7-fold increase in the fraction of total protein that is engaged in pre-rRNA synthesis (α_p). Overall, these factors amount to an 800-fold increase in favour of *E. coli*.

The relative rates of RNA chain elongation and peptide chain elongation were discussed above (section 3.2.3) in the context of coupled transcription/translation. The molecular basis for the slower RNA chain elongation rate in *M. tuberculosis* is not known. It is uncertain whether the enzyme itself is responsible for the slower rate or whether the DNA is packed in such a way that the rate of transcription is diminished. It is recognized (see for example Robinow & Kellenberger 1994) that the bacterial nucleoid is very dynamic. Thus, when the cell is in an active metabolic state, the nucleoid undergoes variations in shape, chiefly in response to continuous activities of transcription. A factor for inversion stimulation (FIS), which has been identified in *E. coli* as a bacterial nucleoid-associated protein, is now believed to modulate chromosomal dynamics during bacterial growth (Schneider *et al.* 1997). The dynamics of changing the shape of the mycobacterial nucleoid and its influence on transcription are matters essential to the cell division cycle. Similarly, it is not known

whether the rate of peptide bond formation is slower in mycobacteria because of the intrinsic activity of the peptidyl transferase centre of the ribosome or because of the availability of tRNA, transcription factors, etc. It is interesting to note, however, that the components of some inducible responses, such as the heat-shock response, are capable of very rapid synthesis (P. Draper, unpublished work 1992), suggesting that there is no intrinsic limitation to the rate of peptide bond formation.

The 15.7-fold increase in the fraction of total protein engaged in pre-rRNA synthesis arises because of differences in the number of *rrn* operons per cell. *M. tuberculosis* has a single *rrn* operon which is replicated after a significant B period and after approximately two-thirds of the *C* period has elapsed. In contrast, *E. coli* has seven *rrn* operons per genome; all seven operons are located near to the origin of replication and DNA replication begins soon after cell division ($B \approx 0$). Hence, there are 14 operons functioning throughout most of the cells life.

Thus, we are able to draw several general conclusions as to why *M. tuberculosis* grows slowly: it has a slower rate of rRNA elongation, a slower rate of peptide chain elongation and a smaller fraction of total protein engaged in RNA synthesis. However, the molecular bases for these differences remain to be elucidated.

4 Bacterial dormancy

The environments to which bacteria are exposed in nature are quite different from those in which they are usually grown in the laboratory. In fact, exposure to optimal growth conditions (for example 37°C, physiological pH and an excess of nutrients) must be rare. In order to survive in unfavourable conditions, non-spore-forming bacteria must be able to persist for long periods of time without cell division occurring. Very often such cells are undetectable and unable to form colonies even when plated onto a suitable growth medium, but divide and become detectable when environmental conditions become favourable (for review see Kaprelyants *et al.* 1993).

It has become apparent, in recent years, that the transition of bacteria from a growing to a dormant state is an active process involving the synthesis of specific 'dormancy' proteins. Thus, in a process which we can think of as analogous to programmed cell death, bacteria can enter a state of programmed dormancy which, by definition, is reversible. In this section we will describe briefly what is known about programmed dormancy in eubacteria, and then discuss the relevance of these ideas to understanding dormancy in mycobacteria.

4.1 Bacterial adaptation to starvation

Bacterial dormancy has usually been studied by investigating the response to nutrient starvation. Non-sporulating bacteria respond to lack of nutrients by activating a variety of molecular survival mechanisms. These mechanisms contribute to survival by:

1 increasing the capacity to scavenge rare nutrients;

2 the use of alternative substrates so that dormancy is delayed;

3 the degradation of intracellular macromolecules as energy and monomer sources; and

4 the protection of a minimal complement of macromolecules so that a dormant phenotype can be maintained.

Although starvation is often used synonymously with stationary phase, this may be misleading. Stationary phase is caused by exhaustion of a single nutrient in the presence of adequate amounts of other nutrients; the nature of the limiting nutrient can have a profound effect on the physiological response (Nystrom *et al.* 1990; Kjelleberg *et al.* 1993).

When bacteria are faced with exhaustion of a particular nutrient, they increase their potential for scavenging that nutrient either by using additional sources of the nutrient or by acquiring a higher affinity for it. Thus, there are defined genetic responses for the enhanced assimilation of carbon (Matin 1992), iron (Bagg & Neilands 1987; see also Chapter 14), phosphate (Wanner 1987) and nitrogen (Kustu *et al.* 1996). Starving bacteria must generate a minimal level of energy to remain viable. Some bacteria have specific stores of polycarbon molecules for

this purpose; for example polyhydroxybutyric acid is used by a number of species as an alternative energy source during starvation (Sierra & Gibbons 1962; Dawes 1976). Degradation of RNA is also used for the generation of energy during nutrient starvation (Dawes 1976); however, in some species at least, while ribosomes are lost, they continue to exist in large excess compared to the demand for protein synthesis during starvation (Flardh *et al.* 1992).

Starvation results in a programmed pattern of gene regulation aimed at promoting bacterial survival. During carbon starvation two general types of genes are induced, those which require cyclic adenosine monophosphate (cAMP) for their induction (called *cst* genes), and those which are cAMP independent (called *pex* genes). Regulation of Cst proteins is involved in enhancing the carbon-scavenging capacity of the bacterium (Matin *et al.* 1989). In addition to cAMP, other 'global' regulators of the starvation response are guanosine diphosphate (GDP) and guanosine triphosphate (GTP) derivatives which carry a pyrophosphate group on the 3′-hydroxyl position of the ribose to give ppGpp and pppGpp (or (p)ppGpp). During amino acid starvation (p)ppGpp accumulates. (p)ppGpp may exert its effect by inhibiting transcription of RNA by, for example, binding to RNA polymerase (for review see Condon *et al.* 1995).

In addition to the *cst* and *pex* genes, and the gene responsible for the accumulation of (p)ppGpp (*relA*), a number of other genes have been associated with the starvation response. For example, the gene *rpoS* (Hengge-Aronis 1993) encoding the putative σ factor σs (KatF) is important because KatF is thought to be involved in the induction of a number of starvation proteins, including some heat-shock and oxidation proteins (McCann *et al.* 1991). Another σ factor, σ^{32}, which is also involved in the heat-shock response, increases during starvation along with the heat-shock proteins DnaK, GroEL and HtpG (Jenkins *et al.* 1991). This involvement of heat-shock proteins is interesting because such proteins are involved in protecting other essential proteins during environmental stress. Thus, their involvement in the starvation response suggests that one of the major roles of protein synthesis during entry into starvation is to protect a minimal requirement of proteins necessary for survival.

4.2 Models of mycobacterial dormancy

Although dormancy is thought to be a crucial component of mycobacterial virulence, it has proved extremely difficult to study. The reason for this is obvious: by definition, dormant bacteria are not dividing and hence cannot be grown in the laboratory. It is only by reversing the dormant state that it is possible to detect the existence of previously dormant bacteria. Therefore, in clinical terms, we are only aware of dormant bacteria in the patients lung when the bacteria leave the dormant state and cause relapse. A similar situation exists with the major animal model of dormant tuberculosis, the so-called 'Cornell model' (McCune *et al.* 1956, 1966). In this model *M. tuberculosis*-infected mice are treated with the tuberculocidal drugs isoniazid and pyrazinamide (NB, in more modern variations, rifampin is often used). After 3 months of treatment it is no longer possible to culture *M. tuberculosis* from the organs of these mice. However, if, after this treatment the mice are simply left for several months, a significant percentage of them will relapse with *M. tuberculosis* infection indicating that even though the tissues appeared to be devoid of viable *M. tuberculosis*, dormant bacteria must have remained. Interestingly, Mitchison and colleagues (de Wit *et al.* 1995) have used modern polymerase chain reaction (PCR)-based methods to detect mycobacterial DNA in mice following drug treatment in the Cornell model. They found that even when no bacteria can be cultured from mouse tissue, DNA equivalent to $\approx 5 \times 10^5$ bacilli can be detected in lung and spleen; however, it is not clear whether this DNA represents dead bacilli, free extracellular DNA, or dormant *M. tuberculosis*. Thus, it appears that in mice, effective chemotherapy achieves a latent infection rather than total sterilization (Grosset 1978).

Although the Cornell model provides evidence of a dormant state in mice, it has not as yet been exploited for understanding the physiological status of dormant bacilli. An *in vitro* system has recently been developed in which it is possible to investigate the molecular and physiological basis of a shift down from rapid to severely restricted or completely repressed growth of *M. tuberculosis* (Wayne 1994; Wayne & Hayes 1996). In this system, *M. tuberculosis* is grown in sealed containers in deep liquid culture, creating a temporal oxygen gradient. It is possible to remove non-replicating bacteria from such cultures and carry out molecular and physiological analysis of these bacteria. Although these studies are relatively recent, a number of interesting findings have started to emerge. First, following a gradual shift down to anaerobiosis *M. tuberculosis* are no longer killed by exposure to rifampicin or isoniazid; however, metronidazole, a drug which is specific for anaerobes and which has no effect on aerobically growing *M. tuberculosis* is bactericidal for such organisms (Wayne & Sramek 1994). This is particularly interesting given the finding from genome sequencing of *M. tuberculosis* that the organism contains a number of genes which are usually associated with anaerobic bacteria (Cole *et al.* 1998; see also Chapter 5). It therefore appears that although *M. tuberculosis* is usually considered to be an obligate aerobe, it may have acquired the ability to enter into an anaerobic survival form when oxygen becomes depleted. These studies are important because the gradual adaptation of *M. tuberculosis* to conditions which are unfavourable for growth by, for example, the gradual depletion of oxygen is probably analogous to the situation which occurs in necrotic tissue where foci of infection become sealed off from normal nutrient supplies. Thus, such models may help us to understand the molecular mechanisms of survival which *M. tuberculosis* can employ.

Other studies of *M. tuberculosis* in stationary phase have revealed evidence of a programmed dormancy response. For example, Yuan *et al.* (1996) found that at least seven proteins were increased in synthesis following entry into stationary phase. One of these, which becomes the predominant stationary-phase protein was identified as α-crystallin-like. This is a small heat-shock protein which probably functions by protecting other essential proteins from degradation. Similarly DeMaio *et al.* (1996) have identified a σ factor encoding gene, *sigF*. The *sigF* gene appears not to be expressed in exponentially growing mycobacteria but is strongly induced in stationary phase and in cultures undergoing nutrient starvation and environmental stress. Preliminary evidence indicates that SigF is present in slow-growing but not rapidly growing mycobacteria, suggesting that it may contribute to survival and persistence of pathogenic mycobacteria.

5 Conclusion

Pathogenic mycobacteria such as *M. tuberculosis* and *M. leprae* are able to survive and grow within host cells such as macrophages (Chapter 19). The macrophage not only provides the intracellular mycobacteria with nutrients, but it potentially imposes considerable environmental stress; thus, the ability to withstand or evade such stress is essential for mycobacteria to survive *in vivo*. The doubling time for *M. tuberculosis* growing inside macrophages is ≈ 24 h, which is similar to the time often found for balanced growth *in vitro* (see Table 11.7). The data presented in Table 11.7 reveal that the mathematical models developed to relate the synthesis of DNA, RNA and proteins during fast and ultra-fast growth also apply to slow-growing mycobacteria. Therefore current ideas about the bacterial cell division cycle should also apply to the growth of *M. tuberculosis*. We infer that *M. tuberculosis* has not developed mechanisms that promote fast or ultra-fast growth.

In addition to growing slowly within macrophages, *M. tuberculosis* also has the ability to enter into prolonged states of dormancy. It is clear from other genera, and from preliminary studies on *M. tuberculosis* itself, that this is a precisely regulated and programmed procedure which maximizes the ability of the organism to survive when conditions are unfavourable to growth.

It seems likely that, over the next few years, we will learn much more about the programmed dormancy response of *M. tuberculosis*; hopefully this knowledge will ultimately lead towards novel agents capable of specifically targeting persisting, dormant tubercle bacilli in the tissues of long-term infected patients.

6 Acknowledgements

We thank our colleague Dr I. D. J. Burdett for his advice and helpful discussion, and Dr G. A. Ellard for permitting us to include his unpublished data on RNA and DNA levels in mycobacteria. This work is supported as part of the European Commission Science Research and Development Programme (contract number ERBIC 18CT 9720253).

7 References

Arnstein, H.R.V. & Cox, R.A. (1992) *Protein Biosynthesis: in Focus*. Oxford: Oxford University Press.

Baess, I. (1984) Determination and re-examination of genome sizes and base ratios on deoxyribonucleic acid from mycobacteria. *Acta Pathologica et Microbiologica et Immunologica Scandinavica, Section B* **92**, 209–211.

Bagg, A. & Neilands, J. (1987) Molecular mechanism of regulation of siderophore-mediated iron assimilation. *Microbiological Reviews* **51**, 509–518.

Bercovier, H., Kafri, O. & Sela, S. (1986) Mycobacteria possess a surprisingly small number of ribosomal RNA genes in relation to the size of their genome. *Biochemical and Biophysical Research Communications* **136**, 1136–1141.

Bi, E. & Lutkenhaus, J. (1990) Analysis of *ftsZ* mutations that confer resistance to the cell division inhibitor SulA (sfiA). *Journal of Bacteriology* **172**, 5602–5609.

Bottger, E.C., Teske, A., Kirschner, P. *et al.* (1992) Disseminated '*Mycobacterium genavense*' infection in patients with AIDS. *Lancet* **340**, 76–80.

Bremer, H. (1975) Parameters affecting the rate of synthesis of ribosomes and RNA polymerase in bacteria. *Journal of Theoretical Biology* **53**, 115–124.

Bremer, H. & Churchward, G. (1977) An examination of the Cooper Helmstetter theory of DNA replication and its underlying assumptions. *Journal of Theoretical Biology* **69**, 645–654.

Bremer, H. & Dennis, P.P. (1987) Modulation of chemical composition and other parameters of the cell growth rate. In: *Escherichia coli and Salmonella typhimurium: Cellular and Molecular Biology* (eds F. C. Neidhardt, J. L. Ingraham, K. B. Low, B. Magasanik, M. Schaechter & H. E. Umbarger). Washington, DC: American Society for Microbiology, pp. 1527–1542.

Brennan, P.J. & Nikaido, H. (1995) The envelope of mycobacteria. *Annual Review of Biochemistry* **64**, 29–63.

Cole, S.T., Brosch, R., Parkhill, J. *et al.* (1998) Deciphering the biology of *Mycobacterium tuberculosis* from the complete genome sequence. *Nature* **393**, 537–544.

Condon, C., French, S., Squires, C. & Squires, C.L. (1993) Depletion of functional ribosomal RNA operons in *Escherichia coli* causes increased expression of the remaining intact copies. *EMBO Journal* **12**, 4305–4315.

Condon, C., Squires, C. & Squires, C.L. (1995) Control of rRNA transcription in *Escherichia coli*. *Microbiological Reviews* **59**, 623–624.

Cooper, S. (1991a) *Bacterial Growth and Division: Biochemistry and Regulation of Prokaryotic and Eukaryotic Division Cycles*. London: Academic Press.

Cooper, S. (1991b) Synthesis of the cell surface during the division cycle of rod-shaped, gram-negative bacteria. *Microbiological Reviews* **55**, 649–674.

Cooper, S. & Helmstetter, C.E. (1968) Chromosome replication and the division cycle of *Escherichia coli* B/r. *Journal of Molecular Biology* **31**, 519–540.

Coyle, M.B., Carlson, L.C., Wallis, C.K. *et al.* (1992) Laboratory aspects of '*Mycobacterium genavense*', a proposed species isolated from AIDS patients. *Journal of Clinical Microbiology* **30**, 3206–3212.

Daffé, M. & Draper, P. (1998) The envelope layers of mycobacteria with reference to their pathogenicity. *Advances in Microbial Physiology* **39**, 131–203.

Dai, K., Mukherjee, A., Xu, Y. & Lutkenhaus, J. (1994) Mutations in *ftsZ* that confer resistance to SulA affect the interaction of *ftsZ* with GTP. *Journal of Bacteriology* **175**, 130–136.

Davidson, L.A., Draper, P. & Minnikin, D.E. (1982) Studies on the mycolic acids from the walls of *Mycobacterium microti*. *Journal of General Microbiology* **128**, 823–828.

Dawes, E.A. (1976) Endogenous metabolism and the survival of starved prokaryotes. In: *The Survival of Vegetative Microbes*, 26th edn. (eds T. R. G. Gray & J. R. Postgate). Cambridge: Cambridge University Press, pp. 19–53.

DeMaio, J., Zhang, Y., Young, D.B. & Bishai, W.R. (1996) A stationary-phase stress-response sigma factor from *Mycobacterium tuberculosis*. *Proceedings of the National Academy of Sciences of the USA* **93**, 2790–2794.

Domenech, P., Menendez, M.C. & Garcia, M.J. (1994) Restriction fragment length polymorphism of 16S rRNA genes in the differentiation of fast-growing mycobacterial species. *FEMS Microbiology Letters* **116**, 19–21.

Doukhan, L., Predich, M., Nair, G. *et al.* (1995) Genomic organisation of the mycobacterial sigma gene cluster. *Gene* **165**, 67–70.

Ellard, G.A. (1991) The use of RNA/DNA ratio measurements to assess rifampicin-induced growth inhibition of *Escherichia coli*. *Journal of Antimicrobial Chemotherapy* **28**, 347–355.

Ellwood, M. & Nomura, M. (1982) Chromosomal locations of the genes for rRNA in *Escherichia coli* K-12. *Journal of Bacteriology* **149**, 458–468.

Erickson, H.P. (1997) *FtsZ*, a tubulin homologue in prokaryote cell division. *Trends in Cell Biology* **7**, 362–367.

Estrada-G, I.C.E., Lamb, F.I., Colston, M.J. & Cox, R.A. (1988) Partial nucleotide sequence of 16S ribosomal RNA isolated from armadillo-grown *Mycobacterium leprae*. *Journal of General Microbiology* **134**, 1449–1453.

Flardh, K., Cohen, P. & Kjelleberg, S. (1992) Ribosomes exist in large excess over the apparent demand for protein synthesis during starvation in marine *Vibrio* sp. strain CCUG 15956. *Journal of Bacteriology* **174**, 6780–6788.

Fsihi, H., De Riossi, E., Salazar, L. *et al.* (1996) Gene arrangement and organisation in a ~76 kb fragment encompassing the *oriC* region of the chromosome of *Mycobacterium leprae*. *Microbiology* **142**, 3147–3161.

Gonzalez-y-Merchand, J.A., Colston, M.J. & Cox, R.A. (1996) The rRNA operons of *Mycobacterium smegmatis* and *Mycobacterium tuberculosis*: comparison of promoter elements and of neighbouring upstream genes. *Microbiology* **142**, 667–672.

Gonzalez-y-Merchand, J.A., Colston, M.J. & Cox, R.A. (1998) Roles of multiple promoters in transcription of ribosomal DNA: effects of growth conditions on precursor rRNA synthesis in mycobacteria. *Journal of Bacteriology* **180**, 5756–5761.

Gonzalez-y-Merchand, J.A., Garcia, M.J., Gonzalez-Rico, S., Colston, M.J. & Cox, R.A. (1997) Strategies used by pathogenic and non-pathogenic mycobacteria to synthesize rRNA. *Journal of Bacteriology* **179**, 6949–6958.

Grosset, J. (1978) The sterilizing value of rifampicin and pyrazinamide in experimental short-course chemotherapy. *Bulletin of the International Union of Tuberculosis* **53**, 5–12.

Harshey, R.M. & Ramakrishnan, T. (1977) Rate of ribonucleic acid chain growth in *Mycobacterium tuberculosis* H37Rv. *Journal of Bacteriology* **129**, 616–212.

Helmstetter, C.E. (1987) Timing of synthetic activities in the cell cycle. In: *Escherichia coli and Salmonella typhimurium: Cellular and Molecular Biology* (eds F. C. Neidhart, J. L. Ingraham, K. B. Low, B. Magasanik, M. Schaechter & H. E. Umbarger). Washington, DC: American Society for Microbiology, pp. 1594–1605.

Helmstetter, C.E. & Cooper, S. (1968) DNA synthesis during the division cycle of rapidly growing *E.coli* B/r. *Journal of Molecular Biology* **31**, 507–518.

Hengge-Aronis, R. (1993) Survival or hunger and stress: the role of *rpoS* in stationary phase gene regulation in *Escherichia coli*. *Cell* **72**, 165–168.

Hiriyanna, K.T. & Ramakrishnan, T. (1986) Deoxyribonucleic acid replication time in *Mycobacterium tuberculosis* H37 Rv. *Archives of Microbiology* **144**, 105–109.

Honore, N.T., Bergh, S., Chanteau, S. *et al.* (1993)

Nucleotide sequence of the first cosmid from the *Mycobacteirum leprae* genome project: structure and function of the *Rif-Str* region. *Molecular Microbiology* **7**, 207–214.

Huang, J., Cao, C. & Lutkenhaus, J. (1996) Interaction between FtsZ and inhibitors of cell division. *Journal of Bacteriology* **178**, 5080–5085.

Huisman, O. & D'Ari, R. (1981) An inducible DNA replication cell division coupling mechanism in *E. coli*. *Nature* **290**, 797–799.

Jarlier, V. & Nikaido, H. (1994) Mycobacterial cell wall: structure and role in natural resistance to antibiotics. *FEMS Microbiological Letters* **123**, 11–18.

Jenkins, D., Auger, E. & Matin, A. (1991) Role of RpoH, a heat shock regulator protein in *Escherichia coli* carbon starvation protein synthesis and survival. *Journal of Bacteriology* **173**, 1992–1996.

Ji, Y.-E., Colston, M.J. & Cox, R.A. (1994a) Nucleotide sequence and secondary structures of precursor 16S rRNA of slow-growing mycobacteria. *Microbiology* **140**, 123–132.

Ji, Y.-E., Kempsell, K., Colston, M.J. & Cox, R.A. (1994b) Nucleotide sequences of the spacer-1, spacer-2 and trailer regions of the *rrn* operons and secondary structure of precursor 23S rRNAs and precursor 5S rRNAs of slow-growing mycobacteria. *Microbiology* **140**, 1763–1773.

Kaprelyants, A.S., Gottschal, J.C. & Kell, D.B. (1993) Dormancy in non-sporulating bacteria. *FEMS Microbiology Reviews* **104**, 271–286.

Kempsell, K.E., Ji, Y.-E., Estrada-G., Colston, M.J. & Cox, R.A. (1992) The nucleotide sequence of the promoter, 16S rRNA and spacer region of the ribosomal RNA operon of *Mycobacterium tuberculosis* and comparison with *Mycobacterium leprae* precursor rRNA. *Journal of General Microbiology* **138**, 1717–1727.

Kjelleberg, S., Albertson, N., Flardh, K. *et al.* (1993) How do non-differentiating bacteria adapt to starvation? *Antonie Van Leeuwenhoek* **63**, 333–341.

Klann, K.G., Belanger, A.E., Abanes-de Mello, A., Lee, J.Y. & Hatfull, G.F. (1998) Characterization of the *dnaG* locus in *Mycobacterium smegmatis* reveals linkage of DNA replication and cell division. *Journal of Bacteriology* **180**, 65–72.

Koch, A.L. (1979) Microbial growth in low concentrations of nutrients. In: *Strategies of Microbial Life in Extreme Environments* (ed. M. Shilo). Weinheim: Verlag Chemie, pp. 261–269.

Koch, A.L. (1987) The variability and individuality of the bacterium. In: *Escherichia coli and Salmonella typhimurium: Cellular and Molecular Biology* (eds F. C. Neidhart, J. L. Ingraham, K. B. Low, B. Magasanik, M. Schaechter & H. E. Umbarger). Washington, DC, American Society of Microbiology, pp. 1606–1614.

Koch, A.L. (1997) Microbial physiology and ecology of slow growth. *Microbiology and Molecular Biology Reviews* **61**, 305–318.

Kustu, S., Sei, K. & Keener, J. (1996) Nitrogen regulation in enteric bacteriology. In: *Regulation of Gene Expression — 25 Years On*. Cambridge: Cambridge University Press, pp. 139–154.

Levy, L. (1976) Studies of the mouse footpad technique for the cultivation of *Mycobacterium leprae* 3. Doubling time during logarithmic multiplication. *Leprosy Review* **47**, 103–106.

Lutkenhaus, J.F. (1983) Coupling of DNA replication and cell division *sulB* is an allele of *Ftsz*. *Journal of Bacteriology* **154**, 1339–1346.

Lutkenhaus, J. & Addinall, S.G. (1997) Bacterial cell division and the Z ring. *Annual Review of Biochemistry* **66**, 93–116.

Maaløe, O. & Kjeldgaard, N.O. (1966) *Control of Macromolecular Synthesis: A Study of DNA, RNA and Protein Synthesis in Bacteria*. New York: W. A. Benjamin.

Mangan, J.A., Sole, K.M., Mitchison, D.A. & Butcher, P.D. (1997) An effective method of RNA extraction from bacteria refractory to disruption, including mycobacteria. *Nucleic Acids Research* **25**, 675–676.

Matin, A. (1992) Physiology, molecular biology and applications of the bacterial starvation response. *Society for Applied Bacteriology Symposium Series* **21**, 49S–57S.

Matin, A., Auger, E., Blum, P. & Schultz, J. (1989) Genetic basis of starvation survival in non-differentiating bacteria. *Annual Review of Microbiology* **43**, 293–316.

McCann, M., Kidwell, J. & Matin, A. (1991) The putative sigma factor KatF has a central role in the development of starvation-mediated general resistance in *Escherichia coli*. *Journal of Bacteriology* **173**, 4188–4194.

McCune, R.M., Feldman, F.M., Lambert, H.P. & McDermott, W. (1966) Microbial persistence. I. The capacity of tubercle bacilli to survive sterilization in mouse tissues. *Journal of Experimental Medicine* **123**, 445–468.

McCune, R.M., Tompsett, R. & McDermott, W. (1956) The fate of *Mycobacterium tuberculosis* in mouse tissues as determined by microbial enumeration techniques. *Journal of Experimental Medicine* **104**, 763–803.

Minnikin, D.E. (1982) Lipids: complex lipids, their chemistry, biosynthesis and roles. In: *The Biology of the Mycobacteria*, Vol. 1. *Physiology, Identification and Classification* (eds C. Ratledge & J. Stanford). London: Academic Press, pp. 95–184.

Nystrom, T., Flardh, K. & Kjelleberg, S. (1990) Response of multiple-nutrient starvation in marine *vibrio* sp. strain CCUG 15956. *Journal of Bacteriology* **172**, 7085–7098.

Philipp, W.J., Poulet, S., Eiglmeier, K. *et al.* (1996) An integrated map of the genome of the tubercle bacillus, *Mycobacterium tuberculosis* H37Rv, and comparison with *Mycobacterium leprae*. *Proceedings of the National Academy of Sciences of the USA* **93**, 3132–3137.

Predich, M., Doukhan, L., Nair, G. & Smith, I. (1995) Characterisation of RNA polymerase and two sigma-factor genes from *Mycobacterium smegmatis*. *Molecular Microbiology* **15**, 355–366.

Qin, M.-H., Madiraju, M.V.V.S., Zachariah, S. & Rajagopalan, M. (1997) Characterisation of the *oriC* region of *Mycobacterium smegmatis*. *Journal of Bacteriology* **179**, 6311–6317.

Robinow, R. & Kellenberger, E. (1994) The bacterial nucleoid revisted. *Microbiological Reviews* **58**, 211–232.

Salazar, L., Fsihi, H., de Rossi, E. *et al.* (1996) Organisation of the origins of replication of the chromosomes of *Mycobacterium smegmatis*, *Mycobacterium leprae* and *Mycobacterium tuberculosis* and isolation of a functional origin from *M. smegmatis*. *Molecular Microbiology* **20**, 283–293.

Schaechter, M., Maaløe, O. & Kjeldgaard, N.O. (1958) Dependency on medium and temperature of cell size and chemical composition during balanced growth of *Salmonella typhimurium*. *Journal of General Microbiology* **19**, 592–606.

Schaechter, M., Williamson, J.P., Hood, J.R. & Jun, Koch, A.L. (1962) Growth, cell and nuclear divisions in some bacteria. *Journal of General Microbiology* **29**, 421–434.

Schneider, R., Travers, A. & Muskhelishvili, G. (1997) FIS modulates growth phase-dependent topological transitions of DNA in *Escherichia coli*. *Molecular Microbiology* **26**, 519–530.

Senaratne, R., Mobasheri, H., Papavinasasundaram, K.G. *et al.* (1998) Expression of a gene for a porin-like protein of the OmpA family from *Mycobacterium tuberculosis* H37Rv. *Journal of Bacteriology* **180**, 3541–3547.

Sierra, G. & Gibbons, N. (1962) Role and oxidation pathway of poly-β-hydroxybutyric acid in *Micrococcus halodenitrificans*. *Canadian Journal of Microbiology* **8**, 255–269.

Smith, P.G. & Moss, A.R. (1994) Epidemiology of tuberculosis. In: *Tuberculosis; Pathogenesis, Protection and Control* (ed. B. R. Bloom). Washington DC: American Society for Microbiology, pp. 47–59.

Suzuki, Y., Nagata, A., Ono, Y. & Yamada, I. (1988) Complete nucleotide sequence of the 16S rRNA gene of *Mycobacterium bovis* BCG. *Journal of Bacteriology* **170**, 2886–2889.

Trias, J. & Benz, R. (1994) Permeability of the cell wall of *Mycobacterium smegmatis*. *Molecular Microbiology* **14**, 283–290.

Wanner, B. (1987) Phosphate regulation of gene expression in *Escherichia coli*. In: *Escherichia coli and Salmonella typhimurium: Cellular and Molecular Biology* (eds F. C. Neidhart, J. L. Ingraham, K. B. Low, B. Magasanik, M. Schaechter & H. E. Umbarger). Washington, DC, American Society of Microbiology, 1326–1333.

Waters, M.F.R., Rees, R.J.W., McDougall, A.C. & Weddell, A.G.M. (1974) Ten years of dapsone in lepromatous leprosy: clinical, bacteriological and histological assessment and the finding of viable leprosy bacilli. *Leprosy Review* **45**, 288–298.

Wayne, L.G. (1994) Dormancy of *Mycobacterium tuberculosis* and latency of disease. *European Journal of Clinical Microbiology and Infectious Diseases* **13**, 908–914.

Wayne, L.G. & Hayes, L.G. (1996) An *in vitro* model for sequential study of shiftdown of *Mycobacterium tuberculosis* through two stages of nonreplicating persistence. *Infection and Immunity* **64**, 2062–2069.

Wayne, L.G. & Kubica, G.P. (1986) The mycobacteria. In: *Bergey's Manual of Systematic Bacteriology*, Vol 2 (eds P. H. A. Sneath, N. S. Mair, M. E. Sharpe, J. G. Holt). London: Williams and Wilkins, pp. 1435–1457.

Wayne, L.G. & Sramek, H.A. (1994) Metronidazole is bactericidal to dormant cells of *Mycobacterium tuberculosis*. *Antimicrobial Agents and Chemotherapy* **38**, 2054–2058.

Winder, F.G. & Rooney, S.A. (1970) Effects of nitrogenous components of the medium on the carbohydrate and nucleic acid content of *Mycobacterium tuberculosis* BCG. *Journal of General Microbiology* **63**, 29–39.

de Wit, D., Wootton, M. & Mitchison, D.A. (1995) The bacterial DNA content of mouse organs in the Cornell model of dormant tuberculosis. *Tubercle and Lung Disease* **76**, 555–562.

Yuan, Y., Crane, D.D. & Barry,C.E., III (1996) Stationary phase-associated protein expression in *Mycobacterium tuberculosis*: function of the mycobacterial α-cystallin homolog. *Journal of Bacteriology* **178**, 4484–4492.

Chapter 12 / Cell wall: physical structure and permeability

JUN LIU, CLIFTON E. BARRY, III & HIROSHI NIKAIDO

1 Introduction

A broad resistance to various antibiotics and chemotherapeutic agents is a common feature of mycobacteria (Jarlier & Nikaido 1994). The molecular composition and structural features of the mycobacterial cell envelope are thought to confer low permeability and thereby contribute to drug resistance. The cell wall of mycobacteria also plays a significant role in pathogenicity, appears to be responsible for properties such as acid fastness (Barksdale & Kim 1977), and is involved in the immunological reactions of the host to mycobacteria (see Chapter 13). In spite of decades of work elucidating the chemical structures of various cell-wall components, until recently little was known about how these molecules are actually arranged. Yet it is the physical organization that controls the influx of solutes, such as nutrients and drugs, and also affects accessibility of the immune factors. Over the last few years, considerable progress has been made in determining the physical organization of mycobacterial cell wall, and now we have a better understanding of structure–function relationships in constituents of the mycobacterial cell envelope. This chapter will focus on these advances, particularly on our current understanding of the ability of the mycobacterial cell wall to function as an effective permeability barrier.

2 Architecture of the mycobacterial cell envelope

The chemistry of mycobacterial cell-wall components is the subject of Chapter 13 and will only be briefly summarized here. The cell-wall skeleton is composed of three covalently linked substructures: peptidoglycan, arabinogalactan (AG) and mycolic acids. The mycobacterial peptidoglycan belongs to one of the most common types found in bacteria with two exceptional features. The muramic acid is *N*-glycolylated instead of the more typical *N*-acetylation, and the crosslinks include bonds between two residues of diaminopimelic acid as well as between diaminopimelic acid and D-alanine. The peptidoglycan is linked to AG via a phosphodiester bridge. The non-reducing termini of the AG polysaccharide consist of branched penta-arabinose units, about two-thirds of which are esterified each with four mycolic acid residues. The most distinctive feature of the cell wall is that up to 60% of its weight is occupied by lipids including mycolic acids. In addition to lipids of the covalently linked skeleton, several types of 'extractable lipids', including trehalose-containing glycolipids, 'phenolic glycolipids' (PGLs) containing phenolphthiocerol, phthiocerol esters such as phthiocerol dimycocerosate, and glycopeptidolipids (GPLs) may be present.

2.1 Features of ultrastructure

Most electron-microscopic techniques available today have been applied to mycobacterial cells (reviewed by Brennan & Draper 1994). Early studies using transmission electron microscopy of thin sections revealed, outside the plasma membrane, a cell wall of a tripartite structure, consisting of an inner layer stained moderately densely, a wide electron-translucent middle layer, and an outer electron-dense layer (Imaeda *et al.* 1968; Barksdale & Kim 1977). Later studies by Rastogi *et al.* (1986) confirmed the triple layer structure and showed that the outer layer was stained strongly by Ruthenium Red, suggesting, in the authors' interpretation, that the outer layer was composed of a polysaccharide. Traditional fixation and embedding procedures, however, are known to extract lipids but a recent study using freeze-substitution, a protocol that minimizes this extraction, also produced images in accord with earlier ones (Paul & Beveridge 1992).

There has been a natural tendency to equate each ultrastructual 'layer' with a physical continuum in the lateral direction, i.e. a layer composed of a single unique material, an assumption that may not be always valid. The innermost, moderately electron-dense layer probably contains the peptidoglycan. Its appearance is consistent with the known staining properties of this molecule, which contains carboxyl groups (of diaminopimelic acid and D-alanine) that bind metal ions (Beveridge & Murray 1980). The wide, electron-transparent, middle layer appears to be the hydrophobic domain of the cell wall, and has been thought to correspond to the mycolyl-AG by many workers. This layer has a thickness of 9–10 nm and is much thicker than the 4–4.5-nm-deep,

electron-transparent layer present in the cytoplasmic membrane. Its transparency to electrons is usually explained by the extremely hydrophobic nature of mycolyl-AG, which presumably excludes the electron-dense heavy metal salts such as uranyl acetate. The outermost, relatively electron-dense layer varies in thickness (from negligible to massive), electron density and appearance (fibrillar, granular or homogeneous), depending on species, growth conditions, and preparation methods for microscopy. This putative polysaccharide-containing layer may contain negatively charged groups as it is stained intensely with Ruthenium Red. However, it is impossible to tell, from electron microscopy, whether an independent polysaccharide 'layer' exists outside the cell wall proper, or the anionic constituents are an integral component of the cell-wall complex (see below).

Barksdale and Kim (1977) also examined the mycobacterial surface structure by negative staining, freeze-etching and freeze-fracture. It is not easy to correlate the various 'layers' observed in these studies to the tripartite structure seen in transmission micrographs. It seems, however, that the material of varying depth on the surface, L_1, which was thought to be made up of GPL (Barksdale & Kim 1977), lies in the outermost area and represents an outward extension of the outer electron-dense layer. The cell wall proper, corresponding to the classical tripartite structure, is bordered by hydrophilic surfaces on both sides, and the L_2 and L_3 'layers' of Barksdale and Kim appear to correspond to these surfaces.

Recent analysis of material released from mycobacterial cell surface by shaking with glass beads, carried out by Daffé and associates (Ortalo-Magné *et al.* 1995; Lemassu *et al.* 1996), showed that a significant amount (corresponding to 2–3% of the dry weight of *Mycobacterium tuberculosis* cells) of protein–polysaccharide mixture was released. Polysaccharide comprised 30–60% of this material, and its major constituent was an $\alpha(1\rightarrow4)$ glucan. Since the presence of such a 'glycogen-like' polysaccharide is expected to produce a highly hydrophilic cell surface, this observation was surprising in view of the well-known hydrophobicity of mycobacterial cell surface. However, this particular glucan is poorly soluble in water, and collects at the interphase when chloroform/methanol/water two-phase partitioning procedure is applied. It is not known what chemical or physical features of this glucan produce this unexpected behaviour. It is also unclear whether these surface polysaccharides contribute to the production of any of the ultrastructural layers observed, especially the 'electron-transparent zones' seen sometimes around the mycobacterial cells.

M. tuberculosis apparently secretes, into the medium, a number of typically cytosolic enzymes such as superoxide dismutase and glutamine synthetase, and shaking with glass beads in the presence of Tween 80 releases several enzymes from the surface layers of nonpathogens such as *M. smegmatis* (Raynaud *et al.* 1998). Although the mechanism of secretion or export to surface is not known, other Gram-positive bacteria are known to export typically cytosolic enzymes, such as glycolytic enzymes, in large amounts to cell surface (Panchioli & Fischetti 1992).

2.2 Physical organization of lipids in the mycobacterial cell wall

In 1982, Minnikin proposed a model for the structure of mycobacterial cell wall in which mycolic acid hydrocarbon chains were packed side by side in a direction perpendicular to the plane of the cell surface. It was also proposed that this mycolic acid-containing inner leaflet was covered by an outer leaflet composed of extractable lipids. Thus, the complete structure formed an asymmetric bilayer (Minnikin 1982) (Fig. 12.1).

When this model was proposed, it was mainly based on chemical principles deduced from studies on conventional lipids and no direct evidence existed. A major argument against this model was that such an orientation would require that AG be located directly exterior to peptidoglycan, either at the border of the electron-transparent and electron-dense layers or as part of the electron-dense layer. It was suggested that such an arrangement was unlikely considering the size ($M_r \approx 30\,000$) and branched nature of AG. Thus, many workers thought that the electron-

transparent, middle layer consisted of AG in which mycolate hydrocarbon chains were enmeshed in more or less random orientations (Draper 1982). Alternatively, it was proposed that AG protrudes into the electron transparent layer from the peptidoglycan in a direction perpendicular to the cell surface (McNeil & Brennan 1991). The model further proposed that the AG acts as a scaffold from which mycolic acids are extended in a direction parallel to the cell surface.

A key question in these models was the packing and orientation of the lipids, particularly mycolic acids, in the cell wall, a question that was not resolved experimentally. An X-ray diffraction study of the cell wall purified from *M. chelonae* showed that most of the hydrocarbon chains in the cell wall were tightly packed in a parallel, quasi-crystalline array and in a direction perpendicular to the cell surface (Nikaido *et al.* 1993)(Fig. 12.2). This study provided direct experimental evidence in favour of the model proposed by Minnikin.

The bilayer model appears to fit with many observations. First, freeze-etching and freeze-fracture electron microscopy of mycobacteria showed two distinct cleavage planes in the cell envelope (Barksdale & Kim 1977); the inner one corresponds to the plasma membrane, the outer plane is presumably within the cell wall, an observation consistent with a bilayer-type cell wall. Second, it is known that different classes of specific antigenic glycolipids are located on the cell surface, and that mycolyl-AG is located outside of the peptidoglycan layer. Third, the transparent layer of the cell wall under electron microscopy is about twice as thick as that of the plasma membrane, as mentioned above, in agreement with the predicted length of hydrocarbon chains across the cell wall (50–60 carbon atoms from the meromycolate chains in the inner leaflet, and 14–19 carbon atoms from short-chain fatty acids in the outer leaflet, see Fig. 12.1) compared with that in plasma membrane (32–36 carbon atoms). Fourth, treatment of the atypical *M. avium* with sublethal concentrations of isoniazid sufficient to inhibit the synthesis of mycolic acids results in the loss of this transparent layer and the appearance of a more

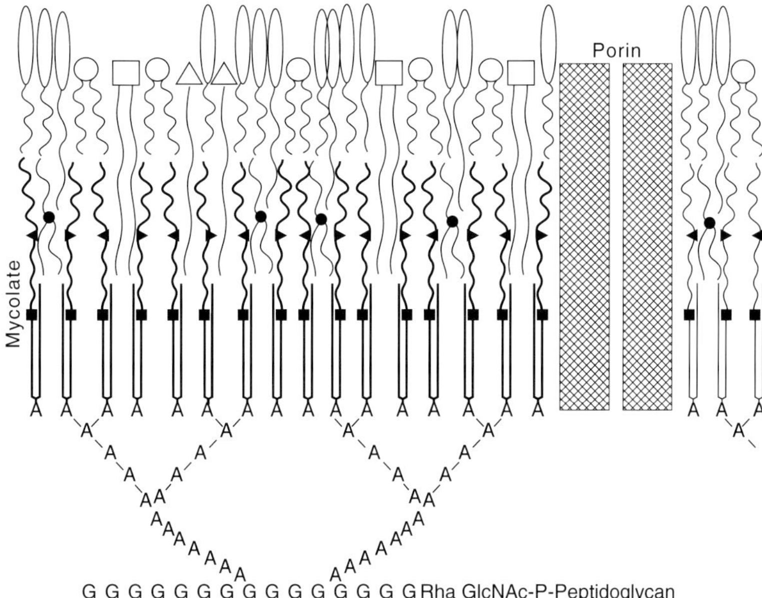

Fig. 12.1 Modified Minnikin model of the mycobacterial cell wall. At the bottom, the arabinogalactan—peptidoglycan complex is shown, with the sugars represented as follows: A, L-arabinose; G, D-galactose; Rha, L-rhamnose; and GlcNAc, *N*-acetyl-D-glucosamine. In the bilayer structure, the mycolate residues (shown in thicker lines), covalently linked to arabinose residues, are thought to produce the inner leaflet. The proximal position (solid squares) of the longer, meromycolate branch may be either double bond or cyclopropane, each of which can be either *cis* or *trans*. The distal position (solid triangles) of the meromycolate branch can be a double bond, cyclopropane, or oxygen-containing structure. The innermost part of the structure is expected to be the least fluid, the fluidity increasing as one moves toward the surface. This fluidity gradient is shown by the increasing waviness of the hydrocarbon chains. Triacylglycerols, with three short fatty acid chains connected to the glycerol residue (solid circles), are assumed to fill the space created by the unequal length of the two branches of a mycolate residue. The outer leaflet is thought to be composed of different types of extractable lipids in different species, some containing short acyl chains, and others chains of intermediate length.

densely staining loosely organized cell periphery (Mdluli *et al.* 1998).

The bilayer model of the mycobacterial cell wall must fulfil several requirements. First, the model predicts that the amount of mycolic acid present in each cell must be sufficient to cover the cell surface area. Calculations based on the amount of mycolic acid present in a known amount of *M. bovis* bacille Calmette–Guérin (BCG) cells and the cross-sectional area of each mycolic acid chain indicate that sufficient mycolic acid is indeed present to cover the entire surface of a mycobacterial cell (Nikaido *et al.* 1993). Second, the monolayer arrangement of the mycolic acids will create a large hydrophobic surface and thus demands that the cell wall contain enough other lipids to form the outer leaflet. Since there are two parallel branches in a single mycolic acid molecule, the extractable lipids should contain approximately twice as many fatty acid residues as mycolic acids. In addition, the two branches of mycolic acid are not equal in length and thus some mechanisms are needed to accommodate the uneven surface of the inner leaflet. Analysis of a pure preparation of cell wall from *M. chelonae* showed that it contained a large amount of saponifiable lipids with C14 to C19 fatty acids (E. Y. Rosenberg and H. Nikaido, unpublished results); the molar ratio between short-chain fatty acids and mycolic acids was found to be $2.3 \pm 0.6 : 1$.

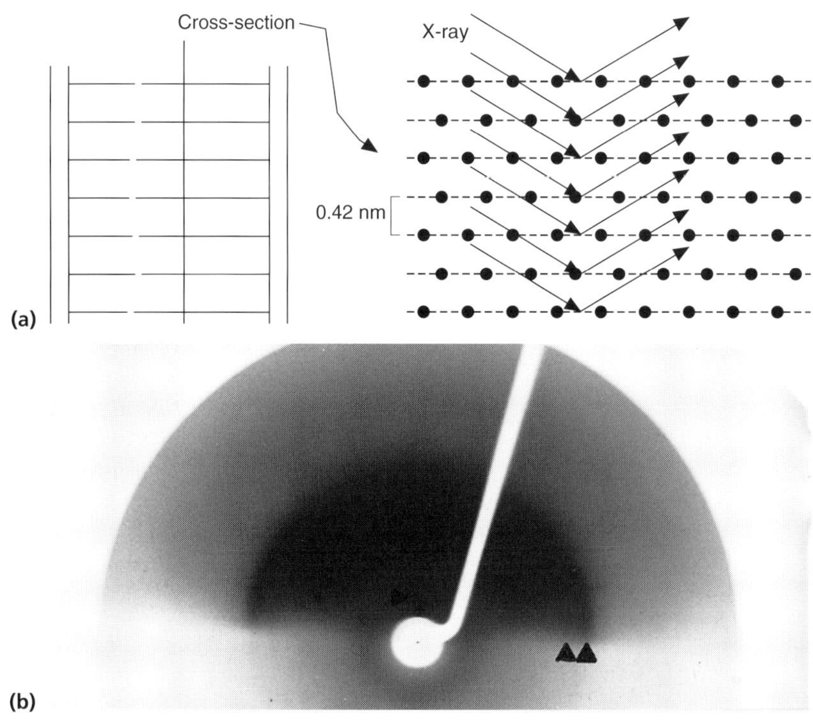

Fig. 12.2 X-Ray diffraction by partially orientated specimen of *M. chelonae* cell wall. (a) According to the Minnikin bilayer model (see Fig. 12.1), the hydrocarbon chains of cell-wall lipids should be orientated perpendicular to the plane of cell surface. Thus, when the cell walls are partially orientated by centrifugation onto a flat glass plate (in the horizontal direction in this figure), the hydrocarbon chains are expected to be aligned in the vertical direction. When an X-ray is directed in a direction perpendicular to the plane of the paper, the regular, narrow spacing of 0.42 nm between tightly packed, paracrystalline hydrocarbon chains should produce a wide-angle diffraction in the horizontal direction. Longer repeated distances reflecting the thickness of cell wall and internal 'layers' within the cell wall will produce narrow-angle diffraction in the vertical direction. (b) The results obtained with the partially orientated specimen of *Mycobacterium chelonae* cell wall were indeed as predicted. Note that the predominant diffraction in the horizontal direction is relatively sharp and corresponds to 0.42 nm, suggesting a crystalline structure. Just inside of these diffractions, a weaker, more diffuse, 0.45 nm diffraction presumably coming from the fluid hydrocarbons of the outer leaflet lipids is also seen. (Part (b) of this figure is from Nikaido *et al.* 1993, with permission.)

Interestingly, a large fraction (about 95%) of the short-chain fatty acids existed as components of triacylglycerols (triglycerides), a lipid species that has been known to exist in the mycobacterial cell in unidentified locations. Since triacylglycerols are apolar they may fill the gap (about 30 carbons long) between the short and long branches of mycolic acids with the acyl chains of the triacylglycerols extending in opposite directions. In some species, they may also fill the gap between the meromycolate arm of α-mycolate and that of other classes of mycolates, which differ 15–19 carbon atoms in length. Ortalo-Magné *et al.* (1996) recently showed that triacylglycerols can be removed from the surface of various mycobacterial species by gentle shaking with glass beads. This result is certainly consistent with the notion that a large amount of triacylglycerols exists within the cell-wall structure, and that these are not confined to the cytosol. Another novel finding in both these studies is the presence of phosphatidylinositol mannosides (PIMs) in mycobacterial cell wall. PIMs were previously thought to be present only in the plasma membrane. Mycobacterial cell walls have generally been assumed to be devoid of lipids containing conventional fatty acids, including glycerophospholipids (Minnikin 1982; Brennan 1988). This is primarily because earlier studies have usually been carried out with whole cells and any component already known to be present in the plasma membrane was automatically assumed to come from that source.

Another major criticism of the lipid bilayer model has been that the close packing of mycolic acid hydrocarbon chains would be difficult since they are covalently linked to AG, a macromolecular polysaccharide, as mentioned above. However, our current knowledge of the AG structure has solved this problem (Daffé *et al.* 1990; McNeil *et al.* 1991). The AG polymer contains ≈ 100 sugar residues, 60–70 of which are L-arabinose (Ara) and 30–40 of which are D-galactose (Gal), all these Ara and Gal residues are in the furanose form. The galactan backbone of AG is made up of Gal*f* units, linked via alternating 1→5 and 1→6 linkages. To this backbone are connected side chains made up of Ara*f* units, the majority of which are 1→5-linked. Both the galactan main chain and the arabinan side chain are constructed so as to allow maximum freedom of movement between the sugar residues, which is likely to facilitate the lateral packing of mycolic acid chains. Furthermore, the recent isolation of a terminal Gal_{25} macromotif devoid of any arabinosyl branching (Besra *et al.* 1995) suggests that the arabinan chains are anchored fairly close to the reducing end of the galactan, which itself is linked to peptidoglycan via linker disaccharide phosphate. These results are consistent with an arrangement in which mycolic acids extend upwards from AG to produce the lipid-filled core of the cell wall and interact with the extractable lipids (see Fig. 12.1), but are inconsistent with earlier models that proposed long upward protrusions of AG to which mycolate chains were attached in either random or horizontal orientation.

In terms of details of the bilayer model, more studies are needed, especially on the organization of the outer leaflet. The presence of large amounts of triacylglycerol can explain, to a large extent, the problems created by the unequal lengths of the two branches of mycolic acid. Most of the extractable lipids from the mycobacterial cell wall, such as GPLs, PGLs, lipooligosaccharides (LOSs), and PIMs, are strong antigens (Brennan 1988). These are therefore likely to be located on the cell surface; in other words, likely to be part of the cell-wall outer leaflet. However, neither their precise location nor their orientation has been experimentally established. This is a serious problem, especially because some of these extractable lipid species do not have the shape and properties expected for a typical bilayer lipid (Fig. 12.3). For example, a GPL molecule has a head group with a large cross-section but a single hydrocarbon chain of smaller cross-section, a shape that would lead to the formation of micelles and fibrillar structures if it were not inserted into a pre-existing bilayer composed of other lipids (Brennan & Nikaido 1995). This would explain the presence of fibrillar structure on the surface of some species of mycobacteria (Draper 1974), or the L_1 layer of Barksdale and Kim (1977), but leaves open the question of the extent of contribution of GPL to form the bilayer structure. Similarly, some PGL molecules do not contain markedly hydrophilic head groups and are expected to exist as amorphous lipid droplets. This also may explain the presence of amorphous, thick, electron-transparent, 'foamy' layers on the surface of organisms such as *M. leprae* (Gaylord & Brennan 1987), but again presents problems as to the role of PGL in bilayer organization.

2.3 Fluidity of the lipid domain of the mycobacterial cell wall

How does the lipid-bilayer model of the cell wall explain its extremely low permeability to various antibiotics and chemotherapeutic agents? The asymmetric nature of the cell-wall bilayer reminds us of the structure of the outer membrane of Gram-negative bacteria. The Gram-negative outer membrane also forms an asymmetric lipid bilayer

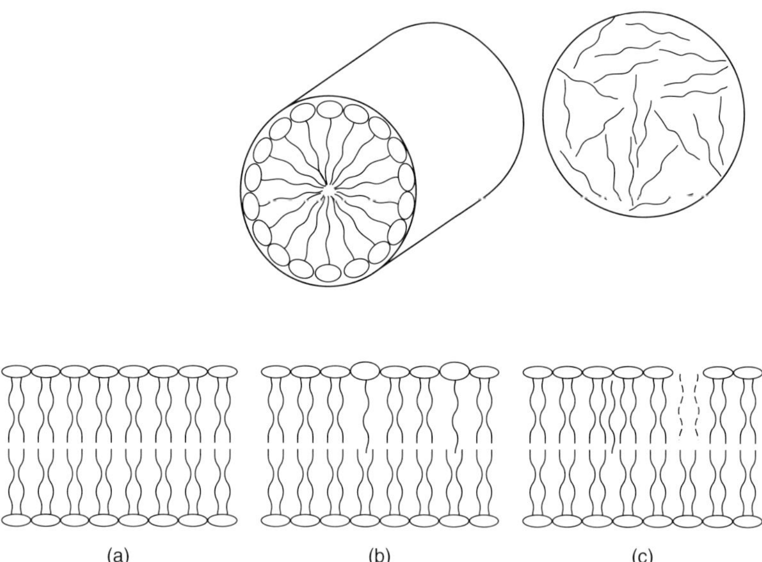

(a)   (b)   (c)

Fig. 12.3 Possible arrangement of various types of lipids in an aqueous environment. (a) Typical bilayer lipids such as glycerophospholipids have head groups with a cross-sectional area close to that of the two hydrocarbon chains. These lipids usually form stable bilayers. (b) Lipid with a large head group and a single hydrocarbon chain, such as glycopeptidolipids, may become inserted into bilayers composed of other lipids (below), but they cannot form bilayers by themselves. They tend to form either spherical aggregates (micelles) or fibrillar structure (above), which is an extension of micellar assembly in one direction. (c) Apolar lipids such as triacylglycerols or phenolic glycolipids do not contain strongly hydrophilic head groups. They may become inserted into the bilayer (below) either at its interior (continuous line) or perhaps even at a location close to surface (broken lines). However, when an excess of these lipids are produced, they are likely to exist as 'oil droplets' with little organization (above).

(reviewed by Nikaido & Vaara 1985); in this case the outer leaflet is composed exclusively of lipopolysaccharide (LPS), whereas the inner leaflet contains only phospholipids. LPS contains six to seven saturated fatty acid chains that pack tightly against each other, decreasing the mobility of the chains and the fluidity of the lipid interior. These factors contribute to the low permeability of the outer membrane bilayer, estimated to be about 50- to 100-fold lower than in the 'typical glycerophospholipid bilayer' (Plésiat & Nikaido 1992).

Consideration of the structure of mycolic acids led us to a similar conclusion regarding the mycobacterial cell wall. Mycolic acid residues have some distinctive features: (i) they are long-chain fatty acids, with a long 'meromycolate' branch of 40–60 carbons and a short branch of 22–24 carbon atoms; (ii) in addition to their extraordinary lengths, mycolic acids contain very few double bonds or cyclopropane groups: the short branch is always saturated and the longer branch has only two positions that can be either a double bond, cyclopropane group or an oxygen-containing group. In the bilayer model of the mycobacterial cell wall, the mycolic acid-containing inner leaflet is expected to have very low fluidity, since lipids containing longer hydrocarbon chains and fewer double bonds tend to become more tightly packed (Lewis & McElhaney 1991). The outer leaflet, containing lipids with shorter fatty acids, should have the usual high fluidity associated with such lipids found in other cells. Thus, a steep gradient of fluidity is likely to exist across the thickness of the mycobacterial cell wall. This is similar to the Gram-negative bacterial outer membrane except that the fluidity gradient in the mycobacterial cell wall has an opposite orientation.

The phase behaviour of a mycolate-containing lipid, trehalose dimycolate, originally known as 'cord factor', was studied earlier by Durand *et al.* (1979a,b). These authors demonstrated, by monolayer studies, that a cord factor containing C_{80} di-*cis*-unsaturated mycolic acids formed a dense, presumably paracrystalline, structure at room temperature (Durand *et al.* 1979a). They have also shown, by differential scanning calorimetry (DSC), that the trehalose dimycolate underwent a cooperative thermal transition with the melting temperature at 46–47°C (Durand *et al.* 1979b). Trehalose dimycolates are only minor components of mycobacterial envelopes so that information obtained on the phase behaviour of these lipids is not directly applicable to the bulk of the AG-bound mycolic acid in the mycobacterial cell wall. However, the evidence presented in these studies did suggest the tendency of mycolic acid hydrocarbons to produce tightly packed, parallel arrays with high melting temperatures.

The extraordinarily low fluidity of the bulk of mycolic acid in the mycobacterial cell wall has been demonstrated by a recent study (Liu *et al.* 1995). DSC was applied directly on purified cell walls isolated from *M. chelonae* by a method that minimized the contamination by plasma membranes. Most of the proteins were removed by extensive protease treatment. DSC analysis of such preparations showed major cooperative thermal transitions with the highest melting temperature around 60°C, suggesting that a significant portion of the lipids existed in a structure of extremely low fluidity in the growing cells. The major components responsible for this high-temperature thermal transition were mycolic acids (see below). Spin-labelled fatty acid probes were inserted into the outer leaflet of the cell wall to examine the local fluidity within the cell-wall structure. Electron-spin resonance spectra of these probes showed considerable fluidity in the outer leaflet, which decreased with increasing depth of probe insertion. These results are thus fully compatible with the prediction of the asymmetric lipid-bilayer model of mycobacterial cell wall, showing the presence of large, highly organized lipid domain composed presumably of mycolic acid, and of a somewhat more fluid outer leaflet.

Cell walls from many other mycobacterial species, including *M. tuberculosis* H37Rv, *M. avium*, *M. terrae*, *M. smegmatis*, *M. chelonae*, *M. vaccae* and *M. aurum*, showed similar phase behaviour to that of *M. chelonae* (Fig. 12.4) (Liu *et al.* 1996a). They all melted at high temperatures (60–70°C), suggesting that this is a general property of mycobacteria. Similar thermal transitions were also observed in whole cells and in

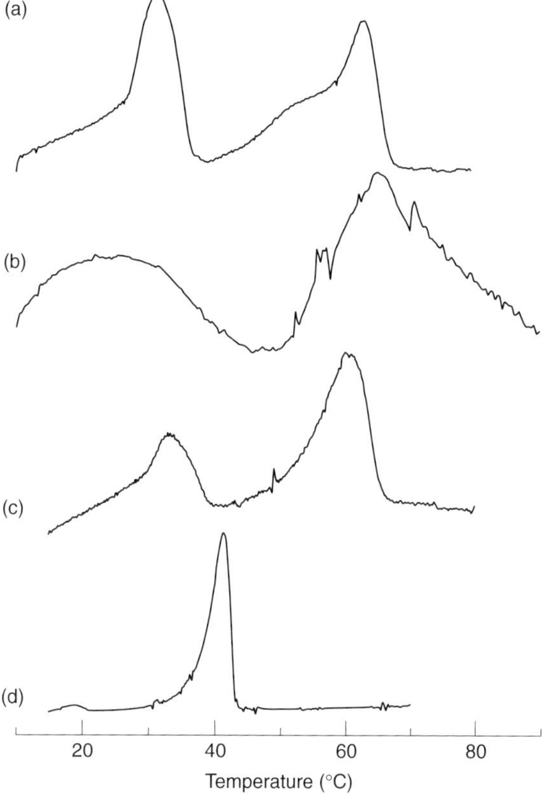

Fig. 12.4 Thermal transitions in *M. tuberculosis* H37Rv. The tightly packed, paracrystalline array of hydrocarbon chains in mycobacterial cell wall undergo a cooperative thermal transition ('melting') when the temperature is gradually raised. This can be observed by the absorption of heat at the melting temperature in the differential scanning calorimeter scan (a peak in the figure). The experiments were performed by using purified, trypsin-treated cell wall (a), whole cells (b), cell wall from which much of extractable lipids were removed with Triton X-114 (c), and methyl esters of mycolates (d), of *M. tuberculosis* H37Rv. (From Liu *et al.* 1996, with permission.)

cell walls from which much of the extractable lipids were removed, suggesting that the major components melting at these temperatures are indeed the mycolic acids.

At least two structural features of mycolic acids appeared to be responsible for the thermal phase transition behaviour of the cell walls. The first is the overall length of the hydrocarbon chains. Thus, the corynebacterial cell walls, containing corynemycolates of 32–38 carbon atoms, which melted at 36–39°C, in a striking contrast to mycobacterial cell walls, which contained mycolates of 74–80 carbon atoms and melted in a correspondingly higher temperature range. The second feature involved a more subtle structural alteration, the ratio of *trans/cis* configuration in the proximal position of meromycolate branch of α-mycolic acids. As mentioned above, mycolic acids are composed of two aligned branches. The shorter chain, containing typically 24–26 carbon atoms, is always without any double bond or cyclopropane group. The longer meromycolate chain contains 40–60 carbon atoms and has two positions that are typically functionalized. The distal position among these two (more than 35 carbon atoms away from the carboxyl end) can be a double bond (or cyclopropane), methoxy, keto or carboxylic ester, and this modification produces, respectively, α-, methoxy-, keto- or wax ester-mycolate. Some species also produce α′-mycolates in which the meromycolate chains are truncated at this position. In many species of mycobacteria, α-mycolates are the most abundant species. The proximal position (about 20 carbons away from the carboxyl end) can be either a double bond or cyclopropane. Interestingly, the *cis* structure of the double bond or cyclopropane at this position is frequently converted into a *trans* double bond or cyclopropane with concomitant introduction of an adjacent methyl branch (Minnikin 1982).

A comparison of the melting temperatures of cell walls from different species of mycobacteria revealed that species containing a significant fraction of the proximal-*trans* α-mycolates (*M. chelonae*, *M. smegmatis* and *M. terrae*) showed higher melting temperatures than those containing smaller amounts of proximal-

trans-α-mycolates (*M. vaccae* and *M. aurum*). This is consistent with our knowledge that *trans* structures are more compatible with close lateral packing of hydrocarbon chains than are *cis* structures, and that the *cis*-to-*trans* conversion raises the thermal transition temperature of short-chain fatty acids (Lewis & McElhaney 1991). In this case, however, the result was far from obvious, as the *cis*-to-*trans* conversion in mycolates is accompanied by the addition of a methyl branch next to the double bond or cyclopropane as mentioned above, and the presence of a methyl branch is known, in some cases, to lower the melting temperature of lipid bilayers appreciably (Lewis & McElhaney 1991). More direct evidence came from studies of purified individual mycolate species. Each of the individual mycolic acid subclasses from *M. tuberculosis* H37Rv (keto and methoxy) and *M. avium* A5 (keto and wax ester) were purified and the ratios of *trans*- to *cis*-cyclopropane at the proximal position were determined by NMR. When the melting temperatures of individual mycolate species were compared, a striking linear relationship was found between the amount of the *trans*-cyclopropane structure and the observed melting temperature (Fig. 12.5) (Liu *et al.* 1996a).

Another factor that may contribute to the cell-wall fluidity was discovered in a study by George *et al.* (1995). The *cma2* gene, whose protein product catalyses the introduction of a *cis*-cyclopropane at the proximal position of the meromycolate chain, was cloned from *M. tuberculosis*. Expression of this gene in *M. smegmatis* resulted in the cyclopropanation of about 30% of the proximal double bonds in the α-mycolate. DSC of detergent-extracted cell walls and purified mycolate species showed that such modification raised the melting temperature by 3°C. In contrast, cyclopropanation of the distal double bond in meromycolate chain did not affect its melting temperature. Thus, introduction of even a *cis*-cyclopropane in the proximal position of meromycolate actually increases the melting temperature. This result explained, at least partly, the unusually high transition temperatures of cell walls of *M. tuberculosis* and *M. avium* (Liu *et al.* 1996a). In spite of their low content of *trans*-α-mycolates, the cell walls from

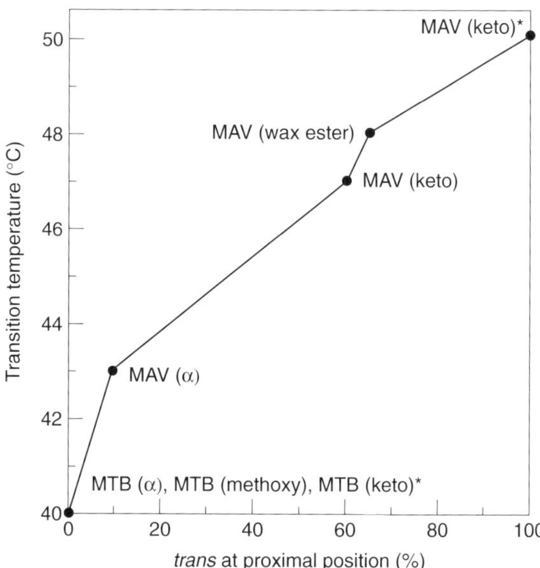

Fig. 12.5 Melting temperature of methylmycolates and fraction of *trans* structure at the proximal position. Methyl esters of mycolates were obtained from *M. tuberculosis* H37Rv (MTB) or from *M. avium* A5 (MAV). They were separated into subclasses (keto, methoxy, and wax ester), and for each subclass the melting behaviour was determined by differential scanning calorimetry and the fraction containing the *trans*-cyclopropane at the proximal position was determined by ^{1}H-NMR. Each preparation showed only one major thermal transition. The methyl ester of ketomycolates from *M. tuberculosis*, however, showed two separate transitions at 40.5 and 49.5°C, suggesting that lateral phase separation resulted in separate melting of all-*cis* and all-*trans* compounds at these two temperatures (shown with asterisks). (From Liu *et al.* 1996, with permission.)

these two species actually melted at higher temperatures than some of the fast-growers. Since all mycolate species in *M. tuberculosis* and *M. avium* contain cyclopropane groups at the proximal position of meromycolate chain, whereas other fast-growing mycobacterial species contain double bonds, this is likely to be the major factor that decreases the fluidity of the cell wall of the slow-growers, including the former two species.

Still other factors may regulate mycobacterial cell-wall fluidity. In a recent study by Yuan *et al.* (1997),

methoxymycolic acid synthetase-1 (MMAS-1), an enzyme encoded in the gene cluster responsible for the biosynthesis of methoxymycolates, was expressed in *M. tuberculosis*. Expression of this enzyme resulted in an increase in the amount of mycolates containing *trans*-cyclopropane at the proximal position (from 4.4 to 25% of total mycolates). This effect was limited to oxygenated types of mycolates (keto-, methoxy-), whereas α-mycolates were not affected. DSC of the intact cells showed that the organism expressing MMAS-1 had a thermal transition that was 7°C lower than the control organisms, which was rather unexpected considering the increase in the amount of proximal *trans*-cyclopropane in the former. However, overexpression of MMAS-1 also caused a significant increase in the fraction of ketomycolates (from 13 to 29% of total mycolates) and a decrease in that of methoxymycolates (from 36 to 12%), although the total amount of α-mycolates changed only little (from 51 to 59%). In addition, mycolates containing *trans* double bonds also appeared in the recombinant organism (about 11% of total mycolates). There are two possible explanations for the unusual transition temperature. First, the introduction of double bond in the proximal position previously occupied by a cyclopropane group in wild-type organisms may decrease the cell-wall fluidity, as discussed above. However, it is difficult to reconcile the large decrease of melting temperature and the small increase in the amount of double bond. The second explanation is that the relative ratio of different species of mycolates may be critical for maintaining the proper cell-wall structure. Thus, the large increase and decrease in the amounts of ketomycolate and methoxymycolate, respectively, in the cell wall of recombinant organisms resulted in the decrease of cell-wall fluidity. This may reflect changes in packing between individual species of mycolic acids, or changes in interaction between lipids in the inner and outer leaflet of the cell wall due to the availability of hydrogen bonding sites in ketomycolate.

The identification of the genes responsible for the biosynthesis of methoxy and ketomycolates (Yuan & Barry 1996) has permitted a direct assessment of the role of ketomycolate in cell-wall structure. By over-expressing MMA-3, the *O*-methyltransferase which converts hydroxymycolate to methoxymycolate, isogenic strains of *M. tuberculosis* were created which contain only methoxymycolate (Y. Yuan and C. E. Barry, unpublished observations). These strains are identical to wild type in uptake of hydrophobic antibiotics, suggesting that the distal position is not involved in formation of the major permeability barrier. On the other hand these strains are seriously compromised in their ability to grow at lower or elevated growth temperatures and in their ability to take up glucose. In addition these strains are severely impaired for growth in macrophages, in spite of a normal appearance of the electron transparent area by electron microscopy. These results suggest a role for ketomycolate in proper assembly and function of the outer leaflet of the cell wall of *M. tuberculosis*.

There are several monolayer studies concerning the interaction of trehalose-dimycolate with phospholipids. Almog and Mannella (1997) observed that minimal compressibility and maximal packing density of mixed cord factor–phosphatidylinositol monolayers occurred at 0.5–0.7 mol fraction of cord factor, suggesting that interdigitation may take place between the acyl chains of the phospholipid and those of cord factor resulting in a realignment of the mycolic acid chains to a more condensed structure. However, earlier studies of mixed monolayers of cord factor and phospholipids having a choline head group (Durand *et al.* 1979b; Crowe *et al.* 1994) reported monolayer expansion at small fractions of cord factor. It is difficult to assess the significance of these data in considering the organization of cell wall, especially because choline-containing phospholipids have not been found in the mycobacterial cell wall.

3 Permeability of the mycobacterial cell wall

As in the outer membrane of Gram-negative bacteria, solutes traverse the mycobacterial cell wall through multiple pathways. Small, hydrophilic solutes can diffuse through porin channels, whereas lipophilic solutes may not be favoured for passage through the water channels formed by porins, and may diffuse

through lipid bilayers. Hints on the pathway used may be obtained by determining solute penetration rates under various conditions. First, if the solutes traverse the cell wall by dissolving into the lipid phase, the permeation rates of lipophilic molecules should have a positive correlation with their oil/water or octanol/water partition coefficients. Second, penetration rates of solutes through the lipid pathway usually show high temperature coefficients, because fluidity of the lipid interior is highly temperature dependent.

3.1 Permeability to hydrophilic solutes

Much of the knowledge on the hydrophilic pathway has been obtained by studying cell-wall permeability to β-lactams. The permeability of mycobacterial cell wall to cephalosporins was first determined in *M. chelonae* (Jarlier & Nikaido 1990). This was done by utilizing the Zimmermann–Rosselet method, previously utilized in studies of Gram-negative outer membrane permeability (Nikaido *et al.* 1983), with some modification to circumvent technical problems due to the aggregation of mycobacterial cells. The rate of hydrolysis of cephalosporins by intact mycobacterial cells was measured, and the cell-wall permeability coefficient was calculated by assuming that drug molecules first diffuse through the cell wall (following Fick's first law of diffusion) and then are hydrolysed by periplasmic β-lactamase (following Michaelis–Menten kinetics). One major requirement of this method is that a rapid enzymatic hydrolysis of β-lactam molecules must occur in the periplasmic space, i.e. the space between the cell wall and the cytoplasmic membrane. Indeed, *M. chelonae* produced constitutively sufficient activity of β-lactamase, without any leakage of enzyme into the medium (Jarlier & Nikaido 1990). The permeability of *M. chelonae* cell wall measured by this approach was very low, for example 10×10^{-8} and 2.3×10^{-8} cm/s for cephaloridine and cefazolin, respectively. These values were about three orders of magnitude lower than that of *Escherichia coli* outer membrane and 10 times lower than that of the notoriously impermeable *Pseudomonas aeruginosa* outer membrane (Fig. 12.6).

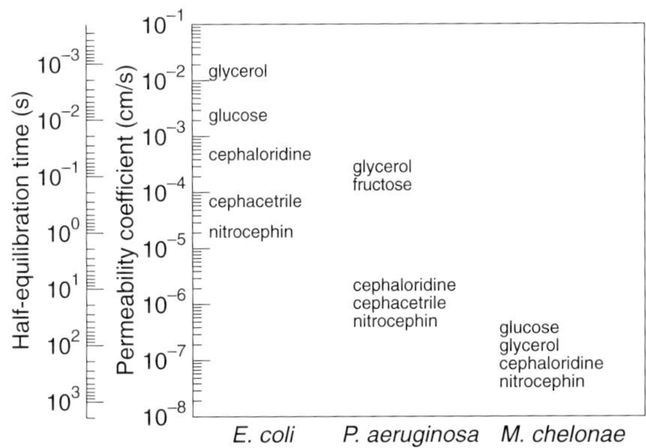

Fig. 12.6 Permeability of *M. chelonae* cell wall to hydrophilic solutes compared to those of *Escherichia coli* and *Pseudomonas aeruginosa*. Permeability coefficients to cephalosporins and nutrients were determined as described in text. Those to nutrients are minimal estimates. Permeability coefficients of outer membranes from two Gram-negative bacteria are shown for comparison. (From Jarlier and Nikaido 1990, with permission.)

Trias *et al.* (1992) made a major contribution to the field by identifying a 59-kDa porin protein from detergent extracts of *M. chelonae* cell wall by fractionation using a proteoliposome reconstitution assay for pore-forming activity. This study explained the extremely low permeability of the *M. chelonae* cell wall to β-lactams (Trias *et al.* 1992; Trias & Benz 1993). First, unlike the porins of *E. coli*, the 59-kDa porin protein is a minor component of the cell wall, accounting for less than 2% of the proteins in the cell wall. Second, although the size of the 59-kDa porin channel (estimated to be 2.2 nm in diameter) seems to be about twice the size of the *E. coli* OmpF porin channel (1.0 nm in diameter), the mycobacterial porin had a specific activity at least 20-fold less than that of the *E. coli* porin. Finally, the porin protein in *M. chelonae* has point negative charges at the channel mouth and tends to exclude anionic solutes, which is consistent with the observation that cephaloridine, a zwitterion, diffused much more rapidly through this channel than monoanionic cephalosporins (Jarlier & Nikaido 1990). It is also known that the temperature coefficient of cephalosporin influx across the mycobacterial cell wall was not

high (Jarlier & Nikaido 1990), and this is also consistent with the notion that the major penetration pathway of cephalosporins across the cell wall is porin mediated.

In *Nocardia*, which is closely related to *Mycobacterium*, Riess *et al.* (1998) have recently purified a cell-wall porin to homogeneity. Very interestingly, this protein behaves as a 87-kDa protein when it is subjected to sodium dodecyl sulphate polyacrylamide gel electrophoresis (SDS-PAGE) without heat denaturation, but after treatment in SDS at 100°C it dissociates into smaller subunit(s) of about 20 kDa. Draper's laboratory has expressed, in *E. coli*, an *ompA* homologue found in the *M. tuberculosis* genome, and showed that the protein had a low level of pore-forming activity (Senaratne *et al.* 1998). Antibody against this protein reacted with a protein in *M. tuberculosis*, but it is not clear how much of the porin activity in *M. tuberculosis* cell wall is due to the activity of this protein.

Permeability of the cell wall to β-lactams has also been measured in *M. smegmatis* (Trias & Benz 1994) and *M. tuberculosis* (Chambers *et al.* 1995). Interestingly, the cell-wall permeability of these two species was about an order of magnitude higher than that in *M. chelonae*. The presence of a porin was also shown in *M. smegmatis* (Trias & Benz 1994). The estimated pore diameter of the *M. smegmatis* porin, around 3 nm, was significantly larger than that of *M. chelonae*. Thus, the difference in permeability between *M. smegmatis* and *M. chelonae* may be the result of differences in the properties of their porins.

Small, hydrophilic, nutrient molecules, such as glucose, glycerol, and amino acids, are also likely to diffuse through porin channels. Permeability of the *M. chelonae* cell wall to these nutrients was measured in a similar way by assuming that once the nutrients penetrated through the cell wall they were actively transported through the cytoplasmic membrane (Jarlier & Nikaido 1990). The latter process is governed by the Michaelis–Menten equation and plays a role that is equivalent to the periplasmic β-lactamase in the original Zimmermann–Rosselet assay. Since the kinetic constants of the active transporters are not known, the permeability coefficients obtained by this method are minimal estimates. The permeability of *M. chelonae* cell wall to these small nutrients was again about five orders of magnitude lower than that of *E. coli* outer membrane (Fig. 12.6). Similarly, the cell wall of *M. smegmatis* and *M. phlei* appeared to have much higher permeability to small nutrients than *M. chelonae* (Jarlier & Nikaido 1990), and once again this difference may be explained by the properties of porins in these species of mycobacteria.

Isoniazid and aminoglycosides are hydrophilic molecules showing significant activities against mycobacteria. Isoniazid is a small molecule that is essentially uncharged at neutral pH, and could use the porin pathway to cross the cell wall. An early study showed that isoniazid was accumulated by mycobacterial cells to up to 50 times its extracellular concentration (Beggs & Jenne 1970). There are large differences in susceptibility to isoniazid among mycobacterial species: *M. tuberculosis* and *M. bovis* are extremely sensitive to isoniazid, whereas *M. avium–intracellulare* complex (MAC), as well as the fast-growers, are resistant to its action. It was often argued that the very low hydrophilic permeability of the cell wall of *M. avium* group was at least partially responsible for its resistance. Rastogi and Goh (1990) showed the efficacy of isoniazid against MAC was increased by converting it into a hydrophobic molecule by addition of a palmitoyl tail. It is likely that the modified isoniazid penetrates more through the lipid bilayer. However, this modification lowered the isoniazid MIC in *M. avium* only to the values seen in rapidly growing mycobacteria, and the most important factor that makes *M. avium* resistant appears to be unrelated to the permeability barrier. Indeed, *M. avium* and *M. tuberculosis* showed a similar degree of crypticity in the KatG-mediated modification of isoniazid, intact cells of both species catalysing this conversion at rates three- to fourfold slower than the rates seen with extracts (Mdluli *et al.* 1998).

Aminoglycosides are larger than isoniazid but their sizes are well within the exclusion limits of mycobacterial porins. Their positive charges should also help in the penetration through the porin channel. Thus, in most mycobacterial species, aminoglycosides

should have no difficulty in penetrating through the porin channels. Streptomycin uptake by *M. tuberculosis* has been shown to occur in two phases, as in other bacteria (Beggs & Williams 1971): an initial rapid phase followed by a slow but eventually massive influx of the drug across the cytoplasmic membrane (Davis 1987). From the foregoing, it seems likely that aminoglycosides have already penetrated across the cell wall during the first phase, but this is not easily detected by the usual methods of measurement because only a small amount of the drug may be located within the cell wall and periplasm, and the drug may diffuse out of these compartments during the washing process. The accumulation data therefore are not easy to interpret. Some mycobacterial species, such as *M. avium*, *M. gordonae* and *M. szulgai,* are naturally resistant to streptomycin, yet their ribosomal protein, S12, appears to be of the antibiotic-susceptible type (Honoré & Cole 1994). In these species the drug may have more difficulty in crossing the plasma membrane.

In Gram-negative bacteria, aminoglycosides have been proposed to use a 'self-promoted pathway' to cross the outer membrane barrier (Hancock 1984). These drugs, being polycations, bind to the negatively charged surface of the outer membrane and are thought to traverse this barrier by essentially disorganizing it. Whether similar interactions occur in the mycobacterial cell wall is unknown but it is not likely that the mycobacterial cell surface carries as dense an array of negative charges as does the surface of some Gram-negative bacteria. Polymyxin, another polycatonic antibiotic, weakly inhibits some species of mycobacteria (David & Rastogi 1985).

3.2 Permeability to hydrophobic solutes

Plésiat and Nikaido (1992) quantitatively measured the permeability of the outer membrane of *Salmonella typhimurium* to highly hydrophobic steroid probes. This was done by coupling the influx of 3-oxosteroids with their subsequent oxidation catalysed by a dehydrogenase. The permeability coefficient of the outer membrane containing wild-type LPS to uncharged steroids was about 10^{-5} cm/s, about 100 times lower than that of a typical biological membrane, and the diffusion appeared to occur mainly through the lipid bilayer domains of the outer membrane. The presence of LPS exclusively in the outer leaflet of Gram-negative outer membrane bilayer decreases its fluidity and, consequently, limits its permeability. Indeed, the permeation rates of steroid probes were markedly increased (up to 100 times) when the LPS leaflet was perturbed either by adding polycations, such as deacyl polymyxin B, or by introducing mutations leading to the production of deep rough LPS (thereby producing a mixed phospholipid–LPS outer leaflet).

Considering the extremely low fluidity of the inner leaflet of the mycobacterial cell-wall bilayer, it can be predicted that the permeability of the cell wall to lipophilic solutes will be even lower than that of the outer membrane of Gram-negative bacteria. Based on the data presented in a study by Sedlaczek *et al.* (1994), which showed that the side chain of β-sitosterol can be cleaved by a strain *Mycobacterium* sp. NRRL MB 3683 to yield androst-1-ene-3,17-dione and androsta-1,4-diene-3,17-dione, we can calculate the cell-wall permeability to β-sitosterol, another steroid, to be $\approx 3 \times 10^{-9}$ cm/s. As anticipated, this is between 3 and 4 orders of magnitude lower than the permeability of Gram-negative outer membrane to steroids.

The lipid bilayer of the mycobacterial cell wall thus has an unusually low permeability. Yet the lipid pathway of uptake is preferred by some lipophilic antibiotics, presumably because the porin pathway is even less efficient for them. Agents such as rifamycins, tetracyclines, macrolides and fluoroquinolones, all relatively lipophilic molecules, thus appear to utilize the lipid bilayer pathway to a significant extent to traverse the mycobacterial cell wall. *M. tuberculosis* is usually susceptible to rifampicin and most of the resistant strains have mutations in the RNA polymerase (Cole 1996). Organisms of the *M. avium* complex, and several fast-growers such as *M. chelonae*, *M. smegmatis* and *M. fortuitum*, are naturally resistant to this antibiotic. Hui *et al.* (1977) showed that strains of

M. intracellulare and *M. smegmatis* containing a susceptible RNA polymerase were resistant to rifampicin, a result suggesting that impaired permeability or active efflux of rifampicin was likely to be the main cause for their natural resistance. The uptake of rifamycins by *M. avium* appeared to be low (David *et al.* 1987) but it is not clear whether this is due to the permeability barrier or the effect of efflux. Interestingly, more hydrophobic derivatives of rifamycins, such as rifabutin, rifapentine, CGP-7040, KRM-1648 and T9, show higher activities against mycobacteria, including *M. avium*, *M. tuberculosis* and *M. leprae* (Heifets *et al.* 1990). A similar positive correlation between lipophilicity and efficacy against mycobacteria was found among other classes of drugs, including tetracycline (Wallace *et al.* 1979; Swenson *et al.* 1982; Gelber 1987), fluoroquinolones (Franzblau & White 1990; the data are analysed by Connell & Nikaido 1994) and macrolides (Fernandes *et al.* 1982; Gorzynski *et al.* 1989; Brown *et al.* 1992). These results suggest that the agents of this type mainly cross the cell wall through its lipid domain. They suggest also that the efficacy is correlated with the rate of diffusion. It is important to realize that such a correlation would hold even when the main mechanism of resistance is efflux, as more rapid influx should be able to compete successfully against the transporter-catalysed, active efflux, which is a saturable process.

As discussed above, mycolyl-AG plays a critical role in determining the fluidity of the cell wall. Thus, agents that inhibit the biosynthesis of mycolyl-AG are expected to increase fluidity of the lipid domain and, as a result, increase the permeability of traditional antimicrobial drugs and their efficacy in a synergistic manner. Ethambutol inhibits an arabinosyl transferase (Belanger *et al.* 1996) and this results in a decrease in the amount of mycolic acids bound to the cell wall (Takayama *et al.* 1979; Takayama & Kilburn 1989). It has been shown that ethambutol synergistically increases the antimycobacterial activities of many other drugs (Rastogi *et al.* 1990). We have, indeed, observed that addition of subinhibitory concentrations of ethambutol resulted in a significant increase in cell wall permeability to the lipophilic probes, chenodeoxycholate and tetracycline (J. Liu and H. Nikaido, unpublished observations).

3.3 The role of the cell-wall barrier in drug resistance of mycobacteria

It has been traditionally assumed that the mycobacterial cell wall plays an important role in the natural resistance of mycobacteria to many antibacterial agents. Early evidence included the observation that addition of detergents to the medium substantially lowered the MIC of antimycobacterial drugs (Hui *et al.* 1977; Mizuguchi *et al.* 1983). The evidence presented in this chapter clearly shows that the extremely low permeability of the mycobacterial cell wall is one major factor in the intrinsic resistance of mycobacteria (see also Chapter 15). However, quantitative calculations show that even the low permeability (experimentally determined as 23×10^{-9} and 3×10^{-9} cm/s for cefazolin and β-sitosterol, respectively, see above) cannot totally exclude the agents. Thus, the concentration of the drug inside the cell wall is expected to reach 50% of the external concentration in 10–60 min under these conditions (for calculations, see Nikaido *et al.* 1983). The influx is thus not instantaneous, yet these half-equilibration times are relatively short in comparison with the long doubling times of the slow growing mycobacteria. We can therefore anticipate that the production of clinically significant levels of resistance would usually require the participation of an *additional* resistance mechanism, such as the enzymatic inactivation or the active efflux of the agents.

The synergism between the cell-wall barrier and the hydrolysis of drug molecules has been shown in the case of β-lactams. Most species of mycobacteria produce β-lactamase (Amicosante *et al.* 1990; Zhang *et al.* 1992; Chambers *et al.* 1995). Mathematical analysis (Jarlier *et al.* 1991; Jarlier & Nikaido 1994) shows that even though the permeability of cephalosporins in *M. chelonae* is extremely low, it will only lower the periplasmic drug concentration by less than 0.5% if it is acting alone. However, in the presence of β-lactamase, the periplasmic concentration of cephaloridine was decreased to 0.2% of the

external concentration. This does not mean that the cell-wall barrier is not important; rather, it shows that the low permeability of the mycobacterial cell wall is necessary for this high degree of resistance. Because the very effective permeability barrier drastically decreases the rate of entry of β-lactam molecules into the cell, β-lactamase can hydrolyse most of these incoming molecules even though its activity in *M. chelonae* is only 20% of that in *E. coli* producing the TEM-type enzyme. In contrast, in *E. coli*, the local drug concentration can be decreased only by about 20% by these two mechanisms, although β-lactamase is much more active than in *M. chelonae*.

The knowledge that the β-lactam resistance of mycobacteria requires synergy between cell-wall barrier and the periplasmic β-lactamase suggests that mycobacteria can be made susceptible to these agents if their β-lactamase can be inhibited. Indeed, Chambers *et al.* (1995) showed that inhibition of *M. tuberculosis* β-lactamase by clavulanic acid or sulbactam does make this species quite susceptible to clinically achievable concentrations of β-lactams such as ampicillin and amoxycillin. This combination therapy may thus be useful in the treatment of infections caused by drug-resistant *M. tuberculosis*.

Other small molecules such as tetracyclines, chloramphenicol and fluoroquinolones, for which inactivating mechanisms have not been described in mycobacteria, should also penetrate the cell envelope and reach their targets. This is because the half-equilibration time across the cell wall is relatively short in comparison with the generation time of the organism, as described above. Indeed, norfloxacin accumulated in the cells and reached a steady-state concentration within 10 min in several species of mycobacteria including the slow-growers *M. tuberculosis* and *M. avium* (Liu *et al.* 1996a). Similar situations, involving a relatively rapid drug influx in comparison with the generation times of the bacteria, are seen in Gram-negative bacteria, where active efflux, often catalysed by multidrug efflux pumps, provides a synergistic mechanism that works together with the outer membrane barrier (Nikaido 1996). Indeed, a multidrug efflux pump has also been discovered in mycobacteria (Liu *et al.* 1996b; Takiff *et al.* 1996): the *lfrA* gene was cloned from chromosomal DNA of a quinolone-resistant strain of *M. smegmatis* and subsequent overexpression of this gene conferred low-level resistance to fluoroquinolones. Biochemical evidence suggested that LfrA protein catalyses the active efflux of several fluoroquinolones as well as ethidium bromide, acriflavine and some quaternary ammonium compounds. The existence of efflux pump(s) conferring high-level resistance to fluoroquinolones in *M. smegmatis* has also been suggested, although these have yet to be characterized (Banerjee *et al.* 1996). In some cases, efflux is mediated by more traditional efflux pumps of narrower specificity, such as that encoded by tetracycline efflux gene in *M. fortuitum* (Pang *et al.* 1994).

Genes coding for drug efflux are extremely prevalent among bacteria and some of these pumps have an extraordinarily broad specificity and appear to catalyse the efflux of any compound containing a moderately lipophilic segment (Nikaido 1996). It therefore seems likely that many of the resistance phenomena in mycobacteria, which have so far been ascribed to the barrier properties of the cell wall, are actually the result of a synergistic collaboration between the cell-wall barrier and an active efflux process. We do not, however, completely understand the mechanism of this synergistic efflux process. In order to be effective in an organism with a surface permeability barrier, the efflux process must have a way to bypass the surface barrier (see Thanassi *et al.* 1995). With Gram-negative bacteria, specialized outer membrane channels for drug efflux are known (Nikaido 1996). Such channels or pathways are not yet known in mycobacteria and it is currently unclear how the efflux process can produce such high levels of resistance.

3.4 An example of the unresolved questions: *Mycobacterium avium–intracellulare* complex

There are significant differences in drug susceptibility among mycobacteria and many species are more resistant to traditional antimycobacterial agents than *M. tuberculosis*. An extreme case is MAC organisms, which are resistant to almost all of the available antimycobacterial drugs. Unlike multiple drug resistance of *M. tuberculosis* that is usually associated with genetic mutations as the result of inadequate therapy, multiple drug resistance is an intrinsic property of MAC. It is thus tempting to assume that the permeability barrier of the cell envelope may play at least some role in this resistance phenotype. Possibly their porins have very low permeability, but this idea is inconsistent with the isoniazid crypticity data described above. The relatively high melting temperature of the cell-wall lipids of *M. avium* (Liu *et al.* 1996a) suggested that its cell wall has a lower fluidity and thus a lower permeability to lipophilic agents. However, this explanation is not entirely satisfactory because the melting temperature of *M. avium* cell-wall lipids is only about 3°C higher than that of *M. tuberculosis* (Liu *et al.* 1996a). Although it is conceivable that the cell-wall structure of MAC is fundamentally different from those of other mycobacterial species, this is unlikely if we consider that the major cell-wall components, i.e. peptidoglycan, AG and mycolic acids, are similar. Some strains of MAC do, however, contain large amounts of GPL as the major component of the cell-surface lipids (Draper 1974; Barrow *et al.* 1980). GPL is abundant in MAC of the SmT (smooth–transparent) colony type but is decreased in the SmD (smooth–domed) colony type and is essentially absent in the Rg (rough) colony types (Barrow & Brennan 1982; reviewed by Aspinall *et al.* 1995). Very interestingly, SmT morphotypes are more resistant to lipophilic antibiotics such as rifampicin and minocycline, than SmD and Rg types (Rastogi *et al.* 1981; Tsukamura *et al.* 1989). (Although a survey of clinical isolates by Tsukamura *et al.* (1989) showed that SmT strains were more resistant also to hydrophilic antibiotics, such as aminoglycosides, this could be a fortuitous correlation.) Indeed, the rates of entry of lipophilic probes, chenodeoxycholate and norfloxacin, were much higher in Rg and SmD colony types than in the SmT colony type derived from the same *M. avium* parent (H. Nikaido and C. E. Barry, III, unpublished observations). Similarly, inhibition of GPL biosynthesis by 3-fluorophenylalanine enhanced the susceptibility of *M. avium* to lipophilic agents such as rifampicin and fluoroquinolones, but not to a hydrophilic agent isoniazid (Rastogi *et al.* 1990).

These data suggest that the presence of GPL is somehow involved in the construction of an effective permeability barrier. However, it is not easy to propose how GPL could be organized to contribute to the barrier. As mentioned already, a fraction of GPL is likely to be inserted into the outer leaflet of the cell wall. An excess of GPL is likely to form micellar or fibrillar aggregates because of the large size of its head group but such an arrangement will not produce an effective permeability barrier. Perhaps the complex head groups of some types of GPL can interact tightly with each other to produce an effective barrier, but currently it is not even possible to propose speculative models on such an arrangement.

Looking back, our current understanding of the barrier properties of the mycobacterial cell wall seems to be somewhat analogous to the situation that was prevailing more than 20 years ago on the properties of Gram-negative outer membrane. At that time we knew that the outer membrane was an effective barrier but knew little about how that effective barrier was constructed. What we knew was the broadest outline of the overall organization and that any perturbation of the final structure, either by physical treatment such as addition of EDTA or polycations, or by genetic mutations, led to the drastic permeabilization of the structure. Similarly, we now know that the mycobacterial cell wall is a very effective barrier to solute penetration, which can be functionally perturbed by a number of agents and genetic changes, very often through mechanisms that we cannot explain in a rigorous manner. Thus, the current situation is not entirely satisfying but our past success in understanding the structure and its functional implications of Gram-negative outer membrane gives us hope that similar progress may be made in the immediate future in the study of mycobacterial cell wall.

4 Acknowledgements

J. Liu was supported in part by University of California AIDS Research Fellowship, and studies in the laboratory of H. Nikaido were supported by a grant from US Public Health Service (AI-09644).

5 References

Almog, R. & Mannella, C.A. (1997) Molecular packing of cord factor and its interaction with phosphatidylinositol in mixed monolyers. *Biophysical Journal* **71**, 3311–3319.

Amicosante, G., Franceschini, N., Segatore, B. *et al.* (1990) Characterization of a β-lactamase produced in *Mycobacterium fortuitum* D316. *Biochemical Journal* **271**, 729–734.

Aspinall, G.O., Chatterjee, D. & Brennan, P.J. (1995) The variable surface glycolipids of mycobacteria: Structures, synthesis of epitopes, and biological properties. *Advances in Carbohydrate Chemistry and Biochemistry* **51**, 169–242.

Banerjee, S.K., Bhatt, K., Rana, S., Misra, P. & Chakraborti, P.K. (1996) Involvement of an efflux system in mediating high level of fluoroquinolone resistance in *Mycobacterium smegmatis*. *Biochemical Biophysical Research Communications* **226**, 362–368.

Barksdale, L. & Kim, K.S. (1977) *Mycobacterium*. *Bacteriological Review* **41**, 217–372.

Barrow, W.W. & Brennan, P.J. (1982) Isolation in high frequency of rough variants of *Mycobacterium intracellulare* lacking C-mycoside glycopeptidolipid antigens. *Journal of Bacteriology* **150**, 381–384.

Barrow, W.W., Ullom, B.P. & Brennan, P.J. (1980) Peptidoglycolipid nature of the superficial cell wall sheath of smooth-colony-forming mycobacteria. *Journal of Bacteriology* **144**, 814–822.

Beggs, W.H. & Jenne, J.W. (1970) Capacity of tubercle bacilli for isoniazid accumulation. *American Review of Respiratory Diseases* **102**, 92–96.

Beggs, W.H. & Williams, N.E. (1971) Streptomycin uptake by *Mycobacterium tuberculosis*. *Applied Microbiology* **21**, 751–753.

Belanger, A.E., Besra, G.S., Ford, M.E *et al.* (1996) The *embAB* genes of *Mycobacterium avium* encode an arabinosyl transferase involved in cell wall arabinan biosynthesis that is the target for the antimycobacterial drug ethambutol. *Proceedings of the National Academy of Sciences of the USA* **93**, 11919–11924.

Besra, G.S., Khoo, K.-H., McNeil, M.R., Dell, A., Morris, H.R. & Brennan, P.J. (1995) A new interpretation of the structure of the mycolyl-arabinogalactan complex of *Mycobacterium tuberculosis* as revealed through characterization of oligoglycosylalditol fragments by fast-atom bombardment mass spetrometry and ^{1}H nuclear magnetic resonance spectroscopy. *Biochemistry* **34**, 4257–4266.

Beveridge, T.J. & Murray, R.G.E. (1980) Sites of metal

deposition in the cell wall of *Bacillus subtilis*. *Journal of Bacteriology*. **141**, 876–887.

Brennan, P.J. (1988) Mycobacterium and other actinomycetes. In: *Microbiol Lipids*, Vol. I (eds C. Ratledge & S. G. Wilkinson). London: Academic Press, pp. 203–298.

Brennan, P.J. & Draper, P. (1994) Ultrastructure of *Mycobacterium tuberculosis*. In: *Tuberculosis: Pathogenesis, Protection and Control* (ed. B. R. Bloom). Washington, DC: American Society for Microbiology, pp. 271–284.

Brennan, P.J. & Nikaido, H. (1995) The envelope of mycobacteria. *Annual Review of Biochemistry* **64**, 29–63.

Brown, B.A., Wallace, R.J. Jr, Onyi, G.O., De Rosas, V., Wallace, R.J., III (1992) Activities of four macrolides, including clarithromycin, against *Mycobacterium fortuitum, Mycobacterium chelonae*, and *M. chelonae*-like organisms. *Antimicrobial Agents and Chemotherapy* **36**, 180–184.

Chambers, H.F., Moreau, D., Yajko, D. *et al.* (1995) Can penicillins and other β-lactam antibiotics be used to treat tuberculosis? *Antimicrobial Agents and Chemotherapy* **39**, 2620–2624.

Cole, S.T. (1996) Rifamycin resistance in mycobacteria. *Research in Microbiology* **147**, 48–52.

Connell, N.D. & Nikaido, H. (1994) Membrane permeability and transport in *Mycobacterium tuberculosis*. In: *Tuberculosis: Pathogenesis, Protection and Control* (ed. B. R. Bloom). Washington, DC: American Society for Microbiology, pp. 333–351.

Crowe, L.M., Spargo, B.J., Ioneda, T., Beaman, B.L. & Crowe, J.H. (1994) Interaction of cord factor (α,α′-trehalose-6,6-dimycolate) with phospholipids. *Biochimica et Biophysica Acta* **1194**, 53–60.

Daffé, M., Brennan, P.J. & McNeil, M. (1990) Predominant structural features of the cell wall arabinogalactan of *Mycobacterium tuberculosis* as revealed through characterization of oligoglycosylalditol fragments by gas chromatography/mass spectrometry and by ^{1}H and ^{13}C NMR analysis. *Journal of Biological Chemistry* **265**, 6734–6743.

David, H.L., Clavel-Seres, S., Clement, F. & Goh, K.S. (1987) Uptake of selected antibacterial agents in *Mycobacterium avium*. *Zentralbatt für Bakteriologie Mikrobiologie Hyg [a]* **265**, 385–392.

David, H.L. & Rastogi, N. (1985) Antibacterial action of colistin (polymyxin E) against *Mycobacterium aurum*. *Antimicrobial Agents and Chemotherapy* **27**, 701–707.

Davis, B.D. (1987) Mechanism of bactericidal action of aminoglycosides. *Microbiological Reviews* **51**, 341–350.

Draper, P. (1974) The mycoside capsule of *Mycobacterium avium* 357. *Journal of General Microbiology* **83**, 431–433.

Draper, P. (1982) The anatomy of mycobacteria. In: *The Biology of the Mycobacteria*, Vol. 1 (eds C. Ratledge & J. Stanford). London: Academic Press, pp. 9–52.

Durand, E., Gillois, M., Tocanne, J.H. & Lannéelle, G. (1979b) Property and activity of mycoloyl esters of methyl glucoside and trehalose. *European Journal of Biochemistry* **94**, 109–118.

Durand, E., Welby, M., Laneelle, G. & Tocanne, J.F. (1979a) Phase behaviour of cord factor and related bacterial glycolipid toxins: a monolayer study. *European Journal of Biochemistry* **93**, 103–112.

Fernandes, P.B., Hardy, D.J., McDaniel, D., Hanson, C.W. & Swanson, R.N. (1982) *In vitro* and *in vivo* activities of clarithromycin against *Mycobacterium avium*. *Antimicrobial Agents and Chemotherapy* **33**, 1531–1534.

Franzblau, S.G. & White, K.E. (1990) Comparative *in vitro* activities of 20 fluoroquinolones against *Mycobacterium leprae*. *Antimicrobial Agents and Chemotherapy* **36**, 180–184.

Gaylord, H. & Brennan, P.J. (1987) Leprosy and the leprosy bacillus: Recent developments in characterization of antigens and immunology of the disease. *Annual Review of Microbiology* **41**, 645–675.

Gelber, R.H. (1987) Activity of minocycline in *Mycobacterium leprae*-infected mice. *Journal of Infectious Diseases* **156**, 236–239.

George, K.M., Yuan, Y., Sherman, D.R. & Barry, C.E. III (1995) The biosynthesis of cyclopropanated mycolic acids in *Mycobacterium tuberculosis*. *Journal of Biological Chemistry* **270**, 27292–27298.

Gorzynski, E.A., Gutman, S.I. & Allen, W. (1989) Comparative antimycobacterial activities of difloxacin, enoxacin, pefloxicin, reference fluoroquinolones, and a new macrolide, clarithromycin. *Antimicrobial Agents and Chemotherapy* **33**, 591–592.

Hancock, R.E.W. (1984) Alterations in outer membrane permeability. *Annual Review of Microbiology* **38**, 237–264.

Heifets, L.B., Lindholm-levy, P.J. & Flory, M.A. (1990) Bactericidal activity *in vitro* of various rifamycins against *Mycobacterium avium* and *Mycobacterium tuberculosis*. *American Review of Respiratory Diseases* **141**, 626–630.

Honoré, N. & Cole, S.T. (1994) Streptomycin resistance in mycobacteria. *Antimicrobial Agents and Chemotherapy* **38**, 238–242.

Hui, J., Gordon, N. & Kajkioka, R. (1977) Permeability barrier to rifampicin in mycobacteria. *Antimicrobial Agents and Chemotherapy* **11**, 773–779.

Imaeda, T., Kanetsuna, F. & Galindo, B. (1968) Ultrastructure of cell walls of genus *Mycobacterium*. *Journal of Ultrastructural Research* **25**, 46–63.

Jarlier, V., Gutmann, L. & Nikaido, H. (1991) Interplay of cell wall barrier and β-lactamase activity determines high resistance to β-lactam antibiotics in *Mycobacterium chelonae*. *Antimicrobial Agents and Chemotherapy* **35**, 1937–1939.

Jarlier, V. & Nikaido, H. (1990) Permeability barrier to hydrophilic solutes in *Mycobacterium chelonae*. *Journal of Bacteriology* **172**, 1418–1423.

Jarlier, V. & Nikaido, H. (1994) Mycobacterial cell wall: Structure and role in natural resistance to antibiotics. *FEMS Micriobiological Letters* **123**, 11–18.

Lemassu, A., Ortalo-Magné, A., Bardou, F., Silve, G., Lanéelle, M.-A. & Daffé, M. (1996) Extracellular and surface-exposed polysaccharides of non-tuberculous mycobacteria. *Microbiology* **142**, 1513–1520.

Lewis, R.N.A.H. & McElhaney, R.N. (1991) The mesomorphic phase behavior of lipid bilayers. In: *The Structure of Biological Membranes* (ed. P. Yeagle). Boca Raton: CRC Press, pp. 73–155.

Liu, J., Barry, C.E. III, Besra, G.S. & Nikaido, H. (1996a) Mycolic acid structure determines the fluidity of the mycobacterial cell wall. *Journal of Biological Chemistry* **271**, 29545–29551.

Liu, J., Rosenberg, E.Y. & Nikaido, H. (1995) Fluidity of the lipid domain of cell wall from *Mycobacterium chelonae*. *Proceedings of the National Academy of Sciences of the USA* **92**, 11254–11258.

Liu, J., Takiff, T.E. & Nikaido, H. (1996b) Active efflux of fluoroquinolones in *Mycobacterium smegmatis* mediated by LfrA, a multidrug efflux pump. *Journal of Bacteriology* **178**, 3791–3795.

McNeil, M.R. & Brennan, P.J. (1991) Structure, function and biogenesis of the cell envelope of mycobacteria in relation to bacterial physiology, pathogenesis and drug resistance; some thoughts and possibilities arising from recent structural information. *Research in Microbiology* **8**, 451–463.

McNeil, M.R., Daffé, M. & Brennan, P.J. (1991) Location of the mycolyl ester substituents in the cell walls of mycobacteria. *Journal of Biological Chemistry* **266**, 13217–13223.

Mdluli, K., Swanson, J., Fischer, E., Lee, R.E. & Barry, C.E. III (1998) Mechanisms involved in the intrinsic isoniazid resistance of *Mycobacterium avium*. *Molecular Microbiology* **27**, 1223–1233.

Minnikin, D.E. (1982) Lipids: complex lipids, their chemistry, biosynthesis, and roles. In: *The Biology of the Mycobacteria*, Vol. 1 (eds C. Ratledge & J. Stanford). London: Academic Press, pp. 95–184.

Mizuguchi, Y., Udou, T. & Yamada, T. (1983) Mechanism of antibiotic resistance in *Mycobacterium intracellulare*. *Microbiology and Immunology* **27**, 425–431.

Nikaido, H. (1996) Multidrug efflux pumps of Gram-negative bacteria. *Journal of Bacteriology* **178**, 5853–5859.

Nikaido, H., Kim, S.-H. & Rosenberg, E.Y. (1993) Physical organization of lipids in the cell wall of *Myobacterium chelonae*. *Molecular Microbiology* **8**, 1025–1030.

Nikaido, H., Rosenberg, E.Y. & Foulds, J. (1983) Porin channels in *Escherichia coli*: Studies with β-lactams in intact cells. *Journal of Bacteriology* **153**, 232–240.

Nikaido, H. & Vaara, M. (1985) Molecular basis of bacterial outer membrane permeability. *Microbiological Reviews* **49**, 1–32.

Ortalo-Magné, A., Dupont, M.-A., Lemassu, A., Andersen, A.B., Gounon, P. & Daffé, M. (1995) Molecular composition of the outermost capsular material of the tubercle bacillus. *Microbiology* **141**, 1609–1620.

Ortalo-Magné, A., Lemassu, A., Lanéelle, M.A. *et al.* (1996) Identification of the surface-exposed lipids on the cell envelopes of *Mycobacterium tuberculosis* and other mycobacterial species. *Journal of Bacteriology* **178**, 456–461.

Panchioli, V. & Fischetti, V.A. (1992) A major surface protein on group A streptococci is a glyceraldehyde-3-phosphate dehydrogenase with multiple binding activity. *Journal of Experimental Medicine* **176**, 415–426.

Pang, Y., Brown, B.A., Steingrube, V.A., Wallace, R.J. Jr & Roberts, M.C. (1994) Tetracycline resistance determinants in *Mycobacterium* and *Streptomyces* species. *Antimicrobial Agents and Chemotherapy* **38**, 1408–1412.

Paul, T.R. & Beveridge, T.R. (1992) Reevaluation of envelope profiles and cytoplasmic ultrastructure of mycobacteria processed by conventional embedding and freeze-substitution protocols. *Journal of Bacteriology* **174**, 6508–6517.

Plésiat, P. & Nikaido, H. (1992) Outer membranes of Gram-negative bacteria are permeable to steroid probes. *Molecular Microbiology* **6**, 1323–1333.

Rastogi, N. & Goh, K.S. (1990) Action of 1-isonicotinyl-2-palmitoyl hydrazine against the *Mycobacterium avium* complex and enhancement of its activity by *m*-fluorophenylalanine. *Antimicrobial Agents and Chemotherapy* **34**, 2061–2064.

Rastogi, N., Fréhel, C. & David, H.L. (1986) Triple-layered structure of the mycobacterial cell wall: evidence for the existence of a polysaccharide-rich outer layer in 18 mycobacterial species. *Current Microbiology* **13**, 237–242.

Rastogi, N., Frehel, C., Ryter, A., Ohayon, H., Lesourd, M. & David, H.L. (1981) Multiple drug resistance in *Mycobacterium avium*: is the wall architecture responsible for the exclusion of antimicrobial agents? *Antimicrobial Agents and Chemotherapy* **20**, 666–677.

Rastogi, N., Goh, K.S. & David, H.L. (1990) Enhancement of drug susceptibility of *Mycobacterium avium* by inhibitors of cell envelope synthesis *Antimicrobial Agents and Chemotherapy* **34**, 759–764.

Raynaud, C., Etienne, G., Peyron, P., Lanéelle, M.-A. & Daffé, M. (1998) Extracellular enzyme activities potentially involved in the pathogenicity of *Mycobacterium tuberculosis*. *Microbiology* **144**, 577–587.

Riess, F.G., Lichtinger, T., Cseh, R. *et al.* (1998) The cell wall porin of *Nocardia farcinica*: biochemical identification of the channel-forming protein and biophysical characterization of the channel properties. *Molecular Microbiology* **29**, 139–150.

Sedlaczek, L., Górminski, B.M. & Lisowska, K. (1994) Effect of inhibitors of cell envelope synthesis of β-sitosterol side chain degradation by *Mycobacterium* sp. NRRL MB 3683. *Journal of Basic Microbiology* **34**, 387–399.

Senaratne, R.H., Mobasheri, H., Papavinasasundaram, K.G., Jenner, P., Lee, E.J. & Draper, P. (1998) Expression of a gene for a porin-like protein of the OmpA family from *Mycobacterium tuberculosis* H37Rv. *Journal of Bacteriology* **180**, 3541–3547.

Swenson, J.M., Thornsberry, C. & Silcox, V.A. (1982) Rapidly growing mycobacteria: testing of susceptibility to 34 antimicrobial agents by broth microdilution *Antimicrobial Agents and Chemotherapy* **16**, 611–614.

Takayama, K. & Kilburn, J.O. (1989) Inhibition of synthesis of arabinogalactan by ethambutol in *Mycobacterium smegmatis*. *Antimicrobial Agents and Chemotherapy* **33**, 1493–1499.

Takayama, K., Armstrong, E.L., Kunugi, K.A. & Kilburn, J.O. (1979) Inhibition by ethambutol of mycolic acid transfer into the cell wall of *Mycobacterium smegmatis*. *Antimicrobial Agents and Chemotherapy* **16**, 240–242.

Takiff, H.E., Cimino, M., Musso, M.C. *et al.* (1996) Efflux pump of the proton antiporter family confers low-level fluoroquinolone resistance in *Mycobacterium smegmatis*. *Proceedings of the National Academy of Sciences of the USA* **93**, 362–366.

Thanassi, D.G., Suh, G.S.B. & Nikaido, H. (1995) The role of outer membrane barrier in efflux-mediated tetracycline resistance in *Escherichia coli*. *Journal of Bacteriology* **177**, 998–1007.

Trias, J. & Benz, R. (1993) Characterization of the channel formed by the mycobacterial porin in lipid bilayer membranes. Demonstration of voltage gating and of negative point charges at the channel mouth. *Journal of Biological Chemistry* **268**, 6234–6240.

Trias, J. & Benz, R. (1994) Permeability of the cell wall of *Mycobacterium smegmatis*. *Molecular Microbiology* **14**, 283–290.

Trias, J., Jarlier, V. & Benz, R. (1992) Porins in the cell wall of mycobacteria. *Science* **258**, 1479–1481.

Tsukamura, M., Mizuno, S. & Murayama, A. (1989) Different correlations of drug susceptibilities to colonial morphology in *Mycobacterium avium* complex strains. *Microbiology and Immunology* **33**, 1001–1011.

Wallace, R.J. Jr, Dalovision, J.R. & Pankey, G.A. (1979) Disk diffusion testing of susceptibility of *Mycobacterium fortuitum* and *Mycobacterium chelonei* to antibacterial agents. *Antimicrobial Agents and Chemotherapy* **22**, 186–192.

Yuan, Y. & Barry, C.E., III (1996) A common mechanism for the biosynthesis of methoxy and cyclopropyl mycolic acids in *Mycobacterium tuberculosis*. *Proceedings of the National Academy of Sciences of the USA* **93**, 12828–12833.

Yuan, Y., Crane, D.C., Musser, J.M., Sreevatsan, S. & Barry, C.E. III (1997) MMAS-I, the branch point between *cis*- and *trans*-cyclopropane-containing oxygenated mycolates in *Mycobacterium tuberculosis*. *Journal of Biological Chemistry* **272**, 10841–10849.

Zhang, Y., Wallace, R.J. Jr, Steingrube, V.A. *et al.* (1992) Isoelectric focusing patterns of β-lactamases in the rapidly growing mycobacteia. *Tubercle and Lung Disease* **73**, 337–344.

Chapter 13 / The cell-wall core of *Mycobacterium*: structure, biogenesis and genetics

ALAIN R. BAULARD, GURDYAL S. BESRA & PATRICK J. BRENNAN

1 Introduction

The most definitive recent review of the envelope of mycobacteria is that of Daffé and Draper (1998). This follows the direction of Barksdale and Kim (1977), Goren and Brennan (1979), Draper (1982), and Brennan and Draper (1994) in discussing interpretations derived from early and more recent ultrastructural studies. The most refreshing aspect of the review by Daffé and Draper (1998) is the evidence, discussed in considerable detail, of a capsule-like surface, such that the envelope of *Mycobacterium tuberculosis* possesses a form of glycocalyx composed in the main of glucans, mannans and arabinogalactan. The consequences of this capsular environment in our comprehension of electron micrographs of the mycobacterial cell wall is discussed in detail by these authors, as is also the consequences for phagocytosis of *M. tuberculosis* and its fate in the intracellular environment.

The evidence for a cell-wall skeleton or core (see Plate 2, between pp. 102 and 103) underpinning any capsule-like segment and the various non-covalently-associated glycolipids, phospholipids and proteins emerged from studies reviewed by Lederer (1975), Petit and Lederer (1984), and Minnikin (1982). This theme was further reviewed by Brennan and Nikaido (1995) and again comes up in the context of Chapter 4 of this volume. Our present-day understanding of the structure and biosynthesis of the cell-wall skeleton—the mycolylarabinogalactan–peptidoglycan (mAGP) complex—of mycobacterial cell walls is the topic of the present review. At this juncture, we will not discuss the structure or biosynthesis of lipoarabinomannan (LAM), an important immunomodulator in tuberculosis and leprosy. Chatterjee and Khoo (1998) have recently reviewed the relevant structural and biological aspects, and Besra *et al.* (1997) have recently described some elegant experiments leading to a fundamental understanding of the biochemical origins of the mannan portion of LAM. Much of the discussion below on the origins of the arabinan of arabinogalactan (AG) also applies to the arabinan portion of LAM.

2 Structure of mycolylarabinogalactan–peptidoglycan

A spate of intensive investigations from 1950 to 1970, resumed again in the late 1980s due to major technical developments in analytical chemistry, allowed the definition of the insoluble cell-wall matrix as a crosslinked peptidoglycan (PG) linked to AG, esterified at the distal ends to the mycolic acids (Misaki & Yukawa 1966; Lederer 1975; Petit & Lederer 1984; Daffé *et al.* 1990; McNeil & Brennan 1991). Historically, PG is thought to consist of alternating units of *N*-acetylglucosamine (GlcNAc) and a modified muramic acid (Mur) (Adam *et al.* 1969; Petit & Lederer 1984). AG was known to be attached to a proportion of the muramic acids via phosphodiester linkage (see Lederer 1975). The tetrapeptide side chains of PG consist of L-alanyl-D-isoglutaminyl-mesodiaminopimelyl-D-alanine (L-Ala-D-iGln-A$_2$pm-D-Ala) (Petit *et al.* 1969) with the Gln being further amidated (Weitzerbin-Falszpan *et al.* 1970). This type of PG (Alaγ) (Schleifer & Kandler 1972) is one of the most common found in bacteria. However, mycobacterial PG differs in two ways: some or all of the *N*-acetyl functions on the Mur residues are further oxidized to glycolic acid (MurNGly) (Adam *et al.* 1969), and the crosslinks include those between two A$_2$pm residues and between A$_2$pm and D-Ala (Ghuysen 1968). There are many unknowns: What are the proportions of MurNGly vs. MurNAc and A$_2$pm-A$_2$pm vs. A$_2$pm-D-Ala linkage? The accepted structure of the peptidoglycan unit of the cell wall of *M. tuberculosis* is shown in Fig. 13.1.

It was known in the 1950s that the major cell-wall polysaccharide is a branched-chain AG with the arabinose (Ara) residues forming the reducing termini. A structural formula, which proved incorrect, was proposed consisting of repeating 11–16 residues. There was much uncertainty about the structure of the galactan, i.e. whether it was 1→4-linked Gal*p* or 1→5-linked Gal*f* (Vilkas *et al.* 1973). More recently, we have confirmed that the polymer is unique not only in its elemental sugars, but, unlike most bacterial polysaccharides (Anderson & Unger 1983), it lacks repeating units, comprised instead of a few distinct structural motifs. Partial depolymerization of the per-*O*-alkylated AG and analysis of the generated oligomers by gas chromatography-mass spectrometry (GC-MS) and fast-atom bombardment (FAB)-MS (McNeil *et al.* 1987, 1990, 1991; Daffé *et al.* 1990, 1991) established that:

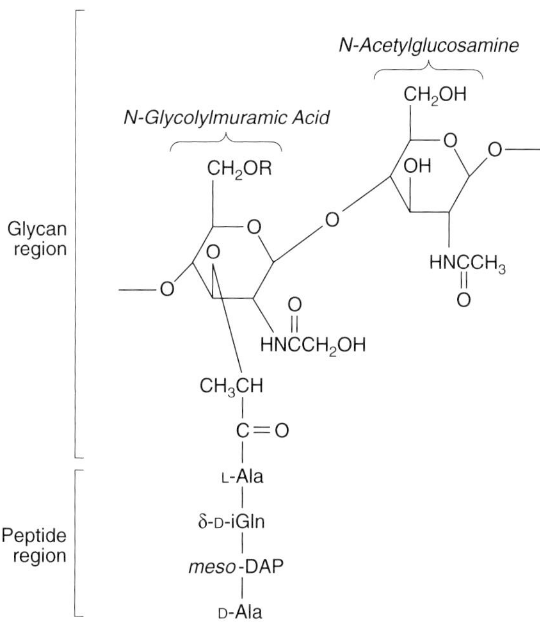

Fig. 13.1 Structure of the basic peptidoglycan unit of mycobacterial cell walls.

1 Ara and Gal are in the furanose form;

2 the non-reducing termini of arabinan consist of the hexasugar motif [β-D-Ara*f*-(1→2)-α-D-Ara*f*]$_2$-3,5-α-D-Ara*f*-(1→5)-α-D-Ara*f*;

3 the majority of the arabinan chains consist of 5-linked α-D-Ara*f* with branching introduced by 3,5-α-D-Ara*f*;

4 the arabinan chains are attached to C-5 of some of the 6-linked Gal*f*, and there are 2–3 such arabinan chains;

5 the galactan consists of linear alternating 5- and 6-linked β-D-Gal*f*;

6 the galactan region of AG is linked to the C-6 of some of the MurNGly residues of PG via a special diglycosyl-P bridge, α-L-Rha*p*-(1→3)-α-D-GlcNAc-(1→P);

7 the mycolic acids are located in clusters of four on the terminal hexa-arabinofuranoside, but only two-thirds of these are mycolated.

More recently, we (Besra *et al.* 1995) obtained oligosaccharide fragments containing up to 26

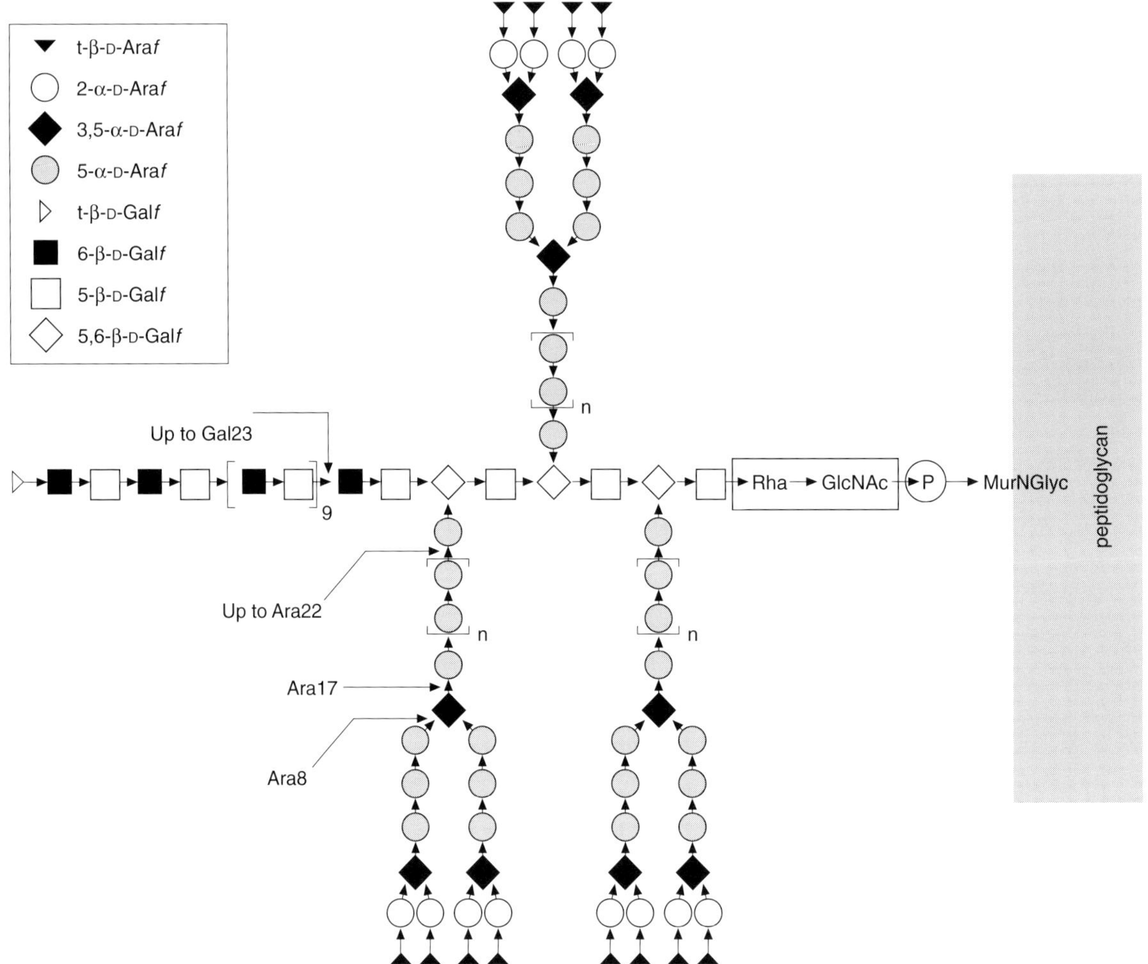

Fig. 13.2 A model of the cell-wall core, the arabinogalactan-peptidoglycan (AGP) complex. The mycolic acids in ester linkage to the terminal *D*-Ara*f* units are not shown.

residues from which molecular weights and alkylation patterns were determined by FAB-MS. The extended non-reducing ends of the arabinan were shown to consist of a tricosarabinoside ('23-mer'), with three such units attached to the galactan unit. The galactan was also isolated and was found to consist of 23 Gal residues of the repeating linear structure, $[\beta\text{-}D\text{-}Galf\text{-}(1\rightarrow5)\text{-}\beta\text{-}D\text{-}Galf\text{-}(1\rightarrow6)]_n$, devoid of branching, thereby demonstrating that the points of attachment of the arabinan chains are close to the reducing end of galactan, itself linked to PG via the linker disaccharide-P. Figure 13.2 represents one such structural unit without the attached mycolic acid residues. The entire flat structure of the PG–linkage unit (LU)–galactan–arabinan–mycolyl complex (mAGP) unit (with just one arabinan chain) is shown in Fig. 13.3.

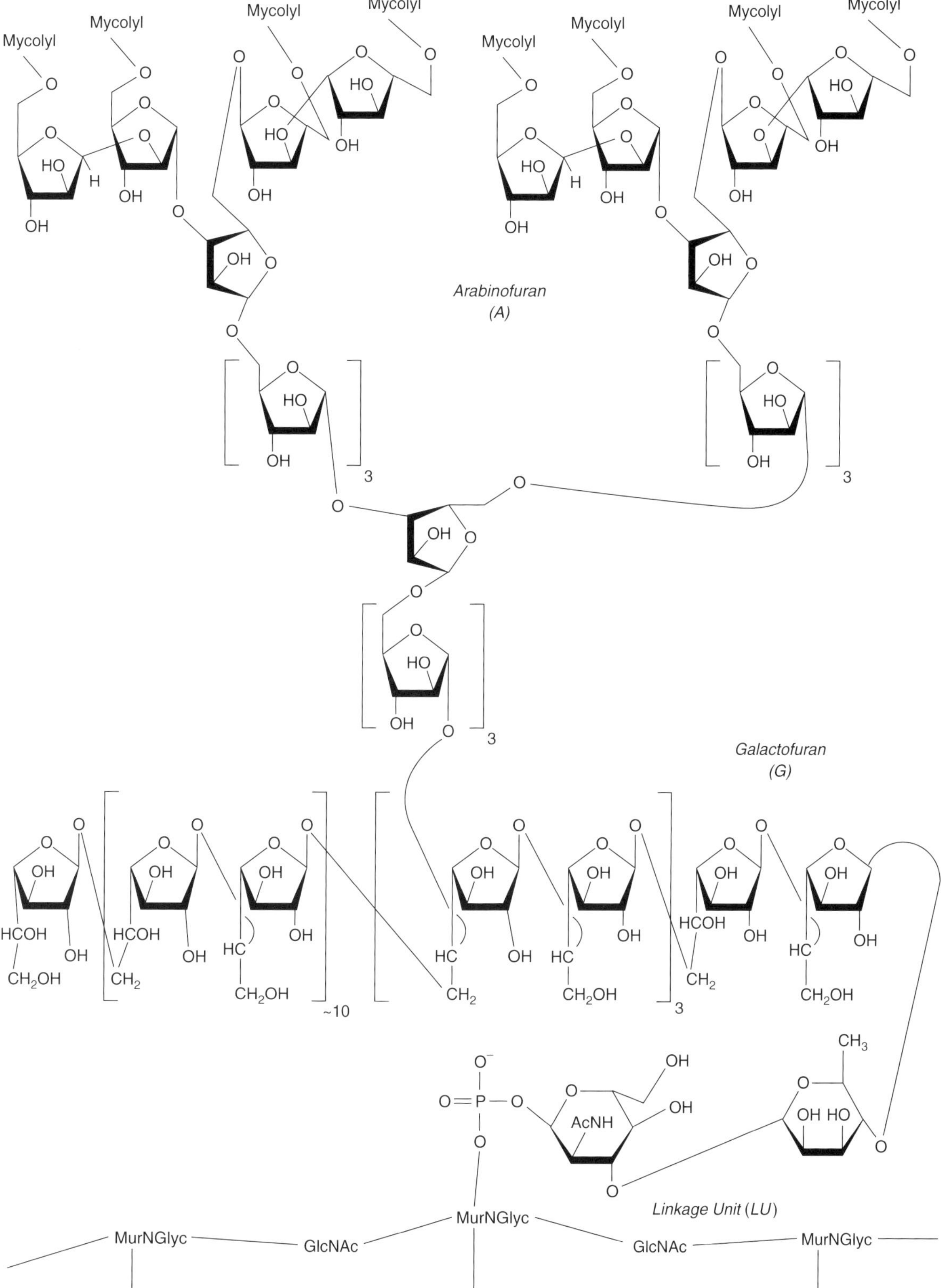

Fig. 13.3 Full structure of the mycolylarabinogalactan-peptidoglycan (mAGP) complex of mycobacterial cell walls.

3 Biosynthesis of the peptidoglycan of mycobacterial cell-wall skeleton

In light of the structural idiosyncracies of mycobacterial PG and inherent resistance of mycobacteria to the β-lactam antibiotics, it is surprising that peptidoglycan synthesis has not been examined in appreciable detail. Both Petit *et al.* (1970) and Takayama *et al.* (1970) demonstrated that exposure of *M. smegmatis* to D-cycloserine resulted in the accumulation of UDP-MurNGly-L-Ala-D-iGln-A$_2$pm, and Eun *et al.* (1978) later demonstrated that exposure to vancomycin resulted in the accumulation of the Park nucleotide equivalent, UDP-MurNGly-L-Ala-D-iGln-A$_2$pm-D-Ala-D-Ala. Based on this limited information, a variation of the well-known pathway of PG biosynthesis in *Escherichia coli* (van Heijenoort 1996) is proposed for *M. tuberculosis* (Fig. 13.4). In *E. coli*, the genes coding for precursor synthesis are found in two major clusters designated *mra* and *mrb*. Most of the individual genes are present in *M. tuberculosis* H37Rv (Fig. 13.5 and Table 13.1), and, hence, the outlined pathway is the expected one. Based on the genome sequence recently published by the Sanger Center (Accession No. AL 123456; see Chapter 5), we have localized on a circular map the various putative genes implicated in PG biosynthesis (Fig. 13.5). The *mra* cluster equivalent to the one present in *E. coli* has been identified in the *M. tuberculosis* H37Rv genome. The putative mycobacterial *Mur* genes are present in the same order as in *E. coli*. Even if not grouped in the same cluster, *rfe*, *murA* and *murI* are localized on a relatively compact locus.

As in *E. coli*, the mycobacterial *mra* cluster also contains some putative genes involved in the latter steps of whole PG synthesis and in the cell-division process. A good representation of the genes for class A and class B high-molecular-weight and low-molecular-weight penicillin-binding proteins (PBPs) has already been found in the *M. tuberculosis* and/or *M. leprae* databases (e.g. pbp1, *ponB*, *pbp1c*, *ponA*, pbp2, pbp3/ftsI, pbp7). This genetic evidence is also supported by biochemical evidence, mostly involving binding to radiolabelled β-lactams, of the presence of high- and low-molecular-weight PBPs (Fattorini *et al.* 1992; Chambers *et al.* 1995; Mukherjee *et al.* 1996).

Table 13.1 Genes responsible for peptidoglycan synthesis in *M. tuberculosis*.

Name	E. C. no.	Sanger no.	PID no.	Putative function in *M. tuberculosis*
1 *murA (murZ)*	2.5.1.7	Rv1315	1235980	UDP-*N*-acetylglucosamine 1-carboxyvinyltransferase
2 *murB*	1.1.1.158	Rv0482	1709183	UDP-*N*-acetylmuramate dehydrogenase
3 *murC*	6.3.2.8	Rv2152c	2104326	UDP-*N*-acetylmuramate-alanine ligase
4 *murD*	6.3.2.9	Rv2155c	2104323	UDP-*N*-acetylmuramylalanine-D-glutamate ligase
5 *murI*	5.1.1.3	Rv1338	1419041	Glutamate racemase
same as			1340104	
6 *murE*	6.3.2.13	Rv2158c	2104320	UDP-*N*-acetylmuramyl-tripeptide synthetase
7 *murF*	6.3.2.15	Rv2157c	2104321	UDP-MurNac-pentapeptide synthetase
8 *ddlA*	6.3.2.4	Rv2981c	1694850	D-Alanine:D-alanine ligase A
9 *alr*	5.1.1.1	Rv3423c	1449364	Alanine racemase
10 *murX (mraY)*	2.7.8.13	Rv2156c	2104322	Phospho-*N*-acetylmuramyl-pentapeptide transferase
11 *murG*	2.4.1.-	Rv2153c	2104325	UDP-*N*-acetylglucosamine-*N*-acetylmuramyl-(pentapeptide) pyrophosphoryl-decaprenol-*N*-acetylglucosamine transferase
12 *ftsW*		Rv2154c	2493593	Cell-division protein
12 *ftsQ*		Rv2151c	2104327	Cell-division protein
12 *ftsZ*		Rv2150c	2104328	Cell-division protein

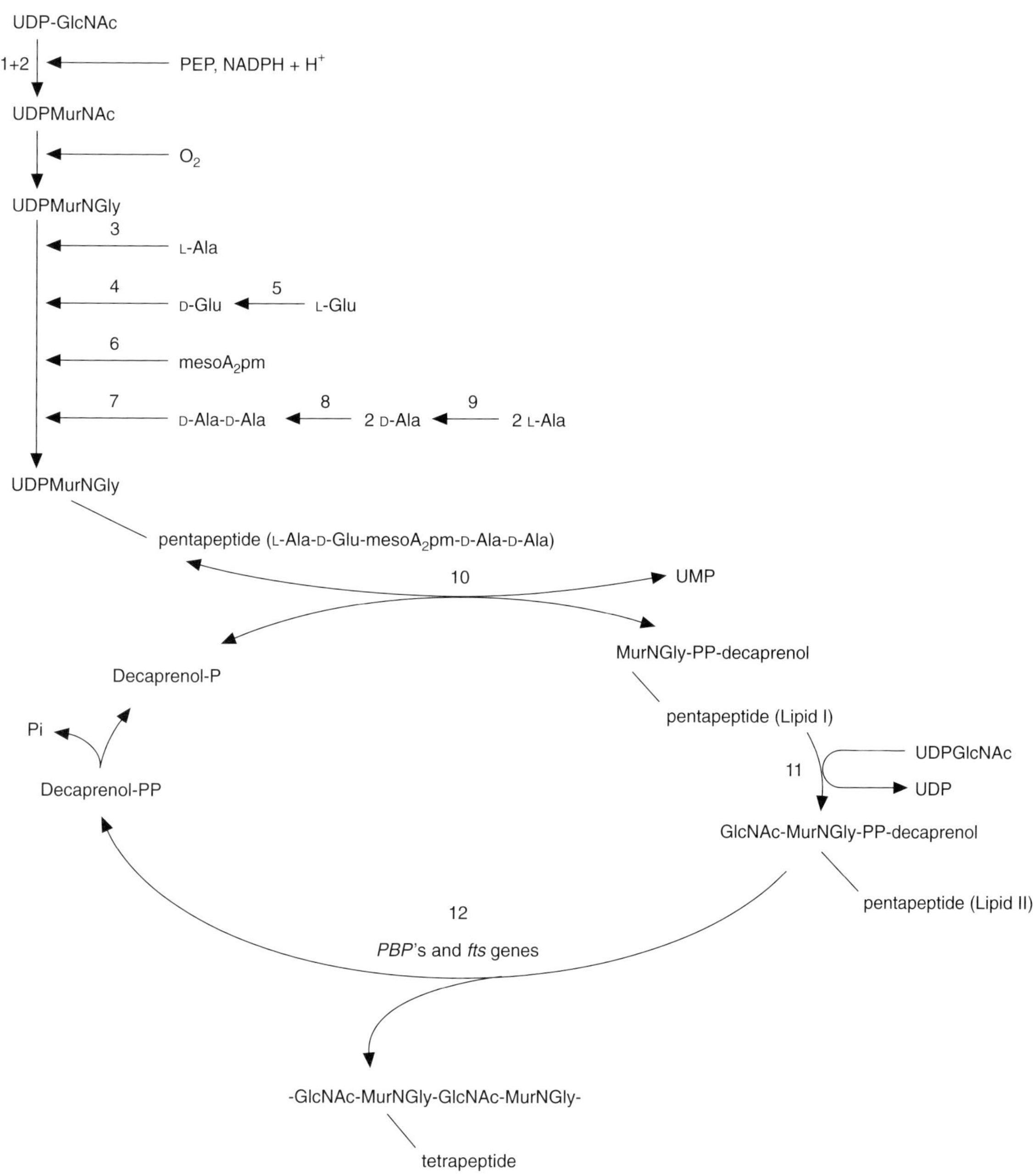

Fig. 13.4 The likely pathway of synthesis of peptidoglycan in mycobacteria. See Table 13.1 for names of associated enzymes and genes.

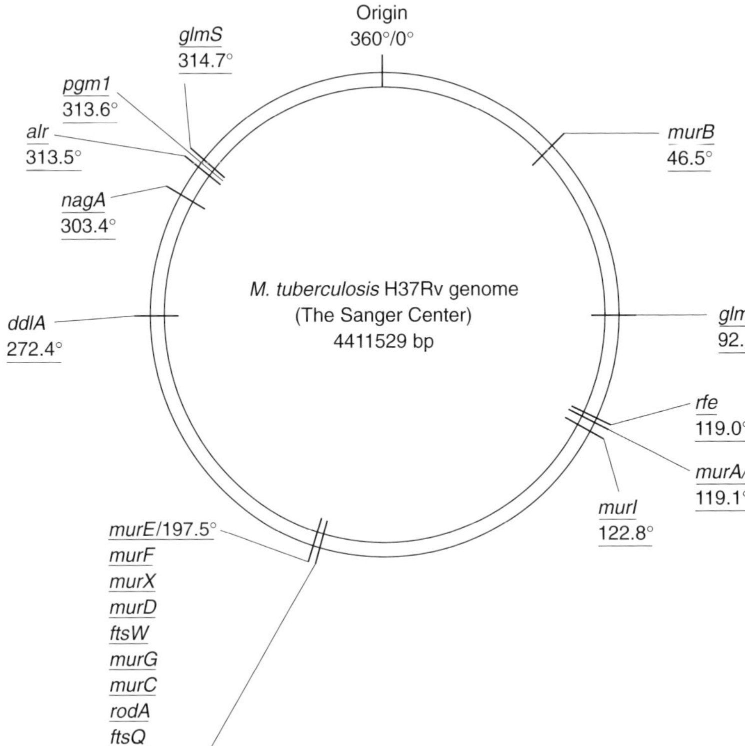

Fig. 13.5 Circular map of the *Mycobacterium tuberculosis* genome with the location of the genes thought to be involved in peptidoglycan synthesis. The origin of the *M. tuberculosis* H37Rv chromosome has been defined by the Sanger Center (see Chapter 5).

4 Biosynthesis of the linkage unit, galactan and arabinan of arabinogalactan

The LU represents the 'bull's eye', the 'Achilles heel', in terms of new drug development, since the whole of the mAG complex is attached to PG via this unit (see Plate 2 and Figs 13.2 and 13.3). In addition, the Araf and Galf residues of AG provide attractive drug targets due to their xenobiotic status in humans. We now believe that the synthesis of the LU, galactan and arabinan is a coupled, sequential event, and, like PG synthesis, the relevant synthetic steps are divided into cytoplasmic, membranous and extramembranous phases (Fig. 13.6).

4.1 Cytoplasmic phase of arabinogalactan–linkage unit synthesis

The immediate donors of the GlcNAc-1-P and Rha units of LU have been identified as UDP-GlcNAc and dTDP-Rha (Mikušová *et al.* 1996), respectively, and both of these originate in the bacterial cytoplasm (Shibaev 1986; Singh & Hogan 1994). We do not regard decaprenyl-P-P-GlcNAc as a donor *per se*; decaprenyl phosphate is the recipient of the GlcNAc-1-P unit donated by UDP-GlcNAc, and the product is the progenitor of much of cell-wall core biosynthesis. Likewise, UDP-Galf is the immediate donor of the Galf units of galactan (Nassau *et al.* 1996; Koplin *et al.* 1997). The Araf units of arabinan obviously originate in the pentose phosphate pathway/hexose monophosphate shunt, but the exact mechanism is not known (Scherman *et al.* 1995, 1996). However, it is

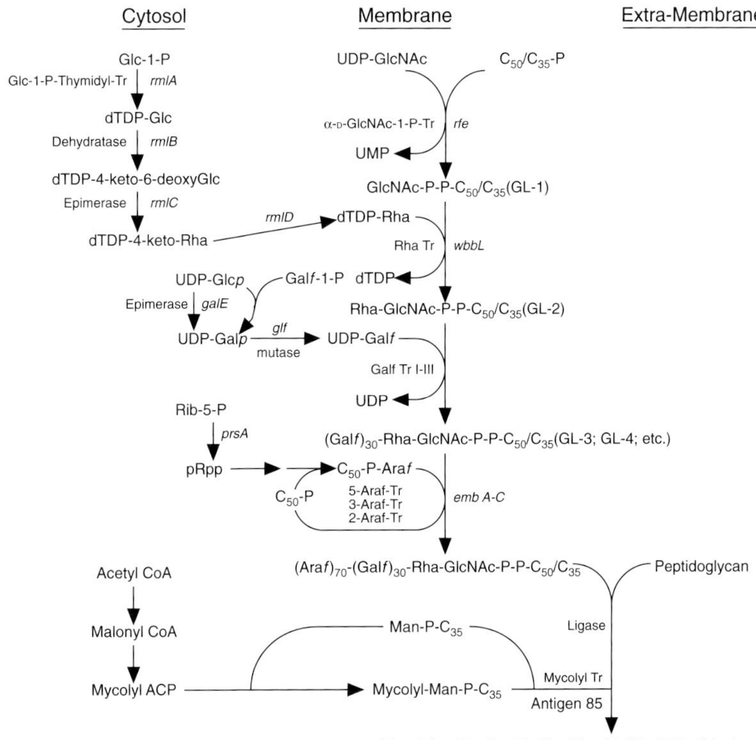

Fig. 13.6 Pathway for the synthesis of the mAGP complex of mycobacterial cell walls. Adapted from the original speculative pathway. Tr, transferase. (From McNeil & Brennan, 1991.)

clear that the immediate donor of the polymerized Ara*f* is the lipid-linked decaprenyl-P-Ara*f* (Fig. 13.7), not a nucleotide sugar (Wolucka *et al.* 1994).

4.1.1 *Enzymology and genetics of linkage unit [Rha-(1→3)GlcNAc-1-P] synthesis*

The sugar donors of the GlcNAc-1-P and Rha units of the LU are UDP-GlcNAc (Mikušová *et al.* 1996), as used widely in the synthesis of analogous LUs bridging cell-wall teichoic acids and the PG of many Gram-positive bacteria (Archibald *et al.* 1993) and dTDP-Rha*p*, as commonly used in bacterial polysaccharide synthesis (Shibaev 1986).

The latest information derived from the *M. tuberculosis* and *M. leprae* genome projects predicts that the biosynthesis of UDP-GlcNAc proceeds exactly as described for *E. coli* and other organisms (Shibaev 1986) (Table 13.2 and Fig. 13.8), the exception being

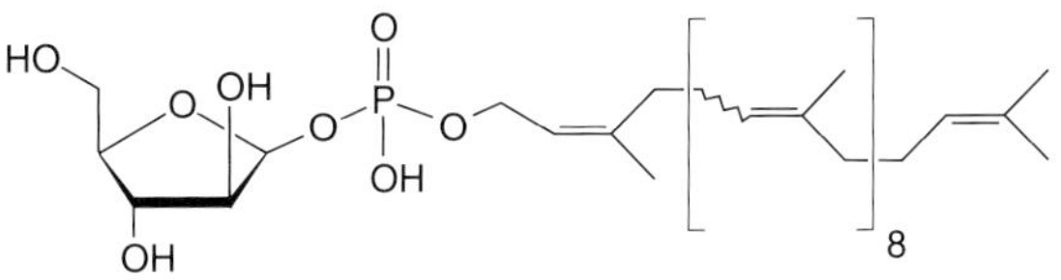

Fig. 13.7 Structure of the C_{50}-P-Ara*f* of mycobacteria, the central donor of Ara*f* units of arabinogalactan (AG) and lipoarabinomannan (LAM).

that decaprenyl-P, rather than undecaprenyl-P, is probably utilized by mycobacteria. However, due to relatively low amino acid sequence conservation of this family of enzymes, it is most likely that putative mycobacterial genes will show higher similarity with other mycobacterial genes than with genes from other genera. Thus, only detailed biochemical analy-

Table 13.2 Genes responsible for synthesis of polyprenol-P-P GlcNAc of *M. tuberculosis*.

	Name	E. C. no.	Sanger no.	PID no.	Putative function in *M. tuberculosis*
1	*glmS*	2.6.1.16	Rv3436c	2388656	1-Glutamine: D-fructose-6-phosphate amidotransferase
		same as		2104357	
2	*nagB*	5.3.1.10	NH[a]		Glucosamine-6-phosphate isomerase
3	?	2.3.1.4			Glucosamine-6-phosphate *N*-acetyltransferase
4	*nagA*	3.5.1.25	Rv3332	2894242	*N*-acetylglucosamine-6-phosphate deacetylase
5	*mrsA* (*pgml*)	5.4.2.3	Rv3441c (WH)[b]	2104362	Phosphoacetylglucosamine mutase
6	?	5.4.2.-			Phosphoglucosamine mutase
7	*glmU*	2.7.7.23	Rv1018c	1870010	UDP-*N*-acetylglucosamine pyrophosphorylase
7	*glmU*[c]	2.3.1.-		1870010	*N*-Acetylglucosamine-1-phosphate acetyltransferase
8	*rfe*	2.4.1.-	Rv1302	1322426	Decaprenyl-phosphate α-*N*-acetylglucosaminyltransferase
9	*murA* (*murZ*)	2.5.1.7	Rv1315	1235980	UDP-*N*-acetylglucosamine 1-carboxyvinyltransferase

[a] (NH): No homologous gene detected in the *M. tuberculosis* H37Rv.
[b] (WH): Weak homology.
[c] *glmU* is a bifunctional enzyme (Mengin-Lecreulx and van Heijenoort, 1994).

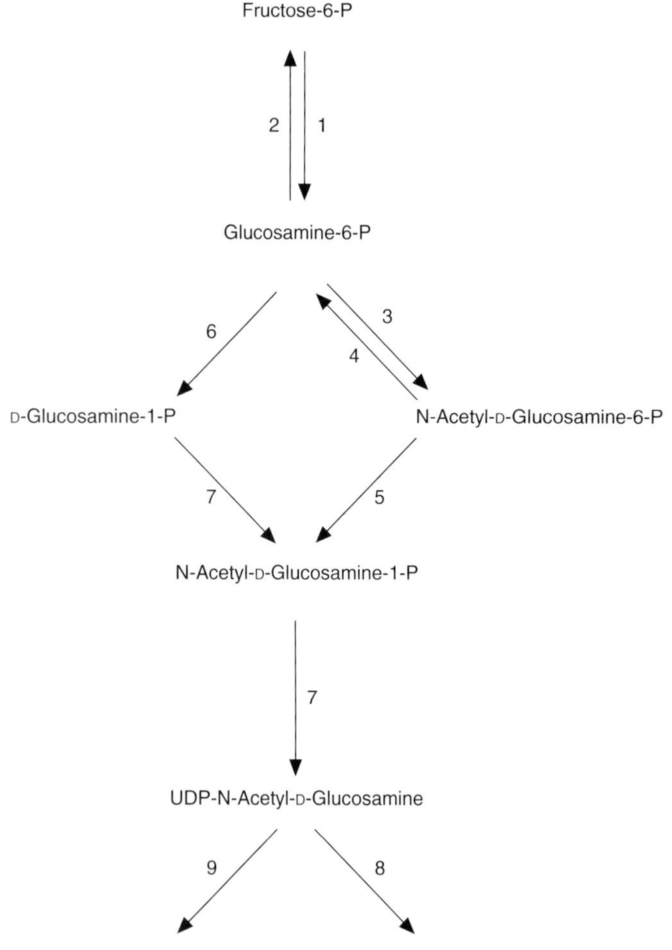

Fig. 13.8 Likely pathway for the synthesis of C_{50}-P-P-GlcNAc, the acceptor for cell-wall core synthesis.

ses will answer the function of the majority of the putative proteins annotated in the database. Thus, it has been difficult to try to identify genes involved in LU biosynthesis based on homology with corresponding or related genes from other bacteria.

As opposed to the situation in *E. coli*, *glmS* and the genes encoding the bifunctional *glmU* are not clustered in the *M. tuberculosis* genome. The genes (see Fig. 13.6) responsible for converting Glc-1-P to dTDP-Rha*p*, i.e. *rmlA*, *rmlB*, *rmlC* and *rmlD*, previously designated *rfb* A–D, are found as a cluster in *E. coli* where they are responsible for the synthesis of an unusual O antigen with branched Rha*p* residues (Liu & Reeves 1994; Stevenson *et al.* 1994; Reeves *et al.* 1996). With the exception of *rmlB/C*, in the case of *M. tuberculosis*, they are found in markedly divergent areas of the genome (Table 13.3). *RmlA* (Rv 0334), the α-D-Glc-1-P thymidyltransferase (see Fig. 13.6) has been cloned and shown to encode the expected enzymatic activity (Ma *et al.* 1997).

The *rmlB* and *rmlC* genes (Table 13.3 and Fig. 13.6) were identified independently of the genome project, in the course of our studies of the genetic fingerprinting of *M. tuberculosis* isolates from Korea (T. Y. Lee *et al.* 1997). These are directly linked to each other. Both of these have been cloned and expressed, and the appropriate enzymatic activity, dTDP-α-D-

Glc-4,6-dehydratase and dTDP-4-keto-6-deoxy-D-Glc-3,5-epimerase, attributed to them (McNeil 1999).

4.1.2 *Origins of uridine diphosphate–Galf*

UDP-Gal*p*, the direct precursor of UDP-Gal*f*, has long been known to arise from UDP-Glc*p* catalysed by UDP-Gal*p* eperimase, the *galE* gene product (Shibaev 1986). The corresponding enzyme from *M. smegmatis* was partially purified and sequenced (Weston *et al.* 1998), which allowed for identification of the corresponding gene in the *M. tuberculosis* H37Rv/Sanger database. In addition, the two enzymes of the galactose salvage pathway, galactokinase and UDP-Glc-Gal-1-P uridyl transferase, are present in *M. smegmatis* (Weston *et al.* 1998). The enzyme responsible for the conversion of UDP-Gal*p* to UDP-Gal*f* (Table 13.3), known as UDP-Gal*p* mutase, encoded by the *glf* gene, was first studied in considerable detail in *E. coli* (Stevenson *et al.* 1994) and, later, *Klebsiella pneumonia* (Koplin *et al.* 1997), allowing the recognition, sequencing and expression of the *M. tuberculosis* version (Lee *et al.* 1996; Nassau *et al.* 1996). The mycobacterial *glf* gene is annotated Rv3809c in the H37Rv database. All indications are that the complex, molecular conversion of UDP-Gal*p* to UDP-Gal*f* is the same in all of these organisms and requires

Table 13.3 Known genes involved in synthesis of the polyprenol-P-P-LU-AG complex of *M. tuberculosis*.

Name	Sanger no.	Putative function in *M. tuberculosis*
rmlA	Rv0334	Glucose-1-phosphate thymidylyl-transferase
rmlB	Rv3464	dTDP-glucose 4,6-dehydratase (V)
rmlC	Rv3465	dTDP-4-dehydrorhamnose 3,5-epimerase (V)
rmlD	Rv3266c	dTDP-4-dehydrorhamnose reductase (NP)
rfe	Rv1302	Decaprenyl-phosphate α-*N*-acetylglucosaminyltransferase
wbbL	Rv3265c	dTDP-rhamnosyl transferase
glf	Rv3809	UDP-galactopyranose mutase
embA	Rv3794	involved in arabinogalactan synthesis
embB	Rv3795	involved in arabinogalactan synthesis
embC	Rv3793	involved in arabinogalactan synthesis
embR	Rv1267c	regulator of embAB genes (AfsR/DndI/RedD family)
fbpA	Rv3804c	antigen 85A, mycolyltransferase
fbpB	Rv1886c	antigen 85B, mycolyltransferase
fbpC2	Rv0129c	antigen 85C, mycolyltransferase

either flavine adenine dinucleotide (FAD) or reduced nicotinamide adenine dinucleotide phosphate (NADPH).

4.1.3 Origins of the Araf units of arabinogalactan and lipoarabinomannan

The discovery of decaprenyl (C_{50})-P-Ara*f* (Wolucka *et al.* 1994), the development of simple Ara*f*-containing tri- and diglycoside acceptors (R. E. Lee *et al.* 1997), and the presence of avid endogenous acceptors of [14C]Ara*f* donated by C_{50}-P-[14C]Ara*f* in cell walls and membranes from disrupted mycobacteria allowed us to demonstrate that C_{50}-P-Ara*f* was the direct precursor of apparently all of the Ara*f* units of mAGP and LAM (Xin *et al.* 1997). The question of the origin of C_{50}-P-Ara*f* itself has led to intriguing experiments and speculation by McNeil and colleagues (Scherman *et al.* 1996) (Fig. 13.9). In eukaryotes and prokaryotes, polyprenyl-P-linked sugars invariably arise in the corresponding nucleotide sugar. However, concerted searches for Ara*f*-containing

nucleotide sugars had failed to confirm reports of their existence (Singh & Hogan 1994). Neither have nucleotide ribofuranosides been reported, although mycobacteria do contain ribans (Mikušová *et al.* 1995). Yet C_{50}-P-D-Rib*f* has been found in *M. smegmatis* (Wolucka & de Hoffman 1995). The evidence for the presence of C_{50}-P-Ara*f* and C_{50}-P-Rib*f* and the corresponding arabinans and ribans combined with evidence that Ara*f* arises in the non-oxidative pentose phosphate pathway (Scherman *et al.* 1995), suggest that the ribose of ribofuran and the arabinose of arabinofuran may not arise in a nucleotide precursor but in 1-P-ribosyl-P-P. McNeil and colleagues (Scherman *et al.* 1996; McNeil 1999) have incubated 1-P-[14C]ribosyl-P-P with cell-free extracts of *M. smegmatis* and demonstrated the formation of both C_{50}-P-[14C]Rib*f* and C_{50}-P-[14C]Ara*f*, and the direct intermediate in the formation of C_{50}-P-[14C]Rib*f* was apparently C_{50}-P-[14C]Rib-5-P. It is assumed, although not yet proven, that C_{50}-P-[14C]Ara*f* arises in an analogous fashion by way of epimerization of P-Rib-P-P to P-Ara-P-P (Fig. 13.9).

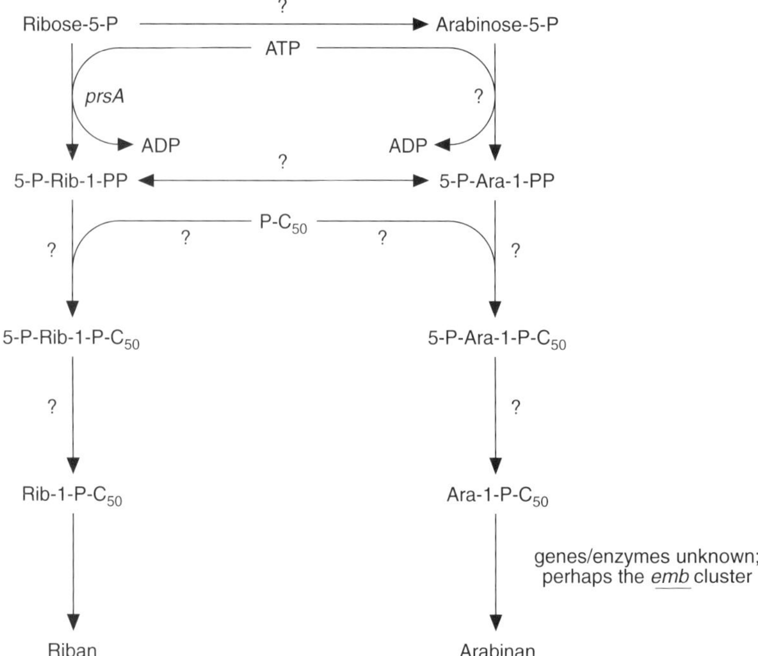

Fig. 13.9 Possible pathway for the synthesis of C_{50}-P-Ara*f*.

4.2 Membrane and extracytoplasmic membrane phase of arabinogalactan–linkage unit synthesis: polymerization, enzymology and genetics of the glycosyl transferases

4.2.1 Definition of the mechanism of galactan and arabinan polymerization

Initial clues as to the course of biosynthesis of the entire LU–AG complex came from the use of cytoplasmic membrane preparations of *M. smegmatis* and *M. tuberculosis* (Mikušová *et al.* 1996), although we do not yet know whether the appropriate glycosyl transferases are membranous, cytoplasmic or operate at the interphase of the two. Membranes from *M. smegmatis* and *M. tuberculosis* catalysed the incorporation of radioactivity from UDP-[^{14}C]GlcNAc into two glycolipids (GL-1 and GL-2) which were mild-acid labile and mild-alkali stable, features consistent with polyprenol-based glycolipids. When tunicamycin was added, a dramatic inhibition of incorporation was observed, suggesting that the initial step in the synthesis of mycobacterial cell wall AG involves formation of a polyprenol-P-P-GlcNAc unit (GL-1), probably either C_{35}-P or C_{50}-P. Incorporation of [^{14}C]Rha from dTDP-[^{14}C]Rha took place only into GL-2, suggesting that GL-2 was a polyprenol-P-P-GlcNAc-Rha, which was then confirmed chemically (Mikušová *et al.* 1996). Addition of a cell-wall-membrane enzyme preparation, prepared by Percoll density gradient centrifugation, resulted in the emergence of more polar glycolipids, GL-3 and GL-4 (see Fig. 13.6). The inclusion of UDP-[^{14}C]Gal*p* resulted in exclusive labelling of GL-3 (trisaccharide) and GL-4 (tetrasaccharide), indicating stepwise growth of the galactan chain on the polyprenol-P-P-GlcNAc-Rha unit, one galactosyl unit at a time. Glycosyl linkage analysis of [^{14}C]Gal-labelled glycolipids 3 and 4 demonstrated the presence of terminal (t) t-Gal*f* and 5-linked Gal*f*, consistent with the proposed structure of AG and suggesting the presence of the UDP-Gal*p* mutase (*glf*) within the preparations converting UDP-Gal*p* to UDP-Gal*f*. More recently (Mikušová *et al.* 1998), we demonstrated that thoroughly washed membrane preparations lost much of the ability to convert UDP-Gal*p* to UDP-Gal*f*, confirming that the enzyme was primarily cytosolic. We have now successfully solubilized the polymerized product resulting from labelling with either UDP-[^{14}C]GlcNAc or UDP-[^{14}C]Gal*p*, and analysis pointed to the emergence of even higher intermediates, GL-5, -6, and so on, eventually resulting in a polymer which possessed the characteristics of mild acid-lability, mild-alkali stability, solubility in an extremely polar organic solvent and exclusion from Bio-Gel P-100, all suggesting a highly polymerized lipid-linked version of GL-1–4 (Besra & Brennan 1997a,b) (see Fig. 13.6). The polymer was found to consist of 35–50 residues, and glycosyl linkage analysis produced t-Gal*f*, 5-linked Gal*f*, 6-linked Gal*f*, and 5,6-linked Gal*f*, suggesting that it contained the full alternating 5- and 6-linked linear galactan components of mycobacteria cell walls with a modicum of branching (exemplified by the presence of 5,6-linked Gal*f* in the methylation products). Moreover, [^{14}C]Ara*f* donated by synthetic C_{50}-P-[^{14}C]Ara*f*, was incorporated into this same polymer characterized by solubility in polar lipid solvents and mild-acid lability and hence was probably lipid linked. Moreover, all of the Ara*f* units as found in both AG and LAM (i.e. 2-, 5-, 3,5-linked) were labelled (Xin *et al.* 1997), indicating synthesis of the entire arabinan while still linked to the polyprenyl-P carrier (although this preliminary evidence does not rule out the possibility of further processing of the arabinan chains of AG and/or LAM, presumably closer to final maturation).

4.2.2 Paucity of information on the responsible glycosyl transferases

The older, historical evidence (Takayama & Kilburn 1989) for a connection between the antituberculosis action of ethambutol (EMB), an effective antituberculosis agent, and arabinan synthesis, combined with the discovery of C_{50}-P-Ara*f* (Wolucka *et al.* 1994), the chemical synthesis of C_{50}-P-[^{14}C]Ara*f* (Lee *et al.* 1995) and arabinosyl group acceptors (Lee *et al.* 1997), and the consequent development of a basic arabinosyl transfer assay (R.E. Lee *et al.* 1995, 1997), all provided us with the opportunity to explore the possibility that the arabinosyltransferases involved in synthesis of the arabinan chains of AG and LAM could be the target of EMB action. A clear outcome of these experiments was the demonstration that all of the *de novo* synthesized arabinan was soluble in polar lipid solvents, whereas there was little appreciable synthesis of insoluble (i.e. PG-bound) arabinan, again indicating that the majority of the newly synthesized arabinan was lipid-linked but that the *in vitro* cell-free systems were ineffective in catalysing the transfer of the LU-GA polymer from its polyprenyl-P carrier to peptidoglycan, a problem that has plagued others investigating the enzymatic and genetic basis of the final ligase step in teichoic acid–LU–PG synthesis in Gram-positive bacteria (Archibald *et al.* 1993).

Following on the earlier work of Takayama and Kilburn (1989), we demonstrated that *in vivo* EMB primarily inhibited the synthesis of the arabinan of AG, while inhibition of the corresponding arabinan components of LAM occurred later, suggesting a secondary target and separate pathways for AG and LAM biosynthesis (Mikušová *et al.* 1995). Moreover, while the synthesis of the arabinans of AG was normal in a spontaneous, EMB-resistant *M. smegmatis* mutant, the addition of EMB to a culture of this strain resulted in partial inhibition of the synthesis of the arabinan of LAM (but not of AG), resulting in the emergence of novel, truncated forms of LAM (Khoo *et al.* 1996). Truncation in the LAM structure was subsequently demonstrated as primarily a consequence of selective and partial inhibition of the synthesis of the linear Ara$_4$ motif, which constitutes a substantial portion of the arabinan termini of LAM but not of AG (Besra *et al.* 1995; Khoo *et al.* 1996). EMB also inhibited arabinosyl transferase activity in the C$_{50}$-P-Ara$f$ assay to a residual activity of 40% at 50 µg/mL (Lee *et al.* 1995). This evidence suggests the existence of an array of arabinosyl transferases with various degrees of susceptibility to EMB, and engaged exclusively in the synthesis of the arabinan of LAM.

Through the use of target overexpression by a plasmid vector as a selection tool, Inamine and col-leagues (Belanger *et al.* 1996) have recently cloned the *M. avium emb* region, which contains three open reading frames, *embR, embA, embB*, which render the otherwise susceptible *M. smegmatis* resistant to EMB. The C$_{50}$ P-Ara$f$ assay demonstrated that *embA* and *embB* are associated with high-level EMB-resistant arabinosyl transferase activity and *embR* appears to modulate their level of expression. Thus, it would appear that *embA* and *embB* encode the drug target of EMB, and these products represent the putative arabinosyl transferases associated with AG synthesis (see Fig. 13.3). However, as *embA* and *B* have not yet been expressed, purified and shown to be active arabinosyl transferases, the possibility still remains that they are only part of a complex of enzymes required for AG synthesis. Nevertheless, the predicted amino acid sequences of embA and B predict several transmembrane loops, all in accord with membrane-associated arabinosyl transferases.

5 Biosynthesis of mycolic acids

Takayama *et al.* (1980; Takayama & Qureshi 1979, 1984) investigated mycolic acid biosynthesis in the late 1970s and proposed a pathway involving four stages.

1 Synthesis of C$_{24}$–C$_{26}$ straight-chain saturated fatty acids to provide the short alkyl chain.

2 Synthesis of C$_{40}$–C$_{60}$ meromycolic acids to provide the main carbon backbone.

3 Modification of this backbone to introduce other functional groups. For instance, the introduction of the distal (Yuan *et al.* 1995) and proximal (George *et al.* 1995) cyclopropane rings has now been ascribed to two genes, *cma*1 (Rv 3392c) and *cma*2 (Rv 0503c), with homology to the cyclopropane fatty acid synthase from *E. coli* (Wang *et al.* 1992).

4 The final mycolic acid condensation step.

The original cell-free system (Takayama & Qureshi 1979; Takayama *et al.* 1980) did not produce complete mycolic acids, but did provide evidence for an elongation system leading to meromycolates. Lacave *et al.* (1990) developed a cell-free system capable of synthesizing whole mycolic acids from [^{14}C]acetate using a particulate/cell-wall enzymatic fraction, work which was later extended (Wheeler *et al.* 1993) to examine the role of putative intermediates involved in mycolic acid biosynthesis. Much of the earlier pathway proposed by Takayama has recently been supported by the elegant work of Barry and colleagues (Mdluli *et al.* 1998) (Fig. 13.10). Most of our present-day understanding of mycolic acid synthesis is based on the enormous body of older (Winder 1982) and recent (Banerjee *et al.* 1994; Quemard *et al.* 1995; Barry & Mdluli 1996) work devoted to the mechanism of isoniazid (INH) action and the genetic and enzymatic basis of resistance to it.

The association between catalase/peroxidase, INH susceptibility/resistance and mycolic acid biosynthesis has been debated for years (Winder 1982). It is now known that INH susceptibility/resistance are mediated by the *kat*G gene which encodes a 80-kDa protein containing haem and structural motifs characteristic of several bacterial catalase/peroxidases (Winder 1982; Zhang *et al.* 1992). The *M. tuberculosis kat*G gene restored sensitivity to INH in a resistant mutant of *M. smegmatis*, and deletion of the gene resulted in INH resistance in *M. tuberculosis* (Zhang *et al.* 1992; Heym *et al.* 1993). Clearly, INH is a pro-drug that requires the *kat*G product for activation, and it has been proposed that AhpC, a homologue of the thioredoxin-dependent alkyl hydroperoxide reductase, can interact directly with activated INH, compensating for the loss of KatG peroxidase in INH resistance (D.R. Sherman *et al.* 1996).

Over 30 years ago, one of us (P.J.B.) helped to show that the primary effect of INH is on mycolic acid synthesis (Winder 1982). Recently, Banerjee *et al.* (1994) isolated a novel gene, *inhA*, which, through point mutations within the 5′-regulatory region, conferred resistance to both INH and ethionamide (ETH). The InhA protein has now been shown to catalyse the NADH-specific reduction of long-chain (C_{12}–C_{24}) 2-*trans*-enoyl acyl carrier protein (ACP) intermediates (Quemard *et al.* 1995) involved in fatty acid elongation consistent with its involvement in the early stages of mycolic acid biosynthesis. This important work served to reopen the debate on the polymorphic nature of INH targets and mechanism of resis-

tance and provided a clearer understanding of the biochemistry of the chain-elongation phases of mycolic acid synthesis. For instance, Barry and Mdluli (1996) recently again proposed that, in *M. tuberculosis*, INH specifically inhibits the insertion of a Δ5 double bond into a C_{24} fatty acid, and thus that the target in the case of *M. tuberculosis* was a Δ5-desaturase rather than a β-keto enoyl reductase, the latter may be the primary target in *M. smegmatis*. Although Takayama originally suggested a Δ5-desaturase as the molecular target of INH, he also suggested that INH may be involved in some aspects of the elongation of C_{30}–C_{56} meromycolates based on the inhibitory effects of INH on the synthesis of fatty acids and hydroxy lipids in a cell-free preparation of *M. tuberculosis* H37Ra (Takayama & Qureshi 1979; Takayama *et al.* 1980). The addition of NADH and NADPH appeared to neutralize the inhibitory effects of INH in this cell-free system, consistent with the nature of the InhA protein (Banerjee *et al.* 1994; Quemard *et al.* 1995). Thus, apparently contrasting views and information on the action of INH, also prevalent in the 1960s and 1970s, can be reconciled by either invoking difference in action in different mycobacteria, or pleiotropic effects. Nevertheless, the interest in INH has precipitated a deeper understanding of mycolic acid synthesis, as discussed above.

6 Predicted final steps in core cell-wall biosynthesis

Our recent contributions have been to the definition of the terminal stages of the mycolic acid pathway. The discovery of mycolyl phospholipid (Myc-PL) (C_{35}-P-Man-mycolyl) (Besra *et al.* 1994) was significant in that it now appears to be the mycolyl-group carrier and donor. Through some fortunate circumstances, we have also isolated a mycolyltransferase that in a cell-free system catalyses the transfer and exchange of mycolic acids from trehalose monomycolate, and apparently Myc-PL, to acceptors (Belisle *et al.* 1997). The *N*-terminal amino acid sequence of the purified mycolyl transferase was similar to that of the antigen 85 (Ag85) complex. When the three proteins comprising the Ag85

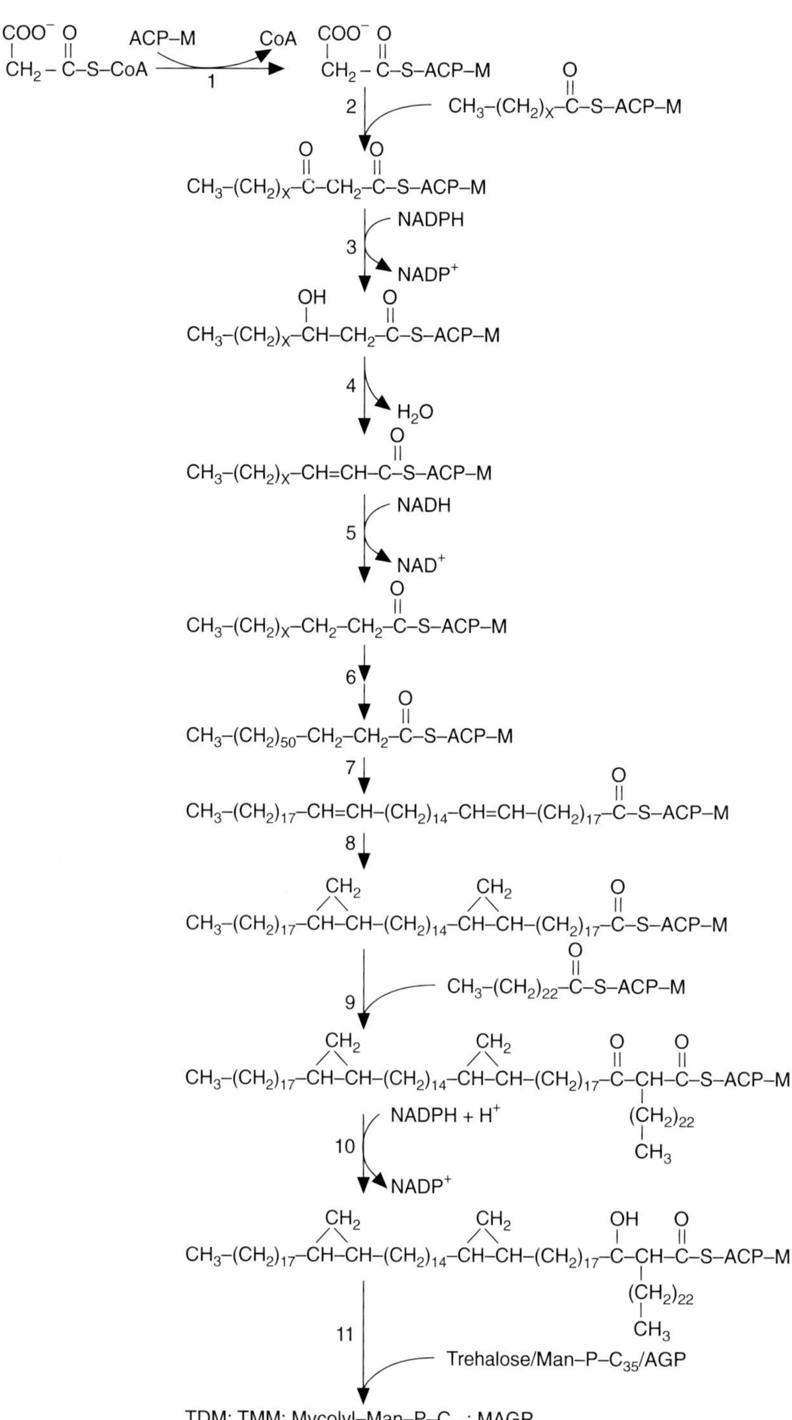

Fig. 13.10 Possible pathway for α-mycolic acid synthesis and deposition in cell wall. (1) FabD, a malonyl-CoA-ACP transacylase. (2) KAS, β-ketoacyl ACP synthase. (3) KR, β-ketoacyl ACP reductase. (4) DE, β-hydroxyacyl ACP dehydratase. (5) EA, enoyl ACP reductase. (6) Many repetitions of steps 2–5. (7) Possible desaturation as discussed by Barry *et al.* (1998). (8) Introduction of cyclopropane groups as discussed by Barry *et al.* (1998). (9) Addition of the α-chain to form the 3-keto mycolate intermediate. (10) Reduction of the 3-keto group. (11) Transfer of mature mycolates from one carrier to either trehalose or a polyprenyl-P-Man or the final AGP complex. (Adapted from Barry *et al.* 1998.)

complex of *M. tuberculosis* were purified to homogeneity and assessed for mycolyl transferase activity, all three exhibited substantial activity. Then, analysis of the full sequences of the Ag85 complex revealed the presence of a highly conserved region homologous to the region surrounding the catalytic site (GXSXG) of the human carboxyesterase D. Site-directed mutagenesis of the putatively equivalent Ser of the recombinant 85C gene (Ser125 Ala) resulted in loss of all mycolyl transferase activity. Thus, one can speculate (Fig. 13.10) on the role of these multiple antigens, now perhaps isoenzymes, in the final stages of mycolic acid deposition.

7 Future prospects

The practical value of an intensive investigation of core cell-wall synthesis lies in its potential for the discovery of new drugs for the treatment of tuberculosis and mycobacterioses in general and, in particular, to help counter the reality of global pockets of multi-drug-resistant tuberculosis. The rationale is that aspects of core synthesis are the focal point of many of the first- and second-line antituberculosis drugs such as INH, ethionamide (ETH), ethambutol, D-cycloserine (D-CS), pyrazinamide (PZA) and probably the thioureas, thiolactomycins (TLM) and thiosemicarbazones. All indications are that the cell-wall core, as distinct from many of the accoutrements that vary with species, isolate and serotype, is essential for bacterial viability. Thus, many of the reactions and assays described in the course of the development of the pathways presented above, and many of the enzymes now cloned, expressed and available in quantity are amenable to high-throughput screens for new antagonists and for study of structure/ activity relationships. While disruption of the synthesis of LAM, a molecule that in whole or part is considered as dispensable, should not undermine the viability of *M. tuberculosis* and other mycobacteria, nevertheless the pathogenesis of the organism should be compromised in light of recent knowledge of the role of LAM in phagocytosis and subsequent survival of *M. tuberculosis* in the host cell.

A more basic reason for exploring the biosynthesis of the mycobacterial cell wall is the fundamental worth of the endeavour. The early work of a host of giant chemists (R. Anderson, E. Lederer, Jean and Cecile Asselineau, T. Kotani, Y. Yamamura, M. Goren, C. E. Ballou) pointed to the extraordinary chemical composition of *Mycobacterium* cell surface. Our more recent chemical endeavors extend this discovery process, and, now with the unfolding of pathways and underlying enzymology and genetics, our awe is heightened. Much has been accomplished but much remains to be done. Recognition of the genes encoding at least six different galactosyl transferases and arabinosyl transferases as well as those involved in final ligation of the AG–LU complex to PG will be one of the more challenging tasks in light of the absence of models with comparable activities. This approach may require a return to classical style enzyme purification, perhaps with the aid of photoaffinity probes.

Another challenging task is the creation of appropriate mutants in order to begin an analysis of the role the various cell-wall entities in viability, intracellular life and disease induction aspects of *M. tuberculosis*. *M. tuberculosis* has been notoriously recalcitrant to a concerted approach to the generation of cell-wall mutants such as the creation of temperature-sensitive, conditional and auxotrophic variants, or the application of transposition mutagenesis or conditional antisense technology. These tools, or alternatively, creative chance, which has helped enormously in the past, will define future directions.

8 Acknowledgements

Work conducted in the authors' laboratories was supported by grants NIH, NIAID AI-18357, NIH, NIAID U19 AI-38087, and NIH, NIAID AI-35220. The authors wish to acknowledge the help of Dr Michael McNeil in providing unpublished data, Dr Sebabrata Mahapatra in providing information on the *mra* and *mrb* gene clusters and Christian Rittner for his information on genes responsible for the synthesis of UDP-GlcNAc. We thank Jack Hurdelbrink for help with preparing the figures and Marilyn Hein for preparing the manuscript.

9 References

Adam, A., Petit, J.F., Weitzerbin-Falszpan, J. *et al.* (1969) L'acide N-glycolylmuramique, constituant des parois de *Mycobacterium smegmatis*: Identification par spectrométrie de masse. *FEBS Letters* **4**, 87–92.

Anderson, L. & Unger, F.M. (1983) *Bacterial Lipopolysaccharides*. ACS Symposium Series 231. Washington, DC, American Chemical Society.

Archibald, A.R., Hancock, I.C. & Haywood, C.R. (1993) *Cell Wall Structure, Synthesis and Turnover in Bacillus subtilis and Other Gram-Positive Bacteria* (eds A. L. Sonensheim, J. A. Hoch & R. Losick). Washington, DC: American Society for Microbiology, pp. 381–410.

Banerjee, A., Dubnau, E., Quemard, A. *et al.* (1994) *inh*A, a gene encoding a target for isoniazid and ethionamide in *Mycobacterium tuberculosis. Science* **263**, 227–230.

Barksdale, L. & Kim, K.-S. (1977) *Mycobacterium. Bacteriological Reviews* **41**, 217–372.

Barry, C.E., III, Lee, R.E., Mdluli, K. *et al.* (1998) Mycolic acids: structure, biosynthesis and physiological functions. *Progress in Lipid Research* **37**, 143–179.

Barry, C.E., III & Mdluli, K. (1996) Drug sensitivity and environmental adaptation of mycobacterial cell wall components. *Trends in Microbiology* **4**, 275–281.

Belanger, A.E., Besra, G.S., Ford, M.E. *et al.* (1996) The *embAB* genes of *Mycobacterium avium* encode an arabinosyl transferase involved in cell wall arabinan biosynthesis that is the target for the antimycobacterial drug ethambutol. *Proceedings of the National Academy of Sciences of the USA* **93**, 11919–11924.

Belisle, J.T., Vissa, V.D., Sievert, T. *et al.* (1997) Role of the major antigen of *Mycobacterium tuberculosis* in cell wall biogenesis. *Science* **276**, 1420–1422.

Besra, G.S. & Brennan, P.J. (1997a) The mycobacterial cell envelope: a target for novel drugs against tuberculosis. *Journal of Pharmacy and Pharmacology* **149** (Suppl. 1), 25–30.

Besra, G.S. & Brennan, P.J. (1997b) The mycobacterial cell wall: biosynthesis of arabinogalactan and lipo-arabinomannan. *Biochemical Society Transactions* **25**, 845–850.

Besra, G.S., Khoo, K.-H., McNeil, M.R. *et al.* (1995) A new interpretation of the structure of the mycolyl-arabinogalactan complex of *Mycobacterium tuberculosis* as revealed through characterization of oligoglycosylalditol fragments by fast-atom bombardment mass spectrometry and ^{1}H nuclear magnetic resonance spectroscopy. *Biochemistry* **34**, 4257–4266.

Besra, G.S., Morehouse, C.B., Rittner, C. *et al.* (1997) Biosynthesis of mycobacterial lipoarabinomannan. *Journal of Biological Chemistry* **272**, 18460–18466.

Besra, G.S., Sievert, T., Lee, R.E. *et al.* (1994) Identification of the apparent transmembrane carrier in mycolic acid synthesis. *Proceedings of the National Academy of Sciences of the USA* **91**, 12735–12739.

Brennan, P.J. & Draper, P. (1994) Ultrastructure of *Mycobacterium tuberculosis*. In: *Tuberculosis: Pathogenesis, Protection and Control* (ed. B. R. Bloom). Washington, DC: American Society for Microbiology, pp. 271–284.

Brennan, P.J. & Nikaido, H. (1995) The envelope of mycobacteria. *Annual Review of Biochemistry* **64**, 29–63.

Chambers, H.F., Moreau, D., Yajho, D. *et al.* (1995) Can penicillins and other β-lactam antibiotics be used to treat tuberculosis? *Antimicrobial Agents and Chemotherapy* **39**, 2620–2624.

Chatterjee, D. & Khoo, K.-H. (1998) Mycobacterial lipoarabinomannan: an extraordinary lipoheteroglycan with profound physiological effects. *Glycobiology* **8**, 113–120.

Daffé, M., Brennan, P.J. & McNeil, M.R. (1990) Predominant structural features of the cell wall arabinogalactan of *Mycobacterium tuberculosis* as revealed through the characterization of oligoglycosyl alditol fragments by gas-chromatography and by ^{1}H and ^{13}C-NMR analyses. *Journal of Biological Chemistry* **265**, 6734–6743.

Daffé, M. & Draper, P. (1998) The envelope layers of mycobacteria with reference to their pathogenicity. *Advances in Microbial Research* **39**, 131–203.

Draper, P. (1982) The anatomy of mycobacteria. In: *The Biology of the Mycobacteria*, Vol. 1 (eds C. Ratledge & J. Stanford). London: Academic Press, pp. 9–52.

Eun, H.-M., Yado, A. & Petit, J.-F. (1978) D-D-Carboxypeptidase activity of membrane fragments of *Mycobacterium smegmatis. European Journal of Biochemistry* **86**, 97–103.

Fattorini, L., Orefici, G., Jin, S.H. *et al.* (1992) Resistance to β-lactams in *Mycobacterium fortuitum. Antimicrobial Agents and Chemotherapy* **36**, 1068–1072.

George, K.M., Yuan, Y., Sherman, D.R. *et al.* (1995) The biosynthesis of cyclopropanated mycolic acids in *Mycobacterium tuberculosis*. Identification and functional analysis of CMAS-2. *Journal of Biological Chemistry* **270**, 27292–27298.

Ghuysen, J.M. (1968) Use of bacteriolytic enzymes in determination of wall structure and their role in cell metabolism. *Bacteriological Reviews* **32**, 425–464.

Goren, M.B. & Brennan, P.J. (1979) Mycobacterial lipids: chemistry and biological activities. In: *Tuberculosis* (ed. G. P. Youmans). Philadelphia: W.B. Saunders, pp. 63–193.

Heym, B., Zhang, Y., Poulet, S. *et al.* (1993) Characterization of the *kat*G gene encoding a catalase-peroxidase required for the isoniazid susceptibility of *Mycobacterium tuberculosis. Journal of Bacteriology* **175**, 4255–4259.

Khoo, K.-H., Douglas, E., Azadi, P. *et al.* (1996) Truncated structural variants of lipoarabinomannan in ethambutol drug-resistant strains of *Mycobacterium smegmatis*: inhibition of arabinan biosynthesis by ethambutol. *Journal of Biological Chemistry* **271**, 28682–28690.

Koplin, R., Brisson, J.R. & Whitfield, C. (1997) UDP-galactofuranose precursor required for formation of the lipopolysaccharide *O* antigen of *Klebsiella pneumoniae* serotype 1 is synthesized by the product of the rfbD$_{KP01}$ gene. *Journal of Biological Chemistry* **272**, 4121–4128.

Lacave, C., Laneelle, M.A. & Laneelle, G. (1990) Mycolic acid synthesis by *Mycobacterium aurum* cell-free extracts. *Biochimica et Biophysica Acta* **1042**, 315–323.

Lederer, E. (1975) Cell walls of mycobacteria and related organisms; chemistry and immunostimulant properties. *Molecular Cell Biochemistry* **7**, 87–104.

Lee, R.E., Brennan, P.J. & Besra, G.S. (1997) Mycobacterial arabinan biosynthesis: the use of synthetic arabinoside acceptors in the development of an arabinosyl transfer assay. *Glycobiology* **7**, 1121–1128.

Lee, R.E., Mikušová, K., Brennan, P.J. *et al.* (1995) Synthesis of the mycobacterial arabinose donor -D-arabinofuranosyl-1-monophosphoryldecaprenol, development of a basic arabinosyl-transferase assay, and identification of ethambutol as an arabinosyl transferase inhibitor. *Journal of the American Chemistry Society* **117**, 11829–11832.

Lee, R.E., Monsey, D., Weston, A. *et al.* (1996) Enzymatic synthesis UDP-galactofuranose and assays for UDP-galactopyranosyl mutase based on HPLC and on linked enzymatic production of NADH. *Analytical Biochemistry* **242**, 1–7.

Lee, T.Y., Lee, T.J., Belisle, J.T. *et al.* (1997) A novel repeat sequence specific to *Mycobacterium tuberculosis* complex and its implications. *Tubercle and Lung Disease* **78**, 13–19.

Liu, D. & Reeves, P.R. (1994) *Escherichia coli* K12 regains its *O* antigen. *Microbiology* **140**, 49–57.

Ma, Y., Mills, J.A., Belisle, J.T. *et al.* (1997) Drug targeting rhamnose biosynthesis in mycobacteria: determination of the pathway for rhamnose biosynthesis in mycobacteria and cloning, sequencing, expressing and determining the genetic organization of the *Mycobacterium tuberculosis* gene encoding -D-glucose-1-phosphate thymidylyltransferase. *Microbiology* **143**, 937–945.

McNeil, M. (1999) Arabinogalactan in mycobacteria: structure, biosynthesis, and genetics. In: *Genetics of Bacterial Polysaccharides* (ed. J.B. Goldberg). Washington, DC: CRC Press, pp. 207–223.

McNeil, M.R. & Brennan, P.J. (1991) Structure, function, and biogenesis of the cell envelope of mycobacteria in relation to bacterial physiology, pathogenesis, and drug resistance; some thoughts and possibilities arising from recent structural information. *Research Microbiology* **142**, 451–463.

McNeil, M., Daffé, M. & Brennan, P.J. (1990) Evidence for the nature of the link between the arabinogalactan and peptidoglycan components of mycobacterial cell walls. *Journal of Biological Chemistry* **265**, 18200–18206.

McNeil, M., Daffé, M. & Brennan, P.J. (1991) Location of the mycolyl ester substituent in the cell walls of mycobacteria. *Journal of Biological Chemistry* **266**, 13217–13223.

McNeil, M., Wallner, S.J., Hunter, S.W. *et al.* (1987) Demonstration that the galactosyl and arabinosyl residues in the cell wall arabinogalactan of *Mycobacterium leprae* and *Mycobacterium tuberculosis* are furanoid. *Carbohydrate Research* **166**, 299–308.

Mdluli, K., Slayden, R.A., Zhu, Y. *et al.* (1998) Inhibition of a *Mycobacterium tuberculosis* β-ketoacyl ACP synthase by isoniazid. *Science* **280**, 1607–1610.

Mengin-Lecreulx, D. & van Heijenoort, J. (1994) Copurification of glucosamine-1-phosphate acetyltransferase and N-acetylglucosamine-1-phosphate uridyltransferase activities of *Escherichia coli*: characterization of the *glmU* gene product as a bifunctional enzyme catalyzing two subsequent steps in the pathway for UDP-N-acetylglucosamine synthesis. *Journal of Bacteriology* **176**, 5788–5795.

Mikušová, K., Mikus, M., Besra, G.S. *et al.* (1996) Biosynthesis of the linkage region of the mycobacterial cell wall. *Journal of Biological Chemistry* **271**, 7820–7828.

Mikušová, K., Morehouse, C., Besra, G.S. *et al.* (1999) Biosynthesis of the galactan of mycobacterial cell walls. Submitted for publication.

Mikušová, K., Slayden, R.A., Besra, G.S. *et al.* (1995) Biogenesis of the mycobacterial cell wall and the site of action of ethambutol. *Antimicrobial Agents and Chemotherapy* **39**, 2484–2489.

Minnikin, D.E. (1982) Lipids: complex lipids, their chemistry, biosynthesis and roles. In: *The Biology of Mycobacteria* (eds C. Ratledge & J. Stanford). London: Academic Press, pp. 95–184.

Misaki, A. & Yukawa, S. (1966) Studies on cell walls of mycobacteria. II. Constitution of polysaccharides from BCG cell walls. *Journal of Biochemistry* **59**, 511–520.

Mukherjee, T., Basu, D., Mahapatra, S. *et al.* (1996) Biochemical characterization of the 49 kDa penicillin-binding protein of *Mycobacterium smegmatis*. *Biochemical Journal* **320**, 197–200.

Nassau, P.M., Martin, S.L., Brown, R.E. *et al.* (1996) Galactofuranose biosynthesis in *Escherichia coli* K12: identification and cloning of UDP-galactopyranose mutase. *Journal of Bacteriology* **178**, 1047–1052.

Petit, J.F., Adam, A., Weitzerbin-Falszpan, J. *et al.* (1969) Chemical structure of the cell wall of *Mycobacterium smegmatis*. I. Isolation and partial characterization of the peptidoglycan. *Biochemical and Biophysical Research Communications* **35**, 478–485.

Petit, J.F., Adam, A., Weitzerbin-Falszpan, J. *et al.* (1970) Isolation of UDP-N-glycolylmuramyl-(Ala, Glu, DAP) from *Mycobacterium phlei. FEBS Letters* **6**, 55–57.

Petit, J.F. & Lederer, E. (1984) The structure of the mycobacterial cell wall. In: *The Mycobacteria, a Sourcebook* (eds G. P. Kubica & L. G. Wayne). New York: Marcel Dekker, pp. 301–322.

Quemard, A., Sacchettini, J.C., Dessen, A. *et al.* (1995) Enzymatic characterization of the target for isoniazid in *Mycobacterium tuberculosis. Biochemistry* **34**, 8235–8241.

Reeves, P.R., Hobbs, M., Valvano, M.A. *et al.* (1996) Bacterial polysaccharide synthesis and gene nomenclature. *Trends in Microbiology* **4**, 495–503.

Scherman, M., Kalbe-Bournonville, L., Bush, D. *et al.* (1996) Polyprenylphosphate-pentoses in mycobacteria are synthesized from 5-phosphoribose pyrophosphate. *Journal of Biological Chemistry* **271**, 29652–29658.

Scherman, M., Weston, A., Duncan, K. *et al.* (1995) Biosynthetic origin of mycobacterial cell wall arabinosyl residues. *Journal of Bacteriology* **177**, 7125–7130.

Schleifer, K.H. & Kandler, O. (1972) Peptidoglycan types of bacterial cell walls and their taxonomic implications. *Bacteriological Reviews* **36**, 407–477.

Sherman, D.R., Mdluli, K., Hickey, M.I. *et al.* (1996) Compensatory *ahp*C gene expression in isoniazid-resistant *Mycobacterium tuberculosis. Science* **272**, 1641–1643.

Shibaev, V.N. (1986) Biosynthesis of bacterial polysaccharide chains composed of repeating units. *Advances in Carbohydrate Chemistry and Biochemistry* **44**, 277–339.

Singh, S. & Hogan, S.E. (1994) Isolation and characterization of sugar nucleotides from *Mycobacterium smegmatis. Microbios* **77**, 217–222.

Stevenson, G., Neal, B., Liu, D. *et al.* (1994) Structure of the O antigen of *Escherichia coli* K-12 and the sequence of its *rfb* gene cluster. *Journal of Bacteriology* **176**, 4144–4156.

Takayama, K. & Kilburn, J.O. (1989) Inhibition of synthesis of arabinogalactan by ethambutol in *Mycobacterium smegmatis. Antimicrobial Agents and Chemotherapy* **33**, 1493–1499.

Takayama, K. & Qureshi, N. (1979) Effects of isoniazid on the synthesis of fatty acids and hydroxy lipids in a cell-free preparation of *Mycobacterium tuberculosis* strain H37Ra. In: *Proceedings of the Fourteenth US–Japan Tuberculosis Research Conference.* US–Japan Cooperative Medical Science Program, NIH/NIAID, Bethesda, Maryland, pp. 168–186.

Takayama, K. & Qureshi, N. (1984) Structure and synthesis of lipids. In: *The Mycobacteria, a Sourcebook* (eds G. P. Kubica & L. G. Wayne). New York: Marcel Dekker, pp. 315–344.

Takayama, K., David, H.L., Wang, L. *et al.* (1970) Isolation and characterization of uridine diphosphate *N*-glycolylmuramyl-L-alanyl-γ-D-glutamyl-meso-α–α′-diaminopimelic acid from *Mycobacterium tuberculosis. Biochemical and Biophysical Research Communications* **39**, 7–12.

Takayama, K., Qureshi, N. & Davidson, L.A. (1980) Studies of the effects of isoniazid on the biosynthesis of nonmycolic C_{30}-C_{56} fatty acids by *Mycobacterium tuberculosis* H37Ra. In: *Proceedings of the of the Fifteenth US–Japan Tuberculosis Research Conference.* US–Japan Cooperative Medical Science Program, NIH/NIAID, Bethesda, Maryland, pp. 71–95.

van Heijenoort, J. (1996) *Murein Synthesis in Escherichia coli and Salmonella: Cellular and Molecular Biology,* Vol. 1, 2nd edn (eds F. C. Neidhardt *et al.*). Washington, DC: American Society for Microbiology, pp. 1025–1034.

Vilkas, E., Amar, C., Markovits, J. *et al.* (1973) Occurrence of a galactofuranose disaccharide in immunoadjuvant fractions of *Mycobacterium tuberculosis* (cell walls and wax D). *Biochimica et Biophysica Acta* **297**, 423–435.

Wang, A.Y., Grogan, D.W. & Cronan, J.E. Jr (1992) Cyclopropane fatty acid synthase of *Escherichia coli*: deduced amino acid sequence, purification, and studies on the enzyme active site. *Biochemistry* **31**, 11020–11028.

Weitzerbin-Falszpan, J., Das, B.C., Azuma, I. *et al.* (1970) Isolation and mass spectrometric identification of the peptide subunits of mycobacterial cell walls. *Biochemistry and Biophysics Research Communications* **40**, 57–63.

Weston, A., Stern, R.J., Lee, R.E. *et al.* (1998) The biosynthetic origin of the mycobacterial cell wall galacto-furanosyl residues. *Tubercle and Lung Disease* **78**, 123–131.

Wheeler, P.R., Besra, G.S., Minnikin, D.E. *et al.* (1993) Stimulation of mycolic acid biosynthesis by incorporation of *cis*-tetracos-5-enoic acid in a cell-free preparation from *Mycobacterium smegmatis. Biochimica et Biophysica Acta* **1167**, 182–188.

Winder, F.G. (1982) Mode of action of the antimycobacterial agents and associated aspects of the molecular biology of mycobacteria. In: *The Biology of the Mycobacteria,* Vol. 1 (eds C. Ratledge & J. Stanford). London: Academic Press, pp. 354–441.

Wolucka, B.A. & de Hoffman, E. (1995) The presence of β-D-ribosyl-L-monophosphodecaprenol in mycobacteria. *Journal of Biological Chemistry* **270**, 20151–20155.

Wolucka, B.A., McNeil, M.R., de Hoffmann, E. *et al.* (1994) Recognition of the lipid intermediate for arabinogalactan/arabinomannan biosynthesis and its relation to the mode of action of ethambutol on mycobacteria. *Journal of Biological Chemistry* **269**, 23328–23335.

Xin, Y., Lee, R.E., Scherman, M.S. *et al.* (1997) Characterization of the *in vitro* synthesized arabinan of mycobacterial cell walls. *Biochimica et Biophysica Acta* **1335**, 231–234.

Yuan, Y., Lee, R.E., Besra, G.S. *et al.* (1995) Identification of a gene involved in the biosynthesis of cyclopropanated mycolic acids in *Mycobacterium tuberculosis. Proceedings of the National Academy of Sciences of the USA* **92**, 6630–6634.

Zhang, Y., Heym, B., Allen, B. *et al.* (1992) The catalase-peroxidase gene and isoniazid resistance of *Mycobacterium tuberculosis. Nature* **358**, 591–593.

Chapter 14 / Iron metabolism

COLIN RATLEDGE

1 Introduction

1.1 The problem with iron

Microorganisms, like all living cells except perhaps for a few species of *Lactobacillus*, require iron for growth. Iron is used in a number of enzymes; it is also involved in the formation of the haem nucleus that occurs in oxygen-carrying and electron-carrying molecules such as the cytochromes, as well as in a variety of other non-haem, iron proteins that are also involved in energy production.

Iron exists in two principal forms: Fe(II), the ferrous ion, and Fe(III), the ferric ion. In aerobic systems, iron is mainly in the oxidized, Fe(III), form, although there are obvious exceptions to this such as haemoglobin in which iron is always in the reduced, ferrous state even when oxygen is attached to it forming oxyhaemoglobin. The oscillatory reduction–oxidation (redox) reactions of iron forms the basis of electron transport through the chain of cytochromes and this is therefore an important aspect of its biochemistry.

The main physical differences between the two forms of iron is that ferrous ions are generally soluble in aqueous systems but ferric ions are not: the solubility product for ferric hydroxide is $\approx 10^{-38}\,\mathrm{mol/L}$ so that at pH7, when the concentration of OH^- and H^+ are both $10^{-7}\,\mathrm{mol/L}$, it can be calculated from eqns 1 and 2 that the concentration of free Fe(III) is $10^{-17}\,\mathrm{mol/L}$. This is too low a concentration to be biologically useful and therefore complex systems have evolved for solubilizing iron, for taking it from the environment or from food for its subsequent transport and storage in cells and tissues. This applies equally to animals, plants and microorganisms. It is only at low pH values that the solubility of Fe(III) become appreciable: for example at pH3, the solubility of free Fe^{3+} is now $0.1\,\mu\mathrm{mol/L}$. (From eqn 2 below, the concentration of OH^- ions at pH3 is 10^{-11}, thus $[OH^-]^3$ is 10^{-33} making $[Fe^{3+}] = 10^{-7}$ mol/L.) At this concentration, iron is probably just sufficiently available for it to be acquired by the cells without the need for any specific solubilizing agent, but, of course, only a few microorganisms are able to grow at pH3 or less. For mycobacteria, the lowest pH at which growth can occur appears to be about 5.5 and this is too high for iron to have an appreciable solubility.

$$Fe(OH)_3 \Leftrightarrow Fe^{3+} + 3OH \qquad (1)$$

$$[Fe^{3+}] = \frac{[Fe(OH)_3]}{[OH^-]^3} = \frac{10^{-38}}{(10^{-7})^3} \qquad (2)$$

Thus, a basic problem for all organisms is the acquisition of iron from the environment (or diet if it is multicellular) with its ultimate incorporation into the various iron-requiring proteins of the cell. This process has to be achieved for cells to grow.

For microorganisms at least two major scenarios for iron assimilation are feasible (see Fig. 14.1). A microbe, whether present in the soil or water as a non-pathogen, or present within an animal host as a pathogen, intracellular symbiont or parasite, may acquire iron by producing its own specific iron-solubilizing agent, known as a siderophore. This is then transported with the iron into the cell following a well-defined path of uptake, iron removal and finally incorporation of iron into the microbial proteins. This is depicted in general terms in Fig. 14.1(a). The alternative route, which has become evident over the past few years, is that some microorganisms have the ability to interact directly with a source of iron — which may be either inorganic iron or protein-bound iron. The iron source is then reduced at the surface of the organism and the ensuing Fe(II) is then directly transported into the cell. This is shown in outline in Fig. 14.1(b).

Both these processes occur in mycobacteria though the non-siderophore mediated route is more speculative and might only occur with one or two species. The emphasis of this review is therefore on the siderophore-mediated route as laid out in Fig. 14.1(a). The alternative route, as shown in Fig. 14.1(b), is discussed mainly in section 7.

1.2 Sources of iron

For microorganisms growing in laboratory culture media, iron, if added at all, may exist in a variety of complexes but usually ferric phosphate is formed and this becomes dispersed in the medium possibly as a mixture of colloidal ferric phosphate and ferric hydroxide. Such forms of iron may have molecular

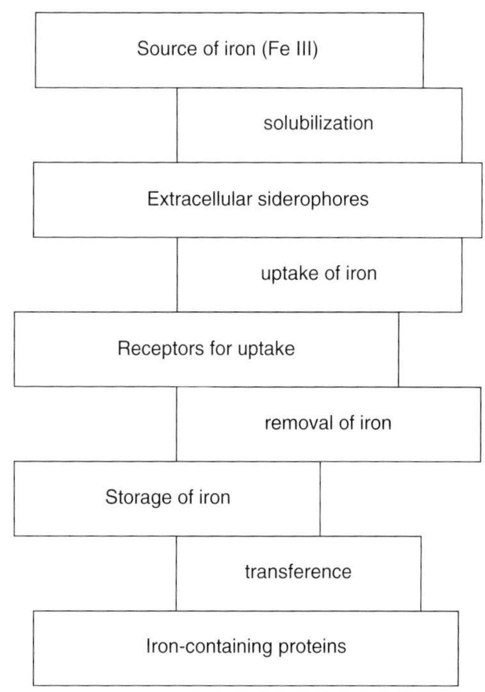

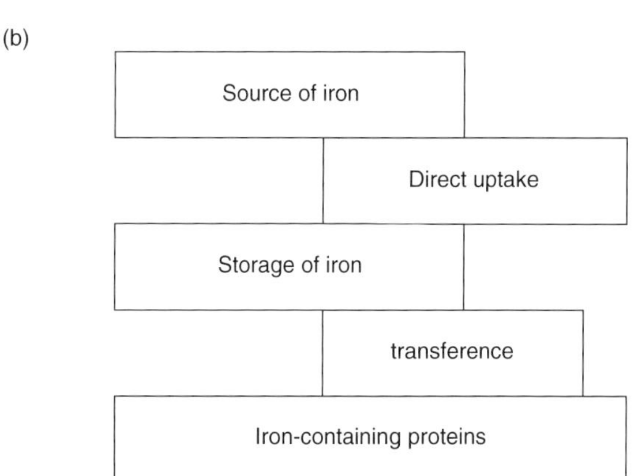

Fig. 14.1 Outlines of the stages of iron assimilation by microorganisms. (a) The siderophore-mediated route of iron solubilization and uptake. (b) The direct, non-siderophore mediated, route of uptake.

sizes of about 40000 Da but are too large to be acquired directly. Solubilization of the iron is therefore necessary by either of the mechanisms given in Fig. 14.1. The addition of metal-solubilizing agents, such as ethylene diamine tetraacetic acid (EDTA), probably docs little to help the overall uptake process but agents, such as citric acid, can form ferric citrate complexes when the molar ratio of iron to citric acid is about 1:20 (Spiro *et al.* 1967a,b) and these can probably be taken up by most microorganisms, including mycobacteria, as a directly assimilatable form of iron (see also section 2.5).

For pathogenic microorganisms, the situation is more complex as they must now acquire iron from one or more of the main iron-containing molecules of the host. Haem, arising from lysis of haemoglobin or other haemproteins by microbial infections can be used directly by some microorganisms and even perhaps by *Mycobacterium haemophilum* but, in general, mycobacteria, which are not known as haemolytic bacteria, acquire their iron probably from either transferrin or ferritin.

Transferrin is a globular glycoprotein with a molecular size of about 80 kDa. It functions to transport iron in the blood stream to all the tissues of the body. It possesses only two chelating centres for Fe(III) which have slightly different binding constants for iron though both are $\approx 10^{20}$ (Aisen 1998). Importantly, transferrin is never fully saturated with iron and usually is only about one-third saturated. This leaves a 'spare' capacity for it to bind iron should it become necessary to do so. For example, any microbial infection of the blood would quickly lead to haemolysis and release of iron but transferrin would then remove this from circulation thereby limiting the bacterial growth.

This process of the withholding of iron from the infecting bacteria has been called 'nutritional immunity' (Kochan 1973, 1976) although there is no participation of the immune system in this process. Simply, the infected body increases its synthesis of transferrin to ensure that as much iron as possible, which might otherwise become available to the bacterial infection, is removed from circulation. Work by Douvas *et al.* (1993) has shown that apotransferrin (i.e. without any iron at all) can arrest the growth of

M. avium within human macrophages and was judged to be a crucial factor in checking the proliferation of the bacteria *in vivo*. A full and detailed review of the structure, role and function of transferrin has been recently provided by Aisen (1998) and this should be consulted for further information. A related molecule, lactoferrin, occurs in secretions of the body and functions in an analogous manner to transferrin.

Transferrin is not an iron storage protein. This role is fulfilled in animals by ferritin which also occurs in a wide range of organisms including both plants and microorganisms. All ferritins are made up of 24 protein subunits forming a hollow sphere which can accommodate over 4000 atoms of iron as an extended inorganic, polymeric matrix (Harrison *et al.* 1998). The mechanism for the acquisition of iron by ferritin has recently been reviewed by Chasteen (1998). It is considered that iron enters the ferritin shell as Fe^{2+} which is then oxidized to Fe^{3+} by ferroxidase located on one of the polypeptide chains. The mobilization of iron from ferritin in animal cells is considered to be by a reductive process but it can also be accomplished by direct removal of Fe(III) by a number of bacterial siderophores, including those from mycobacteria, without the participation of any enzyme.

Thus, an important first step in the acquisition of iron by pathogenic mycobacteria will be to obtain iron probably from ferritin. The ability to use the iron within transferrin is probably less significant as transferrin is predominantly a circulating iron-containing protein with a relatively low (0.6 atoms) iron content per molecule. Ferritin, on the other hand, is ubiquitous in all cell types of animals and presents a large pool of iron for any pathogen to access. All that is required to remove the iron would be a strong chelating agent, i.e. siderophore, to be produced.

1.3 Mycobacteria and iron

Iron has long been known to be required by mycobacteria (Sauton 1912). The necessity for pathogens to be able to acquire iron from their hosts in order to become infectious is known to be an essential component of virulence but clearly is not the cause of virulence itself. Many reviews have been written on this topic though for the sake of brevity only some of the key ones need be listed here: Sigel and Sigel (1998), Guerinot (1994); Winkelmann *et al.* (1987), Bagg & Neilands (1987), and Bullen & Griffiths (1987). A study by Dhople *et al.* (1996) provides direct evidence for the key role of iron in the development of pathogenicity by *M. avium* in mice and confirms, possibly for the first time, what many researchers have already taken to be the case but without any direct proof arising from whole animal experiments, that the availability of iron to a mycobacterial infection is a major determinant for the development of pathogenicity.

The role of iron in the metabolism of mycobacteria was first systematically investigated by Winder and colleagues (one of whom was the present author) from the late 1950s up to the 1980s. An early key finding was that when *M. smegmatis*, used as the principal model organism, was grown with a deliberate deficiency of iron in the medium, the cells contained about 30% of the iron that cells had when grown with 20-fold increase of iron: 64 µg/g cell dry wt as opposed to 224 µg/g (Winder and O'Hara 1966). A number of metabolic consequences of iron deficiency were noted which included a decrease in the DNA/protein ratio, increases in several enzymes including DNA polymerase and not surprisingly a decrease in activity of a number of iron-containing enzymes (Winder & O'Hara 1964) (see Ratledge 1976 for a synopsis of this early work).

Porphyrins and cytochromes are also affected by the limitation of iron. However, not all components are equally disadvantaged by iron deficient growth conditons: flavoproteins and cytochrome c are strongly conserved in *M. smegmatis* (McCready & Ratledge 1978; McCready 1980) suggesting that respiratory components are maintained at the expense of other iron-containing proteins when cells are deprived of iron.

Porphyrins, as the precursors of the haem nucleus, are also adversely affected by iron and, in mycobacteria, their synthesis is strongly repressed when iron is limiting growth: for example, in *M. smegmatis* coproporphyrin III is less than 25 µmol/g dry wt under iron-deficient conditions but is over 200 µmol/g in cells grown with a sufficiency of iron (McCready & Ratledge 1978; McCready 1980). The absence of porphyrins in iron deficiently grown cells then accounts for the very pale ('anaemic') appearance of cells that are otherwise slightly tan in colour, a colour due principally to the porphyrins rather than carotenoids which are the principal pigments of the more orange mycobacteria. This low content of porphyrin in cells, which has also been noted in *M. avium* (R. Barclay and C. Ratledge, unpublished observations 1985), indicates that when cells are undergoing a transition from iron deficiency to iron sufficiency (the 'famine to feast' syndrome) they will be unable to synthesize haems as the porphyrin precursors are absent. Therefore, there is a necessity to hold the iron in some storage form whilst the porphyrin synthetic pathway is de-repressed and porphyrins once more become available for haem synthesis. This may account for some of the unusual features of iron metabolism in mycobacteria and, in particular, explain the necessity for an intracellular iron storage molecule in the form of mycobactin (see sections 4 and 7). The repression of porphyrin synthesis under low iron conditions does not though apparently occur in *Escherichia coli* but more information on this topic may be illuminating about how bacteria adjust their metabolism to take account of low iron states.

Although it is relatively easy to create iron deficient growth conditions in the laboratory and to study the elaboration of various iron-sequestering molecules, it is less clear what is the status of iron in infected tissues. With pathogenic mycobacteria, the macrophage provides the immediate environment for the bacilli as the body begins its task of attempting to eradicate the infection. The macrophage though is one of the main iron-storage cell types of the human body (Finch & Huebers 1982) but iron is withheld from the bacteria by the functioning of a complex series of homeostatic reactions which are still not yet fully understood. Iron plays both a role in preventing pathogen multiplication: it may assist in the generation of toxic oxygen metabolites that are used to check and kill mycobacteria (see Lepper & Wilks 1988; Dussurget *et al.* 1996; Lundrigan *et al.* 1997) as well as serving to stimulate the growth of the bacteria (see sections 2 and 3). The withholding of iron from the bacteria has been noted by several groups (see Alvarez-Hernandez *et al.* 1989; Lepper & Wilks 1988) where inflammatory macrophages have an impaired release of iron possibly involving the participation of the cytokines, interleukin or interferon-γ (Byrd and Horwitz 1993). Iron itself within human monocytes can decrease the release of tumour necrosis factor (TNF) which, in turn, allows the monocytes to differentiate into macrophages so that the proliferation of *M. tuberculosis* is restricted (Byrd 1997). Thus, the very process of infection with the bacilli attempting to gain iron from the host tissues appears to set in train a sequence of events that leads to the active suppression of their growth. The key element in this host defence system is iron. Infected macrophages are quickly converted into an iron-limited state. Douvas *et al.* (1994) was able to show that growth of *M. avium* in cultured human macrophages was enhanced by the addition of iron to the cultures. However, serum also had to be added to achieve this growth-promoting effect and transferrin was implicated as a possible accessory factor. It was not certain though whether the mycobacteria were removing the iron directly by binding to the transferrin+iron complex (=holotransferrin) or were using siderophores to achieve this acquisition. The situation was further complicated by the triacylglycerols present in the added serum also contributing to growth. The conclusion was nevertheless reached that iron must be a limiting nutrient for the mycobacteria within the macrophage otherwise addition of it would not have caused an acceleration of growth. Transferrin can clearly gain access to macrophages and to the ensuing phagosome (Clemens & Horwitz 1996). A mechanism of iron acquisition from this molecule or from ferritin is then necessary as the pH within the macrophage/phagosome is not sufficiently acidic to cause spontaneous dissociation of iron from transferrin (Crowle *et al.* 1991) with the pH within phagosomes containing *M. avium* having been determined as 6.2–6.3 (Sturgill-Koszycki *et al.* 1994; Oh and Staubinger 1996). The possible mechanism for iron acquisition from transferrin is discussed below and is summarized in section 7.

Further details concerning the resistance of the macrophage are discussed in Chapter 19 which explains the key role of Nramp (natural resistance

associated macrophage protein) which can confer resistance of animals to mycobacterial infections (see also Supek *et al.* 1997). A clear picture of the early events is now emerging with iron withdrawal forming the crux of the host's defence system.

2 Siderophores

The microbial acquisition of iron from external sources of iron, whether inorganic or organic (see section 1.2), requires, in most cases, the participation of an extracellular solubilizing agent. Such agents are known as siderophores and numerous examples of several different types have been described over many years. Useful compilations of data concerning the siderophores have been assembled in the monographs by Winkelmann (1991), Winkelmann *et al.* (1987) and Sigel and Sigel (1998) with pertinent recent reviews by Guerinot (1994) and Neilands (1995). Without exception, all siderophores are produced in greatly increased quantities during iron-deficient growth; clearly an indication of the necessity of scavenging all traces of iron from the immediate milieu of the microorganism.

2.1 Salicylic acid

Although salicylic acid (2-hydroxybenzoic acid), or in some cases 6-methysalicylic acid, has long been known as an extracellular product in mycobacteria whose presence is increased some 40-fold by cultivation of the bacteria in iron-deficient conditions (Ratledge & Winder 1962, 1966) (Fig. 14.2), its role as an iron-solubilizing agent has had to be rejected. Initially, even though it was established that [55]Fe-salicylic acid could be readily taken up into mycobacteria suggesting a facile route of iron acquisition (Ratledge & Marshall 1972), it quickly became clear that ferric salicylate could not function *in vivo* in competition with ions such as phosphate which very quickly produced the highly insoluble ferric phosphate (Ratledge *et al.* 1974). Thus, although there continue to be claims that salicylate can act as a siderophore (see, for example, Visca *et al.* 1993), it is nevertheless clear that both in laboratory culture media, which almost invariably contain phosphate ions, as well as in the macrophage/phagosome environment of the pathogenic mycobacteria where phosphate is also ubiquitous (Barclay & Wheeler 1989), salicylate cannot function as a siderophore.

Two immediate questions then arise: if salicylate is not the siderophore of mycobacteria, what is; and secondly, what is the function of salicylate? The answer to the first question is given in the subsequent section (2.2). The answer to the second question is less certain: although salicylate is a direct precursor of

mycobactin (see Fig. 14.6, p. 271) studies with salicylate auxotrophs (Ratledge & Hall 1972) established that mycobactin (and more recently also carboxymycobactin; A. Tadepalli and C. Ratledge, unpublished observations 1998) cannot satisfy the growth requirement of such mutants. Therefore a separate role of salicylate, besides acting as a mycobactin precursor, must exist. This is discussed in further detail in sections 5 and 7.

2.2 Exochelins

Once salicylate had been eliminated as a siderophore for the acquisition and uptake of iron, the search began for the presence of such molecules. Macham and Ratledge (1975) showed that culture filtrates of both *M. smegmatis* and *M. bovis* bacille Calmette–Guérin (BCG) when grown deficient in iron contained a substance or substances that could hold ^{55}Fe in solution and which could pass through a dialysis membrane indicating a molecular size of less than 10 000 Da. These materials were called the exochelins.

Almost immediately, it was appreciated that there were two types of exochelin (Table 14.1): those from *M. smegmatis* and other non-pathogens were water-

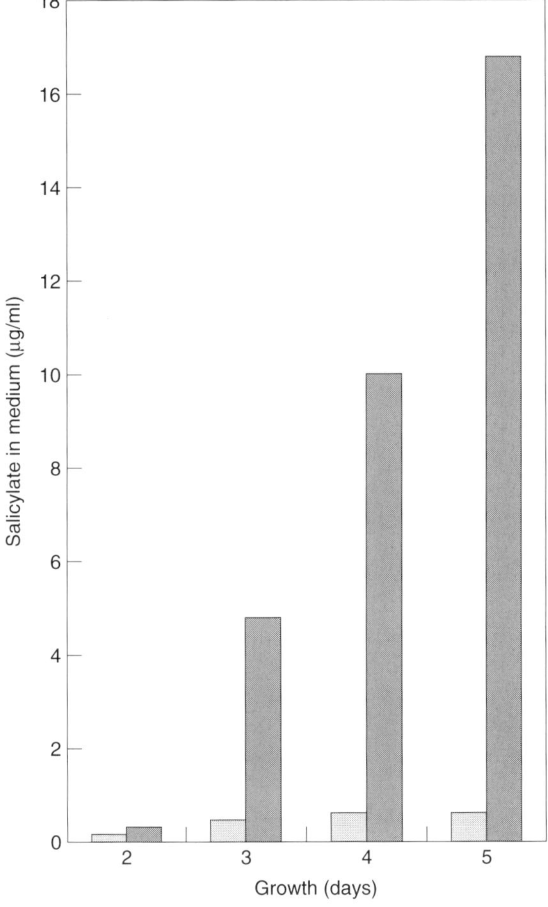

Fig. 14.2 Accumulation of salicylic acid during iron deficient growth of *Mycobacterium smegmatis*. Light bars, iron-sufficiently grown cells; dark bars, iron-deficiently grown cells. (From Ratledge and Winder, 1962.)

Table 14.1 Occurrence of extracellular siderophores in mycobacteria.

Type 1
Water-soluble, non-solvent extractable entities—the
exochelins
Identified in:
  M. smegmatis, M. neoaurum, M. vaccae, armadillo-derived
  Mycobacterium (ADM)

Type 2
Chloroform extractable entities—the
carboxymycobactins
Indentified in:
Pathogens
  M. africanum, M. avium, M. bovis, BCG, *M. intracellulare,*
  M. paratuberculosis (mycobactin-dependent),
  M. scrofulaceum, M. triviale, M. tuberculosis H37Rv and
  H37Ra, *M. xenopi.*
Non-pathogens
  M. smegmatis and *M. neoaurum.*

soluble and could not be extracted into any organic solvent including ethanol whereas the exochelins from the pathogenic mycobacteria were extractable, when converted into their ferric complexes, into chloroform (Macham & Ratledge 1975). It was also established that the two types of siderophore had different modes of uptake: that from the non-pathogens was taken up by an active transport system that was readily inhibited by energy poisons and uncouplers of oxidative phosphorylation whereas the chloroform-soluble siderophores were taken up by an inhibitor-insensitive route, probably by facilitated diffusion and not involving the direct input of energy (Stephenson & Ratledge 1979, 1980). The water-soluble exochelins could not be taken up by the pathogens but the chloroform-soluble ones were taken up by the non-pathogens.

Water-soluble exochelins have been recognized in *M. smegmatis* (Macham *et al.* 1977), *M. neoaurum* (Hall & Ratledge 1987) and *M. vaccae* (Messenger *et al.* 1986). It is likely that other related molecules will be found in the other non-pathogens.

The chloroform-soluble exochelins, whose structure has now been recognized as a variation on mycobactin, are described in detail in the next section (2.3). The term 'exochelin' is now no longer applicable to such molecules and the term 'carboxymycobactin' should be used as a more accurate descriptor. 'Exochelin' should therefore restricted to the water-soluble siderophores.

The structures of exochelins *per se* were not resolved for 20 years. The structure of major exochelin from *M. smegmatis*, first described by Macham and Ratledge (1975), was not established until the work of Sharman *et al.* (1995a). The structure of this exochelin is given in Fig. 14.3. It is an ornithinyl siderophore containing three hydroxamate groups that provide the chelating centre for iron. The siderophore from *M. neoaurum*, exochelin MN (Fig. 14.4) is different and has several unusual structural features not least of which is a β-hydroxyhistidine residue that has only been found so far in one other siderophore. The chelating centre is thus provided by the imidazole ring and the adjacent hydroxyl group and by the two hydroxamate residues.

The principal siderophore from *M. smegmatis*, known as exochelin MS, is produced at up to 150 µg/mL, depending on the degree of iron defi-

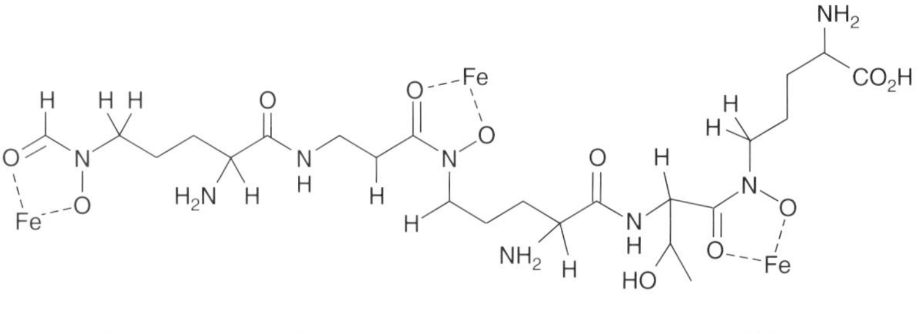

formyl-D-ornithinel-β-alanine-D-ornithine2-D-*allo* threonine-L-ornithine3

Fig. 14.3 Structure of exochelin MS, the extracellular siderophore from *M. smegmatis*. (From Sharman *et al.* 1995a.)

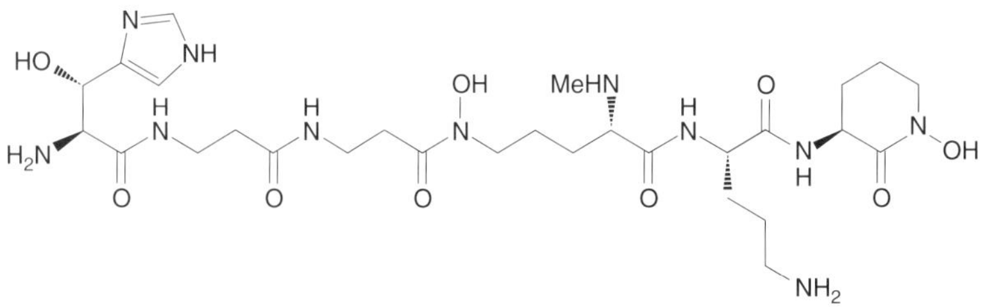

L-*threo*-β-hydroxy histidine-β-alanine-β-alanine-L-α methyl ornithine-L-ornithine-L-(cyclo)ornithine

Fig. 14.4 Structure of exochelin MN, the extracellular siderophore from *M. neoaurum*. (From Sharman *et al.* 1995b.)

ciency created in the culture medium (Ratledge & Ewing 1996). Ferri-exochelin MS is taken up as the intact molecule (as shown by double labelling studies with ^{55}Fe- and ^{3}H-labelled exochelin) by an active transport process that at low concentrations is inhibitable by a number of energy poisons (Stephenson & Ratledge 1979). It is produced in a growth-related manner and can readily solubilize iron from insoluble sources such as ferric phosphate and ferric hydroxide as well as from ferritin (Macham *et al.* 1975, 1977). It thus fulfils all the criteria needed for a siderophore: it can solubilize iron from inorganic and organic sources, it is readily taken up by iron-deficiently grown cells as well as iron-sufficiently grown ones (Stephenson and Ratledge 1979). However, when the concentration of ferric exochelin was increased in uptake studies, a second, non-saturable, non-inhibitable process became evident. From this it was inferred that the first process involved direct uptake of iron into the cell whereas the second and much slower process probably represented a mechanism for storing excess iron within the cells, probably by transfer to mycobactin, prior to the resynthesis of porphyrins and other proteins needed for the acceptance of iron (see section 7 and Fig. 14.7, p. 279).

The mechanism of uptake of the exochelins is discussed further in section 3.

2.3 Exochelin-mediated iron uptake into *Mycobacterium leprae*

Although the extracellular siderophores from *M. leprae* are unknown, Hall *et al.* (1983) and Hall and Ratledge (1987) showed that effective iron uptake in *M. leprae*, taken from armadillo livers, was achieved using the exochelin from *M. neoaurum* (see Fig. 14.4 for structure). The exochelins from *M. vaccae* and *M. smegmatis* (which are possibly equivalent) were ineffective as was the carboxymycobactin from BCG. An exochelin from an armadillo-derived mycobacterium (ADM), as a possible commensal organism living along with *M. leprae* in armadillos (Portaels *et al.* 1985), was also found to be as equally effective as exochelin MN for the uptake of iron into *M. leprae* (Hall & Ratledge 1987). (On the basis of preliminary chromatographic evidence, the exochelins of *M. neoaurum* and the ADM species may be equivalent (Hall & Ratledge 1987).) The unusual structure of exochelin MN (see Fig. 14.4) with its β-hydroxyhistidine residue clearly provides sufficient novelty for its specific recognition and uptake by *M. leprae.*

As the exochelin from *M. neoaurum* is a water-soluble siderophore (see Table 14.1), it is puzzling why an apparent pathogen, such as *M. leprae*, is able to transport iron in this form as this was not seen for the uptake of exochelin MS into *M. bovis* or *M. intracellulare* (Stephenson & Ratledge 1980). Further work on this would obviously be beneficial though the present results may indicate that *M. leprae* is not a pathogen in the same sense as is *M. tuberculosis*. Leprosy certainly is not the killing disease that tuberculosis is and thus *M. leprae* may be an intracellular parasite and not strictly a pathogen; hence, it possesses a non-pathogenic, siderophore-mediated uptake system. As pointed out above, the exochelins can abstract iron from ferritin and can also reverse the bacteriostasis of serum towards *M. smegmatis* which may be sufficient to enable *M. leprae* to obtain iron *in vivo*.

2.4 Exochelin biosynthesis and its regulation

Now that the structure of the major exochelin of *M. smegmatis* is established (see Fig. 14.3), the first steps towards understanding its biosynthesis are taking place. Exochelin MS is essentially a pentapeptide and as such could be presumed to be synthesized, as are other small peptides, by a non-ribosomal pathway. The genes for the synthesis of a number of siderophores from other microorganisms have been described (see Winkelmann *et al.* 1987; Sigel & Sigel 1998). From the recent work of Stachelhaus and Marahiel (1995) it could be presumed that the component amino acids (three ornithines, *allo*-threonine and β-alanine) are linked together using a prosthetic group such as 4′-phosphopantetheine on a protein template and possibly involving the amino acids being activated as their acyl-adenosine monophosphate

(AMP) derivatives. Probably the hydroxylation of the ornithine residues would precede peptide formation as appears to occur in other ornithinyl siderophores, such as ferrichrome (Leong & Winkelmann 1998). Similarly, racemization of L-ornithine to the D configuration would probably precede peptide formation.

The group of Jacobs was the first to address the problem of exochelin MS biosynthesis and identified a 4.3-kb fragment containing four genes named as *fxuA*, *fxuB*, *fxuC* and *fxbA* (Fiss *et al*. 1994). The proteins produced by the first three genes corresponded with ferrienterobactin transport permeases of *E. coli*: FepG, FepC and FepD, respectively. The fourth gene, *fxbA*, was the only one, however, which coded for a synthetic function: the *N*-formyl transferase which would add a formyl group to the N terminus of the exochelin. More recent work by Byers and colleagues has indicated that the multienzyme system used for exochelin biosynthesis could, in fact, be coded for by one single gene with a single multifunctional protein, possibly in excess of 230 kDa (Zhu *et al*. 1998). (Other work by the Jacobs group would confirm this and suggest perhaps an even larger polyfunctional protein (W. R. Jacobs, personal communication 1997).) The gene identified as part of the exochelin transport system has been termed *exiT* (= **ex**ochelin **i**n **t**ransport), and from its base sequence, Zhu *et al*. (1998) were able to deduce that this transport gene was a member of the ABC (=**a**denosine triphosphate (ATP) **b**inding **c**assette) superfamily of transport proteins and consequently concluded that ExiT served to transport exochelin after its synthesis into the extracellular environment.

The regulatory control over siderophore biosynthesis has been extensively examined in *E. coli* and other bacteria for many years (see Braun *et al*. 1998, for a recent review). The key protein is known as Fur (**f**erric **u**ptake **r**egulatory protein) which complexes with any free Fe(II) within the cell and, by binding to a specific site on the DNA (the Fur promoter box), it prevents the transcription of a number of iron-regulated genes including those both for the biosynthesis of the siderophores and its subsequent uptake. The number of genes repressed by Fe(II)–Fur is continually being added to as work on this topic continues apace. Fur is also involved in responses to oxidative stress, acid tolerance, and even in sugar metabolism and toxin synthesis (see Dussurget *et al*. 1996; Crosa 1997; Braun *et al*. 1998). Two *fur* genes (*furA* and *furB*) have been recognized in the genome of *M. tuberculosis* (Cole *et al*. 1998) indicating a consistent pattern of regulation of siderophore biosynthesis possibly occurs in microorganisms.

In the absence of free Fe(II), Fur is inactive and therefore under iron-deficient conditions, a complete change in cell metabolism occurs: not only is the siderophore now synthesized along with its attendant uptake proteins but the whole biochemistry of the cells shifts to take account of the stresses that iron deprivation will cause. Many of these consequences can now be linked to earlier observations of iron stress in mycobacteria and other organisms (see section 1.3).

The work of Issar Smith and colleagues (see also Chapter 4) has identified an IdeR protein in *M. tuberculosis* and several other mycobacterial species that is a homologue to the protein (DtxR) found in *Corynebacterium diphtheriae* that represses synthesis of the diphtheria toxic protein (Doukhan *et al*. 1995; Schmitt *et al*. 1995; Dussurget *et al*. 1996). As DtxR has a role similar to Fur in *E. coli*, the IdeR protein of mycobacteria can be deduced to have a similar role to Fur itself. IdeR regulates the synthesis of exochelin and mycobactin in *M. smegmatis*, although a second regulator appears to be also needed for full repression of exochelin and mycobactin synthesis (Dussurget *et al*. 1996). If the analogy with the Fur protein of *E. coli* holds true, then IdeR-Fe(II) as encoded by the *ideR* gene, will prevent transcription of the exochelin and mycobactin synthesis genes together with those for their uptake and utilization.

This is presently an area of considerable activity with at least three research groups (those of W. Jacobs, I. Smith and B. R. Byers) now attempting to bring the understanding of iron metabolism in mycobacteria somewhere towards the level of comprehension of the *E. coli* system. A summary of the relationship of the genetic information to the biochemical functions is given in Table 14.2. It must though be pointed out that this will rapidly become out of date as results continue to be published.

Table 14.2 Summary of the current understanding between genes, proteins and exochelin-mediated iron metabolism in *M. smegmatis*.

DNA	Regulatory genes	Siderophore biosynthesis	Siderophore transport	Ferric siderophore uptake	Others
Gene description	*IdeR*	*fxbA, fxbB, fxbC* etc	*ExiT*	*fxuA, fxuB, fxuC*	?
Protein	IdeR	*N*-formyl peptide transferase synthetase	*ExiT*	Iron-regulated envelope proteins?	?
Product	Fe(II)-IdeR (repressor for whole operon)	Exochelin	Exochelin export	Exochelin receptor; exochelin uptake	Carboxymycobactin, mycobactin

2.5 Carboxymycobactins

The early work on the siderophores from the pathogenic mycobacteria and the elucidation of their properties and function preceded their structural determination by many years. It was, however, established almost from the outset that carboxymycobactin contained a salicyloyl moiety in common with mycobactin itself (Macham *et al.* 1975) and this was later confirmed by Barclay and Ratledge (1983) for *M. intracellulare*. Somewhat confusingly therefore all the earlier papers refer to the 'exochelins' of these mycobacteria whereas it is now clear that they are an entirely different type of molecule.

The structure of the carboxymycobactins is given in Fig. 14.5. They are variations on the structure of mycobactin (section 4) in which the long alkyl chain (R$_1$) of the intracellular molecule is now a shorter carboxylic acid (Lane *et al.* 1995). This change in structure converts the strongly lipophilic mycobactin from its close association with the cell envelope into a water-soluble, extracellular molecule that becomes soluble in chloroform when converted into its Fe(III) complex. Independently, Gobin *et al.* (1995) and Wong *et al.* (1996) have also determined the structure of the carboxymycobactins from *M. tuberculosis* and *M. avium* but have considered that in both cases the terminal carboxylic acid was a methyl ester (–CO.OCH$_3$). However, these researchers did not have recourse to NMR spectroscopy which had been

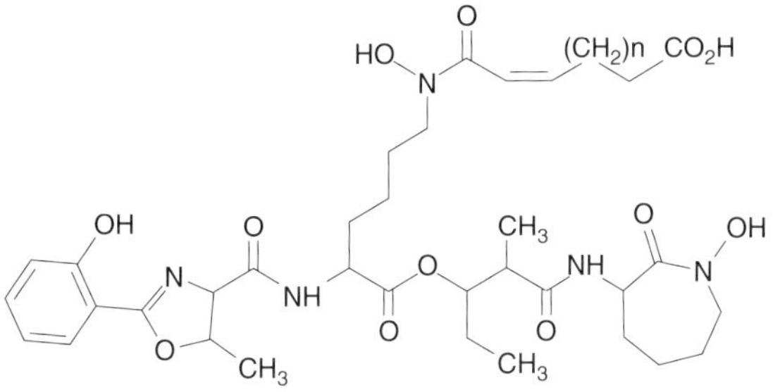

Fig. 14.5 Structure of the carboxymycobactin from *M. avium*, *M. bovis* bacille Calmette–Guérin (BCG) and *M. tuberculosis* (from Lane *et al.* 1995) with *n*=2–9. Related molecules have been reported from *M. smegmatis* (Ratledge & Ewing, 1996; Lane *et al.* 1998) and, as their methyl esters from *M. tuberculosis* (Gobin *et al.* 1995), *M. avium* (Wong *et al.* 1996).

used by Lane *et al.* (1995) to verify the presence of a carboxylic group and not an ester. It is possible that the methyl esters could arise during the late growth phase of the organisms which had been the source of the material for the work of Gobin *et al.* (1995) and Wong *et al.* (1996), whereas the culture filtrates used by Lane *et al.* (1995) had been taken from actively growing cultures of *M. bovis* BCG, *M. avium* and *M. tuberculosis* itself, and would have represented the stage at which iron uptake was likely to have been at its most rapid.

Carboxymycobactins have been recognized (Barclay & Ratledge 1983, 1988) in the following species (see also Table 14.1): *M. tuberculosis* H37Ra and H37Rv and in fresh clinical isolates of *M. tuberculosis* (nine out of nine isolates), *M. africanum* (six out of seven isolates) BCG (one out of one), *M. intracellulare* (13 out of 13), *M. scrofulaceum* (two out of two), *M. avium* (25 out of 25 including 13 strains that were dependent on mycobactin from growth), *M. paratuberculosis* (13 out of 13 mycobactin-requiring strains), *M. triviale* (one out of one) and *M. xenopi* (one out of one). None was found in the only strain of *M. microti* that was examined which also failed to produce a detectable mycobactin. The finding that the mycobactin-dependent strains of *M. avium* and *M. paratuberculosis* (i.e. those organisms that would not grow in laboratory media without mycobactin supplementation) could produce carboxymycobactin when grown in the presence of added mycobactin, could perhaps suggest a simple conversion of one into the other. However, the work of Macham *et al.* (1975) using [^{14}Csalicyloyl]carboxymycobactin failed to detect any significant conversion into mycobactin during growth of BCG. Moreover, the amounts of carboxymycobactin recovered from culture filtrates were greater than the amount of mycobactin that had been added. However, at the time that these studies were carried out the relationship of the 'exochelin' to mycobactin was not evident. Nevertheless the production of carboxymycobactin by *M. paratuberculosis* goes some way to explain how this pathogen (the causative agent of Johne's disease in cattle) may acquire its iron *in vivo* but it does not entirely explain why it is dependent on either carboxymycobactin or mycobactin (see Barclay & Ratledge 1983) for growth in laboratory medium. It is possible that the genes for carboxymycobactin/ mycobactin synthesis are strongly repressed during isolation but on presentation of mycobactin as a growth supplement this, in some way, then leads to production of carboxymycobactin but not by conversion from mycobactin.

Whilst the water-soluble, peptido-exochelins have not been recognized in the pathogenic mycobacteria in spite of detailed investigations (M. Ewing and C. Ratledge, unpublished observations 1989–96),

the carboxymycobactins have been found in the saprophytes, albeit in very low concentrations. In *M. smegmatis*, carboxymycobactin is probably at most only 10% of the total iron-binding capacity of the combined siderophores and can be even less than 1% (Ratledge & Ewing 1996). The structure of the molecule parallels the equivalent mycobactin S structure (Fig. 14.6) with a short 3 unsaturated acyl chain at R$_1$ of mycobactin (Lane *et al.* 1998). Small amounts of a carboxymycobactin have also been recognized in culture filtrates of iron-deficiently grown *M. neoaurum* (T. E. Lee and C. Ratledge, unpublished observations 1995) suggesting perhaps that the carboxymycobactins may be ubiquitous amongst all mycobacteria—pathogen and non-pathogen alike— but more work on this is necessary before this can be concluded with any certainty.

The ability of the carboxymycobactin to solubilize iron from a number of sources is well established. Macham *et al.* (1975) showed that carboxymycobactin from *M. bovis* BCG (then referred to as exochelin MB) could remove iron from ferritin without the need for any ancillary enzyme system. Carboxymycobactin could also reverse the tuberculostatic action of serum on the growth of BCG and of *M. smegmatis*, although interestingly the exochelin of *M. smegmatis* was not able to work with *M. bovis* (Macham *et al.* 1975). This specificity was then confirmed by showing that *M. smegmatis* could utilize both its own exochelin and the carboxymycobactin of *M. bovis* and *M. intracellulare* whereas neither of the latter species could utilize exochelin MS (Stephenson & Ratledge 1980). Thus, carboxymycobactins, being ubiquitous siderophores, are taken up by all mycobacteria but the exochelins, being confined to the non-pathogens, are only taken up by this group of mycobacteria.

The role of the carboxymycobactins as the most likely means of procuring iron for mycobacteria within host tissues is strengthened by the further observations of Barclay and Ratledge (1986a) showing that both *M. avium* and *M. paratuberculosis* would grow in serum-containing medium, which was otherwise inhibitory, when either carboxymycobactin or mycobactin was added. However, for this

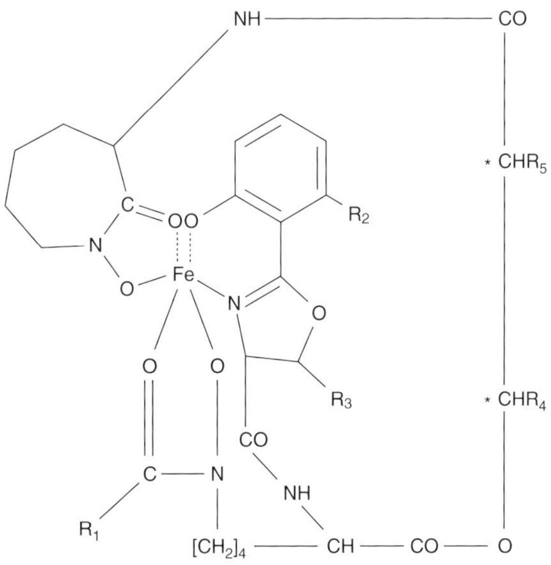

Fig. 14.6 General structure of the ferrimycobactins. References: 1, Snow (1970); 2, Barclay *et al.* (1985); 3, McCullough and Merkal (1982); 4, Ratledge and Snow (1974). The side-chains at R_1 are usually alkenyl groups with a *cis* double bond at C2; there is usually a variety of chain lengths, only the major ones are given.
*There may be up to six chiral centres (see Snow 1970) but only the ones associated with R_4 and R_5 are variable between different mycobactins.
†Indicates two distinct mycobactins are produced by the same strain.
‡These mycobactins were considered to be equivalent.
**The oxazole ring of the mycobactins (i.e. that adjoining the aromatic ring) contains an additional double bond and is thus an oxazoline ring (see ref. 4).

Organism	Mycobactin	Substituents					
		R_1	R_2	R_3	R_4	R_5	Ref.
M. aurum	A	13Δ	CH_3	H	CH_3	H	1
M. fortuitum	F†	$17,11\Delta$	H	CH_3	CH_3	H	1
M. fortuitum	H†	$19,17\Delta$	CH_3	CH_3	CH_3	H	1
M. marinum	M†	1	H	CH_3	$C_{17}H_{35}$	CH_3	1
M. marinum	N†	2	H	CH_3	$C_{17}H_{35}$	CH_3	1
M. phlei	P	$17cis\Delta$	CH_3	H	C_2H_5	CH_3	1
M. terrae	R	19Δ	H	H	C_2H_5	CH_3	1
M. smegmatis	S	$17,15cis\Delta$	H	H	CH_3	H	1
M. tuberculosis	T	19Δ	H	H	CH_3	H	1
M. avium	Av	$\Delta2$ alkenyl	H	H	$C_{10}H_{23}$?	CH_3	2
M. intracellulare	Av‡	$\Delta2$ alkenyl	H	CH_3	satd alkyl	CH_3	2
M. scrofulaceum	Av‡	alkenyl	H	H	satd alkyl	CH_3	2
M. paratuberculosis	Av‡	$\Delta3$ alkenyl	H	CH_3	satd alkyl	CH_3	2
M. paratuberculosis	J	15Δ	H	H	isopropyl	CH_3	3
Nocardia asteroides	NA**	1	H	CH_3	satd alkyl	CH_3	4

effect to be manifested, the cultures had to be pre-grown under iron-deficient conditions, presumably to induce the other components that were needed for iron assimilation. Whilst it was presumed both in this work and in the earlier studies of Kochan *et al.* (1971) and Macham *et al.* (1975) that the serum was causing bacteriostasis because of the iron-withholding nature

of transferrin within it, this has only been directly confirmed by the recent work of Gobin and Horwitz (1996). Purified carboxymycobactin from *M. tuberculosis* could rapidly remove iron from transferrin whether its iron saturation was 95% or 40%. Carboxymycobactin also removed iron from lactoferrin as well as from ferritin. Again, as previously observed, these events were non-enzyme mediated; in other words, once the carboxymycobactin is liberated by an infecting mycobacterium in a tissue, it will be able to solubilize iron from transferrin (or lactoferrin) or ferritin and thereby make the iron accessible to the bacilli. The iron-withholding action of transferrin is therefore nullified by carboxymycobactin. The interactions of carboxymycobactin and mycobactin with transferrin were equally rapid (Gobin & Horwitz 1996) indicating that the length of the R_1 alkyl or acyl chain (Figs 14.5 and 14.6) is irrelevant in this process. The stability constant of the chelating centres (which are obviously equal for these two similar molecules) is probably a key determinant for iron acquisition. The removal of iron from transferrin and lactoferrin, however, is to be expected in view of the considerable differences in binding constants between transferrin (and lactoferrin) and carboxymycobactin (and mycobactin): the K_s values are approx 10^{20} and 10^{36} (see Snow 1970).

2.6 Utilization of other siderophores including xenosiderophores

The major extracellular siderophores of the mycobacteria are the exochelins and carboxymycobactins. Other siderophores can be used even though these are not synthesized by the bacteria. Messenger and Ratledge (1982) showed that *M. smegmatis* was able to take up ferric citrate in a system independent of the exochelin route and was not affected by any metabolic inhibitor. The system, unlike a similar one in *E. coli*, was constitutive and did not require cells to be grown in the presence of citrate for its induction. Subsequent studies by Matzanke *et al.* (1997) indicated that the ferric ion was directly transferred from citrate to mycobactin (section 4) within the cells. A simple exchange of iron, not requiring participation of any reductive process, was suggested, as the chelating strength of mycobactin is far in excess of that of citrate. Whilst this may well be the ultimate fate of iron from ferric citrate, Messenger and Ratledge (1982) had suggested that the route of uptake may be more complex, as iron uptake was not in competition with iron uptake from ferric salicylate

which had been shown to involve mycobactin directly. Moreover, mycobactin was not an essential component of ferric citrate uptake as Messenger *et al.* (1986) showed that this process also readily occurred in a strain of *M. vaccae* that was devoid of mycobactin (Hall & Ratledge 1984, 1986). Perhaps in this instance direct incorporation of iron, when released from the citrate, into the cells was being achieved (Fig. 14.7). Two genes have been reported in the genome sequence of *M. tuberculosis* (Cole *et al.* 1998) as being equivalent to FeIII-dicitrate transporters in *E. coli*, *fecB* and *fecB2*, indicating that this route is also of importance for the acquisition of iron in the mycobacteria.

Uptake of 'xenosiderophores', i.e. siderophores produced by foreign (non-mycobacterial) microorganisms, can also occur. Unpublished work from the author's laboratory had indicated that ferrirhodotorulic acid could act as an iron source for *M. smegmatis* (see Wheeler & Ratledge 1994) and recent work by Matzanke *et al.* (1997) has shown that not only can rhodotorulic acid (a siderophore from yeasts) be used by *M. smegmatis* and *M. fortuitum* but also ferricrocin (from *Aspergillus viridi-nutans*), serratiochelin (from *Serratia marcescens*) and myxochelin (from a *Myxobacterium*). Some stimulation of growth was found with ferrioxamine B and rhizoferrin but not with aerobactin or enterobactin. With ferricrocin, the rate of uptake of iron was less than 10% of the rates of uptake observed with ferri-exochelin (Stephenson & Ratledge 1979) though the system was still inhibited by several respiratory poisons indicating that, like the ferric exochelin transport process, this too was an active uptake system. The process was deduced not to involve mycobactin but, in view of its very slow uptake rate, conclusions advanced for its uptake must be viewed cautiously as the most likely route of uptake of ferriocrocin is via the exochelin route which, at low concentrations, does not involve mycobactin. Genes corresponding to uptake systems for vibriobactin (from *Vibrio* spp.) and ferripyochelin (from *Pseudomonas aeruginosa*) have been found in the genome of *M. tuberculosis* (Cole *et al.* 1998) but not for the other siderophores mentioned above.

3 Receptors for siderophores: iron-regulated envelope proteins

As indicated in section 2.4, some of genes and the corresponding proteins are now being identified for exochelin synthesis and its transport. With the active transport of any solute, a receptor protein for recognizing the material is usually necessary. Uptake of microbial siderophores is well known to involve such receptors proteins (van der Helm 1998). Attempts to recognize equivalent receptors for the mycobacterial siderophores have been made using isolated envelope (wall+membrane) fractions from the cells—the so-called iron-regulated envelope proteins (IREPs). Table 14.3 summarizes the present state of knowledge of these proteins. Prominent in almost all species examined so far is a 29-kDa protein. This was first recognized in envelope preparations from iron-deficiently grown *M. smegmatis* (Hall *et al.* 1987). When antibodies were raised to this protein they inhibited ferri-exochelin uptake into whole cells. Antibodies raised to three other IREPs failed to inhibit iron uptake. The 29-kDa protein has now been isolated from *M. smegmatis* (Dover & Ratledge 1996) using affinity chromatography with ferriexochelin as the binding ligand. Although it has been subsequently partially sequenced (G. Nixon, unpublished observations 1998), its consensus DNA sequence did not correspond to any of the currently published gene sequences for exochelin transport (Fiss *et al.* 1994; Zhu *et al.* 1998).

A possible similar group of iron-regulated proteins (IRPs) have been isolated from whole cells of *M. tuberculosis* but without any prior separation of the envelope fraction (Calder & Horwitz 1998). Seven such proteins were identified of which three (15, 24 and 29 kDa) were expressed in cells grown in high iron conditions and four were expressed under low iron growth conditions with molecular sizes of 10, 13, 23 and 28 kDa. The gene for the first protein, *irp10*, was found to be next to a gene, *mtaA*, coding for a metal-transporting adenosine triphosphatase (ATPase) and, although no evidence for the corresponding protein, MtaA, at 72 kDa being iron regulated was presented, it was nevertheless suggested that the two proteins, Irp10 and MtaA, functioned as a two-component metal-transport system in *M. tuberculosis*. Further work on this system could clearly help to explain some of the aspects of iron uptake in *M. tuberculosis* particularly now that the complete genome sequence of *M. tuberculosis* is available (Cole *et al.* 1998). The genome sequence indicates nine metal-transporting ATPases to be present as well as one for Cd^{2+} transport, two for Mg^{2+} and three for K^+.

IREPs have also been examined in mycobacteria grown *in vivo* and *in vitro*. Sritharan and Ratledge (1990) showed the presence of IREPs of 180, 29, 21 and 14 kDa in the *in vitro*, iron-deficiently grown cells of *M. avium*. The same organism, when grown and recovered from C57 mice, still produced the same proteins (Table 14.3) in the wall fraction of the envelope. When the organism was grown iron sufficiently *in vitro*, these proteins were no longer in evidence. The clear inference is that the IREPs, induced by iron deficiency *in vitro*, must have been induced by similar conditions existing *in vivo*. Further work is needed to verify that the proteins isolated from *in vitro* and *in vivo* are in fact the same though the present evidence strongly supports this proposition.

During this investigation of IREPs in *M. avium* (and also *M. leprae*), Sritharan and Ratledge (1990) observed the presence of two high-molecular-weight proteins at 240 and 250 kDa in the animal-grown bacteria. Such proteins had been previously recognized both in *M. smegmatis* and in *M. avium* but only when they were grown iron sufficiently. These were termed the high-iron proteins (HIPs).

It would therefore appear that mycobacteria recovered from infected animals, are simultaneously both iron deficient and iron sufficient. This apparent paradox, however, is explicable. It has been suggested (Sritharan & Ratledge 1990) that during the initial stages of infection the bacteria are probably deprived of iron by the iron-withholding action of transferrin (sections 1.2 and 1.3). The bacteria then respond by expressing all the genes involved in iron sequestration and, very quickly, the carboxymycobactin (the extracellular siderophore of the pathogens) gains the iron from the surrounding material. The cells do not continue to express the iron-regulated genes as iron is now available and will

Table 14.3 Occurrence of iron-regulated envelope proteins (IREPs) and high iron proteins (HIPs) as analysed by SDS-PAGE in mycobacteria grown *in vitro* and *in vivo*.[a]

Organism	HIPs of:		IREPs of:							
	250 kDa	240 kDa	180 kDa	120 kDa	84 kDa	29 kDa	25 kDa	21 kDa	14 kDa	11 kDa
M. smegmatis	*	*	+	–	+	+	+	–	+	–
M. neoaurum	*	*	+	*	–	+	–	+	+	–
ADM 8563	?	?	–	–	–	+	–	–	+	+
M. avium (*in vitro*)	*	*	+	–	–	+	–	+	+	–
M. avium (*in vivo*)	✓	✓	✓	–	–	✓	–	✓	✓	–
M. leprae (*in vivo*)	✓	✓	–	–	–	✓	–	✓	✓	–

[a] The results are derived from work by Sritharan and Ratledge (1990).
+ and – denote the presence or absence of the IREP in iron deficiently-grown cells.
* Presence of protein only in iron-sufficient cells.
? Very faint band, uncertain presence.
✓ Present *in vivo* where iron status is uncertain.
ADM, armadillo-derived mycobacterium.

inevitably lead to their repression. The *in vivo* state is unlike the contrived *in vitro* state where iron is not available to the cells at any stage of their growth because it has been deliberately removed; *in vivo* iron is present but needs to be continuously sequestered. Hence, the infecting bacteria must oscillate between iron deprivation and iron sufficiency: they are on the cusp of iron availability. Consequently, such cells must express components both of a high-iron state (i.e. the HIPs) and of a low-iron state (i.e. the IREPs) thereby explaining how both sets of proteins were recognized in *M. avium* and *M. leprae* recovered from animal infections.

4 Mycobactins: the intracellular siderophore

4.1 Structures

Mycobactins were amongst the very first bacterial siderophores to be described (Francis *et al.* 1949, 1953). The elegant work of Snow and colleagues in the 1950s and 1960s elucidated the structures of 10 mycobactins from different mycobacteria (see Fig. 14.6). Since this work, the structures of only two further mycobactins have been reported: mycobactin J from *M. paratuberculosis* (McCullough & Merkal 1982) and mycobactin Av from the three closely species *M. avium*, *M. intracellular* and *M. scrofulaceum*

(Barclay *et al.* 1985). The structure of a related molecule from *Nocardia asteroides,* termed a nocobactin, has been reported (Ratledge & Snow 1974).

As mycobactin J is produced and is available commercially, it should be pointed out that it is unique in having an isopropyl [$CH_3.CH(CH_3)$-] grouping at the R_4 position—see Fig. 14.6. Barclay *et al.* (1985) who also examined the same strain of *M. paratuberculosis* failed to find this mycobactin and instead found a mycobactin of Av type suggesting that *M. paratuberculosis* was closely related to *M. avium*. Mycobactins of the Av type, nevertheless, are unusual in having two alkyl chains in the molecule instead of the normal one. The origins of mycobactin J remain a mystery.

Although the structural determinations of new mycobactins have been limited, this has not prevented them being detected in almost every single mycobacterium that has been examined. They are also present as nocobactins in some species of *Nocardia* and *Rhodococcus* (see Ratledge 1984). Simple methods have been devised for promoting the formation of mycobactin in both solid media (Hall & Ratledge 1982) and liquid media (Barclay *et al.* 1992; Bosne *et al.* 1993). As mycobactins are useful chemotaxonomic markers amongst the mycobacteria (Snow 1970; Hall & Ratledge 1984), rapid methods for their separation and identification (but not structural elucidation) have been developed including high-performance liquid chromatography (HPLC) (Hall 1986; Hall & Ratledge 1984), high-performance thin layer chromatography (HPTLC) (Hall & Ratledge 1984; Barclay *et al.* 1992) and thin layer chromatography (TLC) (Bosne *et al.* 1993; Leite *et al.* 1995). Thus, the predictions of Snow (1970) that mycobactins, because of their subtle variations in structures between species, would prove useful chemotaxonomic markers has been more than adequately borne out by the numerous publications mentioned above and also others that have involved applications to the rapid identification of clinical mycobacteria (Hall & Ratledge 1985a,b; Barclay & Ratledge 1988; Bosne & Levy-Frebault 1992; Leite *et al.* 1995).

4.2 Biosynthesis of mycobactin and carboxymycobactin

Although very little work has been carried out on the biosynthesis of mycobactin *per se*, the biosynthetic origins of the molecule are nevertheless clear. The aromatic moiety, salicylic acid, is synthesized via the shikimic acid pathway with chorismic acid, isochorismic acid and finally 2,3-dihydro-2,3-dihydroxybenzoic acid being the intermediates (Marshall & Ratledge 1971, 1972). In those mycobactins where 6-methylsalicylate occurs (i.e. mycobactins A, H and P—see Fig. 14.6), this is synthesized by the polyketide route and not by the shikimic acid route (Snow 1970). The remainder of the molecule is probably assembled, as is the case with exochelin (see section 2.4), by a series of non-ribosomal, peptide synthetases involving serine (or sometimes threonine), N^6-hydroxylysine (two molecules), a short-chain (C_4 or C_5) β-hydroxyacid and a long-chain fatty acid. Eight or nine genes that are involved in mycobactin or carboxymycobactin biosynthesis have now been recognized in *M. tuberculosis* (Cole *et al.* 1998) and, on the basis of the putative identity of one of these genes, salicylate is probably incorporated into the molecule via its activation as salicyloyl-AMP.

As mycobactin and carboxymycobactin are not interconvertible, they probably arise from a common precursor, a desacylmycobactin, that is the molecule lacking the long alkyl side-chain (see Fig. 14.6). The

final step leading to mycobactin formation would then be via reaction of this intermediate with a fatty acyl-CoA ester. As fatty acyl-CoA esters occur as an even numbered series of acids, this then explains the related series of alkyl groups on mycobactin increasing by two C atoms (see Fig. 14.6). However, with carboxymycobactin the final acylation donor is much less clear. First, as the terminal group is a carboxylic acid, the acyl donor should be a dicarboxylic acid. Second, as the length of the acyl chain on carboxymycobactin increases by unit numbers, the donor cannot be a fatty acid simply being converted to a dicarboxylic acid as this would give a series of acyl chains increasing by two C atoms. Instead, the donor has been suggested (Lane *et al.* 1998) as probably being a long chain fatty acid or derivative that is cleaved to give one or perhaps two diacyl fatty acids: thus C_{12} gives $2 \times C_6$, C_{14} gives $2 \times C_7$, etc. The biochemistry of such a cleavage reaction is though far from clear.

4.3 Are mycobactins essential?

Mycobactins are not ubiquitous amongst mycobacteria. *M. paratuberculosis* and some strains of *M. avium* require mycobactin for growth in the laboratory (Matthews *et al.* 1976; Lambrecht & Collins 1993) suggesting that it fulfils some essential function. However, carboxymycobactin (section 2.5), but not the water-soluble exochelins, will also act as an alternative growth factor and is not converted into mycobactin (Barclay & Ratledge 1983). Thus, strictly, mycobactin *per se* is not an obligate nutrient for these species though a specific means of iron solubilization and transport obviously is required.

A mycobacterium that grows well in laboratory media and yet is devoid of mycobactin is *M. vaccae* (Hall & Ratledge 1984). The organism shows some impairment of growth only if it is grown in competition with a mycobactin-producing species and then, when iron is suddenly presented to this mixed culture, *M. vaccae* is disadvantaged as it fails to acquire sufficient iron for subsequent growth (Hall & Ratledge 1986). On its own, this organism is able to acquire its iron by using the exochelin-mediated route (Messenger *et al.* 1986) but has no apparent means of storing a surfeit of it. *M. microti* also appears to be devoid of mycobactin (Barclay & Ratledge 1988) and although it only grows slowly in laboratory media, its growth is not enhanced by mycobactin or any other mycobacterial siderophore (M. Ewing and C. Ratledge, unpublished observations 1996).

The presence of mycobactin in *M. leprae* is uncertain. As this organism cannot be grown in axenic culture, it must be obtained from infected tissues, usually armadillo livers or mouse footpads. Kato (1985) examined *M. leprae* taken from the former source and, after careful extraction, failed to detect mycobactin either directly or by its action to promote the growth of a mycobactin-requiring strain of *M. paratuberculosis*. However, Dhople and Osborne (1988) following an almost identical protocol succeeded in producing growth of *M. paratuberculosis* using a chloroform extract of a leprosy-infected armadillo liver. The presence of mycobactin in the *in vivo M. leprae* was therefore considered highly likely as uninfected livers gave no growth stimulation of *M. paratuberculosis*.

Of major concern both in this work and with mycobactin-competent mycobacteria, is whether mycobactin is indeed produced during *in vivo* growth. Lambrecht and Collins (1993) concluded that neither *M. tuberculosis* nor *M. avium* recovered from infected animals contained any mycobactin, again using growth of *M. paratuberculosis* as an assay method.

All the above studies beg the question of whether mycobactin fulfils an essential role in iron metabolism as: (i) some mycobacteria clearly do grow both in animals and in laboratory media without it; and (ii) what would be the minimal amount of mycobactin to sustain growth *in vivo* when the amount of mycobactin that occurs in cultures grown iron sufficiently in the laboratory is very small indeed. Mycobactin is only produced if cultures are iron deficient and it is far from proven that *in vivo* cultures are in this state. Evidence from the presence IREPs (see section 3) suggests that mycobacteria *in vivo* are probably on the cusp of iron deficiency and may be repressed for mycobactin biosynthesis by the time they are recovered from infected tissues. The only reliable method to ascertain if mycobactin is essential for *in vivo* growth is to delete one of the mycobactin-synthesizing genes (together with a

similar experiment deleting one of the genes for carboxymycobactin biosynthesis) from a pathogen and then determine if the pathogenicity of such a mutant have been affected.

A possible scenario for the role of mycobactin is that it is an essential component in the transfer of iron from the envelope, where it is located (Ratledge *et al.* 1982), into the cytoplasm. Only a few molecules of mycobactin per cell may be needed but without them, iron assimilation may not take place (see also section 5).

The alternative model to using mycobactin for iron assimilation is to suggest, as Lambrecht and Collins (1993) have done, that iron assimilation *in vivo* may occur by direct binding of mycobacteria to transferrin or lactoferrin. This is already established with other bacteria including a number of pathogens: *Neisseria* spp., *Haemophilus*, *Moraxella catarrhalis*, *Actinobacillus pleuropneumonia* and *Porphyromonas gingwalis* (see Byers & Arceneaux 1998). In most cases, the iron is removed from transferrin or lactoferrin but in some cases haem itself may be used. Although receptors for neither transferrin or lactoferrin have yet been recognized in mycobacteria, an intrinsic part of this uptake procedure (see Fig. 14.1b) is the direct removal of iron from the molecule by reduction. Such a novel reductase has recently been identified in *M. paratuberculosis* (Homuth *et al.* 1998). Necessarily, the reductase is extracellular and it could function not only with transferrin but also with ferritin and ferric ammonium citrate. It had an M_r of 17 000 and its gross amino acid composition was also reported. Significantly, antibodies raised to the reductase were able to detect the enzyme in *M. paratuberculosis* recovered from infected bovine tissue thus signifying that this enzyme was fulfilling a role in the *in vivo* state.

It will therefore be of clear interest to determine whether related extracellular ferric reductases may be found in *M. tuberculosis* and other pathogens as this would go a considerable way to explain the mechanism of iron acquisition in the *in vivo* state: is it via the siderophore-mediated route (see Fig. 14.1a) or the direct acquisition pathway (see Fig. 14.1b)?

Mycobactin, however, is able to fulfil a definite role

in those mycobacteria that possess it. Although it can be argued (see Wheeler & Ratledge 1994) that the amounts of mycobactin found in mycobacteria grown in the laboratory are due to nothing more than the contrived continuous withholding of iron from the cells, nevertheless it must be supposed that some mycobactin will be produced when the cells grow *in vivo*. (It is hardly likely that mycobactin biosynthesis only occurs in mycobacteria grown in laboratory media.) Thus, when iron is suddenly presented to mycobacteria grown under iron-limiting conditions, the iron is taken up quickly (within 3–4 min) by the cells and they become visibly red due to the formation of ferric mycobactin (McCready & Ratledge 1979). The mycobactin appears therefore to act as a store of iron. From the work of Matzanke *et al.* (1997) it would appear that mycobactin is a short-term storage molecule for iron before it is passed on into the cell (see section 7).

As mycobacteria repress the synthesis of porphyrins during iron deprivation (see Table 14.1), a shift up in metabolism must occur before the iron can be transferred into porphyrins and apo-proteins. Therefore the hypothesis is that whilst porphyrin and protein synthesis are being de-repressed, the iron is held within mycobactin in readiness for the shift up in iron metabolism. Mycobactin safeguards the cells against a sudden iron overload, which would be a highly dangerous event, and at the same time holds the iron in a form that can gradually be used as the demand for iron increases. The 'feast and famine' scenario of bacteria nutrition (Koch 1971) is thus helped by the presence of key storage molecules such as mycobactin for iron. In the *in vivo* situation, the extremes of iron deprivation and supply are likely to be avoided but nevertheless surges in the supply of iron to the cells can be anticipated to occur, particularly if a molecule such as ferritin, with over 4000 atoms of iron, was being degraded within a tissue.

5 Release of iron from mycobactin and the other siderophores

Although mycobactin (and carboxymycobactin which has the same chelating centre) and the exo-

chelins have relatively high stability constants for iron (about 10^{36} and 10^{25}, respectively), iron can be removed from them by reduction to Fe(II). A ferrimycobactin ferric reductase was first reported by Ratledge (1971) in anaerobic extracts of *M. smegmatis* in which reduced nicotinamide adenine dinucleotide (NADH) or reduced NAD phosphate (NADPH) was used as the external reductant. Brown and Ratledge (1975a) showed that the reaction rate was increased almost 10-fold if EDTA was included in the assay mixture to prevent the rapid reoxidation of the ferrous ions and McCready and Ratledge (1979) found that salicylate could substitute for EDTA in the same reaction thereby suggesting a possible role for salicylate as an intracellular chelator of ferrous ions.

The ferric reductase had a K_m value for NADH of 1.75 mmol/L and less than 4 µmol/L for ferrimycobactin. It was strongly inhibited by thiol reactive reagents, $HgCl_2$ and *N*-ethylmaleamide as well as *p*-chloromercuribenzoate and iodoacetate (McCready & Ratledge 1979). There was no significant hydrolysis of the mycobactin indicating that the reduction was occurring without changing the iron-chelating properties of the molecule. However, as ferrimycobactin was also reduced by extracts from other microorganisms, it appeared that the enzyme nvolved could simply be any dehydrogenase of broad specificity. Indeed this 'ferric reductase' of *M. smegmatis* could not only reduce ferrimycobactin as well as ferriexochelin, it also reduced the iron in ferriferrioxamine B (Desferal) and ferric ammonium citrate. Further work to purify the enzyme was therefore not pursued at that time though it is now evident that ferric reductases are as ubiquitous and probably are the sole route by which iron can be removed from the ferric siderophores (Leong & Winkelmann 1998).

Although ferricarboxymycobactin was not examined as a substrate for the ferric reductase of *M. smegmatis* (McCready & Ratledge 1979), it would be expected that it too would be as readily reducible as ferrimycobactin itself.

The fate of iron after leaving its siderophore is discussed in section 7.

6 Role of bacterioferritin

Iron-storage proteins can be divided into two somewhat related groups: the ferritins, which are found in both eukaryotes and prokaryotes, and the haem-containing bacterioferritins found in the true bacteria (i.e. eubacteria) and fungi (Harrison *et al.* 1998). The occurrence of bacterioferritin in mycobacteria has been described (Brooks *et al.* 1991; Inglis *et al.* 1994; Pessolani *et al.* 1994) and its presence is already helping to clarify key aspects of iron metabolism.

Bacterioferritins occur in a large variety of microorganisms, Gram-positive and Gram-negative bacteria, micro-algae, fungi and others (Chasteen 1998). They differ from the ferritins by containing between eight and 12 haem groups (although 12 is the usual number) and these are bound between two polypeptides which go to make up a 24-mer molecule. The molecular size of the protein subunit in *M. leprae* is 18.2 kDa (Pessolani *et al.* 1994) and the total molecular mass for the whole molecule was found to be 380 kDa. The number of atoms of iron contained within the protein shell was between 1000 and 4000. (These numbers and molecular sizes do not strictly accord with a 24-mer macromolecule but instead suggest that the bacterioferritin of *M. leprae* could be a 20-mer, although this remains to be resolved.)

The presence of bacterioferritin in most if not all mycobacteria is therefore to be expected. Indeed, Pessolani *et al.* (1994) were able to recover sufficient of the molecule from *M. leprae,* isolated from infected armadillo livers and spleens, to be able to partially characterize the molecule and to suggest that it must play a crucial role in iron metabolism in the *in vivo* mycobacteria in view of its abundance. However, the essentiality of bacterioferritin for growth of mycobacteria in macrophages, and thus in infected tissues, is by no means certain as Denoel *et al.* (1997) reported that the survival of *Brucella melitensis* in human macrophages was unaffected when a bacterioferritin-deletion mutant of it was also placed in macrophages. A similar experiment, perhaps using *M. tuberculosis,* would therefore be illuminating to see if indeed bacterioferritin is essential for pathogenicity and

in vivo survival. The gene sequence for the bacterioferritin was determined by Pessolani *et al.* (1994), who were then able to identify equivalent gene sequences for bacterioferritin in several mycobacterial species: *M. tuberculosis, M. avium, M. paratuberculosis, M. intracellulare* and *M. scrofulaceum* but not in *M. smegmatis* or *M. xenopi* but this may only indicate a genetic diversity and not a lack of bacterioferritin. Indeed, the work of Matzanke *et al.* (1997) discussed in the next paragraph indicates that *M. smegmatis* (and *M. fortuitum*) does have a bacterioferritin. The complete gene sequence for the bacterioferritin from *M. avium* has been independently published by Inglis *et al.* (1994) and two genes, *bfrA* and *bfrB*, for bacterioferritin are now sequenced in the genome of *M. tuberculosis* (Cole *et al.* 1998). The finding of the *bfr* gene in *M. paratuberculosis* confirms the earlier report of this gene sequence by Brooks *et al.* (1991).

The role of bacterioferritin has been partially clarified by the recent work of Matzanke *et al.* (1997) using *M. smegmatis* and *M. fortuitum*. These workers, by following the course of iron uptake from ferric citrate using Mossbauer spectroscopy, were able to deduce the likely course of events as:

ferric citrate → ferric mycobactin → Fe(II) →
  bacterioferritin

The movement of iron from mycobactin to bacterioferritin was slow, indicating that it was not a spontaneous transfer. It is tempting therefore to suggest that transfer of iron from mycobactin to bacterioferritin may take place across the cytoplasmic membrane by an enzyme-mediated process (see section 7).

For the movement of iron out of bacterioferritin, where it is stored as the crystalline ferric oxide/hydroxide core (Chasteen 1998), it is supposed that this will be by a reductase or reductant but very little information is yet available on this subject. Ferredoxin, which has been proposed as a possible (cytosolic) storage molecule in *M. smegmatis* (Kikuchi *et al.* 1994), may possibly play a role but this will probably be only in iron replete or iron overloaded cells.

7 The big picture

The multitude of components involved in iron uptake by microorganisms is extremely large. Hopefully with respect to mycobacteria, most of the major pieces of the jigsaw are now known and it is up to the interested researcher to try to assemble these into a coherent picture.

As stated in section 1 (see Fig. 14.1), two possible scenarios for iron uptake can be envisaged: (i) the siderophore-mediated route of iron solubilization; and (ii) the direct contact route. Both routes may occur in mycobacteria. The key aspects of these routes, and the establishment of their essentiality, will, however, only be resolved when the genes coding for the individual components have been deleted and the ensuing mutants examined for changes in their pathogenicity, survival, growth and virulence in the host tissue or preferably whole animal. Until such experiments are undertaken, we can only speculate as to what may be 'the big picture' (Fig. 14.7).

The siderophore-mediated route, as typified by exochelin in *M. smegmatis*, is probably the best understood but, in comparison with bacteria such as *E. coli*, our knowledge is still rudimentary and the interpretations of data still subject to much uncertainty. The receptor protein for exochelin uptake (Rep in Fig. 14.7) is probably the 29-kDa protein described in section 3. The other uptake proteins (FxuA, B, C, etc.) are those described by Fiss *et al.* (1994) (see section 2.4). The route of uptake of carboxymycobactin, the extracellular siderophore of the pathogenic mycobacteria, is uncertain particularly as this is by facilitated diffusion and not therefore involving a direct input of energy. Uptake of ferric citrate and perhaps other xenosiderophores (Matzanke *et al.* 1997) may involve transfer into mycobactin (see sections 2.6 and 4).

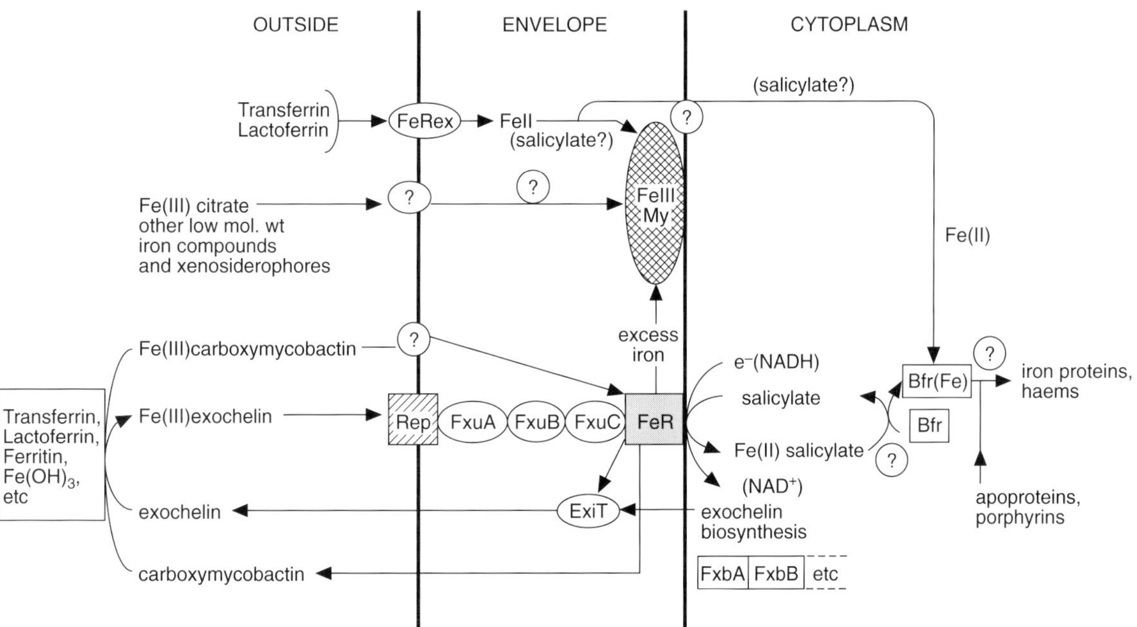

Fig. 14.7 Possible mechanisms of iron transport in mycobacteria, see text for details. FeRex, ferric reductase extracellular; FeR, ferric reductase; My, mycobactin; Rep, receptor protein for exochelin; FxuA, FxuB, etc., ferriexochelin uptake proteins; ExiT, exochelin transport protein; FxbA, FxbB, etc., ferriexochelin biosynthesis proteins; Bfr, bacterioferritin; ? indicates unknown mechanism.

For all siderophore-mediated routes of iron uptake, it is thought that the final step for iron to pass into the cytoplasm will be via a ferric reductase (FeR in Fig. 14.7) which will reduce Fe(III), carried in through the envelope, to Fe(II). This may involve the participation of salicylic acid (McCready & Ratledge 1979). Ferrosalicylate may then be the transfer system for its incorporation into bacterioferritin, the likely long-term storage protein for iron within the cytoplasm of the cells. For the incorporation of iron into porphyrins and apoproteins, this possibly involves ferric reductase again and possibly also salicylate.

Mycobactin is undoubtedly of importance for short-term iron storage in the cell envelope and may serve to regulate the iron flux through the cytoplasmic membrane. Iron overloading is prevented by rapid transfer (within 1–2 min) of excess iron from exochelins or carboxymycobactin into mycobactin, although whether mycobactin is strictly essential remains an open question (see section 4.3). Mycobactin-requiring mycobacteria, such as *M. paratuberculosis,* will grow if carboxymycobactin is provided to them and, as this molecule is not convertible into mycobactin, mycobactin *per se* cannot be an absolute necessity.

For the non-siderophore mediated route of uptake, the key presence of an extracellular ferric reductase (FeRex in Fig. 14.7) (Homuth *et al.* 1998) would indicate that direct acquisition of iron from host tissue sources such as transferrin, lactoferrin or even ferritin, is possible. An additional receptor protein for these molecules may be needed but FeRex itself could be both receptor and reductase. The fate of the Fe(II) from this source is completely unknown as yet but might involve its direct uptake into the cytoplasm (possibly involving once more salicylate as an appropriate carrier) or into mycobactin.

8 Inhibitors of iron metabolism

The only reason for studying mycobacteria is to be able to kill them. The eradication of tuberculosis and leprosy is a major objective of world health care. The paucity of antimycobacterial agents and the necessity of using multidrug regimes for the treatment of tuberculosis and related diseases means that there is

considerable urgency for the discovery of new agents, particularly when there is probably very little prospect of the adventitious discovery of new antibiotics over the next 10, and maybe even 20, years. Iron metabolism is such a vital aspect of bacterial growth that inhibition of this process appears an attractive proposition for the development of new chemotherapeutic drugs particularly as the mechanism involves siderophores and their attendant proteins that are not found in animals.

One such drug may, however, already exist. *p*-Aminosalicylic acid (PAS) has been known for over 50 years as an antituberculosis drug (Lehmann 1946) but it has usually been considered as an antifolate agent due to its closeness to *p*-aminobenzoic acid which is a component of folic acid itself. There have always been numerous inconsistencies with this proposal (see Winder 1982) not least of which is that PAS is not effective against other bacteria and that other known antifolates, such as the sulphonamides, are ineffective against mycobacteria. It therefore appears a more likely proposition that PAS acts as a salicylate analogue and interferes, in some way, with iron metabolism (Ratledge & Brown 1972; Brown & Ratledge 1975a,b). PAS is not an inhibitor of mycobactin biosynthesis but significantly PAS-treated cells behave biochemically as if they are iron deprived even in the presence of a surfeit of iron. Furthermore, a mycobactin auxotroph of *M. smegmatis* was 50 times more sensitive to PAS than the wild-type strain and more recent findings have shown that a salicylate auxotroph of the same organism has an equally increased sensitivity to PAS (A. Tadepalli, P. D. Ayling and C. Ratledge, unpublished observations 1998). Although Brown and Ratledge (1975b) could find no evidence for PAS acting at the level of ferrimycobactin ferric reductase (see Fig. 14.7 and section 5), the current theory for the involvement of bacterioferritin in mycobacterial metabolism suggests other possible sites where PAS could be inhibitory to iron acquisition. As PAS is inhibitory to most mycobacteria, but not to other bacteria, there has to be some unique aspect of mycobacterial metabolism not found elsewhere that must be affected and the prime candidate has to be an aspect of salicylate metabolism involved in iron assimilation. Until more details of the process are understood, it is still premature to speculate further as to what reaction may be affected. However, as a large number of variations of PAS structure have already been synthesized (see Winder 1964) it possibly is unrewarding at this stage to synthesize further analogues but, when the exact site of action of PAS is identified, the present powers of combinatorial chemistry could be usefully addressed to synthesizing analogues acting against this target.

Another useful approach that has already been tried to block iron uptake, is to use metal analogues of siderophores. For example, Rogers *et al.* (1980, 1982) showed that the scandium and indium complexes of enterochelin (=enterobactin) were inhibitory to *E. coli*. Although a number of similar metal analogues of mycobactin and carboxymycobactin (then called exochelin) were prepared (Barclay & Ratledge 1986b) and tested against *M. intracellulare* and *M. tuberculosis*, none of them was particularly effective and, indeed, they were sometimes positively beneficial for growth. The problem with this strategy is that the stability of siderophores for iron is always higher than for other metals so that alternative complexes will always exchange their metal for iron and thus be changed from a potential antagonist of iron to a potential growth-promoting substance, a fact that Snow (1970) had already advanced in his study of the mycobactins (see also Snow & White 1969).

The final concept that offers some promise is to use the siderophores as a means of introducing a lethal agent into the cell that otherwise would not gain access: the so-called 'Trojan Horse' concept. this novel mechanism of drug delivery has been pioneered by M. J. Miller, and a number of papers and reviews on this topic have been published (for reviews see Miller 1989; Miller & Malouin 1993, 1994). To achieve success with mycobacteria, Hu and Miller (1997) have achieved the total synthesis of mycobactin S, a notable achievement in its own right. They have then found, somewhat surprisingly in view of the earlier work of Snow (1970), that at a relatively high concentration of 12.5 μg/mL it was able to cause greater than 99% inhibition of the growth of *M. tuberculosis* H37Rv. As mycobactin S differs from mycobactin T (see Fig. 14.6) in only one chiral centre (the one at the R$_4$ sub-

stituent), it was surmised by Hu and Miller (1997) that this was sufficient to explain the complete inhibition of growth. Further studies on synthetic analogues of mycobactin have recently been reported (Xu & Miller 1998) with one analogue being able to inhibit *M. tuberculosis* with a minimum inhibitory concentration (MIC) value of less than 0.2 μg/mL. Related work has also been reported by Bosne-David *et al.* (1997) in which *M. aurum* was inhibited by a synthetic catechol siderophore, FR160, and by high concentrations of mycobactin J and mycobactin S. Mycobactin A (from *M. aurum*) was growth-promoting at similar high concentration but ferrioxamine (Desferal) was without effect.

The conceptual idea of using mycobactin, or one of the other mycobacterial siderophores, to smuggle a toxic agent into the bacteria that otherwise could not gain access is an attractive proposition (see *et al.* 1984, 1989; Ohi *et al.* 1986; Silley *et al.* 1990; Watanabe *et al.* 1987; Miller 1989; Miller & Malouin *et al.* 1993, 1994). An obvious candidate for a toxic agent would be one of the β-lactam antibiotics as there is no reason why mycobacteria should not be inhibited by penicillin or cephalosporin except that these antibiotics appear not to be able to be taken up by mycobacteria (see Chapters 12 and 15). Ghosh and Miller (1993) have already synthesized a number of such β-lactam-siderophore conjugates using the citrate-based siderophores of aerobactin, schizokinen, arthrobactin and nannochelin (from *Myxobacterium nannocystis*). These proved to be effective bacteriostatic agents against *E. coli* but were not toxic and merely delayed the onset of growth. Further work along these lines will therefore be awaited with keen interest to see if, indeed, the siderophores of mycobacteria can find practical applications that will lead to the defeat of tuberculosis and the other mycobacterial disease.

9 References

Aisen, P. (1998) Transferrin, the transferrin receptor and the uptake of iron by cells. In: *Metal Ions in Biological Systems*, Vol. 35, *Iron Transport and Storage in Microorganisms, Plants and Animals*. (eds A. Sigel, & H. Sigel). New York: Marcel Dekker, pp. 585–631.

Alvarez-Hernandez, X., Liceaga, J., McKay, I. & Brock, J.H. (1989) Induction of hypoferrenica and modulation of macrophage iron metabolism by tumor necrosis factor. *Laboratory Investigation* **61**, 319–322.

Bagg, A. & Neilands, J.B. (1987) Molecular mechanism of siderophore-mediated iron assimilation. *Microbiological Reviews* **54**, 509–530.

Barclay, R. & Ratledge, C. (1983) Iron-binding compounds of *Mycobacterium avium, M. intracellulare, M. scrofulaceum* and mycobactin-dependent *M. paratuberculosis* and *M. avium. Journal of Bacteriology* **153**, 1138–1146.

Barclay, R. & Ratledge, C. (1986a) Participation of iron in the growth inhibition of pathogenic strain of *Mycobacterium avium* and *M. paratuberculosis* in serum. *Zentralbatt für Bakteriologie, Mikrobiologie und Hygiene* **A 262**, 189–194.

Barclay, R. & Ratledge, C. (1986b) Metal analogues of mycobactin and exochelin fail to act as effective antimycobacterial agents. *Zentralbatt für Bakteriologie, Mikrobiologie und Hygiene* **A 264**, 203–207.

Barclay, R. & Ratledge, C. (1988) Mycobactins and exochelins of *Mycobacterium tuberculosis, M. bovis, M. africanum* and other related strains. *Journal of General Microbiology* **134**, 771–776.

Barclay, R. & Wheeler, P.R. (1989) Metabolism of mycobacteria in tissues. In: *The Biology of the Mycobacteria,* Vol. 3 (eds C. Ratledge, J. Stanford & J. M. Grange). London: Academic Press, pp. 37–196.

Barclay, R., Ewing, D.E. & Ratledge, C. (1985) Isolation, identification and structural analysis of the mycobactins of *Mycobacterium avium, M. intracellulare, M. scrofulaceum* and *M. paratuberculosis. Journal of Bacteriology* **164**, 896–905.

Barclay, R., Furst, V. & Smith, I. (1992) A simple and rapid method for the detection and identification of mycobacteria using mycobactin. *Journal of Medical Microbiology* **37**, 286–290.

Basker, M.J., Edmondson, R.A., Knott, S.J., Ponsford, R.J., Slocombe, B. & White, S.J. (1984) *In vitro* antibacterial properties of BRL 36650, a novel 6-substituted penicillin. *Antimicrobial Agents and Chemotherapy* **26**, 734–740.

Basker, M.J., Frydrych, C.H., Harrington, F.P. & Milner, P.H. (1989) Antibacterial activity of catecholic piperacillin analogues. *Journal of Antibiotics* **42**, 1328–1330.

Bosne, S. & Levy-Frebault, V.V. (1992) Mycobactin analysis as an aid for the identification of *Mycobacterium fortuitum* and *M. chelonae* subspecies. *Journal of Clinical Microbiology* **30**, 1225–1231.

Bosne, S., Papa, F. & Clavel-Sérès and Rastogi, N. (1993) A simple and reliable EDOA method for mycobactin production in mycobacteria: optimal conditions and use in mycobacterial speciation. *Current Microbiology* **26**, 353–358.

Bosne-David, S., Bricard, L., Ramiandrasoa, F., DeRoussent, A., Kunesch, A. & Andremont, A. (1997) Evaluation of growth promotion and inhibition from mycobactin and non-mycobacterial siderophores (desferrioxamine and FR160) in *Mycobacterium aurum. Antimicrobial Agents and Chemotherapy* **41**, 1837–1839.

Braun, V., Hantke, K. & Köster, W. (1998) Bacterial iron transport: mechanisms, genetics and regulation. In: *Metal Ions in Biological Systems*, Vol. 35, *Iron Transport and Storage in Microorganisms, Plants and Animals* (eds A. Sigel & H. Sigel). New York: Marcel Dekker, pp. 67–145.

Brooks, B.W., Young, N.M., Watson, D.C., Robertson, R.H., Sugden, E., Nielsen, K.H. & Becker, S.A. (1991) *Mycobacterium paratuberculosis* antigen D; characterization and evidence that it is a bacterioferritin. *Journal of Clinical Microbiology* **29**, 1652–1658.

Brown, K.A. & Ratledge, C. (1975a) Iron transport in *Mycobacterium smegmatis*: ferrimycobactin reductase (NAD (P)H: ferrimycobactin oxidoreductase), the enzyme releasing iron from its carrier. *FEBS Letters* **53**, 262–266.

Brown, K.A. & Ratledge, C. (1975b) The effect of *P*-aminosalicylic acid on iron transport and assimilation in mycobacteria. *Biochimica et Biophysica Acta* **385**, 207–220.

Bullen, J.J. & Griffiths, E. (1987) *Iron and Infection: Molecular, Physiological and Clinical Aspects.* Chichester: John Wiley & Sons.

Byers, R.B. & Arceneaux, J.E.L. (1998) Microbial iron transport: iron acquisition by pathogenic microorganisms. In: *Metal Ions in Biological Systems*, Vol. 35, *Iron Transport and Storage in Microorganisms, Plants and Animals* (eds A. Sigel & H. Sigel). New York: Marcel Dekker, pp. 37–66.

Byrd, T. (1997) Tumor necrosis factor (TNF) promotes growth of virulent *Mycobacterium tuberculosis* in human monocytes: iron mediated growth suppression is correlated with decreased release of TNF from iron-treated infected monocytes. *Journal of Clinical Investigations* **99**, 2518–2529.

Byrd, T.F. & Horwitz, M.A. (1993) Regulation of transferrin receptor expression and ferritin content in human mononuclear phagocytes. *Journal of Clinical Investigation* **91**, 969–976.

Calder, K.M. & Horwitz, M.A. (1998) Identification of iron-regulated proteins of *Mycobacterium tuberculosis* and cloning of tandem genes encoding a low-induced protein and a metal transporting ATPase with similarities to two-component metal transport systems. *Microbial Pathogen* **24**, 133–143.

Chasteen, N.D. (1998) Ferritin. Uptake, storage and release of iron. In: *Metal Ions in Biological Systems*, Vol. 35, *Iron Transport and Storage in Microorganisms, Plants and Animals* (eds A. Sigel & H. Sigel). New York: Marcel Dekker, pp. 479–514.

Clemens, D.L. & Horwitz, M.A. (1996) The *Mycobacterium tuberculosis* phagosome interacts with early endosomes and is accessible to exogenously administered transferrin. *Journal of Experimental Medicine* **184**, 1349–1355.

Cole, S.T. & 41 others (1998) Deciphering the biology of *Mycobacterium tuberculosis* from the complete sequence. *Nature* **393**, 537–544.

Crosa, J.H. (1997) Signal transduction and trancriptional and posttranscriptional control of iron-regulated genes in bacteria. *Microbiological Molecular Biological Review* **61**, 319–336.

Crowle, A., Dahl, R., Ross, E. & May, M. (1991) Evidence that vesicles containing living virulent *M. tuberculosis* or *M. avium* in cultured human macrophages are not acidic. *Infection and Immunity* **59**, 1823–1831.

Denoel, P.A., Crawford, R.M., Zygmunt, M.S., Tibor, A., Weynauts, V.E., Godfreid, F., Hoover, D.L. & Letesson, J.-J. (1997) Survival of a bacterioferritin deletion mutant of *Brucella melitensis* 16M in human monocyte-derived macrophages. *Infection and Immunit* **65**, 4337–4340.

Dhople, A.M., Ibanez, M.A. & Poirer, T.C. (1996) Role of iron in the pathogenesis of *Mycobacterium avium* infection in mice. *Microbios* **87**, 77–87.

Dhople, A.M. & Osborne, L.J. (1988) Presence of mycobactin-like substance in *Mycobacterium leprae*. *Indian Journal of Leprosy* **60**, 348–359.

Doukhan, L., Predich, M., Nair, G.L., Dussurget, O., Mandic-Mulec, I., Cole, S.T., Smith, D.R. & Smith, I. (1995) Genomic organization of the mycobacterial sigma gene cluster. *Gene* **165**, 67–70.

Douvas, G.S., May, M.H. & Crowle, A.J. (1993) Transferrin, iron, and serum lipids enhance or inhibit *Mycobacterium avium* replication in human macrophages. *Journal of Infection and Disease* **167**, 857–864.

Douvas, G.S., May, M.H., Pearson, J.R., Lain, E., Miller, L. & Tsuchida, N. (1994) Hypertriglyceridemic serum, very low density lipoprotein, and iron enhance *Mycobacterium avium* replication in human macrophages. *Journal of Infectious Diseases* **170**, 1248–1255.

Dover, L.G. & Ratledge, C. (1996) Identification of a 29 kDa protein in the envelope of *Mycobacterium smegmatis* as a putative ferri-exochelin receptor. *Microbiology* **142**, 1521–1530.

Dussurget, O., Rodriguez, M. & Smith, I. (1996) An *ideR* mutant of *Mycobacterium smegmatis* has derepressed siderophore production and an altered oxidative stress response. *Molecular Microbiology* **22**, 535–544.

Finch, C.A. & Huebers, H. (1982) Perspectives in iron metabolism. *New England Journal of Medicine* **306**, 1520–1528.

Fiss, E.H., Yu, S.Y. & Jacobs, W.R. (1994) Identification of genes involved in the sequestration of iron in mycobacteria: the ferric exochelin biosynthetic and uptake pathways. *Molecular Microbiology* **14**, 557–569.

Francis, J., Macturk, H.M., Madinaveita, J. & Snow, G.A. (1953) Mycobactin, a growth factor for *Mycobacterium johnei*. I. Isolation from *Mycobacterium phlei*. *Biochemical Journal* **55**, 596–607.

Francis, J., Madinaveita, J., Macturk, H.M. & Snow, G.A. (1949) Isolation from acid-fast bacteria of a growth-factor for *Mycobacterium johnei* and a precursor of phthicol. *Nature* **163**, 365.

Ghosh, A. & Miller, M.J. (1993) Synthesis of novel citrate-based siderophores and siderophore-β-lactam conjugates. Iron transport-mediated drug delivery systems. *Journal of Organic Chemistry* **58**, 7652–7659.

Gobin, J. & Horwitz, M.A. (1996) Exochelins of *Mycobacterium tuberculosis* remove iron from human iron-binding proteins and donate iron to mycobactins in the *M. tuberculosis* cell wall. *Journal of Experimental Medicine* **183**, 1527–1532.

Gobin, J., Moore, C.H., Reeve, J.R., Wong, D.K., Gibson, B.W. & Horwitz, M.A. (1995) Iron acquisition by *Mycobacterium tuberculosis*: isolation and characterization of a family of iron-binding exochelins. *Proceedings of the National Academy of Sciences of the USA* **92**, 5189–5193.

Guerinot, M.L. (1994) Microbial iron transport. *Annual Review of Microbiology* **48**, 743–772.

Hall, R.M. (1986) Mycobactins: how to obtain them and how to employ them as chemotaxonomic characters for the mycobacteria and related organisms. *Actinomycetes* **19**, 92–106.

Hall, R.M. & Ratledge, C. (1982) A simple method for the production of mycobactin, the lipid-soluble siderophore from mycobacteria. *FEMS Microbiological Letters* **15**, 133–136.

Hall, R.M. & Ratledge, C. (1984) Mycobactins as chemotaxonomic characters for some rapidly growing mycobacteria. *Journal of General Microbiology* **130**, 1883–1892.

Hall, R.M. & Ratledge, C. (1985a) Equivalence of mycobactins from *Mycobacterium senegalense, M. farcinogenes* and *M. fortuitum*. *Journal of General Microbiology* **131**, 1691–1696.

Hall, R.M. & Ratledge, C. (1985b) Mycobactins in the classification and identification of armadillo-derived mycobacteria. *FEMS Microbiological Letters* **28**, 243–247.

Hall, R.M. & Ratledge, C. (1986) Mycobactin and the competition for iron between two species of mycobacterium, *M. neoaurum* and *M. vaccae*. *Journal of General Microbiology* **132**, 839–843.

Hall, R.M. & Ratledge, C. (1987) Exochelin-mediated iron acquisition by the leprosy bacillus, *Mycobacterium leprae*. *Journal of General Microbiology* **133**, 193–199.

Hall, R.M., Sritharan, M., Messenger, A.J.M. & Ratledge, C. (1987) Iron transport in *Mycobacterium smegmatis*: occurrence of iron-regulated envelope proteins as potential receptors for iron uptake. *Journal of General Microbiology* **133**, 2107–2114.

Hall, R.M., Wheeler, P.R. & Ratledge, C. (1983) Exochelin-mediated iron uptake into *Mycobacterium leprae. International Journal of Leprosy* **51**, 490–494.

Harrison, P.M., Hempstead, P.C., Artymiuk, P.J. & Andrews, S.C. (1998) Structure–function relationships in the ferritins. In: *Metal Ions in Biological Systems*, Vol. 35, *Iron Transport and Storage in Microorganisms, Plants and Animals* (eds A. Sigel & H. Sigel). New York: Marcel Dekker, pp. 435–477.

Homuth, M., Valentin-Weigand, P., Rohle, M. & Gerlach, G.-F. (1998) Identification and characterization of a novel extracellular ferric reductase from *Mycobacterium paratuberculosis. Infection and Immunity* **66**, 710–716.

Hu, J. & Miller, M.J. (1997) Total synthesis of a mycobactin S, a siderophore and growth promoter of *Mycobacterium smegmatis*, and determination of its growth inhibitory activity against *Mycobacterium tuberculosis. Journal of the American Chemical Society* **119**, 3462–3468.

Inglis, N.F., Stevenson, K., Hosie, A.H. & Sharp, J.M. (1994) Complete sequence of the gene encoding the bacterioferritin subunit of *Mycobacterium avium* subspecies *silvaticum. Gene* **150**, 205–206.

Kato, L. (1985) Absence of mycobactin in *Mycobacterium leprae*; probably a microbe dependent microorganism implications. *International Journal of Leprosy* **57**, 58–70.

Kikuchi, S., Fukumoto, M. & Takahashi, H. (1994) Iron storage in *Mycobacterium smegmatis* grown under iron-sufficient and iron-overload conditions. *Bioscience, Biotechnology and Biochemistry* **58**, 885–888.

Koch, A.L. (1971) The adaptive responses of *Escherichia coli* to a feast and famine existence. *Advances in Microbial Physiology* **6**, 147–217.

Kochan, I. (1973) The role of iron in bacterial infections with special consideration to host–tubercle bacillus interaction. *Current Topics in Microbiological Immunology* **60**, 1–30.

Kochan, I. (1976) Role of iron in the regulation of nutritional immunity. *Bioorganic Chemistry* **2**, 55–57.

Kochan, I., Pellis, N.R. & Golden, C.A. (1971) Mechanisms of tuberculostasis in mammalian serum. III. Neutralization of serum tuberculostasis by mycobactin. *Infection and Immunity* **3**, 553–558.

Lambrecht, R.S. & Collins, M.T. (1993) Inability to detect mycobactin in mycobacteria-infected tissues suggests an alternative iron acquisition mechanism by mycobacteria *in vivo. Microbial Pathogenesis* **14**, 229–238.

Lane, S.J., Marshall, P.S., Upton, R.J. & Ratledge, C. (1998) Isolation and characterization of carboxymycobactins as the second extracellular siderophores in *Mycobacterium smegmatis. Biometals* **11**, 13–20.

Lane, S.J., Marshall, P.S., Upton, R.J., Ratledge, C. & Ewing, M. (1995) Novel extracellular mycobactins, the carboxymycobactins from *Mycobacterium avium. Tetrahedron Letters* **36**, 4129–4132.

Lehmann, J. (1946) *Para*-aminosalicylic acid in the treatment of tuberculosis. *Lancet* **i**, 15–16.

Leite, C.Q.F., Barreto, A.M.W. & Leite, S.R.A. (1995) Thin-layer chromatography of mycobactins and mycolic acid for the identification of clinical mycobacteria. *Revista Microbiologica, Sao Paulo* **26**, 192–199.

Leong, S.A. & Winkelmann, G. (1998) Molecular biology of iron transport in fungi. van der Helm, D. (1998) The physical chemistry of bacterial outer-membrane siderophore receptor proteins In: *Metal Ions in Biological Systems*, Vol. 35, *Iron Transport and Storage in Microorganisms, Plants and Animals* (eds A. Sigel & H. Sigel). New York: Marcel Dekker, pp. 147–186.

Lepper, A.W.D. & Wilks, C.R. (1988) Intracellular iron storage and the pathogenesis of paratuberculosis. Comparative studies with other mycobacterial, parasitic or infectious conditions of veterinary importance. *Journal of Comparative Pathology* **98**, 31–53.

Lundrigan, M.D., Arceneaux, J.E.L., Zhu, W. & Byers, B.R. (1997) Enhanced hydrogen-peroxide sensitivity and altered stress protein expression in iron-starved *Mycobacterium smegmatis. Biometals* **10**, 215–225.

Macham, L.P. & Ratledge, C. (1975) A new group of water-soluble iron-binding compounds from mycobacteria: the exochelin. *Journal of General Microbiology* **89**, 379–382.

Macham, L.P., Ratledge, C. & Nocton, J.C. (1975) Extracellular iron acquisition by mycobacteria: role of the exochelins and evidence against the participation of mycobactin. *Infection and Immunity* **12**, 1242–1251.

Macham, L.P., Stephenson, M.C. & Ratledge, C. (1977) Iron transport in *Mycobacterium smegmatis*: the isolation, purification and function of exochelin MS. *Journal of General Microbiology* **101**, 41–49.

Marshall, B.J. & Ratledge, C. (1971) Conversion of chorismic acid and iso-chorismic acid to salicylate by cell-free extracts of *Mycobacterium smegmatis. Biochimica et Biophysica Acta* **230**, 643–654.

Marshall, B.J. & Ratledge, C. (1972) Salicylic acid biosynthesis and its control in *Mycobacterium smegmatis. Biochimica et Biophysica Acta* **264**, 106–116.

Matthews, P.R.J., McDiarmid, A., Collins, P. & Brown, A. (1976) The dependence of some strains of *Mycobacterium avium* on mycobactin for initial and subsequent growth. *Journal of Medicine Microbiology* **11**, 53–57.

Matzanke, B.F., Böhnke, R., Möllmann, U., Reissbrodt, R., Schünemann, V. & Trantwein, A.X. (1997) Iron uptake and intracellular metal transfer in mycobacteria mediated by xenosiderophores. *Biometals* **10**, 193–203.

McCready, K.A. (1980) Studies in iron metabolism in *Mycobacterium smegmatis, M. avium* and other mycobacteria. PhD Thesis, University of Hull, UK.

McCready, K.A. & Ratledge, C. (1978) Amounts of iron, haem and related compounds in *Mycobacterium smegmatis* grown in various concentrations of iron. *Biochemical Society Transactions* **6**, 421–423.

McCready, K.A. & Ratledge, C. (1979) Ferrimycobactin

reductase activity from *Mycobacterium smegmatis*. *Journal of General Microbiology* **113**, 67–72.

McCullough, W.G. & Merkal, R.S. (1982) Structure of mycobactin J. *Current Microbiology* **7**, 337–341.

Messenger, A.J.M., Hall, R.M. & Ratledge, C. (1986) Iron uptake processes in *Mycobacterium vaccae* R877R, a mycobacterium lacking mycobactin. *Journal of General Microbiology* **132**, 845–852.

Messenger, A.J.M. & Ratledge, C. (1982) Iron transport in *Mycobacterium smegmatis*: uptake of iron from ferric citrate. *Journal of Bacteriology* **149**, 131–135.

Miller, M.J. (1989) Syntheses and therapeutic potential of hydroxamic acid based siderophores and analogues. *Chemical Reviews* **89**, 1563–1579.

Miller, M.J. & Malouin, F. (1993) Microbial iron chelators as drug delivery agents: the rational design and synthesis of siderophore-drug conjugates. *Accounts in Chemistry Research* **26**, 241–249.

Miller, M.J. & Malouin, F. (1994) Siderophore-mediated drug delivery: the design, synthesis, and study of siderophore-antibiotics and anti-fungal conjugates. In: *The Development of Iron Chelators for Clinical Use* (eds R. J. Bergeron & G. M. Brittenham). Boca Raton: CRC Press, pp. 275–333.

Neilands, J.B. (1995) Siderophores: structure and function of microbial iron transport compounds. *Journal of Biological Chemistry* **270**, 26723–26726.

Oh, Y.K. & Staubinger, R.M. (1996) Intracellular fate of *Mycobacterium avium*: use of dual-label spectrofluorometry to investigate the influence of bacterial viability and opsonization on phagosomal pH and phagosome–lysosome interaction. *Infection and Immunity* **64**, 319–325.

Ohi, N., Aoki, B., Shinozaki, T. *et al.* (1986) Semisynthetic β-lactam antibiotics. I. Synthesis and antibacterial activity of new ureidopenicillin derivatives having catechol moieties. *Journal of Antibiotics* **39**, 230–241.

Pessolani, M.C.V., Smith, D.R., Rivoire, B. *et al.* (1994) Purification, characterization, gene sequence, and significance of a bacterioferritin from *Mycobacterium leprae*. *Journal of Experimental Medicine* **180**, 319–327.

Portaels, F., DeRidder, K. & Pattyn, S.R. (1985) Cultivatable mycobacteria isolated from organs of armadillo uninoculated and inoculated *Mycobacterium leprae*. *Annals of Microbiology (Institute Pasteur)* **136A**, 181–190.

Ratledge, C. (1971) Transport of iron by mycobactin in *Mycobacterium smegmatis*. *Biochemical and Biophysical Research Communications* **45**, 856–861.

Ratledge, C. (1976) The physiology of the mycobacteria. *Advances in Microbial Physiology* **13**, 115–244.

Ratledge, C. (1984) Metabolism of iron and other metals by mycobacteria. In: *The Mycobacteria, a Sourcebook, Part A* (eds G. P. Kubica & L. G. Wayne). New York: Marcel Dekker, pp. 603–627.

Ratledge, C. & Brown, K.A. (1972) Inhibition of mycobactin formation in *Mycobacterium smegmatis* by *p*-aminosalicylate. A new proposal for the mode of action of *p*-aminosalicylate. *American Review of Respiratory Diseases* **106**, 774–776.

Ratledge, C. & Ewing, M. (1996) The occurrence of carboxymycobactin, the siderophore of pathogenic mycobacteria, as a second extracellular siderophore in *Mycobacterium smegmatis*. *Microbiology* **142**, 2207–2212.

Ratledge, C. & Hall, M.J. (1972) Isolation and properties of auxotrophic mutants of *Mycobacterium smegmatis* requiring salicylic acid or mycobactin. *Journal of General Microbiology* **72**, 143–150.

Ratledge, C., Macham, L.P., Brown, K.A. & Marshall, B.J. (1974) Iron transport in *Mycobacterium smegmatis*: a restricted role for salicylic acid in the extracellular environment. *Biochimica et Biophysica Acta* **372**, 39–51.

Ratledge, C. & Marshall, B.J. (1972) Iron transport in *Mycobacterium smegmatis*: the role of mycobactin. *Biochimica et Biophysica Acta* **279**, 58–74.

Ratledge, C. & Snow, G.A. (1974) Isolation and structure of nocobactin NA, a lipid-soluble iron-binding compound from *Nocardia asteroides*. *Biochemical Journal* **139**, 407–413.

Ratledge, C. & Winder, F.G. (1962) The accumulation of salicylic acid by mycobacteria during growth on iron-deficient medium. *Biochemical Journal* **84**, 501–506.

Ratledge, C. & Winder, F.G. (1966) Biosynthesis and utilization of aromatic compounds by *Mycobacterium smegmatis* with particular reference to the origin of salicylic acid. *Biochemical Journal* **101**, 274–283.

Ratledge, C., Patel, P.V. & Mundy, J. (1982) Iron transport in *Mycobacterium smegmatis*: the location of mycobactin by electron microscopy. *Journal of General Microbiology* **128**, 1559–1565.

Rogers, H.J., Synge, C. & Woods, V.E. (1980) Antibacterial effect of scandium and indium complexes of enterochelin on *Klebsiella pneumoniae*. *Antimicrobial Agents and Chemotherapy* **18**, 63–68.

Rogers, H.J., Woods, V.E. & Synge, C. (1982) Antibacterial effect of scandium and indium complexes of enterochelin on *Escherichia coli*. *Journal of General Microbiology* **128**, 2389–2394.

Sauton, B. (1912) Sur la nutrition minérale du bacille tuberculeux. *Compte rendu hebdomadaire séances de l'Académie des Sciences (Paris)* **155**, 860–861.

Schmitt, M.P., Predich, M., Doukhan, L., Smith, I. & Holmes, R.K. (1995) Characterization of an iron-dependent regulatory protein (IdeR) of *Mycobacterium tuberculosis* as a functional homolog of the diphtheria toxin repressor (DtxR) from *Corynebacterium diphtheriae*. *Infection and Immunity* **63**, 4284–4289.

Sharman, G.J., Williams, D.H., Ewing, D.F. & Ratledge, C. (1995a) Isolation, purification and structure of exochelin MS, the extracellular siderophore from *Mycobacterium smegmatis*. *Biochemical Journal* **305**, 187–196.

Sharman, G.J., Williams, D.H., Ewing, D.F. & Ratledge, C. (1995b) Determination of the structure of exochelin MN, the extracellular siderophore from *Mycobacterium neoaurum. Chemistry and Biology* **2**, 553–561.

Sigel, A. & Sigel, H. (eds) (1998) *Metal Ions in Biological Systems*, Vol. 35, *Iron Transport and Storage in Microorganisms, Plants and Animals.* New York: Marcel Dekker.

Silley, P., Griffiths, J.W., Monsey, D. & Harrison, A.M. (1990) Mode of action of GR69153, a novel catechol-substituted cephalosporin, and its interaction with the *tonB*-dependent iron transport system. *Antimicrobial Agents and Chemotherapy* **34**, 1806–1808.

Snow, G.A. (1970) Mycobactins: iron-chelating growth factors from mycobacteria. *Bacteriological Reviews* **34**, 99–125.

Snow, G.W. & White, A.J. (1969) Chemical and biological properties of mycobactins isolated from various mycobacteria. *Biochemical Journal* **115**, 1031–1045.

Spiro, T.G., Bates, G. & Saltman, P. (1967b) The hydrolytic polymerization of ferric citrate. II. The influence of excess citrate. *Journal of the American Chemical Society* **89**, 5559–5562.

Spiro, T.G., Pape, L. & Saltman, P. (1967a) The hydrolytic polymerization of ferric citrate. I. The chemistry of the polymer. *Journal of the American Chemical Society* **89**, 5555–5559.

Sritharan, M. & Ratledge, C. (1990) Iron-regulated envelope proteins of mycobacteria grown *in vitro* and their occurrence in *Mycobacterium leprae* grown *in vivo. Biology of Metals* **2**, 203–208.

Stachelhaus, T. & Marahiel, M.A. (1995) Modular structure of genes encoding multifunctional peptide synthetases required for non-ribosomal peptide synthesis. *FEMS Microbiology Letters* **125**, 3–14.

Stephenson, M.C. & Ratledge, C. (1979) Iron transport in *Mycobacterium smegmatis*: uptake of iron from ferriexochelin. *Journal of General Microbiology* **110**, 193–202.

Stephenson, M.C. & Ratledge, C. (1980) Specificity of exochelins for iron transport in three species of mycobacteria. *Journal of General Microbiology* **116**, 521–523.

Sturgill-Koszycki, S., Schlesinger, P.H., Chakraborty, P. *et al.* (1994) Lack of acidification of *Mycobacterium* phagosomes produced by exclusion of the vesicular proton-ATPase. *Science* **263**, 678–681.

Supek, F., Supekova, L., Nelson, H. & Nelson, N. (1997) Function of metal-ion homeostasis in the cell division cycle, mitochondrial protein processing, sensitivity to mycobacterial infection and brain function. *Journal of Experimental Biology* **200**, 321–330.

van der Helm, D. (1998) The physical chemistry of bacterial outer-membrane siderophore receptor proteins In: *Metal Ions in Biological Systems*, Vol. 35, *Iron Transport and Storage in Microorganisms, Plants and Animals* (eds A. Sigel & H. Sigel). New York: Marcel Dekker, pp. 335–401.

Visca, P., Ciervo, A., Sanfilippo, V. & Orsi, N. (1993) Iron-regulated salicylate synthesis by *Pseudomonas* spp. *Journal of General Microbiology* **139**, 1995–2001.

Watanabe, N.-A., Nagasu, T., Katsu, K. & Kitoh, K. (1987) E-0702, a new cephalosporin, is incorporated into *Escherichia coli* cells via the *tonB*-dependent iron transport system. *Antimicrobial Agents and Chemotherapy* **31**, 497–504.

Wheeler, P.R. & Ratledge, C. (1994) Metabolism of *Mycobacterium tuberculosis*. In: *Tuberculosis: Pathogenesis, Protection, and Control* (ed. B. R. Bloom). Washington, DC: American Society for Microbiology, pp. 353–385.

Winder, F.G. (1964) The antibacterial action of streptomycin, isoniazid and PAS. In: *Chemotherapy of Tuberculosis* (ed. V.C. Barry). London: Butterworths, pp. 111–149.

Winder, F.G. (1982) Mode of action of the antimycobacterial agents and associated aspects of the molecular biology of the mycobacteria. In: *The Biology of the Mycobacteria*, Vol. 1 (eds C. Ratledge & J. Stanford). London: Academic Press, pp. 353–438.

Winder, F.G. & O'Hara, C. (1964) Effects of iron deficiency and of zinc deficiency on the activities of some enzymes in *Mycobacterium smegmatis. Biochemical Journal* **90**, 122–126.

Winder, F.G. & O'Hara, C. (1966) Levels of iron and zinc in *Mycobacterium smegmatis* grown under conditions of trace metal limitation. *Biochemical Journal* **100**, 38P–39P.

Winkelmann, G. (ed.) (1991) *CRC Handbook of Microbial Iron Chelates*. Boca Raton: CRC Press.

Winkelmann, G., van der Helm, D. & Neilands, J.B.(eds) (1987) *Iron Transport in Microbes, Plants and Animals.* Weinheim: VCH.

Wong, D.K., Gobin, J., Horwitz, M.A. & Gibson, B.W. (1996) Characterization of exochelins of *Mycobacterium avium*: evidence for saturated and unsaturated acid for acid and ester forms. *Journal of Bacteriology* **178**, 6394–6398.

Xu, Y. & Miller, M.J. (1998) Total synthesis of mycobactin analogues as potent antimycobacterial agents using a minimal protecting group strategy. *Journal of Organic Chemistry* **63**, 4314–4322.

Zhu, W., Arceneaux, J.E.L., Beggs, M.L., Byers, R.B., Eisenach, K. & Lundrigan, M.D. (1998) Exochelin genes in *Mycobacterium smegmatis*: identification of an ABC transporter and two non-ribosomal peptide synthetase genes. *Molecular Microbiology* **29**, 629–639.

Chapter 15 / Antibiotics* and antibiotic resistance in mycobacteria

VERA WEBB & JULIAN DAVIES

1 Introduction

Mycobacteria have, unquestionably, played a significant role in human disease throughout history and such infections have been a considerable force in the social history of humankind. The availability of effective antibiotic therapy in the late 1940s and early 1950s had a significant impact on morbidity and mortality due to *Mycobacterium tuberculosis* infection and had dramatic social implications in the industrialized and developing worlds. Regrettably, mycobacteria have presented unusual problems during and since the antibiotic era (since 1945). The early and dramatic success of streptomycin in the treatment of tuberculosis, followed by the introduction of other highly effective drugs (isoniazid, pyrazinamide, ethambutol, rifampin) suggested that effective, worldwide control of mycobacterial infections would be guaranteed. However, the development of antibiotic resistance intervened in this anticipated success. The result has been initial encouragement followed by subsequent (and recent) treatment failure on a widespread basis with the unanticipated re-emergence of tuberculosis and other mycobacterial diseases in industrialized nations. Resistance to

*In this review, 'antibiotic' refers to any 'antimicrobial' agent, whether it be naturally occurring or synthetic.

antimycobacterial drugs has played a key role in this revanche although other factors such as changes in social behaviour, and failures in the delivery and maintenance of effective health care have contributed. At present, there is no antibiotic treatment that is free of the problem of resistance development. Recent reviews of antibiotic resistance in mycobacteria have been published by Jarlier and Nikaido (1994), Musser (1995), Blanchard (1996), Nolan (1997) and Rattan *et al.* (1998).

In this review, we will focus primarily on antitubercular agents but will include information on treatments for leprosy, *M. avium* complex (MAC) disease and other ailments caused by mycobacterial species in humans. While we have tried to put the problem of antibiotic resistance in mycobacteria in historical context, we have devoted most of our attention to reports published between 1994 and 1998.

2 The adversaries

Other chapters in this volume discuss the structure and physiology of mycobacteria and the ways in which these unique characteristics affect antibiotic susceptibility. The mycobacteria are divided into two types, namely the fast-growers and the slow-growers. These two groups have different virulence/pathogenicity profiles and different susceptibilities to a range of antimicrobial agents; hence, different drug therapies are recommended and different types of resistance have developed.

A frequently used minimal definition of multidrug-resistant (Mdr) *M. tuberculosis*, is resistance to, at the least, both rifampicin and isoniazid. Strains have been isolated from many sites in the United States (Bifani *et al.* 1996) that are resistant to rifampicin and isoniazid plus pyrazinamide and streptomycin. In a recent study of new cases of tuberculosis in the United States, it was found that about 13% were resistant to at least one of the five front-line drugs (Bloch *et al.* 1994). Outbreaks of Mdr tuberculosis have been reported in Italy, France, England, Spain, Argentina and Mozambique (Caugant *et al.* 1995; Nolan 1997).

In general, mycobacteria are intrinsically resistant to many of the commonly used antibiotics; however, there is considerable variation in the susceptibility of mycobacteria to different antibiotics. For example, *M. tuberculosis* is quite resistant to many drugs that are active against other mycobacterial strains, but shows almost unique and exquisite susceptibility to isoniazid, an agent which is ineffective against other mycobacteria, such as *M. smegmatis*. Similarly, *M. tuberculosis* and *M. bovis* are susceptible to another front-line drug, rifampicin, while *M. smegmatis* and MAC are recalcitrant. The genetic basis for this variation is unclear, although it has been the subject of much discussion (Deretic *et al.* 1996). These differences in susceptibility have several biochemical explanations, such as:

1 the composition and structure of the mycobacterial cell wall preventing the uptake of many types of molecules;

2 the existence of multisubstrate efflux pumps;

3 unique metabolic functions that create specific drug susceptibilities; and

4 chromosomally encoded resistance to certain classes of antibiotics (β-lactams, aminoglycosides).

While *M. tuberculosis* and other mycobacteria do possess chromosomal genes that encode resistance by drug inactivation to certain classes of antibiotics, there is no evidence to suggest that antibiotic resistance in mycobacteria is the result of anything other than mutational alteration of the appropriate chromosomal gene. Multidrug resistance is the result of the accumulation of independent mutations and many such mutations have been isolated and examined (see below). Furthermore, the fact that there is no evidence for acquired, plasmid-mediated, antibiotic resistance in mycobacteria sets them apart from all other bacterial pathogens.

3 The weapons and how they act

Table 15.1 summarizes the drugs that are currently available for the treatment of mycobacterial infections.

3.1 Antitubercular agents: the front-line drugs

Today, the standard 'short-course' therapy for tuberculosis involves the treatment of patients with a four-drug combination of rifampicin, isoniazid, ethambutol, and pyrazinamide for 2 months, followed by treatment with a combination of rifampicin and isoniazid for an additional 4 months. While multidrug therapy appears unusual when compared to the treatment of other infectious diseases such as 'strep throat' or gonorrhea, this approach has been the rule for tuberculosis since the 1950s. Streptomycin is not part of this 'short course', but it is still widely used and considered one of the front-line drugs. When first used against tuberculosis in the 1940s, streptomycin was administered alone, but it was commonly found that within 3 months 80% of patients were harbouring streptomycin-resistant organisms. Grosset has recently reviewed the history of chemotherapy for cavitary pulmonary tuberculosis (Grosset 1996) and notes that, given the rapid development of streptomycin resistance, there was no hope of curing tuberculosis with any single antimicrobial agent. Subsequent treatment schemes indicated that suppression of resistant mutants was possible if patients were cotreated with *p*-aminosalicylic acid (PAS) together with isoniazid, and so by the mid-1950s combination therapy became the rule. Because a small number of drug-sensitive organisms survived even this three-drug regimen, 18–24 months of treatment were necessary for the immune

Table 15.1 Antibiotics effective against mycobacteria.

Site of action	Antibiotic	Mode of resistance
Inhibitors of cell-wall synthesis	Isoniazid[a] (ethionamide) Ethambutol[a,b] D-cycloserine	*inh*A, *kat*G (mycolic acid synthesis) *emb*B D-alanine racemase
Inhibitors of nucleic acid synthesis 　RNA 　DNA	 Rifampicin[a] (and derivatives) Fluoroquinolones[b] (and derivatives)	 β-subunit of RNA polymerase (*rpo*B) A-subunit of topoisomerase I (*gyr*A)
Inhibitors of protein synthesis	Clarithromycin[b] Azithromycin[b] Streptomycin Capreomycin Viomycin Tuberactinomycin Kanamycin Amikacin[b] Tobramycin Hygromycin Apramycin	All mutations in rRNA or r-protein binding
Unknown	Pyrazinamide[a] Clofazimine[b]	*pnc*A (pyrazinamidase) Formation of superoxide radicals?

[a] Denotes front-line antitubercular drug.
[b] Denotes anti-*M. avium* complex drug.

system to eliminate the persisting infection. Relapse rates, however, were still at about 10%.

A radical change in treatment occurred with the introduction of rifampicin in the 1960s. This drug was effective against both actively multiplying organisms and resting *M. tuberculosis* that were responsible for post-therapy relapse. Rifampicin in combination with streptomycin and isoniazid effected the cure of almost 100% of patients in only 9 months, i.e. half the time of the previous three-drug combination therapy. In the 1980s with the rediscovery of pyrazinamide (PZA) and its inclusion in the combination therapy, treatment time was further reduced to 6 months.

3.1.1 *Streptomycin and other aminoglycosides*

Streptomycin, discovered in Waksman's laboratory in 1944, was the first product of a deliberate programme to screen the biological activity of secondary metabolites of actinomycetes. *M. tuberculosis* (also an actinomycete!) was highly susceptible to streptomycin when tested clinically. By 1946, streptomycin was into the market, an amazingly short period by today's standards. The discovery of streptomycin was significant in two respects: it was the first effective treatment for tuberculosis and an ugly lawsuit led to the legal partitioning of the Nobel prize (awarded to Waksman). The success of streptomycin was phenomenal but it was marred by two therapeutic problems: toxicity and resistance. The long-term dosage required for a tuberculosis cure led many sufferers to lose their hearing, and the chances of resistance developing were very high. In spite of its efficacy, streptomycin is no longer the drug of choice for tuberculosis.

Aminoglycosides (AGs), of which streptomycin is a member, are broad-spectrum pseudosaccharide antibiotics, composed of three or more cyclitol units. They bind to the 30S subunit of the ribosome and interfere with the transition from the initiation complex to the elongation complex; they also disrupt the decoding process. In most bacterial pathogens, resistance to aminoglycosides is due to the acquisition of aminoglycoside-inactivating enzymes (Shaw *et al*. 1993), and target-site mutations are rare. Most fast-growing bacteria have multiple copies of the rRNA genes, and because resistance is genetically recessive to antibiotic sensitivity, only rare mutations, in the gene for protein S12, have been isolated under normal situations with such bacteria. Dominant rRNA mutants can be established in strains carrying multiple copies of the resistant allele (on a plasmid). Mutations altering one or more ribosomal proteins have been identified in laboratory studies and rarely in clinical isolates of pathogens such as *Neisseria gonorrhoeae*. However, because the slow-growing mycobacteria possess only single copies of the rRNA genes, resistance to streptomycin and other AGs can arise by mutational alteration of either 16S rRNA or ribosomal protein S12. Mutations in these ribosomal components are the only types of aminoglycoside resistance found clinically in mycobacteria (Finken

et al. 1993). Recent work by Puglisi and coworkers (Fourmy *et al*. 1996) has defined the interaction between paromomycin (an aminoglycoside that had a brief period as a potential antitubercular agent) and its rRNA target in precise molecular terms. Further analysis of this interaction in concert with increasing knowledge of the working of the ribosome in translation should define the action of this class of drugs.

Streptomycin-resistant mutants were the first functional alterations to be defined in a ribosomal target. A brief history of streptomycin is described in Table 15.2. Laboratory studies with *E. coli* identified mutational alteration of ribosomal protein S12 as the principal basis of streptomycin resistance and dependence (some of the clinical isolates are actually dependent on streptomycin for growth!); rRNA changes were not suspected for the reasons described above. However, in 1993, almost 50 years after the isolation of the first streptomycin-resistant *M. tuberculosis*, resistance was shown to be due to either an altered S12 ribosomal protein or a single base substitution at one of several bases in the decoding site of the 16S rRNA (Fig. 15.1a). Significant genetic and biochemical analyses of streptomycin resistance have been carried out; essentially, the resistant ribosomes

Table 15.2 A brief history of streptomycin.

1944	Streptomycin the first bacterially-produced antibiotic discovered in soil-screening programme (Waksman) Streptomycin proposed for the treatment of Gram-negative infections and also for TB
1947	Streptomycin became generally available for treatment of TB High level resistance to streptomycin developed during prolonged treatment
1952	Streptomycin used to demonstrate polarity of bacterial conjugation (Hayes)
1955	Multiply drug-resistant *Shigella dysenteriae* (including streptomycin) isolated in Japan
1959	Streptomycin resistance transferable between Gram-negative bacteria Streptomycin shown to inhibit bacterial protein synthesis
1960	Notion of endosymbiosis supported by streptomycin inhibition of chloroplast function
1964	Streptomycin shown to cause mistranslation of genetic code Streptomycin resistance in *Escherichia coli* due to alterations in bacterial ribosomes
1968	Streptomycin inactivation due to R factor-encoded enzymes
1969	Streptomycin resistance in *E. coli* due to amino acid substitutions in ribosomal protein S12
1994	Streptomycin resistance in *M. tuberculosis* due to mutations in ribosomal protein S12 or in 16S rRNA

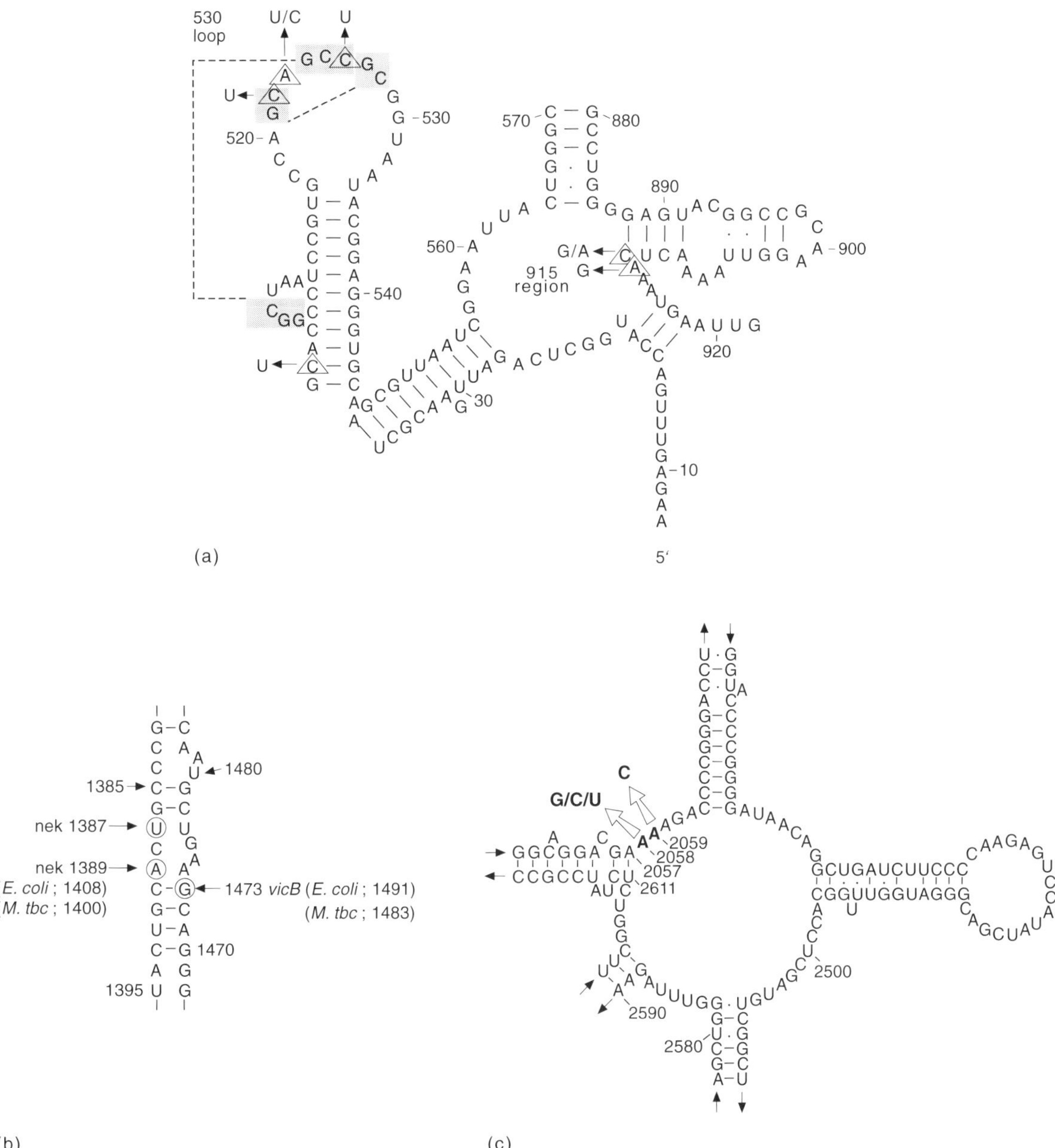

Fig. 15.1 Ribosome mutations leading to antibiotic resistance. (a) Molecular basis of streptomycin resistance is caused by mutations in the decoding loop of the 16S rRNA molecule. Arrows indicate mutations associated with resistance to streptomycin in *Mycobacterium tuberculosis*. (b) Molecular basis of resistance to aminoglycoside (kanamycin) and peptide (viomycin) antibiotics located in the 16S rRNA. (c) Molecular basis of resistance to macrolide antibiotics (clarithromycin) located in the central loop of domain V of 23S rRNA.

either fail to bind or bind the drug weakly (Sander *et al.* 1996).

In general, the aminoglycosides (apart from streptomycin) are rarely used in tuberculosis treatment because the (necessarily) long treatment period leads to significant side-effects. An exception to this is found in Japan where kanamycin has been used for the treatment of tuberculosis. Interestingly, mutational alteration of the ribosome rarely gives rise to kanamycin resistance in the laboratory. However, resistance in *M. smegmatis* is due to an A→G transition at position 1389 (equivalent to 1408 of *Escherichia coli*) in the 16S rRNA; high-level resistant *M. tuberculosis* strains had an A→G transition at position 1400 (equivalent to A1389 in *M. smegmatis*; see Fig. 15.1b) (Taniguchi *et al.* 1997). The A→G transition at position 1400 has been reported to confer high-level resistance in a majority of clinical isolates of *M. tuberculosis*, *M. abscessus* and *M. chelonae* resistant to amikacin, kanamycin and other 2-deoxystreptamine aminoglycoside antibiotics (Alangaden *et al.* 1998; Prammananan *et al.* 1998).

While clinical resistance to streptomycin in *M. tuberculosis* is mediated by target-site mutation (see above), the presence of aminoglycoside-modifying enzymes in the genus as a whole has been widely reported (Mitsuhashi *et al.* 1977; Udou *et al.* 1986, 1987). More recently Aínsa *et al.* (1996, 1997), using DNA hybridization, have identified genes encoding aminoglycoside 2′-*N*-acetyltransferase aac(2′) in all mycobacterial species tested, including *M. tuberculosis*. The aac2′ enzyme confers resistance to gentamicin, tobramycin, dibekacin, netilmicin, 2′-*N*-ethyl-netilmicin and 6′-*N*-ethyl-netilmicin. When *M. smegmatis* mc²155 was transformed with plasmids carrying the *aac(2′)-lb* gene from *M. fortuitum* or the *aac(2′)-ld* from *M. smegmatis* mc²155, the minimum inhibitory concentrations (MICs) for these aminoglycosides increased by two- to 60-fold. Aínsa *et al.* (1997) concluded from their phylogenetic analysis of the *aac(2′)* genes, that every mycobacterial species appears to have a specific aminoglycoside 2′-*N*-acetyltransferase. Nonetheless, the function of these enzymes remains unclear. When the *aac(2′)* gene was disrupted in *M. smegmatis* mc²155, the culture became

more susceptible to lysozyme treatment, leading to the suggestion that the aac(2′) enzyme may be involved with the acetylation of cell-wall components. This idea is supported by an earlier observation (Udou *et al.* 1989) that amino sugars and derivatives of coenzyme A inhibited aminoglycoside acetyl transferase activity from cultures of *M. fortuitum*.

3.1.2 Rifamycin

Rifamycins are a class of lipophilic compounds (ansamycins) that diffuse rapidly across the hydrophobic mycobacterial cell envelope. The most familiar member of this group is rifampicin. As noted above, rifampicin is one of the front-line drugs for tuberculosis treatment and is also used frequently to treat other mycobacterial diseases such as leprosy. Rifampicin inhibits bacterial DNA-dependent RNA polymerases. RNA polymerase is a complex enzyme which exists in two forms: the core enzyme (consisting of two α-subunits, and the β- and β′-subunits), which catalyses transcription elongation, and the holoenzyme (core enzyme plus one of many of the σ subunits), which recognizes specific promoters and catalyses transcription initiation (see Chapter 4). Mutations in the gene encoding the β-subunit of RNA polymerase (*rpoB*) give high-level resistance to rifampicin, and greater than 90% of all resistant mycobacteria fall into this category. These missense mutations occur in a short region of 27 codons near the centre of *rpoB* and consist predominantly of point mutations (Fig. 15.2) (Cole 1994). The appearance of resistant strains as a result of *rpoB* mutations is quite common in multidrug-resistant strains of *M. tuberculosis* and had been considered an effective surrogate marker for multiple drug resistance before the development of fast and reliable assays based on the polymerase chain reaction (PCR) for resistant genotypes. Cole notes that unlike *E. coli*, the majority of mutations from clinical isolates of *M. tuberculosis* do not appear to confer a significant growth disadvantage (Cole 1996).

Natural resistance to rifampicin has been reported for *M. avium–intracellulare* and *M. smegmatis* which are resistant to >25 μg/mL. Genetic analysis of the *rpoB*

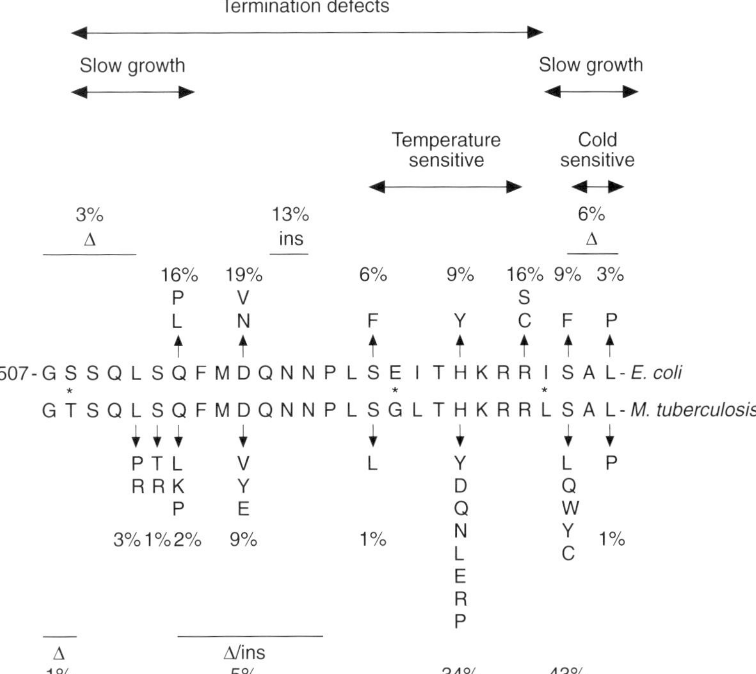

Fig. 15.2. Structure of the rifampicin-resistance locus from the rpoB genes of *Escherichia coli* and *M. tuberculosis*, showing the positions of individual mutations, their frequencies, and for *E. coli*, the associated phenotypes. Amino acid differences between *E. coli* and *M. tuberculosis* are shown as asterisks and insertions/deletions (ins/Δ) are represented by horizontal bars. (From Cole, 1994.)

genes from these organisms indicates that they are of the susceptible type. This intrinsic resistance is thought to be due to a property of the cell envelope of these species, since inclusion of the detergent Tween 80 in the growth media renders these organisms susceptible to rifampicin. However, other explanations for natural resistance among fast-growing mycobacteria are possible. Dabbs *et al.* (1995) have reported the inactivation of rifampicin by ribosylation at the C23 position or removal of a side chain at the C3 position of the drug in several species of fast-growing mycobacteria. Furthermore, Cole (1996) has reported that about 5% of clinically isolated rifampicin resistant *M. tuberculosis* strains have no mutation in their *rpoB* gene and notes that some type of inactivation mechanism in these strains cannot be ruled out.

The semisynthetic rifampicin analogue, rifabutin, has been used alone as a prophylactic for MAC disease in acquired immune deficiency syndrome (AIDS) patients. While it appears to have the same therapeutic and pharmacokinetic advantages as its parent, the MICs against MAC for this compound are two to 16 times lower than rifampicin (Kunin 1996). There have been reports that rifabutin has some activity against certain rifampicin-resistant *M. tuberculosis* strains. Cole (1996) reported that a subset of rifampicin-resistant mutants with the substitutions Leu511Pro, Asp516Tyr, Asp516Val or Ser522Leu, remained slightly susceptible to rifabutin. However, since cross-resistance to the rifamycins has been demonstrated, it is essential to exclude the possibility of *M. tuberculosis* infection before prescribing rifabutin as a prophylactic for MAC disease.

3.1.3 Isoniazid

Isoniazid (isonicotinic acid hydrazide, INH) was first described in 1912, but not until 1952 was it known to be effective in the treatment of *M. tuberculosis*. Today it is the most commonly prescribed antimycobacterial drug for prophylaxis and treatment. Both *M. tubercu-*

losis and *M. bovis* are extremely susceptible to the drug (MICs 0.02–0.2 µg/mL) while other mycobacterial species, such as *M. leprae* and *M. avium*, are less susceptible (MICs 5–50 µg/mL) and enteric bacteria like *E. coli* are resistant to at least 500 µg/mL. Recent studies of the oxidative stress pathways of the mycobacteria have uncovered an explanation for the exquisite sensitivity of *M. tuberculosis* to isoniazid. A number of groups have shown that *M. tuberculosis*, unlike most mycobacteria, lacks a functional *oxyR* gene which encodes a transcriptional regulator that stimulates expression of the *ahpC* gene during times of oxidative stress (Deretic *et al.* 1995, 1996, 1997; Sherman *et al.* 1996; Zhang *et al.* 1996; Dhandayuthapani *et al.* 1997; see also Chapter 4). The *ahpC* gene encodes the small, 22-kDa subunit of alkyl hydroperoxide reductase that is part of the oxygen detoxification system.

Shortly after the inclusion of INH in the treatment for tuberculosis, studies on specimens taken from patients treated with it reported the loss of the characteristic acid-fast property. This observation led to the suggestion that the drug may interfere with the synthesis or incorporation of mycolic acids or other cell surface components. Subsequent studies by Takayama and coworkers demonstrated that INH specifically inhibits the biosynthesis of unsaturated fatty acids greater than 26 carbons in length (Takayama *et al.* 1975; Davidson and Takayama 1979). However, the story of the antitubercular activity of isoniazid was far from solved.

As with other antibiotics, resistance to isoniazid developed soon after its introduction into clinical use. The earliest reports of resistance recorded the loss of catalase activity (reviewed in Sacchettini & Blanchard 1996). At first glance it is difficult to see how this mechanism of resistance is related to the inhibition of fatty acid synthases, but recent genetic and biochemical studies of INH-resistant mycobacteria have reported how the two biochemical effects might be linked and eventually lead to INH resistance. Making use of a spontaneous INH-resistant mutant of *M. smegmatis*, Jacobs and coworkers have cloned and sequenced the *inhA* locus which confers INH resistance to *M. bovis* when transferred on a mul-

ticopy plasmid. A single amino acid change at residue 94 converting a serine to an alanine is sufficient to confer INH resistance in *M. smegmatis*. Amino acid sequence alignment analysis indicated that the 29-kDa InhA protein had significant homology to the EnvM protein of *E. coli*, a protein thought to be involved in fatty acid biosynthesis (Banerjee *et al.* 1994). PCR amplification of genomic DNA allowed the cloning and overexpression of the *inhA* gene from *M. tuberculosis* in *E. coli*. The InhA protein has been characterized biochemically and found to be an enoyl-acyl carrier protein (ACP) reductase, which catalyses the reduced nicotinamide adenine dinucleotide (NADH)-dependent reduction of fatty 2-*trans*-enoyl thioesters of either CoA or ACP involved in fatty acid synthesis (Quemard *et al.* 1995). A recent survey of spontaneous *M. smegmatis* INH-resistant mutants showed that at least half are defective in NADH dehydrogenase (Miesel *et al.* 1998). X-ray crystallographic and mass spectrographic studies of the *M. tuberculosis* enoyl-ACP reductase (Rozwarski *et al.* 1998) have demonstrated the covalent attachment of the activated form of isoniazid to the nicotinamide ring of NAD bound within the active site of the enzyme.

In other studies, Cole and coworkers have characterized a series of clinically isolated INH-resistant *M. tuberculosis* which have either deletion or missense mutations in the *katG* gene encoding catalase–peroxidase (Zhang *et al.* 1992; Heym *et al.* 1995). Recent epidemiological studies indicate that 50–75% of INH-resistant isolates are *katG* mutants (Musser *et al.* 1996; Haas *et al.* 1997). The current hypothesis is that isoniazid is a prodrug which is oxidized by the catalase–peroxidase to an active compound which then binds to the enoyl-ACP reductase (Sacchettini and Blanchard 1996). Thus, resistance to isoniazid arises either when mutations in the *katG* gene prevent the conversion of INH to the active drug or when mutations occur in the target structural gene, *inhA*. However, there are some problems with this theory. First, hundreds of isolates of isoniazid-resistant *M. tuberculosis* have been screened: polymorphisms have been identified in the *inhA* structural gene, yet none has conferred resistance when moved to a susceptible host (Mdluli *et al.* 1996). Second, how does *M. tuberculosis* survive the oxidatively stressful environment of the macrophage without a functional catalase–peroxidase?

Biochemical and genetic analyses reported from a number of groups begin to address these questions. In experiments comparing the fatty acid profiles of iso-niazid-treated cultures of *M. tuberculosis* and *M. smegmatis,* it was found that fatty acid chains of 24- and 26-carbons accumulated on a 12-kDa ACP (AcpM) in *M. tuberculosis*, while in the *M. smegmatis* cultures only short-chain fatty acids accumulated (Mdluli *et al.* 1996; Mdluli *et al.* 1998), suggesting that isoniazid affects a target further along in mycolic acid biosyn-thesis in *M. tuberculosis*. In addition, a protein species isolated from isoniazid-treated *M. tuberculosis* was shown to consist of a covalent complex of INH, AcpM and a β-ketoacyl ACP synthase (KasA) (Mdluli *et al.* 1998). Amino acid-altering mutations in the KasA protein have been identified from INH-resistant clini-cal isolates (Mdluli *et al.* 1998). These results suggest that activated isoniazid may target other enzymes as well as enoyl-ACP reductase in *M. tuberculosis* (Mdluli *et al.* 1996). This hypothesis is further supported by genetic experiments in which transformation with multicopy plasmids containing the *inhA* genes con-ferred INH resistance on *M. smegmatis* but not on *M. tuberculosis* (Mdluli *et al.* 1996).

Nucleotide sequence analysis of isoniazid-resistant *katG* mutants of *M. tuberculosis* by a number of differ-ent groups has revealed mutations in the promoter region of the *ahpC* gene which give compensatory gene expression (Dhandayuthapani *et al.* 1996; Sherman *et al.* 1996; Wilson & Collins 1996; Kelley *et al.* 1997) (see section 4.2). Thus, *katG* mutants appear to counteract the loss/reduction of catalase–peroxidase activity by hyperexpression of the alkyl hydroperoxidase encoded by *ahpC*. The mutations are seen only in *katG* mutants and are unable to confer INH resistance when transferred into sensitive *M. bovis* strains (Sherman *et al.* 1996). These observa-tions are confirmed in a recent study investigating the relationship between overexpression of alkyl hydroperoxide reductase and virulence (Heym *et al.* 1997). However, others suggest that mutations in the

upstream region of *ahpC* alone can contribute to INH resistance (Dhandayuthapani *et al.* 1996; Wilson & Collins 1996; Sreevatsan *et al.* 1997a).

The relationship between *katG* mutants and com-pensatory mutations in the *ahpC* has been shown to be quite complex. An analysis of the *ahpC* region from clinical isolates of isoniazid-resistant *M. tubercu-losis* found a paucity of *ahpC* promoter mutations in *katG* codon 315 mutants (Sreevatsan *et al.* 1997a). The most common *katG* mutations occur within codon 315 (Haas *et al.* 1997) and have been reported to confer high-level INH resistance plus a 20-fold decrease in catalase–peroxidase activity (Heym *et al.* 1995). One explanation might be that different amino acid substitutions at codon 315 have different effects on either the catalase or the peroxidase activity of the enzyme and would thus have dif-ferent requirements for compensatory mutations (Sreevatsan *et al.* 1997a). Finally, *M. tuberculosis* strains producing catalase–peroxidase are more virulent than isogenic *katG*-deficient mutants, in immunocompetent mice; however, the loss of catalase–peroxidase activity is less important when immunodeficient mice, unable to produce activated macrophages, are infected (Heym *et al.* 1997).

3.1.4 Ethambutol

Ethambutol (EMB), also a synthetic drug, has been known for its antimycobacterial activity since it was initially described in 1961. The critical target for ethambutol lies in the pathway for the biosynthesis of cell-wall components. Mycobacterial cell walls are comprised of a framework of covalently attached mycolic acids, arabinan, galactan and peptidoglycan, known as the mycolylarabinogalactan–peptidoglycan complex (reviewed by Brennan &Nikaido 1995; see also Chapters 12 and 13). Studies in the late 1980s defined the synthesis of the arabinan component of arabinogalactan as the target. It was further demon-strated that treatment of ethambutol-susceptible cul-tures of *M. smegmatis* inhibited arabinogalactan and arabinomannan synthesis and resulted in the accu-mulation of decaprenyl phosphoarabinose (DPA), an intermediate in arabinan biosynthesis (Wolucka *et al.*

1994). Subsequently, a cell-free assay using [14C]-labelled DPA was developed to demonstrate that ethambutol specifically inhibits arabinosyl transferase activity (Lee *et al*. 1995).

Utilizing the DPA assay and a recombinant strain that was drug resistant due to target overexpression on a plasmid vector, a 9.8-kb region of *M. avium* DNA encoding the *embAB* and *embR* genes was cloned and characterized (Belanger *et al*. 1996). The *emb* operon of *M. smegmatis*, M. *tuberculosis* and *M. leprae* has also been characterized recently (Telenti *et al*. 1997). A plasmid containing a 9-kb DNA fragment from a high-level EMB-resistant strain conferred resistance when transformed into *M. smegmatis* mc2155. The *M. smegmatis* genes were then used to identify the *M. tuberculosis* homologue. The *emb* operon of *M. smegmatis, M. tuberculosis*, and *M. leprae* is composed of three structural genes, *embC, embA, embB*, each ≈ 3200 bp in length. These genes represent examples of gene duplication, encoding proteins with amino acid similarities of 61–68%. Telenti and coworkers noted that the topology of the EmbCAB proteins suggests that they are integral membrane proteins with 12 transmembrane domains (Telenti *et al*. 1997). In contrast, the *emb* operon of *M. avium* comprises only two structural genes, *embA* and *embB*, homologues of the genes found in the other mycobacterial species. In addition, the *M. avium* operon includes a regulatory gene *embR*, which has N-terminal amino acid sequence homology to AfsR and Dnr I, transcriptional activators for operons encoding the biosynthesis of secondary metabolites from *Streptomyces* (Belanger *et al*. 1996). Using the biosynthesis of secondary metabolites as a model, these authors note that the synthesis of these complex molecules always occurs from the condensation of single identical starter units and that mycobacterial arabinan may be considered in a similar fashion, since this homopolysaccharide is generated from multiple D-arabinofuranose units. Interestingly, Telenti and coworkers reported no sequence similarities among the *embR* genes of *M. smegmatis, M. tuberculosis*, and *M. leprae*.

Recently, the *embB* genes from 85 EMB-resistant and 33 EMB-susceptible strains of *M. tuberculosis* were analysed and mutations at codon 306 were found to have a critical role in resistance to ethambutol (Sreevatsan *et al*. 1997c). Over 60% of the EMB-resistant organisms had a Met306Leu, Met306Val or Met306Ile replacement. About 30% of the EMB-resistant isolates had no change in the *embB* gene suggesting that there are probably other targets for ethambutol activity. Several reports from Brennan's group indicate that ethambutol has a differential effect on arabinogalactan and lipoarabinomannan synthesis in *M. smegmatis* (Deng *et al*. 1995; Mikusova *et al*. 1995). It has been suggested that EMB may exert its effect by inhibiting any one of a number of arabinosyltransferases involved in mycobacteria cell-wall biosynthesis (Khoo *et al*. 1996).

3.1.5 Pyrazinamide

Pyrazinamide (PZA), a synthetic analogue of nicotinamide, was discovered in 1952, but its unique effect in accelerating antimycobacterial therapy when used in combination with isoniazid and rifampicin was discovered only in the 1980s. These observations led to a reduction in the duration of tuberculosis therapy from 9 to 6 months and made PZA one of the front-line drugs in the treatment of tuberculosis. Pyrazinamide is unique among antimycobacterial drugs because it is effective against the semidormant bacterial population persisting in low-pH environments (pH4.8–5.6). It is also effective in acute-inflammation sites and within the phagosomes of infected macrophages (Mitchison 1985). However, despite almost 50 years of use, its mode of action is still not completely understood. Resistance to PZA has been correlated with the loss of pyrazinamidase (PZAase) activity. The current model is that pyrazinamide is a prodrug that must be activated or converted by bacterial PZAase to pyrazinoic acid, which then attacks a target not yet identified in the mycobacterial cell. PZAase is a nicotinamidase, an enzyme involved in the conversion of nicotinamide to nicotinic acid and found in most species of bacteria. The PZAase gene (*pncA*) of *M. tuberculosis* has been cloned by PCR, using degenerate primers based on the sequence of the *pncA* gene from *E. coli* (Scorpio &

Zhang 1996). Sequence analysis revealed that the *M. tuberculosis pncA* gene (558 bases) encoded a protein of 186 amino acids with a 35% identity to the *E. coli* nicotinamidase.

The conventional PZA-susceptibility tests are both time-consuming (2–6 weeks) and unreliable. In two recent studies 10–17% of the PZA-resistant isolates analysed were false or misidentified, i.e. they had no mutations in the *pncA* structural gene and when retested proved to be sensitive to PZA (Scorpio & Zhang 1996; Scorpio *et al.* 1997). With the nucleotide sequence now available, DNA-based tests to identify a PZA-sensitive phenotype have been developed to study the epidemiology of resistance and the structure–function mechanisms of resistance. Prior to the characterization of the *pncA* gene, susceptibility to PZA was rarely included in epidemiological studies of multidrug-resistant *M. tuberculosis*. For example, the analysis of PZA resistance was not included in the analysis of the origin and spread of an Mdr clone of *M. tuberculosis* in New York City (Bifani *et al.* 1996). With the *pncA* sequence available, Sreevatsan and coworkers reanalysed the Bifani Mdr strains and demonstrated that PZA resistance arose once in a W strain that was already resistant to rifampicin, isoniazid and streptomycin (Sreevatsan *et al.* 1997b). Unlike mutations in the *rpoB*, *rspL* and *katG* genes, which are confined to a small region of the gene, mutations in *pncA* have been found dispersed throughout the gene (Scorpio *et al.* 1997; Sreevatsan *et al.* 1997b). These studies analysed over 80 PZA-resistant isolates and located mutations from nucleotide –11 to nucleotide +518, which included amino acid substitutions in residue 5 through residue 171. Clearly the crystal structure of the enzyme would be useful in understanding the interactions between the enzyme and its substrate.

These studies of PZA-resistant *M. tuberculosis* suggest that mutations in the *pncA* gene are the primary mechanism of resistance to this drug. However, it is still unclear what the target of pyrazinoic acid, the metabolite of PZA, might be. After eliminating false-resistants, Scorpio and coworkers identified only one strain that had no mutation in the *pncA* gene and which had low-level resistance to PZA

(MIC 200–300 µg/mL as opposed to >900 µg/mL for high-level resistance). They also reported that after repeated attempts they were unable to obtain mutants resistant to pyrazinoic acid on 7H11 agar. They suggest that mycobacteria with mutations in this target may not be viable *in vitro* in normal medium or *in vivo* in patients (Scorpio *et al.* 1997). The target must be unique to mycobacteria, since the conversion of PZA to pyrazinoic acid by pyrazinamidase/nicotinamidase does not appear to have deleterious effects in other bacteria.

3.2 Other antibiotics used for treatment of mycobacterial infections

3.2.1 Inhibitors of nucleic acid biosynthesis: the fluoroquinolones

Nucleic acid metabolism has attracted much attention as a potential target for antimicrobial drugs; the strategy of hitting at the 'heart' of the microbe seemed the most likely to lead to effective bactericidal agents. Unfortunately, the ubiquity of DNA and RNA and the failure to identify discriminating target differences in the biosynthetic enzymes made this search unproductive until recently, when the fluoroquinolones (FQs, a class of synthetic drugs with no known natural analogues) were introduced. However, resistance mechanisms were not long in appearing, and the FQs, instead of being the 'superdrug' needed, are already limited in their efficacy by widespread resistance.

The molecular basis of FQ activity has been recently reviewed (Drlica & Zhao 1997). Nalidixic acid, the prototype quinolone antibiotic, was discovered in 1962 and had limited therapeutic use for urinary tract infections caused by Gram-negative bacteria. Resistance developed by mutation, and the Nal[R] phenotype proved to be the first useful marker for *gyrA*, the gene encoding the A subunit of DNA gyrase (topoisomerase II). The development of high-level resistance to nalidixic acid occurred solely by this type of mutation, and plasmid-determined resistance has never been reliably identified. This is not surprising, given that nalidixic acid is a purely synthetic chemical and no natural analogue has been identified.

In the late 1970s the first FQ antimicrobials were introduced; these proved to be vastly superior to nalidixic acid and remain among the most potent antimicrobial agents known. A number FQ antibiotics have now been introduced as anti-infectives and most show good broad-spectrum activity. The FQs show moderate *in vitro* activity (MIC 0.25–0.5 µg/mL) against *M. tuberculosis* (Garcia-Rodriguez & Garcia 1993), and sparfloxacin has been shown to be the most effective (Lalande *et al.* 1993), followed by ciprofloxacin and ofloxacin. Clinically, FQs have excellent bactericidal activity and achieve effective serum, tissue and intracellular levels following oral administration (reviewed by Alangaden & Lerner 1997). In addition FQs produce few adverse effects.

In laboratory studies, mutations to high-level resistance occurred at relatively low frequency, and genetic studies identified a number of different targets associated with DNA replication. As with nalidixic acid, the A subunit of DNA gyrase (topoisomerase II) is the principal target, but other targets involved in DNA replication (such as topoisomerase IV) which give the FQ-resistant phenotype have been identified in different bacterial species. In clinical use, resistance to the newer FQs has been found to develop quite rapidly as a result of one or more mutations. The *gyrA* gene of *M. tuberculosis* has been cloned, sequenced and used to characterize FQ-resistant mutants (Takiff *et al.* 1994). The mutations within codons analogous to those described in other FQ-resistant bacterial species were found in all strains resistant to ciprofloxacin concentrations greater than 2 µg/mL (Takiff *et al.* 1994). Similar results were found when a well-characterized collection of multi-drug resistant (Mdr) tuberculosis strains was examined for FQ resistance. Eleven of the 13 isolates possessed *gyrA* alleles normally associated with FQ resistance in other bacteria (Xu *et al.* 1996). Recently Alangaden and Lerner (1997) reviewed clinical reports that examine the use of FQs in the treatment of tuberculosis, leprosy and MAC disease. They conclude that when a FQ is the sole active agent in a multidrug therapy, resistance to FQs emerges in the same manner as during mono-therapy. They further suggest that FQs be used only when effective alternative drugs are not available.

Since mycobacterial resistance to the FQs occurs by *gyrA* mutations, it can be assumed that DNA gyrase is the primary target. However, in the absence of additional biochemical studies, one cannot rule out the possibility that FQs interact with other components of the DNA replication system to inhibit mycobacteria. The biochemical mechanism of action of FQs is not completely resolved: the drugs have been shown to interact with the DNA–topoisomerase complex. Novobiocin and coumermycin, well-known inhibitors of the topoisomerase B subunit (encoded by the *gyrB* gene), are active against mycobacteria, but to our knowledge they have not been used clinically for mycobacterial disease treatment.

Active efflux of the FQs has also been found to be an important mechanism leading to resistance in many bacterial genera. For example, the *norA* mutation identifies a multidrug-resistance system in *S. aureus*. A gene *lfrA*, encoding a putative proton antiporter efflux pump from *M. smegmatis*, has been cloned. When overexpressed on a multicopy plasmid, ciprofloxacin resistance is conferred on an FQ-sensitive strain of *M. smegmatis* (Takiff *et al.* 1996). The *lfrA* gene is homologous to the *qacA* gene of *Staphylococcus aureus*. Interestingly, Takiff and coworkers found that increased expression of *lfrA* augments the appearance of subsequent mutations to higher levels of FQ resistance. Thus, it is likely that resistance to FQ antibiotics is due to multiple mutations leading to both increased efflux and alteration of components of the DNA synthetic apparatus. Recently, Doran *et al.* (1997) reported the presence in *M. tuberculosis* of the *efpA* gene that encodes a putative efflux protein with homology to the QacA transporters family.

3.2.2 Other protein-synthesis inhibitors

Peptide antibiotics

The peptide antibiotics tuberactinomycin, viomycin and capreomycin have been employed for tuberculosis treatment, and the latter is a reserve drug in North America. These translation inhibitors interfere with the functions of both ribosomal subunits during early stages of peptide bond formation. It has been found that single and double mutations in 16S or 23S rRNA generate a resistance phenotype in *M. smegmatis*. Once again, such mutations are difficult to obtain in the laboratory with *E. coli*, usually at low frequency (10^{-9}); however, in *M. tuberculosis* they are isolated as single-step mutants to resistance and are often cross-resistant to the aminoglycosides (see Fig. 15.1b) (Taniguchi *et al.* 1997). Nucleic acid sequence analysis of PCR-amplified rDNA fragments from resistant strains of *M. smegmatis* showed that a limited number of base substitutions were involved. Similar alterations were found in resistant clinical isolates of *M. tuberculosis*. The nephrotoxicity of the tuberactinomycin class of antibiotics restricts their use, but in general, the frequency of mutation(s) to resistance is relatively rare.

Tetracyclines

Tetracyclines inhibit protein synthesis by blocking the binding of aminoacyl tRNAs to the A-site of the ribosome. The two most common mechanisms leading to tetracycline resistance are active efflux of the drug from the cell or the protection of the ribosomal target. Tetracyclines have been rarely applied to the treatment of diseases caused by the slow-growing mycobacteria (*M. tuberculosis* and *M. leprae*); however, they have occasionally been used to treat soft tissue or bone infections caused by the fast-growing mycobacteria *M. fortuitum* and *M. chelonae*. Despite this infrequent use, resistance has been described. Pang *et al.* (1994) analysed human mycobacterial isolates that were found in mixed culture with streptomycetes. Many of the strains carried the genes for either the TetL or TetK determinants (active efflux) or the OtrA (ribosome protection) and OtrB (active efflux) resistance alleles. A surprising finding was the presence of the *otr* genes (typically found in streptomycete strains which produce oxytetracycline) in mycobacteria and, coincidentally, of TetL or TetK (typically found in bacterial pathogens) in streptomycetes. These results suggest that tetracycline resistance determinants are readily exchanged with mycobacteria by an unknown

mechanism (Pang *et al.* 1994). However, E. Paget and J. E. Davies (unpublished observations 1998) found that *M. smegmatis* strains often possess a gene for the TetM determinant (ribosome protection 1998), apparently included in an element of the Tn*916* type. More recently, the presence of the genes encoding the TetL and TetK determinants has been demonstrated in a number of clinical isolates *M. avium* and *M. intracelluare* (Doran *et al.* 1997).

Macrolides

The newer macrolides (azithromycin and clarithromycin) have good activity against mycobacteria and are recommended for treatment of disseminated MAC disease. The use of clarithromycin against *M. avium* infections has recently been reviewed (Heifets 1996). Briefly, monotherapy with clarithromycin resulted in elimination of bacteremia in almost all patients with disseminated infection; however, a relapse of bacteremia in patients who survive long enough to reach this event inevitably followed. The relapses of bacteremia were caused by the multiplication of pre-existing mutants. Resistance occurs by mutations at A2058 in the central loop of domain V of the 23S rRNA (see Fig. 15.1c). Surprisingly, multiple-resistance mutations at the same site were isolated from single patients (Meier *et al.* 1996). Thus, although the frequency of spontaneous clarithromycin resistance is of the order of 10^{-9}, the microbial load during infection allows for the possibility of such rare mutations in a species with only a single copy of the rRNA gene. Therefore, a major problem in the treatment of MAC disease is the selection of companion drugs to be used in combination with clarithromycin (or azithromycin; cross-resistance to azithromycin has been confirmed). A recent randomized open-label clinical trial found that the inclusion of ethambutol with clarithromycin and clofazimine reduced relapses and the emergence of clarithromycin resistance (Dubé *et al.* 1997).

Clarithromycin has been tested against *M. tuberculosis* in *in vitro* assays and in mouse models by a number of groups (Luna-Herrera *et al.* 1995; Truffot-Pernot *et al.* 1995; Rastogi *et al.* 1996) with the same conclusion: it is inactive against this organism! However, a recent study on *M. tuberculosis* infection of human macrophages reported a synergistic effect on reducing the MIC when clarithromycin was used in combination with pyrazinamide (Mor & Esfandiari 1997).

3.2.3 Other cell-wall synthesis inhibitors

D-Cycloserine

The antimycobacterial activity of D-cycloserine (DCS) has been known for many years but it has not been routinely incorporated into standard therapeutic regimens because of its neurotoxic effects. Nonetheless, DCS, an analogue of alanine and a potent inhibitor of D-alanine racemase and ligase, is recommended for tuberculosis therapy under certain circumstances. Early studies showed that DCS-resistant mutants of *M. tuberculosis* could be obtained, but the biochemical mechanism was not elucidated. Recently, in detailed studies using *M. smegmatis*, Cáceres *et al.* (1997) demonstrated that one mechanism of resistance was due to overexpression of the D-alanine racemase gene and that such a mechanism was likely to be operative in *M. tuberculosis*. It will be of interest to see if other classes of DCS-resistant mutants can be obtained. The authors emphasize the specificity of the bacterial target of DCS and noted that additional studies to obtain DCS analogues could lead to the development of useful antimycobacterial drugs.

β-Lactams

As mentioned above, β-lactam antibiotics are ineffective against most mycobacteria, including *M. tuberculosis*. This intrinsic resistance is due to the combined effects of the permeability barrier of the mycobacterial cell wall and to a lesser degree to chromosomally encoded β-lactamases (reviewed in Jarlier & Nikaido 1994: see Chapter 12). *M. tuberculosis* possesses at least four penicillin-binding proteins (PBPs) of ≈94, 82, 52 and 37 kDa, and they bind therapeutically achievable concentrations of ampicillin, amoxicillin and imipenem. The three largest of these are the critical targets (Chambers *et al.* 1995). Permeability studies indicated that the rate of penetration of β-lactam antibiotics to these targets was not sufficient to account for resistance. Furthermore, the inclusion of clavulanic acid (a β-lactamase inhibitor) in the assay could reverse resistance (Chambers *et al.* 1995). An endogenous β-lactamase of *M. tuberculosis* has been cloned and characterized (Hackbarth *et al.* 1997). Substrate profiles and amino acid sequence analysis have shown it to be a typical class A enzyme with 60% homology to class A β-lactamases from other actinomycetes. This study also identified the nucleotide sequence of a gene from the *M. tuberculosis* genome project database encoding a class C β-lactamase. More recently, Voladri and coworkers report the expression and characterization of the class A β-lactamase of *M. tuberculosis* which accounts for almost 90% of the postchromatofocusing β-lactamase activity (Voladri *et al.* 1998).

3.2.4 Other targets

Folic acid biosynthesis

Folic acid is involved in the transfer of one-carbon groups utilized in the synthesis and metabolism of amino acids such as methionine and glycine and in nucleotide precursors adenine, guanine and thymine. Folic acid is converted in two reduction steps to tetrahydrofolate (THF), which serves as the intermediate carrier of hydroxymethyl, formyl or methyl groups in a large number of enzymatic reactions in which one-carbon groups are transferred from one metabolite to another or are interconverted. The synthetic antibacterial agents, sulphonamides and trimethoprim, inhibit specific steps in the biosynthesis of THF. The current state of resistance to sulphonamides and trimethoprim in major bacterial pathogens and the mechanisms involved have recently been reviewed (Huovinen *et al.* 1995).

The sulphonamides were first discovered in 1932 and introduced into clinical practice in the late 1930s.

They have a wide spectrum of antibacterial activity and have been used principally in urinary tract infections due to the Enterobacteriaceae, in respiratory tract infections due to *Streptococcus pneumonia* and *Haemophilus influenzae*, in skin infections due to *Staphylococcus aureus*, and in gastrointestinal tract infections due to *E. coli* and *Shigella* spp. The enzyme dihydropteroate synthase (DHPS) catalyses the formation of dihydropteric acid, the immediate precursor of dihydrofolic acid, and is the target of sulphonamides. These drugs are structural analogues of *p*-aminobenzoic acid, the normal substrate of DHPS, and act as competitive inhibitors for this enzyme.

Sulphonamides have been little used against mycobacterial infections in general, although dapsone and its derivatives remain important drugs for the amelioration of leprosy. Thus, one would not expect to find sulphonamide resistance in this group of bacteria. The mechanisms of resistance to this class of synthetic drug commonly involve mutation to overproduction of the target enzyme, dihydropteroate synthase, or modification of the enzyme structure. In Gram-negative bacteria, sulphonamide resistance is usually plasmid determined in the form of a novel dihydropteroate synthase gene that encodes an enzyme refractory to the inhibitor. It was all the more surprising therefore to find a sulR gene (*sul3*) in *M. fortuitum* associated with a defective integron (Martin *et al.* 1990). Integrons are the naturally occurring expression systems that have been shown to be responsible for much of the acquired antibiotic resistance in the Enterobacteriaceae (Recchia & Hall 1995). The demonstration of such a structure (albeit lacking a complete integrase gene) in mycobacteria is surprising and may indicate that the mycobacteria participate in horizontal acquisition and dissemination of antibiotic-resistance genes in natural populations.

Trimethoprim (TMP) was first introduced in 1962 and since 1968 has been used (often in combination with sulphonamides for a supposed synergistic activity) in numerous clinical situations. Like the sulphonamides, trimethoprim has a wide spectrum of activity and low cost. The enzyme dihydrofolate reductase (DHFR) is essential in all living cells. Trimethoprim is a structural analogue of dihydrofolic acid and acts as a competitive inhibitor of the reductase. The human DHFR is naturally resistant to trimethoprim, which is the basis for its use as an antibacterial agent. Mycobacteria are poorly inhibited by TMP. A new DHFR inhibitor, epiroprim, when used alone has somewhat better activity than TMP against MAC; however, a synergistic effect was seen when epiroprim was used in combination with dapsone (Locher *et al.* 1996).

Clofazimine

Clofazimine, a phenazine compound used initially to treat leprosy, has been useful in treating MAC disease. Its antimicrobial activity results from the generation of superoxide radicals. Resistance is probably due to induction of protective mechanisms against superoxide damage, such as superoxide dismutase, catalase and carotenoid pigments (Warek & Falkinham 1996).

4 Other factors leading to resistance

4.1 Hypermutability

At this point it is worth considering the generation of antibiotic-resistance mutations in mycobacteria. For many resistance alleles, the frequency of spontaneous mutations in laboratory studies is relatively low, especially with reference to functions involved in replication (*gyrA*), transcription (*rpoB*) or translation (*rpsL*, *rrnA*, *rrnB*); such mutations occur generally at 10^{-7} per generation (or less) in bacteria. However, given the rapidity with which mycobacterial infections become recalcitrant to antibiotics, it is suggested that mutations to antibiotic resistance occur at a more significant rate *in vivo* (Davies 1998). The DNA repair systems of mycobacteria have not been well studied; the *mutT* locus encodes a hydrolase that removes the mutagenic base 8-oxo-dGTP (Taddei *et al.* 1997a,b). Are successful intracellular bacterial pathogens more likely to be deficient in this function?

4.2 Compensatory mutations

The development of compensatory mutations leading to isoniazid resistance in clinical isolates of *M. tuberculosis* has been observed (see section 3.1.3). However, compensatory mutations in targets for other antibiotics are also possible and have been identified in *Salmonella typhimurium*. When components of essential macromolecular biosynthetic complexes (such as ribosomes) are altered by mutation, as in antibiotic-resistant strains, the mutants are frequently growth defective, which may lead to reduction in virulence of the antibiotic-resistant strains as compared to the drug-sensitive wild type. It has been shown recently (Schrag *et al.* 1997; Björkman *et al.* 1998) that compensatory mutations appear and restore the 'fitness' of the pathogen. Björkman *et al.* (1998) analysed the appearance of virulent streptomycin-resistant *S. typhimurium* in mouse infection studies. While the majority of the infecting organisms did not survive, rapid selection of fast-growing (surviving) strains took place: these strains had acquired additional mutations (sometimes intracodon) compensating the growth defect without altering the resistant phenotype. Interestingly, the compensatory mutations, when isolated from their cognate resistance mutation, reduce the survival fitness of the host. This suggests that antibiotic-resistant microbial pathogens are probably multiple mutants: the initial mutation to generate the resistance phenotype, accompanied by the compensating mutation(s) that maintain full virulence characteristics.

5 Conclusions and afterthoughts

Whether one is considering antibiotic susceptibility or resistance, the mycobacteria are a special case. Given the nature of tuberculosis, the extended duration of treatment leads to the inevitable appearance of resistant strains during the course of infection and is the major problem of tuberculosis therapy. Thus, in the half-century since streptomycin was discovered, effective treatment has been complicated by the frequent appearance of multiple drug resistance. The treatment of antibiotic-resistant tuberculosis is probably going to be the principal consideration for the foreseeable future; this means that multidrug therapy and complete compliance must be the norm. Unfortunately, afflicted persons often have a primary defect in their immune response due to AIDS, drug abuse, malnutrition or some other deficiency. Seventy years ago tuberculosis was considered a death sentence; and still today, an immunosuppressed person infected with multidrug-resistant *M. tuberculosis* has virtually no chance of reprieve.

Because of the unique physiology and biochemistry of the mycobacteria, there is no shortage of potential drug targets. What is surprising is that no effective new agent for tuberculosis treatment has been discovered for more than 25 years! Perhaps this is because in the industrialized world tuberculosis is no longer a major economic and social concern, in spite of the advent of Mdr tuberculosis. Is it too much to expect the development of a new generation of antitubercular drugs that are rapid-acting and effective on the intracellular pathogen in its dormant state? We would like to argue for more concerted research efforts on the treatment of dormant intracellular pathogens. This may not be the dominant type of infection at the moment, but as the population ages steadily, treatments for this disease in persons with reduced immunocompetence are likely to become more and more needed in the next century.

It is indeed fortunate that the mycobacteria are defective (or reduced) in sexual activity, since the non-appearance of transmissible drug resistance in tuberculosis and related mycobacterial diseases has made treatment uncomplicated by the promiscuous transmission of antibiotic-resistance genes and virulence factors. In retrospect, the pharmaceutical industry should not have given up the search for antimycobacterial drugs, and in prospect, the search should begin again. In spite of the difficulties in dealing with mycobacterial drugs, diagnostic methodology is now reaching the point where a few infecting microbes can be identified reliably and rapidly, thus permitting the early institution of treatment. However, in addition to new methods of diagnosis and surveillance, increased studies of mycobacterial infection and the ways in which they become refractory to drug treatment are essential to maintaining some form of therapeutic truce or parity with ever-present mycobacterial infection.

6 Acknowledgements

We thank the Canadian Bacterial Diseases Network and the National Science and Engineering Council of Canada for support.

7 References

Aínsa, J.A., Martín, C., Gicquel, B. & Gómez-Lus, R. (1996) Characterization of the chromosomal aminoglycoside 2′-*N*-acetyltransferase gene from *Mycobacterium fortuitum*. *Antimicrobial Agents and Chemotherapy* **40**, 2350–2355.

Aínsa, J., Pérez, E., Pelicic, V., Berthet, F.-X., Gicquel, B. & Martín, C. (1997) Aminoglycoside 2′-*N*-acetyltransferase genes are universally present in mycobacteria: characterization of the *aac (2′)-Ic* gene from *Mycobacterium tuberculosis* and the *aac (2′) -Id* gene from *Mycobacterium smegmatis*. *Molecular Microbiology* **24**, 431–441.

Alangaden, G.J., Kreiswirth, B.N., Aouad, A. *et al.* (1998) Mechanism of resistance to amikacin and kanamycin in *Mycobacterium tuberculosis*. *Antimicrobial Agents and Chemotherapy* **42**, 1295–1297.

Alangaden, G.J. & Lerner, S.A. (1997) The clinical use of fluoroquinolones for the treatment of mycobacterial diseases. *Clinical Infection and Disease* **25**, 1213–1221.

Banerjee, A., Dubnau, E., Quemard, A. *et al.* (1994) *inhA*, a gene encoding a target for isoniazid and ethionamide in *Mycobacterium tuberculosis*. *Science* **263**, 227–230.

Belanger, A.E., Besra, G.S., Ford, M.E. *et al.* (1996) The *embAB* genes of *Mycobacterium avium* encode an arabinosyl transferase involved in cell wall arabinan biosynthesis that is the target for the antimycobacterial drug ethambutol. *Proceedings of the National Academy of Sciences of the USA* **93**, 11919–11924.

Bifani, P.J., Plikaytis, B.B., Kappur, V. *et al.* (1996) Origin and interstate spread of a New York City multidrug-resistant *Mycobacterium tuberculosis* clone family. *Journal of the American Medicical Association* **275**, 452–457.

Björkman, J., Hughes, D. & Andersson, D.I. (1998) Virulence of antibiotic resistant *Salmonella typhimurium*. *Proceedings of the National Academy of Sciences of the USA* **95**, 3949–3953.

Blanchard, J.S. (1996) Molecular mechanisms of drug resistance in *Mycobacterium tuberculosis*. *Annual Review of Biochemistry* **65**, 215–239.

Bloch, A.B., Cauthen, G.M., Onorato, I.M. *et al.* (1994) Nationwide survey of drug-resistant tuberculosis in the United States. *Journal of the American Medical Association* **271**, 665–671.

Brennan, P.J. & Nikaido, H. (1995) The envelope of mycobacteria. *Annual Review of Biochemistry* **64**, 29–63.

Cáceres, N.E., Harris, N.B., Wellehan, J.F., Feng, Z., Kapur, V. & Garletta, R.G. (1997) Overexpression of the D-alanine racemase gene confers resistance to D-cycloserine in *Mycobacterium smegmatis*. *Journal of Bacteriology* **179**, 5049–5055.

Caugant, D.A., Sandven, P., Eng, J., Jeque, J.T. & Tønjm, T. (1995) Detection of rifampin resistance among isolates of *Mycobacterium tuberculosis* from Mozambique. *Microbial Drug Resistance* **1**, 321–326.

Chambers, H.F., Moreau, D., Yajko, D. *et al.* (1995) Can penicillins and other β-lactam antibiotics be used to treat tuberculosis? *Antimicrobial Agents and Chemotherapy* **39**, 2620–2624.

Cole, S.T. (1994) *Mycobacterium tuberculosis*: drug-resistance mechanisms. *Trends in Microbiology* **2**, 411–415.

Cole, S.T. (1996) Rifamycin resistance in mycobacteria. *Research Microbiology* **147**, 48–52.

Dabbs, E.R., Yazawa, K., Mikami, Y., Miyaji, M., Morisaki, N., Iwasaki, S. & Furihata, K. (1995) Ribosylation by mycobacterial strains as a new mechanism of rifampin inactivation. *Antimicrobial Agents and Chemotherapy* **39**, 1007–1009.

Davidson, L.A. & Takayama, K. (1979) Isoniazid inhibition of the synthesis of monosaturated long-chain fatty acids in *Mycobacterium tuberculosis*. *Antimicrobial Agents and Chemotherapy* **16**, 104–105.

Davies, J. (1998) Antibiotic resistance in mycobacteria. In: *Genetics and Tuberculosis* (eds D. J. Chadwick & G. Cardew), Novartis Foundation Symposium 217. Chichester: John Wiley & Sons, pp. 195–208.

Deng, L., Mikusova, K., Robuck, K.G., Scherman, M., Brennan, P.J. & McNeil, M.R. (1995) Recognition of multiple effects of ethambutol on metabolism of mycobacterial cell envelope. *Antimicrobial Agents and Chemotherapy* **39**, 694–701.

Deretic, V., Pagán-Ramos, E., Zhang, Y., Dhandayuthapani, S. & Via, L.E. (1996) The extreme sensitivity of *Mycobacterium tuberculosis* to the front-line antituberculosis drug isoniazid. *Nature Biotechnology* **14**, 1557–1561.

Deretic, V., Philipp, W., Dhandayuthapani, S. *et al.* (1995) *Mycobacterium tuberculosis* is a natural mutant with an inactivated oxidative-stress regulatory gene: implications for sensitivity to isoniazid. *Molecular Microbiology* **17**, 889–900.

Deretic, V., Song, J. & Pagán-Ramos, E. (1997) Loss of oxyR in *Mycobacterium tuberculosis*. *Trends in Microbiology* **5**, 367–371.

Dhandayuthapani, S., Mudd, M. & Deretic, V. (1997) Interactions of OxyR with the promoter region of the *oxyR* and *ahpC* genes from *Mycobacterium leprae* and *Mycobacterium tuberculosis*. *Journal of Bacteriology* **179**, 2401–2409.

Dhandayuthapani, S., Zhang, Y., Mudd, M.H. & Deretic, V. (1996) Oxidative stress response and its role in sensitivity to isoniazid in mycobacteria: characterization and inducibility of *ahpC* by peroxides in *Mycobacterium smegmatis* and lack of expression in *M. aurum* and *M. tuberculosis*. *Journal of Bacteriology* **178**, 3641–3649.

Doran, J.L., Pang, Y., Mduli, K.E. *et al.* (1997) *Mycobacterium tuberculosis efp*A encodes an efflux protein of the QacA transporter family. *Clinical and Diagnostic Laboratory Immunology* **4**, 23–32.

Drlica, K. & Zhao, X. (1997) DNA gyrase, topoisomerase IV, and the 4-quinolones. *Microbiological Molecular Biological Review* **61**, 377–392.

Dubé, M.P., Sattler, F.R., Torriani, F.J. *et al.* (1997) A randomized evaluation of ethambutol for prevention of relapse and drug resistance during treatment of *Mycobacterium avium* complex bacteremia with clarithromycin-based combination therapy. *Journal of Infectious Diseases* **176**, 1225–1232.

Finken, M., Kirschner, P., Meier, A., Wrede, A. & Böttger, E.C. (1993) Molecular basis of streptomycin resistance in *Mycobacterium tuberculosis*: alterations of the ribosomal protein S12 gene and point mutations within a functional 16S ribosomal RNA pseudoknot. *Molecular Microbiology* **9**, 1239–1246.

Fourmy, D., Recht, M.I., Blanchard, S.C. & Puglisi, J.D. (1996) Structure of the A site of *Escherichia coli* 16S ribosomal RNA complexed with an aminoglycoside antibiotic. *Science* **274**, 1367–1371.

Garcia-Rodriguez, J.A. & Garcia, A.C.G. (1993) *In-vitro* activities of quinolones against mycobacteria. *Antimicrobial Agents and Chemotherapy* **32**, 797–808.

Grosset, J. (1996) Current problems with tuberculosis treatment. *Research Microbiology* **147**, 10–16.

Haas, W.H., Schilke, K., Brand, J. *et al.* (1997) Molecular analysis of *katG* gene mutations in strains of *Mycobacterium tuberculosis* complex from Africa. *Antimicrobial Agents and Chemotherapy* **41**, 1601–1603.

Hackbarth, C.J., Unsal, I. & Chambers, H.F. (1997) Cloning and sequence analysis of a class A β-lactamase from *Mycobacterium tuberculosis* H37Ra. *Antimicrobial Agents and Chemotherapy* **41**, 1182–1185.

Heifets, L.B. (1996) Clarithromycin against *Mycobacterium avium* complex infections. *Tubercle and Lung Disease* **77**, 19–26.

Heym, B., Alzari, P.M., Honoré, N. & Cole, S.T. (1995) Missense mutations in the catalase-peroxidase gene, *katG*, are associated with isoniazid resistance in *Mycobacterium tuberculosis*. *Molecular Microbiology* **15**, 235–245.

Heym, B., Stavropuolos, E., Honoré, N. *et al.* (1997) Effects of overexpression of the alkyl hydroperoxide reductase AhpC on the virulence and isoniazid resistance of *Mycobacterium tuberculosis*. *Infection and Immunity* **65**, 1395–1401.

Huovinen, P., Sundström, L., Swedberg, G. & Sköld, O. (1995) Trimethoprim and sulfonamide resistance. *Antimicrobial Agents and Chemotherapy* **39**, 279–289.

Jarlier, V. & Nikaido, H. (1994) Mycobacterial cell wall: structure and role in natural resistance to antibiotics. *FEMS Microbiological Letters* **123**, 11–18.

Kelley, C.L., Rouse, D.A. & Morris, S.L. (1997) Analysis of *ahpC* gene mutations in isoniazid-resistant clinical isolates of *Mycobacterium tuberculosis*. *Antimicrobial Agents and Chemotherapy* **41**, 2057–2058.

Khoo, K.-H., Douglas, E., Asadi, P. *et al.* (1996) Truncated structural variants of lipoarabinomannan in ethambutol drug-resistant strains of *Mycobacterium smegmatis*: inhibition of arabinan biosynthesis by ethambutol. *Journal of Biological Chemistry* **271**, 28682–28690.

Kunin, C.M. (1996) Antimicrobial activity of rifabutin. *Clinical Infection and Disease* **22** (Suppl. 1), S3–S14.

Lalande, V., Truffot-Pernot, A., Paccaly-Moulin, A. & Grosset, J. (1993) Powerful activity of sparfloxacin (AT 4140) against *Mycobacterium tuberculosis* in mice. *Antimicrobial Agents and Chemotherapy* **37**, 407–413.

Lee, R.L., Mikusova, K., Brennan, P.J. & Besra, G.S. (1995) Synthesis of the mycobacterial arabinose donor β-D-arabinofuranosyl-1-monophosphonyldecaprenol, development of a basic arabinosyl-transferase assay, and identification of ethambutol as an arabinosyl transferase inhibitor. *Journal of the American Chemical Society* **117**, 11829–11832.

Locher, H.H., Schlunegger, H., Hartman, P.G., Angehrn, P. & Then, R.L. (1996) Antibacterial activities of epiroprim, a new dihydrofolate reductase inhibitor, alone and in combination with dapsone. *Antimicrobial Agents and Chemotherapy* **40**, 1376–1381.

Luna-Herrera, J., Reddy, V.M., Daneluzzi, D. & Gangadharam, P.R. (1995) Antituberculosis activity of clarithromycin. *Antimicrobial Agents and Chemotherapy* **39**, 2692–2695.

Martin, C., Timm, J., Rauzier, J., Gomez-Lus, R., Davies, J. & Gicquel, B. (1990) Transposition of an antibiotic resistance element in mycobacteria. *Nature* **345**, 739–743.

Mdluli, K., Sherman, D.R., Hickey, M.J., *et al.* (1996) Biochemical and genetic data suggest that InhA is not the primary target for activated isoniazid in *Mycobacterium tuberculosis*. *Journal of Infection and Disease* **174**, 1085–1090.

Mdluli, K., Slayden, R.A., Zhu, Y. *et al.* (1998) Inhibition of a *Mycobacterium tuberculosis* β-ketoacyl ACP synthase by isoniazid. *Science* **280**, 1607–1610.

Meier, A., Heifets, L., Wallace, R.J. Jr, Zhang, Y., Brown, B.A., Sander, P. & Böttger, E.C. (1996) Molecular mechanisms of clarithromycin resistance in *Mycobacterium avium*: observation of multiple 23S rRNA mutations in a clonal population. *Journal of Infection and Disease* **174**, 354–360.

Miesel, L., Weisbrod, T.R., Marcinkeviciene, J.A., Bittman, R. & Jacobs, W.R. Jr (1998) NADH dehydrogenase defects confer isoniazid resistance and conditional lethality in *Mycobacterium smegmatis*. *Journal of Bacteriology* **180**, 2459–2467.

Mikusova, K., Slayden, R.A., Besra, G.S. & Brennan, P.J. (1995) Biogenesis of the mycobacterial cell wall and the site of action of ethambutol. *Antimicrobial Agents and Chemotherapy* **39**, 2484–2489.

Mitchison, D.A. (1985) The action of antituberculosis drugs in short-course chemotherapy. *Tubercle* **66**, 219–225.

Mitsuhashi, S., Tanaka, T., Kawabe, H. & Umezawa, H. (1977) Biochemical mechanism of kanamycin resistance in *Mycobacterium tuberculosis*. *Microbiological Immunology* **21**, 325–327.

Mor, N. & Esfandiari, A. (1997) Synergistic activities of clarithromycin and pyrazinamide against *Mycobacterium tuberculosis* in human macrophages. *Antimicrobial Agents and Chemotherapy* **41**, 2035–2036.

Musser, J.M. (1995) Antimicrobial agent resistance in mycobacteria: molecular genetic insights. *Clinical Microbiological Review* **8**, 469–514.

Musser, J.M., Kapur, V., Williams, D.L., Kreiswirth, B.N., van Soolingen, D. & van Embden, J.D.A. (1996) Characterization of the catalase-peroxidase gene (*katG*) and *inhA* locus in isoniazid-resistant and -susceptible strains of *Mycobacterium tuberculosis* by automated DNA sequencing: restricted array of mutations associated with drug resistance. *Journal of Infection and Disease* **173**, 196–202.

Nolan, C.M. (1997) Nosocomial multidrug-resistant tuberculosis—global spread of the third epidemic. *Journal of Infection and Disease* **176**, 748–751.

Pang, Y., Brown, B.A., Steingrube, V.A., Wallace, R.J., Jr & Roberts, M.C. (1994) Tetracycline resistance determinants in *Mycobacterium* and *Streptomyces* species. *Antimicrobial Agents and Chemotherapy* **38**, 1408–1412.

Prammananan, T., Sander, P., Brown, B.A. *et al.* (1998) A single 16S RNA substitution is responsible for resistance to amikacin and other 2-deoxystreptamine aminoglycosides in *Mycobacterium abscessus* and *Mycobacterium chelonae. Journal of Infection and Disease* **177**, 1573–1581.

Quemard, A., Sacchettini, J.C., Dessen, A., Jacobs, W.R. Jr & Blanchard, J.S. (1995) Enzymatic characterization of the target for isoniazid in *Mycobacterium tuberculosis*. *Biochemistry* **34**, 8235–8241.

Rastogi, N., Labrousse, V. & Goh, K.S. (1996) *In vitro* activities of fourteen antimicrobial agents against drug susceptible and resistant clinical isolates of *Mycobacterium tuberculosis* and comparative intracellular activities against the virulent H37Rv strain in human macrophages. *Current Microbiology* **33**, 167–175.

Rattan, A., Kalia, A. & Ahmad, N. (1998) Multidrug-resistant *Mycobacterium tuberculosis*: molecular perspectives. *Emerging Infectious Diseases* **4**, 195–209.

Recchia, G.D. & Hall, R.M. (1995) Gene cassettes: a new class of mobile element. *Microbiology* **141**, 3015–3027.

Rozwarski, D.A., Grant, G.A., Barton, D.H.R., Jacobs, W.R. Jr & Sacchettini, J.C. (1998) Modification of the NADH of the isoniazid target (InhA) from *Mycobacterium tuberculosis*. *Science* **279**, 98–102.

Sacchettini, J.C. & Blanchard, J.S. (1996) The structure and function of the isoniazid target in *M. tuberculosis*. *Research Microbiology* **147**, 36–43.

Sander, P., Meier, A. & Böttger, E.C. (1996) Ribosomal drug resistance in mycobacteria. *Research Microbiology* **147**, 59–67.

Schrag, S.J., Perrot, V. & Levin, B.R. (1997) Adaptation to the fitness costs of antibiotic resistance in *Escherichia coli*. *Proceedings of the Royal Society of London B* **264**, 1287–1291.

Scorpio, A., Lindholm-Levy, P., Heifets, L. *et al.* (1997) Characterization of *pncA* mutations in pyrazinamide-resistant *Mycobacterium tuberculosis*. *Antimicrobial Agents and Chemotherapy* **41**, 540–543.

Scorpio, A. & Zhang, Y. (1996) Mutations in *pncA*, a gene encoding pyrazinamidase/nicotinamidase, cause resistance to the antituberculous drug pyrazinamide in tubercle bacillus. *Nature Medicine* **2**, 662–667.

Shaw, K.J., Rather, P.N., Hare, R.S. & Miller, G.H. (1993) Molecular genetics of aminoglycoside resistance genes and familial relationships of the aminoglycoside-modifying enzymes. *Microbiological Reviews* **57**, 138–163.

Sherman, D.R., Mdluli, K., Hickey, M.J. *et al.* (1996) Compensatory *ahpC* gene expression in isoniazid-resistant *Mycobacterium tuberculosis*. *Science* **272**, 1641–1643.

Sreevatsan, S., Pan, X., Zhang, Y., Deretic, V. & Musser, J.M. (1997a) Analysis of the *oxyR-ahpC* region in isoniazid-resistant and -susceptible *Mycobacterium tuberculosis* complex organisms recovered from diseased humans and animals in diverse localities. *Antimicrobial Agents and Chemotherapy* **41**, 600–606.

Sreevatsan, S., Pan, X., Zhang, Y., Kreiswirth, B.N. & Musser, J.M. (1997b) Mutations associated with pyrazinamide resistance in *pncA* of *Mycobacterium tuberculosis* complex organisms. *Antimicrobial Agents and Chemotherapy* **41**, 636–640.

Sreevatsan, S., Stockbauer, K.E., Pan, X. *et al.* (1997c) Ethambutol resistance in *Mycobacterium tuberculosis*: critical role of *embB* mutations. *Antimicrobial Agents and Chemotherapy* **41**, 1677–1681.

Taddei, F., Hayakawa, H., Bouton, M.-F. *et al.* (1997a) Counteraction by MutT protein of transcriptional errors caused by oxidative damage. *Science* **278**, 128–130.

Taddei, F., Radman, M., Maynard-Smith, J., Toupance, B., Gouyon, P.H. & Godelle, B. (1997b) Role of mutator alleles in adaptive evolution. *Nature* **387**, 700–702.

Takayama, K., Schnoes, H.K., Armstrong, E.L. & Boyle, R.W. (1975) Site of inhibitory action of isoniazid in the synthesis of mycolic acids in *Mycobacterium tuberculosis*. *Journal of Lipid Research* **16**, 308–317.

Takiff, H.E., Cimino, M., Musso, M.C. *et al.* (1996) Efflux pump of the proton antiporter family confers low-level fluoroquinolone resistance in *Mycobacterium smegmatis*. *Proceedings of the National Academy of Sciences of the USA* **93**, 362–366.

Takiff, H.E., Salazar, L., Guerrero, C. *et al.* (1994) Cloning and nucleotide sequence of *Mycobacterium tuberculosis gyrA* and *gyrB* genes and detection of quinolone resistance mutations. *Antimicrobial Agents and Chemotherapy* **38**, 773–780.

Taniguchi, H., Chang, B., Abe, C., Nikaido, Y., Mizuguchi, Y. & Yoshida, S.-I. (1997) Molecular analysis of kanamycin and viomycin resistance in *Mycobacterium smegmatis* by use of the conjugation system. *Journal of Bacteriology* **179**, 4795–4801.

Telenti, A., Philipp, W.J., Sreevatsan, S. *et al.* (1997) The *emb* operon, a gene cluster of *Mycobacterium tuberculosis* involved in resistance to ethambutol. *Nature Medicine* **3**, 567–570.

Truffot-Pernot, C., Lounis, N., Grosset, J.H. & Ji, B. (1995) Clarithromycin is inactive against *Mycobacterium tuberculosis*. *Antimicrobial Agents and Chemotherapy* **39**, 2827–2828.

Udou, T., Mizuguchi, Y. & Wallace, R.J. (1987) Patterns and distribution of aminoglycoside-acetylating enzymes in rapidly growing mycobacteria. *American Review of Respiratory Diseases* **136**, 338–343.

Udou, T., Mizuguchi, Y. & Wallace, R.J. Jr (1989) Does aminoglycoside-acetyltransferase in rapidly growing mycobacteria have a metabolic function in addition to aminoglycoside inactivation? *FEMS Microbiological Letters* **57**, 227–230.

Udou, T., Mizuguchi, Y. & Yamada, T. (1986) Biochemical mechanisms of antibiotic resistance in a clinical isolate of *Mycobacterium fortuitum*. Presence of beta-lactamase and aminoglycoside-acetyltransferase and possible participation of altered drug transport on resistance mechanism. *American Review of Respiratory Diseases* **133**, 653–657.

Voladri, R.K.R., Lakey, D.L., Hennigan, S.H., Menzies, B.E., Edwards, K.M. & Kernodle, D.S. (1998) Recombinant expression and characterization of the major β-lactamase of *Mycobacterium tuberculosis*. *Antimicrobial Agents and Chemotherapy* **42**, 1375–1381.

Warek, U. & Falkinham, J.O. (1996) Action of clofazimine on the *Mycobacterium avium* complex. *Research Microbiology* **147**, 43–48.

Wilson, T.M. & Collins, D.M. (1996) *ahpC*, a gene involved in isoniazid resistance of the *Mycobacterium tuberculosis* complex. *Molecular Microbiology* **19**, 1025–1034.

Wolucka, B.A., McNeil, M.R., Hoffmann, E.D., Chojnacki, T. & Brennan, P.J. (1994) Recognition of the lipid intermediate for arabinogalactan/arabinomannan biosynthesis and its relation to the mode of action of ethambutol on mycobacteria. *Journal of Biological Chemistry* **269**, 23328–23335.

Xu, C., Kreiswirth, B.N., Sreevatsan, S., Musser, J.M. & Drlica, K. (1996) Fluoroquinolone resistance associated with specific gyrase mutations in clinical isolates of multidrug-resistant *Mycobacterium tuberculosis*. *Journal of Infection and Disease* **174**, 1127–1130.

Zhang, Y., Dhandayuthapani, S. & Deretic, V. (1996) Molecular basis for the exquisite sensitivity of *Mycobacterium tuberculosis* to isoniazid. *Proceedings of the National Academy of Sciences of the USA* **93**, 13212–13216.

Zhang, Y., Heym, B., Allen, B., Young, D. & Cole, S. (1992) The catalase-peroxidase gene and isoniazid resistance of *Mycobacterium tuberculosis*. *Nature* **358**, 591–593.

Chapter 16 / Immunotherapy for mycobacterial diseases

JOHN L. STANFORD & GRAHAM A.W. ROOK

1 Introduction

Prior to the development of effective chemotherapy for leprosy and tuberculosis, many attempts at immunotherapy were made, some of which appear to have been highly successful if judged by the published evidence. However, the concepts of Jenner and his successors confused the issue, and lack of understanding of immunity led to hit and miss results. With the success of chemotherapy, interest in 'stimulating the phagocytes' declined. Today, with the potential decline in effectiveness of chemotherapy resulting from the selection of drug-resistant strains, poor prescription and the tendency of patients not to comply with the long regimens required, immunotherapy deserves reconsideration.

Although immunoprophylaxis and immunotherapy have separate uses and have separate problems, they are intimately related. Despite there being separate chapters in this book on immunity and vaccines, it would be difficult to give full consideration to immunotherapy without consideration of these topics. By immunoprophylaxis we refer to the use of a vaccine to modify immunity to prevent the establishment of infection, and by immunotherapy we refer to the use of substances modulating the immune system in a way that will contribute to overcoming disease due to infection, whether this be latent or actual. These substances may be steroidal or non-steroidal immunoregulatory drugs, purified cytokines, or bacterial products exerting immunoregulatory activity. With mycobacteria we have a further complication in that at least some of them have the capacity to remain apparently dormant in the tissues following infection, and to activate to cause clinical disease at some later date, when the immune status favours it. This survival of persistors (Grange 1992; Parrish *et al.* 1998) confers a condition of latency and offers a further opportunity for immunological intervention, that of preventive immunotherapy given to overcome persisters and remove latency rather than to treat overt disease. Thus, the topic can be considered in three parts:

1 vaccination of the uninfected—immunoprophylaxis;

2 eradication of live bacilli persisting in the tissues but not causing disease—preventive immunotherapy;

3 immunotherapy for active disease.

Reagents of four types are available for evaluation at the moment. The only one in worldwide general use is bacille Calmette–Guérin (BCG). The others include BCG plus an additive, or a replacement for BCG. Also possible is the direct injection of purified cytokines, and other adjuvants or immunomodulating substances, the use of which will not be considered further here.

1.1 Immunity and mycobacterial disease

Only the simplest review of this most complicated subject will be given here. Fundamental to immunity is the interaction between the brain, the endocrine system and the immune system itself: psycho-neuro-endocrino-immunology. The purpose of immunity is to maintain a stable balance between external and internal milieu; to adequately respond to external antigens and allergens, and to modify responses to determinants that these share with self antigens. Together, these encompass almost all human disease. Figure 16.1 illustrates some aspects of this fascinating relationship. At this stage in our knowledge the maturation patterns of T-helper (CD4+) lymphocytes, their production of different series' of interrelating cytokines, their effects on macrophages, antibody production and B-lymphocyte function, and their influence on killer cells, cytotoxic and CD8+ T cells, appear central to the whole process.

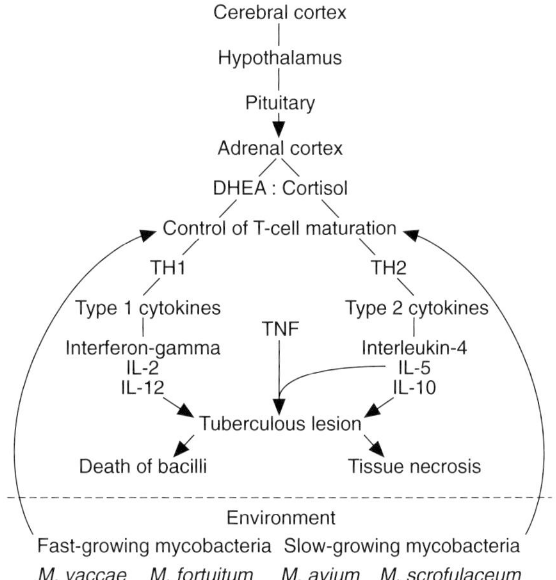

Fig. 16.1 An outline of psycho-neuro-endocrino-immunology and interaction with environmental mycobacteria. TH1 and TH2 are two pathways of T lymphocyte maturation, each results in production of different sets of cytokines. DHEA, dehydroepiandrosterone; IL, interleukins of the numbers indicated; TNF, tumour necrosis factor alpha.

Although Fig. 16.1 is constructed with mycobacteria being given the key roles, other influences include virus infections, parasitic infestations, and tobacco smoking. In coordination with human genetics, the figure illustrates the chief mechanism of homoeostasis, and a major determining factor in prevention, or susceptibility to, cancer, cardiovascular disease, the major psychoses and the autoimmune and other degenerative diseases. Undoubtedly the same system is involved in ontogeny and the process of ageing.

1.2 Environmental influences

Unlike some other infectious diseases, mycobacterioses occur superimposed upon the immunological experience of contact with closely related environmental species (Paul *et al.* 1975a). Mycobacteria do not make up part of the normal flora of the body in its generally accepted sense, but they are so common in many situations that daily exposure to some of them is the experience of most of us (Rook & Stanford 1998). Indeed, it can be argued that these 'harmless' environmental species are the most important of all mycobacteria. After all, only about 10% of people worldwide develop clinical mycobacterial disease at any time in their lives, and much of the successful protection afforded to the remaining 90% originates from immune responses to these environmental organisms. Such organisms are not entirely harmless, many of them give rise to occasional cases of disease, the opportunist mycobacterial infections.

Influences on vaccination and autoimmunity

Some species can also exert a more insidious effect, pre-empting vaccination with BCG (Stanford *et al.* 1981), undermining immunity and giving the individual a secondary susceptibility to more pathogenic species. Environmental mycobacteria may also be important because they possess antigenic epitopes shared with those of human tissues, and present them together with their potent adjuvants. These may have very wide-reaching consequences, modulating our responsiveness to ourselves in ways that may lead to protection from, or development of, a whole range of diseases not usually associated with a mycobacterial aetiology. These include the autoimmune diseases from rheumatoid arthritis to schizophrenia, cancer and atherosclerosis.

The mechanism of contact

Contact with environmental mycobacteria may be through the skin. Although it is doubtful if they can pass though intact skin, they can certainly enter through abrasions, and this is the chief mode of invasion with some opportunists. The lipophilic nature of mycobacteria may mean that they can persist on the skin surface for some considerable time and that anti-

gens released may be absorbed. It is quite unknown whether this is an important means of contacting mycobacterial antigens. The percutaneous method of tuberculin testing depends on this mechanism (Backman *et al.* 1984).

Animal experiments

Some experiments have been carried out in which mice were injected subcutaneously with small numbers of a variety of mycobacterial species, and three consequences have been observed. Species pathogenic for mice may establish a local lesion, spread to the draining lymph node and disseminate. Other species result in a small local granuloma with an enlarged draining lymph node before resolving completely. Still others cause no macroscopic lesion at all. Biopsy of injection sites in the first two situations shows acid-fast bacilli persisting for some days or weeks with an infiltrate of macrophages and lymphocytes. In the third situation, acid-fast bacilli disappeared within 24h (Shepard *et al.* 1980). It may be that one or other of these situations, depending on the species invading, is frequently experienced by everyone.

1.3 The process of infection

Inhalation

Many mycobacteria are undoubtedly inhaled and this is the predominant route of infection for most pathogenic species, either lodging in the turbinates of the nose, or passing through the trachea and bronchi to the alveoli. Experiments suggest that most inhaled particles are too large to reach the periphery of the lung, become entrapped in bronchial mucus and follow the mucus flow to the pharynx where they are swallowed. Clumps of two or three bacilli may reach the alveolar passages where they are taken up by alveolar macrophages. How mycobacteria are handled in this situation when they are not of pathogenic species is quite unknown, as is the part they may play in predetermining the fate of subsequent challenges with pathogens as a result of local immunity and in the establishment of diseases such as asthma (Grange *et al.* 1994).

Ingestion

Many mycobacteria are ingested, either in food or drink, or in the mucus flow coming from the trachea. This is a much less important route for successful invasion by pathogens, although it was, and in some countries perhaps still is, the major route of infection for bovine tubercle bacilli sometimes present in large numbers in infected milk. Studies of sewage from tuberculosis sanatoria showed large numbers of living tubercle bacilli to be present. This suggests that the very large numbers of bacilli brought up in the sputum and swallowed by pulmonary tuberculosis patients pass straight through the gut. Whether secretory IgA plays any part in preventing mycobacteria attaching to villi etc., or influences their uptake by M cells in the crypts of Lieberkühn between the villi of the small intestine is unknown. Certainly tuberculosis patients have detectable levels in their serum of mycobacterium-specific IgA, and it is highly likely that a reasonable proportion of this attaches to secretory factor. The fate of ingested environmental mycobacterial species is even less well-known, and sampling of gut content by M cells may be an important means in maintaining immune mechanisms to mycobacteria, and regulating responsiveness to self. Some unpublished studies of skin testing with reagents prepared from environmental mycobacteria in Gurkha soldiers rotating for 2-year stints in Surrey, UK, Hong Kong and Nepal, showed that the pattern of responses at the end of 2 years in one place were different from the pattern of their responses after 2 years in a different place. The same individuals passing from station to station both lost previous sensitizations and acquired new ones, showing that skin-test memory can be for less than 2 years (D. Jolliffe, personal communication 1981). Reactivity is retained or lost presumably as a result of regulatory responses dependant upon environmental contact, and the M cells could be major mediators of this.

The pharynx

Other important sites where the immune system regularly meets mycobacteria are the pharynx and tonsils. In some cases this may associate with tonsillar enlargement, in others there may be spread to draining cervical lymph nodes resulting in mycobacterial cervical lymphadenitis. In the UK the principal causes of this are *Mycobacterium avium*, *M. intracellulare*, *M. malmoense* and *M. scrofulaceum*. Many less potentially pathogenic species must frequently pass the tonsils leaving an immunological message but no clinical lesion.

Bacterial cultures of tonsillectomy specimens have shown environmental rapid-growing mycobacteria such as *M. fortuitum* to be present most commonly in specimens obtained in the winter months (Stewart *et al.* 1970). Further studies showed that children acquire more skin-test positivity to *M. avium* in the winter months than in summer months, and that this is associated with outdoor games. Recent environmental studies of farmland have shown seasonal changes in frequency of mycobacteria, and an influence on their distribution when conventional farming methods are replaced by organic methods (Donoghue *et al.* 1997). Further evidence for this came from skin-test studies of Burmese villagers living in the region where the major trial of BCG vaccination against leprosy was carried out. It was found that those drawing their drinking water from one well had different responses from those taking water from a different well in the same village (M. J. Shield, personal communication 1978). Thus, there is seasonal variation in contact with mycobacteria, and small local changes influence the distribution of species. The overall effect of such differences may influence the distribution of mycobacterial disease (and perhaps many other diseases) by changing the resistance or susceptibility of the local population.

1.4 Infection leading to disease

Mycobacterial diseases start with the bacilli being ingested, breathed in or rubbed into abrasions, and their subsequent phagocytosis. Primarily intracellular, some infections such as with *M. leprae* almost always remain so, although as disease progresses and tissue damage increases, a greater proportion of bacilli become extracellular and may multiply in necrotic tissue. The assumed rate of bacillary replication in leprosy based on doubling times of stainable acid-fast bacilli in experimentally infected animals suggests that this is so very slow that necrotic tissue is going to have been cleared up before there is time for any significant degree of extracellular replication. This may not be the case, however, if unconventional means of replication occur with *M. leprae*. Tubercle bacilli ostensibly multiply much faster than leprosy bacilli, much more readily appear outside of cells, and undoubtedly achieve their greatest replication rate in extracellular situations such as the internal surfaces of air-filled cavities.

The entry of tubercle bacilli

The major portal of entry for tubercle bacilli leading to development of disease is by inhalation into the lower respiratory tract. Bundles of two or three bacilli travel in the central air stream to the alveolar passages where they are engulfed by wandering phagocytic cells and carried back through the alveolar membrane into the tissues. Larger bundles of bacilli inhaled tend to fall out of the central air stream and are trapped in the surface mucus. The mucus flow carries them to the pharynx from whence they are swallowed. Bacilli surviving the stomach acid may be phagocytosed by the M cells in the crypts of Lieberkühn in the small intestine, but establishment of tuberculous disease in the small intestine is infrequent, though still living bacilli appear in the faeces.

The entry of leprosy bacilli

The portal of entry for leprosy bacilli may be the same as that for tuberculosis but with bacilli having to be carried from the site of entry to skin or peripheral nerves before disease becomes established. A more likely site of first invasion by leprosy bacilli is through the nasal mucosa where the earliest lesions are often discovered. Normally, nasal secretions are swallowed,

just as are those of the lower respiratory tract, and thus leprosy bacilli must also reach the gut-associated lymphoid tissues. Although leprosy bacilli are not known to cause lesions in the gut itself, chronic lepromatous patients often have enlarged spleens that may be packed with leprosy bacilli. The mechanism for immune tolerance that seems to be an essential part of the establishment of multibacillary leprosy may be through absorbing immunosuppressive substances from nasal lesions (Nye *et al.* 1986), or may be an unfortunate result of the homoeostatic mechanisms associated with tolerance induced via the gut.

Portals of entry for Mycobacterium marinum *and* Mycobacterium ulcerans

Mycobacteria are not thought capable of invading the intact skin but are well known to enter through minor abrasions, probably being rapidly phagocytosed by polymorphs and macrophages. Both *M. marinum* infection and *M. ulcerans* infection may start from such sites, and lupus vulgaris may follow tubercle bacilli entering by this method.

1.5 Progression of infection

In leprosy, where the very earliest lesions may be in the nasal mucosa, the developing infection may be completely asymptomatic. Similarly, in tuberculosis, the very earliest lesions, often in the periphery at the apex of an upper lobe of the lung, are quite symptom-free. Many cases of either disease probably abort at this stage, which may be the starting point for latency. Others progress locally and in the draining lymph nodes, and still others rapidly disseminate to a number of sites.

Although there is little doubt that *M. marinum* enters through abrasions and swimming-pool granulomas (Linnell & Norden 1954; Collins *et al.* 1985) start from the point of inoculation, it is possible that *M. ulcerans* is inhaled or ingested, enters the bloodstream and sets up Buruli ulcers at sites of trauma (*loci minoris resistentiae*). There is also some evidence suggesting that *M. ulcerans* can persist and give rise to disease months or even years after infection.

Evidence for this comes from patients developing disease long after they have left endemic regions, and from a second peak of incidence of the disease in old age when patients are no longer in regular contact with the environment where it is thought the bacilli have their natural reservoir. Once disease starts either at the site of infection or of deposit from the circulation, at least some of the phagocytic cells must be sessile or are immobilized as an early effect of their infection. Subsequently other cells may transfer bacilli to the local lymph nodes or even further from the infected site.

Whatever the portal of entry, unless the individual is new-born or severely immunosuppressed, the potential invader must meet an immune response. This might have been initiated by a previous contact with the pathogen, by environmental contact, or by vaccination. This immune response can be either protective, or that of immunopathology increasing the likelihood of developing disease.

2 Vaccination of the uninfected

2.1 Possible times for prophylactic vaccination

It is probably not realistic to use vaccination to defend against the initial step of phagocytosis and the death of a few cells, but subsequent stages of disease development offer opportunities for successful vaccination against clinical manifestation. In the best situation, the initially infected phagocytes should release T-cell attractant cytokines such as interleukin 1 (IL-1), and stimulate local production of tumour necrosis factor alpha (TNF-α). Amongst the T helper (Th) cells should be some that recognize mycobacterial antigens presented on the macrophage surface. These cells begin to multiply in a draining lymph node and to release cytokines controlling macrophage function such as interferon gamma (IFN-γ) and IL-2. When the infecting mycobacteria are potent pathogens such as leprosy or tubercle bacilli or *M. ulcerans*, and the challenge dose is large, they may interfere with immune functions by the secretion of soluble bacterial substances, and these may need to be inacti-

vated by complexing with antibodies. The immunosuppressive substance of *M. ulcerans* is probably the same as its toxin (Krieg *et al.* 1974; Pimsler *et al.* 1988), now shown to be a macrolide (P. Small: presented at the International Conference on Buruli Ulcer; Yamoussoukro, Cote d'Ivoire; July 6–9, 1998), but the immunoregulatory substances of tubercle and leprosy bacilli are better defined by their activities than by their biochemistry, and may be proteins. Comprehensive immunity comprises both cell mediated and humoral aspects, the latter of which have been largely disregarded since the 1930s.

Successful immune response should lead to the death of all invading bacilli, and repair of the locally damaged tissue leaving no detectable scar, and an enhanced immune memory. It is probable that a proportion of recipients of BCG vaccine achieve this condition since skin-test responses to tuberculin appear to wax and wane without lesions being detectable, perhaps reflecting silent challenges overcome. In other recipients of BCG, in whom immune mechanisms must be subtly different, the bacilli are not killed but are walled off with fibrous tissue at the point of invasion and are thus prevented from spreading to other sites. Post-mortem studies of such people disclose tiny healed foci in the lungs from which living tubercle bacilli may be recovered. Such persons have latent infection, capable of activation under immune suppressive conditions. It is very surprising that bacilli should apparently remain viable for years in such sites, since organisms in cultures left at 37°C soon die off in laboratory conditions. A possible explanation is that metabolites harmful to bacilli are removed in the tissues, whereas they build up in culture conditions. It is probable that mycobacteria, in common with some other Gram-positive genera, have the genetic capacity to switch on mechanisms giving protection during resting phases, and to switch them off again when conditions change.

Less effective immune responses which still give a degree of protection occur. In these, T-helper cells that recognize mycobacterial antigens are few and take time to produce effective numbers and for sufficient antibody to be produced to neutralize secreted bacterial products. This can then result in larger local lesions often spreading to local lymph nodes. In such situations the local response to bacillary invasion is different. Infected macrophages die with live bacilli still within them, as occurs with *M. ulcerans* where macrophage death is due to the macrolide toxin (P. Small: presented at the International Conference on Buruli Ulcer; Yamoussoukro, Cote d'Ivoire; July 6–9, 1998). Small capillaries are damaged by circulating TNF-α in the presence of interleukins released by Th2 cells responding to messengers of tissue damage such as stress proteins, rather than to the invading bacteria. As true antibacterial mechanisms develop, the disease progress that has started is aborted, but live bacilli may become walled off by fibrous tissue laid down around small areas of necrosis, again giving rise to latency. The classic example of this response to challenge is the Ghon focus, in which routine X-ray of a child in contact with an index case of tuberculosis shows a small opacity in the peripheral lung field with streaks of lymphatics marking drainage to an enlarged hilar lymph node. In most cases such foci heal without progressing to clinical disease. This subclinical infection in childhood can be detected in the adult as a calcified lesion 4–5 mm across in the periphery of the lung, together with some calcified spicules in a hylar lymph node. Similar involvement of lymphatics is seen in the 'sporotrichoid' spread of *M. marinum* from the initial skin lesion. Interestingly, swimming-pool or fish-tank granuloma is usually seen in previous recipients of BCG vaccine, typically young Europeans.

2.2 Vaccination with bacille Calmette–Guérin

This consists of intradermal injection of a small number of living tubercle bacilli rendered of low virulence by the process of attenuation. This was achieved some 70 years ago by taking a strain of bovine tubercle bacillus from a naturally infected calf, and passaging it on media containing bile salts for more than 10 years. A bovine strain was selected since it was thought that such organisms were less pathogenic for humans (Grange *et al.* 1983). At the end of the period of attenuation the bacilli were shown to have lost their virulence for calves and guinea pigs, and were used initially as an oral vaccine for children in France. Since that time the vaccine has experienced both successes and set-backs. When it is effective it provides long-lasting protection from leprosy and tuberculosis of up to 80% following a single dose. When it is unsuccessful it may give no protection from tuberculosis, or even increase susceptibility (Stanford *et al.* 1981). In recent times it has been shown to be protective in some situations against leprosy, where it is not effective against tuberculosis. Table 16.1 shows the results of several of the trials of BCG carried out against tuberculosis or leprosy (see also Chapter 17).

2.3 Where and when bacille Calmette–Guérin fails

Unfortunately, vaccination with BCG is not always followed by any degree of protective immunity, and in fact there is more than a theoretical possibility that susceptibility rather than protective immunity can be enhanced. The immunological process of developing tuberculosis appears to require a combination of both Th1 and Th2 mechanisms. Under these conditions tissue necrosis occurs around cells expressing mycobacterial antigens, probably as a result of TNF

Table 16.1 Results of some of the trials of vaccination with BCG against tuberculosis or leprosy.

Vaccination against tuberculosis		Vaccination against leprosy	
United Kingdom	78%	Uganda	80%
Alabama & Georgia	14%	Burma	18%
South India	<0%[a]	New Guinea	50%
South India	15%[b]	South India	30%[c]
Malawi	<0%	Malawi	49%

 The results shown are percentages of protection in comparison with randomized, non-vaccinated control groups.

[a] First results at 5-year follow-up.

[b] Results of 10-year follow-up.

[c] Results obtained in retrospect from the tuberculosis trial.

< Indicates that BCG apparently increased susceptibility.

becoming toxic in the presence of Th2 cytokines (Hernandez-Pando *et al.* 1997). The combination may also result in most of the circulating cortisone being in the active cortisol form, continually promoting Th2 mechanisms and compounding tissue necrotizing effects. This is illustrated by the massive necrotic response that may follow BCG immunotherapy for malignancies, and the development of arthritis similar to Poncet's syndrome. Once a Th2 response has developed to mycobacterial antigens, BCG enhances it, making the situation worse, increasing susceptibility and probably reducing bacteriostatic mechanisms that may be holding bacilli in a state of persistence in those with latent tuberculosis. Thus, BCG seems incapable of conferring protection to those with latent disease, or those who have developed Th2 mechanisms in response to excessive contact with certain environmental mycobacteria. Environmental species incriminated in this situation include *M. scrofulaceum*, *M. kansasii* and *M. intracellulare*, which were probably responsible for the failures of BCG in Burma and South India (Bechelli *et al.* 1974; Tuberculosis Prevention Trial Madras 1980). How could such persons be given protective immunity?

2.4 Bacille Calmette–Guérin and the prevention of other diseases

A number of studies have shown neonatal BCG vaccination to be effective in preventing childhood leukaemias and solid tumours (Davignon *et al.* 1970; Crispin and Rosenthal 1976), but this only seems to have occurred where BCG is also effective against tuberculosis. Where BCG fails to protect from tuberculosis, it also fails to protect from these cancers (Grange and Stanford 1990).

Recent work in Japan has shown that BCG vaccination resulting in long-lasting tuberculin positivity is associated with a reduced incidence of childhood allergies (Shirakawa *et al.* 1997). It is postulated that these effects and that on childhood cancer are due to enhanced TH1 responsiveness to stress proteins. On the other hand, the induction of a rheumatoid arthritis-like condition in BCG immunotherapy (Torisu *et al.* 1978) for leukaemias in Japan, and for carcinoma *in situ* in the United Kingdom, suggests enhanced TH2 responses (antibody) to these stress proteins. Many of these proteins have been conserved throughout evolution so that the structure and sequence of amino acids in some of the stress proteins of humans are very similar (60% or more) to the heat shock proteins of bacteria, including mycobacteria. They appear to have the same 'chaperone' functions in both humans and mycobacteria and have similar molecular weights. Thus, the 65 kDa heat shock protein of BCG is analogous to the 60 kDa stress protein of humans.

2.5 Summary

Prophylactic vaccination with BCG is successful when the individual being vaccinated has an immune system primed to make a Th1 response which can be maximized, resulting in rapid recognition of invading mycobacteria by their group I, common mycobacterial antigens (Stanford *et al.* 1981). When Th2 mechanisms predominate, the requirement is for preventive immunotherapy and for this BCG is inadequate.

3 Preventive immunotherapy

The problem of overcoming the prejudicial effects of excessive contact with slow-growing mycobacteria in the environment to enable protective immunity to be imposed may be the same as that of overcoming latent infection with tubercle bacilli. In either situation preventive immunotherapy is required since the existent immune status has to be changed before protection can be provided. Although undoubtedly simpler than the truth, the concepts of Th1 and Th2 (Mosmann *et al.* 1986) offer a means of expression of the requirement of which there are two parts. First, the switching off of Th2 mechanisms, and perhaps separately and secondarily switching on of Th1 mechanisms. The question as to whether these are in reality a single process or two separate ones has recently been addressed in experimental systems.

3.1 Switching off of Th2 and promoting Th1

That mycobacteria possess both capacities has been shown in experimental animal vaccination systems with the fast-growing species *M. vaccae*, and has been shown to occur in many clinical situations in humans (see below). In mouse studies production of IL-5 and IgE, markers of Th2 mechanisms in allergic responses, have been shown to be completely switched off by injection of a killed suspension of *M. vaccae*, a process abrogated when the allergen (ovalbumin) is injected along with the *M. vaccae*. On the other hand, induction of IFN-γ is increased by injection of the combination (Wang & Rook 1998).

One thing is certain, BCG does not cause a switch in Th1/Th2 predominance, but enhances whichever exists at the time that the vaccine is injected. Thus, sometimes it induces protective immunity and sometimes it does not. Indeed it may even increase susceptibility. The killed suspension of *M. vaccae* was originally designed as an additive to BCG to try to make it induce protection, whatever the immune status of the individual receiving an injection of the combination (Stanford 1991a). Thus, the combination should act to enhance protective immunity in those in whom BCG would be expected to be effective alone, and to induce protection in those in whom BCG alone would be ineffective (Ganapati *et al.* 1989; Stanford *et al.* 1989). Thus, the combination containing the non-specific adjuvant of cell-mediated immunity in BCG and the specific Th1 adjuvant properties of *M. vaccae* should act both as an immunotherapeutic agent for those previously uninfected with mycobacteria, and as a preventive immunotherapeutic in those with latent disease. Persuasive indirect evidence supports this contention. This includes evidence of enhanced cellular-immune recognition of group i, common mycobacterial antigens, persistent for at least 8–10 years in the children of Iranian leprosy patients (Stanford *et al.* 1989), and reduced post-vaccination response sizes to skin tests with tuberculin (Stanford & Eshetu Lemma 1983).

Subsequent studies have shown that a suspension of 10^8 killed *M. vaccae* given alone may be equally effective as when 10^7 *M. vaccae* are injected mixed with BCG (Stanford 1991a; Bottasso *et al.* 1998).

Clinical trials of preventive immunotherapy wait to be done, but the theoretical structure is in place and preliminary evidence supports the likelihood of its success.

3.2 Summary

Latent disease and enhanced susceptibility due to the induction of Th2 predominance by excessive contact with environmental, slow-growing mycobacteria both require the same corrective mechanism— switching off of Th2 and enhancement of Th1. BCG cannot achieve this but, when combined with a powerful Th1 adjuvant such as killed *M. vaccae* or when the latter is used alone, preventive immunotherapy should be achievable.

4 Immunotherapy for active disease

4.1 History of immunotherapy based on tubercle bacilli

The breakthrough, as in so many aspects of tuberculosis, was made by Robert Koch (1890) who introduced a treatment for tuberculosis apparently successful in a proportion of patients. This was based on repeated injections of increasing amounts of his 'brown fluid', now known to be tuberculin. Although dangerous, and even lethal in some patients (Anonymous 1890), Koch showed that changing the immune response to tubercle bacilli could be beneficial to the patient and this discovery sowed the seeds for immunotherapy. Almost simultaneously, Coley (1893) introduced immunotherapy for cancer using bacterial preparations. Both Koch's and Coley's methods appear to have achieved the same immunological end—the maximization of Th2. In the case of tuberculosis, this led to necrosis and walling off of infected tissue and, in the case of tumours, it led to massive necrosis of the malignant tissues. Forty years later Coley noticed that his 'toxins' had been more effective in the New York of the 1890s than they were in 1930. It has been suggested that this was because latent tuberculosis, with its association with Th2, was commoner at the earlier date (Starnes 1992).

Many others followed Koch, and the literature between 1890 and 1930 is littered with papers describing success with such attempts. All, or almost all, depended on maximizing necrosis to isolate and slow down bacillary replication by exaggerating the Th2 mechanism. Among the early papers on tuberculosis, however, there are some that are less easily explained. The benefits of Spahlinger's serum therapy seem unassailable (Macassey & Saleeby 1934). Very advanced cases of incontrovertible tuberculosis are described to respond and apparently achieve cure after repeated injections of these sera. Spahlinger's use of sera raised to heat-stressed tubercle bacilli (Spahlinger 1922) for 'surgical' tuberculosis was prophetic of the explosion of interest in stress, or heat-shock proteins that started in the 1980s (Young 1992). It is beginning to seem that regulation of immune responses to these shared proteins may be fundamental to protection from and susceptibility to many diseases, and may also be a key feature in their treatment.

Serum therapies were extensively investigated between 1910 and the development of chemotherapy in the late 1940s, although they received a severe check prior to this with the development of the dogma that antibodies had no part to play in the treatment of tuberculosis. Like most dogma, this has turned out to be an oversimplification and interest has recently rekindled in the part that antibodies may play in prevention of tuberculosis in infancy and childhood.

The modern proposal of using thalidomide in tuberculosis and to control the erythema nodosum of leprosy is based on control of TNF production (Moreira *et al.* 1993). A careful reading of the observations of Sechehaye on the use of Umkeloabo, or Steven's cure, in the early years of this century suggests that this so-called quack medicine was indeed doing the same thing (Sechehaye, *c.* 1916). Thus,

both the use of therapeutic vaccines and of immuno-manipulative drugs was born at the beginning of this century rather than at its end!

The major difference between Koch's treatment and the modern use of killed *M. vaccae* (SRL-172) is that the immunological aims are different. Whereas Koch attempted to maximize the mechanism of immunopathology, modern immunotherapy aims to switch this off and replace it with effective antibacterial mechanisms (Stanford 1991b). Thus, the conceptual framework for immunotherapy of active tuberculosis is the same as that for preventive immunotherapy of latent disease.

4.2 Immunotherapy with various mycobacterial species

The first person to develop a non-pathogenic mycobacterial species to treat tuberculosis was Friedmann. He had described the *Schildkröten-tuberkelbazillus* (now called *M. chelonei*), originally isolated from a diseased turtle in Berlin Zoo (Friedmann 1903). The pathology of the disease in the turtle was strikingly similar to that in humans, leading Friedmann to the reasonable conclusion that his bacillus possessed all the pathogenic attributes of the human tubercle bacillus except virulence for humans. Thus, he thought that it would provide an ideal vaccine and immunotherapeutic. He developed this as an intradermal injection of living bacilli that he called 'Anningzochen' (Friedmann 1958), which is still available, although now apparently in an autoclaved form. Like *M. vaccae*, as described below, Anningzochen had activities affecting many non-mycobacterial diseases.

4.3 Historical immunotherapy for leprosy

Koch made available some of his 'brown fluid' for use in leprosy patients as early as 1891 and this was investigated in a small number of leprosy patients in London (Abraham 1890), with claims for an effect. Subsequently a variety of mycobacteria and other organisms isolated from leprous tissues and the clothes, etc. of leprosy patients were used in attempted immunotherapy for the disease. Carried out prior to an understanding of the immunopathological spectrum of leprosy, these early studies are difficult to evaluate today. The first scientific investigation of immunotherapy for leprosy was carried out with a material called 'Antigen Marianin'. This consisted of a suspension of killed organisms of a species first isolated from leprous tissues and initially called *M. marianum* (Blanc *et al.* 1953). Subsequently, this name was considered confusingly similar to the earlier *M. marinum* and was replaced by the later synonym *M. scrofulaceum*. In some hands Antigen Marianin appeared to be highly successful (Ondoua *et al.* 1964) but in others it did little or nothing (Beckett 1958). Unfortunately the severe local reactions it induced in many patients, whether or not accompanied by clinical benefit, led to its use being dropped. In retrospect Antigen Marianin appears to have worked like Koch's tuberculin and enhanced Th2 mechanisms and the death of tissues containing leprosy bacilli. It is interesting to note that skin-test positivity to scrofulin (a new tuberculin) from contact with *M. scrofulaceum* in the environment appears to have undermined the vaccinating capacity of BCG in the Burmese trial area, perhaps by priming for a Th2 response to BCG (Shield & Stanford 1982; Shield 1983).

5 The development and use of *Mycobacterium vaccae*

This rapidly growing environmental organism was first recognized as of potential immunoregulatory value in the early 1970s, following its isolation from Ugandan soil (Stanford & Paul 1973) and demonstration, by skin testing, of its natural immunosensitizing capacity (Paul *et al.* 1975a,b; Stanford *et al.* 1978). It was hoped that a killed suspension might be added to BCG vaccine to ensure effective vaccination everywhere, and this may well be the case.

Preclinical investigation

Mouse experiments in London showed that adding live *M. vaccae* to their drinking water in advance of vaccinating with BCG enhanced the protective effect of vaccination (Stanford *et al.* 1981). In the same series of experiments it was shown that replacing *M. vaccae* with *M. scrofulaceum* in the animals' drinking water entirely altered the immune response following BCG vaccination (Rook *et al.* 1981), apparently reducing its protective effect. *M. scrofulaceum* had been identified as being one of the environmental mycobacteria blocking the protective effect of BCG against leprosy in Burma. These were antecedent observations of the subsequently described Th1 and Th2 mechanisms, respectively (Mosmann *et al.* 1986). A second stage in the experiment showed that the same respective results could be achieved by adding killed *M. vaccae* or killed *M. scrofulaceum* to BCG and injecting the combinations.

In studies of experimental *M. ulcerans* infection in the mouse, prior injection of killed *M. vaccae* could modulate the progress of subsequent disease. The effects of different doses of killed *M. vaccae* in the mouse were titrated using a simple model based on the known dose of live *M. leprae* required for cellular immune recognition being 10^6. Mice vaccinated a month previously with 10^5, 10^6, 10^7, 10^8 or 10^9 *M. vaccae* were injected with 10^4 *M. leprae* and 3 months later they were footpad tested with leprosin A, a soluble skin-test reagent prepared from leprosy bacilli harvested from the tissues of experimentally infected nine-banded armadillos. Maximal footpad swelling 24h later was recorded from the animals vaccinated with 10^7 *M. vaccae*. Recent studies have confirmed this as the optimal dose and have elaborated and partially explained the basic observations (Hernandez-Pando *et al.* 1997).

Preliminary clinical evaluation

Two different types of investigative reagents were prepared from a Ugandan stain of *M. vaccae* R877R, NCTC 11659, grown on solidified Sauton's medium; a skin-test reagent, vaccin, by the new tuberculin method (Shield *et al.* 1977), and a suspension of the organism in 66 mmol/L borate-buffered saline at pH8.0. Early batches of this reagent were sterilized by irradiation, and later batches by autoclaving. These reagents were used in the preliminary investigations in humans.

Vaccine was used as one of four new tuberculins made from different mycobacterial species in a system of quadruple skin testing. Using this system, not only was it possible recognize persons who had experienced immunological recognition of different species, but to divide people into different responder categories (Lockwood *et al.* 1987; McManus *et al.* 1988). These separated people according to whether they could produce positive skin-test responses to mycobacterial antigens of groups i, ii or iv (Fig. 16.2), and have proved a valuable tool in dissecting the immune response to mycobacteria. Categorization has also been valuable in detecting the immune status of individuals with a range of diseases such as rheumatoid arthritis, human immunodeficiency virus (HIV), psoriasis and the silent seropositive phase of infection with *Trypanosoma cruzi* infestation. A further use of the quadruple skin-test system has been as a surrogate marker of protective immunity before or after vaccination procedures (Stanford & Eshetu Lemma 1983; Stanford *et al.* 1989; Bottasso *et al.* 1998).

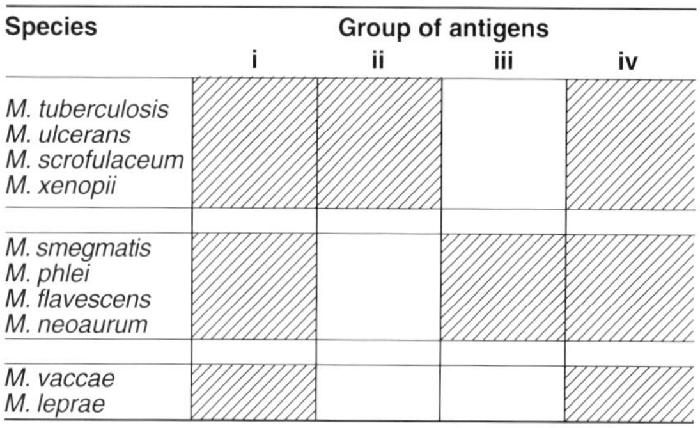

With the discovery that footpad or skin-test positivity to mycobacterial antigens, respectively, in mouse and human, could be by two different pathways indicating different immune situations, investigation of immunotherapy could be undertaken with mixed skin tests. By this means it has been shown that mixtures of new tuberculins prepared from different species can be used to demonstrate two types of regulation (Nye *et al.* 1983). One of these is local, only affecting the response to the mixture and the other works at a distance modulating the response to reagents injected into the skin of the other arm. Reagents made from any fast-growing mycobacterial species can locally regulate responses to slow-growers but only certain fast-growing species can influence distant responses (Morton *et al.* 1984). Amongst such species is *M. vaccae* NCTC 11659, and vaccine derived from it was fractionated to try to separate local and distant immunoregulatory fractions. Most fractions could exert a local effect, but only two fractions contained the distantly active moiety (Nye *et al.* 1986). By extrapolation, if an intradermal injection given on one arm can regulate responses on the opposite arm, why not in the deep tissues at the site of tuberculous lesions?

Species	Group of antigens			
	i	ii	iii	iv
M. tuberculosis M. ulcerans M. scrofulaceum M. xenopii	▨	▨		▨
M. smegmatis M. phlei M. flavescens M. neoaurum	▨		▨	▨
M. vaccae M. leprae	▨			▨

Group i, the common mycobacterial antigens are shared by all mycobacterial species and include the bacterial heat shock proteins, some of which share marked amino-acid sequence homology with the human stress proteins.
Group ii, the slow-grower associated antigens are shared by all slow-growing mycobacterial species, are heat stable and the cause of cross-reactivity between tuberculins.
Group iii, the fast-grower associated antigens are shared by almost all fast-growing mycobacterial species.
Group iv, the species-specific antigens are limited to individual species, and may reflect sub-specific and serotype variation.

Fig. 16.2 Mycobacterial antigens demonstrable in double diffusion tests against rabbit antisera.

Clinical evaluation

1 The first studies were in three volunteers with histories of confirmed or probable past tuberculosis. A man in his 50s treated 20 years earlier for pulmonary tuberculosis was found to have a small, well-regulated response to tuberculin, which, 3 months after an intradermal injection of 10^7 irradiated *M. vaccae*, was unaltered. The two others, in their late 20s with large and florid responses to tuberculin, each received an injection of irradiated *M. vaccae* and 3 months later their responses to tuberculin were small and well-regulated, quite different from how they had started. This too is a model of potential immunotherapeutic influence.

2 A small group of in-patients receiving treatment for pulmonary tuberculosis in the Middlesex Hospital in London, volunteered to receive injections of irradiated *M. vaccae* or saline, and to be investigated by repeat skin testing (Pozniak *et al.* 1987). The results of this investigation is shown in Table 16.2.

3 The next studies were carried out in volunteers amongst chronic leprosy patents in Sanatorio Fontilles in Spain (Stanford *et al.* 1987). Patients with an original diagnosis of multibacillary disease, and unable to respond to Mitsuda lepomin, or leprosin A, were given injections of 10^7, 10^8 or 10^9 *M. vaccae* and a year later they were retested with leprosin A. Although the two lower doses were ineffective, more than a third of those receiving the largest dose developed positivity to leprosin A. This could be improved further so that about half of patients became leprosin A positive by adding a low dose of tuberculin to the irradiated *M. vaccae*. This is shown in Table 16.3. When a number of these patients receiving an injection of *M. vaccae* were retested with lepromin some 10 years later, some of them were now positive by the Mitsuda reaction!

4 Studies were set up to investigate the effect of an injection of *M. vaccae* or saline given to volunteer patients in Kuwait, 1 month after they had started short course chemotherapy for pulmonary tuberculosis (Bahr *et al.* 1990; Stanford *et al.* 1990). Re-cognition of group i, common mycobacterial antigen was assessed by lymphocyte proliferation 2 months later. The best recognition of these antigens was achieved after an injection of 2×10^9 irradiated *M. vaccae* but this reagent gave a more severe local reaction and has not been investigated further. The dose of 10^9 autoclaved *M. vaccae* was better than the irradiated preparation both in this study and in animal experiments, and has been used in all later studies (Table 16.4).

5.1 Phase II trials of immunotherapy for tuberculosis

With different regimens of chemotherapy

This randomized, blind, placebo-controlled trial was performed in Banjul in the Gambia, under the auspices of the Medical Research Council of the UK (Corrah 1994). Patients newly diagnosed with pulmonary tuberculosis with sputum smear or culture positive for acid-fast bacilli were enrolled to receive a single injection of 10^9 irradiation-killed *M. vaccae* NCTC 11659, or saline, 6 weeks after starting a full course of either standard chemotherapy for 18 months, or short-course chemotherapy for 6 months. Standard chemotherapy consisted of daily streptomycin, isoniazid and thiacetazone for 2 months fol-

Table 16.2 Skin-test responses to tuberculin after an injection of 10^9 irradiation-killed *M. vaccae* or saline of patients receiving chemotherapy for pulmonary tuberculosis in London.

No.	Before saline	One month later	No.	Before *M. vaccae*	One month later
1	25 mm (K)[a]	21 mm (K)	1	24 mm (K)	10 mm (L)
2	25 mm (K)	28 mm (K)	2	15 mm (K)	7 mm (L)
			3	20 mm (K)	14 mm (L)
			4	11 mm (K)	8 mm (L)
			5	17 mm (K)	12 mm (L)
			6	28 mm (K)	11 mm (L)

[a] K refers to a 'Koch-type' skin-test reaction with a necrotizing component. L refers to a 'listeria-like' reaction that is soft and lacks a necrotic component. These are outdated terms referring to responses with or without a Th2 component, respectively. The numbers shown are for individual patients receiving either *M. vaccae* or placebo.

Table 16.3 Skin-test responses to antigens of *M. leprae* in leprosin A one to 2 years after injection of various preparations of irradiation-killed *M. vaccae* in volunteers with fully treated multibacillary leprosy in Spain.

Intervention	Response to leprosin A	
Saline	1/29 (3.4%)	
10^7 irradiated *M. vaccae*	2/28 (7.1%)	
10^8 irradiated *M. vaccae*	0/4	
10^9 irradiated *M. vaccae*	10/29 (34.5%)	22/82 (26.8%)*
10^9 irradiated *M. vaccae* + tuberculin	10/21 (47.6%)	

*$p < 0.005$ from saline intervention.

Table 16.4 Results from patients with pulmonary tuberculosis volunteering to enter a study of immunotherapy in Kuwait. Each received an injection of saline or of one of several different preparations of killed *M. vaccae* one month after starting chemotherapy. The numbers of patients are shown who produced lymphoproliferative responses to common mycobacterial antigens in blood samples taken two months after intervention.

Intervention	Response to group i antigen	
Saline	5/49 (11%)	
10^8 irradiated *M. vaccae*	2/8 (25%)	
10^9 irradiated *M. vaccae*	11/38 (29%)	32/80 (40%)*
2×10^9 irradiated *M. vaccae*	9/12 (75%)	
10^9 autoclaved *M. vaccae*	10/22 (45%)	

*$p < 0.0002$ from saline intervention.
Although the dose of 2×10^9 irradiated *M. vaccae* gave the best result, the local response to injection was more severe.

Table 16.5 Results obtained in the Gambian trial of immunotherapy with *M. vaccae* against tuberculosis. Patients were considered cured if their sputum no longer contained tubercle bacilli and if clinical and radiological parameters of their disease improved.

Intervention	*M. vaccae*	Placebo
Standard chemotherapy[a]		
Cured	43/56 (76.8%)	50/76 (65.8%)
Died	4/56 (7.1%)	12/76 (15.8%)
Short-course chemotherapy[b]		
Cured	53/59 (89.8%)	49/60 (81.7%)
Died	2/59 (3.4%)	4/60 (6.7%)
BCG + ve[c] standard chemotherapy		
Cured	16/19 (84.2%)	25/27 (92.6%)
Died	0/19	0/27
BCG + ve short-course chemotherapy		
Cured	19/21 (90.5%)	13/17 (76.5%)
Died	0/21	0/17
BCG-ve[d] standard chemotherapy		
Cured	27/37 (73%)	25/49 (51%)
Died	4/37 (10.8%)	12/49 (24.5%)
BCG-ve short-course chemotherapy		
Cured	34/38 (89.5%)	36/43 (83.7%)
Died	2/38 (5.3%)	4/43 (9.3%)

[a] 18 month regimen based on streptomycin.
[b] Six month regimen based on rifampicin.
[c] Denotes individuals with a scar of past BCG vaccination.
[d] Denotes those without evidence of past BCG vaccination.

lowed by daily isoniazid and thiacetazone for 16 months. Short-course chemotherapy consisted of thrice-weekly rifampicin, isoniazid, pyrazinamide and ethambutol for 2 months and rifampicin and isoniazid for 4 months. It was planned to enrol 150 patients to receive immunotherapy and the same number to receive a placebo. The purpose of the study was to evaluate the practical part that immunotherapy might play in treatment of the disease. Unfortunately, in retrospect, recruitment was stopped after 119 patients had received immunotherapy and 145 had received placebo which meant that, although there were many interesting trends, few of them quite reached statistical significance. One hundred and forty-one patients received standard chemotherapy and 123 received the short course. The most important outcomes of the study are shown in Table 16.5.

After the time of intervention, none of the 85 patients who had a scar of past BCG vaccination had died, whereas 13.2% of those without such a scar had died during the course of chemotherapy. More died amongst those receiving standard chemotherapy (18.6%) than amongst those receiving short-course chemotherapy (7.4%). Many deaths occurred between 6 and 18 months of standard treatment. Since follow-up after the end of chemotherapy was incomplete, it is not known whether this would have uncovered the same death rate amongst those receiving the short-course regimen in the year following its completion. Of those without a BCG scar receiving the intervention and followed up to the end of chemotherapy, fewer died amongst those injected with *M. vaccae* (6/75) compared with the placebo group (16/92). Thus, it seemed that the immunotherapy could replace the advantage of prior BCG vaccination.

This trial was performed concurrently with the studies in Kuwait, described above, in which killing *M. vaccae* with heat (autoclaving) produced a more effective reagent than killing it with irradiation. Subsequently, the trials outlined below employed the autoclaved product.

On patients with drug resistance, chronic relapse or treatment failure

Two randomized, part-blind, placebo-controlled studies were set up in Romania in the cities of Bucharest and Brasov; one to evaluate immunotherapy in newly diagnosed pulmonary tuberculosis (Corlan *et al.* 1997a) and the other in chronic patients relapsing repeatedly or failing chemotherapy (Corlan *et al.* 1997b). All patients received a short course, intermittent (twice-weekly) regimen of streptomycin, isoniazid, rifampicin and pyrazinamide for 2 months and isoniazid and rifampicin for 4 months. Immunotherapy or placebo was given as a single intradermal injection after 1 month of chemotherapy. These studies clearly indicate the advantages of immunotherapy in the treatment of chronic disease and show beneficial trends in patients with newly diagnosed disease, as shown in Tables 16.6a and b.

Table 16.6a Bacteriological, body weight and erythrocyte sedimentation rate results for the two studies carried out in Bucharest and Brasov in Romania.

Intervention:	New cases		Chronic cases	
	M. vaccae	Placebo	*M. vaccae*	Placebo
1 month after intervention				
Microscopy positive	11/88 (13%)	21/100 (21%)	16/53* (30%) $p < 0.02$[a]	25/45* (56%)
Culture positive	14/97 (14%) ($p = 0.08$[a])	26/107 (24%)	25/56 (45%) $p = 0.008$[a]	32/45* (72%)
Body weight (increase)	59.6 ± 9.2 (3.1 kg)	58.7 ± 9.6 (2.6 kg)	59.7± 9.1 (1.3 kg)	55.8 ± 10.5 (1.9 kg)
ESR	27.4 ± 21.4	31.3 ± 23.7	34.9 ± 30.3 $p < 0.02$[b]	44.4 ± 24.6
5 months after intervention				
Microscopy positive	4/85 (5%)	8/96 (8%)	7/47 (11%) $p < 0.02$[a]	14/43 (30%)
Culture positive	2/91 (2%) ($p = 0.06$[a])	9/105 (9%)	10/50 (20%)	12/43 (28%)
Body weight (increase)	60.9 ± 9.5 (4.4 kg)	59.4 ± 9.5 (3.3 kg)	62.3 ± 9.5 (3.9 kg)	57.6 ± 8.7 (3.7 kg)
ESR	16.7 ± 12.1 $p = 0.002$[b]	24.9 ± 19.5	26.0 ± 20.5 $p < 0.02$[b]	38.0 ± 24.3
11 months after intervention				
Microscopy positive	3/89 (3%)	2/93 (2%)	1/39* (3%) $p < 0.02$[a]	7/28 (25%)
Culture positive	2/89 (2%)	2/93 (2%)	1/39* (3%) $p < 0.02$[a]	7/28 (25%)
Body weight (increase)	63.6 ± 9.9 (7.1 kg)	60.5 ± 9.0 (4.4 kg)	64.5 ± 9.6 (6.1 kg)	58.5 ± 9.2 (4.6 kg)
ESR	13.5 ± 10.1 $p < 0.0001$[b]	22.4 ± 18.6	19.9 ± 15.3 $p < 0.001$[b]	41.2 ± 22.7

[a] Two-tailed Fisher's exact test.
[b] Student's *t* test.
ESR, erythrocyte sedimentation rate.

Table 16.6b Radiological results for the studies carried out in Romania.

Intervention	New cases		Chronic cases	
	M. vaccae	Placebo	*M. vaccae*	Placebo
At entry				
Nos with cavities	79/97 (81%)	84/109 (77%)	50/56 (89%)	41/45 (89%)
Cavity surface area (cm^2)	54.6 ± 66.5	43.1 ± 38.6	83.7 ± 99.7	63.7 ± 56.1
Mean lesional score	2.1 ± 0.7	2.1 ± 0.6	2.4 ± 0.6	2.5 ± 0.5
5 months after intervention				
Still with cavities	47/73 (64%)	50/77 (65%)	32/47 (68%)	36/43 (84%)
Cavity surface area (cm^2)	26.2 ± 34.0	31.9 ± 33.4	47.2 ± 56.8	50.6 ± 46.8
Mean lesional score	1.4 ± 0.6	1.6 ± 0.6	1.7 ± 0.7 $p < 0.01$[b]	2.1 ± 0.7
11 months after intervention				
Still with cavities	19/72 (26%)	22/77 (31%)	17/45 (38%) $p < 0.01$[a]	22/32 (69%)
Cavity surface area (cm^2)	28.6 ± 44.0	28.4 ± 38.2	32.7 ± 31.2	45.4 ± 39.4
Mean lesional score	1.2 ± 0.5	1.3 ± 0.6	1.4 ± 0.6 $p < 0.001$[b]	2.0 ± 0.7

[a] Two-tailed Fisher's exact test.
[b] Student's *t* test.

Parameters that best showed the advantages of immunotherapy were clearance of the bacilli from the sputum, regain of body weight, fall in erythrocyte sedimentation rate (ESR) and chest radiology.

On chronic multidrug-resistant tuberculosis

This was an open study to evaluate immunotherapy with multiple doses of autoclaved *M. vaccae* performed in an hospital in Mashad, Iran, at a time when only first-line antitubercular chemotherapy was available (Etemadi *et al.* 1992; Farid *et al.* 1994). Most patients' bacilli were resistant to these drugs. Against an historical cure rate in such patients of no more than one in 100 with the chemotherapy available, 11 out of 41 were bacteriologically cured with a combination of chemotherapy and up to four doses of immunotherapy. An interesting finding was that the four successes after a single dose of immunotherapy occurred in patients with a shorter history of chemotherapy (14.5 ± 2.6 months) than in the seven successes achieved after three or four doses (81.1 ± 37.1 months; $p < 0.01$). Chemotherapy was continued for 6 months after their sputum became negative and none relapsed during the 18 months after finishing treatment. Table 16.7 shows the results obtained in those successfully cured compared with the treatment failures.

With grossly inadequate chemotherapy

To evaluate immunotherapy in routine practice in the poor conditions frequently found in Africa, this randomized, placebo-controlled, part-blind study of single-dose immunotherapy with *M. vaccae* was carried out at the Infectious Diseases Hospital in Kano, Nigeria (Onyebujoh *et al.* 1995). At that time only streptomycin was available for most of the year, and sometimes isoniazid. Other drugs had to bought by patients at high prices from private dispensaries. Often these purchased drugs were fake, out of date and had been stored at high ambient temperatures. Against this background, the immunotherapy was very effective, with the placebo group doing no better than would have been expected without any treatment given at all, as shown in Table 16.8a.

A further important aspect of this study was that nearly 20% of patients were found to be seropositive

Table 16.7 Results obtained in 41 patients with multidrug-resistant tuberculosis receiving a combination of up to four injections of *M. vaccae* in Mashad, Iran.

	Successes		Failures
Numbers	11		15
Age	40.5 ± 13.1 years	ns	37.1 ± 13.0 years
Duration of disease	58 ± 46 months	ns	69 ± 48 months
Body weight increase	5.1 ± 3.6 kg	$p < 0.005$	1.6 ± 1.9 kg
Fall in ESR	32.1 ± 19.7 mm	$p < 0.001$	6.5 ± 9.2 mm

ESR, erythrocyte sedimentation rate; ns, not statistically significant.
Four out of 41 were cured after one dose of *M. vaccae*, three more after three doses and four more after four doses. The intervals between doses were 6 months, 2 months and 2 months. During the course of the study seven patients died and eight were lost to follow-up.

Table 16.8a Results of follow-ups of newly diagnosed tuberculosis patients, 20 days and 10–14 months after receiving an injection of autoclaved *M. vaccae* or saline 1–3 weeks after starting very inadequate courses of antituberculosis drugs in Kano, Nigeria.

Intervention	*M. vaccae*		Placebo
20 days after intervention			
Positive microscopy	20/75 (26.6%)	$p < 0.00001$	53/65 (82%)
Body weight increase	2.9 kg	$p < 0.0001$	0.55 kg
Fall in ESR	25.4 mm	$p < 0.001$	4.0 mm
10–14 months after intervention			
Positive microscopy	11/33 (33%)	$p < 0.00002$	22/26 (84.6%)
Body weight increase	7.91 kg	$p < 0.003$	2.04 kg
Fall in ESR	42.0 mm	$p < 0.001$	15.0 mm
Survival	34/34 (100%)	$p < 0.00001$	28/47 (59.6%)

by two enzyme-linked immunosorbent assay (ELISA) tests and western blotting to HIV-1. Table 16.8b shows the results for these HIV-seropositive patients, and it can be seen that the immmunotherapy for tuberculosis was equally effective whatever the HIV status (Stanford *et al.* 1993; Onyebujoh *et al.* 1995).

5.2 Assessment of immunotherapy

As a result of the studies outlined above, a number of parameters have been identified for assessment of the effects of immunotherapy.

Bacteriological

The disappearance of viable tubercle bacilli from the sputum is spoken of as the 'gold standard' for efficacy of chemotherapy and is imposed on immunotherapy,

Table 16.8b Results for patients seropositive for HIV-1 in Kano.

Intervention	*M. vaccae*	Placebo
At entry		
Numbers	8	9
Lymphadenopathy	5/8	7/9
Positive microscopy	8/8	9/9
Starting weight	44.75 kg	49.5 kg
ESR	69.8 mm	92.5 mm
10–14 months after intervention		
Survival	8/8	3/9
Lymphadenopathy	0/8	3/3
Positive microscopy	1/8	3/3
Body weight increase	5.5 kg	0 kg
ESR	25.25 mm	54.5 mm

ESR, erythrocyte sedimentation rate.

probably wrongly. As a measure of immunotherapeutic activity it may be less efficient. Not that immunotherapy does not also aim to sterilize the sputum, the mechanisms of its action may be slower, yet more thorough, than those of antitubercular drugs. Whereas the majority of antitubercular chemotherapeutic agents act directly on the bacilli to kill them or prevent their replication whether they be within or without tissue cells, immunotherapy works on the hosts immunity and is likely to be most effective against intracellular organisms. However, that does not mean that necrotic or cavitating lesions lie totally outside the influence of immunotherapy since the mechanisms of necrosis are based on the presence of a Th2 component that is switched off by immunotherapy. Once the propensity for further necrosis stops, tissue repair commences, bacilli are phagocytosed and become susceptible to immune mechanisms. Thus, speed of sputum conversion is not the ideal measure of immunotherapeutic efficacy, but nevertheless effects are seen, as shown in Tables 16.6a and 16.8a.

Erythrocyte sedimentation rate

This crude and non-specific indicator of disease depends on the concentration of proteins in the plasma. As shown in Tables 16.6a, 16.7 and 16.8, it too has been a useful indicator of immunotherapeutic benefit, returning to normal values faster than in patients receiving placebo or chemotherapy alone.

Regain of body weight

Marked loss of body weight is a frequent, but by no means invariable, component of clinical tuberculosis, thought to be due to a toxic effect of circulating TNF (cachectic factor) in the presence of Th2 cytokines. Recovery of weight in such individuals normally takes place steadily over the period of chemotherapy. The rate of weight regain is enhanced in those receiving immunotherapy, as shown in Tables 16.6a, 16.7 and 16.8, and by Mayo (1999).

Radiological improvement

This is a somewhat difficult parameter to measure, but several studies have shown faster resolution of cavities and opacities in immunotherapy recipients (see Table 16.6b), seemingly with less residual fibrosis. This is likely to reflect regulation of IL-6 activity in the lung.

Resolution of symptoms

Although somewhat difficult to quantify, it has been the observation of clinicians that patients receiving immunotherapy show rapid symptomatic relief and may be able to leave hospital earlier and return to work earlier (Onyebujoh *et al.* 1995). This phenomenon is likely to be due to the reduction in toxicity of circulating TNF associated with switching off of the production of Th2 cytokines. Thus, fever resolves faster, and cough and chest pain may follow the same pattern.

5.3 Phase III good clinical practice trial in newly diagnosed tuberculosis

Only one phase III trial of immunotherapy for tuberculosis has been completed so far and that was carried out to the modern standard of good clinical practice (GCP). It has been run in Durban, South Africa. All patients enrolled were smear-positive and culture-positive for *M. tuberculosis*, and those analysed were infected with bacilli fully sensitive to rifampicin, isoniazid, ethambutol and pyrazinamide with which they were treated. Intervention with an injection of 10^9 autoclaved *M. vaccae* or buffered saline was given after 8 days of chemotherapy. About a third of the patients were seropositive for HIV-1, and results were analysed for HIV-positive and -negative patients both separately and together.

In this study in which patients were hospitalized for the first 8 weeks of their chemotherapy, during which compliance with treatment was very high, the primary and secondary endpoints measured at the end of the 8 weeks showed no obvious advantage for the addition of immunotherapy (Durban

Immunotherapy Trial Group 1999). Acid-fast bacilli persisted just as long in the sputum, body weight was regained equally well in both groups and ESR fell at similar rates in both groups. Radiological assessment after 2 months of treatment showed no differences between the study intervention groups.

5.4 Subsequent phase I/II trial in Uganda

The results were recently released (June 1998) of this American trial run to the standards of GCP in Uganda. It was performed on patients with sputum smear and culture positive, newly diagnosed pulmonary tuberculosis, selected to be HIV seronegative, and to be infected with drug-susceptible bacilli.

The results show a significant reduction ($p<0.03$) in sputum culture positivity 28 days after administration of *M. vaccae*. The immunotherapy also resulted in significant radiological improvement ($p<0.02$) after 6 months (end of chemotherapy) and after 1 year ($p<0.05$).

5.5 Further evaluation of immunotherapy for tuberculosis

The studies and trials described above show that the addition of immunotherapy to a full course of chemotherapy measurably improves the treatment for those with newly diagnosed pulmonary tuberculosis caused by fully drug-sensitive bacilli in some countries but perhaps not in others. Nevertheless, analyses of sera from the Durban trial show that significant changes in antibody levels have taken place in those receiving the immunotherapy. Other studies show that when conditions are not optimal, for example when the infecting bacilli are drug resistant, when the patient has repeatedly relapsed, or failed to be cured, and when good treatment is either not available or not taken, then immunotherapy plays an important role.

Drug-resistant tuberculosis

Studies in individual patients show that many who have repeatedly failed to respond to the available

chemotherapy, can be bacteriologically cured by repeated doses of immunotherapy. In a study of such patients in Vietnam, 10 out of 11 were successfully treated with chemotherapy plus up to 12 doses of *M. vaccae* given at 2-month intervals.

Fluorescence flow cytometry to determine the proportion of cells that can make IL-2, IL-4 or IFN shows significant anomalies that are corrected by immunotherapy. Prior to immunotherapy, such patients have an higher percentage than normal of both CD4+ and CD8+ cells producing the Th2 cytokine IL-4, and a lower percentage than normal of cells producing the Th1 cytokines IL-2 and IFN. This situation is corrected after one or two injections of *M. vaccae*, as shown in Table 16.9a.

Abbreviated chemotherapy

Because of poor prescription, lack of drugs, transport difficulties or patient non-compliance, chemotherapy can fail even when the bacilli are fully drug sensitive. Under these conditions, common in the developing world, immunotherapy offers major advantages as shown in the Kano study. In order to investigate this further, studies have been carried out in Vietnam in which their usual streptomycin-based, 9-month treatment regimen has been compared with the same regimen shorted to 6 months and 4 months, with an injection of *M. vaccae* given on the first day of treatment. These studies have been followed for up to 18 months after the end of chemotherapy and no differences were found in rates of treatment success. It is hoped to carry out a phase III trial to confirm this. If successful, then immunotherapy should enable a marked reduction in regimen with all the major benefits that this would have.

5.6 Immunotherapy for opportunist mycobacterial infections

Infections with opportunist mycobacteria in HIV-seronegative patients may affect the lymph nodes, the skin or the lungs. Many opportunist species are naturally resistant to the usual antitubercular drugs, and often respond poorly to macrolides and quinolones.

Table 16.9 Fluorescence activated cytometry for CD4+ and CD8+ T lymphocytes producing IFN-γ, IL-2 and IL-4 in peripheral blood: effect of injection of heat-killed *M. vaccae*.

(a) Samples from a patient with multidrug-resistant tuberculosis

	Initial sample	After 2 weeks	After 2 months	After 3 months
CD4+				
IFN-γ	4.6%	9.6%	36%	39%
IL-2	3%	5.5%	14.5%	15%
IL-4	6.1%	0.6%	0.8%	0.6%
CD8+				
IFN-γ	21%	14%	52%	50%
IL-2	1.6%	0.6%	1.1%	1%
IL-4	1.4%	1.1%	2.3%	2.1%

(b) Two patients with opportunist mycobacterial disease of the lungs

Patient 1 — Wegener's granulomatosis + M. intracellulare *infection*

	Initial sample	After 1 month	After 2 months	After 3 months
CD4+				
IFN-γ	2.0%	7.9%	20%	28%
IL-2	4.1%	11%	8.9%	15%
IL-4	6.7%	4.6%	3.8%	2.1%
CD8+				
IFN-γ	6%	23%	25%	59%
IL-2	1.3%	3.3%	1.4%	6.8%
IL-4	3.7%	4.1%	2.7%	1.8%

Patient 2 — asthma + M. xenopi *infection*

	Initial sample	After 2 weeks	After 2 months	After 3 months
CD4+				
IFN-γ	38%	48%	54%	41%
IL-2	22%	21%	20%	22%
IL-4	9.5%	4.2%	2.2%	1.8%
CD8+				
IFN-γ	13%	20%	43%	17%
IL-2	0.2%	5.9%	4.5%	6.2%
IL-4	4.1%	0.9%	1.6%	1.1%

Injections of 10^9 heat-killed *M. vaccae* were given after taking each sample. Percentages are shown of CD4+ and CD8+ T-cells producing IFN-γ, IL-2 or IL-4, after stimulation with phorbol myristic acid, mitomycin-C and monensin. (The data for this table was provided by Navin Thapa, PhD student with JLS.)

Cervical lymphadenitis of children

This disease occurring in children usually aged between 1 and 6 years who have not received BCG vaccination, is frequently caused by *M. avium*, *M. intracellulare*, *M. malmoense* or *M. scrofulaceum*. The disease is usually treated by surgical removal, with or without antimycobacterial chemotherapy. Sometimes the disease recurs after surgery, fails to respond to chemotherapy or is in a situation where surgical removal would be difficult. The few cases of this type that have been treated with a single dose of *M. vaccae* have resolved very well, and a trial of immunotherapy for this disease is about to start.

Swimming-pool or fish-tank granuloma

This superficial infection with *M. marinum* usually responds to a short course of broad-spectrum antibiotics without the need for immunotherapy.

Opportunist mycobacterial infections of the lungs

These occur almost exclusively in patients over 40 years of age, and are often associated with some particular exposure to heavily contaminated aerosols, some debilitating condition or long-term use of steroids. A proportion, however, occur in seemingly healthy individuals without any predisposing condition or previous lung damage. The mycobacterial species involved in the UK include *M. kansasii*, *M. intracellulare*, *M. malmoense*, *M. xenopi* and *M. avium* in this descending order of frequency. Except for *M. kansasii*, which responds to regimens including rifampicin and ethambutol, the others are often very difficult to cure. Two trials of combinations of drugs, with or without four doses of immunotherapy with *M. vaccae* at 2-month intervals, are currently being conducted by the British Thoracic Society. Anecdotal cases have responded well to immunotherapy, and cytokine studies show this to be associated with switching off Th2 and enhancing Th1 cytokine production, as shown in Table 16.9b.

5.7 Immunotherapy for leprosy

Leprosy is the only mycobacterial disease in recent times that has received immunotherapeutic attention from several research groups, notably from Bombay (Deo *et al.* 1983), Delhi (Chaudhuri *et al.* 1983), Caracas (Convit *et al.* 1982) and our group in London (Stanford 1994). All four groups have developed approaches based on whole-cell preparations of different mycobacteria. Those in Bombay have developed an organism called the ICRC (Indian Cancer Research Centre) bacillus, which is a strain of *M. intracellulare* grown on a special medium originally developed for eucaryote cell culture. The group in Delhi have selected a similar organism called *Mycobacterium 'W'*, also a strain of *M. intracellulare*, which they selected on the basis of lymphoproliferation studies. The Venezuelan group have used armadillo-grown *M. leprae*, with or without the addition of BCG vaccine, and our group has used *M. vaccae*. Whereas the Indian researchers have used Mitsuda lepromin conversion as a marker of potential immunotherapeutic activity, the South Americans have used a soluble extract of leprosy bacilli producing a reaction similar to the Fernandez reaction, and we have used tuberculin-type skin-test conversion to leprosin A (see Table 16.3).

Skin-test reagents used in leprosy

Mitsuda lepromin is prepared from autoclaved leprosy bacilli extracted from biopsies of heavily bacilliferous tissue from untreated lepromatous leprosy patients or from experimentally infected nine-banded armadillos (*Dasypus novemcinctus*). Injection of this material into the skin produces two types of reaction, one maximal at 48 h known as the Fernandez reaction, which is positive in many individuals who may or may not have ever contacted leprosy patients, and the other is a small granuloma appearing at the injection site 2–4 weeks later. This is called the Mitsuda reaction which is positive in paucibacillary (TT-BT) leprosy patients and negative in multibacillary (BT-LL) patients, and in healthy persons may be an indicator of previous contact with leprosy bacilli.

The soluble antigen used by the Venezuelan group (Convit *et al.* 1975) is a filtrate of a Mitsuda-type lepromin prepared from heat-killed bacilli extracted from the tissues of experimentally infected eight-banded armadillos (*Dasypus sabanicola*). When injected into the skin this material induces a 48-h Fernandez reaction, and only rarely a Mitsuda reaction.

As mentioned above, leprosin A prepared in London by R. J. W. Rees and colleagues (Shield *et al.* 1982) is a sonicated preparation of irradiation-killed leprosy bacilli extracted from the tissues of experimentally infected *Dasypus novemcinctus*. This material, prepared as a new tuberculin, produces a tuberculin-type skin reaction at 72 h, largely but not exclusively to its content of group i, common mycobacterial antigens.

The effects of immunotherapy in leprosy

Within a little time the effects of immunotherapy appear to have been the same with each of the preparations in these days of multiple drug therapy (MDT) for leprosy. Following immunotherapy there is an enhanced clearance of bacilli from the tissues of multibacillary patients which may be of value in preventing recurrent erythema nodosum leprosum (ENL) reactions (Ramu *et al.* 1998). There is induction of immune responsiveness to antigens of *M. leprae* in lepromatous patients, seemingly to both group i, common mycobacterial and group iv, species-specific antigens (Stanford *et al.* 1987). Both the occurrence and severity of ENL are reduced. These would have been very valuable contributions in the days of dapsone monotherapy but now with MDT, the benefit is chiefly for the small proportion of patients troubled by ENL reactions. Results obtained in long-term-treated leprosy patients are described below. Immunotherapy has not been properly investigated in paucibacillary patients.

5.8 Immunotherapy for Buruli ulcer

Histopathological aspects

Although immunotherapy has not been tried yet for this disease, it remains an important and interesting potential use. Classically, the disease starts as a small subcutaneous nodule slowly enlarging as the supporting trabeculae for the subcutaneous fat are progressively destroyed by local invasion by *M. ulcerans* and production of its lipid toxin (Krieg *et al.* 1974). The central area of skin ulcerates and necrotic fat loaded with bacilli is released, leaving a 'collar stud' ulcer, deeply undermined, with a rolled edge exposing deep fascia at its base as shown in Fig. 16.3. The site of active infection and tissue destruction leads to a steady increase in size of the lesion (MacCallum *et al.* 1948). A point is reached, remaining ill-understood, at which the character of the disease changes from extending necrosis to granuloma formation and healing with massive fibrosis.

Immunosuppressive activities

During the necrotic phase there is no apparent immune recognition of the disease, probably because of the local immunosuppressive activity of the toxin on antigen-presenting cells (Pimsler *et al.* 1988).

There is no local infiltration of macrophages and lymphocytes, and frequently no enlargement of draining lymph nodes. There is also a generalized suppressive effect on T lymphocytes such that pre-existing tuberculin positivity is suppressed. With the change in histopathology to granuloma formation marked infiltration of lymphocytes and macrophages occurs around the lesion with rapid disappearance of invading bacilli, re-expression of tuberculin positivity and development of skin-test positivity to the new tuberculin prepared from *M. ulcerans* known as Burulin (Stanford *et al.* 1975). These changes are shown in the proposed spectrum for Buruli ulcer presented in Fig. 16.4.

Treatment

Although *M. ulcerans* is susceptible to a number of antitubercular drugs *in vitro* (Stanford *et al.* 1973), these are clinically ineffective probably because the bacilli are in necrotic fat with a poor, or absent blood supply. Treatment is almost exclusively surgical, which is very effective for early small lesions, but much less cost-effective, or clinically effective in advanced disease.

A disease of low mortality but high morbidity, its natural progress to immunologically mediated spontaneous recovery strongly suggests that im-

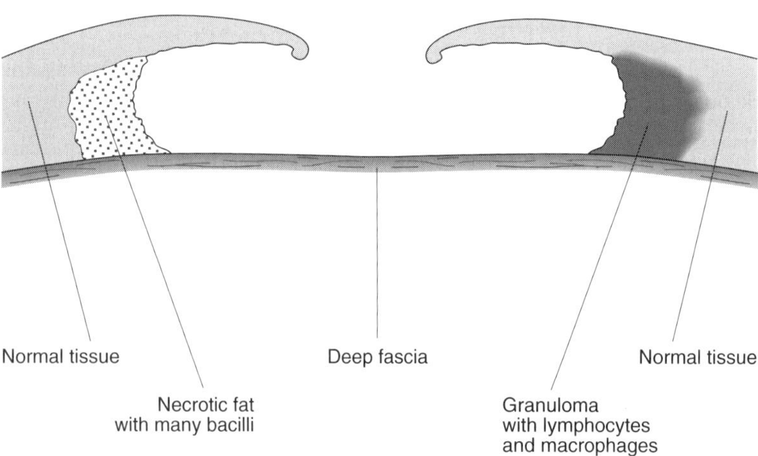

Fig. 16.3 Diagram of a buruli ulcer in the necrotizing and early ulcerative phase (left side) and in the granulomatous, healing phase (right side). The tissues are shown of approximately natural size for a well-nourished person, but the undermined edges of the ulcer would normally collapse on to the deep fascia.

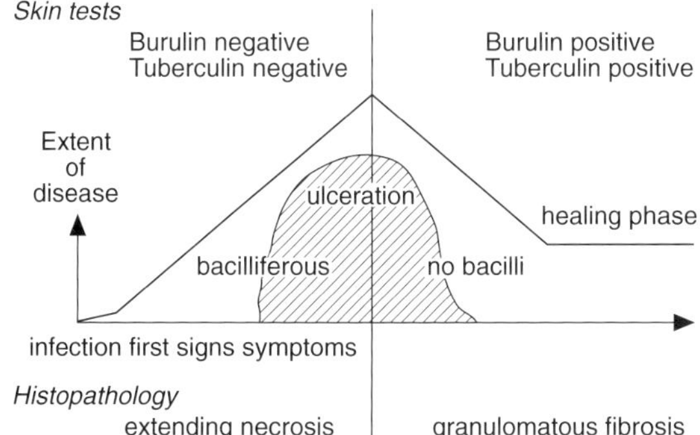

Fig. 16.4 A proposed spectrum for Buruli ulcer (*M. ulcerans* infection).

munotherapy should lead to earlier recovery. As with other mycobacterial diseases, promotion of Th1 with reintroduction of immune recognition of group i, common mycobacterial antigens is likely to be required, possibly also needing development of anti-toxin. Early work suggests that the lipid toxin can be neutralized by antibodies, at least in experimental animal models (P. Small and K. George, personal communication 1998).

6 Immunotherapy with *Mycobacterium vaccae* for other diseases

6.1 For cancer

Coley's work

Earlier in this chapter, Coley's work on developing a bacterium-based immunotherapy for cancer was mentioned as concurrent with that of Koch on tuberculosis. Coley developed his concepts from the observation that severe erysipelas occurring in patients with malignant sarcomas could lead to massive necrosis of the tumours and cure in some cases (Coley 1893). His treatment consisted of repeated injections of extracts of group A streptococci and of a Gram-negative bacillus. In a proportion of patients with sarcomas he achieved remarkable results with massive necrosis of the tumours by a reaction analo-

gous to the Koch phenomenon dependent on exquisite local toxicity of TNF in the presence of Th2 cytokines. His treatment was much less successful in treating carcinomas and became less successful with time so that a higher proportion of patients could be cured in the 1890s than could be achieved at the end of his professional life in the 1930s. It has been suggested that this was because Koch-type responsiveness to tubercle bacilli was a necessary primer for the efficacy of Coley's toxins and this had become much less common in New York patients over this time period (Starnes 1992).

A shift in conceptual framework: Mycobacterium vaccae *and cancer*

There are a number of reports claiming (Davignon *et al.* 1970; Crispin & Rosenthal 1976) or denying (Kinlen & Pike 1971; Skegg 1975) that BCG vaccination in childhood can reduce the incidence of juvenile leukaemias and solid tumours. Grange and Stanford (1990) recognized that this effect of BCG was limited to those areas where BCG also provided effective protection from tuberculosis. Subsequently it was suggested that the mode of action required might be the promotion of apoptosis by the same mechanism as that of immunotherapy for mycobacterial infections (Grange *et al.* 1995). Just as the mechanism of immunotherapy for tuberculosis with *M. vaccae* is the opposite of that of Koch's method, immunotherapy for cancer might be the opposite of Coley's. If this is true then injections of *M. vaccae* should have immunotherapeutic efficacy against cancers, providing there are shared antigens that could focus the apoptosis. Such antigens could well be the stress proteins, notably the 65-kDa hsp of mycobacteria and the 60-kDa stress protein of humans. Preliminary trials of repeated injections of killed *M. vaccae* show that the hypothesis is true, at least in principle, and successful phase I/II studies have been completed in stage IV malignant melanoma (Baban 1998) and in symptomatic non-small-cell carcinoma of the lung (M. O'Brien, personal communication 1999). A phase III trial of *M. vaccae* against the latter is about to begin.

6.2 For psoriasis

Psoriasis is generally spoken of as a disease of unknown aetiology in which exaggerated Th1 responses take place in the skin resulting in scattered erythematous squamous lesions histologically showing keratinocyte proliferation, vascular changes and dermo-epidermal inflammatory infiltrates. Studies of lymphoproliferation show that psoriasis patients have defective responses to group i, common mycobacterial antigens (Bay *et al.* 1998)—in common with patients with tuberculosis, multibacillary leprosy, HIV infection (Khoo *et al.* 1996), rheumatoid arthritis (Bahr *et al.* 1988) and the silent seropositive phase of infestation with *Trypanosoma cruzi*, Chagas' disease (Bottasso *et al.* 1994).

In the course of studies of the effects of immunotherapy with *M. vaccae* for ENL in an Indian leprosy patient with concurrent psoriasis, it was noticed that the immunotherapy had a marked beneficial effect on the psoriasis lesions. This observation was backed with a small study of *M. vaccae* in psoriasis patients without leprosy, again with striking improvement (Ramu *et al.* 1990). Subsequently a number of unpublished studies confirmed the observations and a randomized, blind trial of immunotherapy for chronic plaque psoriasis has been carried out in Argentina (Lehrer *et al.* 1998). The results show striking, and long-lasting beneficial effects.

6.3 For autoimmune and allergic diseases

Evidence from chronic leprosy

Evidence is steadily building up for beneficial effects of enhancement of Th1, and reduction of Th2, and the consequent regulation of responses to stress proteins that follows injection of *M. vaccae*, in patients with diseases associated with autoimmunity, and in some forms of allergy. Much of this work followed investigation of immunotherapy with *M. vaccae* in bacteriologically negative, long-term treated, leprosy patients in Iran (Stanford 1994). Such patients may be troubled with a number of chronic sequelae associated with autoimmune vascular lesions induced by recurring reactional episodes. These include arthritis, neuritis, vasculitis and anterior uveitis. Such patients have grossly raised levels of IgG antibody to the 65-kDa hsp, and probably to other stress proteins. Following injection of 10^9 heat-killed *M. vaccae*, blood flow to the fingertips was found to improve with an increase towards normal of their fingertip skin temperature. Anterior uveitis was found to resolve and the levels of antibody to the 65-kDa hsp were halved (Rafi 1995).

Autoimmune diseases

Many diseases are now thought to be due to autoimmunity and in many others, autoimmune phenomena play a part. In many cases autoimmunity is the cause of underlying vascular damage. Thus the damage to the arterial intima due to bacterially induced antibody binding to stressed cells expressing the 60-kDa stress protein is now thought to be the fundamental lesion on which atheromatous plaque is deposited (Wick *et al.* 1995). Endarteritis has been recognized as the first step in the development of rheumatoid arthritic lesions since the early years of the twentieth century (Bannatyne 1906) and several studies of adjuvant arthritis in animals have shown injections of *M. vaccae* to improve or suppress the disease (Thompson *et al.* 1991). Anecdotal experience in patients with psoriatic arthropathy or early rheumatoid disease have shown marked clinical improvement lasting between 9 and 12 months after a single injection. Evidence has been building up for the last 80 years that the major psychoses, schizophrenia and acute depression may be immunologically mediated through autoimmune mechanisms (Amital & Shoenfeld 1993). Antibodies from a proportion of schizophrenic patients bind to parts of the 65-kDa hsp of BCG and also to extracts from certain neuronal cell lines (Kilidirias *et al.* 1992).

Allergic diseases

Such diseases are known to be associated with increased Th2 mechanisms, IL-5 production and IgE production. As such, regulation of Th2 by

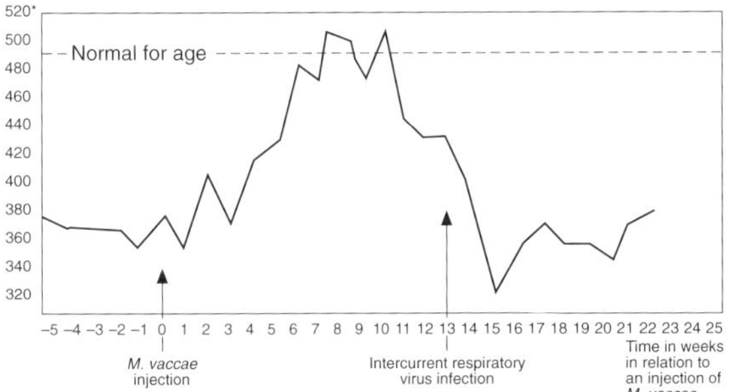

Peak expiratory air flow in litres per minute.

The patient suffered asthmatic attacks without marked seasonal variation over 15 years. These were triggered by house dust and cats and controlled with daily low-dose inhaled steroid and salbutamol.

Fig. 16.5 Peak flow values for an 18-year-old girl with moderate asthma before and after an injection of *M. vaccae*.

immunotherapy with *M. vaccae* would be a logical approach to treatment. So far, allergic bronchospasm has been investigated, first as anecdotal cases, then as part of hayfever induced by allergens of the grass *Phleum pratense* (Hopkin *et al.* 1998), and a current study is addressing perennial asthma. Figure 16.5 shows the changes in peak flow readings in an 18-year-old volunteer with asthma related to house dust mites and cat fur. A single injection of *M. vaccae* resulted in a return to normal peak flow readings for 3 months until the effects were undermined by a viral infection.

7 Conclusion

Immunotherapy remains a widely disregarded means of treatment for many diseases. Largely eclipsed by the overwhelming enthusiasm for antibiotics and antivirals, with the increasing problems of bacterial and viral resistance, attention has to be drawn again to the body's capacity to overcome infection when properly primed. Similarly, with increasing understanding of the basic mechanisms of cancer, autoimmunity and allergy, the treatment of symptoms needs to be reinforced by correction of the underlying immunological problems. In these days of molecular

and genetic engineering, it seems strange to revert to the 'nineteenth century' approach of whole-cell products. Nevertheless, the killed suspension of whole *M. vaccae* with its combination of Th1 adjuvant activity and antigenic content, and excellent safety record, is going to be difficult to replace even with modern technology.

8 References

Abraham, P.T. (1890) Leprosy and Professor Koch's treatment. *Lancet* **2**, 1300.

Amital, H. & Shoenfeld, Y. (1993) Autoimmunity and schizophrenia: an epiphenomenon or an etiology? *Israel Journal of Medical Sciences* **29**, 594–598.

Anonymous. (1890) Professor Koch's remedy for tuberculosis; Austria. *British Medical Journal* **2**, 1490.

Baban, B. (1998) The evaluation of a heat-killed suspension of *Mycobacterium vaccae* as an immunomodulating agent in the treatment of cancer. PhD Thesis, University of London.

Backman, A., Pirila, V., Forstrom, L., Heiskala, M., Nurmela, T., Tala, E. & Uotila, K. (1984) A new method for testing tuberculin skin reactivity—chamber test. *Tubercle* **65**, 279–283.

Bahr, G.M., Sattar, M.I., Stanford, J.L. *et al.* (1988) HLA-DR and tuberculin tests in rheumatoid arthritis and tuberculosis. *Annals of Rheumatic Diseases* **48**, 63–68.

Bahr, G.M., Shaaban, M.A., Gabriel, M. *et al.* (1990) Improved immunotherapy for pulmonary tuberculosis with *Mycobacterium vaccae*. *Tubercle* **71**, 259–266.

Bannatyne, G.A. (1906) *Rheumatoid Arthritis. Its Pathology, Morbid Anatomy and Treatment*, 4th edn. Bristol: John Wright and Co.,

Bay, M.-L., Lehrer, A., Bressanelli, A., Morini, J., Bottasso, O. & Stanford, J. (1998) Psoriasis patients have T-cells with reduced responsiveness to common mycobacterial antigens. *FEMS Immunology and Medical Microbiology* **21**, 65–70.

Bechelli, L.M., Lwin, K., Garbajosa, P.G. *et al.* (1974) BCG vaccination of children against leprosy: nine-year findings of the controlled WHO trial in Burma. *Bulletin of the World Health Organization* **51**, 93–99.

Beckett, D.W. (1958) A trial of Antigen Marianum as an adjunct to DDS in the treatment of lepromatous leprosy. *Leprosy Review* **29**, 209–214.

Blanc, M., Prost, M.-T. & Sister Marie-Suzanne. (1953) Influence de l'injection d'une mycobacterium isole d'un cas de lèpre (souche Chauvire) sur la reaction de Mitsuda. *Bulletin de la Société de Pathologie Exotique* **46**, 1009–1014.

Bottasso, O.A., Ingledew, N., Keni, M. *et al.* (1994) Cellular immune response to common mycobacterial antigens in subjects seropositive for *Trypanosoma cruzi*. *Lancet* **344**, 1540–1541.

Bottasso, O., Merlin, V., Cannon, L., *et al.* (1998) Studies of vaccination of persons in close contact with leprosy patients in Argentina. *Vaccine* **16**, 1166–1171.

Chaudhuri, S., Fotedar, A. & Talwar, G.P. (1983) Lepromin conversion in repeatedly negative BL/LL patients after immunization with autoclaved *Mycobacterium w*. *International Journal of Leprosy* **51**, 159–168.

Coley, W.B. (1893) The treatment of malignant tumors by repeated inoculations of erysipelas; with a report of ten original cases. *American Journal of Medical Science* **105**, 487–511.

Collins, C.H., Grange, J.M., Noble, W.C. & Yates, M.D. (1985) *Mycobacterium marinum* infections in man. *Journal of Hygiene* **94**, 135–149.

Convit, J., Aranzazu, N., Ulrich, M., Pinardi, M.E., Reyes, O. & Alvarado, J. (1982) Immunotherapy with a mixture of *Mycobacterium leprae* and BCG in different forms of leprosy and in Mitsuda negative contacts. *International Journal of Leprosy* **50**, 415–424.

Convit, J., Pinardi, M.E., Avila, J.L. & Aranzazu, N. (1975) Specificity of the 48-hour reaction to Mitsuda antigen. *Bulletin of the World Health Organization* **52**, 187–191.

Corlan, E., Marica, C., Macavei, C., Stanford, J.L. & Stanford, C.A. (1997a) Immunotherapy with *Mycobacterium vaccae*. (1) In the treatment of newly diagnosed pulmonary tuberculosis in Romania. *Respiratory Medicine* **91**, 13–19.

Corlan, E., Marica, C., Macavei, C., Stanford, J.L. & Stanford, C.A. (1997b) Immunotherapy with *Mycobacterium vaccae*. (2) In the treatment of chronic or relapsed tuberculosis in Romania. *Respiratory Medicine* **91**, 21–29.

Corrah, P.T. (1994) Studies of tuberculosis in The Gambia. PhD Thesis, The Open University.

Crispin, R.G. & Rosenthal, S.R. (1976) BCG vaccination and cancer mortality. *Cancer Immunology and Immunotherapy* **1**, 139–142.

Davignon, L., Lemonde, P., Robillard, P. & Frappier, A. (1970) BCG vaccination and leukaemia mortality. *Lancet* **2**, 638.

Deo, M.G., Bapat, C.V. & Bhalerao, V. (1983) Antileprosy potentials of ICRC vaccine. A study of patients and healthy volunteers. *International Journal of Leprosy* **51**, 540–549.

Donoghue, H.D., Overend, E. & Stanford, J.L. (1997) A longitudinal study of environmental mycobacteria on a farm in south-west England. *Journal of Applied Microbiology* **82**, 57–67.

Durban Immunotherapy Trial Group (1999) Randomised controlled trial of immunotherapy with *Mycobacterium vaccae* in newly diagnosed pulmonary tuberculosis patients in South Africa. *Lancet* in press.

Etemadi, A., Farid, R. & Stanford, J.L. (1992) Immunotherapy for drug-resistant tuberculosis (letter). *Lancet* **340**, 360–361.

Farid, R., Etemadi, A., Mehvar, M., Stanford, J.L., Dowlati, Y. & Velayati, A.A. (1994) *Mycobacterium vaccae* immunotherapy in the treatment of multi-drug-resistant tuberculosis: a preliminary report. *Iranian Journal of Medical Sciences* **19**, 37–39.

Friedmann, F.F. (1903) Spontane Lungentuberkulöse bei Schildkröten und die Stellung des Tuberkelbazillus im System. *Zeitschrift für Tuberkulöse* **4**, 439–457.

Friedmann, F.F. (1958) *The Friedmann Preparation: Anningzochin*. Ronnenberg, Hannover: Laves Arzneimittel GmbH.

Ganapati, R.P., Revankar, C.R., Lockwood, D.N.P. *et al.* (1989) A pilot study of three potential vaccines for leprosy in Bombay. *International Journal of Leprosy* **57**, 33–37.

Grange, J.M. (1992) The mystery of the mycobacterial persister. *Tubercle and Lung Disease* **73**, 249–251.

Grange, J.M. & Stanford, J.L. (1990) BCG vaccination and cancer. *Tubercle* **71**, 61–64.

Grange, J.M., Gibson, J., Osborn, T.W., Collins, C.H. & Yates, M.D. (1983) What is BCG? *Tubercle* **64**, 129–139.

Grange, J.M., Stanford, J.L. & Rook, G.A.W. (1995) Tuberculosis and cancer: parallels in host responses and therapeutic approaches? *Lancet* **345**, 1350–1352.

Grange, J.M., Stanford, J.L., Rook, G.A.W. & Wright, P. (1994) Role of viral infections in the inception of childhood asthma and allergies (letter). *Thorax* **50**, 701.

Hernandez-Pando, R., Pavon, L., Arriaga, K., Orozco, H., Madrid-Marina, V. & Rook, G. (1997) Pathogenesis of tuberculosis in mice exposed to low and high doses of an environmental mycobacterial saprophyte before infection. *Infection and Immunity* **65**, 3317–3327.

Hopkin, J.M., Shaldon, S., Ferrie, B. *et al.* (1998) Mycobacterial immunisation in grass pollen asthma and rhinitis. *Thorax* **53** (suppl. 4), S63.

Khoo, S.H., Wilkins, E.G.L., Fraser, I. & Stanford, J.L. (1996) Multiple skin testing of HIV-infected persons with new tuberculins. *Thorax* **51**, 932–935.

Kilidirias, K., Latov, N., Strauss, D.H. *et al.* (1992) Antibodies to the 60 kDa heat shock protein in patients with schizophrenia. *Lancet* **340**, 569–572.

Kinlen, L.J. & Pike, M.C. (1971) BCG and leukaemia, evidence of vital statistics. *Lancet* **2**, 398–402.

Koch, R. (1890) An address on bacteriological research delivered before the International Medical Congress, held in Berlin, August 1890. *British Medical Journal* **2**, 380–383.

Krieg, R.E., Hockmeyer, W.T. & Connor, D.H. (1974) Toxin of *M. ulcerans*: production and effects in guinea pig skin. *Archives of Dermatology* **110**, 783–788.

Lehrer, A., Bressanelli, A., Wachsmann, V. *et al.* (1998)

Immunotherapy with *Mycobacterium vaccae* in the treatment of psoriasis. *FEMS Immunology and Medical Microbiology* **21**, 71–77.

Linnell, F. & Norden, A. (1954) *Mycobacterium balnei*: new acid-fast bacillus occurring in swimming pools and capable of producing skin lesions in humans. *Acta Tuberculosea Scandinavica* **33** (Suppl.) 1–34.

Lockwood, D.N.J., McManus, I.C., Stanford, J.L., Thomas, A., Abeyagunawardana, D.V.P. (1987) Three types of human response to mycobacterial antigen: population studies in India and Sri Lanka. *European Journal of Respiratory Diseases* **71**, 348–355.

Macassey, L. & Saleeby, C.W. (1934) Spahlinger contra tuberculosis 1908–34: an international tribute. eds L. Macassey & C. W. Saleeby. London: John Bale and Sons and Danielsson.

MacCallum, P., Tolhurst, J.C., Buckle, G. & Sissons, H.A. (1948) A new mycobacterial infection in man. *Journal of Pathology and Bacteriology* **60**, 93–122.

Mayo, R.E.P. (1999) *Mycobacterium vaccae* immunotherapy for tuberculosis in rural Kwazulu, South Africa. MD Thesis, University of London.

McManus, I.C., Lockwood, D.N.J., Stanford, J.L., Shaaban, M.A., Abdul Ati, M. & Bahr, G.M. (1988) Recognition of a category of responders to group ii, slow-grower associated antigens amongst Kuwaiti senior school children. *Tubercle* **69**, 275–281.

Moreira, A.L., Sampaio, E.P., Zmuidzinas, A., Frindt, P., Smith, K.A. & Kaplan, G. (1993) Thalidomide exerts its inhibitory action on tumor necrosis factor alpha by enhancing mRNA degradation. *Journal of Experimental Medicine* **177**, 1675–1680.

Morton, A., Nye, P.M., Rook, G.A.W., Samuel, N. & Stanford, J.L. (1984) A further investigation of skin-test responsiveness and suppression in leprosy patients and healthy school children in Nepal. *Leprosy Review* **55**, 273–281.

Mosmann, T.R., Cherwinski, H., Bond, M.W., Giedlin, M.A. & Coffman, R.L. (1986) Two types of murine helper T cell clone. (1) Definition according to profiles of lymphokine activities and secreted proteins. *Journal of Immunology* **136**, 2348–2357.

Nye, P.M., Price, J.E., Revankar, C.R., Rook, G.A.W. & Stanford, J.L. (1983) The demonstration of two types of suppressor mechanism in leprosy patients and their contacts by quadruple skin-testing with mycobacterial reagent mixtures. *Leprosy Review* **54**, 9–18.

Nye, P.M., Stanford, J.L., Rook, G.A.W. *et al.* (1986) Suppressor determinants of mycobacteria and their potential relevance to leprosy. *Leprosy Review* **57**, 147–157.

Ondoua, P., Prost, M.-T., Sister M. de la Trinite (1964) Clinical and immunological results obtained with the Marianin antigen after more than 10 years of therapeutic use. *Leprosy Review* **35**, 297–303.

Onyebujoh, P., Abdulmumini, T., Robinson, S., Rook, G.A.W. & Stanford, J.L. (1995) Immunotherapy for tuberculosis in African conditions. *Respiratory Medicine* **89**, 199–207.

Parrish, N.M., Dick, J.D. & Bishai, W.R. (1998) Mechanisms of latency in *Mycobacterium tuberculosis. Trends in Microbiology* **6**, 107–112.

Paul, R.C., Stanford, J.L. & Carswell, J.W. (1975b) Multiple skin testing in leprosy. *Journal of Hygiene (Cambridge)* **75**, 57–68.

Paul, R.C., Stanford, J.L., Misljenovic, O. & Lefering, J. (1975a) Multiple skin testing of Kenyan schoolchildren with a series of new tuberculins. *Journal of Hygiene (Cambridge)* **75**, 303–313.

Pimsler, M., Sponsler, T.A. & Meyers, W.M. (1988) Immunosuppressive properties of the soluble toxin from *Mycobacterium ulcerans. Journal of Infectious Diseases* **157**, 577–580.

Pozniak, A., Stanford, J.L., Johnson, N.M. & Rook, G.A.W. (1987) Preliminary studies of immunotherapy of tuberculosis in man. Proceedings of the International Tuberculosis Congress, Singapore, 1986. *Bulletin of the International Union Against Tuberculosis* **62**, 39–40.

Rafi, A. (1995) Molecular and immunological studies on 'fully treated' long-term leprosy patients. PhD Thesis, University of London.

Ramu, G., Grange, J. & Stanford, J. (1998) Evaluation of adjunct immunotherapy with killed *Mycobacterium vaccae* in the treatment of multibacillary leprosy and erythema nodosum leprosum. In: *Clinical Mycobacteriology* (ed. M. Casal). Barcelona Prous Science.

Ramu, G., Prema, G.D., Balakrishnan, S., Shanker Narayan, N.P. & Stanford, J.L. (1990) A preliminary report on the immunotherapy of psoriasis. *Indian Medical Gazette* **124**, 381–382.

Rook, G.A.W., Bahr, G.M. & Stanford, J.L. (1981) The effect of two distinct forms of cell-mediated response to mycobacteria on the protective efficacy of BCG. *Tubercle* **62**, 63–68.

Rook, G.A.W. & Stanford, J.L. (1998) Give us this day our daily germs. *Immunology Today* **19**, 113–116.

Sechehaye, A. (*c.* 1916) *The Treatment of Pulmonary and Surgical Tuberculosis with Umckaloabo (Stevens' Cure).* London: B. Fraser.

Shepard, C.C., Minagawa, F., Van Landingham, R. & Walker, L. (1980) Foot-pad enlargement as a measure of induced immunity to *Mycobacterium leprae. International Journal of Leprosy* **48**, 371–381.

Shield, M.J. (1983) The importance of immunologically effective contact with environmental mycobacteria. In: *The Biology of the Mycobacteria*, Vol. 2 (eds C. Ratledge & J. L. Stanford). London: Academic Press.

Shield, M.J. & Stanford, J.L. (1982) The epidemiological evaluation in Burma, of the skin test reagent LRA6; a cell-free extract from armadillo-derived *Mycobacterium leprae*. Part 2: Close contacts and non-contacts of bacilliferous leprosy patients. *International Journal of Leprosy* **50**, 446–454.

Shield, M.J., Stanford, J.L., Garbajosa, G., Draper, P. & Rees, R.J.W. (1982) The epidemiological evaluation in Burma, of the skin test reagent LRA6; a cell-free extract from armadillo-derived *Mycobacterium leprae*. Part 1: Leprosy patients. *International Journal of Leprosy* **50**, 436–445.

Shield, M.J., Stanford, J.L., Paul, R.C. & Carswell, J.W. (1977) Multiple skin testing of tuberculosis patients with a range of new tuberculins, and a comparison with leprosy and *Mycobacterium ulcerans* infection. *Journal of Hygiene (Cambridge)* **78**, 331–348.

Shirakawa, T., Enomoto, T., Shimazu, S. & Hopkin, J.M. (1997) The inverse association between tuberculin responses and atopic disorder. *Science* **275**, 77–79.

Skegg, D.C.G. (1975) BCG vaccination and the incidence of lymphomas and leukaemia. *International Journal of Cancer* **21**, 18–21.

Spahlinger, H. (1922) Note on the treatment of tuberculosis. *Lancet* **i**, 5–8.

Stanford, J.L. (1991a) Improving on BCG. *Acta Pathologica, Microbiologica et Immunologica Scandinavica* **99**, 103–113.

Stanford, J.L. (1991b) Koch's phenomenon: can it be corrected? *Tubercle* **72**, 13–20.

Stanford, J.L. (1994) The history and future of vaccination and immunotherapy for leprosy. *Tropical and Geographical Medicine* **46**, 93–107.

Stanford, J.L. & Eshetu Lemma (1983) The use of a sonicate preparation of *Mycobacterium tuberculosis* (new tuberculin) in the assessment of BCG vaccination. *Tubercle* **64**, 275–282.

Stanford, J.L. & Paul, R.C. (1973) A preliminary report on some studies of environmental mycobacteria. *Annales de la Société Belge de Médecine Tropiques* **53**, 389–393.

Stanford, J.L., Bahr, G.M., Rook, G.A.W. *et al.* (1990) Immunotherapy with *Mycobacterium vaccae* as an adjunct to chemotherapy in the treatment of pulmonary tuberculosis. *Tubercle* **71**, 87–93.

Stanford, J.L., de Terencio Las Aguas, J., Torres, P., Gervasioni, B. & Ravioli, R. (1987) Studies on the effects of a potential immunotherapeutic agent in leprosy patients. *Quaderni di Cooperazione Sanitaria* **7**, 201–206.

Stanford, J.L., Hutt, M.S.R., Phillips, I. & Revill, W.D.L. (1973) Antibiotic treatment in *Mycobacterium ulcerans* infection. *Ain Shams Medical Journal* **25** (Suppl.), 258–:261.

Stanford, J.L., Onyebujoh, P.C., Rook, G.A.W., Grange, J.M. & Pozniak, A. (1993) Old plague, new plague and a treatment for both? (letter) *AIDS* **7**, 1275–1277.

Stanford, J.L., Revill, W.D. & Gunthorpe, W.J. (1975) The production and preliminary investigation of Burulin, a new skin test reagent for *Mycobacterium ulcerans* infection. *Journal of Hygiene (Cambridge)* **74**, 7–16.

Stanford, J.L., Shield, M.J., Rook, G.A.W. (1978) *Mycobacterium leprae*, other mycobacteria, and a possible vaccine. In: *Proceedings of the XI International Leprosy Congress, Mexico City* (ed. F. Latapi). Excerpta Medica, 102–107.

Stanford, J.L., Shield, M.J. & Rook, G.A.W. (1981) How environmental mycobacteria may predetermine the protective efficacy of BCG. *Tubercle* **62**, 55–62.

Stanford, J.L., Stanford, C.A., Ghazi Saidi, K. *et al.* (1989) Vaccination and skin test studies on the children of leprosy patients. *International Journal of Leprosy* **57**, 38–44.

Starnes, C.O. (1992) Coley's toxins in perspective. *Nature* **357**, 11–12.

Stewart, C.J., Dixon, J.M.S. & Curtis, B.A. (1970) Isolation of mycobacteria from tonsils, naso-pharyngeal secretions and lymph nodes in East Anglia. *Tubercle* **51**, 178–183.

Thompson, S.J., Butcher, P.D., Patel, V.K.R. *et al.* (1991) Modulation of pristane-induced arthritis by mycobacterial antigens. *Autoimmunity* **11**, 35–43.

Torisu, M., Miyahara, T., Shinohara, N., Ohsato, K. & Sonozaki, H. (1978) A new side-effect of BCG immunotherapy; BCG-induced arthritis in man. *Cancer Immunology and Immunotherapy* **5**, 77.

Tuberculosis Prevention Trial Madras. (1980) Trial of BCG vaccines in South India for tuberculosis prevention. *Indian Journal of Medical Research* **72** (Suppl.), 1–74.

Wang, C.C. & Rook, G.A.W. (1998) Inhibition of an established allergic response to ovalbumin in BALB/c mice by killed *Mycobacterium vaccae*. *Immunology* **93**, 307–313.

Wick, G., Schett, G., Amberger, A., Kleindienst, R. & Xu, Q. (1995) Is atherosclerosis an immunologically mediated disease? *Immunology Today* **16**, 27–33.

Young, D.B. (1992) Heat-shock proteins: immunity and autoimmunity. *Current Opinion in Immunology* **4**, 396–400.

Chapter 17 / Vaccines

DOUGLAS B. LOWRIE

1 Introduction

Imagine this as an epitaph: 'There were new vaccines; they could not be evaluated.' The irony would be wasted on the three million people dying of tuberculosis every year if, through lack of political will, the epitaph actually comes to be written. Indeed the disappointment for us all would far exceed that from actually evaluating a range of traditional-style, non-recombinant, whole-organism vaccines against leprosy (Fine 1996) and finding them to be all about equally protective. Molecular biology holds such promise, surely we can do better.

The reality today is that there is a buzz of excitement as the biotechnology revolution generates novel candidate vaccines (for recent reviews see Bloom & Fine 1994; Collins 1994; Kaufmann & Hess 1997; Orme 1997), but this is rightly tempered by an awareness that the hurdles (clinical, political and economic) that must be overcome before there can be a proper clinical appraisal of any one of them are daunting. The more innovative the vaccine, the

greater the caution. However, we cannot place our hopes solely on the alternative approach of discovering new antimycobacterial drugs. New drugs alone cannot be the answer because there will always be newly acquired multidrug resistance. Outstanding vaccine candidates must also be properly evaluated somehow.

Presuming that an emerging champion vaccine candidate will clear the hurdles, the question becomes: how do we define and then find the outstanding competitors?

The tasks that we require new vaccines to successfully accomplish immunologically are neither simple nor yet understood. They can be simply stated: we want vaccines to protect against future new infection (prevention) and to treat existing disease (therapy) so that there is less pathology, fewer relapses and fewer antimycobacterial drugs are needed. Less obviously, we would also like a vaccine to eliminate the dormant persistent *M. tuberculosis* infection that produces most of the active disease currently seen in formerly high prevalence countries such as those of the

West. However, for both prevention and therapy we need vaccines to selectively promote beneficial and not harmful kinds of immune response and our knowledge of how to do this, in anything other than an empirical fashion for either tuberculosis or leprosy, is still only partial. Studies of the effect of bacille Calmette–Guérin (BCG) vaccine or of exposure to a variety of different mycobacterial species have given abundant evidence that this is achievable. BCG can prevent both tuberculosis and leprosy (Fine 1995). Hence, in the context of prevention the issue is whether we can devise something that is better than BCG. It is in this area of prevention that this chapter is focused. In contrast, the use of mycobacteria or mycobacterial antigen for immunotherapy of mycobacterial disease does not have such an encouraging background, frequently being either ineffective or generating adverse reactions. Nevertheless, recent advances suggest that therapeutic vaccines may have a role in future and this is dealt with in Chapter 16. Furthermore, there is actually rather little demand for new preventative vaccines against leprosy, not least because BCG works so well (Klatser *et al.* 1996; Young 1996) (although the emergence of rifampicin resistance in leprosy may change this attitude), hence the scope of this chapter is essentially restricted to tuberculosis.

2   Environmental mycobacteria and bacille Calmette–Guérin failure

The immunological cross-reactions between the various mycobacteria that cause disease with BCG and the ubiquitous non-pathogenic mycobacteria in the environment, raise paradoxes that we have not resolved and that threaten to undermine new vaccines (Fine 1995). BCG vaccination at birth, prior to sensitization to any other mycobacteria, confers substantial protection against the relatively non-infectious childhood forms of tuberculosis (Rodrigues *et al.* 1993; Colditz *et al.* 1995) but protection probably declines with time after vaccination. Vaccination given later gives protection that ranges from high (e.g. almost 80% in the UK) (Hart & Sutherland 1977) to zero (e.g. in South India) (Tripathy 1979)

against the predominantly pulmonary and infectious adult forms. The effect in the UK became undetectable 15 years later. Exposure and sensitization to environmental mycobacteria (and *Mycobacteria tuberculosis*) is relatively light in the UK so that declining protection probably reflects declining immunological memory, but it is not known if repeating the BCG vaccination would make a difference. In contrast in South India, and in other regions where BCG efficacy against tuberculosis is low, exposure and sensitization to environmental mycobacteria is intensive. Indeed, it was the prediction that environmental mycobacteria might interfere with BCG that led to the siting of the major controlled clinical trial in South India that duly failed to show any protective effect (Tripathy 1979). Even repeating BCG vaccination still had no effect in Malawi, which is another location where there is extensive environmental sensitization and the vaccine fails, although paradoxically it did enhance protection against leprosy (Fine & Smith 1996). It is striking that environmental mycobacteria also appear to interfere with the efficacy of BCG vaccination in cattle (Fine 1995). The consensus view is that exposure to environmental mycobacteria is likely to be a far more important determinant of vaccine efficacy than any contribution by genetic factors in either host or pathogen. However, it is still not clear whether sensitization by the environment can confer full protection (equal to the maximum effect of BCG), thereby leaving no room for BCG to add anything, or if the immune system becomes subverted so that protective responses can no longer be evoked by BCG, perhaps even enhancing harmful responses (Stanford & Rook 1983). The distinction is important because it could profoundly affect the whole approach to finding something better than BCG.

2.1   Protection by environmental mycobacteria

Ninety per cent of people who become infected with *M. tuberculosis* do not develop the disease; BCG can protect up to 80% of the remainder. Hence, if it is true that sensitization to environmental mycobacteria is equal to the effect of BCG, then we are looking for a new vaccine that will do what BCG does plus protect the remaining 20% of susceptible people or 2% of the total population. The balance of evidence from animal models is in favour of this interpretation, that environmental mycobacteria are protective (Palmer & Long 1966; Edwards *et al.* 1982; Orme & Collins 1984; Herbert *et al.* 1994; Kamala *et al.* 1996). Furthermore, the incidence of tuberculosis in the control, non-vaccinated, population in South India was much lower than expected (Tripathy 1979) and *in vitro* tests of peripheral blood monocyte antimyco-bacterial function before and after vaccination were consistent with pre-existing protection (Cheng *et al.* 1993). If this is so, we can only speculate on the prospects for protecting the remaining 2% of the total population. Even following a patient's recovery from tuberculosis there remains a significant chance of exogenous reinfection despite the naturally acquired immunity (Rich 1951; ten Dam 1984). Can novel vaccines achieve in these individuals what infection with either BCG or *Mycobacterium tuberculosis* itself can not? Perhaps, but only if the immune response is not inherently defective. In that case, a better vaccine might be obtained by genetically modifying ('engineering') BCG, or some other non-pathogenic mycobacterium, to be more immunogenic by includ-ing antigens unique to *M. tuberculosis*, such as crip-pled virulence determinants (Jacobs *et al.* 1987). Alternatively, the response might be made more pro-tective by a vaccine that changes the type of response obtained, for example by including adjuvants or cytokines (Ottenhoff *et al.* 1997), rather than changing the antigens. In either case, a major problem is how to demonstrate clinically that a new vaccine is better than BCG; very large numbers of vaccination candidates would be needed for there to be a chance.

The view that the failure of BCG to protect against tuberculosis is due to pre-existing protection having taken place by exposure to environmental mycobac-teria does not sit comfortably with simultaneous observations that BCG protects against leprosy (both multibacillary lepromatous and paucibacillary tuber-culoid forms) in the same populations (Fine & Smith 1996). There is so much cross-protection between mycobacteria that it might seem inherently unlikely that environmental sensitization that protects against tuberculosis would not protect against such a feeble-growing organism as the leprosy bacillus. However, it has been difficult to establish if there is direct cross-protection between leprosy and tuberculosis (Glaziou *et al.* 1993; Seghal & Jain 1995). Lepromatous leprosy might be associated with increased susceptibility to tuberculosis, even though in tuberculoid leprosy the immune response may give some protection against tuberculosis. If there is little cross-protection between leprosy and tuberculosis then the absence of protec-tion by the environment against leprosy, despite the presence of protection by the environment against tuberculosis, is less surprising.

2.2   Immune subversion by environmental mycobacteria

Actual interference by environmental mycobacteria has been demonstrated in animals, although dose, route and timing of exposure can all be critical (Brown *et al.* 1985). Immunization with a high dose of the killed, environmental mycobacterium *M. vaccae* can decrease the resistance of mice to tubercu-losis even though a low dose enhances resistance (Hernandez-Pando *et al.* 1997). Hence, it is not ex-cluded that intensive environmental sensitization can set the immune response into a pathway that is not fully protective against tuberculosis and is resis-tant to correction by BCG. In this case, again, a better vaccine would be one that can switch the protective responses back on, by changing the way the antigens are seen by the immune system rather than neces-sarily changing the antigens seen. It has recently been similarly argued that BCG is normally used at a dose that is too high to elicit an optimum protective response (Bretscher 1994; Bretscher *et al.* 1996). Although it is improbable that a low dose would overcome environmental sensitization, it might be better than a high dose if the vaccine is given before such sensitization has occurred. If the idea that environmental mycobacteria abrogate or divert the protective response is correct, there is in principle a much better prospect of demonstrating that a new vaccine that overcomes this is better than BCG; the 80% of susceptible subjects not being protected by BCG are available for protection by the new vaccine.

The kind of immunological correction needed might be provided by, for example, switching between the expression of predominantly antibody and predominantly cellular immunity. These fundamentally different immune responses are now known to be modulated by T lymphocytes that secrete signalling molecules (cytokines) that either favour development of cellular immunity and discourage antibody immunity (known as T-helper type 1, or Th1 cells; or more broadly as the type 1 response) or secrete cytokines that favour antibody and discourage cellular immunity (T-helper type 2, Th2 or type 2 response). Although, once established, these response types tend to be self-sustaining through cytokine feedback, appropriate intervention can cause the cytokine profile and the balance of the responses to shift from one type to the other. Thus, a Th2 response can be switched into a Th1 response in mice (Raz *et al*. 1996; Ottenhoff *et al*. 1997). A vaccine that is designed to do this might have a therapeutic role in tuberculosis if the Th2 response contributes to disease progression (Rook & Hernandez-Pando 1994). Such vaccines should therefore also be clinically evaluated as adjuncts to chemotherapy. Possible precedent for this exists in the killed *M. vaccae* vaccine that elicits a response in which CD8$^+$ T cells with specific cytotoxicity and type 1 cytokine profile predominate (Skinner *et al*. 1997). This vaccine is currently being evaluated for a therapeutic application (Stanford & Stanford 1994; Onyebujoh *et al*. 1995), although initial reports from a controlled trial are not encouraging.

3   Features required of a replacement vaccine

3.1   General

Whatever the mechanism whereby environmental sensitization reduces BCG efficacy against tuberculo-

sis, the primary requirement of a replacement for BCG is that it should be significantly more efficacious in those same environments. The best bet for achieving this would be to aim to have a vaccine that produces a protective response in infancy that is both (i) intrinsically stronger and (ii) longer lasting than that normally induced by BCG or tuberculosis infection. If it is to be used beyond infancy, or as a therapeutic vaccine, it should probably also be capable of redirecting a pre-existing non-protective response. Since disease is usually a consequence of regrowth of the bacteria at secondary sites of infection after dissemination, often after long periods of quiescence, a vaccine that minimizes persistence and reactivation of dormant bacteria is also desirable.

3.2   Cross-protection

There are secondary attributes of BCG that we would like to retain. For example BCG gives significant protection against diseases due to atypical mycobacteria, as indeed does immunization by infection with tuberculosis (Horsburgh *et al*. 1996). Disease due to atypical mycobacteria are occurring with increasing frequency in the general population in Western countries, perhaps partly as a result of decreasing use of BCG (Trnka *et al*. 1994; Romanus *et al*. 1995). BCG also gives a legacy of protection against both *M. avium* and tuberculosis that is seen as human immunodeficiency virus (HIV) infection progresses into acquired immune deficiency syndome (AIDS) (Grange 1994), important activities to retain in a replacement vaccine. Such vaccines should also obviously differ from BCG in being incapable of multiplying to cause disease ('BCG-osis') as the profound cellular immunodeficiency develops (Felten & Leichsenring 1995; O'Brien *et al*. 1995; Marsh *et al*. 1997). The protection given by BCG against leprosy has already been mentioned; it would clearly be advantageous also to retain this property in any replacement (Fine 1996).

3.3   Diagnostic compatibility

To achieve a vaccine that does not interfere with the subsequent use of diagnostic skin hypersensitivity tests is also highly desirable. There is the same requirement in both human and domestic animals to retain immunoreactivity as a diagnostic tool (Collins 1994; Cirillo *et al*. 1995). BCG vaccination does interfere with skin testing and, although the case against BCG may have been overrated (Smith 1970), a fractionated or subunit vaccine approach is most appropriate to avoid this. There is no evidence that specific antigens of *M. tuberculosis* are necessary for protection, so by using only some of the cross-reactive antigens in a vaccine the remainder, together with any specific ones, are available for skin testing. There was initially doubt that a subunit vaccine comprising only a handful of antigens could give a level of protection equal to that obtained with the whole bacterium but these are being dispelled. Indeed it can be argued that the subunit approach has the potential to be superior to the live vaccine because of the removal of immunosuppressive effects of mycobacterial infection (Lagrange 1984; Reiner 1994; Kaye 1995; Maes *et al*. 1996). The cell-wall glycolipids may be particularly important modulators of the immune response to whole mycobacteria (Chatterjee *et al*. 1992; Barrow *et al*. 1993; Bradbury & Moreno 1993); the subunit approach allows removal of undesired modulatory glycolipids in addition to the replacement of beneficial adjuvanting ones, perhaps as synthetic analogues.

4 Animal models

Having defined what we want, albeit in general terms, we can turn to the fruits of molecular biology and assess the current progress. This has to be done in the context of laboratory animal models of tuberculosis, since no vaccine has yet entered human trials. There has been much debate about the value of using different models (Wiegeshaus *et al*. 1971; Smith 1985; Harboe *et al*. 1996a) because different animals differ widely in their susceptibility to natural and experimental infection and in the pathology of the disease (Soltys 1952; Francis 1958).

There is agreement that the protective effects of vaccination are mediated by accelerated expression of acquired cell-mediated immunity. Antibodies are not protective. Even in non-vaccinated humans, this response is usually able to control the infection at the primary focus in the lung, but not always before the infection has disseminated to other organs. It is this disseminated infection that subsequently causes disease in about 10% of the people initially infected (Balasubramanian *et al*. 1994). The disease is usually extrapulmonary in children and predominantly confined initially to the upper lobes of the lung in adults. The accelerated cellular response in vaccinated individuals prevents disease by preventing this dissemination (Balasubramanian *et al*. 1994); a replacement for BCG should have the same property. Aerosol infections of laboratory animals can be used to assess this directly and reflect the natural route of transmission in humans. These are difficult and expensive to do safely. However, any model which demonstrates earlier arrest of bacterial growth followed by decline in bacterial viability at initial sites of infection is likely to be a good guide and this can include challenge by injection. Nevertheless, results with the same vaccine in different models can differ significantly (Wiegeshaus *et al*. 1971), so a combination of test models is most appropriate (Harboe *et al*. 1996a). Mice and guinea pigs are preferred. Mice are rather resistant but convenient for immunological analyses and economical, whereas guinea pigs are excessively susceptible but develop granulomas similar to humans. Although tuberculosis and leprosy are essentially diseases due to immunopathology, the target for preventative vaccines is to reduce bacterial survival. Reduction of pathology is secondary and potentially independent. The reverse might be true for therapeutic vaccines that are to be given as adjuncts to chemotherapy but, unless reduction of immunopathology is the objective, our concern in the model is primarily with the fate of the bacteria, not the animal.

Murine models in which vaccination against dormant infection can be assessed have been developed (Lecoeur *et al*. 1981; deWit *et al*. 1995) and a monkey, *Macaca fascicularis*, which has a higher resistance to tuberculosis than the rhesus, may also be suitable (Walsh *et al*. 1996). Reactivation may often require liquefaction of the caseous material that comprises the old lesion; rabbits may be the best model for this aspect (Dannenberg & Sugimoto 1976; Converse *et al.* 1996). The kind of immune response needed to eliminate persistent bacteria and the kind needed to contain initial infection may be different. Furthermore, if the antigens produced by actively multiplying bacteria are different to those from dormant bacteria then the antigens required in the vaccine would also be different.

5 Vaccine approaches

5.1 Recombinant mycobacteria

The recombinant mycobacterium is the most difficult approach and is primarily motivated by the desire to use BCG as a vehicle to express antigens cloned from unrelated pathogens (Matsuo *et al.* 1990; Cirillo *et al.* 1995; Honda *et al.* 1995; Matsumoto *et al.* 1996). Such a multivalent vaccine might protect against a range of diseases. Although doubts about the stability of long-term expression of foreign genes remain, the technical feasibility is demonstrated by the integration and expression of foreign proteins in BCG (Kong & Kunimoto 1995; Norman *et al.* 1995) and by the immunological activity of foreign proteins in BCG (Kameoka *et al.* 1994; Langermann *et al.* 1994; Honda *et al.* 1995; Wada *et al.* 1996; Lagranderie *et al.* 1997). Most interestingly, BCG has been engineered to secrete murine cytokines that can change the nature of the immune response (Kong & Kunimoto 1995; Murray *et al.* 1996; Marshall *et al.* 1997), although it is not clear how acceptable these would be for use in a prophylactic vaccine. Auxotrophic mutants of BCG have been generated with a view to having a vaccine vehicle that is safer to deploy than the original strains now that we have entered the AIDS era with its attendant increased risk of live vaccines multiplying to cause disease in severely immunodeficient people. The nutritional impairment in such mutants severely decreases their ability to survive and multiply in macrophages while permitting bacterial metabolism and antigen production to continue long enough to generate effective

immune responses (McAdam *et al.* 1995; Bange *et al.* 1996). Methods have also recently been developed that will permit precisely targeted alterations to be made to the genomes of BCG and *M. tuberculosis* with reasonable efficiency (Balasubramanian *et al.* 1996; Pelicic *et al.* 1997).

5.1.1 *Fast-growing mycobacteria*

Mycobacterium smegmatis has been used as a more convenient, faster growing model host in which to express antigens from the slow-growers yet retain features of mycobacterial post-translational antigen modification (Garbe *et al.* 1993). The enhancement of protective immunogenicity against mycobacterial infections by expressing BCG or *M. leprae* antigens in *M. smegmatis* has been demonstrated (Falcone *et al.* 1995; Baumgart *et al.* 1996). Expression of a secreted protein (Mig) that is only produced by *M. avium* inside macrophages has been shown to enhance the intracellular survival of *M. smegmatis* (Plum *et al.* 1997) and so has expression of the gene for thioredoxin-thioredoxin reductase of *M. leprae* in *M. smegmatis* (Wieles *et al.* 1997). *M. vaccae* should also be an attractive model but, surprisingly, expression of a 19-kDa glycosylated antigen from *M. tuberculosis* in *M. vaccae* eliminated the protective efficacy of this vaccine candidate in mice, despite eliciting a type 1 cellular immune response (Abou-Zeid *et al.* 1997) and despite enhancing virulence by complementation in defective mutants (Lathigra *et al.* 1996).

5.1.2 *Slow-growing mycobacteria*

Virulence in tubercle bacilli seems to be associated more with differences in ability to grow *in vivo* than with differences in production of toxic factors. Nevertheless, the potential of the general vaccine approach of including virulence genes in non-pathogenic mycobacteria deserves further exploration. The attenuation of BCG is now known to be associated with the deletion of major segments of the *M. bovis* genome (Harboe *et al.* 1996b; Mahairas *et al.* 1996), including a 9.5-kb segment that encodes, among other proteins, a major protective antigen (ESAT-6, see below). This deletion also results in impaired capacity of the mycobacterium to up-regulate its stress protein responses. Whether reintroduction of elements from this segment into BCG will give a better vaccine remains to be seen. Another deletion, of a 10.7-kb sequence which includes the gene encoding antigen MPT64, occurs in BCG Pasteur and its derived substrains and not others (Harboe *et al.* 1996b; Mahairas *et al.* 1996) but there is little evidence that BCG strains with and without this segment differ in their protective efficacy (Brewer & Colditz 1995). *M. microti*, a natural pathogen of field voles, is equivalent to BCG as a vaccine against tuberculosis in humans (Hart & Sutherland 1977) and has been extensively used in Czechoslovakia (Sula & Radkovsky 1976). It is a slow-grower that might be more suitable than BCG for engineering because it is unlikely to have suffered major DNA deletions.

5.2   Other recombinant vehicles

Efforts to use recombinant vaccinia and other pox viruses as vectors of mycobacterial genes to protect against mycobacterial disease have so far had only limited success (Lyons *et al.* 1990; Baumgart *et al.* 1996), perhaps because of residual activity of viral proteins that interfere with immune responses (Smith 1993; Symons *et al.* 1995). In the most recent study, although other antigens did not give protection, vaccinia vectors expressing the 38-kDa antigen, now known as PstS-1 (formerly PhoS, pab, Ag78 or Ag5), or indeed expressing the 19-kDa antigen that abrogated protection when expressed in *M. vaccae*, protected mice from *M. tuberculosis* challenge given 2 weeks later (Zhu *et al.* 1997a). *Streptomyces lividans* (Colston 1990) and the more distantly related bacteria, *Escherichia coli* (Colston 1990) and *Salmonella typhimurium* (Wiegeshaus & Smith 1989), have also been briefly tested as potential vaccine vehicles for mycobacterial antigen (the 65-kDa heat-shock protein (hsp65)) but failed to generate protective responses.

5.3   Protein subunit vaccines

5.3.1   Secreted proteins

The subunit approach, originally aimed at using purified proteins or other chemically homogeneous fractions directly as a vaccine, has evolved since the initial focus on antigens that are present in whole bacterial extracts and give good antibody responses. It was realized that antigens released by multiplying bacteria may be among the most important for protection and these would not necessarily be either present in extracts or be good antibody inducers. Culture filtrates that are collected during the early phase of mycobacterial growth and before bacterial lysis has occurred contain a large number of proteins (Andersen *et al.* 1991b; Fifis *et al.* 1991; Nagai *et al.* 1991; Pal & Horwitz 1992; Wiker *et al.* 1992; Gulle *et al.* 1995). Novel antigens are present, to which T cells from purified protein derivative (PPD)-sensitive people and BCG-vaccinated mice give strong specific cellular responses (Collins *et al.* 1988). Responses to culture filtrate antigens appeared early during infection in mice (Andersen *et al.* 1991b) and in humans (Boesen *et al.* 1995), suggesting that some of these antigens at least were related to, and probably identical to, those released by live mycobacteria during the early stages of infection. It is possible that other antigens, released by bacteria growing inside mammalian cells but not in bacteriological culture media, remain to be discovered since the environments are very different, and radiolabelling studies indicate that the pattern of protein synthesis is also very different (Lee & Horwitz 1995; Monahan *et al.* 1996).

Protection in animal models

Despite the caveat that antigens secreted *in vivo* may differ from those secreted *in vitro*, antigens in early culture filtrates have been found to elicit protective immunity when administered with suitable adjuvant to guinea pigs (Pal & Horwitz 1992) and mice (Hubbard *et al.* 1992; Andersen 1994b). Several of the individual purified proteins, when adsorbed to nitrocellulose particles and injected twice subcutaneously, gave significant, although modest, protection to mice against an aerosol challenge with *M. tuberculosis* (Hubbard *et al.* 1992). A series of subcutaneous injections of a crude preparation in incomplete Freund's adjuvant was found to give a significant degree of protection to guinea pigs challenged 3 weeks later by aerosol with virulent bacteria (Pal & Horwitz 1992); bacterial multiplication in lungs and spleens was about 10-fold lower 5–9 weeks after infection. A similar degree of protection, equivalent to that obtained with live BCG, was found in lungs and spleens of mice that had received a series of subcutaneous injections of a similar preparation in dimethyldioctadecylammonium chloride adjuvant when they were challenged intravenously 3 weeks later (Andersen 1994b); bacterial multiplication in spleens and lungs was about 10-fold less after 2 weeks.

Some of the culture filtrate antigens responsible for protective effects have been identified. Five of the most abundant proteins were purified and corresponded to known antigens, antigens 85A, 85B, superoxide dismutase, MPT51 and MPT63. They were tested separately and in combinations in an adjuvant formulation for protection of guinea pigs against the development of the tuberculosis pathology that follows aerosol challenge (Horwitz *et al.* 1995). Hsp70 purified from the growth medium of heat-shocked cultures was also included in the study. The combinations were protective whilst, separately, hsp70 had some protective effect and antigen 85B was as effective as the mixtures. Hsp70 is currently unique in having apparently given strong protection in mice when they were immunized with purified native protein in Freund's incomplete adjuvant (Dhiman & Khuller 1997). It seems likely that many more of the individual antigens in filtrates will be found to be protective when they are administered with appropriate adjuvants. The problem with applying the protein-plus-adjuvant approach in practical vaccines is the lack of suitable adjuvants. No combinations have yet been shown to give long-lasting protection and they tend to give strong antibody and type 2 cytokine responses (as does BCG), whereas

antigen-specific T cells producing type 1 cytokines and displaying a cytotoxic phenotype are needed for protection (Orme 1995; Bonato *et al.* 1998).

Immunogenicity in humans

A similar group of secreted proteins, including antigen 85, superoxide dismutase and MPB64 plus the 38-kDa antigen PstS-1 (now known to be one of a group of related antigens (Lefevre *et al.* 1997)) had been identified as prominent antigens by others (Andersen *et al.* 1991a). T cells from tuberculosis patients and from individuals who had been infected without developing the disease, as revealed by the presence of delayed-type hypersensitivity to PPD of *M. tuberculosis*, also responded strongly to crude early culture filtrate. Nevertheless, responses to any one of 12 characterized filtrate antigens were found to be infrequent (Boesen *et al.* 1995). It is notable therefore that fractionation of the filtrate to identify the antigens that were seen by T cells early during infection in mice revealed, in addition to the antigen 85 complex, a novel 6-kDa antigen (ESAT-6) (Andersen 1994c; Andersen *et al.* 1995; Sorensen *et al.* 1995). T cells from most tuberculosis patients responded to antigen(s) of about this size in fractionated filtrate (Boesen *et al.* 1995) suggesting that it could be an important ingredient for a vaccine. Similar results were obtained with guinea pigs (Haslov *et al.* 1995) and with cattle infected with *M. bovis* (Pollock & Andersen 1997); response to ESAT-6 was prominent, although in guinea pigs MPB64 was recognized more often than antigen 85 in contrast to findings in mice. Others have found that tuberculosis patients and their contacts frequently have T cells that recognize the secreted antigens MPB64 and MPB70 in addition to antigen 85 (Roche *et al.* 1994). Prominent type 1 T cell responses to low molecular weight fractions of short-term culture filtrate were also found in humans after BCG vaccination (Zhu *et al.* 1997a); since ESAT-6 is absent from BCG it is evident that other antigens of similar molecular weight remain to be defined. The prominence and early appearance (Boesen *et al.* 1995; Haslov *et al.* 1995; Pollock & Andersen 1997) of responses to antigens of low molecular weight may to some extent be an experimental artefact, since screening is done with antigen fractions standardized by weight, not number of molecules, so that for example 10-fold more molecules of 6 kDa than of 60 kDa are present in assays.

5.3.2   Stress proteins

A variety of antigens that are not prominent in early culture filtrate, but may be released later, are also frequently recognized by T cells from tuberculosis patients (Havlir *et al.* 1991). These include 19-kDa antigen (Oftung *et al.* 1987) hsp10 (Barnes *et al.* 1992; Mehra *et al.* 1992; Mendezsamperio *et al.* 1995), PstS-1, hsp65 and hsp70 (Mendezsamperio *et al.* 1995). The heat-shock (or stress) proteins in particular have generated interest as potential protective antigens, largely on the basis of their prominence in the immune response in mice and humans at both antibody and T-cell level (Kaufmann 1990; Young 1990; Barnes *et al.* 1992; Lee *et al.* 1992; Mehra *et al.* 1996). Furthermore, hsp65 appears to be produced in increased amounts by intracellular tubercle bacilli (Lee & Horwitz 1995), presumably in response to the stress of intracellular existence. Also a 17-kDa antigen that is a homologue of the α-crystallin stress protein becomes a dominant protein under such conditions (Yuan *et al.* 1996). Both hsp65 and hsp10 proteins decreased growth of *M. leprae* in mice when they were given in incomplete Freund's adjuvant (Gelber *et al.* 1994). It is intriguing that the α-crystallin homologue and hsp70, hsp65 and hsp10 have been identified as prominent immunogens in mycobacterial ribosomal preparations (Tantimavanich *et al.* 1993; Matsuoka *et al.* 1997) that represent one of the more successful of the earlier tuberculosis subunit vaccines (Youmans & Youmans 1965).

Thus, it seems likely that antigens, such as stress proteins, that are produced late in infection could also be good targets for a protective immune response, particularly if the development of bacterial persistence and dormancy are to be overcome, and even more so if there is to be a therapeutic vaccine. However, stress proteins are highly conserved proteins with immunological cross-reactivity with homologues not only in other microorganisms, including gut flora, but also in mammalian cells (Young 1990). Background cross-reactivity to these proteins may serve a normal function, potentiating immune responses by alerting the immune system to stressed or damaged cells (Young & Elliott 1989; Nomoto & Yoshikai 1991; Dicesare *et al.* 1992; Murray & Young 1992; Denagel & Pierce 1993). If so, there could be additional advantages to including them in subunit vaccines, but against this must be set the possibility that recognition of microbial stress proteins could potentiate autoimmune reactions (Young 1990).

There are alternatives to leaving these major antigens out of vaccines: the epitopes involved in autoimmunity might be readily engineered out of the DNA (if they differ sufficiently from the protective epitopes) (Anderton *et al.* 1994; van Eden *et al.* 1995; Prakken *et al.* 1996); the formulation and delivery might be selected to direct the response away from a harmful and towards a beneficial type. Indeed, T cells that recognize mycobacterial heat-shock proteins can either protect against or potentiate autoimmune disease, depending on their phenotype (Billingham *et al.* 1990; Cohen 1991; Anderton *et al.* 1994; Anderton & VanEden 1996; Birk *et al.* 1996; Cohen & Elias 1996; Kingston *et al.* 1996; Adams *et al.* 1997). We recently showed that DNA vaccination with mycobacterial hsp65 protected Lewis rats against induction of adjuvant arthritis (Ragno *et al.* 1997). This may be the best approach in the long run because, to the extent that an acquired immune response depends on recognition of 'altered self' rather than 'foreign', essentially all antigens may have the potential for autoimmune involvement.

Something that has not yet been taken into account in the push for a BCG replacement is the possibility that natural isolates of *M. tuberculosis* vary in their expression of key antigens. Precedent for this is seen in the absence of the 19-kDa antigen from some strains (Lathigra *et al.* 1996). The molecular probe techniques that enable the diversity of field isolates to be defined make it now possible to distinguish between heterogeneity of antigen availability and heterogeneity of response to antigen in the clinical contexts where vaccination is proposed. There would be no merit in including in a new vaccine an antigen that is not expressed by the locally prevailing strains of *M. tuberculosis* even though possible differences in responsiveness in the target populations are at present a more obvious concern.

5.4   DNA vaccination

Perhaps the most remarkable recent development has been the realization that individual mycobacterial protein antigens can generate protection to a degree similar to that obtained with live BCG vaccine. Findings with purified proteins and adjuvants were summarized above, but the strongest evidence for this has come from studies in which mycobacterial DNA is taken out of the bacteria and expressed directly within mouse cells. Initially this was shown with the gene encoding hsp65 of *M. leprae* expressed from a retroviral vector in a mouse macrophage cell line that was used as the vaccine (Silva & Lowrie 1994). This was followed by the finding that protection could also be generated by direct DNA vaccination (Lowrie *et al.* 1994; Tascon *et al.* 1996) and that a range of different mycobacterial antigens can also be protective when they are expressed from plasmid DNA, including 36-kDa proline-rich antigen (Tascon *et al.* 1996), Ag85 (Huygen *et al.* 1996), hsp70 and ESAT-6 (Lowrie *et al.* 1997b), PstS-1 (Zhu *et al.* 1997b) and MPT83 (R. E. Tascon, D.B. Lowrie, R.G. Hewinson and M. J. Colston, unpublished observations 1998). The essence of this approach is that plasmid DNA encoding the antigen of interest is introduced directly into the normal tissues and cells of the body which then synthesize the antigen under the control of appropriate viral or eukaryotic promoters. This approach offers many attractions (Hassett & Whitton 1996; Kumar & Sercarz 1996; Robertson 1996; Siegrist & Lambert 1996; Donnelly *et al.* 1997). For example, it avoids the problems of purifying proteins without denaturing them and offers a way of rapidly testing different combinations and mixtures, selecting the best epitopes, omitting or modifying unwanted epitopes, targeting different antigen-presenting pathways, incorporating DNA encoding cytokines, etc. All this can be done with almost no concern for vector constraints.

The simplicity of production, stability of the product and extreme versatility have made it a very attractive tool in vaccine development in general. Its application for protection against HIV, influenza, herpes and malaria in humans is already undergoing clinical trials. At the present state of the art, DNA vaccination is rather inefficient. Although nanogram amounts of plasmid coated on gold particles and fired by a compressed-gas gun into the skin can be enough to generate protection in some infection models (Haynes *et al.* 1994; Johnston & Tang 1994), by direct intramuscular injections, of the kind being tested in humans, tens and even hundreds of micrograms of DNA are needed, even in mice (Davis *et al.* 1994; Babiuk *et al.* 1995). However, efficient oral delivery systems are under development (Darji *et al.* 1997; Jones *et al.* 1997; Lowrie 1998). This could make oral DNA vaccines very attractive for deployment against diseases in developing countries; this would include tuberculosis since oral BCG is already established as clinically effective (Editorial 1960). Furthermore, tests in mice suggest that DNA vaccination of newborn infants could raise good immune responses (Barrios *et al.* 1996; Bot *et al.* 1997). DNA vaccines have an inherent adjuvanticity that can play a strong part in biasing the immune response towards type 1 cytokine production (Halpern *et al.* 1996; Krieg 1996) and this, as we will see below, is exactly the kind that is needed for tuberculosis. There are safety concerns with DNA vaccination that are as yet unresolved, although these are not so great as to prevent phase I clinical trials that include healthy volunteers, and so far there are no reports of adverse reactions (Robertson 1996).

5.5   Best antigens

At this stage it is not possible to say precisely which are the best antigens to use in a practical vaccine, irrespective of whether this is to be in the form of a DNA vaccine, proteins or peptides together with adjuvant, or live recombinant vectors. However, in general terms, they are those that are released in largest amount by the intracellular mycobacteria, that carry multiple or promiscuous epitopes and that are presented on major histocompatibility complex (MHC) class I and MHC class II (for immunological help). Those proven or reported to have at least some protective effect as DNA vaccines are hsp65, hsp70, 36-kDa PRA and ESAT-6 (Lowrie *et al.* 1997b), MPT83 (R. E. Tascon, D.B. Lowrie, R.G. Hewinson and M. J. Colston, unpublished observations 1998), PstS-1 (Haslov *et al.* 1990; Zhu *et al.* 1996) and Ag85A (Huygen *et al.* 1996). The number of potential candidate antigens is large (Young *et al.* 1992; Andersen 1994a; Andersen & Brennan 1994) and more are being described all the time. Some good candidates might be identified from current efforts to identify proteins that are produced in increased amounts by intracellular mycobacteria (Lee & Horwitz 1995; Monahan *et al.* 1996; Russell *et al.* 1996) or that carry secretory signals (Lim *et al.* 1995). It is probable that others will be found by screening mycobacterial genome libraries as DNA vaccines in either totally random fashion (Barry *et al.* 1995) or focused on putative genes identified through sequencing the whole genome (Cole *et al.* 1998).

6   Correlates of protection

If we know how protective immunity works, not only is the rational design of vaccines facilitated, we can also develop rapid tests to monitor whether or not vaccination has conferred protection. This could greatly facilitate clinical trials. Unfortunately it has proved difficult to demonstrate acquired protective antimycobacterial immunity in humans, short of measuring impact on the disease (Cheng *et al.* 1993; Warwick Davies *et al.* 1994; Boom 1996). The best we can do at present is to measure both antigen-specific generation of interferon-γ (IFN-γ) and antigen-specific cytotoxicity. IFN-γ is essential (Cooper *et al.* 1993; Flynn *et al.* 1993), probably for activation of macrophage antimycobacterial action. Cytotoxicity is associated with protection in mice (Flynn *et al.* 1992;

Kaufmann 1993; Gupta *et al.* 1995; Tsukaguchi *et al.* 1995; Bonato *et al.* 1998), possibly because the process that kills the host cells also kills the mycobacteria.

6.1   Interferon-γ, cytotoxicity and CD8+ lymphocytes

From the point of view of vaccine design, it has become apparent that a good way of generating both of these responses is to present the antigen to the immune system via the intracellular (endogenous) pathway and hence in association with MHC class I for recognition by CD8+ T cells. Thus, protection resulted when protein was expressed from a retroviral promoter in tumour cells (Silva & Lowrie 1994) or bone marrow cells (Silva *et al.* 1995), or from the promoters of cytomegalovirus immediate-early gene (Huygen *et al.* 1996; Tascon *et al.* 1996; Zhu *et al.* 1997b) or murine hydroxymethylglutaryl-CoA reductase (Tascon *et al.* 1996) after injection of plasmid DNA (DNA vaccination) into muscles or skin (Lowrie *et al.* 1998), or when pox viruses generated the antigen intracellularly (Zhu *et al.* 1997a) or when intravenously injected cationic liposomes carried the antigen into cell cytosol (Lowrie *et al.* 1997a). All of these procedures resembled BCG vaccination in that they gave protection against challenge with virulent *M. tuberculosis* and the plasmid DNA, tumour cell and liposome systems all generated splenic T-cell populations in which CD8+ and CD4+ cells were about equally increased, whereas injection of protein alone or in Freund's incomplete adjuvant did not protect and only increased CD4+ cells selectively (Silva *et al.* 1996; and C. L. Silva, D. B. Lowrie, R. E. Tascon, V. M. F. Lima, V. L. D. Bonato, unpublished work 1998). Dead and live recombinant *M. vaccae* also evoked these type 1 responses (AbouZeid *et al.* 1997; Skinner *et al.* 1997). Analysis of the functions of purified T-cell subpopulations or large numbers of T-cell clones obtained from spleens after tumour-hsp65 or plasmid-hsp65 immunization or after *M. tuberculosis* infection showed that CD8+ cells had the greatest antimycobacterial activities *in vitro* and *in vivo* (Silva *et al.* 1996; Bonato *et al.* 1998). Adoptive transfer of protective immunity to naïve mice by transfer of T-cell clones showed that the most protective T cells were those that produced IFN-γ and had the greatest cytolytic activity against infected macrophages. It is not clear that cytotoxicity as such contributes to protection in mice since genetic deletions of the known mechanisms of cytotoxicity had no effect on immunity (Cooper *et al.* 1997; Laochumroonvorapong *et al.* 1997); hence, some undefined function that is associated with the cytotoxic phenotype might be responsible.

6.2   Interleukin 4 and CD44[+] lymphocytes

A striking difference between the immune response to DNA vaccination and the immune response to either BCG or *M. tuberculosis* infection is that DNA induces almost entirely a protective CD44[hi] type 1 cytokine response, whereas the mycobacterial infections have a major component of non-cytotoxic/CD44[lo] T cells that produce a type 2 cytokine (interleukin 4) response that is associated with antibody production (Bonato *et al.* 1998). Clones of this type 2 phenotype obtained from mice infected with *M. tuberculosis* have virtually no protective capacity in adoptive transfer experiments and their role in the disease is unclear: do they primarily serve the purposes of the host or of the infectious agent? Since the type 1 and type 2 responses are mutually antagonistic, the type 2 is thought to be there to downregulate the type 1 response in order to limit the tissue pathology on which the disease process depends (Yamamura *et al.* 1992; Sander *et al.* 1995; Hernandez-Pando *et al.* 1996). However, on this basis, it may also impede bacterial killing, since this also seems to depend on the cytotoxicity (and IFN-γ) of type 1 responding cells. Hence, the type 1 response that follows DNA vaccination may be preferable to the type 0 (mixed type 1 and type 2) response that follows BCG vaccination.

6.3   Switching the immune response

The hypothesis that inappropriate type1/type 2 switching of the immune response might underly the failure of BCG in some human populations (Rook *et al.* 1993) has already been alluded to. Exposure to environmental mycobacteria might give a predominantly type 2 response to common mycobacterial antigens that stays that way despite subsequent BCG vaccination. Whilst it is not clear what causes the mixed response seen in mycobacterial infection, the type 1 and type 2 responses are thought to tend to be self-sustaining, through a combination of positive and negative feedback by cytokines in the local environment (Croft *et al.* 1994; Seder & Paul 1994). It may be relevant therefore that DNA vaccination can reportedly switch an existing type 2 response into a type 1 response that is stable to subsequent boosting by exogenous antigen (Raz *et al.* 1996). On that basis, DNA might work where BCG fails, and could even have a therapeutic role. These possibilities are at present pure speculation but are amenable to experimentation in animal models.

7   Duration of protection

Protection by endogenous hsp65 is transient except after immunization with plasmid DNA. After a series of intramuscular plasmid injections, protection equal to the effect of BCG persisted undiminished for at least 8 months (Silva *et al.* 1996; C. L. Silva, D. B. Lowrie, R. E. Tascon, V. M. F. Lima, V. L. D. Bonato, unpublished work 1998). Analysis of the splenic T-cell populations indicated that this was due to a progressive increase during the same period in numbers of cells with the hsp65-specific memory/activated phenotype (CD44[hi]/CD8[+]) that produced IFN-γ and antigen-specific cytotoxicity; the other immunization procedures did not achieve this. Plasmid DNA does not integrate into host-cell DNA but persists episomally and can express encoded antigen for the lifetime of the injected mouse (Wolffe *et al.* 1992). It may be relevant that the persistence of viable BCG has been argued to make a significant contribution to the persistence of protection after BCG vaccination, although it appears not to be essential (Lefford & McGregor 1974). Presumably, the continued synthesis and presentation of endogenous hsp65 antigen may account for the persistent protection after DNA vaccination. This potentially important observation requires confirmation in other hands and with other antigens.

8 Projections to the future

The answer to the question—can we devise a subunit vaccine for use in humans that is as good as BCG?—now seems likely to be, yes! But we need to know first which antigens to use, what delivery system, and what immune response we want evoke in people. This will require substantial animal work and analysis of responses in human vaccinees.

8.1 Oral delivery

It could be very helpful to attain a subunit vaccine that is 'only' as good as BCG. Antigens that are not included in the vaccine will remain available for use in diagnostic tests for infection (this has hitherto been a major obstacle to deployment of BCG in some countries, including the United States). More importantly, oral subunit vaccines now seem likely to be effective and innocuous (Lowrie 1998) and are highly desirable for use in developing countries. BCG is effective as an oral vaccine but is not acceptable because of side-effects, particularly lymphadenitis. An oral DNA vaccine (for example) with equal efficacy and no side-effects would be a useful advance.

8.2 Bias towards type 1 responses

The likelihood of attaining a vaccine that is more protective than BCG is difficult to assess. It is obviously a goal worth striving for, since BCG efficacy ranges from less than 80% down to zero. A rational approach demands that we should first understand why BCG efficacy varies. Most likely it is due to immunization by environmental mycobacteria. This either (i) confers protection equal to that attainable by BCG, or (ii) subverts the immune response so that BCG cannot evoke protection. It is surprisingly difficult to distinguish between these two alternatives in humans. If environmental mycobacteria have a protective effect equal to BCG this reduces the likelihood

of finding (or demonstrating) that a new vaccine is better. The window, from 80% to 100% protection, is rather small. In contrast, if the immune response is subverted by environmental mycobacteria there is a much better prospect. Although animal models usually show protection by environmental mycobacteria, they have never been properly set up to test the alternative hypothesis. This could be done, by establishing a non-protective response (predominantly antibody or type 2) and trying to convert it into a protective response (predominantly a type 1 cellular response) by vaccinating. Precedent exists in *Leishmania* models. Tuberculosis and BCG both establish responses with a major type 2 component and a case can be made for biasing tuberculosis vaccine responses towards type 1 (CD8, cytotoxicity and IFN-γ). New ways of doing this (adjuvanting) are being developed.

8.3 Targeting dormant bacteria

It would be highly desirable to have a vaccine that would eliminate the dormant persistent bacteria that are already established in as much as a third of the world's population. Whether such a vaccine (or drug, or vaccine/drug combination) would ever be administered on an appropriate global scale is highly questionable, even though these persistent bacteria can account for most of the infectious disease that is seen in adults. However, there may be a reasonable prospect of developing and using a vaccine that would prevent the establishment of the persistent bacteria in the first place. The target antigens and the immune parameters needed to achieve this may be quite different from those needed to deal with the actively multiplying bacteria that predominate at the early stages of infection. They remain to be identified in animal models.

9 References

Abou-Zeid, C., Gares, M.P., Inwald, J. *et al.* (1997) Induction of a type 1 immune response to a recombinant antigen from *Mycobacterium tuberculosis* expressed in *Mycobacterium vaccae*. *Infection and Immunity* **65**, 1856–1862.

Adams, E., Basten, A., Rodda, S. & Britton, W.J. (1997)
Human T-cell clones to the 70-kilodalton heat shock
protein of *Mycobacterium leprae* define Mycobacterium-
specific epitopes rather than shared epitopes. *Infection and
Immunity* **65**, 1061–1070.

Andersen, A.B. (1994a) *Mycobacterium tuberculosis*
proteins—structure, function, and immunological
relevance. *Danish Medical Bulletin* **41**, 205–215.

Andersen, A.B. & Brennan, P. (1994) Proteins and antigens
of *Mycobacterium tuberculosis.* In: *Tuberculosis: Pathogenesis,
Protection, and Control* (ed. B. R. Bloom), pp. 307–332.
American Society for Microbiology, Washington, D.C.

Andersen, P. (1994b) Effective vaccination of mice against
Mycobacterium tuberculosis infection with a soluble mixture
of secreted mycobacterial proteins. *Infection and Immunity*
62, 2536–2544.

Andersen, P. (1994c) The T cell response to secreted
antigens of *Mycobacterium tuberculosis. Immunobiology* **191**,
537–547.

Andersen, P., Andersen, A.B., Sorensen, A.L. & Nagai, S.
(1995) Recall of long-lived immunity to *Mycobacterium
tuberculosis* infection in mice. *Journal of Immunology* **154**,
3359–3372.

Andersen, P., Askgaard, D., Ljungqvist, L., Bennedsen, J. &
Heron, I. (1991a) Proteins released from *Mycobacterium
tuberculosis* during growth. *Infection and Immunity* **59**,
1905–1910.

Andersen, P., Askgaard, D., Ljungqvist, L., Bentzon, M.W. &
Heron, I. (1991b) T-cell proliferative response to antigens
secreted by *Mycobacterium tuberculosis. Infection and
Immunity* **59**, 1558–1563.

Anderton, S.M., van der Zee, R., Noordzij, A. & van Eden,
W. (1994) Differential mycobacterial 65-kDa heat shock
protein T cell epitope recognition after adjuvant arthritis-
inducing or protective immunization protocols. *Journal of
Immunology* **152**, 3656–3664.

Anderton, S.M. & van Eden, W. (1996) T-lymphocyte
recognition of hsp 60 in experimental arthritis. In: *Stress
Proteins in Medicine* (eds W. van Eden, & D. B. Young).
New York: Marcel Dekker, pp. 73–91.

Babiuk, L.A., Lewis, P.J., Cox, G., van Drunen Littel-van
den Hurk, S., Baca-Estrada, M. & Tikoo, S.K. (1995) DNA
immunization with bovine herpesvirus-1 genes. *Annals of
the New York Academy of Sciences* **772**, 47–63.

Balasubramanian, V., Pavelka, M.S., Bardarov, S.S. *et al.*
(1996) Allelic exchange in *Mycobacterium tuberculosis* with
long linear recombination substrates. *Journal of
Bacteriology* **178**, 273–279.

Balasubramanian, V., Wiegeshaus, E.H. & Smith, D.W.
(1994) Mycobacterial infection in guinea pigs.
Immunobiology **191**, 395–401.

Bange, F.C., Brown, A.M. & Jacobs, W.R. (1996) Leucine
auxotrophy restricts growth of *Mycobacterium bovis* BCG
in macrophages. *Infection and Immunity* **64**, 1794–1799.

Barnes, P.F., Mehra, V., Rivoire, B. *et al.* (1992)
Immunoreactivity of a 10-kDa-antigen of *Mycobacterium
tuberculosis. Journal of Immunology* **148**, 1835–1840.

Barrios, C., Brawand, P., Berney, M., Brandt, C., Lambert,
P.H. & Siegrist, C.A. (1996) Neonatal and early life
immune responses to various forms of vaccine antigens
qualitatively differ from adult responses: Predominance
of a Th2-biased pattern which persists after adult
boosting. *European Journal of Immunology* **26**, 1489–
1496.

Barrow, W.W., Carvalho de Sousa, J.P., Davis, T.L., Wright,
E.L., Bachelet, M. & Rastogi, N. (1993)
Immunomodulation of human peripheral blood
mononuclear cell functions by defined lipid fractions of
Mycobacterium avium. Infection and Immunity **61**,
5286–5293.

Barry, M.A., Lai, W.C. & Johnston, S.A. (1995) Protection
against mycoplasma infection using expression-library
immunization. *Nature* **377**, 632–635.

Baumgart, K.W., McKenzie, K.R., Radford, A.J., Ramshaw,
I. & Britton, W.J. (1996) Immunogenicity and protection
studies with recombinant mycobacteria and vaccinia
vectors coexpressing the 18-kilodalton protein of
Mycobacterium leprae. Infection and Immunity **64**,
2274–2281.

Billingham, M.E.J., Carney, S., Butler, R. & Colston, M.J.
(1990) A mycobacterial 65-kDa heat shock protein
induces antigen-specific suppression of adjuvant arthritis,
but is not itself arthritogenic. *Journal of Experimental
Medicine* **171**, 339–344.

Birk, O.S., Douek, D.C., Elias, D. *et al.* (1996) A role of
Hsp60 in autoimmune diabetes: analysis in a transgenic
model. *Proceedings of the National Academy of Sciences of the
USA* **93**, 1032–1037.

Bloom, B.R. & Fine, P.E.M. (1994) The BCG experience:
implications for future vaccines against tuberculosis. In:
Tuberculosis: Pathogenesis, Protection, and Control (ed. B. R.
Bloom). Washington, DC: American Society for
Microbiology, pp. 531–557

Boesen, H., Jensen, B.N., Wilcke, T. & Andersen, P. (1995)
Human T-cell responses to secreted antigen fractions of
Mycobacterium tuberculosis. Infection and Immunity **63**,
1491–1497.

Bonato, V.L.D., Lima, V.M.F., Tascon, R.E., Lowrie, D.B. &
Silva, C.L. (1998) Identification and characterization of
protective T cells in hsp65 DNA vaccinated and
Mycobacterium tuberculosis infected mice. *Infection and
Immunity* **66**, 169–175.

Boom, W.H. (1996) The role of T-cell subsets in
Mycobacterium tuberculosis infection. *Infectious Agents and
Disease: Reviews, Issues and Commentary* **5**, 73–81.

Bot, A., Antohi, S., Bot, S., Garcia-Sastre, A. & Bona, C.
(1997) Induction of humoral and cellular immunity
against influenza virus by immunization of newborn
mice with a plasmid bearing a hemagglutinin gene.
International Immunology **9**, 1641–1650.

Bradbury, M.G. & Moreno, C. (1993) Effect of lipoarabinomannan and mycobacteria on tumour necrosis factor production by different populations of murine macrophages. *Clinical and Experimental Immunology* **94**, 57–63.

Bretscher, P.A. (1994) Prospects for low dose BCG vaccination against tuberculosis. *Immunobiology* **191**, 548–554.

Bretscher, P.A., Menon, J.N. & Ogunremi, O. (1996) Towards a strategy of universally efficacious vaccination against pathogens uniquely susceptible to cell-mediated attack. *Journal of Biotechnology* **44**, 1–4.

Brewer, T.F. & Colditz, G.A. (1995) Relationship between bacille Calmette–Guérin (BCG) strains and the efficacy of BCG vaccine in the prevention of tuberculosis. *Clinical Infectious Diseases* **20**, 126–135.

Brown, C.A., Brown, I.N. & Swinburne, S. (1985) The effect of oral *Mycobacterium vaccae* on subsequent responses to BCG vaccination. *Tubercle* **66**, 251–260.

Chatterjee, D., Roberts, A.D., Lowell, K., Brennan, P.J. & Orme, I.M. (1992) Structural basis of capacity of lipoarabinomannan to induce secretion of tumor necrosis factor. *Infection and Immunity* **60**, 1249–1253.

Cheng, S.H., Walker, K.B., Lowrie, D.B., *et al.* (1993) Monocyte antimycobacterial activity before and after *Mycobacterium bovis* BCG vaccination in Chingleput, India, and London, United Kingdom. *Infection and Immunity* **61**, 4501–4503.

Cirillo, J.D., Stover, C.K., Bloom, B.R., Jacobs, W.R. & Barletta, R.G. (1995) Bacterial vaccine vectors and bacillus Calmette–Guérin. *Clinics in Infectious Disease* **20**, 1001–1009.

Cohen, I.R. (1991) Autoimmunity to chaperonins in the pathogenesis of arthritis and diabetes. *Annual Review of Immunology* **9**, 567–589.

Cohen, I.R. & Elias, D. (1996) Immunity to 60 kDa heat shock protein in autoimmune diabetes. *Diabetes Nutrition and Metabolism* **9**, 229–232.

Colditz, G.A., Berkey, C.S., Mosteller, F. *et al.* (1995) The efficacy of bacillus Calmette–Guérin vaccination of newborns and infants in the prevention of tuberculosis: Meta-analyses of the published literature, *Pediatrics* **96**, 29–35.

Cole, S.T., Brosch, R., Parkhill, J. *et al.* (1998) Deciphering the biology of *Mycobacterium tuberculosis* from the complete genome sequence. *Nature* **393**, 537–544.

Collins, F.M. (1994) The immune response to mycobacterial infection: Development of new vaccines. *Veterinary Microbiology* **40**, 95–110.

Collins, F.M., Lamb, J.R. & Young, D.B. (1988) Biological activity of protein antigens isolated from *Mycobacterium tuberculosis* culture filtrate. *Infection and Immunity* **56**, 1260–1266.

Colston, M.J. (1990) Protective immunity against mycobacterial infections: investigating cloned antigens.

In: *Molecular Biology of the Mycobacteria* (ed. J. McFadden). London: Surrey University Press, pp. 69–76.

Converse, P.J., Dannenberg, A.M., Estep, J.E. *et al.* (1996) Cavitary tuberculosis produced in rabbits by aerosolized virulent tubercle bacilli. *Infection and Immunity* **64**, 4776–4787.

Cooper, A.M., Dalton, D.K., Stewart, T.A., Griffin, J.P., Russell, D.G. & Orme, I.M. (1993) Disseminated tuberculosis in interferon gamma gene-disrupted mice. *Journal of Experimental Medicine* **178**, 2243–2247.

Cooper, A.M., D'Souza, C., Frank, A.A. & Orme, I.M. (1997) The course of *Mycobacterium tuberculosis* infection in the lungs of mice lacking expression of either perforin- or granzyme-mediated cytolytic mechanisms. *Infection and Immunity* **65**, 1317–1320.

Croft, M., Carter, L., Swain, S.L. & Dutton, R.W. (1994) Generation of polarized antigen-specific CD8 effector populations: reciprocal action of interleukin (IL)-4 and IL-12 in promoting type 2 versus type 1 cytokine profiles. *Journal of Experimental Medicine* **180**, 1715–1728.

ten Dam, H.G. (1984) Research on BCG vaccination. *Advances in Tuberculosis Research* **21**, 79–106.

Dannenberg, A.M. & Sugimoto, M. (1976) Liquefaction of caseous foci in tuberculosis. *American Review of Respiratory Diseases* **113**, 257–259.

Darji, A., Guzman, C.A., Gerstel, B. *et al.* (1997) Oral somatic transgene vaccination using attenuated *S. typhimurium*. *Cell* **91**, 765–775.

Davis, H.L., Michel, M.L., Mancini, M., Schleef, M. & Whalen, R.G. (1994) Direct gene transfer in skeletal muscle: Plasmid DNA-based immunization against the hepatitis B virus surface antigen. *Vaccine* **12**, 1503–1509.

Denagel, D.C. & Pierce, S.K. (1993) Heat shock proteins in immune responses. *Critical Reviews in Immunology* **13**, 71–81.

de Wit, D., Wootton, M., Dhillon, J. & Mitchison, D.A. (1995) The bacterial DNA content of mouse organs in the Cornell model of dormant tuberculosis. *Tubercle and Lung Disease* **76**, 555–562.

Dhiman, N. & Khuller, G.K. (1997) Immunoprophylactic properties of 71-kDa cell wall-associated protein antigen of *Mycobacterium tuberculosis* H37Ra. *Medical Microbiology and Immunology* **186**, 45–51.

Dicesare, S., Poccia, F., Mastino, A. & Colizzi, V. (1992) Surface expressed heat-shock proteins by stressed or Human immunodeficiency virus (HIV)-infected lymphoid cells represent the target for antibody-dependent cellular cytotoxicity. *Immunology* **76**, 341–343.

Donnelly, J.J., Ulmer, J.B., Shiver, J.W. & Liu, M.A. (1997) DNA vaccines. *Annual Review of Immunology* **15**, 617–648.

Editorial. (1960) Oral BCG vaccination. *Tubercle* **41**, 302–303.

Edwards, K.L., Goodrich, J.M., Muller, D., Pollack, A., Ziegler, J.E. & Smith, D.W. (1982) Infection with *Mycobacterium avium/intracellulare* and the protective effects of bacille Calmette–Guérin. *Journal of Infectious Diseases* **145**, 733–741.

Falcone, V., Bassey, E., Jacobs, W. & Collins, F. (1995) The immunogenicity of recombinant *Mycobacterium smegmatis* bearing BCG genes. *Microbiology (UK)* **141**, 1239–1245.

Felten, M.K. & Leichsenring, M. (1995) Use of BCG in high prevalence areas for HIV. *Tropical Medicine and Parasitology* **46**, 69–71.

Fifis, T., Costopoulos, C., Radford, A.J., Bacic, A. & Wood, P.R. (1991) Purification and characterization of major antigens from a *Mycobacterium bovis* culture filtrate. *Infection and Immunity* **59**, 800–807.

Fine, P.E.M. (1995) Variation in protection by BCG: Implications of and for heterologous immunity. *Lancet* **346**, 1339–1345.

Fine, P.E.M. (1996) Primary prevention of leprosy. *International Journal of Leprosy* **64**, S44–S49.

Fine, P.E.M. & Smith, P.G. (1996) Vaccination against leprosy—the view from 1996. *Leprosy Review* **67**, 249–252.

Flynn, J.L., Chan, J., Triebold, K.J., Dalton, D.K., Stewart, T.A. & Bloom, B.R. (1993) An essential role for interferon gamma in resistance to *Mycobacterium tuberculosis* infection. *Journal of Experimental Medicine* **178**, 2249–2254.

Flynn, J.L., Goldstein, M.M., Triebold, K.J., Koller, B. & Bloom, B.R. (1992) Major histocompatibility complex Class-I-restricted T-cells are required for resistance to *Mycobacterium tuberculosis* infection. *Proceedings of the National Academy of Sciences of the USA* **89**, 12013–12017.

Francis, J. (1958) *Tuberculosis in Animals and Man: a Comparative Pathology*. London: Cassell.

Garbe, T., Harris, D., Vordermeier, M., Lathigra, R., Ivanyi, J. & Young, D. (1993) Expression of the *Mycobacterium tuberculosis* 19-kilodalton antigen in *Mycobacterium smegmatis*—immunological analysis and evidence of glycosylation. *Infection and Immunity* **61**, 260–267.

Gelber, R.H., Mehra, V., Bloom, B. *et al.* (1994) Vaccination with pure *Mycobacterium leprae* proteins inhibits *M. leprae* multiplication in mouse footpads. *Infection and Immunity* **62**, 4250–4255.

Glaziou, P., Cartel, J.-L., Moulia-Pelat, J.-P. *et al.* (1993) Tuberculosis in leprosy patients detected between 1902 and 1991 in French Polynesia. *International Journal of Leprosy* **61**, 199–1993.

Grange, J.M. (1994) Is the incidence of AIDS-associated *Mycobacterium avium/intracellulare* disease affected by previous exposure to BCG. *M tuberculosis* or environmental mycobacteria? *Tuberculosis and Lung Disease* **75**, 234–236.

Gulle, H., Fray, L.M., Gormley, E.P., Murray, A. & Moriarty, K.M. (1995) Responses of bovine T cells to fractionated lysate and culture filtrate proteins of *Mycobacterium bovis* BCG. *Veterinary Immunology and Immunopathology* **48**, 183–190.

Gupta, L.K., Bhatnagar, S., Ram, G.C. & Bansal, M.P. (1995) Role of cytotoxic lymphocytes in immunity against bovine tuberculosis. *Medical Sciences Research* **23**, 535–538.

Halpern, M.D., Kurlander, R.J. & Pisetsky, D.S. (1996) Bacterial DNA induces murine interferon-gamma production by stimulation of interleukin-12 and tumor necrosis factor-alpha. *Cellular Immunology* **167**, 72–78.

Harboe, M., Andersen, P., Colston, M.J. *et al.* (1996a) Vaccines against tuberculosis: Report of the Expert Panel IX, European Commission COST/STD Initiative. *Vaccine* **14**, 701–716.

Harboe, M., Oettinger, T., Wiker, H.G., Rosenkrands, I. & Andersen, P. (1996b) Evidence for occurrence of the ESAT-6 protein in *Mycobacterium tuberculosis* and virulent *Mycobacterium bovis* and for its absence in *Mycobacterium bovis* BCG. *Infection and Immunity* **64**, 16–22.

Hart, P.D'A. & Sutherland, I. (1977) BCG and vole bacillus vaccines in the prevention of tuberculosis in adolescence and early adult life: Final report to the Medical Research Council. *British Medical Journal* **2**, 293–295.

Haslov, K., Andersen, A.B., Ljungqvist, L. & Bentzon, M.W. (1990) Comparison of the immunological activity of 5 defined antigens from *Mycobacterium tuberculosis* in 7 inbred guinea pig strains. *Scandinavian Journal of Immunology* **31**, 503–514.

Haslov, K., Andersen, A., Nagai, S., Gottschau, A., Sorensen, T. & Andersen, P. (1995) Guinea pig cellular immune responses to proteins secreted by *Mycobacterium tuberculosis*. *Infection and Immunity* **63**, 804–810.

Hassett, D.E. & Whitton, J.L. (1996) DNA immunization. *Trends in Microbiology* **4**, 307–312.

Havlir, D.V., Wallis, R.S., Boom, W.H., Daniel, T.M., Chervenak, K. & Ellner, J.J. (1991) Human immune response to *Mycobacterium tuberculosis* antigens. *Infection and Immunity* **59**, 665–670.

Haynes, J.R., Fuller, D.H., Eisenbraun, M.D., Ford, M.J. & Pertmer, T.M. (1994) Accell (R) particle-mediated DNA immunization elicits humoral, cytotoxic, and protective immune responses. *AIDS Research and Human Retroviruses* **10**, S43–S45.

Herbert, D., Paramasivan, C.N. & Prabhakar, R. (1994) Protective response in guinea-pigs exposed to *Mycobacterium avium intracellulare, M. scrofulaceum*, BCG & South Indian isolates of *M. tuberculosis*. *Indian Journal of Medical Research* **99**, 1–7.

Hernandez-Pando, R., Orozcoe, H., Sampieri, A. *et al.* (1996) Correlation between the kinetics of Th1/Th2 cells and pathology in a murine model of experimental pulmonary tuberculosis. *Immunology* **89**, 26–33.

Hernandez-Pando, R., Pavon, L., Arriaga, K., Orozco, H., Madrid-Marina, V. & Rook, G. (1997) Pathogenesis of tuberculosis in mice exposed to low and high doses of an environmental mycobacterial saprophyte before infection. *Infection and Immunity* **65**, 3317–3327.

Honda, M., Matsuo, K., Nakasone, T. *et al.* (1995) Protective immune responses induced by secretion of a chimeric soluble protein from a recombinant *Mycobacterium bovis* bacillus Calmette–Guérin vector candidate vaccine for human immunodeficiency virus type 1 in small animals. *Proceedings of the National Academy of Sciences of the USA* **92**, 10693–10697.

Horsburgh, C.R., Hanson, D.L., Jones, J.L. & Thompson, S.E. (1996) Protection from *Mycobacterium avium* complex disease in human immunodeficiency virus-infected persons with a history of tuberculosis. *Journal of Infectious Diseases* **174**, 1212–1217.

Horwitz, M.A., Lee, B.W.E., Dillon, B.J. & Harth, G. (1995) Protective immunity against tuberculosis induced by vaccination with major extracellular proteins of *Mycobacterium tuberculosis*. *Proceedings of the National Academy of Sciences of the USA* **92**, 1530–1534.

Hubbard, R.D., Flory, C.M. & Collins, F.M. (1992) Immunization of mice with mycobacterial culture filtrate proteins. *Clinical and Experimental Immunology* **87**, 94–98.

Huygen, K., Content, J., Denis, O. *et al.* (1996) Immunogenicity and protective efficacy of a tuberculosis DNA vaccine. *Nature Medicine* **2**, 893–898.

Jacobs, W.R., Tuckman, M. & Bloom, B.R. (1987) Introduction of foreign DNA into mycobacteria using a shuttle phasmid. *Nature* **327**, 532–535.

Johnston, S.A. & Tang, D.C. (1994) Gene gun transfection of animal cells and genetic immunization. *Methods in Cell Biology* **43**, 353–365.

Jones, D.H., Corris, S., McDonald, S., Clegg, J.C.S. & Farrar, G.H. (1997) Poly(DL-lactide-co-glycolide) -encapsulated plasmid DNA elicits systemic and mucosal antibody responses to encoded protein after oral administration. *Vaccine* **15**, 814–817.

Kamala, T., Paramasivan, C.N., Herbert, D., Venkatesan, P. & Prabhakar, R. (1996) Immune response and modulation of immune response induced in the guinea-pigs by *Mycobacterium avium* complex (MAC) and *M. fortuitum* complex isolates from different sources in the south Indian BCG trial area. *Indian Journal of Medical Research* **103**, 201–211.

Kameoka, M., Nishino, Y. & Matsuo, K. (1994) Cytotoxic T lymphocyte response in mice induced by a recombinant BCG vaccination which produces an extracellular alpha antigen that fused with the human immunodeficiency virus type 1 envelope immunodominant domain in the V3 loop. *Vaccine* **12**, 153–158.

Kaufmann, S.H.E. (1990) Heat shock proteins and the immune response. *Immunology Today* **11**, 129–136.

Kaufmann, S.H.E. (1993) Immunity to intracellular bacteria. *Annual Review of Immunology* **11**, 129–163.

Kaufmann, S.H.E. & Hess, J. (1997) Rational design of

antituberculosis vaccines: Impact of antigen display and vaccine localization. *Biologicals* **25**, 169.

Kaye, P.M. (1995) Costimulation and the regulation of antimicrobial immunity. *Immunology Today* **16**, 423–427.

Kingston, A.E., Hicks, C.A., Colston, M.J. & Billingham, M.E.J. (1996) A 71-kD heat shock protein (hsp) from *Mycobacterium tuberculosis* has modulatory effects on experimental rat arthritis. *Clinical and Experimental Immunology* **103**, 77–82.

Klatser, P.R., Britton, W.J., Cole, Naafs, B., Gupte, M.D. & Young, D. (1996) Immunoprophylaxis against leprosy; The case for improved vaccines against leprosy — Discussion. *International Journal of Leprosy and Mycobacterial Disease* **64**, S79.

Kong, D.Q. & Kunimoto, D.Y. (1995) Secretion of human interleukin 2 by recombinant *Mycobacterium bovis* BCG. *Infection and Immunity* **63**, 799–803.

Krieg, A.M. (1996) An innate immune defense mechanism based on the recognition of CpG motifs in microbial DNA. *Journal of Laboratory and Clinical Medicine* **128**, 128–133.

Kumar, V. & Sercarz, E. (1996) Genetic vaccination: the advantages of going naked. *Nature Medicine* **2**, 857–859.

Lagranderie, M., Balazuc, A.M., Gicquel, B. & Gheorghiu, M. (1997) Oral immunization with recombinant *Mycobacterium bovis* BCG Simian immunodeficiency virus nef induces local and systemic cytotoxic T-lymphocyte responses in mice. *Journal of Virology* **71**, 2303–2309.

Lagrange, P.H. (1984) Cell-mediated immunity and delayed-type hypersensitivity. In: *The Mycobacteria, a Sourcebook* (eds G.P. Kubica, & L.G. Wayne). New York: Marcel Dekker, pp. 681–720.

Langermann, S., Palaszynski, S.R., Burlein, J.E. *et al.* (1994) Protective humoral response against pneumococcal infection in mice elicited by recombinant bacille Calmette–Guérin vaccines expressing pneumococcal surface protein A. *Journal of Experimental Medicine* **180**, 2277–2286.

Laochumroonvorapong, P., Wang, J., Liu, C.C. *et al.* (1997) Perforin, a cytotoxic molecule which mediates cell necrosis, is not required for the early control of mycobacterial infection in mice. *Infection and Immunity* **65**, 127–132.

Lathigra, R., Zhang, Y., Hill, M., Garcia, M.J., Jackett, P.S. & Ivanyi, J. (1996) Lack of production of the 19-kDa glycolipoprotein in certain strains of *Mycobacterium tuberculosis*. *Research in Microbiology* **147**, 237–249.

Lecoeur, H., Lagrange, P.H., Truffot, C. & Grosset, J. (1981) Microbial persistence and tuberculosis hypersensitivity after chemotherapy of the mouse experimental tuberculosis. *Annales d'Immununologie (Paris)* **132D**, 237–248.

Lee, B.Y., Hefta, S.A. & Brennan, P.J. (1992) Characterization of the major membrane protein of virulent *Mycobacterium tuberculosis*. *Infection and Immunity* **60**, 2066–2074.

Lee, B.Y. & Horwitz, M.A. (1995) Identification of macrophage and stress-induced proteins of *Mycobacterium tuberculosis*. *Journal of Clinical Investigation* **96**, 245–249.

Lefevre, P., Braibant, M., De Wit, L. *et al.* (1997) Three different putative phosphate transport receptors are encoded by the *Mycobacterium tuberculosis* genome and are present at the surface of *Mycobacterium bovis* BCG. *Journal of Bacteriology* **179**, 2900–2906.

Lefford, M.J. & McGregor, D.D. (1974) Immunological memory in tuberculosis. I. Influence of persisting viable organisms. *Cellular Immunology* **14**, 417–428.

Lim, E.M., Rauzier, J., Timm, J. *et al.* (1995) Identification of *Mycobacterium tuberculosis* DNA sequences encoding exported proteins by using phoA gene fusions. *Journal of Bacteriology* **177**, 59–65.

Lowrie, D.B. (1998) DNA vaccination exploits normal biology. *Nature Medicine* **4**, 147–148.

Lowrie, D.B., Colston, M.J., Tascon, R.E. & Silva, C.L. (1997a) DNA encoding individual mycobacterial antigens protects mice against tuberculosis. In: *Vaccines 97: Molecular Approaches to the Control of Infectious Diseases* (eds F. Brown, D. Burton, P. Doherty, J. Mekalanos & E. Norrby): Cold Spring Harbor: Cold Spring Harbor Laboratory Press, pp. 163–166.

Lowrie, D.B., Silva, C.L., Colston, M.J., Ragno, S. & Tascon, R.E. (1997b) Protection against tuberculosis by a plasmid DNA vaccine. *Vaccine* **15**, 834–838.

Lowrie, D.B., Silva, C.L. & Tascon, R.E. (1998) Progress towards a new tuberculosis vaccine. *Biodrugs* **10**, 201–213.

Lowrie, D., Tascon, R.E., Colston. M.J. & Silva, C.L. (1994) Towards a DNA vaccine against tuberculosis. *Vaccine* **12**, 1537–1540.

Lyons, J., Sinos, C., Destree, A. *et al.* (1990) Expression of *Mycobacterium tuberculosis* and *Mycobacterium leprae* proteins by vaccinia virus. *Infection and Immunity* **58**, 4089–4098.

Maes, H.H., Causse, J.E. & Maes, R.F. (1996) Mycobacterial infections: are the observed enigmas and paradoxes explained by immunosuppression and immunodeficiency? *Medical Hypotheses* **46**, 163–171.

Mahairas, G.G., Sabo, P.J., Hickey, M.J., Singh, D.C. & Stover, C.K. (1996) Molecular analysis of genetic differences between *Mycobacterium bovis* BCG and virulent *M. bovis*. *Journal of Bacteriology* **178**, 1274–1282.

Marsh, B.J., von Reyn, C.F., Edwards, J. *et al.* (1997) The risks and benefits of childhood bacille Calmette–Guérin immunization among adults with AIDS. *AIDS* **11**, 669–672.

Marshall, B.G., Chambers, M.A., Wangoo, A., Shaw, R.J. & Young, D.B. (1997) Production of tumor necrosis factor and nitric oxide by macrophages infected with live and dead mycobacteria and their suppression by an interleukin-10-secreting recombinant. *Infection and Immunity* **65**, 1931–1935.

Matsumoto, S., Tamaki, M., Yukitake, H. *et al.* (1996) A stable *Escherichia coli*–mycobacteria shuttle vector 'pSO246' in *Mycobacterium bovis* BCG. *FEMS Microbiology Letters* **135**, 237–243.

Matsuo, K., Yamaguchi, R., Yamazaki, A. *et al.* (1990) Establishment of a foreign antigen secretion system in mycobacteria. *Infection and Immunity* **58**, 4049–4054.

Matsuoka, M., Nomaguchi, H., Yukitake, H. *et al.* (1997) Inhibition of multiplication of *Mycobacterium leprae* in mouse foot pads by immunization with ribosomal fraction and culture filtrate from *Mycobacterium bovis* BCG. *Vaccine* **15**, 1214–1217.

McAdam, R.A., Weisbrod, T.R., Martin, J. *et al.* (1995) *In vivo* growth characteristics of leucine and methionine auxotrophic mutants of *Mycobacterium bovis* BCG generated by transposon mutagenesis. *Infection and Immunity* **63**, 1004–1012.

Mehra, V., Bloom, B.R., Bajardi, A.C. *et al.* (1992) A major T-cell antigen of *Mycobacterium leprae* is a 10-kD heat-shock cognate protein. *Journal of Experimental Medicine* **175**, 275–284.

Mehra, V., Gong, J.H., Iyer, D. *et al.* (1996) Immune response to recombinant mycobacterial proteins in patients with tuberculosis infection and disease. *Journal of Infectious Diseases* **174**, 431–434.

Mendez-Samperio, P., Gonzalez-Garcia, L., Pineda-Fragoso, P.R. & Ramos-Sanchez, E. (1995) Specificity of T cells in human resistance to *Mycobacterium tuberculosis* infection. *Cellular Immunology* **162**, 194–201.

Monahan, I.M., Banerjee, D.K., Jacobs, W.R. & Butcher, P.D. (1996) Gene expression of *Mycobacterium bovis* BCG during interaction with macrophages. *Third International Conference on the Pathogenesis of Mycobacterial Diseases, Stockholm, Sweden*. Stockholm: Swedish Institute for Infectious Disease Control, p. 43.

Murray, P.J. & Young, R.A. (1992) Stress and immunological recognition in host–pathogen interactions. *Journal of Bacteriology* **174**, 4193–4196.

Murray, P.J., Aldovini, A. & Young, R.A. (1996) Manipulation and potentiation of antimycobacterial immunity using recombinant bacille Calmette–Guérin strains that secrete cytokines. *Proceedings of the National Academy of Sciences of the USA* **93**, 934–939.

Nagai, S., Wiker, H.G., Harboe, M. & Kinomoto, M. (1991) Isolation and partial characterization of major protein antigens in the culture fluid of *Mycobacterium tuberculosis*. *Infection and Immunity* **59**, 372–382.

Nomoto, K. & Yoshikai, Y. (1991) Heat-shock proteins and immunopathology—regulatory role of heat-shock protein-specific T-cells. *Springer Seminars in Immunopathology* **13**, 63–80.

Norman, E., Dellagostin, O.A., McFadden, J. & Dale, J.W.

(1995) Gene replacement by homologous recombination in *Mycobacterium bovis* BCG. *Molecular Microbiology* **16**, 755–760.

O'Brien, K.L., Ruff, A.J., Louis, M.A. *et al.* (1995) Bacillus Calmette–Guérin complications in children born to HIV-1-infected women with a review of the literature. *Pediatrics* **95**, 414–418.

Oftung, F., Mustafa, A.S., Husson, R., Young, R.A. & Godal, T. (1987) Human T cell clones recognise two abundant *Mycobacterium tuberculosis* protein antigens expressed in *Escherichia coli. Journal of Immunology* **138**, 927–931.

Onyebujoh, P.C., Abdulmumini, T., Robinson, S., Rook, G.A.W. & Stanford, J.L. (1995) Immunotherapy with *Mycobacterium vaccae* as an addition to chemotherapy for the treatment of pulmonary tuberculosis under difficult conditions in Africa. *Respiratory Medicine* **89**, 199–207.

Orme, I.M. (1995) Prospects for new vaccines against tuberculosis. *Trends in Microbiology* **3**, 401–404.

Orme, I.M. (1997) Progress in the development of new vaccines against tuberculosis. *International Journal of Tuberculosis and Lung Disease* **1**, 95–100.

Orme, I.M. & Collins, F.M. (1984) Efficacy of *Mycobacterium bovis* BCG vaccination in mice undergoing prior pulmonary infection with atypical mycobacteria. *Infection and Immunity* **44**, 28–32.

Ottenhoff, T.H., Spierings, E., Nibbering, P.H. & de Jong, R. (1997) Modulation of protective and pathological immunity in mycobacterial infections. *International Archives of Allergy and Immunology* **113**, 400–408.

Pal, P.G. & Horwitz, M.A. (1992) Immunization with extracellular proteins of *Mycobacterium tuberculosis* induces cell-mediated immune responses and substantial protective immunity in a guinea pig model of pulmonary tuberculosis. *Infection and Immunity* **60**, 4781–4792.

Palmer, C.E. & Long, M.W. (1966) Effects of infection with atypical mycobacteria on BCG vaccination and tuberculosis. *American Review of Respiratory Diseases* **94**, 553–568.

Pelicic, V., Jackson, M., Reyrat, J.M., Jacobs, W.R., Gicquel, B. & Guilhot, C. (1997) Efficient allelic exchange and transposon mutagenesis in *Mycobacterium tuberculosis. Proceedings of the National Academy of Sciences of the USA* **94**, 10955–10960.

Plum, G., Brenden, M., Clark-Curtiss, J.E. & Pulverer, G. (1997) Cloning, sequencing, and expression of the *mig* gene of *Mycobacterium avium*, which codes for a secreted macrophage-induced protein. *Infection and Immunity* **65**, 4548–4557.

Pollock, J.M. & Andersen, P. (1997) Predominant recognition of the ESAT-6 protein in the first phase of infection with *Mycobacterium bovis* in cattle. *Infection and Immunity* **65**, 2587–2592.

Prakken, A.B.J., van Eden, W., Rijkers, G.T. *et al.* (1996) Autoreactivity to human heat-shock protein 60 predicts disease remission in oligoarticular juvenile rheumatoid arthritis. *Arthritis and Rheumatism* **39**, 1826–1832.

Ragno, S., Colston, M.J., Lowrie, D.B., Winrow, V.R., Blake, D.R. & Tascon, R. (1997) Immunization with naked DNA encoding mycobacterial hsp65 protects rats from adjuvant arthritis. *Arthritis and Rheumatism* **40**, 277–283.

Raz, E., Tighe, H., Sato, Y. *et al.* (1996) Preferential induction of a Th-1 immune response and inhibition of specific IgE antibody formation by plasmid DNA immunization. *Proceedings of the National Academy of Sciences of the USA* **93**, 5141–5145.

Reiner, N.E. (1994) Altered cell signaling and mononuclear phagocyte deactivation during intracellular infection. *Immunology Today* **15**, 374–381.

Rich, A.R. (1951) *The Pathogenesis of Tuberculosis*, 2nd edn. Blackwell Scientific Publications, Oxford.

Robertson, J.S. (1996) The potential of DNA vaccines. *Journal of Biotechnology in Healthcare* **3**, 239–247.

Roche, P.W., Triccas, J.A., Avery, D.T., Fifis, T., Billmanjacobe, H. & Britton, W.J. (1994) Differential T cell responses to mycobacteria-secreted proteins distinguish vaccination with bacille Calmette–Guérin from infection with *Mycobacterium tuberculosis. Journal of Infectious Diseases* **170**, 1326–1330.

Rodrigues, L.C., Diwan, V.K. & Wheeler, J.G. (1993) Protective effect of BCG against tuberculous meningitis and miliary tuberculosis: a meta-analysis. *International Journal of Epidemiology* **22**, 1154–1158.

Romanus, V., Hallander, H.O., Wahlen, P., Olinder Nielsen, A.M., Magnusson, P.H. & Juhlin, I. (1995) Atypical mycobacteria in extrapulmonary disease among children. Incidence in Sweden from 1969 to 1990, related to changing BCG-vaccination coverage. *Tuberculosis and Lung Disease* **76**, 300–310.

Rook, G.A.W. & Hernandez-Pando, R. (1994) T cell helper types and endocrines in the regulation of tissue-damaging mechanisms in tuberculosis. *Immunobiology* **191**, 478–492.

Rook, G.A.W., Onyebujoh, P. & Stanford, J.L. (1993) TH1/TH2: switching and loss of CD4+ T-cells in chronic infections—an immunoendocrinological hypothesis not exclusive to HIV. *Immunological Today* **14**, 568–569.

Russell, W.R., Pollock, J.M., Butcher, P.D. & Hewinson, R.G. (1996) The detection of *M. bovis* antigens expressed inside macrophages. *Proceedings of the Second International Conference on Mycobacterium bovis 'Tuberculosis in Wildlife and Domestic Animals', Otago Conference* (eds F. Griffin, & G. de Lisle). Series no. 3. F. Dunedin: University of Otago Press, pp. 135–138.

Sander, B., Skansen-Saphir, U., Damm, O., Hakansson, L., Andersson, J. & Andersson, U. (1995) Sequential production of Th1 and Th2 cytokines in response to live bacillus Calmette–Guérin. *Immunology* **86**, 512–518.

Seder, R.A. & Paul, W.E. (1994) Acquisition of lymphokine-producing phenotype by CD4+ T cells. *Annual Review of Immunology* **12**, 635–673.

Seghal, V.N. & Jain, S. (1995) Concomitant occurrence of leprosy and tuberculosis: leprosy vaccine, a myth or reality? *Leprosy Review* **66**, 179–181.

Siegrist, C.A. & Lambert, P.H. (1996) DNA vaccines: What can we expect? *Infectious Agents and Disease: Reviews, Issues and Commentary* **5**, 55–59.

Silva, C.L. & Lowrie, D.B. (1994) A single mycobacterial protein (hsp65) expressed by a transgenic antigen-presenting cell vaccinates mice against tuberculosis. *Immunology* **82**, 244–248.

Silva, C.L., Pietro, R.L.R., Januario, A. *et al.* (1995) Protection against tuberculosis by bone marrow cells expressing mycobacterial hsp65. *Immunology* **86**, 519–524.

Silva, C.L., Silva, M.F., Pietro, R.C.L.R. & Lowrie, D.B. (1996) Characterization of T cells that confer a high degree of protective immunity against tuberculosis in mice after vaccination with tumor cells expressing mycobacterial hsp65. *Infection and Immunity* **64**, 2400–2407.

Skinner, M.A., Yuan, S., Prestidge, R., Chuk, D., Watson, J.D. & Tan, P.L.J. (1997) Immunization with heat-killed *Mycobacterium vaccae* stimulates CD8 (+) cytotoxic T cells specific for macrophages infected with *Mycobacterium tuberculosis*. *Infection and Immunity* **65**, 4525–4530.

Smith, D.W. (1970) Why not vaccinate against tuberculosis? *Annals of Internal Medicine* **72**, 419–422.

Smith, D.W. (1985) Protective effect of BCG in experimental tuberculosis. *Advances in Tuberculosis Research* **22**, 1–97.

Smith, G.L. (1993) Vaccinia virus glycoproteins and immune evasion. The sixteenth Fleming Lecture. *Journal of General Virology* **74**, 1725–1740.

Soltys, M.A. (1952) *Tubercle Bacillus and Laboratory Methods in Tuberculosis*. London: Livingstone.

Sorensen, A.L., Nagai, S., Houen, G., Andersen, P. & Andersen, A.B. (1995) Purification and characterization of a low-molecular-mass T-cell antigen secreted by *Mycobacterium tuberculosis*. *Infection and Immunity* **63**, 1710–1717.

Stanford, J.L. & Rook, G.A.W. (1983) Environmental mycobacteria and immunization with BCG. *Medical Microbiology* **2**, 43–69.

Stanford, J.L. & Stanford, C.A. (1994) Immunotherapy of tuberculosis with *Mycobacterium vaccae* NCTC 11659. *Immunobiology* **191**, 555–563.

Sula, L. & Radkovsky, J. (1976) Protective effects of *M. microti* vaccine against tuberculosis. *Journal of Hygiene, Epidemiology, Microbiology and Immunology* **20**, 1–6.

Symons, J.A., Alcami, A. & Smith, G.L. (1995) Vaccinia virus encodes a soluble type I interferon receptor of novel structure and broad species specificity. *Cell* **81**, 551–560.

Tantimavanich, S., Nagai, S., Nomaguchi, H., Kinomoto, M., Ohara, N. & Yamada, T. (1993) Immunological properties of ribosomal proteins from *Mycobacterium bovis* BCG. *Infection and Immunity* **61**, 4005–4007.

Tascon, R.E., Colston, M.J., Ragno, S., Stavropoulos, E., Gregory, D. & Lowrie, D.B. (1996) Vaccination against tuberculosis by DNA injection. *Nature Medicine* **2**, 888–892.

Tripathy, S.P. (1979) Trial of BCG vaccines in South India for tuberculosis prevention: First Report. *Bulletin of the World Health Organization* **57**, 819–827.

Trnka, L., Dankova, D. & Svandova, E. (1994) Six years' experience with the discontinuation of BCG vaccination. 4. Protective effect of BCG vaccination against the *Mycobacterium avium/intracellulare* complex. *Tuberculosis and Lung Disease* **75**, 348–352.

Tsukaguchi, K., Balaji, K.N. & Boom, W.H. (1995) CD4 (+) alpha beta T cell and gamma delta T cell responses to *Mycobacterium tuberculosis*—Similarities and differences in antigen recognition, cytotoxic effector function, and cytokine production. *Journal of Immunology* **154**, 1786–1796.

van Eden, W., Anderton, S.M., van der Zee, R., Prakken, A.B.J. & Rijkers, G.T. (1995) Specific immunity as a critical factor in the control of autoimmune arthritis: The example of hsp60 as an ancillary and protective autoantigen. *Scandinavian Journal of Rheumatology* **101**, 141–145.

Wada, N., Ohara, N., Kameoka, M. *et al.* (1996) Long-lasting immune response induced by recombinant Bacillus Calmette–Guérin (BCG) secretion system. *Scandinavian Journal of Immunology* **43**, 202–209.

Walsh, G.P., Tan, E.V., de la Cruz, E.C. *et al.* (1996) The Philippine cynomolgus monkey (*Macaca fascicularis*) provides a new non-human primate model of tuberculosis that resembles human disease. *Nature Medicine* **2**, 430–436.

Warwick-Davies, J., Dhillon, J.L.O.B., Andrew, P.W. & Lowrie, D.B. (1994) Apparent killing of *Mycobacterium tuberculosis* by cytokine-activated human monocytes can be an artefact of a cytotoxic effect on the monocytes. *Clinical and Experimental Immunology* **96**, 214–217.

Wiegeshaus, E.H., Harding, G., McMurray, D., Grover, A.A. & Smith, D.W. (1971) A co-operative evaluation of test systems used to assay tuberculosis vaccines. *Bulletin of the World Health Organization* **45**, 543–550.

Wiegeshaus, E.H. & Smith, D.W. (1989) Evaluation of the protective potency of new tuberculosis vaccines. *Reviews of Infectious Diseases* **11S**, 484–490.

Wieles, B., Ottenhoff, T.H.M., Steenwijk, T.M., Franken, K.L.M.C., de Vries, R.R.P. & Langermans, J.A.M. (1997) Increased intracellular survival of *Mycobacterium smegmatis* containing the *Mycobacterium leprae* thioredoxin-thioredoxin reductase gene. *Infection and Immunity* **65**, 2537–2541.

Wiker, H.G., Nagai, S., Harboe, M. & Ljungqvist, L. (1992) A family of cross-reacting proteins secreted by *Mycobacterium tuberculosis*. *Scandinavian Journal of Immunology* **36**, 307–319.

Wolffe, J.A., Ludtke, J.J., Acsadi, G., Williams, P. & Jani, A. (1992) Long-term persistence of plasmid DNA and foreign gene expression in mouse muscle. *Human Molecular Genetics* **1**, 363–369.

Yamamura, M., Wang, X.H., Ohmen, J.D. *et al.* (1992) Cytokine patterns of immunologically mediated tissue damage. *Journal of Immunology* **149**, 1470–1475.

Youmans, A.S. & Youmans, G.P. (1965) Immunogenic activity of a ribosomal fraction obtained from *Mycobacterium tuberculosis*. *Journal of Bacteriology* **89**, 1291–1298.

Young, D. (1996) Immunoprophylaxis of leprosy—lessons from the TB program. *International Journal of Leprosy and Other Mycobacterial Diseases* **64**, S75–S76.

Young, D.B., Kaufmann, S.H.E., Hermans, P.W.M. & Thole, J.E.R. (1992) Mycobacterial protein antigens—A compilation. *Molecular Microbiology* **6**, 133–145.

Young, R.A. (1990) Stress proteins and immunology. *Annual Review of Immunology* **8**, 401–420.

Young, R.A. & Elliott, T.J. (1989) Stress proteins, infection, and immune surveillance. *Cell* **59**, 5–8.

Yuan, Y., Crane, D.D. & Barry, C.E. (1996) Stationary phase-associated protein expression in *Mycobacterium tuberculosis*: Function of the mycobacterial alpha-crystallin homolog. *Journal of Bacteriology* **178**, 4484–4492.

Zhu, X., Venkataprasad, N., Ivanyi, J. & Vordermeier, H.M. (1997a) Vaccination with recombinant vaccinia viruses protects mice against *Mycobacterium tuberculosis* infection. *Immunology* **92**, 6–9.

Zhu, X.J., Venkataprasad, N., Thangaraj, H.S. *et al.* (1997b) Functions and specificity of T cells following nucleic acid vaccination of mice against *Mycobacterium tuberculosis* infection. *Journal of Immunology* **158**, 5921–5926.

Zhu, X., Venkataprasad, N., Thangaraj, H. & Vordermeier, H.M. (1996) Nucleic acid vaccination for protection against *M. tuberculosis*. *Third International Conference on the Pathogenesis of Mycobacterial Infections, Stockholm, Sweden*. Stockholm: Swedish Institute for Infectious Disease Control, p. 25.

Chapter 18 / Mycobacterial antigens

JELLE THOLE, RINY JANSSEN & DOUGLAS YOUNG

1 Introduction

Mycobacterial antigens represent the interface between mycobacteria and the host immune system. They can be studied from the perspective of the immune response, and many researchers have identified and characterized antigens as an approach to development of improved strategies for disease control through vaccination and immunodiagnosis. They can also be studied from the perspective of their functional role for the mycobacterium and, in this sense, may be viewed as lead components in studies aiming to advance our understanding of the physiology and virulence of mycobacterial pathogens. In this chapter we emphasize antigen characterization primarily from the perspective of the microbe.

2 Identification and characterization

Two major strategies have been followed to identify and characterize mycobacterial protein antigens. The first strategy was based on the screening of genomic expression libraries using antibodies and T cells as probes, and this approach has led to the identification of a multitude of protein antigens. Subsequent sequence analysis of the corresponding genes has allowed further characterization and provided clues as to their location and functional role in the mycobacterial cell. The second strategy has focused on the use of biochemical methods to characterize antigens in mycobacterial subcellullar fractions, with a particular emphasis on components present in filtrates prepared from mycobacterial cultures. Amino acid determination of components isolated from such fractions, complemented with reversed genetics, has led to the identification of an additional set of proteins (Young *et al.* 1992; Andersen & Brennan 1994; Thole *et al.* 1995). The biochemical approach has also allowed the identification of non-protein antigens such as glycolipids and phosphate-containing molecules from mycobacteria. An updated list of the best characterized mycobacterial antigens is presented in Table 18.1.

3 Structure and function

The structural and functional analysis of many of these antigens is in progress and has revealed that a broad repertoire of antigens—varying in their structural characteristics, biochemical function and subcellular compartmentalization—can provide targets for recognition by the host immune response.

3.1 Cytoplasmic antigens

3.1.1 Heat-shock protein families

These proteins were identified based on homologies to proteins of the highly conserved heat-shock families, in particular of the hsp60 and hsp70 families

Table 18.1 List of mycobacterial protein antigens.

Code no[b]	Organism	Subunit size (kDa)[c]	Names	MAbs[e]	Function[f]	Immunological characteristics	Genome annotations[g]
1T	*M. tuberculosis*	71	DnaK, CIE Ag63	51A, HAT1, HAT3	Heat-shock protein; role in protein. folding and translocation; >50% sequence identity with *E. coli* DnaK and human *hsp-70*	Antibody response in mouse and humans; proliferative T-cell response in patients and controls; potential target of autoreactivity?	Rv 0350
1B	BCG	70	DnaK				
1L	*M. leprae*	70	DnaK				Cosmid L611
2T	*M. tuberculosis*	65	GroEL, CIE Ag82	HAT5, CBA1, H2.16, TB78*	Heat-shock protein; role in protein folding and translocation; >50% sequence identity with *E. coli* GroEL and human *hsp-60*; functions with GroES. More recently, a second GroEL (Rv 3417c) was described for mycobacterium, that is organized in operon with GroES(Rv 3418c)	Antibody response in mouse and humans; proliferative and cytotoxic T-cell responses in patients and controls; recognized by some γδ T cells; multiple peptide epitopes mapped; autoreactive responses in rodents and humans	Rv 0440
2B	BCG	65	GroEL, 64-kDa antigen, MbaA,				
2L	*M. leprae*	65	GroEL	IIIE9, *IVD8, *IIH9, D5H, IIC8, Y1.2, ML30, CW1F2E8			cosmid B1620
3T	*M. tuberculosis*	38	PhoS, CIE Ag78, Pab, US-Japan Ag5	TB71, *TB72, *HYT28, *HBT 12, *C38.D1, *F67.19, *HAT2	Role as 'binding protein' in phosphate transport; potential signal peptide and lipoprotein consensus sequence; carbohydrate associated with purified protein	*M. tuberculosis* complex-specific antibody response in smear-positive patients; proliferative T-cell response in patients and after BCG vaccination	Rv 0932c
4T	*M. tuberculosis*	23	SodA, CIE Ag62		Superoxide dismutase; >50% sequence identity with *E. coli* SodA and MnSod from human mitochondria	Recognized by monoclonal antibodies. Recognized by T-cells from leprosy patients	Rv 3846
4L	*M. leprae*	28	SodA	F116.5, D2D			
5T	*M. tuberculosis*	12	GroES		Heat-shock protein; role in protein folding and translocation; functions with GroEL; extensive sequence identity with *E. coli* GroES	Recognized by monoclonal antibodies; induces strong proliferative T-cell response Response in tuberculoid leprosy and lepromin tests	Rv 3418c
5B	BCG	12	GroES, MPB57, BCG-a, MCP-I	SA12			
5L	*M. leprae*	14	GroES	CS-01			cosmid B1936
6T	*M. tuberculosis*	40		HBT10*	L-alanine dehydrogenase	Antibody HBT10 distinguishes BCG and *M. tuberculosis*	Rv 2780
8T	*M. tuberculosis*	36			Proline-rich antigen		Rv 1078
8L	*M. leprae*	36	Pra	F47.9*	Proline-rich antigen	Recognized by monoclonal antibodies, and antibodies in patients. Induces proliferative and suppressive responses in patients and infected individuals	cosmid B1306

Continued on p. 358

Table 18.1 Continued

Code no[b]	Organism	Subunit size (kDa)[c]	Names	MAbs[e]	Function[f]	Immunological characteristics	Genome annotations[g]
10T(a)	M. tuberculosis	30/31	CIE Ag85A, P32, MPT44	HYT27	Multigene family encoding three or more closely related major secreted proteins; signal peptide cleaved in mature proteins; fibronectin-binding proteins; function as mycolyl transferase. Also described for M. leprae and BCG	Recognized by MAbs; cross-reactive antibody response in leprosy and tuberculosis patients; proliferative T-cell response in mouse and humans; recognized by T-cells from synovial fluid of rheumatoid arthritis patients	Rv 3804c
10T(b)	M. tuberculosis	30/31	CIE Ag85B, MPT59, α antigen, US-Japan Ag6	HYT27			Rv 1886c
10T(c)	M. tuberculosis	30/31	CIE Ag85C	HYT27			Rv 0129c
11L	M. leprae	28			Possible iron-regulated protein; potential signal sequence	Antibody response in lepromatous leprosy patients	cosmid B1846
12T	M. tuberculosis	23	MPT64	L24,b4,C24,b1	Major secreted protein specific to M. tuberculosis complex		Rv 1980c
13T	M. tuberculosis	19		TB23, HYT6, F29, 47, 21-2H3	Potential signal peptide and lipoprotein consensus sequence; carbohydrate associated with purified protein. Potential role in phosphate uptake.	Antibody response in mouse and humans; proliferative T-cell response in patients and controls	Rv 3763
13B	M. bovis	19					
14L	M. leprae	18	L5*		Member of low-molecular-weight heat-shock protein family	Antibody response in mouse and humans; proliferative T-cell response in patients and controls	cosmid L537
15T	M. tuberculosis	18	MPT70		Major secreted protein		Rv 2875
15B	BCG	18	MPB70		Major secreted antigen	Specific antibody response in M. bovis infection	
16L	M. leprae	15	LSR2, A15		Unknown, contains RGD motif for binding to fibronectin	Antibody and T-cell responses in leprosy patients and healthy contacts, strongly recognized by ENL patients	cosmid B4
28L	M. leprae	35	MMP-1		Membrane-associated protein	Recognized by antibodies in patients	cosmid B2548
31T	M. tuberculosis	34	CIE84		Cytoplasmic protein		Rv 2145c
31L	M. leprae	34	CIE84		Cytoplasmic protein	Recognized by monoclonal antibodies and antibodies in patients	
36T	M. tuberculosis	16		TB68*, F23.49*, F24-2*	Member of low-molecular-weight heat-shock protein family		Rv 2031c

37L	*M. leprae*	12	ML10		Homologous to glutamine amido transferase	Recognized by monoclonal antibodies	
43L	*M. leprae*	45		FAP, 45/47 kDa, MPT32	Secreted antigen, binds to fibronectin. Also found in other mycobacteria	Recognized by antibodies and T cells in infected individuals	
47L	*M. leprae*	18		22 kDa MMPII, Bfr	Bacterioferritin, potentially iron-regulated 'iron-storage' protein	Recognized by antibodies and T cells in infected individuals	cosmid B38
57L	*M. leprae*	49			Bifunctional thioredoxin-reductase in *M. leprae*, in other mycobacteria expressed as 35 and 12 kDa proteins	Recognized by antibodies and T cells in infected individuals	cosmid L222
58L	*M. leprae*	46			Membrane protein; possibly involved in transport of small molecules through membrane	Recognized by antibodies	
59T	*M. tuberculosis*			Des	Acyl-acyl carrier protein desaturase	Recognized by antibodies in patients	Rv 0824c
60T	*M. tuberculosis*	6		ESAT6	Secreted antigen	Recognized by T-cells. Partial protection in animal models	Rv 3875
61T	*M. tuberculosis*	18		MPT63	Secreted antigen	Recognized by antibodies and T-cells	Rv 1926c
62B	*M. bovis*	23		MPB83	Glycosylated lipoprotein homologous to MPB70	Recognized by antibodies in patients	Rv 2873
63T	*M. tuberculosis*				Isocitrate dehydrogenase	Recognized by antibodies in patients	Rv 3339c
64T	*M. tuberculosis*				Malate dehydrogenase	Recognized by antibodies in patients	Rv 1240

[a] Modified from Young *et al*. 1992, Andersen and Brennan 1994, and Thole *et al*. 1995.

[b] An antigen is assigned a code number if (i) its subunit molecular mass is known and (ii) either the complete gene has been cloned or the N-terminal sequence has been established. A letter added after the code number designates *M. tuberculosis* (T), *M. bovis* (B) or *M. leprae* (L).

[c] Estimates of subunit molecular mass are based on those observed during SDS-PAGE, which often differ from those derived by sequence analysis.

[d] CIE numbers refer to the system based on analysis of mycobacterial extracts by CIE.

[e] Only MAbs included in WHO workshops are listed. In addition to the workshops already published (Engers *et al*. 1985, 1986), this includes results of a third workshop completed in May 1991 (Khanolkar-Young *et al*., 1992). Asterisks denote antibodies reported to show species specificity.

[f] In most cases, the proposed function is based on sequence comparison rather than on direct experiments with the mycobacterial proteins.

[g] Data obtained from Sanger database and MycDB.

(Young *et al.* 1988). Heat-shock proteins (hsp) are produced under normal physiological conditions but their synthesis is markedly increased when the cells are exposed to elevated temperature or other stress conditions, such as oxidative radicals (hence their alternative designation as 'stress proteins'). Heat-shock proteins have multiple functions, acting as chaperones to mediate polypeptide folding and assembly into multimers, and facilitating transport across membranes. These functions probably also assist cells to adapt to environmental changes induced by stress conditions by maintaining and protecting proteins from losing their 'functional' structure. Heat-shock proteins perform an essential function in all living cells and the hsp60 and hsp70 families are amongst the most highly conserved proteins in nature. Such molecules were found to represent important antigens from mycobacterial cell extracts, and were particularly prominent during early attempts to generate monoclonal antibodies against mycobacterial proteins in mice (Engers *et al.* 1985, 1986).

Hsp60/GroEL

The hsp60, or GroEL, homologue in mycobacteria is generally referred to as the 65-kDa antigen, and is probably the most extensively studied mycobacterial heat-shock protein. Limiting dilution analysis has shown it to be highly immunodominant (Kaufmann *et al.* 1987) and detailed B-cell and T-cell epitope maps of the *Mycobacterium tuberculosis* and *M. leprae* molecules have been constructed (Thole *et al.* 1990). The high degree of sequence conservation between bacterial and mammalian heat-shock proteins has stimulated discussion as to their possible role in autoimmune responses. A role for hsp60 has indeed been demonstrated in several animal models of autoimmunity, acting as a target of autoreactive T cells or as an immunomodulatory agent capable of altering the course of disease (van Eden 1991). The role of heat-shock proteins in human autoimmune disease is far less clear. Although autoreactive responses have been reported (Res *et al.* 1991) detailed analysis of the specificity of T-cell responses

to heat-shock proteins in individuals infected with mycobacteria indicate a preferential recognition of epitopes that are not shared with the corresponding self protein (Adams *et al.* 1997).

The GroEL protein forms a multimeric complex with a second heat-shock protein, GroES, to provide a 'scaffolding' structure that catalyses proper folding of newly synthesized polypeptides, and genes encoding the two proteins are generally linked on the bacterial chromosome. While a *groEL* gene does form an operon with the *groES* gene in mycobacteria, the 65-kDa antigen is in fact encoded by a second unlinked member of the *groEL* family (Rinke de Wit *et al.* 1992; Kong *et al.* 1993). It is not known why mycobacteria and related streptomycetes (Mazodier *et al.* 1991) require a second GroEL protein when common bacteria such as *Escherichia coli* and *Bacillus subtilis* make do with a single copy. The two genes are relatively divergent—judged by sequence similarity—the two mycobacterial genes are as closely related to *E. coli groEL* as they are to each other. Each GroEL protein might perform some of the functions carried out by the *E. coli* enzyme; alternatively, a second GroEL may have evolved to fulfil some additional purpose specific to the mycobacteria. The relative immunogenicity of the two proteins is also unclear; with only the 65-kDa antigen as yet identified as a prominent immune target. The 10-kDa GroES molecule, like the 65-kDa antigen, has been identified as a major B- and T-cell antigen in human and murine studies of *M. tuberculosis* and *M. leprae* (Barnes *et al.* 1992; Mehra *et al.* 1992; Orme *et al.* 1992).

Hsp70/DnaK

The hsp70 molecule is another heat-shock protein identified as a major antigen in several mycobacteria. The mycobacterial molecule, like other hsp70 homologues such as *E. coli* DnaK, displays adenosine triphosphatase (ATPase) and autophosphorylating activity (Peake *et al.* 1991), and is able to bind to short peptide molecules (Roman *et al.* 1994). Conformational changes associated with the ATPase activity are thought to enable hsp70 to regulate folding of polypeptide substrates. As in other bacteria, the mycobacterial *dnaK* gene is linked to genes encoding additional heat-shock proteins, which are required for regulation of the functional activity of DnaK. In *M. tuberculosis*, the *dnaK* gene forms an operon which includes genes encoding GrpE, DnaJ and a putative regulatory protein termed HspR (Bucca *et al.* 1997). While temperature-dependent induction of heat-shock proteins has been demonstrated by analysis of transcription and protein synthesis in mycobacteria (Mehlert & Young 1989; Mangan *et al.* 1997), the regulatory mechanisms underlying the heat-shock response have not been fully elucidated. Induction of DnaK is probably mediated by changes in a temperature–sensitive interaction involving the HspR protein (P. O'Gaora and D. Young, unpublished observations 1998). Similarly, GroES/EL induction is thought to be mediated by upstream regulatory proteins analogous to those described in *B. subtilis* (Hecker *et al.* 1996). Both hsp60 and hsp70 promoters are commonly used to direct expression of foreign genes in recombinant mycobacterial systems (Aldovini & Young 1991; Stover *et al.* 1991). Some induction of heat-shock protein expression has been reported following uptake of mycobacteria by macrophages (Lee & Horwitz 1995), although a high level of expression is routinely observed in the absence of stress stimuli.

α-crystallin

A third group of heat-shock proteins that have received attention as antigens in the host immune reponse to mycobacteria are the α-crystallin homologues. These molecules comprise a rather heterogenous group of so-called 'small heat shock proteins' that all share 20–30% homology to the α-crystallin lens protein. In mycobacteria, two antigens with molecular weights of 16 kDa (*M. tuberculosis*) and 18 kDa (*M. leprae*) have been characterized as belonging to this category of proteins, and both have been the subject of detailed immunological analysis (Mustafa *et al.* 1986; Booth *et al.* 1988; Verbon *et al.* 1992). A function for the 16-kDa protein in enhancing protein stability, and thus promoting long-term survival of mycobacteria, has recently been proposed. Increased

amounts of the 16-kDa antigen in stationary-phase and anaerobic cultures suggests that these proteins may play a role in survival of 'dormant' mycobacteria which remain quiescent within the host for extended time periods (Yuan *et al.* 1996).

3.1.1 Other cytoplasmic proteins

A variety of other cytoplasmic proteins have been identified as target molecules for the host immune response (see below): 10 kDa (Kuwabara 1975); 28 kDa (Cherayil & Young 1988); alanine dehydrogenase (Andersen *et al.* 1992); 34 kDa (Hermans *et al.* 1995); isocitrate and malate dehydrogenases (Ohman & Ridell 1996). Cytoplasmic antigens whose functions have been investigated in more detail are described below.

Superoxide dismutase

Superoxide dismutases (SOD) are metalloenzymes that form part of the defence against toxic forms of oxygen, produced either endogenously as byproducts of aerobic respiration, or exogenously through the respiratory burst of phagocytic host cells. Whereas the SOD produced by most mycobacteria utilizes Mn^{2+} as cofactor, the predominant SOD expressed by *M. tuberculosis* binds iron in its active site (Thangaraj *et al.* 1989; Zhang *et al.* 1991). A recent study has shown that alteration of a single amino acid residue converts the *M. tuberculosis* enzyme from an Fe-SOD to an Mn-SOD, prompting the suggestion that the unusual metal ion preference might reflect adaptation of *M. tuberculosis* to survival in a manganese-deficient environment (Bunting *et al.* 1998). A second unusual feature of the *M. tuberculosis* Fe-SOD is that, while it has characteristics of a typical cytoplasmic protein, it is also detected in the culture medium (Zhang *et al.* 1991). Similar findings have been reported with other slow-growing mycobacteria (Escuyer *et al.* 1996) and with *Nocardia asteroides* in which the extracellular SOD is thought to contribute to resistance to oxidative killing by host phagocytes (Beaman & Beaman 1990). In the absence of a signal peptide, it is not clear how SOD could be secreted from mycobacteria: potential mechanisms might include leakage from damaged cells, or the action of some unusual export pathway. A second *sod* gene is present in the genome of *M. tuberculosis*. In this case, sequence analysis predicts expression of a surface-exposed lipoprotein form of Cu/Zn-SOD. The crystal structure of the Fe-SOD of *M. tuberculosis* has been established. It is a multimeric protein in which the tight association of four 23-kDa subunits leaves an outward extending loop which is distal from the reactive site and which is not involved in subunit interactions (Cooper *et al.* 1994; Cooper *et al.* 1995). The ability to express recombinant SOD at high levels in a variety of mycobacteria, combined with the reported immunogenicity of this protein, has allowed the development of an efficient epitope delivery system based on the insertion of antigenic determinants in this outward extending loop (Hetzel *et al.*, 1998).

Bacterioferritin

Iron is an essential element for many functions in the bacterial cell. In *M. leprae* an 18-kDa antigen was recently characterized which might function as a bacterioferritin in the storage of iron (Pessolani *et al.* 1994). An 'iron-box' precedes the gene encoding this protein, which is a key component in a native complex of 380 kDa. This complex is proposed to function as a iron depository during conditions of iron deprivation encountered inside host cells, when exochelin- or mycobactin-mediated uptake may not suffice (see also Chapter 14).

Thioredoxin/thioredoxin reductase

Thioredoxin and thioredoxin reductase, together with reduced nicotinamide adenine dinucleotide phosphate (NADPH), form an integrated system that provides electrons for a wide variety of metabolic processes in eukaryotic and prokaryotic cells. Organization of genes encoding the two enzymes varies amongst mycobacteria (Wieles *et al.* 1995). In most mycobacteria they are located separately on the chromosome, whereas in bacteria belonging to the *M. tuberculosis* complex, both genes are found in the same locus overlapping in one nucleotide. *M. leprae* is unique in that both enzymatic functions are encoded by a single gene that specifies a 49-kDa protein—consisting of the thioredoxin reductase linked to the thioredoxin via a peptide spacer. The 49-kDa protein was initially identified as an antigen recognized by serum samples from leprosy patients.

3.2 Cell-wall antigens

3.2.1 *Non-protein antigens*

Cell-wall molecules, such as lipoarabinomannan (LAM), mycolylarabinogalactan and the phenolic glycolipid of *M. leprae* and some members of the *M. tuberculosis* complex, have long been recognized as dominant targets of the antibody response to mycobacteria. More recently, study of 'double-negative T cells' (that lack both the CD4 and the CD8 coligand) has shown that these non-protein molecules are also recognized by the cell-mediated arm of the immune response (Beckman *et al.* 1994; Sieling *et al.* 1995; Tanaka *et al.* 1995). Both mycolic acid and LAM molecules seem to be frequent targets for these T cells which recognize antigens bound to CD1 molecules on the surface of antigen-presenting cells (Moody *et al.* 1997). The importance of non-peptide antigens in the cell-mediated immune response is further highlighted by the discovery that a phosphorylated nucleotide derivative stimulates a dominant subset of T cells present in human blood that express a receptor comprising γ and δ chains, in place of the more common α/β chains (Tanaka *et al.* 1994; Behr *et al.* 1996). The functional role of these different T-cell subsets remains to be clarified, but the ability to respond to non-protein determinants provides an extension to the immune repertoire which may be particularly important in the case of the glycolipid-rich mycobacterial pathogens.

In addition to acting as targets for antibody and T-cell recognition, cell-wall components play an important role in triggering expression of cytokines by infected cells. LAM has been shown to induce, for example, tumour necrosis factor (TNF), an activity that is markedly decreased in the case of the 'mannose-capped' LAM expressed by slow-growing mycobacteria, which is characterized by the presence of mannose rather than arabinose as terminal residues (Chatterjee *et al.* 1992). The non-protein antigens are presumed to function primarily in maintaining the structure, or regulating the permeability, of the mycobacterial cell wall. In the context of understanding the evolution of mycobacterial virulence, it would be interesting to determine whether or not the structural features of these molecules that are essential for their role in microbial viability are the same as those that regulate their potency in cytokine induction. Glycolipids and carbohydrates exposed on the surface of mycobacteria are likely to play an important role in uptake by host cells (Hoppe *et al.* 1997).

3.2.2 Lipoproteins

The immunological significance of lipoproteins as antigens has been recognized through reports that acylation of proteins enhances their immunogenicity: for example, recombinant bacille Calmette–Guérin (BCG) expressing OspA as a lipoprotein protected mice against *Borrelia burgdorferi* infection, whereas no protection was provided when this antigen was expressed as an intracellular or secreted antigen (Stover *et al.* 1991). Four lipoproteins, with molecular masses of 19, 26, 27 and 38 kDa, have been identified in *M. tuberculosis* by detergent phase separation and metabolic labelling (Young & Garbe 1991). From these, the 19-kDa and 38-kDa lipoproteins have been identified as potent immunogens. The genes for both proteins predict a signal peptide and a cysteine motif, characteristic of bacterial lipoproteins. Analysis of the genome sequence of *M. tuberculosis* identifies a further 30–50 open reading frames encoding potential lipoproteins with similar motifs (Cole *et al.* 1998). The 38-kDa protein shares 30% sequence identity with the PstS, or PhoS, periplasmic protein of *E. coli*, suggesting a role in phosphate transport (Andersen & Hansen 1989). Many of the other potential lipoproteins identified by genome analysis (including additional PstS homologues) also resemble proteins involved in nutrient transport in

other bacteria. The 19-kDa antigen shows no significant homology to known proteins and its function remains to be determined (Ashbridge *et al.* 1989). Both the 38-kDa and 19-kDa antigens have homologues in some other slow-growing mycobacteria; and, as in *M. tuberculosis*, multiple *pstS* genes have been identified in *M. intracellulare* (Thangaraj *et al.* 1996).

In addition to acylation, there is evidence of glycosylation in the case of both the 19-kDa and 38-kDa antigens (Herrmann *et al.* 1996). Protein glycosylation, though common in eukaryotic systems, is relatively rare amongst prokaryotes; a series of reports of mycobacterial glycoproteins seeming to represent an exception to this general rule. Glycosylation has been thoroughly characterized in the case of a 45-kDa secreted antigen of *M. tuberculosis* (Dobos *et al.* 1996). The 45-kDa antigen is modified by covalent attachment of short mannose-containing glycans to a series of threonine residues at both ends of the protein. Structural analysis of the 19-kDa antigen demonstrates the presence of similar *O*-linked glycans close to the acylated N terminus, and a further *N*-linked glycan close to the C terminus (M. Ward, C. Abou-Zeid and D. Young, unpublished observations 1998). When threonine residues involved in *O*-linked glycosylation were substituted in a recombinant expression system, the resulting 19-kDa protein was found to be susceptible to protease cleavage, suggesting a model in which glycosylation may act to regulate proteolytic conversion from a cell-associated lipoprotein to a soluble form of the antigen (Herrmann *et al.* 1996). Several other mycobacterial lipoproteins have a similar glycosylation motif close to the N terminus—including the 38-kDa antigen and MPB83, a prominent antigen of *M. bovis* (Hewinson *et al.* 1996)—and may also be capable of release in a soluble form. The *N*-linked glycan on the 19-kDa antigen is a more complex branched chain structure typical of those found in eukaryotic glycoproteins; its function is unknown.

The immunological properties of the 38-kDa and 19-kDa antigens have been extensively documented in terms of antibody and T-cell responses (Vordermeier *et al.* 1992a,b). The 38-kDa protein is one of most promising candidate antigens for the development of a serodiagnostic test against *M. tuberculosis* (Wilkinson *et al.* 1997). Similarly, MPB83 has been used as a target for immunodiagnosis of bovine tuberculosis (Wiker *et al.* 1998). The 19-kDa antigen, while capable of inducing a strong T-cell response, was found to have a detrimental effect when incorporated into a recombinant mycobacterial vaccine (Abou-Zeid *et al.* 1997). The ability of lipoproteins to induce immune responses (particularly, antibody-mediated responses) has been exploited for the development of antigen delivery systems. Expression of heterologous proteins as lipoproteins through the addition of signal sequence and acylation motif derived from the 19-kDa antigen has resulted in the development of various recombinant BCG vaccine candidates (Stover *et al.* 1991, 1993).

3.2.3 Other cell-wall proteins

It is likely that many additional proteins contribute to the potent T-cell stimulation induced by mycobacterial cell-wall preparations, for example the 35-kDa 'major membrane protein' (MMP1) (Winter *et al.* 1995; Triccas *et al.* 1996) and a 46-kDa protein (Oskam *et al.* 1995). Proline-rich repetitive domains identified in several antigens provide an indication of potential association with the cell wall (for example, see Thole *et al.* 1990; Romain *et al.* 1993; Berthet *et al.* 1995), and analysis of genome data suggests that some proteins of *M. tuberculosis* may be anchored to the cell wall by hexapeptide C-terminal anchoring domains characteristic of other Gram-positive bacteria (Jenkinson 1995). Proteins exposed on the outer surface of mycobacteria may play an important role in mediating interaction and uptake via specific host-cell receptors (Arruda *et al.* 1993; Menozzi *et al.* 1996; Rambukkana *et al.* 1997). Antibodies that block such interactions could provide a route to modulation of the course of mycobacterial infection.

3.3 Secreted antigens

Secreted antigens have attracted particular attention as candidate antigens for subunit vaccines. It is rea-soned that, since they are available for processing and presentation to T cells prior to release of cytoplasmic or cell-wall proteins from dead bacteria, they will act as key targets for protective immune responses at the early stages of infection (Andersen *et al.* 1991; Orme *et al.* 1993). Secreted antigens have been extensively characterized and some have been shown to induce protection in animal models (Andersen 1994; Horwitz *et al.* 1995). In addition to the antigens described below, important studies have focused on a desaturase enzyme (Des) (Jackson *et al.* 1997), MPB70 and MPB83 (expressed in large amounts by *M. bovis*) (Hewinson *et al.* 1996; Wiker *et al.* 1998), and MPT63 (Terasaka *et al.* 1989; Manca *et al.* 1997).

Antigen 85 complex

These represent a particularly immunogenic group of secreted proteins that have been found to be widely recognized by antibodies and T cells in infected individuals. The antigen 85 complex consists of three distinct, but highly homologous proteins (85A, B and C), each of ≈ 30 kDa. The three proteins share ≈ 70–80% identity and are encoded by three genes located at separate chromosomal loci (Eiglmeier *et al.* 1993; Rinke de Wit *et al.* 1993). All three molecules are found in the culture medium as well as being present in the cell wall of mycobacteria, but differences exist in the relative amount of each component in either compartment (Wiker *et al.* 1991). MPT51, a 27-kDa protein displaying 40% identity to the antigen 85A, B and C components, may represent a fourth member of this category of proteins (Rinke de Wit *et al.* 1993). Several reports have documented the fibronectin-binding abilities of these molecules but this ability has also been questioned and the biological relevance remains unclear (Abou-Zeid *et al.* 1988; Pessolani & Brennan 1992). More recently, an enzymatic function has been attributed to antigen 85 complex molecules, with the 85B component shown to display mycolyl transferase activity (Belisle *et al.* 1997). Mycolic acids are major components of the mycobacterial cell wall, and the different members of the antigen 85 complex probably represent a set of enzymes with different transferase activities involved in cell-wall biosynthesis. Induction of an immune response to members of the antigen 85 complex using the technique of DNA vaccination has been shown to confer protection against challenge with *M. tuberculosis* in a mouse model (Huygen *et al.* 1996).

ESAT6

Screening for the ability to induce expression of interferon-γ by primed T cells resulted in identification of a 6-kDa protein as a potent early culture filtrate antigen (ESAT6) (Sorensen *et al.* 1995). This protein is of particular interest in that the segment of the chromosome including the gene encoding ESAT6 has been deleted from BCG (Mahairas *et al.* 1996), thus providing an opportunity for the development of a diagnostic tool that could discriminate between BCG vaccination and infection with other mycobacteria such as *M. tuberculosis* (Pollock & Andersen 1997). Interestingly, the deduced gene sequence revealed no obvious secretion signal for ESAT6, perhaps again indicating the existence of some unusual protein export system in mycobacteria. No function has yet been attributed to this molecule.

MPT64

This 24-kDa molecule received considerable attention when it was found to be present mainly in *M. tuberculosis* complex but was absent from the four most commonly used BCG vaccine strains (Harboe *et al.* 1986; Yamaguchi *et al.* 1989; Li *et al.* 1993). Like ESAT6, this antigen therefore has the potential of being used in the development of a diagnostic test to discriminate between BCG vaccination and infection with *M. tuberculosis* (Oettinger *et al.* 1997). The function of MPT64 remains as yet unknown.

Fibronectin attachment protein

A fibronectin attachment protein (FAP), which may function to attach bacteria to fibronectin, has been identified in several mycobacteria. The genes for several mycobacterial homologues have recently been characterized (Wieles *et al.* 1994; Laqueyrerie *et al.* 1995; Schorey *et al.* 1995) and displayed two fibronectin binding sites of 24 or 25 amino acids in length (Schorey *et al.* 1996). An essential role for this protein in the interaction with the host has been proposed. Both antibodies and T cells in infected individuals and patients were shown to recognize this protein.

4 Future advances

The recent completion of the genome sequence of *M. tuberculosis* (Cole *et al.* 1998), together with the anticipated completion of the *M. leprae* sequence, will undoubtedly have a major impact on the future course of antigen research. The availability of sequence data for all of the proteins of *M. tuberculosis* will greatly enhance the power of approaches to antigen identification based on biochemical fractionation. Proteome analysis, for example, in which two-dimensional gel electrophoresis is combined with sensitive mass spectrometry analysis of amino acid sequence 'tags', represents a feasible strategy for a general cataloguing of the expressed protein complement (Sonnenberg & Belisle 1997). Bioinformatic analysis of genome data allows prediction of protein function, post-translational modification, and subcellular location which will facilitate identification of novel cell-wall and secreted antigens. This approach may uncover antigen classes that have been missed by previous biochemical and genetic strategies. A remarkable observation arising from genome analysis is the presence of a large number of open reading frames encoding proteins with conserved glycine and alanine-rich repetitive elements—corresponding to the polymorphic GC-rich sequences (PGRS) used for PGRS strain typing. If these open reading frames are actually expressed, they may represent an important source of antigenic variation between *M. tuberculosis* isolates. The ability to compare sequences between different mycobacterial genomes will greatly facilitate the search for species-specific antigens for use in diagnosis. The identification of genomic regions deleted from BCG represents a clear example (Mahairas *et al.* 1996): in addition to ESAT6 and

MPT64, sequence analysis reveals a series of additional open reading frames that can be screened for the presence of antigens that may be used to improve the current BCG vaccine or for the development of *M. tuberculosis*-specific diagnostic tests. The world of mycobacterial antigens is poised for a rapid expansion over the next decade under the influence of genomics and proteomics.

5 References

Abou-Zeid, C., Gares, M.P., Inwald, J., Janssen, R., Zhang, Y., Young, D.B., Hetzel, C., Lamb, J.R., Baldwin, S.L., Orme, I.M., Yeremeev, V., Nikonenko, B.V. & Apt, A.S. (1997) Induction of a type 1 immune response to a recombinant antigen from *Mycobacterium tuberculosis* expressed in *Mycobacterium vaccae*. *Infection and Immunity* **65**, 1856–1862.

Abou-Zeid, C., Ratliff, T.L., Wiker, H.G., Harboe, M., Bennedsen, J. & Rook, G.A. (1988) Characterization of fibronectin-binding antigens released by *Mycobacterium tuberculosis* and *Mycobacterium bovis* BCG. *Infection and Immunity* **56**, 3046–3051.

Adams, E., Basten, A., Rodda, S. & Britton, W.J. (1997) Human T-cell clones to the 70-kilodalton heat shock protein of *Mycobacterium leprae* define mycobacterium-specific epitopes rather than shared epitopes. *Infection and Immunity* **65**, 1061–1070.

Aldovini, A. & Young, R.A. (1991) Humoral and cell-mediated immune responses to live recombinant BCG-HIV vaccines. *Nature* **351**, 479–482.

Andersen, A.B., Andersen, P. & Ljungqvist, L. (1992) Structure and function of a 40,000-molecular-weight protein antigen of *Mycobacterium tuberculosis*. *Infection and Immunity* **60**, 2317–2323.

Andersen, A.B. & Brennan, P.J. (1994) Proteins and antigens of *Mycobacterium tuberculosis*. In: *Tuberculosis; Pathogenesis, Protection and Control* (ed. B. R. Bloom). Washington, DC: American Society for Microbiology, pp. 307–332.

Andersen, A.B. & Hansen, E.B. (1989) Structure and mapping of antigenic domains of protein antigen b, a 38,000-molecular-weight protein of *Mycobacterium tuberculosis*. *Infection and Immunity* **57**, 2481–2488.

Andersen, P. (1994) Effective vaccination of mice against *Mycobacterium tuberculosis* infection with a soluble mixture of secreted mycobacterial proteins. *Infection and Immunity* **62**, 2536–2544.

Andersen, P., Askgaard, D., Ljungqvist, L., Bentzon, M.W. & Heron, I. (1991) T-cell proliferative response to antigens secreted by *Mycobacterium tuberculosis*. *Infection and Immunity* **59**, 1558–1563.

Arruda, S., Bomfim, G., Knights, R., Huima-Byron, T. & Riley, L.W. (1993) Cloning of an *M. tuberculosis* DNA fragment associated with entry and survival inside cells. *Science* **261**, 1454–1457.

Ashbridge, K.R., Booth, R.J., Watson, J.D. & Lathigra, R.B. (1989) Nucleotide sequence of the 19 kDa antigen gene from *Mycobacterium tuberculosis*. *Nucleic Acids Research* **17**, 1249.

Barnes, P.F., Mehra, V., Rivoire, B., Fong, S.J., Brennan, P.J., Voegtline, M.S., Minden, P., Houghten, R.A., Bloom, B.R. & Modlin, R.L. (1992) Immunoreactivity of a 10-kDa antigen of *Mycobacterium tuberculosis*. *Journal of Immunology* **148**, 1835–1840.

Beaman, L. & Beaman, B.L. (1990) Monoclonal antibodies demonstrate that superoxide dismutase contributes to protection of *Nocardia asteroides* within the intact host. *Infection and Immunity* **58**, 3122–3128.

Beckman, E.M., Porcelli, S.A., Morita, C.T., Behar, S.M., Furlong, S.T. & Brenner, M.B. (1994) Recognition of a lipid antigen by CD1-restricted alpha beta+ T cells. *Nature* **372**, 691–694.

Behr, C., Poupot, R., Peyrat, M.A., Poquet, Y., Constant, P., Dubois, P., Bonneville, M. & Fournie, J.J. (1996) *Plasmodium falciparum* stimuli for human gamma-delta T cells are related to phosphorylated antigens of mycobacteria. *Infection and Immunity* **64**, 2892–2896.

Belisle, J.T., Vissa, V.D., Sievert, T., Takayama, K., Brennan, P.J. & Besra, G.S. (1997) Role of the major antigen of *Mycobacterium tuberculosis* in cell wall biogenesis. *Science* **276**, 1420–1422.

Berthet, F.X., Rauzier, J., Lim, E.M., Philipp, W., Gicquel, B. & Portnoi, D. (1995) Characterization of the *Mycobacterium tuberculosis erp* gene encoding a potential cell surface protein with repetitive structures. *Microbiology* **141**, 2123–2130.

Booth, R.J., Harris, D.P., Love, J.M. & Watson, J.D. (1988) Antigenic proteins of *Mycobacterium leprae*. Complete sequence of the gene for the 18-kDa protein. *Journal of Immunology* **140**, 597–601.

Bucca, G., Hindle, Z. & Smith, C.P. (1997) Regulation of the *dnaK* operon of *Streptomyces coelicolor* A3(2) is governed by HspR, an autoregulatory repressor protein. *Journal of Bacteriology* **179**, 5999–6004.

Bunting, K., Cooper, J.B., Badasso, M.O., Tickle, I.J., Newton, M., Wood, S.P., Zhang, Y. & Young, D. (1998) Engineering a change in metal-ion specificity of the iron-dependent superoxide dismutase from *Mycobacterium tuberculosis*—X-ray structure analysis of site-directed mutants. *European Journal of Biochemistry* **251**, 795–803.

Chatterjee, D., Roberts, A.D., Lowell, K., Brennan, P.J. & Orme, I.M. (1992) Structural basis of capacity of lipoarabinomannan to induce secretion of tumor necrosis factor. *Infection and Immunity* **60**, 1249–1253.

Cherayil, B.J. & Young, R.A. (1988) A 28-kDa protein from

Mycobacterium leprae is a target of the human antibody response in lepromatous leprosy. *Journal of Immunology* **141**, 4370–4375.

Cole, S.T., Brosch, R., Parkhill, J. *et al.* (1998) Deciphering the biology of the tubercule bacillus: The complete genome sequence of *Mycobacterium tuberculosis* H37Rv. *Nature* **393**, 537–544.

Cooper, J.B., Driessen, H.P., Wood, S.P., Zhang, Y. & Young, D. (1994) Crystallization and preliminary X-ray analysis of the superoxide dismutase from *Mycobacterium tuberculosis*. *Journal of Molecular Biology* **235**, 1156–1158.

Cooper, J.B., McIntyre, K., Badasso, M.O., Wood, S.P., Zhang, Y., Garbe, T.R. & Young, D. (1995) X-ray structure analysis of the iron-dependent superoxide dismutase from *Mycobacterium tuberculosis* at 2.0 Angstroms resolution reveals novel dimer–dimer interactions. *Journal of Molecular Biology* **246**, 531–544.

Dobos, K.M., Khoo, K.H., Swiderek, K.M., Brennan, P.J. & Belisle, J.T. (1996) Definition of the full extent of glycosylation of the 45-kilodalton glycoprotein of *Mycobacterium tuberculosis*. *Journal of Bacteriology* **178**, 2498–2506.

Eiglmeier, K., Honore, N., Woods, S.A., Caudron, B. & Cole, S.T. (1993) Use of an ordered cosmid library to deduce the genomic organization of *Mycobacterium leprae*. *Molecular Microbiology* **7**, 197–206.

Engers, H.D. & Workshop Participants (1985) Results of a World Health Organization-sponsored workshop on monoclonal antibodies to *Mycobacterium leprae*. *Infection and Immunity* **48**, 603–605.

Engers, H.D. & Workshop Participants (1986) Results of a World Health Organization-sponsored workshop to characterize antigens recognized by mycobacterium-specific monoclonal antibodies. *Infection and Immunity* **51**, 718–720.

Escuyer, V., Haddad, N., Frehel, C. & Berche, P. (1996) Molecular characterization of a surface-exposed superoxide dismutase of *Mycobacterium avium*. *Microbial Pathogenesis* **20**, 41–55.

Harboe, M., Nagai, S., Patarroyo, M.E., Torres, M.L., Ramirez, C. & Cruz, N. (1986) Properties of proteins MPB64, MPB70, and MPB80 of *Mycobacterium bovis* BCG. *Infection and Immunity* **52**, 293–302.

Hecker, M., Schumann, W. & Volker, U. (1996) Heat-shock and general stress response in *Bacillus subtilis*. *Molecular Microbiology* **19**, 417–428.

Hermans, P.W., Abebe, F., Kuteyi, V.I., Kolk, A.H., Thole, J.E. & Harboe, M. (1995) Molecular and immunological characterization of the highly conserved antigen 84 from *Mycobacterium tuberculosis* and *Mycobacterium leprae*. *Infection and Immunity* **63**, 954–960.

Herrmann, J.L., O'Gaora, P., Gallagher, A., Thole, J.E. & Young, D.B. (1996) Bacterial glycoproteins: a link between glycosylation and proteolytic cleavage of a 19 kDa antigen from *Mycobacterium tuberculosis*. *EMBO Journal* **15**, 3547–3554.

Hetzel, C., Janssen, R., Ely, S., Kristensen, N., Bunting, K., Cooper, J., Lamb, J.R., Young, D.B. & Thole, J.E. (1998) An epitope delivery system using recombinant mycobacteria. *Infection and Immunity* **66**, 3643–3648.

Hewinson, R.G., Michell, S.L., Russell, W.P., McAdam, R.A., Jacobs, W.R., Jr (1996) Molecular characterization of MPT83: a seroreactive antigen of *Mycobacterium tuberculosis* with homology to MPT70. *Scandinavian Journal of Immunology* **43**, 490–499.

Hoppe, H.C., de Wet, B.J., Cywes, C., Daffe, M. & Ehlers, M.R. (1997) Identification of phosphatidylinositol mannoside as a mycobacterial adhesin mediating both direct and opsonic binding to nonphagocytic mammalian cells. *Infection and Immunity* **65**, 3896–3905.

Horwitz, M.A., Lee, B.W., Dillon, B.J. & Harth, G. (1995) Protective immunity against tuberculosis induced by vaccination with major extracellular proteins of *Mycobacterium tuberculosis*. *Proceedings of the National Academy of Sciences of the USA* **92**, 1530–1534.

Huygen, K., Content, J., Denis, O., Montgomery, D.L., Yawman, A.M., Deck, R.R., DeWitt, C.M., Orme, I.M., Baldwin, S., D'Souza, C., Drowart, A., Lozes, E., Vandenbussche, P., Van Vooren, J.P., Liu, M.A. & Ulmer, J.B. (1996) Immunogenicity and protective efficacy of a tuberculosis DNA vaccine. *Nature Medicine* **2**, 893–898.

Jackson, M., Portnoi, D., Catheline, D., Dumail, L., Rauzier, J., Legrand, P. & Gicquel, B. (1997) *Mycobacterium tuberculosis* Des protein: an immunodominant target for the humoral response of tuberculous patients. *Infection and Immunity* **65**, 2883–2889.

Jenkinson, H.F. (1995) Anchorage and release of Gram-positive bacterial cell-surface polypeptides. *Trends in Microbiology* **3**, 333–335.

Kaufmann, S.H., Vath, U., Thole, J.E., van Embden, J.D. & Emmrich, F. (1987) Enumeration of T cells reactive with *Mycobacterium tuberculosis* organisms and specific for the recombinant mycobacterial 64-kDa protein. *European Journal of Immunology* **17**, 351–357.

Kong, T.H., Coates, A.R., Butcher, P.D., Hickman, C.J. & Shinnick, T.M. (1993) *Mycobacterium tuberculosis* expresses two chaperonin-60 homologs. *Proceedings of the National Academy of Sciences of the USA* **90**, 2608–2612.

Kuwabara, S. (1975) Amino acid sequence of tuberculin-active protein from *Mycobacterium tuberculosis*. *Journal of Biological Chemistry* **250**, 2563–2568.

Laqueyrerie, A., Militzer, P., Romain, F., Eiglmeier, K., Cole, S. & Marchal, G. (1995) Cloning, sequencing, and expression of the *apa* gene coding for the *Mycobacterium tuberculosis* 45/47-kilodalton secreted antigen complex. *Infection and Immunity* **63**, 4003–4010.

Lee, B.Y. & Horwitz, M.A. (1995) Identification of

macrophage and stress-induced proteins of *Mycobacterium tuberculosis*. *Journal of Clinical Investigation* **96**, 245–249.

Li, H., Ulstrup, J.C., Jonassen, T.O., Melby, K., Nagai, S. & Harboe, M. (1993) Evidence for absence of the MPB64 gene in some substrains of *Mycobacterium bovis* BCG. *Infection and Immunity* **61**, 1730–1734.

Mahairas, G.G., Sabo, P.J., Hickey, M.J., Singh, D.C. & Stover, C.K. (1996) Molecular analysis of genetic differences between *Mycobacterium bovis* BCG and virulent *M bovis*. *Journal of Bacteriology* **178**, 1274–1282.

Manca, C., Lyashchenko, K., Wiker, H.G., Usai, D., Colangeli, R. & Gennaro, M.L. (1997) Molecular cloning, purification, and serological characterization of MPT63, a novel antigen secreted by *Mycobacterium tuberculosis*. *Infection and Immunity* **65**, 16–23.

Mangan, J.A., Sole, K.M., Mitchison, D.A. & Butcher, P.D. (1997) An effective method of RNA extraction from bacteria refractory to disruption, including mycobacteria. *Nucleic Acids Research* **25**, 675–676.

Mazodier, P., Guglielmi, G., Davies, J. & Thompson, C.J. (1991) Characterization of the *groEL*-like genes in *Streptomyces albus*. *Journal of Bacteriology* **173**, 7382–7386.

Mehlert, A. & Young, D.B. (1989) Biochemical and antigenic characterization of the *Mycobacterium tuberculosis* 71kDa antigen, a member of the 70kDa heat-shock protein family. *Molecular Microbiology* **3**, 125–130.

Mehra, V., Bloom, B.R., Bajardi, A.C., Grisso, C.L., Sieling, P.A., Alland, D., Convit, J., Fan, X.D., Hunter, S.W. & Brennan, P.J. (1992) A major T cell antigen of *Mycobacterium leprae* is a 10-kDa heat-shock cognate protein. *Journal of Experimental Medicine* **175**, 275–284.

Menozzi, F.D., Rouse, J.H., Alavi, M., Laude-Sharp, M., Muller, J., Bischoff, R., Brennan, M.J. & Locht, C. (1996) Identification of a heparin-binding hemagglutinin present in mycobacteria. *Journal of Experimental Medicine* **184**, 993–1001.

Moody, D.B., Reinhold, B.B., Guy, M.R., Beckman, E.M., Frederique, D.E., Furlong, S.T., Ye, S., Reinhold, V.N., Sieling, P.A., Modlin, R.L., Besra, G.S. & Porcelli, S.A. (1997) Structural requirements for glycolipid antigen recognition by CD1b- restricted T cells. *Science* **278**, 283–286.

Mustafa, A.S., Gill, H.K., Nerland, A., Britton, W.J., Mehra, V., Bloom, B.R., Young, R.A. & Godal, T. (1986) Human T-cell clones recognize a major *M. leprae* protein antigen expressed in *E. coli*. *Nature* **319**, 63–66.

Oettinger, T., Holm, A. & Haslov, K. (1997) Characterization of the delayed type hypersensitivity-inducing epitope of MPT64 from *Mycobacterium tuberculosis*. *Scandinavian Journal of Immunology* **45**, 499–503.

Ohman, R. & Ridell, M. (1996) Purification and characterisation of isocitrate dehydrogenase and malate dehydrogenase from *Mycobacterium tuberculosis* and evaluation of their potential as suitable antigens for the serodiagnosis of tuberculosis. *Tubercle and Lung Disease* **77**, 454–461.

Orme, I.M., Andersen, P. & Boom, W.H. (1993) T cell response to *Mycobacterium tuberculosis*. *Journal of Infectious Diseases* **167**, 1481–1497.

Orme, I.M., Miller, E.S., Roberts, A.D., Furney, S.K., Griffin, J.P., Dobos, K.M., Chi, D., Rivoire, B. & Brennan, P.J. (1992) T lymphocytes mediating protection and cellular cytolysis during the course of *Mycobacterium tuberculosis* infection. Evidence for different kinetics and recognition of a wide spectrum of protein antigens. *Journal of Immunology* **148**, 189–196.

Oskam, L., Hartskeerl, R.A., Hermans, C.J., De Wit, M.Y., Jarings, G.H., Nicholls, R.D. & Klatser, P.R. (1995) A 46 kDa integral membrane protein from *Mycobacterium leprae* resembles a number of bacterial and mammalian membrane transport proteins. *Microbiology* **141**, 1963–1968.

Peake, P., Basten, A. & Britton, W.J. (1991) Characterization of the functional properties of the 70-kDa protein of *Mycobacterium bovis*. *Journal of Biological Chemistry* **266**, 20828–20832.

Pessolani, M.C. & Brennan, P.J. (1992) *Mycobacterium leprae* produces extracellular homologs of the antigen 85 complex. *Infection and Immunity* **60**, 4452–4459.

Pessolani, M.C., Smith, D.R., Rivoire, B., McCormick, J., Hefta, S.A., Cole, S.T. & Brennan, P.J. (1994) Purification, characterization, gene sequence, and significance of a bacterioferritin from *Mycobacterium leprae*. *Journal of Experimental Medicine* **180**, 319–327.

Pollock, J.M. & Andersen, P. (1997) The potential of the ESAT-6 antigen secreted by virulent mycobacteria for specific diagnosis of tuberculosis. *Journal of Infectious Diseases* **175**, 1251–1254.

Rambukkana, A., Salzer, J.L., Yurchenco, P.D. & Tuomanen, E.I. (1997) Neural targeting of *Mycobacterium leprae* mediated by the G domain of the laminin-alpha2 chain. *Cell* **88**, 811–821.

Res, P., Thole, J. & de Vries, R. (1991) Heat-shock proteins and autoimmunity in humans. *Springer Seminars in Immunopathology* **13**, 81–98.

Rinke de Wit, T.F., Bekelie, S., Osland, A., Miko, T.L., Hermans, P.W., van Soolingen, D., Drijfhout, J.W., Schoningh, R., Janson, A.A. & Thole, J.E. (1992) Mycobacteria contain two *groEL* genes: the second *Mycobacterium leprae groEL* gene is arranged in an operon with *Groes*. *Molecular Microbiology* **6**, 1995–2007.

Rinke de Wit, T.F., Bekelie, S., Osland, A., Wieles, B., Janson, A.A. & Thole, J.E. (1993) The *Mycobacterium leprae* antigen 85 complex gene family: identification of the genes for the 85A, 85C, and related MPT51 proteins. *Infection and Immunity* **61**, 3642–3647.

Romain, F., Augier, J., Pescher, P. & Marchal, G. (1993)

Isolation of a proline-rich mycobacterial protein eliciting delayed-type hypersensitivity reactions only in guinea pigs immunized with living mycobacteria. *Proceedings of the National Academy of Sciences of the USA* **90**, 5322–5326.

Roman, E., Moreno, C. & Young, D. (1994) Mapping of Hsp70-binding sites on protein antigens. *European Journal of Biochemistry* **222**, 65–73.

Schorey, J.S., Holsti, M.A., Ratliff, T.L., Allen, P.M. & Brown, E.J. (1996) Characterization of the fibronectin-attachment protein of *Mycobacterium avium* reveals a fibronectin-binding motif conserved among mycobacteria. *Molecular Microbiology* **21**, 321–329.

Schorey, J.S., Li, Q., McCourt, D.W., Bong-Mastek, M., Clark-Curtiss, J.E., Ratliff, T.L. & Brown, E.J. (1995) A *Mycobacterium leprae* gene encoding a fibronectin binding protein is used for efficient invasion of epithelial cells and Schwann cells. *Infection and Immunity* **63**, 2652–2657.

Sieling, P.A., Chatterjee, D., Porcelli, S.A., Prigozy, T.I., Mazzaccaro, R.J., Soriano, T., Bloom, B.R., Brenner, M.B., Kronenberg, M. & Brennan, P.J. (1995) CD1-restricted T cell recognition of microbial lipoglycan antigens. *Science* **269**, 227–230.

Sonnenberg, M.G. & Belisle, J.T. (1997) Definition of *Mycobacterium tuberculosis* culture filtrate proteins by two-dimensional polyacrylamide gel electrophoresis, N-terminal amino acid sequencing, and electrospray mass spectrometry. *Infection and Immunity* **65**, 4515–4524.

Sorensen, A.L., Nagai, S., Houen, G., Andersen, P. & Andersen, A.B. (1995) Purification and characterization of a low-molecular-mass T-cell antigen secreted by *Mycobacterium tuberculosis*. *Infection and Immunity* **63**, 1710–1717.

Stover, C.K., Bansal, G.P., Hanson, M.S., Burlein, J.E., Palaszynski, S.R., Young, J.F., Koenig, S., Young, D.B., Sadziene, A. & Barbour, A.G. (1993) Protective immunity elicited by recombinant bacille Calmette–Guérin (BCG) expressing outer surface protein A (OspA) lipoprotein: a candidate Lyme disease vaccine. *Journal of Experimental Medicine* **178**, 197–209.

Stover, C.K., de la Cruz, V.F., Fuerst, T.R., Burlein, J.E., Benson, L.A., Bennett, L.T., Bansal, G.P., Young, J.F., Lee, M.H. & Hatfull, G.F. (1991) New use of BCG for recombinant vaccines. *Nature* **351**, 456–460.

Tanaka, Y., Morita, C.T., Nieves, E., Brenner, M.B. & Bloom, B.R. (1995) Natural and synthetic non-peptide antigens recognized by human gamma delta T cells. *Nature* **375**, 155–158.

Tanaka, Y., Sano, S., De Nieves, E., Libero, G., Rosa, D., Modlin, R.L., Brenner, M.B., Bloom, B.R. & Morita, C.T. (1994) Nonpeptide ligands for human gamma delta T cells. *Proceedings of the National Academy of Sciences of the USA* **91**, 8175–8179.

Terasaka, K., Yamaguchi, R., Matsuo, K., Yamazaki, A.,

Nagai, S. & Yamada, T. (1989) Complete nucleotide sequence of immunogenic protein MPB70 from *Mycobacterium bovis* BCG. *FEMS Microbiological Letters* **49**, 273–276.

Thangaraj, H.S., Bull, T.J., De Smet, K.A., Hill, M.K., Rouse, D.A., Moreno, C. & Ivanyi, J. (1996) Duplication of genes encoding the immunodominant 38 kDa antigen in *Mycobacterium intracellulare*. *FEMS Microbiological Letters* **144**, 235–240.

Thangaraj, H.S., Lamb, F.I., Davis, E.O. & Colston, M.J. (1989) Nucleotide and deduced amino acid sequence of *Mycobacterium leprae* manganese superoxide dismutase. *Nucleic Acids Research* **17**, 8378.

Thole, J.E., Stabel, L.F., Suykerbuyk, M.E., De Wit, M.Y., Klatser, P.R., Kolk, A.H. & Hartskeerl, R.A. (1990) A major immunogenic 36,000-molecular-weight antigen from *Mycobacterium leprae* contains an immunoreactive region of proline-rich repeats. *Infection and Immunity* **58**, 80–87.

Thole, J.E., Wieles, B., Clark-Curtiss, J.E., Ottenhoff, T.H. & de Wit, T.F. (1995) Immunological and functional characterization of *Mycobacterium leprae* protein antigens: an overview. *Molecular Microbiology* **18**, 791–800.

Triccas, J.A., Roche, P.W., Winter, N., Feng, C.G., Butlin, C.R. & Britton, W.J. (1996) A 35-kilodalton protein is a major target of the human immune response to *Mycobacterium leprae*. *Infection and Immunity* **64**, 5171–5177.

van Eden, W. (1991) Heat-shock proteins as immunogenic bacterial antigens with the potential to induce and regulate autoimmune arthritis. *Immunological Review* **121**, 5–28.

Verbon, A., Hartskeerl, R.A., Moreno, C. & Kolk, A.H. (1992) Characterization of B cell epitopes on the 16K antigen of *Mycobacterium tuberculosis*. *Clinical Experimental Immunology* **89**, 395–401.

Vordermeier, H.M., Harris, D.P., Friscia, G., Roman, E., Surcel, H.M., Moreno, C., Pasvol, G. & Ivanyi, J. (1992a) T cell repertoire in tuberculosis: selective anergy to an immunodominant epitope of the 38-kDa antigen in patients with active disease. *European Journal of Immunology* **22**, 2631–2637.

Vordermeier, H.M., Harris, D.P., Mehrotra, P.K., Roman, E., Elsaghier, A., Moreno, C. & Ivanyi, J. (1992b) *M. tuberculosis*-complex specific T-cell stimulation and DTH reactions induced with a peptide from the 38-kDa protein. *Scandinavian Journal of Immunology* **35**, 711–718.

Wieles, B., van Agterveld, M., Janson, A., Clark-Curtiss, J., Rinke de Wit, T., Harboe, M. & Thole, J. (1994) Characterization of a *Mycobacterium leprae* antigen related to the secreted *Mycobacterium tuberculosis* protein MPT32. *Infection and Immunity* **62**, 252–258.

Wieles, B., van Soolingen, D., Holmgren, A., Offringa, R.,

Ottenhoff, T. & Thole, J. (1995) Unique gene organization of thioredoxin and thioredoxin reductase in *Mycobacterium leprae. Molecular Microbiology* **16**, 921–929.

Wiker, H.G., Harboe, M. & Nagai, S. (1991) A localization index for distinction between extracellular and intracellular antigens of *Mycobacterium tuberculosis. Journal of General Microbiology* **137**, 875–884.

Wiker, H.G., Lyashchenko, K.P., Aksoy, A.M., Lightbody, K.A., Pollock, J.M., Komissarenko, S.V., Bobrovnik, S.O., Kolesnikova, I.N., Mykhalsky, L.O., Gennaro, M.L. & Harboe, M. (1998) Immunochemical characterization of the MPB70/80 and MPB83 proteins of *Mycobacterium bovis. Infection and Immunity* **66**, 1445–1452.

Wilkinson, R.J., Haslov, K., Rappuoli, R., Giovannoni, F., Narayanan, P.R., Desai, C.R., Vordermeier, H.M., Paulsen, J., Pasvol, G., Ivanyi, J. & Singh, M. (1997) Evaluation of the recombinant 38-kilodalton antigen of *Mycobacterium tuberculosis* as a potential immunodiagnostic reagent. *Journal of Clinical Microbiology* **35**, 553–557.

Winter, N., Triccas, J.A., Rivoire, B., Pessolani, M.C., Eiglmeier, K., Lim, E.M., Hunter, S.W., Brennan, P.J. & Britton, W.J. (1995) Characterization of the gene encoding the immunodominant 35 kDa protein of *Mycobacterium leprae. Molecular Microbiology* **16**, 865–876.

Yamaguchi, R., Matsuo, K., Yamazaki, A., Abe, C., Nagai, S., Terasaka, K. & Yamada, T. (1989) Cloning and characterization of the gene for immunogenic protein MPB64 of *Mycobacterium bovis* BCG. *Infection and Immunity* **57**, 283–288.

Young, D.B. & Garbe, T.R. (1991) Lipoprotein antigens of *Mycobacterium tuberculosis. Research in Microbiology* **142**, 55–65.

Young, D.B., Kaufmann, S.H., Hermans, P.W. & Thole, J.E. (1992) Mycobacterial protein antigens: a compilation. *Molecular Microbiology* **6**, 133–145.

Young, D., Lathigra, R., Hendrix, R., Sweetser, D. & Young, R.A. (1988) Stress proteins are immune targets in leprosy and tuberculosis. *Proceedings of the National Academy of Sciences of the USA* **85**, 4267–4270.

Yuan, Y., Crane, D.D. & Barry, C.E., III (1996) Stationary phase-associated protein expression in *Mycobacterium tuberculosis*: function of the mycobacterial alpha-crystallin homolog. *Journal of Bacteriology* **178**, 4484–4492.

Zhang, Y., Lathigra, R., Garbe, T., Catty, D. & Young, D. (1991) Genetic analysis of superoxide dismutase, the 23 kilodalton antigen of *Mycobacterium tuberculosis. Molecular Microbiology* **5**, 381–391.

Chapter 19 / *Mycobacterium* and the seduction of the macrophage

DAVID G. RUSSELL

1 Introduction

If any adjective personifies the interaction between pathogenic mycobacteria and the macrophage it is 'enduring'. Although many pathogens are capable of transient infections of macrophages, few show levels of chronic, persistent infection comparable to those attained by *Mycobacterium tuberculosis* and *M. leprae*. All pathogenic mycobacteria reside primarily within the phagocytes of their host and, although many factors influence disease development, it is this pathogen–phagocyte interplay that is central to determination of the outcome of an infection (Russell *et al.* 1997). Despite the usual dominant role of the macrophage *in vivo*, in *in vitro* experiments, macrophages are unable to regulate division of *M. tuberculosis*, and even that of less pathogenic species like *M. avium*. This inability of macrophages to regulate infection in the absence of an immune component is in turn pivotal to the adoption of *M. tuberculosis* as one of the most reliable indicators of coinfection with human immunodeficiency virus (HIV). Macrophages are therefore both the saints and the sinners in this interplay capable of providing a protective haven for the bacilli, or, if activated by an appropriate cellular immune response, an effective regulator of bacterial division and mediator of bacterial death. It is this fine balance that makes chronic infections, like tuberculosis, so interesting with respect to the interplay between phagocyte and pathogen.

This chapter explores our current knowledge in the establishment and maintenance of mycobacterial infections in macrophages and attempts to expand this framework and examine the consequences of macrophage activation on the intracellular compartments in which the bacilli reside. Obviously, the ability of macrophages to regulate infection once activated will have placed strong selective pressure on the bacteria to develop strategies to avoid the induction or consequences of such a response and these strategies are also examined in this chapter which will interface closely with that of Cooper and Orme (Chapter 20) on immune responses.

2 Route of host-cell entry

The primary role of the macrophage in host defence is to circulate throughout the body sequestering foreign, particulate matter and digesting it in its highly developed lysosomal system. To perform this function, these cells are equipped with a range of receptors capable of identifying and binding to a diverse range of ligands. These ligands fall into two basic categories: self and non-self. The self ligands include antibodies, complement components and extracellular matrix proteins which the host produces and constitute a bridge, or opsonin, between the foreign particle and the receptors of the macrophage. The non-self ligands are obviously more diverse and are, for mycobacteria, predominantly carbohydrate in nature. These are recognized by the lectin receptors on the macrophage surface which show a range of specificities to both host-derived and foreign structures.

2.1 Identification of receptor–ligand pairings active during host-cell entry

There are two main principles motivating studies conducted on the identification of ligand–receptor pairings active in the mycobacterium–macrophage interaction. First, that the receptor–ligand pairing(s) involved in host-cell entry may influence directly the outcome of the invasion process, namely the establishment of a productive infection vs. an abortive invasion culminating in microbial death. Second, that the receptor–ligand pairing(s) may influence tissue distribution and the ability of the infection to establish itself within a particular type of phagocyte or host cell.

Early studies on the binding and uptake of *M. tuberculosis*, *M. leprae* and *M. avium* all indicated that the complement receptors, CR1, CR3 and CR4, specific for fragments of C3, the third component of complement, played an active role in invasion (Catanzaro & Wright 1990; Schlesinger *et al.* 1990). C3 is the central component of the alternative arm of the complement cascade which functions in the absence of specific antibody. C3 deposition leads to the combina-

tion with Factor B to form a C3 convertase which amplifies the generation of C3b and its covalent association with the activating surface. C3 is present on the mycobacterial surface in both its C3b and inactive C3bi forms, specific for the CR1 and CR4/CR4 receptors, respectively.

Although the interaction of mycobacteria with these receptors dominates binding to macrophages in the presence of serum, data has also been generated suggesting that binding to CR3 and CR4 can also be observed in the absence of exogenous serum (Schlesinger 1993; Stokes *et al.* 1993). It has been proposed that this interaction procedes via a lectin-like, mannose-binding activity mediated by another ligand binding domain in CR3. These data are controversial and an alternate explanation lies in the known ability of the macrophage to synthesize and secrete complement components, including C3 (Blackwell *et al.* 1985). In this scenario, macrophages are capable of 'localized opsonization' or depositing C3 on the surface of particles directly leading to binding to the complement receptors.

The specificity of deposition of C3, i.e. identification of the C3 acceptor molecule(s), is complex. C3, once cleaved to C3b exposing the thioester bond, will combine with either carbohydrate or amines, yielding either an ester or amide bond, respectively. Although C3 deposition does show differential efficiency, the nature of this interaction suggests that many surface-exposed mycobacterial constituents could act as C3 acceptors. For example, it has been shown that isolated lipoarabinomannan (LAM) and a phenolic glycolipid can both activate the alternative pathway and bind C3 (Schlesinger & Horwitz 1994; Schlesinger *et al.* 1994).

Of greater potential significance is the recent observation that pathogenic mycobacteria, *M. avium* and bacille Calmette–Guérin (BCG) are capable of generating their own unique C3 convertase, thus accelerating C3 deposition on the surface of the bacilli. Schorey *et al.* (1997) extended previous observations that horse serum enhanced uptake of bacteria by identifying the underlying mechanism whereby pathogenic mycobacteria bind the complement protein C2a directly to form an active C3 convertase on the bacterial surface. Under normal circumstances C2a will only form a C3 convertase in the context of the classical, or antibody-mediated, complement cascade in association with C4b. However, a molecule on the surface of the pathogenic, and not free-living, mycobacteria is capable of binding C2a in its proteolytically active form, facilitating cleavage of C3 and its deposition on the bacterial surface. Such behaviour would enhance markedly the ability of the bacteria to bind to and enter host phagocytes.

The exploitation of complement receptors to gain entry to macrophages is a recurring theme in microbial–phagocyte interactions. These receptors have several qualities to recommend themselves. They are high-affinity receptors; they exploit an abundant serum opsonin provided by the host, and they are relatively benign because they trigger minimal microbicidal behaviour from the macrophage. The only requirement for efficient exploitation of this route of entry is a lack of sensitivity to complement mediated lysis, which all mycobacteria species display.

Despite the dominant role occupied by the complement receptors in this interaction mycobacteria can also bind via other host-cell receptors. The mannose–fucose receptor has been implicated in binding of LAM (Schlesinger *et al.* 1994). In addition, the fibronectin-binding activities of both the α-antigen family and the fibronectin attachment protein (FAP) have been shown to mediate adherence using a fibronectin bridge (Schorey *et al.* 1995, 1996). Although this latter binding is of little significance for phagocytes, the FAP protein has been suggested to mediate binding and entry of *M. leprae* into Schwann cells. Of greater potential significance to pulmonary *M. tuberculosis* infections is the recent observation that pulmonary surfactant protein A can also act as an opsonin and enhance uptake of bacteria into alveolar macrophages (Downing *et al.* 1995; Gaynor *et al.* 1995). Binding and phagocytosis requires the glycosylated form of the protein and is inhibitable by mannan and antimannose receptor antibodies.

What should be clear from this section is that these bacilli exploit multiple routes of entry suggesting several tiers of redundancy; a theme recurrent in microbial pathogenesis. It is therefore unlikely that mycobacteria rely on one solitary surface molecule to mediate these interactions. Data that imply the dominant activity of a single surface moiety, such as the mycobacterial 'invasin' homologue (Arruda *et al.* 1993), must be carefully evaluated in the context of the intact bacterium, an approach rendered more accessible by recent advances in homologous recombination procedures in *Mycobacterium* species (see Chapter 1).

2.2 Influence of receptor–ligand pairing on infection and survival

The central question behind these studies is to what extent does the receptor regulate the efficiency of survival of the invading bacilli. This is not a trivial question to answer, but data from a few laboratories have relevance to the issue.

First, observations from Hart and colleagues indicated that entry via members of the Fc receptor family enhanced the fusion of phagosomes containing *M. tuberculosis* with lysosomes preloaded with ferritin (Armstrong & Hart 1975). The polymeric Fc receptor, FcγRII, may be a 'special' case because it has evolved to cooperate with the immune system through binding and internalization of antibody-opsonized particles. FcγRII triggers a superoxide burst even in resting macrophages. Despite the relatively aggressive behaviour of the FcγRII receptor, and the apparent fusion of phagosomes containing mycobacteria with lysosomes there was no marked drop in the viability of the infecting bacilli. This finding has been extended by recent work on *M. avium* by Oh and Straubinger (1996) who demonstrated that antibody opsonization of bacteria did not affect the ability of the bacteria to block the acidification of their phagosomes.

In another recent study on the relative contribution of receptors involved in non-immune uptake of *M. tuberculosis,* Zimmerli *et al.* (1996) blocked independently various receptor families with antibodies or competitive inhibitors and then examined the relative survival of bacteria internalized through the remaining routes. These researchers inhibited selectively the activities of CR1, CR3, CR4 and mannose receptors, and found that although the numbers of internalized bacteria varied with the identity of the receptor(s) blocked, the intracellular survival and replication rates were equivalent. The authors concluded that the route of entry was not the determining feature in bacterial survival but that this occurred subsequent to phagocytosis and was common to all entry pathways examined.

3 Establishment and maintenance of an intracellular infection

Some controversy exists as to whether or not pathogenic mycobacteria species remain intravacuolar following phagocytosis by macrophages. Although the majority of studies report that *M. tuberculosis* bacilli are confined within membranous compartments there are reports to the contrary, most recently by McDonough *et al.* (1993). However, the body of published evidence suggests that in most studies the bacteria appear almost exclusively intravacuolar (Xu *et al.* 1994; Clemens & Horwitz 1995). Mycobacterial infections induce CD8 responses in both humans and mice, indicating that bacterial antigens are presented in context of class I major histocompatibility complex (MHC) antigens. Presentation via class I MHC molecules is more common for cytosolic antigens, and it has been suggested that a cytosolic location of bacilli could explain the apparent anomaly of class I MHC presentation of proteins derived from intravacuolar bacteria. It should be noted, however, that CD8 responses have been readily observed for *Leishmania* infections (Stefani *et al.* 1994), and following inoculation with antigens conjugated to inert, particulate carriers (Oh *et al.* 1997), and neither of these particles escape from the vacuole.

3.1 The limited acidification of vacuoles containing mycobacteria

One of the defining characteristics of the endosomal network of higher eukaryotes is the drop in pH experienced by material passed through the network to lysosome. Although early studies by Sprick (1956) suggested that the vacuoles containing live *M. tuberculosis* were acidic, work by Hart and colleagues in the 1970s produced a substantial body of data indicating that these vacuoles did not fuse with lysosomes (Hart *et al.* 1972; Hart & Armstrong 1974; Hart 1979). A recent version of these experiments is shown in Fig. 19.1. Hart demonstrated an inverse correlation between lysosomal fusion and bacterial viability and suggested that the vacuoles were not acidic. Crowle *et al.* (1991) exploited the relative sequestration of a dinitrophenol-derivatized weak base in vacuoles containing mycobacteria vs. dense lysosomes and demonstrated that the compartments in which the bacteria reside had a higher pH than lysosomes. In 1994, these studies were extended by measurement of the pH of phagosomes containing *M. avium* labelled with *N*-hydroxysuccinimide carboxyfluorescein (Sturgill-Koszycki *et al.* 1994). The vacuoles containing mycobacteria equilibrated to pH6.2–6.3, whilst those containing inert IgG beads, *Leishmania* parasites or yeast cell walls dropped to below pH5. The pH of the vacuoles containing mycobacteria has since been confirmed in an independent study by Oh and Straubinger (1996).

Biochemical analysis of vacuoles, containing *M. avium*, isolated from infected macrophages revealed that the lack of acidification was due to a failure to accumulate proton–adenosine triphosphatase (ATPase) complexes (Sturgill-Koszycki *et al.* 1994). Eukaryote proton–ATPases are multisubunit complexes involved in the acidification of intracellular organelles such as mitochondria, lysosomes, and the trans-Golgi network (Grinstein *et al.* 1992). Some of the subunits are organelle specific, and those involved in lysosomal acidification are known as vATPases, or vacuolar ATPases. Immunoblotting of vacuoles in which mycobacteria reside with antibodies against the E subunit (31 kDa), B subunit (56 kDa) and the 110-kDa transmembrane accessory protein were all negative, whilst the acidic phagosomes containing IgG beads were positive for each protein. Despite the paucity of proton–ATPases, vacuoles containing mycobacteria were positive for lysosome-associated membrane protein (LAMP)-1, a sialic

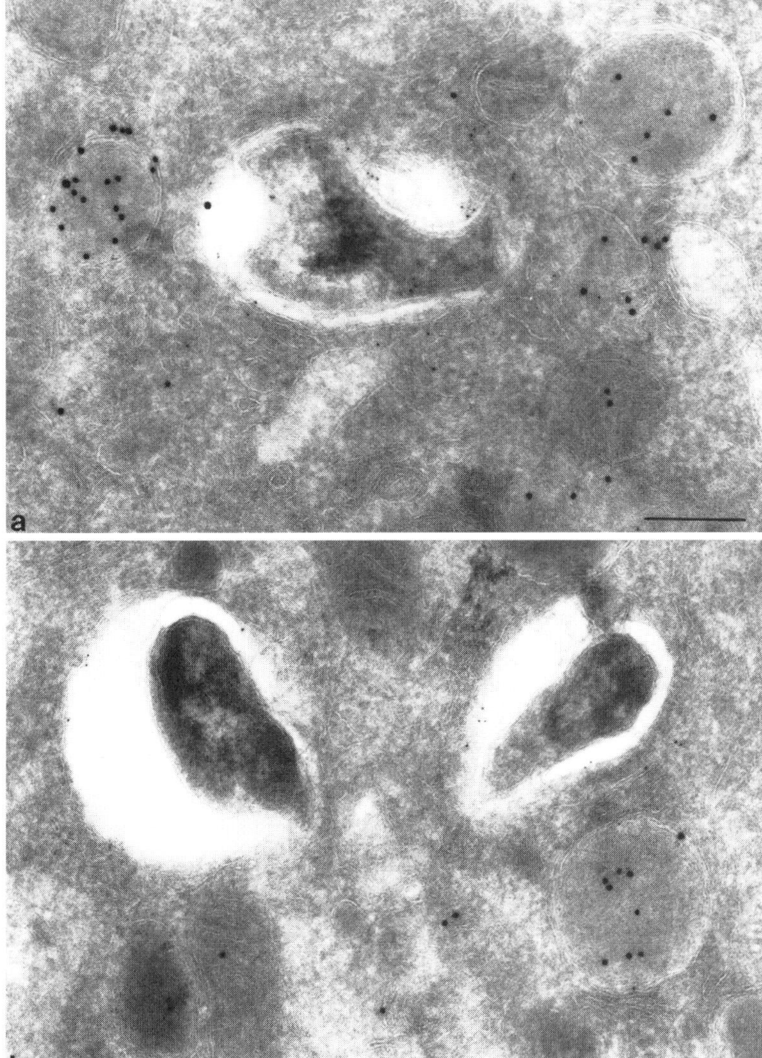

Fig. 19.1 Vacuoles containing mycobacteria do not fuse with lysosomal compartments. Infected cells incubated for 45 min with the fluid-phase marker biotin-dextran (10 kDa) (500 µg/mL). Both samples were probed with streptavidin/rabbit antistreptavidin (15 nm gold-anti-rabbit IgG) and anti-LAMP 1 (5 nm gold-anti-rat IgG). The endocytic marker failed to enter or accumulate within the mycobacterial vacuoles. (a) *M. avium*-infected cells incubated with biotinylated dextran (500 µg/ mL), 4 days postinfection. (b) *M. tuberculosis*-infected cells incubated with biotinylated dextran (500 µg/ mL), 8 days postinfection. Scale bars = 0.25 µm. (From Xu *et al.* 1994.)

acid-rich membrane protein abundant in the late endosomes and lysosomes of mammalian cells (Sturgill-Koszycki *et al.* 1994; Xu *et al.* 1994). The presence of LAMP-1 was also reported in vacuoles containing *M. tuberculosis* by Clemens and Horwitz (1995), although the authors noted that the density of label was less than in neighbouring lysosomes.

These data, in combination with studies from many laboratories reporting the limited access of the vacuoles harbouring mycobacteria to a range of endocytic tracers, all fostered the idea that these vacuoles were sequestered outwith the 'normal' endosomal–lysosomal continuum and were relatively inert.

3.2 Vacuoles in which mycobacteria reside fuse with early endosomal compartments

The first indications that vacuoles containing mycobacteria may interact selectively with early endosomal compartments came from the observations of de Chastellier *et al.* (1995) who reported transient access of horseradish peroxidase to vacuoles inhabited by *M. avium*, and from the immunoelectron microscopy studies of Clemens and Horwitz (1995) who observed the transferrin receptor within vacuoles containing *M. tuberculosis*. The dynamic nature of vacuoles in which either *M. tuberculosis* or *M. avium* reside was demonstrated directly by analysis of the trafficking of the glycosphingolipid, GM1, in infected macrophages (Russell *et al.* 1996). GM1, complexed with biotinylated cholera toxin B subunit, gained access to vacuoles containing mycobacteria within 5 min of uptake and reached steady-state levels after 10–15 min (Fig. 19.2). Delivery to

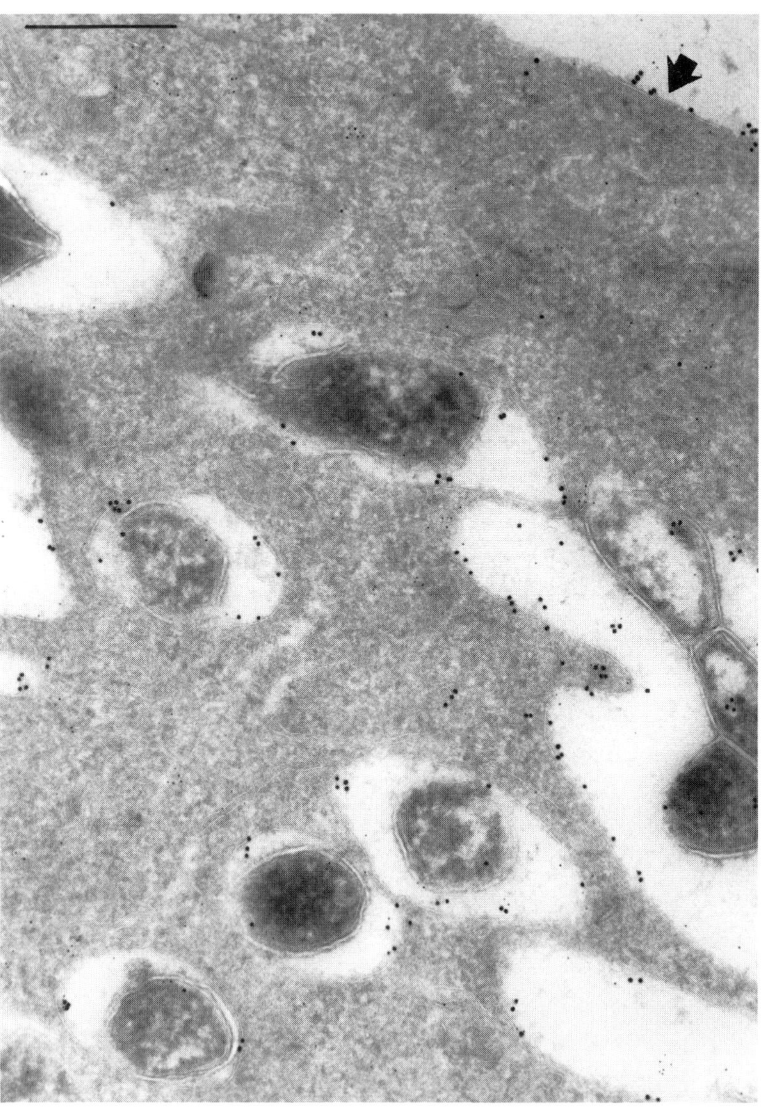

Fig. 19.2 Biotinylated cholera toxin-B subunit (CTx-B) enters vacuoles containing *M. tuberculosis.* Bone-marrow-derived macrophages, infected 72 h previously with *M. tuberculosis,* were incubated in biotinylated Ctx-B (10 μg/mL) for 15 min, washed and placed in prewarmed medium for a 45-min chase period. This section from an *M. tuberculosis*-infected macrophage was probed with streptavidin/antistreptavidin (anti-rabbit IgG-18 nm gold) and anti-LAMP 1 (anti-rat IgG-5 nm gold). Ctx-B can be seen in all the bacterial vacuoles visible in this field. The cell surface membrane is indicated by an arrow. Scale bars = 0.5 μm. (From Russell *et al.* 1996.)

phagolysosomes containing IgG beads was markedly slower.

Direct demonstration of the interaction between early endosomes and vacuoles containing mycobacteria came as a result of the detection of the transferrin receptor in these compartments (Clemens & Horwitz 1995, 1996). Mammalian cells acquire iron through the ability of their transferrin receptors to bind iron-loaded transferrin. The transferrin is internalized into early endosomes, and delivered to sorting endosomes where, at pH6.2–6.3, the iron in the form of Fe^{2+}, comes off the transferrin. Iron-

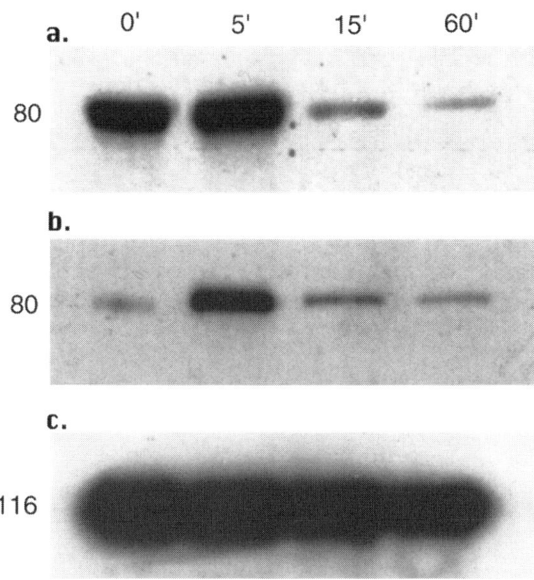

Fig. 19.3 Transferrin trafficking through vacuoles containing mycobacteria. Immunoblots demonstrating the kinetics of transit of digoxigenin–transferrin through vacuoles containing mycobacteria. The gels were run with macrophage homogenates (a) and isolated *M. avium*-containing vacuoles (b and c). Infected macrophages were incubated with digoxigenin–transferrin (10 µg/mL) on ice, then washed and transferred to medium at 37°C for the time period indicated; A and B were probed with antidigoxigenin antibody demonstrating the rate of loss of transferrin from the whole cell (a) vs. the increase, from 0 to 5 min, and decrease, from 5 to 15 min, of transferrin in the vacuoles containing mycobacteria (b). (c) shows these vacuoles probed with ID4B, anti-LAMP 1 antibody, to demonstrate equivalent loading of material from the different time points. (From Sturgill-Koszycki *et al.* 1996.)

depleted transferrin, still bound to its receptor, then recycles back to the cell surface through the recycling endosomal system (Dautry-Varsat *et al.* 1983). At neutral pH on the cell surface, the affinity of the receptor for iron-free transferrin is lessened and the ligand is displaced competitively from the receptor by iron-loaded transferrin. Immunoelectron microscopy studies on *M. tuberculosis*-infected human monocyte-derived macrophages (Clemens & Horwitz 1996) and biochemical studies on *M. avium*-infected murine macrophages (Sturgill-Koszycki *et al.* 1996) both demonstrated that transferrin had access to the vacuoles in which mycobacteria reside. Pulse chase experiments revealed that transferrin passed in and out of the vacuoles containing the mycobacteria showing that the vacuoles were part of a dynamic continuum (Fig. 19.3).

3.3 Route of delivery of lysosomal constituents to vacuoles containing mycobacteria

Despite the data demonstrating delivery of plasma-lemma-derived material to vacuoles containing mycobacteria, the question still remains how were late endosomal/lysosomal proteins trafficked into these vacuoles. Lysosomal hydrolases, such as cathepsins D, B, and L and β-glucuronidase, are delivered from the trans-Golgi network to the endosomal system predominantly through the activity of the cation-independent, mannose 6-phosphate receptor. However, the site(s) of delivery of the enzymes within the endosomal network is the subject of some debate (Ludwig *et al.* 1991). Studies on cathepsin D trafficking indicate that the enzyme is synthesized and glycosylated as a 52–55-kDa inactive precursor protein with an N-terminal pro-sequence (Rijnboutt *et al.* 1992; Delbrueck *et al.* 1994). Removal of the pro-sequence starts prior to delivery to the 'light' endosomal fraction which contain both the 52–55-kDa precursor and the 48-kDa immature form of the enzyme. Once these vesicles differentiate into dense or 'heavy' lysosomal compartments, cathepsin D is processed into its mature, two-chain form of 17- and 31-kDa polypeptides. Immunoblotting of phagosomes containing

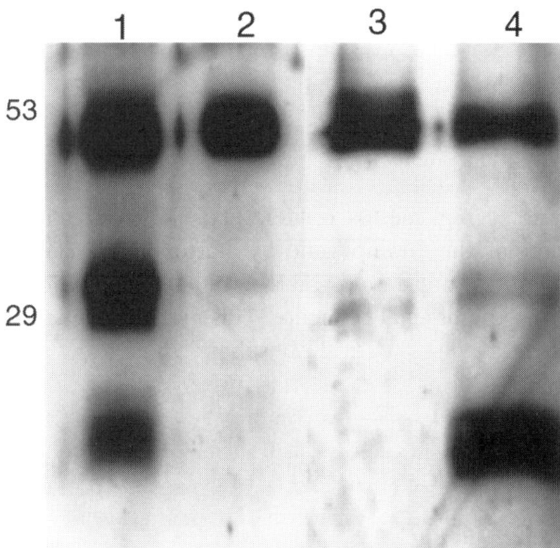

Fig. 19.4 Vacuoles in which mycobacteria reside contain immature cathepsin D. Immunoblot of vacuoles containing *M. avium* and IgG-bead-containing phagosomes with anticathepsin D antibody demonstrating the restricted processing of cathepsin D in the vacuoles with *M. avium*-bacilli, and the subsequent proteolysis of cathepsin D on acidification of isolated vacuoles. Sodium dodecyl sulphate polyacrylamide gel electrophoresis (SDS-PAGE) (12%) gels were run with 60 min IgG-bead-containing phagosomes (lane 1), and *M. avium*-containing phagosomes isolated 9 days (lane 2), and 60 min postinfection (lanes 3 & 4). Lane 4 contains 60 minute *M. avium*-containing phagosomes following incubation in 50 mmol/L acetate buffer, pH4.5, with 0.05% Nonidet P40 at 37°C for 10 min. Vacuole preparations, shown in lanes 3 & 4, were isolated in the absence of proteinase inhibitors. The samples were normalized for protein content prior to electrophoresis. (From Sturgill-Koszycki *et al.* 1996.)

M. avium isolated from macrophages infected 2 h or 9 days previously revealed cathepsin D in its 48-kDa form, while IgG-bead phagolysosomes contained processed, mature enzyme (Sturgill-Koszycki *et al.* 1996). This is shown in Fig. 19.4. When the vacuoles from infected macrophages were acidified in the test tube, the cathepsin D was processed as a consequence of cathepsin B and L activity. These data provide a graphic illustration of the limited hydrolytic capacity of vacuoles containing mycobacteria but, more importantly, they suggest that the acquisition of

cathepsin D, and probably other endosomal constituents, is from the cell's own synthetic pathway and not through fusion with pre-existing, acidified, endosomes.

Intriguingly, characterization of IgG-bead phagosomes shortly after internalization generate similar profiles of pro-cathepsin D, LAMP-1 and minimal proton–ATPase content. Furthermore, IgG-bead phagosomes show a transient interaction with the transferrin-loaded recycling/sorting endosomal network (Sturgill-Koszycki *et al.* 1996). This suggests that the vacuoles in which mycobacteria reside are not chimeric or aberrant compartments generated by the bacilli, but represent early endosomal compartments that are accessible to the sorting/recycling endosomal system. It would appear that both *M. tuberculosis* and *M. avium* have evolved mechanisms to sustain the homotypic fusion of these early endosomal vesicles and yet avoid the acquisition of fusion machinery that would facilitate fusion with later endosomal and lysosomal vesicles which is the normal consequence of phagosome maturation.

The mechanism(s) behind this arrested maturation are unknown. However, investigators have shown that weak bases and inhibitors of proton–ATPase activity, such a bafilomycin A, have profound effects on the retention of material within the sorting and recycling endosomal machinery, suppressing its delivery to lysosomes (Clague *et al.* 1994; van Weert *et al.* 1995). Furthermore, the treatment of cells with lysosomotropic weak bases can modify markedly the fusion capacity of phagosomes. Chloroquine increases phagosome–lysosome fusion (Hart & Young 1978), whilst ammonium chloride, which also effects cytosolic pH, suppresses phagosome–lysosome fusion but enhances phagosome–endosome fusion (Hart & Young 1991). These data indicate an intimate relationship between acidification and the passage of material through to the late endosomal–lysosomal system. An often quoted suggestion was that the ammonia produced by the bacterium's urease blocked vacuolar acidification. A similar proposal was also made regarding the amines produced by the glutamine synthase (Harth *et al.* 1994) and, given the association between acidification and maturation, it is possible that the maintenance of a high pH could block transition to lysosomes. No direct evidence exists for this association, and the recent characterization of urease-negative BCG (Reyrat *et al.* 1996), generated by homologous recombination and deletion of the urease locus, indicated that there was a marginal drop in virulence in mice. However, because the parental BCG strain has only a limited virulence in animals, this result does not disprove conclusively a role for urease in blocking vacuolar maturation. Regardless of the debate surrounding the mechanism, the data suggest that it is the exploitation of a decision point encountered early in phagosome biogenesis that allows mycobacteria to survive in macrophages and prevent fusion of the vacuoles containing the bacilli with lysosomes.

4 Innate resistance at the level of the macrophage: BCGr/Nramp

Resistance to mycobacterial infections at the level of the macrophage is the product of both immune and innate mechanisms. Although the innate mechanisms are of limited efficacy, and in most instances are synergistic with immune mechanisms, they have been shown to limit early bacterial expansion.

The BCG resistance gene was identified as a single, or tightly linked genetic locus on mouse chromosome 1 capable of limiting the intracellular expansion of *Leishmania donovani*, *Salmonella typhimurium* and some *Mycobacterium* spp. (Blackwell *et al.* 1994). The phenotype of BCGr(resistant) vs. BCGs(susceptible) mice differs with species and strain of *Mycobacterium* used, the size of the challenge dose, and the tissue distribution of infection. Electron-microscopical studies on *M. avium*-infected macrophages from BCGr and BCGs mice indicated that there was greater fusion with lysosomes and a drop in bacterial viability in the cells from BCGr mice (de Chastellier *et al.* 1993).

Vidal *et al.* (1993) cloned the gene encoding the BCGr protein, which they called Nramp (natural resistance-associated macrophage protein). The protein appeared to have 10 membrane-spanning domains and showed homology to other eukaryote and prokaryote transporter proteins. The authors suggested that the resistant vs. susceptible phenotype was the product of a single amino acid substitution at residue 105 of Nramp. Expression of this Nramp gene was restricted to cells of reticuloendothelial lineage and macrophage-like cell lines. Further characterization of the Nramp gene by Barton *et al.* (1994) identified a 64 amino acid N-terminal region omitted from the original sequence. This domain contained three protein kinase C phosphorylation sites and a putative Src homology 3-binding domain, suggesting that the protein was capable of responding to signalling cascades. This form of the gene family was the only form found to be expressed in macrophages. More recently, another form of Nramp, Nramp 2, has been found in murine tissues (Gruenheid *et al.* 1995). This gene shows 78% homology with Nramp 1 and is situated on chromosome 15. It appears constitutively expressed at low levels in all tissues examined.

The Nramp gene belongs to a rapidly expanding family of genes that was initially implicated in nitrate transport in plants (Belouchi *et al.* 1995). However, recent data indicated that the primary function of the protein in *Saccharomyces cerevisiae* was manganese transport (Supek *et al.* 1997), and it was suggested that Nramp may function by decreasing superoxide dismutase activity in the bacterium through removal of the Mn^{2+} (or Fe^{2+}) from the active centre of the enzyme. Detailed analysis of the rat Nramp 1 homologue (DCT-1, divalent cation transporter 1) and its functional analysis through expression of cDNA in frog oocytes revealed that the protein was a metal ion transporter of unusually broad specificity capable of transporting Fe, Zn, Mn, Co, Cd, Cu, Ni and Pb ions (Gunshin *et al.* 1997). Transport is proton-coupled and requires a membrane potential. These data add an interesting twist to the observations concerning the limited acidification of vacuoles containing mycobacteria because transporter activity at pH 6.2 was markedly reduced from the level expressed at pH 5.5. In addition, recent immunolocalization of the Nramp 1 gene product revealed that the protein was present in endosomal and phagosomal membranes (Gruenheid *et al.* 1997).

The possible mechanism(s) by which Nramp could limit bacterial growth are multiple and currently unsubstantiated. The depletion of Mn^{2+} or Fe^{2+} could decrease bacterial superoxide dismutase or catalase activity and thus diminish the bacteria's ability to deal with reactive oxygen radicals. Alternately, acquisition of Fe ions by the host macrophage could increase production of hydroxyl radicals and hypervalent iron, both of which are toxic to the bacilli. The putative cytosolic signalling domains could function in transduction of pro-inflammatory cascades (Barton *et al.* (1994). Recent observations by Hackam *et al.* (1998) indicate a direct effect of Nramp1 on phagosome maturation and acidification. This intriguing phenotype was observed through comparison of vacuole pH in BCG-containing phagosomes in macrophages from BCGʳ versus BCGˢ mice.

Although data exist in the mouse model that links this locus to increased susceptibility to a range of intracellular pathogens, including several species of *Mycobacterium*, a comparable correlate has yet to be observed in the human population.

5 Programmed cell death and its role in mycobacterial infections

Programmed cell death, or apoptosis, in mycobacterium-infected macrophages has been the subject of numerous studies in recent literature. Its role in mycobacterial infections is one of the more controversial and confusing areas of current mycobacterial literature.

Although some studies suggest that mycobacterial infections can, under some circumstances, protect macrophages from progressing into programmed cell death (Durrbaum-Landmann *et al.* 1996), there has been more of a consensus in recent reports indicating that all pathogenic mycobacteria, including *M. avium*, will, under most conditions, induce apoptosis in their host cells (Gan *et al.* 1995; Keane *et al.* 1997). All cells are rendered more susceptible to apoptosis by blocking protein synthesis or growth. Therefore it is likely that monocytes and macrophages show a similar, growth-related sensitivity to apoptosis. One recent study showed that granulocyte/monocyte colony-stimulating factor (GM-CSF) will protect

M. avium-infected macrophages from apoptosis and the authors correlated this protection with the increased ability of the macrophages to synthesize plasminogen activator inhibitor type 2 (PAI-2) (Gan *et al.* 1995). PAI-2 is a potent and specific inhibitor of urokinase-type, plasminogen activator which Gan and colleagues suggest is involved in triggering apoptosis in infected macrophages.

Studies of human alveolar macrophages, both in lavage and *in situ*, have demonstrated that *M tuberculosis,* particularly virulent strains such as H37Rv or clinical infections, can be a potent stimulator of apoptosis. Interestingly, alveolar macrophages from infected individuals were more prone to go apoptotic suggesting that the infection could 'prime' the cells to response to apoptotic stimuli (Placido *et al.* 1997). The apoptotic response can be enhanced by exposure to tumour necrosis factor-α (TNF-α), and in *M. avium*-infected macrophages, was also enhanced by coinfection with HIV (Newman *et al.* 1993).

In general, most studies suggest that apoptosis in mycobacterial infections is an antimicrobial response. Remold and colleagues noted that once a macrophage becomes apoptotic, it is recognized by receptors on neighbouring phagocytes and can be ingested by these cells (Fratazzi *et al.* 1997). They postulate that, although mycobacteria can limit the development of the vacuole after phagocytosis, they would have difficulty overcoming the hydrolytic, acidic pH of the lysosome into which an apoptotic macrophage, with its bacterial cargo, would be delivered. The most intriguing observations are those that emerged from the studies of Kaplan and colleagues who demonstrated that the induction of programmed cell death in BCG- or *M. avium*-infected macrophages by exposure to high extracellular ATP or H_2O_2, but not to ligation of Fas, led to death of the microbes (Laochumroonvorapong *et al.* 1996). This work has been followed up by a recent study from Lammas *et al.* (1997) who demonstrated that the mycobactericidal behaviour required expression of purinergic receptors, P2Z, on the host cell. Although the actual mechanism of killing of the bacteria is unknown, the recent results of Silva *et al.* (1996) on the mechanism of cytolysis and mycobacterial death, following lysis of infected macrophages by CD8[+] T cells, may suggest that this type of mycobactericidal behaviour could play a significant role in both innate (apoptosis) and immune (cytotoxic T cells) regulation of infection. Our appreciation of direct killing of intramacrophage bacilli by cytolytic T cells was extended considerably by the recent studies of Stenger and colleagues who demonstrated that granulysin, in combination with perforin, was capable of killing *Mycobacterium* within host macrophages (Stenger *et al.* 1997, 1998).

6 The consequences of macrophage activation

Macrophages, and other professional phagocytes, possess a range of microbiocidal mechanisms that enable them to fulfil their primary function as the major barrier against microbiocidal invasion of the body. This armoury is sufficient to deal with the majority of infections without the host being cognizant of their activity. Many of these mechanisms are expressed in resting macrophages, such as the acidic pH of the lysosome, the lysosomal hydrolases, bactericidal peptides, and a low level of superoxide production on ligation of receptors for the Fc portion of antibodies. However, on activation with IFN-γ and TNF-α, there is a marked up-regulation in production of reactive oxygen intermediates and, in the murine system, the cells express the inducible nitric oxide (NO) synthase (iNOS). Although expression of iNOS has been shown in human macrophages, the signals required for its induction appear more complex.

Recently, much of the attention on killing of intracellular pathogens in macrophages in particular has focused on iNOS, which is capable of producing NO from arginine. NO is active against *Mycobacterium* spp. *in vitro* (O'Brien *et al.* 1994; Rhoades *et al.* 1997), although its efficacy varies with the strain of bacteria. However, infections with a range of intracellular pathogens, including *Listeria, Salmonella, Leishmania, Plasmodium* and *Mycobacterium* spp. all progress in an uncontrolled manner in the absence of iNOS activity in macrophage infections *in vitro* (Flesch & Kaufmann 1991; Chan *et al.* 1992) and in iNOS knockout mice (Wei *et al.* 1995; MacMicking *et al.* 1995). Obviously iNOS fulfills a necessary function in regulation of

these infections but a full appreciation of its mode of action must take into account the cascade of other physiological changes that occur during macrophage activation.

In the case of intracellular pathogens that actively need to maintain their own intracellular compartments, such as *Mycobacterium* spp., it is unclear which event comes first. The death, or compromise of the infecting microbe, or the differentiation of their compartment into an acidic, hydrolytically competent lysosome. If the latter is true, this translocation could drastically alter the environment and could potentiate or enable NO's efficacy. Macrophages activated with interferon-γ (IFN-γ) and lipopolysaccharide (LPS) prior to uptake of *M. avium* were able to acidify the phagosomes containing the bacilli to below pH5.3 within 2–3 h of uptake. However, the rate of acidification was markedly slower than that observed with inert particles suggestive of a 'struggle' between the bacilli and the macrophage (Schaible *et al.* 1998). When these vacuoles were isolated and characterized biochemically they revealed an accumulation of proton–ATPase, and were no longer accessible to transferrin delivered from outside the host cell. This physiological alteration indicated a translocation of the vacuoles from the recycling pathway deeper down the endosomal pathway to acidic, hydrolytic compartments. Analysis of the metabolic activity and viability of *M. avium* in resting macrophages indicated that the bacilli remain static for a period prior to commencement of growth (Sturgill-Koszycki *et al.* 1997). In activated macrophages, the infecting bacteria never established exponential growth and were slowly killed by their host cells. Retention of the vacuoles in which the bacilli reside within the early endosomal machinery appears to require metabolic activity because dead bacilli are internalized into vacuoles that acidify and fuse with lysosomes (S. Sturgill-Koszycki and D.G. Russell, unpublished observations 1994).

Electron-microscopical analysis has indicated that one of the first phenotypic alterations in activated macrophages is the coalescence of vacuoles containing individual *M. avium* bacilli into communal vacuoles with many bacteria (Fig. 19.5). At the time of fusion, the majority of these bacilli show few signs of damage or degradation and there was little change

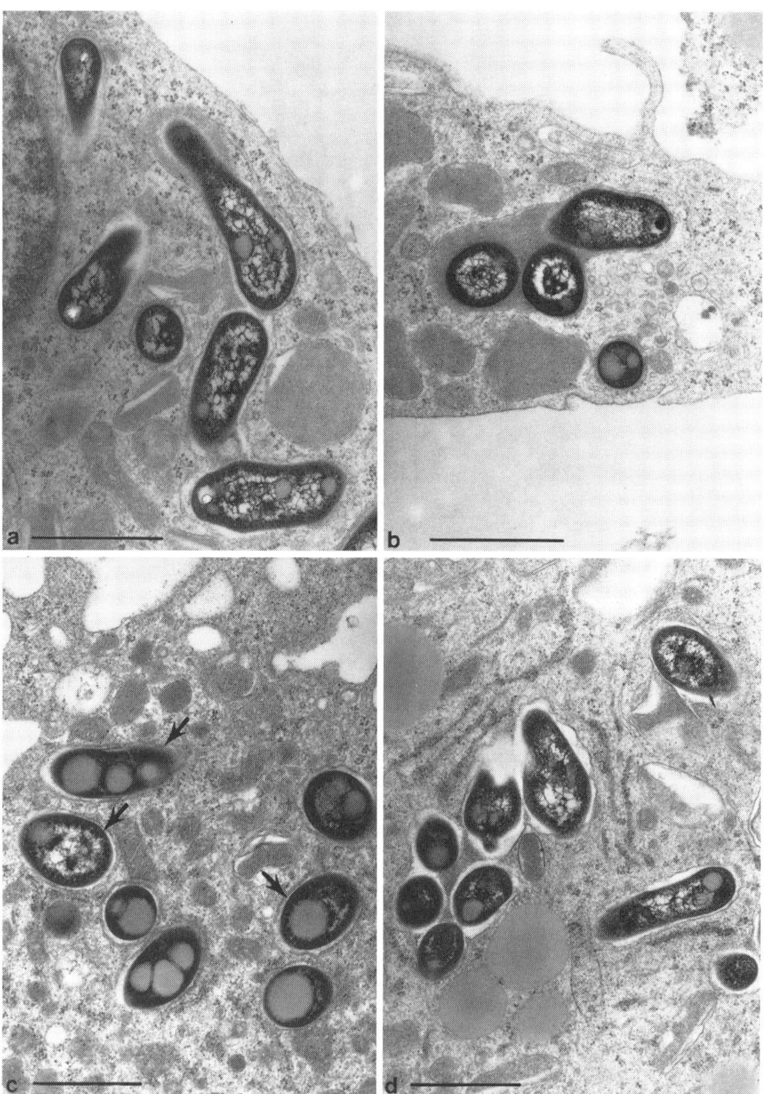

Fig. 19.5. Macrophage activation alters the biology of compartments containing mycobacteria. Electron micrographs of macrophages derived from murine bone marrow and infected with *M. avium* revealing alterations in vacuole macrophage morphology following activation of macrophages with interferon-γ (IFN-γ) and lipopolysaccharide (LPS). (a) Resting macrophages 2 h following infection. The bacilli tend to be sequestered in individual vacuoles which show little evidence of lysosomal fusion. (b) Activated macrophages 2 h following infection. The bacteria are observed more frequently in communal vacuoles that contain dense, lysosomal matrix. Macrophages were activated with IFN-γ (400 U/mL) for 16 h and LPS (500 ng/mL) for 2 h prior to infection. (c) Resting macrophages 5 days postinfection. *M. avium* persist and divide in individual vacuoles. Many of these replicating organisms have prominent ribosomes (arrowed). (d) Activated macrophages 5 days postinfection. Again there is a marked tendency for the bacteria to be in vacuoles containing multiple bacilli. Although there is little obvious degeneration of the bacilli the ribosomes are not as numerous or developed as those seen in (c). Macrophages were activated on day 4 with IFN-γ (400 U/mL) for 16 h and lipopolysaccharide (500 ng/mL) for 2 h prior to processing. (From Schaible *et al.* 1997.)

in bacterial viability measured by colony-forming units (Schaible *et al.* 1997). These observations indicate that the functional transition of these vacuoles to acidic endosomes preceded the drop in microbial viability, suggesting that the acidic lysosomal environment could be integral to the microbicidal mechanisms of the host macrophage.

The functional translocation towards more lysoso-mal compartments precedes any marked drop in microbial viability suggesting that it is the product of an alteration in macrophage physiology, rather than a consequence of microbial death. The lysosomal environment of activated macrophages could potentiate NO toxicity in several ways. Oxidation of NO to nitrite and nitrate will be retarded at acidic pH. NO can combine with superoxide, O_2^-, the production of which is upregulated in activated macrophages, to make peroxynitrite (ONOO⁻). NO can release metal ions, such as Fe²⁺, from metalloproteins which can combine with H_2O_2 to produce ·OH and hypervalent Fe. Furthermore, the action of lysosomal hydrolases on the microbial cell wall will likely expose more targets for oxidative attack. The microbicidal responses of activated macrophages are probably based on the complex interactions of several antimicrobial phenomena and more work is required on the effects of activation on the regulation of intracellular fusion within the endosomal–lysosomal continuum before these interactions can be appreciated.

7 The interface between the infected macrophage and the host's immune system

Considering that the outcome of a productive immune response is death of the infecting bacilli, there is significant pressure on mycobacteria to develop strategies that avoid the induction or consequences of a cellular immune response leading to release of macrophage-activating cytokines. Although the purview of this chapter has been the mycobacterium–macrophage interface, this final section has strayed into some immunology. However, it is difficult to deal with the macrophage as a separate entity given its intimate role as an initiator and effector cell at both ends of the cytokine response; this topic is covered in detail by Cooper and Orme (Chapter 20).

The literature discussing mycobacterial infections is rife with reports of nonresponsiveness and the suppression of cellular immunity. Nash and Douglass (Nash & Douglass 1980) noted that up to 25% of patients with pulmonary tuberculosis, despite their heavy bacterial load, had negative skin tests to purified protein derivative (PPD) of mycobacteria. Other researchers have reported a lack of delayed-type hypersensitivity (DTH) response and decreased *in vitro* proliferation of lymphocytes in mice infected with *Mycobacterium bovis* BCG (Colizzi *et al*. 1984). This effect was mediated by macrophage-derived, soluble factors which could inhibit DNA synthesis and interleukin 2 (IL-2) production by T cells. More recently,

Pancholi *et al*. (1993) demonstrated that human monocytes chronically infected with BCG were defective in their ability to present mycobacterial antigens to autologous T cells, but were able to present antigens added exogenously. These results are in contrast to those of Gercken *et al*. (1994) who reported that the decreased ability of mycobacteria-infected macrophages to process and present exogenous antigens was reflected in their diminished expression of surface MHC class II and accessory molecules.

It is likely that several mechanisms are responsible for the reduced ability of mycobacteria-infected macrophages to induce a cellular immune response. However, in addition to the avoidance mechanisms alluded to above, there are data supporting the active suppression of T-cell responses in the vicinity of infection foci. Sussman and Wadee (1992) reported that the supernatants from CD8 T cells incubated with mycobacterial cell-wall components blocked T-cell blastogenesis. This inhibition was reversed by addition of neutralizing levels of anti-IL-6 antibody. Furthermore, recombinant IL-6 blocked cytokine production by mononuclear cells. These data are likely to be related to the results of VanHeyningen *et al*. (1997) who demonstrated that BCG-infected bone-marrow-derived macrophages were suppressed in their ability to stimulate a T-cell response and that this suppressive effect could be transferred with conditioned medium from infected macrophage cultures to uninfected macrophages (Fig. 19.6). This suppression was observed in both T-cell hybridomas and polyclonal T-cell populations. Attempts to neutralize either the production of the inhibitory factor or its activity with neutralizing antibodies against IL-1, IL-1, IL10, TGF-α and TNF-γ, or blocking prostaglandin and NO production, all failed. In contrast, the inhibitory effect of conditioned medium could be completely removed by immunodepletion of IL-6. Moreover, recombinant IL-6 had a comparable inhibitory effect on antigen-dependent responses of both primary T cells and T-cell hybridomas (VanHeyningen *et al*. 1997). The classical cytokine cascade by which TNF-α upregulates IL-1, which in turn upregulates IL-6, does not appear to be causing the induction of IL-6 in mycobacterium-infected

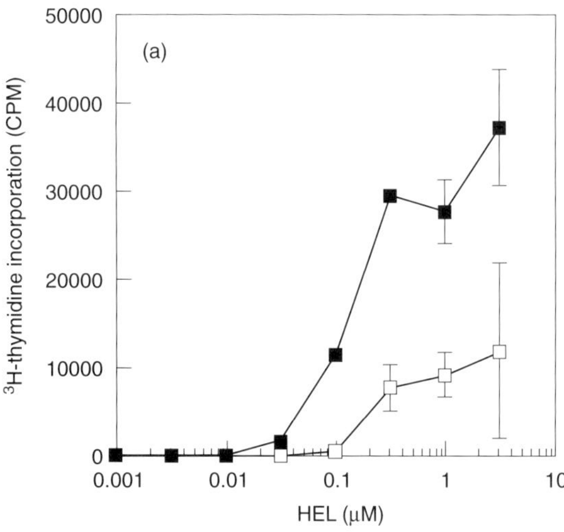

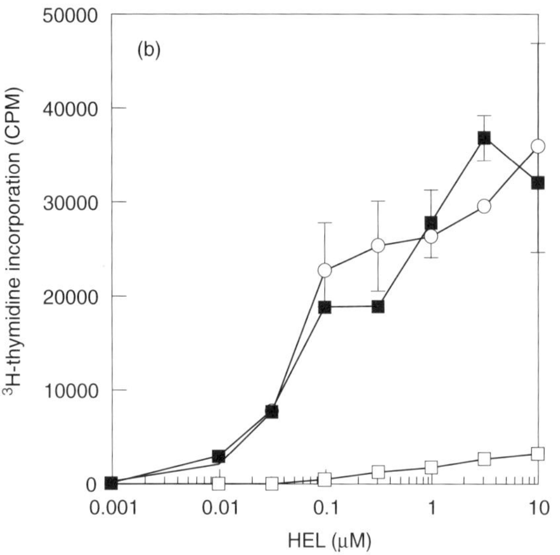

Fig. 19.6 Bacille Calmette–Guérin (BCG)-infected macrophages release a factor that suppresses T-cell responses. (a) Macrophages (5×10^4 cells/well) were infected with *M. bovis* BCG (5×10^5/well; open squares), or left uninfected (solid squares). Two days later, macrophages were pulsed with hen-egg white lysozyme protein (HEL) and the HEL-specific T-cell hybridoma line 3A9 (10^5 cells/well) were added. After 24 h supernatants were collected and assayed for IL-2 by proliferation of the IL-2-dependent cell line, CTLL-2. Results are expressed as mean (± standard deviation) of [^{3}H]thymidine incorporation of triplicate wells. The inhibition is mediated by a soluble factor produced by BCG-infected macrophages. (b) Uninfected macrophages (5×10^4/well) were incubated with HEL and 3A9 T-cell hybridomas (10^5/well) in either control medium (solid squares), infected macrophage culture medium (open squares) or a 1 : 8 dilution of supernatant from *in vitro* grown BCG (open circles). After 24 h, supernatants were collected and assayed for IL-2. Results are expressed as mean (± standard deviation) of [^{3}H]thymidine incorporation of triplicate. (From VanHeyningen *et al.* 1997.)

macrophages because neutralizing antibodies against either IL-1, IL-1, or TNF-α, or all three in combination, had no effect on IL-6 expression.

The primary function of IL-6 is the induction of terminal differentiation of activated B cells to immunoglobulin-producing plasma cells (Tosato *et al.* 1988). The cytokine also regulates acute phase protein synthesis by hepatocytes, is necessary for growth and function of T cells, and is synergistic with IL-3 and GM-CSF in recruitment and differentiation of bone-marrow-derived cells (Hirano 1992). There are several reports detailing induction of IL-6 in *in vivo* infections. Champsi *et al.* (1995) compared the levels of IL-6 induced after infection of mice with *M. avium* over a 5-week period and found IL-6 was produced by cultured spleen cells 2 weeks postinfection and levels were maintained throughout the 5-week infection period. Recently, the levels of IL-6 in bronchoalveolar lavage from patients with active pulmonary tuberculosis were found to be elevated as well (Law *et al.* 1996). These data mesh well with the sustained levels of IL-6 observed after infection of macrophages *in vitro*. However, demonstration that the inhibitory effect of IL-6 in T-cell proliferation assays *in vitro* has relevance to *in vivo* infections requires further experimentation.

Although the mycobacterial cell-wall constituent LAM induces release of IL-6 in macrophages and monocytes, this induction is transient and is thus similar to other proinflammatory stimuli like bacterial LPS. In contrast, the production of IL-6 induced by *M. avium* or BCG infections in macrophages is sustained and suggests that LAM, by itself, is not sufficient. Moreover, infected macrophages induce release of IL-6 in uninfected bystander macrophages. This production correlates with the transfer of cell wall lipids between cells (T. VanHeyningen and D.G. Russell, unpublished observations 1997), and may be comparable to the high-molecular-weight mycobacterial cell-wall fraction used by Sussman and Wadee (1991) to induce CD8 T cells to produce IL-6. Despite an existing study demonstrating the release and vesicular trafficking of LAM in *M. tuberculosis*-infected murine macrophages, the identification, routing and means of transfer of LAM and other cell-wall constituents between infected and uninfected cells remains to be elucidated.

8 Conclusions

Our appreciation of the interplay between the macrophage and pathogenic mycobacteria has increased tremendously over the past few years. Nonetheless, the molecular mechanisms behind the strategies employed by these bacteria to block the normal acidification and maturation of their phagosomes, trigger macrophage apoptosis, and modulate immune responses, and host cell responsiveness, all remain to be delineated. The recent development of transposon mutagenesis, allellic exchange, and the information becoming available from the different genome projects on *M. tuberculosis* and *M. leprae* all herald more rapid progression in the experimental analysis of different facets of this intriguing host cell/pathogen interplay.

9 Acknowledgements

D.G.R. is supported by USPHS grants AI 33348 and HL 55936 and is a recipient of the Burroughs Wellcome Scholar Award in Molecular Parasitology.

10 References

Armstrong, J.A. & Hart, P.D. (1975) Phagosome–lysosome interactions in cultured macrophages infected with virulent tubercle bacilli. Reversal of the usual nonfusion pattern and observations on bacterial survival. *Journal of Experimental Medicine* **142**, 1–16.

Arruda, S., Bomfim, G., Knights, R., Huima-Byron, T. & Riley, L.W. (1993) Cloning of an *M. tuberculosis* DNA fragment associated with entry and survival inside cells. *Science* **261**, 1454–1457.

Barton, C.H., White, J.K., Roach, T.I. & Blackwell, J.M. (1994) NH_2-terminal sequence of macrophage-expressed natural resistance-associated macrophage protein (Nramp) encodes a proline/serine-rich putative Src homology 3-binding domain. *Journal of Experimental Medicine* **179**, 1683–1687.

Belouchi, A., Cellier, M., Kwan, T., Saini, H.S., Leroux, G. & Gros, P. (1995) The macrophage-specific membrane protein Nramp controlling natural resistance to infections in mice has homologues expressed in the root system of plants. *Plant Molecular Biology* **29**, 1181–1196.

Blackwell, J.M., Barton, C.H., White, J.K. *et al.* (1994) Genetic regulation of leishmanial and mycobacterial infections: the Lsh/Ity/Bcg gene story continues. *Immunology Letters* **43**, 99–107.

Blackwell, J.M., Ezekowitz, R.A.B., Roberts, M.B., Channon, J.Y., Sim, R.B. & Gordon, S. (1985) Macrophage complement and lectin-like receptors bind *Leishmania* in the absence of serum. *Journal of Experimental Medicine* **162**, 324–332.

Catanzaro, A. & Wright, S.D. (1990) Binding of *Mycobacterium avium-Mycobacterium intracellulare* to human leukocytes. *Infection and Immunity* **58**, 2951–2956.

Champsi, J., Young, L.S. & Bermudez, L.E. (1995) Production of TNF-alpha, IL-6 and TGF-beta, and expression of receptors for TNF-alpha and IL-6, during murine *Mycobacterium avium* infection. *Immunology* **84**, 549–554.

Chan, J., Xing, Y., Magliozzo, R.S. & Bloom, B.R. (1992) Killing of virulent *Mycobacterium tuberculosis* by reactive nitrogen intermediates produced by activated murine macrophages. *Journal of Experimental Medicine* **175**, 1111–1122.

Clague, M.J., Urbe, S., Aniento, F. & Gruenberg, J. (1994) Vacuolar ATPase activity is required for endosomal carrier vesicle formation. *Journal of Biological Chemistry* **269**, 21–24.

Clemens, D.L. & Horwitz, M.A. (1995) Characterization of the *Mycobacterium tuberculosis* phagosome and evidence that phagosomal maturation is inhibited. *Journal of Experimental Medicine* **181**, 257–270.

Clemens, D.L. & Horwitz, M.A. (1996) The *Mycobacterium tuberculosis* phagosome interacts with early endosomes and is accessible to exogenously administered transferrin. *Journal of Experimental Medicine* **184**, 1349–1355.

Colizzi, V., Ferluga, J., Garreau, J., Malkovsky, M. & Asherson, G.L. (1984) Suppressor cells induced by BCG release non-specific factors *in vitro* which inhibit DNA synthesis and interleukin 2 production. *Immunology* **51**, 65–72.

Crowle, A.J., Dahl, R., Ross, E. & May, M.H. (1991) Evidence that vesicles containing living virulent *Mycobacterium tuberculosis* or *Mycobacterium avium* in cultured human macrophages are not acidic. *Infection and Immunity.* **59**, 1823–1829.

Dautry-Varsat, A., Ciechanover, A. & Lodish, H.F. (1983) pH and the recycling of transferrin during receptor mediated endocytosis. *Proceedings of the National Academy of Sciences of the USA* **80**, 2258–2262.

de Chastellier, C., Frehel, C., Offredo, C. & Skamene, E. (1993) Implication of phagosome–lysosome fusion in restriction of *Mycobacterium avium* growth in bone marrow macrophages from genetically resistant mice. *Infection and Immunity* **61**, 3775–3784.

de Chastellier, C., Lang, T. & Thilo, L. (1995) Phagocytic processing of the macrophage endoparasite, *Mycobacterium avium*, in comparison to phagosomes which contain *Bacillus subtilis* or latex beads. *European Journal of Cell Biology* **68**, 167–182.

Delbrueck, R., Desel, C., von Figura, K. & Hille-Rehfeld, A. (1994) Proteolytic processing of cathepsin D in prelysosomal organelles. *European Journal of Cell Biology* **64**, 7–14.

Downing, J.F., Pasula, R., Wright, J.R., Twigg, H.L. & Martin, W.J. (1995) Surfactant protein A promotes attachment of *Mycobacterium tuberculosis* to alveolar macrophages during infection with human immunodeficiency virus. *Proceedings of the National Academy of Sciences of the USA* **92**, 4848–4852.

Durrbaum-Landmann, I., Gercken, J., Flad, H.D. & Ernst, M. (1996) Effect of *in vitro* infection of human monocytes with low numbers of *Mycobacterium tuberculosis* bacteria on monocyte apoptosis. *Infection and Immunity* **64**, 5384–5389.

Flesch, I.E. & Kaufmann, S.H. (1991) Mechanisms involved in mycobacterial growth inhibition by gamma interferon-activated bone marrow macrophages: role of reactive nitrogen intermediates. *Infection and Immunity* **59**, 3213–3218.

Fratazzi, C., Arbeit, R.D., Carini, C. & Remold, H.G. (1997) Programmed cell death of *Mycobacterium avium* serovar 4-infected human macrophages prevents the mycobacteria from spreading and induces mycobacterial growth inhibition by freshly added uninfected macrophages. *Journal of Immunology* **158**, 4320–4327.

Gan, H., Newman, G.W. & Remold, H.G. (1995) Plasminogen activator inhibitor type 2 prevents programmed cell death of human macrophages infected with *Mycobacterium avium*, serovar 4. *Journal of Immunology* **155**, 1304–1315.

Gaynor, C.D., McCormack, F.X., Voelker, D.R., McGowan, S.E. & Schlesinger, L.S. (1995) Pulmonary surfactant protein A mediates enhanced phagocytosis of *Mycobacterium tuberculosis* by direct interaction with human macrophages. *Journal of Immunology* **155**, 5343–5351.

Gercken, J., Pryjma, J., Ernst, M. & Flad, H.D. (1994) Defective antigen presentation by *Mycobacterium tuberculosis*-infected monocytes. *Infection and Immunity* **62**, 3472–3478.

Grinstein, S., Nanda, A., Lukacs, G. & Rotstein, O. (1992) V-ATPases in phagocytic cells. *Journal of Experimental Biology* **172**, 179–192.

Gruenheid, S., Cellier, M., Vidal, S. & Gros, P. (1995) Identification and characterization of a second mouse Nramp gene. *Genomics* **25**, 514–525.

Gruenheid, S., Pinner, E., Desjardins, M. & Gros, P. (1997) Natural resistance to infection with intracellular pathogens: the Nramp1 protein is recruited to the membrane of the phagosome. *Journal of Experimental Medicine* **185**, 717–730.

Gunshin, H., Mackenzie, B., Berger, U.V. *et al.* (1997) Cloning and characterization of a mammalian proton-coupled metal-ion transporter. *Nature* **388**, 482–488.

Hackam, D.J., Rotstein, O.D., Zhang, W. *et al.* (1998) Host resistance to intracellular infection: mutation of natural resistance-associated macrophage protein 1 (Nramp 1) impairs phagosomal acidification. *Journal of Experimental Medicine* **188**, 351–364.

Hart, P.D. (1979) Phagosome–lysosome fusion in macrophages: a hinge in the intracellular fate of ingested microorganisms? *Frontiers of Biology* **48**, 409–423.

Hart, P.D. & Armstrong, J.A. (1974) Strain virulence and the lysosomal response in macrophages infected with *Mycobacterium tuberculosis*. *Infection and Immunity* **10**, 742–746.

Hart, P.D., Armstrong, J.A., Brown, C.A. & Draper, P. (1972) Ultrastructural study of the behavior of macrophages toward parasitic mycobacteria. *Infection and Immunity* **5**, 803–807.

Hart, P.D. & Young, M.R. (1978) Manipulations of the phagosome–lysosome fusion response in cultured macrophages. Enhancement of fusion by chloroquine and other amines. *Experimental Cell Research* **114**, 486–490.

Hart, P.D. & Young, M.R. (1991) Ammonium chloride, an inhibitor of phagosome–lysosome fusion in macrophages, concurrently induces phagosome–endosome fusion, and opens a novel pathway: studies of a pathogenic mycobacterium and a nonpathogenic yeast. *Journal of Experimental Medicine* **174**, 881–889.

Harth, G., Clemens, D. & Horwitz, M.A. (1994) Glutamine synthetase of *Mycobacterium tuberculosis*: extracellular release and characterization of its enzymatic activity. *Proceedings of the National Academy of Sciences of the USA* **91**, 9342–9346.

Hirano, T. (1992) Interleukin-6 and its relation to inflammation and disease. *Clinical Immunology and Immunopathology* **S60**, 62–83.

Keane, J., Balcewicz-Sablinska, M.K., Remold, H.G. *et al.* (1997) Infection by *Mycobacterium tuberculosis* promotes human alveolar macrophage apoptosis. *Infection and Immunity* **65**, 298–304.

Lammas, D.A., Stober, C., Harvey, C.J., Kendrick, N., Panchalingam, S. & Kumararatne, D.S. (1997) ATP-induced killing of mycobacterium by human macrophages is mediated by purinergic P2Z (P2X7) receptors. *Immunity* **3**, 433–444.

Laochumroonvorapong, P., Paul, S., Elkon, K.B. & Kaplan, G. (1996) H_2O_2 induces monocyte apoptosis and reduces viability of *Mycobacterium avium–M. intracellulare* within cultured human monocytes. *Infection and Immunity* **64**, 452–459.

Law, K., Weiden, M., Harkin, T., Tchou-Wong, K., Chi, C. & Rom, W.N. (1996) Increased release of interleukin 1 beta, interleukin 6 and tumor necrosis factor alpha by bronchoalveolar cells lavaged from involved sites in pulmonary tuberculosis. *American Journal of Respiratory and Critical Care Medicine.* **153**, 799–807.

Ludwig, T., Griffiths, G. & Hoflack, B. (1991) Distribution of newly synthesized lysosomal enzymes in the endocytic pathway of normal rat kidney cells. *Journal of Cell Biology* **115**, 1561–1572.

MacMicking, J.D., Nathan, C., Horn, G. *et al.* (1995) Altered responses to bacterial infection and endocytic shock in mice lacking inducible nitric oxide synthase. *Cell* **81**, 641–646.

McDonough, K., Kress, Y. & Bloom, B.R. (1993) Pathogenesis of tuberculosis: Interactions of *Mycobacterium tuberculosis* with macrophages. *Infection and Immunity.* **61**, 2763–2774.

Nash, D.R. & Douglass, J.E. (1980) Anergy in active pulmonary tuberculosis: a comparison between positive and negative reactors and an evaluation of 5 TU and 250 TU skin test doses. *Chest* **77**, 32–37.

Newman, G.W., Kelley, T.G., Gan, H. *et al.* (1993) Concurrent infection of human macrophages with HIV-1 and *Mycobacterium avium* results in decreased cell viability, increased *M. avium* multiplication and altered cytokine production. *Journal of Immunology* **151**, 2261–2272.

O'Brien, L., Carmichael, J., Lowrie, D.B. & Andrew, P.W. (1994) Strains of *Mycobacterium tuberculosis* differ in susceptibility to reactive nitrogen intermediates *in vitro. Infection and Immunity.* **62**, 5187–5190.

Oh, Y.K. & Straubinger, R.M. (1996) Intracellular fate of *Mycobacterium avium*: use of dual-label spectrofluorometry to investigate the influence of bacterial viability and opsonization on phagosomal pH and phagosome–lysosome interaction. *Infection and Immunity* **64**, 319–325.

Oh, Y.K., Harding, C.V. & Swanson, J.A. (1997) The efficiency of antigen delivery from macrophage phagosomes into cytoplasm for MHC class I-restricted antigen presentation. *Vaccine* **15**, 511–518.

Pancholi, P., Mizra, A., Bardwaj, N. & Steinman, R.M. (1993) Sequestration from immune CD4+ T cells of mycobacteria growing in human macrophages. *Science* **260**, 984–987.

Placido, R., Mancino, G., Amendola, A. *et al.* (1997) Apoptosis of human monocytes/macrophages in *Mycobacterium tuberculosis* infection. *Journal of Pathology* **181**, 31–38.

Reyrat, J.M., Lopez-Ramerez, G., Ofredo, C., Gicquel, B. & Winter, N. (1996) Urease activity does not contribute dramatically to persistence of *Mycobacterium bovis* bacillus Calmette–Guérin. *Infection and Immunity* **64**, 3934–3936.

Rhoades, E.R., Frank, A.A. & Orme, I. (1997) Susceptibility of a panel of virulent strains of *Mycobacterium tuberculosis* to reactive nitrogen intermediates. *Infection and Immunity* **65**, 1189–1195.

Rijnboutt, S., Stoorvogel, W., Geuze, H.J. & Strous, G.J. (1992) Identification of subcellular compartments involved in biosynthetic processing of cathepsin D. *Journal of Biological Chemistry* **267**, 15665–15672.

Russell, D.G., Dant, J. & Sturgill-Koszycki, S. (1996) *Mycobacterium avium* and *Mycobacterium tuberculosis*-containing vacuoles are dynamic, fusion-competent vesicles that are accessible to glycosphingolipids from the host cell plasmalemma. *Journal of Immunology* **156**, 4764–4773.

Russell, D.G., Sturgill-Koszycki, S., VanHeyningen, T., Collins, H. & Schaible, U. (1997) Why intracellular parasitism need not be a degrading experience for *Mycobacterium. Philosophical Transactions of the Royal Society, Series B* **352**, 1303–1310.

Schaible, U.E., Sturgill-Koszycki, S., Schlesinger, P. & Russell, D.G. (1998) Cytokine activation leads to acidification and increase maturation of *Mycobacterium avium*-containing phagosomes in murine macrophages. *Journal of Immunology* **160**, 1290–1296.

Schlesinger, L.S. (1993) Macrophage phagocytosis of virulent but not attenuated strains of *Mycobacterium tuberculosis* is mediated by mannose receptors in addition to complement receptors. *Journal of Immunology* **150**, 2920–2930.

Schlesinger, L.S. & Horwitz, M.A. (1994) A role for natural antibody in the pathogenesis of leprosy: antibody in nonimmune serum mediates C3 fixation to the *Mycobacterium leprae* surface and hence phagocytosis by human mononuclear phagocytes. *Infection and Immunity* **62**, 280–289.

Schlesinger, L.S., Bellinger-Kawahara, C.G., Payne, N.R. & Horwitz, M.A. (1990) Phagocytosis of *Mycobacterium tuberculosis* is mediated by human monocyte complement receptors and complement component C3. *Journal of Immunology* **144**, 2771–2780.

Schlesinger, L.S., Hull, S.R. & Kaufman, T.M. (1994) Binding of the terminal mannosyl units of lipoarabinomannan from a virulent strain of *Mycobacterium tuberculosis* to human macrophages. *Journal of Immunology* **152**, 4070–4079.

Schorey, J.S., Carroll, M.C. & Brown, E.J. (1997) A macrophage invasion mechanism of pathogenic mycobacteria. *Science* **277**, 1091–1093.

Schorey, J.S., Holsti, M.A., Ratliff, T.L., Allen, P.A. & Brown, E.J. (1996) Characterization of the fibronectin attachment protein of *Mycobacterium avium* reveals a fibronectin binding motif conserved among mycobacteria. *Molecular Microbiology* **21**, 321–329.

Schorey, J.S., Li, Q., McCourt, D.W. *et al.* (1995) A *Mycobacterium leprae* gene encoding a fibronectin binding protein is used for efficient invasion of epithelial cells and Schwann cells. *Infection and Immunity* **63**, 2652–2657.

Silva, C.L., Silva, M.F., Pietro, R.C. & Lowrie, D.B. (1996) Characterization of T cells that confer a high degree of protective immunity against tuberculosis in mice after vaccination with tumor cells expressing mycobacterial hsp65. *Infection and Immunity* **64**, 2400–2407.

Sprick, M.G. (1956) Phagocytosis of *M. tuberculosis* and *M. smegmatis* stained with indicator dyes. *American Reviews in Tuberculosis and Pulmonary Diseases* **74**, 552–556.

Stefani, M.M., Muller, I. & Louis, J.A. (1994) *Leishmania major* specific CD8+ T cells are inducers and targets of nitric oxide produced by parasitized macrophages. *European Journal of Immunology* **24**, 746–752.

Stenger, S., Mazzaccaro, R.J., Uyemara, K. *et al.* (1997) Differential effects of cytolytic T cell subsets on intracellular infection. *Science* **276**, 1684–1687.

Stenger, S., Hanson, D.A., Teitelbaum, R. *et al.* (1998) An antimicrobial activity of cytolytic T cells mediated by granulysin. *Science* **282**, 121–125.

Stokes, R.W., Haidl, I.D., Jefferies, W.A. & Speert, D.P. (1993) Mycobacteria–macrophage interactions. Macrophage phenotype determines the nonopsonic binding of *Mycobacterium tuberculosis* to murine macrophages. *Journal of Immunology* **151**, 7067–7076.

Sturgill-Koszycki, S., Haddix, P. & Russell, D.G. (1997) Analysis of the interaction of *Mycobacterium* with macrophages by 2-dimensional-SDS PAGE. *Electrophoresis.* **18**, 2558–2565.

Sturgill-Koszycki, S., Schaible, U.E. & Russell, D.G. (1996) Mycobacterium-containing phagosomes are accessible to early endosomes and reflect a transitional state in normal phagosome biogenesis. *EMBO Journal* **15**, 6960–6968.

Sturgill-Koszycki, S., Schlesinger, P.H., Chakraborty, P. *et al.* (1994) Lack of acidification in *Mycobacterium* phagosomes produced by exclusion of the vesicular proton-ATPase. *Science* **263**, 678–681.

Supek, F., Supekova, L., Nelson, H. & Nelson, N. (1997) Function of metal ion homeostasis in the cell division cycle, mitochondrial protein processing, sensitivity to mycobacterial infection and brain function. *Journal of Experimental Biology* **200**, 321–330.

Sussman, G. & Wadee, A.A. (1991) Production of a suppressor factor by CD8+ lymphocytes activated by mycobacterial components. *Infection and Immunity* **59**, 2825–2835.

Sussman, G. & Wadee, A.A. (1992) Supernatants derived from CD8+ lymphocytes activated by mycobacterial fractions inhibit cytokine production. The role of IL-6. *Biotherapy* **4**, 87–95.

Tosato, G., Seamon, K.B., Goldman, N.D. *et al.* (1988) Identification of a monocyte derived human B cell growth factor as interferon-beta 2 (BSF-2, IL-6). *Science* **239–243**, 502.

VanHeyningen, T., Collins, H.L. & Russell, D.G. (1997) IL-6 produced by macrophages infected with *Mycobacterium* species suppress T cell responses. *Journal of Immunology* **158**, 330–337.

van Weert, A.W.M., Dunn, K.W., Geuze, H.J., Maxfield, F.R. & Stoorvogel, W. (1995) Transport from late endosomes to lysosomes, but not sorting of integral membrane proteins in endosomes, depends on the vacuolar proton pump. *Journal of Cell Biology* **130**, 821–834.

Vidal, S.M., Malo, D., Vogan, K., Skamene, E. & Gros, P. (1993) Natural resistance to infection with intracellular parasites: Isolation of a candidate for Bcg. *Cell* **73**, 469–485.

Wei, X.Q., Charles, I.G., Smith, A. *et al.* (1995) Altered immune responses in mice lacking inducible nitric oxide synthase. *Nature* **375**, 408–411.

Xu, S., Cooper, A., Sturgill-Koszycki, S *et al.* (1994) Intracellular trafficking in *Mycobacterium tuberculosis* and *Mycobacterium avium*-infected macrophages. *Journal of Immunology* **153**, 2568–2578.

Zimmerli, S., Edwards, S. & Ernst, J.D. (1996) Selective receptor blockade during phagocytosis does not alter the survival and growth of *Mycobacterium tuberculosis* in human macrophages. *American Journal of Respiratory Cell and Molecular Biology* **15**, 760–770.

Chapter 20 / Cytokines in immunity to tuberculosis

ANDREA M. COOPER & IAN M. ORME

1 Introduction

The host response to infection with *Mycobacterium tuberculosis* is expressed by T cells and macrophages, but much of the communication between these cells is mediated by soluble protein molecules that modulate cellular activity (cytokines) or cellular attraction/movement (chemokines). Over the past decade an increasingly large number of molecules with such properties have now been identified, many of which appear to be directly involved in the expression of cell-mediated immunity against intracellular bacterial pathogens.

Cytokines are small, potent proteins which primarily act over short distances between cells involved in immune activity. They fall into two broad groups, those derived from antigen-activated lymphocytes and those derived from other activated cells, particularly macrophages. Cytokines can have a variety of functions which may overlap with the functions of other cytokines and the resulting interactive network is very complex.

The macrophage is the primary host cell for mycobacterial infection and is a principal source of many cytokines; it is also the target of many of these molecules and responds rapidly to them. The cytokines which play a primary role in activating macrophages to microbicidal activity are interferon-γ (IFN-γ) and tumour necrosis factor-α (TNF-α). Both these cytokines can be induced by bacterial products acting on macrophages to induce TNF-α and on natural killer (NK) cells to induce IFN-γ. Upon activation, macrophages also express the cytokines interleukin 10 (IL-10) and IL-12. These molecules counteract each other as IL-12 helps in the expression of IFN-γ from lymphocytes and IL-10 down-regulates macrophage activation. Other important macrophage products are IL-1 and IL-6. These proinflammatory cytokines have many functions, some of which overlap. Transforming growth factor-β (TGF-β) is an enigmatic pluripotent cytokine principally involved in wound healing; it can however, strongly down-regulate immune responses.

Of the antigen-activated lymphocyte-derived cytokines, IL-2 and IL-4 are T-cell growth factors which allow the expansion of an antigen-specific response. IL-3, granulocyte/monocyte colony-stimulating factor (GM-CSF) and monocyte colony-stimulating factor (M-CSF) target the haematopoietic tissues and stimulate the production of phagocytic cells.

The chemokines are a more recently described group of molecules which appear to have even more overlapping functions than the cytokines. They are potent inducers of both immune cell activation and chemotaxis and are the principal mediators of cellular accumulation at the site of infection. Two examples of these chemokines are macrophage chemoattractant protein 1 (MCP-1) and RANTES (regulated upon activation, T-cell expressed and secreted). MCP-1 is released by macrophages and cells of the tissue architecture (i.e. endothelial cells) and is a potent activator and recruiter of monocytes. RANTES, a T-cell product, is a potent recruiter of both monocytes and memory T cells.

In the context of tuberculosis infection, at least three cytokine networks can be identified: those which mediate the early response, those involved in the effector response and those necessary to down-regulate the cellular response. In this brief review these pathways will be described, and the way in which cytokine modulation of these pathways may influence therapy of the disease will be discussed.

2 Initial host–parasite interactions

Tuberculosis is caused by the inhalation of bacteria into the peribronchial areas of the lung, or into an alveolus (Wells 1955; Riley & O'Grady 1961). In the first case, bacteria engulfed by macrophages on the surface of the bronchial tree are most likely carried up to the oesophagus and destroyed in the stomach. Alternatively, bacteria may be carried into a lymphatic vessel and hence into lymphoid tissues in the lung, giving rise to initial immunity.

If the bacillus is picked up by macrophages in the surfactant/tissue fluid, where the bacillus may have become opsonized (Schlesinger *et al.* 1990, 1994), the host cell has a tendency to spread and adhere tightly to the alveolar wall. It is the case of who wins: if the bacillus has properties of virulence it will propagate, if not, it is probably destroyed. Under the former conditions, dividing mycobacteria may eventually be able to lyse the host cell and erode through the alveolar basement membrane into the interstitium, establishing a primary site of infection. Due to this local tissue damage, vasoactive amines and prostaglandins are released, resulting in relaxation of the adjacent capillary vessels and the influx of tissue fluid giving rise to swelling of the interstitial spaces. This increased vascular permeability also encourages the influx of macrophages, which then engulf the free bacilli.

This initial interaction with host cells, and the subsequent initiation of various intracellular antigen presentation pathways, can stimulate a large array of cytokine responses. In fact, just the phagocytic event itself can induce the production of a variety of pro-inflammatory cytokines (IL-1, IL-6, etc.), TNF, IL-12 and various colony-stimulating factors (GM-CSF, M-CSF, IL-3, etc.), the latter of which stimulate increased leucocyte production at the bone-marrow level (Moore *et al.* 1990).

Once the bacteria are able to proliferate within the macrophage phagosome, proteins that leak or are actively secreted/exported from the bacillus are removed into an endosomic intracellular pathway leading to presentation of associated peptides in the cleft of major histocompatibility complex (MHC) class II molecules, thus leading to the sensitization of CD4 T cells and the emergence of acquired specific resistance (Orme *et al.* 1993).

Where does this happen? It seems unlikely to us that T-cell sensitization occurs at the primary site of infection. Matching a T cell with the correct receptor shape to a few macrophages sitting in an interstitial space seems equivalent to finding a needle in a haystack. Hence, we suggest that sensitization probably occurs at two separate places: (i) in the bronchial lymphoid tissue, as discussed above; and (ii) in the spleen. In this second scenario, a few bacilli escape the primary focus via the adjacent capillary and are captured by macrophages in this organ, through which T cells regularly permeate. This hypothesis is of course also consistent with the concept of 'haematogenous spread', in which bacilli also erode into lung tissues where the ventilation/perfusion ratio is more favourable. If this occurs, then there is no reason bacteria cannot also get as far as the spleen.

Either way, such a mechanism would allow for the generation of needed protective T cells in a relatively short time (in the mouse, low-dose, aerosol model, perhaps 15–20 days). These cells then migrate to sites of infection, directed to these sites by adhesion molecules on the adjacent blood-vessel surface.

Other host cellular mechanisms may also exist. There is increasing evidence that small antigenic fragments containing pyrophosphate groups may stimulate major subsets of γδ T cells (Constant *et al.* 1994; Tanaka *et al.* 1995), although whether host presentation molecules (MHC or otherwise) are used (or needed) is not known. The role of γδ T cells sensitized in this manner is also unclear; but, in their absence, granulomatous lesions contain substantially more neutrophils (D'Souza *et al.* 1997). This suggested that γδ T cells control/prevent the influx of neutrophils into inflammatory granulomas, or alternatively somehow reduce local tissue injury that would otherwise attract neutrophils. Finally, in humans at least, macrophages appear to be able to also use non-polymorphic CD1 molecules to present non-proteinaceous mycobacterial products such as mycolic acids and lipoarabinomannan (Beckman *et al.* 1994; Sieling *et al.* 1995). These observations were initially made using clones of T cells which did not express the CD4 or CD8 markers normally associated with T cells, but more recent data suggest certain clones of CD8 T cells, as well as a subset of T cells which express CD4 and the NK1.1[+] marker, may also recognize such antigens in a CD1-restricted manner (Stenger *et al.* 1997).

These *in vitro* observations have recently been proposed as a form of protective immunity. However, this seems to us to be rather unlikely, given the location of mycolic acids and lipoarabinomannan in the mycobacterial cell wall: protective immunity would surely have to pre-exist in order to activate macrophage to kill the bacillus and literally dig these materials out. Another possible scenario is that the CD1 pathway is used to clear away these materials if they are released by dead macrophages. This model would also be consistent with the fact that the CD4+NK1.1+ T cell is a potent source of IL-4, which would promote antibody production and clearance to these potentially noxious materials (Bendelac *et al.* 1996).

3 Early cytokine-mediated events in the infected lung

As described above, local vasculature relaxation will permit the influx of tissue fluid into the otherwise extremely narrow lung interstitium, thus allowing the egress of cells from the bloodstream. Why macrophages and T cells, rather than neutrophils, begin to accumulate in the interstitium is probably a reflection of (i) the local chemokine environment, and (ii) differential expression of adhesion molecules on local blood vessels that promote the entry of these cells. In addition, as described above, there is the intriguing possibility that γδ T cells help control this influx.

This specific cellular influx is probably controlled by release of the cytokine TNF-α by infected macrophages which induces local epithelial and endothelial cells to secrete chemokines such as MCP-1. The amount of TNF-α generated seems to be directly related to the degree of cell influx and hence the eventual size of the granuloma (Rhoades *et al.* 1995). Thus, strains that induce a lot of TNF-α secretion tend to have very large granulomas, whereas others that induce much less have much smaller, compact granulomas. Whether this is good or bad for the host in the long run remains unclear.

TNF-α production in the lungs is an example of a 'double-edged sword'. It is clear from animal models in which TNF-α is neutralized by antibodies (Kindler *et al.* 1989) or the gene has been disrupted (Flynn *et al.* 1995) that TNF-α is essential to the early formation of the granuloma and its integrity. However, if TNF-α production continues, then this almost certainly gives rise to caseous necrosis and potential cavitation seen later during the disease process. IL-12 is probably also involved, at least during the early stages of the infection, as suggested by the poor granulomatous response observed in infected mice infused with neutralizing anti-IL-12 antibodies (Cooper *et al.* 1995), or in which IL-12 genes have been disrupted (Cooper *et al.* 1997).

As this process continues, messages for the various chemokines can be detected in lung tissues (Rhoades *et al.* 1995). Some, like MCP-1 and RANTES, seem to have a specific chemoattractant property for CD4 T cells and monocytes (Schall & Bacon 1994). Although still mainly circumstantial, the evidence seems to be pointing to a grand scheme of things whereby the living mycobacterial infection is a potent stimulus for IL-12 production, which in turn helps to drive a chemokine response that recruits the correct type of cellular influx needed to deal with the infection (Fig. 20.1).

In this regard, as the inflammatory response proceeds during the early phases of the infection, the stage is set for a protective T-cell response. Secretion of IL-6 and IL-1 by macrophages will promote cell differentiation and induce expression of IL-2 receptors by incoming T cells (Stein & Singer 1992). Colony-stimulation factors will promote monocyte production by the bone marrow and finally, IL-12 heterodimer production will drive incoming sensitized CD4 T cells to an end-point short-lived IFN-γ-secreting phenotype.

4 Effector mechanisms

The successful expression of acquired specific resistance to tuberculosis infection depends upon the prompt mobilization of a CD4 IFN-γ secreting protective T-cell population (Orme 1987; Orme *et al.* 1993), the function of which is amplified by IL-12 cytokine production by infected macrophages. Hence, in mice in which either the genes for IL-12 or for IFN-γ are disrupted, the animal is unable to prevent the florid growth and rapid dissemination of tubercular infections (Cooper *et al.* 1993, 1997; Flynn *et al.* 1993). The kinetics of the CD4 IFN-γ response parallel the control and containment of primary intravenous infections (Orme *et al.* 1993), and these cells from either immune mice (Orme *et al.* 1992) or memory immune mice (Andersen *et al.* 1995) secrete IFN-γ, *in vitro*, in response to proteins from the early culture filtrate of *M. tuberculosis*. Taken together, these facts support the hypothesis that the culture filtrate proteins contain similar antigens to those which are released by bacteria which are growing *in vivo* (Fig. 20.2).

Although the primary role of IFN-γ is fully accepted, the crucial components of macrophage

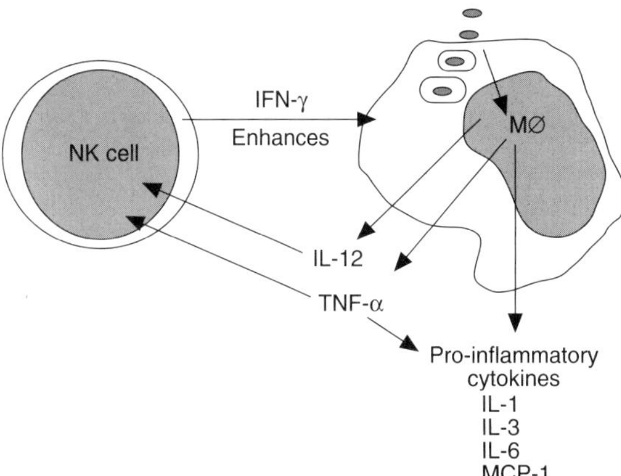

Fig. 20.1 The initial interaction between *Mycobacterium tuberculosis* and the macrophage (MØ) results in the expression of the cytokines interleukin 12 (IL-12) and tumour necrosis factor-α (TNF-α). These then act upon cells of the innate immune system, such as natural killer (NK) cells, which produce the macrophage activating molecule interferon-γ (IFN-γ). Phagocytosis of *M. tuberculosis* also induces the release of the proinflammatory cytokines IL-1, IL-3 and IL-6 and the chemokine MCP-1.

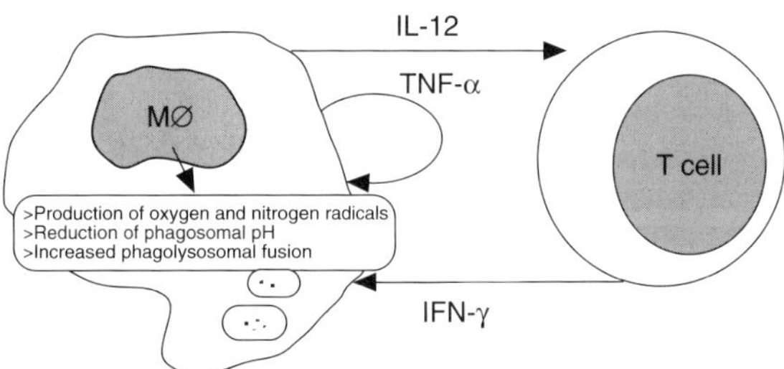

Fig. 20.2 Infected MØ respond in an autocrine manner to the TNF-α induced by the bacteria and to the IFN-γ-derived from antigen-specific CD4 T cells. These cytokines act together to activate the macrophage to a microbiocidal state. The armaments induced include the reduction of the phagosomal pH, increased phagolysosomal fusion and the production of reactive oxygen and nitrogen radicals. The combination of these mechanisms results in a highly toxic environment which can control the growth of the mycobacteria.

activation which lead to the control of bacterial growth still have not been fully elucidated. Again, the study of mice in which microbiocidal pathways have been genetically disrupted by homologous recombination has allowed some progress in this area. For instance, it has recently been demonstrated that IL-12 gene-disrupted mice and IFN-γ gene-disrupted mice are unable to induce expression of the gene for inducible nitric oxide synthase (iNOS) in macrophages, and hence cannot generate toxic nitric oxide molecules (Flynn *et al.* 1993; Cooper *et al.* 1997). In addition, mice which lack the 55-kDa TNF-α receptor are delayed in their expression of iNOS and are more susceptible to disease (Flynn *et al.* 1995). However, although these results and others (Chan *et al.* 1995, 1992) imply a central role for nitric oxide in killing mycobacteria, more recent evidence (Rhoades & Orme 1997) suggests that many clinical isolates of *M. tuberculosis* may be resistant to the effects of nitric oxide.

Under acidic conditions, nitric oxide can react with oxygen radicals to form molecules such as peroxynitrite which are highly toxic to mycobacteria. Hence, one can reason that whether this reaction happens or not depends on the degree of activation of the infected macrophage, mediated by IFN-γ, and hence the degree of proton pump association with the phagosomal membrane (Sturgill-Koszycki *et al.* 1994). How the mycobacterium is able to combat the acidification of the phagosomal environment, which may explain its initial ability to survive and proliferate, remains one of the more interesting questions in the field.

Secretion of toxic oxygen and nitrogen radicals can damage local tissues, and as a result two negative-feedback loops come into play. First, as the growth of the infection is curtailed and bacteria are killed, production, release and presentation of key mycobacterial antigens diminishes, resulting in a progressive loss of IFN secreting protective T cells. Second, macrophages continue to secrete the cytokines IL-10 and TGF-β (Barnes *et al.* 1993), both of which further down-regulate the production of IFN-γ as well as reducing antigen presentation (Fig. 20.3).

These events occur in concert with the walling off

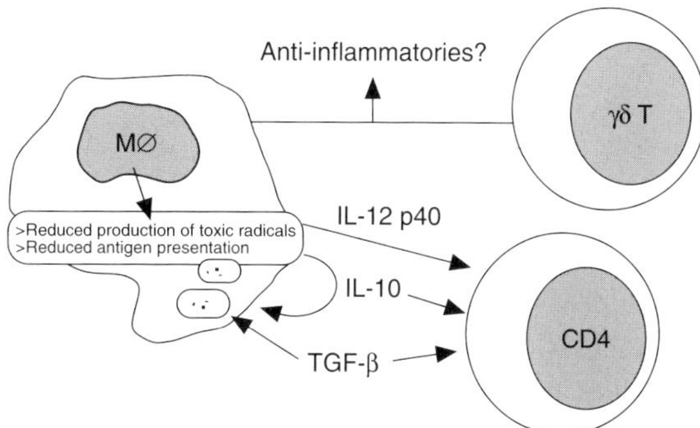

Fig. 20.3 Phagocytosis of *M. tuberculosis* results in the expression of cytokines such as IL-10 and transforming growth factor-β (TGF-β) and the homodimeric form of IL-12 [p40/p40]. These can limit both T-cell and MØ activation resulting in the control of the expanding inflammatory response. Other T cells appear to have a role in the cellular infiltration, as the absence of γδ T cells results in increased neutrophil involvement in the normally mononuclear granuloma.

of the lesion by the continued development of the granulomatous response. In fact, while protective immunity is relatively short-lived, peaking around day 20–23 in the mouse lung infection model, the overall granulomatous response is almost a life-long event, as we have recently demonstrated (Rhoades *et al.* 1997). Moreover, these two mechanisms can be dissociated using the intracellular adhesion molecule 1 (ICAM-1) gene-disrupted mice in which initial protection takes place in the absence of any granulomatous response (Johnson *et al.* 1998).

In mouse strains such as the C57BL/6 that are relatively resistant to tuberculosis, or in bacille Calmette–Guérin (BCG)-vaccinated guinea pigs, the bacteria that survive/escape the phase of protective immunity enter into a chronic state of disease that can last for most of the lifespan of the animal. However, while the bacterial load does not appreciably increase in the lungs, the presence of these bacilli seems to continue to elicit a continuing stimulus/inflammatory signal in the host, resulting in necrosis and extensive fibrosis in the mouse (Rhoades *et al.* 1997), and continuing long consolidation in the guinea pig (McMurray 1994). How this happens and what is the actual basis of the stimulus remains unclear, but it is almost certainly mediated by cytokines and chemokines, raising the possibility of eventual therapeutic control.

5 Other feedback mechanisms

As the IFN-γ-secreting T-cell population gradually disappears, a population of IL-4 T cells emerges (Orme *et al.* 1993). Given the fact that in several infectious disease models the IFN pathway and the IL-4 pathway can cross-modulate each other (Mossman & Coffman 1989), and the fact that experimental mouse models of tuberculosis become chronic as the IL-4 response emerges, we looked at this possibility using IL-4 gene-knockout mice. In this system, however, we obtained no evidence to support the concept that the IL-4 response was in any way interfering with the adequate expression of the protective IFN-γ response (Saunders *et al.* 1999).

Another possible regulatory cytokine loop involves the switch-over by macrophages from the production of the heterodimeric, bioactive IL-12 molecule which contains the p35 and p40 chains, to the production of the homodimeric molecule which contains only the p40 chain. While the role of this molecule remains uncertain, preliminary data may suggest that it down-regulates CD4 T cell activity. Intriguingly, it may also promote CD8 T-cell activity; this is pertinent because it is known that CD8 T cells which are IFN-γ-secreting (Orme *et al.* 1992) and non-cytolytic (Cooper *et al.* 1997; Laochumroonvorapong *et al.* 1997) may play an anti-inflammatory role in the lungs as the CD4 response wanes (or is actively turned off by the absence of IL-12 heterodimer).

6 Cytokine immunotherapy

With the emergence of multidrug-resistant strains of *M. tuberculosis,* there has been increased emphasis placed on the potential therapy of mycobacterial infections using cytokines. Several have been tried in various clinical trials, with varying success.

For example, IFN-γ, which has already been ap-

proved for the treatment of chronic granulomatous disease in humans, has recently been tested in patients with multidrug-resistant tuberculosis with some promising early results (Condos *et al.* 1997; Dr S. Holland, personal communication). Similarly, IL-12 has been used in clinical trials in cancer patients, and despite concerns about its toxicity has been tolerated reasonably well; there has been no attempt to use it in tuberculosis patients as yet, but in individuals suffering from atypical mycobacterial some promising early results have been seen (Dr S. Holland, personal communication).

Perhaps a more promising approach has been the use of IL-2, a cytokine crucial to the clonal expansion of antigen-specific cells, in patients with multidrug-resistant tuberculosis, including human immunodeficiency virus (HIV)-positive patients (Johnson *et al.* 1997) Daily intradermal injections are well tolerated and appear to give rise to clinical improvement.

These data indicate that these therapeutic approaches may have positive, albeit moderate results. Some trials indicate relapses when the therapy is removed, thus it is likely that any cytokine therapy will be prolonged and therefore expensive. Side-effects of long-term therapy, such as autoimmune reactions to the cytokines, are as yet unknown.

Because of this, it may actually be preferable to try to control the inflammation mediated by the continuing presence of persisting bacteria. In this regard, inhibition of TNF-α production by thalidomide results in an increase in weight of HIV-positive multidrug-resistant tuberculosis patients (Tramontona *et al.* 1995).

7 Vaccination enhancement by cytokines

Protective vaccines against non-viral pathogens are few in number. The world-wide use of BCG (an attenuated *M. bovis* strain which is the current anti-tuberculosis vaccine) has had varying efficacy in numerous clinical trials (Rodrigues & Smith 1990; see also Chapters 16 and 17). As a result a concerted effort is now underway to try to develop new vaccines.

It is unlikely that new vaccination protocols will be undertaken in areas where exposure has not already occurred either through previous BCG vaccination or by exposure to environmental mycobacteria (Orme 1997). As such therefore the target of any 'boosting' type of vaccine should be memory T cells. While no experimental data as yet exist regarding this approach, it is nevertheless supported by some rather surprising results from our laboratory.

In those studies (Baldwin *et al.* 1998), we observed that addition of IL-2 to a mixture of filtrate protein antigens and a relatively mild adjuvant protected guinea pigs for over 30 weeks from an aerosol infection with virulent *M. tuberculosis* (controls lived 8–15 weeks). More importantly, the guinea pigs did not develop the caseous necrosis that is the hallmark of this animal model. Since vaccines that did not contain IL-2 did not have much effect, we hypothesize that the presence of IL-2 was necessary to generate the long-lived state of immunity these animals clearly possessed.

8 Future studies

Just as work over the past decade has revealed that the T-cell response to tuberculosis infection is complex (Orme *et al.* 1993), our increasing knowledge of the cytokine/chemokine response shows it to be equally complex. In addition, while it is unlikely that we will discover any more major T-cell subsets, the pace of discovery of new soluble molecules seems unabated.

The 'undiscovered country', if there is one, are the chemokines. As more are identified, and their pleiotropic effects and redundant systems characterized, the outline of their role in infection will remain a considerable scientific challenge. Their promise, however, may be considerable; they help focus the correct cells to the site of infection, they are key molecules in the delayed-type hypersensitivity (DTH) reaction underlying the tuberculin diagnostic test in humans, and their continued production and presence may underlie the chronic inflammation and tissue damage caused by a tuberculosis infection. The development of vaccination or immunotherapuetic protocols which increase chemokine expression early in infection or which decrease chemokine expression during chronic infection would therefore be a plausible target of research endeavours.

9 Acknowledgements

This work was supported by NIH grant AI-40488. We thank our many colleagues in the Mycobacteria Research Laboratories at Colorado State University for their contributions to the ideas expressed above.

10 References

Andersen, P., Andersen, A., Sorenson, A. & Nagai, S. (1995) Recall of long-lived immunity to *Mycobacterium tuberculosis* infection in mice. *Journal of Immunology* **154**, 3359–3372.

Baldwin, S.L., D'Souza, C.D., Roberts, A.D. *et al.* (1998) Evaluation of new vaccines in mouse and guinea pig models of tuberculosis. *Infection and Immunity* **60**, 2951–2959.

Barnes, P., Lu, S., Abrams, J., Wang, E., Yamamura, M. & Modlin, R.L. (1993) Cytokine production at the site of disease in human tuberculosis. *Infection and Immunity* **61**, 3482–3489.

Beckman, E., Porcelli, S., Morita, C., Behar, S.M., Furlong, S.T. & Brenner, M.B. (1994) Recognition of a lipid antigen by CD1 restricted alpha/beta T cells. *Nature* **372**, 691–694.

Bendelac, A., Hunziker, R. & Lantz, O. (1996) Increased interleukin-4 and immunoglobulin E production in transgenic mice overexpressing NK1 T cells. *Journal of Experimental Medicine* **184**, 1285–1295.

Chan, J., Tanaka, K., Carroll, D., Flynn, J. & Bloom, B. (1995) Effects of nitric oxide synthase inhibitors on murine infection with *Mycobacterium tuberculosis*. *Infection and Immunity* **63**, 736–740.

Chan, J., Xing, Y., Magliozzo, R. & Bloom, B. (1992) Killing of virulent *Mycobacterium tuberculosis* by reactive nitrogen imtermediates produced by activated murine macrophages. *Journal of Experimental Medicine* **175**, 1111–1122.

Condos, R., Rom, W. & Schluger, N. (1997) Treatment of multi-drug resistant pulmonary tuberculosis with interferon-gamma via aerosol. *Lancet* **349**, 1513–1515.

Constant, P., Davodeau, F., Peyrat, M. *et al.* (1994) Stimulation of human γδ T cells by non-peptidic mycobacterial ligands. *Science* **264**, 267–269.

Cooper, A.M., D'Souza, C.D., Frank, A.A. & Orme, I.M. (1997) The course of *Mycobacterium tuberculosis* infection in the lungs of mice lacking either perforin or granzyme mediated cytolytic mechanisms. *Infection and Immunity* **65**, 1317–1320.

Cooper, A.M., Dalton, D.K., Stewart, T.A., Griffin, J.P., Russell, D.G. & Orme, I.M. (1993) Disseminated tuberculosis in interferon gamma gene-disrupted mice. *Journal of Experimental Medicine* **178**, 2243–2247.

Cooper, A.M., Magram, J., Ferrante, J. & Orme, I.M. (1997) IL-12 is crucial to the development of protective immunity in mice intravenously infected with *Mycobacterium tuberculosis*. *Journal of Experimental Medicine* **186**, 39–46.

Cooper, A.M., Roberts, A.D., Rhoades, E.R., Callahan, J.E., Getzy, D.M. & Orme, I.M. (1995) The role of interleukin-12 in acquired immunity to *Mycobacterium tuberculosis* infection. *Immunology* **84**, 423–432.

D'Souza, C., Cooper, A., Frank, A., Mazzaccaro, R.J., Bloom, B.R. & Orme, I.M. (1997) An anti-inflammatory role for γδ T lymphocytes in acquired immunity to *Mycobacterium tuberculosis*. *Journal of Immunology* **158**, 1217–1221.

Flynn, J.L., Chan, J., Triebold, K.J., Dalton, D.K., Stewart, T.A. & Bloom, B.R. (1993) An essential role for interferon gamma in resistance to *Mycobacterium tuberculosis* infection. *Journal of Experimental Medicine* **178**, 2249–2254.

Flynn, J.L., Goldstein, M.M., Chan, J. *et al.* (1995) Tumor necrosis factor-alpha is required in the protective immune response against *Mycobacterium tuberculosis* in mice. *Immunity* **2**, 561–572.

Johnson, B.J., Beckker, L.G., Rickman, R. *et al.* (1997) rhuIL-2 adjunctive therapy in multi-drug resistant tuberculosis: a comparison of two treatment regimes. In: *Thirty-Second US–Japan Cooperative Medical Science Program: Tuberculosis-Leprosy Research Conference, Cleveland, OH*.

Johnson, C.M., Cooper, A.M., Frank, A.A. & Orme, I.M. (1998) Dissociation between control and containment of tuberculosis infection in the mouse lung. *Infection and Immunity*, in press.

Kindler, V., Sappino, A.-P., Grau, G., Piguet, P.-F. & Vassalli, P. (1989) The inducing role of tumor necrosis factor in the development of bacterial granulomas during BCG infection. *Cell* **56**, 731–740.

Laochumroonvorapong, P., Wang, J., Liu, C.-C.Y.e.W. *et al.* (1997) Perforin, a cytotoxic molecule which mediates cell necrosis, is not required for the early control of mycobacterial infection in mice. *Infection and Immunity* **65**, 127–132.

McMurray, D. (1994) Guinea pig model of tuberculosis. In: *Tuberculosis: Pathogenesis, Protection and Control* (ed. B. R. Bloom). Washington, D.C: American Society for Micrpbiology, pp. 135–148.

Moore, M., Muench, M., Warren, D. & Laver, J. (1990) Cytokine networks involved in hematopoietic stem cell proliferation and differentiation in molecular control of hematopoiesis. *CIBA Symposium* **148**, 43–52.

Mossman, T. & Coffman, R. (1989) Th1 and Th2 cells: Different patterns of lymphokine secretion lead to different functional properties. *Annual Review of Immunology* **7**, 145–173.

Orme, I.M. (1987) The kinetics of emergence and loss of mediator T lymphocytes acquired in response to infection with *Mycobacterium tuberculosis*. *Journal of Immunology* **138**, 293–298.

Orme, I. (1997) Progress in the development of new vaccines against tuberculosis. *International Journal of Tuberculosis and Lung Disease* **1**, 95–100.

Orme, I., Andersen, P. & Boom, W. (1993) T cell response to *Mycobacterium tuberculosis*. [Review.] *Journal of Infectious Diseases* **167**, 1481–1497.

Orme, I.M., Miller, E.S., Roberts, A.D. *et al.* (1992) T lymphocytes mediating protection and cellular cytolysis during the course of *Mycobacterium tuberculosis* infection. Evidence for different kinetics and recognition of a wide spectrum of protein antigens. *Journal of Immunology* **148**, 189–196.

Orme, I.M., Roberts, A.D., Griffin, J.P. & Abrams, J.S. (1993) Cytokine secretion by CD4 T lymphocytes acquired in response to *Mycobacterium tuberculosis* infection. *Journal of Immunology* **151**, 518–525.

Rhoades, E., Cooper, A. & Orme, I. (1995) Chemokine response in mice infected with *Mycobacterium tuberculosis*. *Infection and Immunity* **63**, 3871–3877.

Rhoades, E., Frank, A. & Orme, I. (1997) Progression of chronic pulmonary tuberculosis in mice aerogenically infected with virulent *Mycobacterium tuberculosis*. *Tubercle and Lung Disease* **78**, 57–67.

Rhoades, E. & Orme, I. (1997) Susceptibility of a panel of virulent strains of *Mycobacterium tuberculosis* to reactive nitrogen intermediates. *Infection and Immunity* **65**, 1189–1195.

Riley, R. & O'Grady, F. (1961) *Airborne Infection: Transmission and Control*. New York: Macmillan.

Rodrigues, L. & Smith, P. (1990) Tuberculosis in developing countries and methods for its control. *Transactions of the Royal Society of Tropical Medicine and Hygiene* **1990**, 739–744.

Saunders, B., Cooper, A.M. & Orme, I.M. (1999) A protective role for interleukin-6 but not interleukin-4 in immunity to *Mycobacterium tuberculosis* infection. Submitted.

Schall, T. & Bacon, K. (1994) Chemokines, leukocyte trafficking and inflammation. *Current Opinion in Immunology* **6**, 865–873.

Schlesinger, L., Bellinger-Kawahara, C., Payne, N. & Horwitz, M.A. (1990) Phagocytosis of *Mycobacterium tuberculosis* is mediated by human monocyte complement receptors and complement component C3. *Journal of Immunology* **144**, 2771–2780.

Schlesinger, L.S., Hull, S.R. & Kaufman, T.M. (1994) Binding of the terminal mannosyl units of lipoarabinomannan from a virulent strain of *Mycobacterium tuberculosis* to human macrophages. *Journal of Immunology* **152**, 4070–4074.

Sieling, P., Chaterjee, D., Porcelli, S. *et al.* (1995) CD1 restricted T cell recognition of microbial lipoglycan antigens. *Science* **269**, 227–230.

Stein, P. & Singer, A. (1992) Similar stimulation requirements of CD4+ and CD8+ primary helper T cells: role of IL-1 and IL-6 in inducing IL-2 secretion and subsequent proliferation. *International Immunology* **3**, 327–335.

Stenger, S., Mazzaccaro, R., Uyemura, K. *et al.* (1997) Differential effects of cytolytic T cell subsets on intracellular infection. *Science* **276**, 1684–1687.

Sturgill-Koszycki, S., Schlesinger, P., Chakraborty, P. *et al.* (1994) Lack of acidification in *Mycobacterium* phagosomes produced by exclusion of the vesicular proton-ATPase. *Science* **263**, 678–681.

Tanaka, Y., Morita, C., Tanaka, Y., Nieves, E., Brenner, M. & Bloom, B. (1995) Natural and synthetic non-peptide antigens recognised by human gamma-delta T cells. *Nature* **375**, 155–157.

Tramontona, J., Utaipat, U., Molloy, A. *et al.* (1995) Thalidomide treatment reduces tumor necrosis factor-alpha production and enhances weight gain in patients with pulmonary tuberculosis. *Molecular Medicine* **1**, 384–397.

Wells, W. (1955) *Airborne Contagion and Air Hygiene.* Cambridge, MA: Harvard University Press.

Index

DATE DUE

GAYLORD

PRINTED IN U.S.A.